INTRODUCTION TO
ELECTRIC
CIRCUITS

INTRODUCTION TO
ELECTRIC
CIRCUITS

SEVENTH EDITION

Herbert W. Jackson
and
Preston A. White III

Prentice Hall, Englewood Cliffs, New Jersey, 07632

Library of Congress Cataloging-in-Publication Data

Jackson, Herbert W.
    Introduction to electric circuits / Herbert W. Jackson, Preston A.
White III. — 7th ed.
        p.   cm.
    Includes index.
    ISBN 0-13-477142-7
    1. Electric circuits.   I. White, Preston A.   II. Title.
TK454.J28   1989
621.319′2—dc19   88-14684
                                                    CIP

Editorial/production supervision: Chris Baumle, Editing, Design & Production, Inc.
Interior design: Anne T. Bonanno
Cover design: Linda J. Den Heyer Rosa
Manufacturing buyer: Robert Anderson

 © 1989, 1986, 1981, 1976, 1970, 1965, 1959 by Prentice-Hall, Inc.
A Division of Simon & Schuster
Englewood Cliffs, New Jersey 07632

Printed in the United States of America

10 9 8 7 6 5 4 3 2 1

ISBN 0-13-477142-7

PRENTICE-HALL INTERNATIONAL (UK) LIMITED, *London*
PRENTICE-HALL OF AUSTRALIA PTY. LIMITED, *Sydney*
PRENTICE-HALL CANADA INC., *Toronto*
PRENTICE-HALL HISPANOAMERICANA, S.A., *Mexico*
PRENTICE-HALL OF INDIA PRIVATE LIMITED, *New Delhi*
PRENTICE-HALL OF JAPAN, INC., *Tokyo*
SIMON & SCHUSTER ASIA PTE. LTD., *Singapore*
EDITORA PRENTICE-HALL DO BRASIL, LTDA., *Rio de Janeiro*

CONTENTS

9 EQUIVALENT-CIRCUIT THEOREMS 210

10 ELECTRICAL MEASUREMENT 248

PART III CAPACITANCE AND INDUCTANCE 281

11 CAPACITANCE 282

12 CAPACITANCE IN DC CIRCUITS 307

13 MAGNETIC CIRCUITS 345

14 INDUCTANCE 383

15 INDUCTANCE IN DC CIRCUITS 397

PART IV ALTERNATING CURRENT 429

16 ALTERNATING CURRENT 430

17 REACTANCE 455

18 PHASORS 471

19 IMPEDANCE 497

20 POWER IN ALTERNATING-CURRENT CIRCUITS 525

PART V IMPEDANCE NETWORKS   557

21 SERIES AND PARALLEL IMPEDANCES   558

22 IMPEDANCE NETWORKS   590

23 RESONANCE   632

24 TRANSFORMERS   657

PREFACE

This book is designed specifically for the first- and second-semester introductory electric circuits course for the electrical/electronics engineering technician/technology program in community colleges, technical institutes, and universities. It assumes only that students have a working knowledge of algebra and elementary trigonometry.

Often students are attracted to technology programs because they are not comfortable with a mathematical derivation as the sole justification of a technical principle, as is customary in engineering science programs. On the other hand, once a principle is thoroughly understood in accurate qualitative terms, a mathematical expression of that principle does make sense. Consequently, all topics in this book are developed in a methodical, step-by-step, cause-and-effect manner geared to the need for technology students to be able to "visualize" why electric circuits behave as they do. There are no "it-can-be-shown-that" formulas to be memorized in this informal presentation style.

Increasing emphasis on circuit analysis techniques in recent years has often cut into classroom time available for developing this appreciation of the fundamental principles of electricity and magnetism. Students will find that the comprehensive, step-by-step development of these all-important topics permits self study to compensate for such time restrictions.

The increasing availability of personal computers and circuit-analysis software increases the temptation to solve network problems by simply responding to the computer program's requests for input data—the modern equivalent of plugging-numbers-into-formulas. Although this may be a great labor-saving technique for *graduates*, I believe that *students* should learn to visualize exactly what is happening as they apply various network simplification tools. Consequently, every network simplification technique is developed from basic principles. Several methods of achieving the same results are explained so that instructors can select the method best suited to a particular course outline, and students have more than one way of visualizing circuit behavior. Computers are discussed as a *tool*, indicating situations where they are better suited to the study of electric circuit analysis than the usual hand-held calculator.

Notwithstanding the emphasis on understanding electric circuit *operation*, all topics included in a typical introductory circuit *analysis* course are covered. The order of presentation is the traditional dc before ac format with dc network analysis introduced as early as possible. However, the contents are packaged into five parts in such a manner that variations (such as dc + ac fundamentals before circuit analysis techniques) can readily be accommodated.

I belong to a generation of technical college instructors who taught this introductory electric circuits course during the 1940s, '50s, and '60s. Many of us came into technical education from a training background in the armed forces. We presided over the evolution of the course from a circuit operation format to the modern circuit analysis emphasis. We are now retiring from active duty to make way for a new generation of technical instructors who may have different ideas about this all-important first course.

I am, therefore, happy to welcome the collaboration of Professor Preston White of the Southern College of Technology in Marietta, Georgia, in preparing the seventh edition. In addition to his many helpful suggestions on content, Preston is the author of the expanded learning aids available with this edition. I am sure that his knowledge and experience will add much to the usefulness of this current edition.

H.W. Jackson
Toronto, Canada.

A BRIEF MESSAGE FROM THE NEW CO-AUTHOR

Having taught electrical circuits courses for many years, I know personally how great a challenge it is to design and conduct a program that will help electrical/electronics engineering technology students be successful in dc and ac circuits studies. These courses are quite difficult for most students, but the principles are the cornerstone of all electrical/electronics technology. It is, therefore, most important to encourage students to develop a deep personal comprehension of this material. Student success in these early circuits courses is critical, since experience has shown that retention of promising students and their perseverance throughout higher-level studies depends on this early success.

This seventh edition is different from the previous edition in some important ways. The technical material has been updated and the emphasis has been changed somewhat to provide the appropriate foundation for the advanced courses found in modern electrical/electronics technology programs. The problems and questions have been re-worked but retain the practical flavor of the past editions. In order to make these chapter-end problems even more useful, they are now coded to point students to the section that contains the information upon which the problem is based. Because it is important to provide students with continuous feedback on their degree of comprehension, Self-Check problems have been added at the end of most sections. Students can test their own knowledge by comparing their solutions to the Self-Check problems, with the annotated solutions included at the back of the book. A chapter-by-chapter glossary is also included to help students build their technical vocabulary.

Besides the improvements to the text itself, a complete set of supplementary materials is now available. By using these materials both instructors and students can tailor the learning environment to meet their needs exactly. The supplementary materials include:

- **Laboratory Investigations Manual.** A collection of 29 laboratory exercises and five appendices. The investigations are designed to be easily adapted to any schedule. They provide practical, step-by-step reinforcement of the text studies and encourage students to refine their hands-on skills. Circuit design and computer literacy projects are included in each investigation.

- **Transparency Master Book.** More than a hundred figures, tables, and graphs taken from the text are reproduced in a format from which large transparencies can be made. Using these transparencies will enhance formal presentations and strongly establish the connection between classroom work and the text.

- **Solutions Manual.** The solution to each end-of-chapter problem is contained in this manual. Also included is the solution to the Preparation part of each of the 29 laboratory investigations.

■ **Software.** A diskette is available that contains BASIC programs from the text and programs that facilitate the solution of simultaneous equations.

It is with great pleasure that I join Herb Jackson in providing this set of educational materials designed to help the new electrical/electronics engineering technology student grasp the fundamental principles that guide the behavior of electrical circuits.

Preston A. White III
Marietta, Georgia

PART I   THE BASIC ELECTRIC CIRCUIT

ELECTRIC CIRCUIT THEORY consists of a few basic laws and a series of mathematical equations which state quite clearly the relationships that must hold true in all electric and electronic circuits. Consequently, much of an introductory electric circuits course deals with the solution of numerical examples. However, developing an understanding of why electric circuits must behave as they do involves much more than entering numbers into a computer program. To be able to apply basic electric circuit principles to more sophisticated electrical and electronic systems, we also need to form clear concepts of the physical basis for electric circuit theory.

The chapters of Part I are devoted to developing these basic concepts in a detailed, step-by-step manner. This material has been included primarily for students who do not have a sound knowledge of the principles of electricity from a prior course in physics.

1

INTRODUCTION

Our standard of living in today's technological society is very much dependent on an abundant, convenient, and economical supply of **energy** in its many forms and on the relative ease with which we can *convert* energy from one form to another. We need energy to heat (or cool) and light our homes, to drive our cars, to run the machinery of industry, and to operate the many household appliances which we take for granted.

Electricity is a form of energy which was no more than a scientific curiosity in the nineteenth century. But once human beings found a means of *converting* vast amounts of natural energy into an electrical form and then *converting* it at will into heat, light, mechanical, or other useful forms of energy, electricity became the very foundation of modern technology.

Even more significant is the ease and precision with which we can *control* electric energy. As a result, electric and electronic control systems have invaded every branch of engineering technology. To be able to work effectively with these systems, we must first investigate the basic principles which govern the behavior of electric circuits. Once we understand why electric circuits must behave as they do, we can then proceed to develop techniques for analyzing the performance of practical electric and electronic circuits as we may encounter them.

1-1 CIRCUIT DIAGRAMS

Two energy conversions take place in the basic **electric circuit** illustrated in Figure 1-1. Chemical energy stored in the battery is converted into electric energy. This energy is conveyed by a pair of copper electric **conductors** to the lamp, which, in turn, converts the electric energy into light (and some heat) energy. An electric **switch** in this circuit

FIGURE 1-1
Pictorial Representation of a
Basic Electric Circuit.

FIGURE 1-2
Schematic Diagram for a
Basic Electric Circuit.

allows us to start and stop the energy conversion at will by interrupting the transmission of electric energy from the energy **source** (battery) to the **load**[†] (lamp).

Since we can show the exact interconnection of the various components of an electric circuit much more clearly with a sketch than with a description in words, we shall follow the customary practice of drawing a **circuit diagram** for each electric circuit we discuss. As we progress to more elaborate circuits, the pictorial representation of Figure 1-1 would have serious limitations. To keep circuit diagrams as simple and as clear as possible, we represent the various circuit elements by standard graphic symbols rather than by pictorial sketches. We refer to a circuit diagram composed of such symbols as a **schematic diagram.** Although there have been some variations in graphic symbols in the past, those shown in Table 1-1 are the accepted standard in North America at present. As we encounter a symbol for the first time in a circuit diagram, we can refer back to Table 1-1 for its interpretation. Figure 1-2 shows the schematic diagram for the basic electric circuit of Figure 1-1.

1-2 THE INTERNATIONAL SYSTEM OF UNITS

In 1883, Lord Kelvin observed, "I often say that when you can measure what you are speaking about, and express it in numbers, you know something about it; but when you cannot measure it, when you cannot express it in numbers, your knowledge is of a meagre and unsatisfactory kind." We can appreciate the significance of this remark when we

---

[†]In talking about electric circuits, we use the general term **load** for any device that receives electric energy and converts it to some other form of energy. In digital electronics terminology, we encounter the term **sink** to describe a device that *drains* energy from an electric circuit.

TABLE 1-1

GRAPHIC SYMBOLS USED IN THIS TEXT (IEEE STD 315-1975)

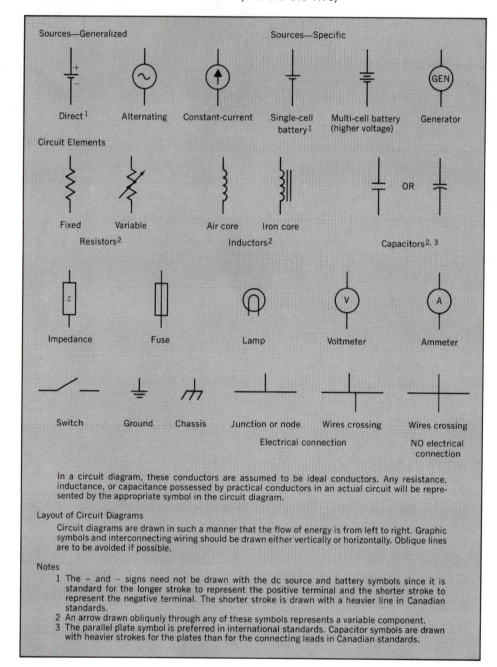

Sources—Generalized

Sources—Specific

Direct [1]   Alternating   Constant-current   Single-cell battery [1]   Multi-cell battery (higher voltage)   Generator

Circuit Elements

Fixed   Variable   Air core   Iron core

Resistors [2]   Inductors [2]   OR   Capacitors [2, 3]

Impedance   Fuse   Lamp   Voltmeter   Ammeter

Switch   Ground   Chassis   Junction or node   Wires crossing   Wires crossing

Electrical connection   NO electrical connection

In a circuit diagram, these conductors are assumed to be ideal conductors. Any resistance, inductance, or capacitance possessed by practical conductors in an actual circuit will be represented by the appropriate symbol in the circuit diagram.

Layout of Circuit Diagrams

Circuit diagrams are drawn in such a manner that the flow of energy is from left to right. Graphic symbols and interconnecting wiring should be drawn either vertically or horizontally. Oblique lines are to be avoided if possible.

Notes

1 The + and − signs need not be drawn with the dc source and battery symbols since it is standard for the longer stroke to represent the positive terminal and the shorter stroke to represent the negative terminal. The shorter stroke is drawn with a heavier line in Canadian standards.

2 An arrow drawn obliquely through any of these symbols represents a variable component.

3 The parallel plate symbol is preferred in international standards. Capacitor symbols are drawn with heavier strokes for the plates than for the connecting leads in Canadian standards.

realize that the scientists of Lord Kelvin's time had no knowledge of atomic structure and the free electron. To apply his philosophy to our study of electric circuits, we must have at our disposal a universally recognized system of measuring units.

The English-speaking world in Lord Kelvin's time still used an awkward and haphazard set of measuring units which included inches, feet, yards, rods, chains, fathoms, pints, bushels, teaspoons, cups, etc., ad infinitum. Most of the rest of the world had already switched to a simple decimal system of units which originated in France—the **metric** system. Almost a hundred years would elapse before North America finally began to adopt an updated version of the metric system for everyday measurement. However, in October 1965, the Institute of Electrical and Electronics Engineers adopted the **International System of Units** proposed at the eleventh General Conference on Weights and Measures held in France in 1960. The SI units[†] represent a rationalized selection of the MKS system of units which has been universally used by scientists for many years. Once the three basic MKS units—meter, kilogram, and second—are defined, we can *derive* all additional units that we might need in mechanics from these base units. By defining one additional base unit—the ampere—we can derive all the units required for electrical measurement as we proceed through this book.

The base unit of **length** or **distance** in the International System of Units is the **meter.** Originally calculated as one ten-millionth (1×10^{-7}) of the distance at sea level from the earth's equator to the pole, the meter is now more accurately defined in terms of the wavelength of the atomic radiation of krypton-86.

The original metric unit of mass—the *gram*—was related to the meter by defining the gram as the mass of one cubic centimeter of pure water at 4°C. The gram was too small a unit for practical purposes. Hence the SI base unit for **mass** is 1000 grams—the **kilogram.** The international standard for the kilogram is a platinum-iridium cylinder preserved by the International Bureau of Weights and Measures at Sévres in France.

The SI unit of **time** is the **second.** Originally defined as 1/86 400 of a mean solar day, we now define the second more accurately in terms of an "atomic clock" based on the atomic radiation of cesium. The second is not a true decimal unit, since the SI system still recognizes *minutes* and *hours* for longer periods of time.

The fourth base SI unit which will extend the system of units to electrical measurement is the **ampere.** In order to relate electrical measurement to mechanical measurement, the ampere is defined in terms of the electromagnetic force between two current-carrying electric conductors. We shall define the ampere when we first encounter it in Section 2-6, but we leave details of the international standard ampere until Section 13-3.

As we derive additional units from these base SI units, we shall want to show the relationship among the units by mathematical equations. Hence we need letter symbols to represent both the quantities being measured and the SI units of measurement. Table 1-2 shows the letter symbols for those base SI units that we use in electricity and electronics.

Occasionally, we may encounter an *abbreviation* of a unit name. For example, electricians often abbreviate *ampere* to *amp*. We do not use such abbreviations as substitutes for the designated **unit symbol** in equations.

[†]The abbreviation SI comes from the original French—Système Internationale d'Unités.

TABLE 1-2

LETTER SYMBOLS FOR SI BASE UNITS			
Quantity	Quantity Symbol	SI Base Unit	Unit Symbol
Length or distance	ℓ s	Meter	m
Mass	m	Kilogram	kg
Time	t	Second	s
Electric current	I	Ampere	A

Note the use of *italic* letters for quantity symbols and roman letters for unit symbols.

EXAMPLE 1-1

When the switch is closed in the basic electric circuit of Figure 1-1, the electric current through the lamp is 0.25 ampere. Express this statement as a mathematical equation.

SOLUTION

$$I = 0.25 \text{ A}$$

1-3 ELECTRONIC CALCULATORS

Following Lord Kelvin's advice, we find that numerical examples using SI units of measurement provide the best means of understanding the behavior of electric and electronic circuits. Although the mathematics required can be kept reasonably simple, the calculations themselves can often be tedious. The examples throughout this book can all be solved by "longhand" techniques. However, the speed and accuracy of a hand-held electronic calculator are well worth the modest investment. For most examples in an introductory electric circuits course, the simplicity and portability of a hand-held calculator are preferable to the sophistication of a personal computer. Hence, solutions of numerical examples in this book assume that students are using a suitable pocket calculator. From time to time, the solutions include hints on calculator techniques.

You should choose a "scientific notation" calculator, preferably with an Eng key which will give answers in powers of 10 corresponding to SI unit prefixes. It does not have to have expensive "programmable" features; but make sure that it can solve exponentials e^x and convert rectangular to polar coordinates → P, and vice versa → R. Some so-called "scientific" calculators do not have these functions.

1-4 NUMERICAL ACCURACY

At first glance, it might seem that an electric current of 12 amperes is exactly the same as a current of 12.0 amperes. However, there is a subtle but significant difference between 12.0 and 12. A current stated as 12 amperes means that the exact value is closer to 12 than it is to either 11 or 13. Possibly the ammeter is calibrated in 1-ampere intervals so that it can be read only to the closest whole number of amperes. A current of 12.0

amperes is closer to 12.0 than it is to 11.9 or to 12.1; consequently, it must have been possible to read the ammeter in this case to the nearest tenth of an ampere. We say that the figure 12 has two significant digits, whereas the figure 12.0 has three significant digits.

An electronic calculator assumes that any numerical quantity we enter is precise, no matter how many digits it contains. Hence, if we enter 34 ÷ 2.3, the calculator will show 14.782608. But if the original quantities are accurate to only two digits, the string of digits displayed by the calculator can only tell us that the answer is closer to 15 than it is to 14. In performing chain calculations with a calculator, we can leave the strings of digits contained in the calculator until the final answer. However, in writing down the *final* answer, we should avoid trying to show more significant digits in the answer than the least accurate of any figures representing measured quantities used in the calculations. If we have to write down the calculator results for an intermediate step in the examples in this book, it is satisfactory to record *four* significant digits.

SELF-CHECK

1-4:1. Suppose we measure the side of a square room carefully and write down the measurement as 7.06 m. Express the area of this room properly.
1-4:2. What is the circumference of a circle which has a radius of 18 mm? Be sure to express the answer to the proper number of significant figures.

1-5 SCIENTIFIC NOTATION

Using the SI *base* units, we will often encounter some very large quantities—such as a radio frequency of 455 000 hertz—and some very small quantities—such as an inductance of 0.000 75 henry. Scientists use powers of 10 to avoid having to write down long strings of zeros.

Given the frequency of a radio signal as 455 000 hertz, if we shift the decimal point three places to the left, it is equivalent to *dividing* the original quantity by 1000. We can restore the original value by then *multiplying* by 1000 or 10^3. Hence,

$$f = 455\,000 \text{ Hz} = 455 \times 1000 \text{ Hz} = 455 \times 10^3 \text{ Hz}$$

Similarly, if we shift the decimal point *five* places to the left, multiplying by 10^5 restores the original magnitude,

$$f = 455\,000 \text{ Hz} = \mathbf{4.55 \times 10^5 \text{ Hz}}$$

Given the inductance of a radio tuning circuit as 0.000 75 henry, if we shift the decimal point four places to the right, it is equivalent to *multiplying* by 10 000. Consequently, to restore the original value, we must *divide* 7.5 by 10 000 or *multiply* by 10^{-4}. Hence,

$$L = 0.000\,75 \text{ H} = \frac{7.5}{10\,000} \text{H} = \mathbf{7.5 \times 10^{-4} \text{ H}}$$

To express a quantity in **scientific notation,** we shift the decimal point until there is *one* significant digit to the left of the decimal point, and then we multiply the result by the appropriate power of 10 to restore the quantity to its original value.

EXAMPLE 1-2

Express 839 000 meters in scientific notation.

SOLUTION

To obtain one digit to the left of the decimal point, we must shift the decimal point *five* places to the left. To restore the quantity to its original value, we must then multiply by 10^5.

$$\therefore 839\,000 \text{ m} = \mathbf{8.39 \times 10^5} \text{ m}$$

EXAMPLE 1-3

Express 1/200 second in scientific notation.

SOLUTION

In a longhand solution, the first step is to divide the fraction out into decimal form.

$$\frac{1}{200} \text{ s} = 0.005 \text{ s}$$

We now shift the decimal point three places to the right and multiply the result by 10^{-3}, obtaining

$$\frac{1}{200} \text{ s} = \mathbf{5.0 \times 10^{-3}} \text{ s}$$

Most scientific-notation calculators automatically combine these two steps.

We can illustrate the advantage of scientific notation for calculator computation by the following example based on the equation for inductive reactance (which we shall derive in a later chapter).

EXAMPLE 1-4

Given $X_L = 2\pi f L$, $f = 455\,000$ Hz, and $L = 0.000\,75$ H, calculate X_L.

SOLUTION

Substituting the numerical values in the equation gives

$$X_L = 2 \times 3.1416 \times 455\,000 \times 0.000\,75 \text{ } \Omega \text{ (ohms)} \tag{1-1}$$

If we enter the quantities in this form into a calculator, we may have trouble keeping track of the number of zeros. However, if we express f and L in scientific notation, we have

$$X_L = 2 \times \pi \times 4.55 \times 10^5 \times 7.5 \times 10^{-4} \ \Omega \qquad (1\text{-}2)$$

Using the scientific-notation feature of the calculator, we enter this equation using the EE key[†] as follows:

$$X_L = 2 \ \boxed{\times} \ \boxed{\pi} \ \boxed{\times} \ 4.55 \ \boxed{EE} \ 5 \ \boxed{\times} \ 7.5 \ \boxed{EE} \ \boxed{+/-} \ 4 \ = \ \mathbf{2144} \ \Omega \qquad (1\text{-}3)$$

Even in a longhand solution, scientific notation speeds up the computation since, to *multiply* powers of 10, we simply *add* their indices algebraically. Hence, Equation (1-2) becomes

$$X_L = 6.2832 \times 4.55 \times 7.5 \times 10 = \mathbf{2144} \ \Omega \qquad (1\text{-}4)$$

SELF-CHECK

1-5:1. Given the equations:

$$X_C = 1/[2 \cdot \pi \cdot f \cdot C]$$
$$\text{and } Z = \sqrt{R^2 + X_C^2}$$

and given the measurements:

$$f = 5 \ 100 \ \text{Hz},$$
$$C = 0.000 \ 000 \ 000 \ 27 \ \text{F}$$
$$\text{and } R = 180 \ 000 \ \Omega$$

(a) Express f, C, and R in scientific notation.
(b) Use your calculator to find X_C. Express X_C in scientific notation using the proper number of significant figures.
(c) Use your calculator to find Z. Express Z in scientific notation using the proper number of significant figures.

1-5:2. Given the equation:

$$f = [2\pi\sqrt{L \times C}]^{-1}$$

and given the measurements:

$$L = 0.159 \ \text{H}$$
$$\text{and } C = 0.000 \ 047 \ \mu\text{F}$$

(a) Express L and C in scientific notation using the proper number of significant figures.
(b) Express f in scientific notation using the proper number of significant figures.

[†] EE key = Enter Exponent. Note that the calculator's permanent memory includes the numerical value for π so that the operator need not remember that $\pi = 3.1416$, as in the longhand solution (or with most personal computer programs).

1-6 SI UNIT PREFIXES

Engineering personnel prefer to use a variation of scientific notation to express numerical quantities clearly. We prefer to tack an appropriate *prefix* onto the *root* unit‡ of measurement to represent the equivalent power of 10. To create decimal multiples or submultiples of root units, SI provides the standard unit prefixes listed in Table 1-3. For engineering purposes, we prefer to shift the decimal point in multiples of *three*. Hence, the prefixes *hecto, deka, deci,* and *centi* are seldom used in working with electric circuits. The other unit prefixes shown in lightface type in Table 1-3 are used with quantities much larger or much smaller than we shall encounter in this book. We sometimes refer to quantities expressed with SI unit prefixes as **engineering notation.**

TABLE 1-3

SI UNIT PREFIXES		
Prefix	Factor by Which the Root Unit Is Multiplied	Letter Symbol
exa	$1\,000\,000\,000\,000\,000\,000 = 10^{18}$	E
peta	$1\,000\,000\,000\,000\,000 = 10^{15}$	P
tera	$1\,000\,000\,000\,000 = 10^{12}$	T
giga	$1\,000\,000\,000 = 10^{9}$	G
mega	$1\,000\,000 = 10^{6}$	M
kilo	$1\,000 = 10^{3}$	k
hecto	$100 = 10^{2}$	h
deka	10	da
deci	$0.1 = 10^{-1}$	d
centi	$0.01 = 10^{-2}$	c
milli	$0.001 = 10^{-3}$	m
micro	$0.000\,001 = 10^{-6}$	μ*
nano	$0.000\,000\,001 = 10^{-9}$	n
pico	$0.000\,000\,000\,001 = 10^{-12}$	p
femto	$0.000\,000\,000\,000\,001 = 10^{-15}$	f
atto	$0.000\,000\,000\,000\,000\,001 = 10^{-18}$	a

*The symbol for micro is the Greek letter μ (mu).

Shifting the decimal point three places at a time means that some numerical values will have more than one digit to the left of the decimal point. Consequently, when using prefixes with root units, SI recommends that we select prefixes that will permit the numerical values of the quantities to be between 0.1 and 1000.

⌐ *EXAMPLE 1-2A*
Express 839 000 meters in a more convenient size of unit.

‡*Root* units have no prefixes. In most cases in SI, root units are also base units. The exception is the base unit for mass—the kilogram—which historically already has the prefix *kilo* and is, therefore, not a root unit.

SOLUTION

Shifting the decimal point *three* places to the left gives a figure of 839, which we must now multiply by 1000 or 10^3 to restore its proper magnitude. Hence,

$$839\,000 \text{ m} = 839 \times 10^3 \text{ m}$$

From Table 1-3, we can replace 10^3 by tacking the prefix *kilo* onto the root unit.

$$\therefore 839\,000 \text{ m} = 839 \times 10^3 \text{ m} = \mathbf{839 \text{ km}} \text{ (kilometers)}$$

The kilometer is 1000 times as large as the base unit, the meter.

EXAMPLE 1-3A

Express 1/200 second in a more appropriate form using SI unit prefixes.

SOLUTION

Again the first step is to divide the fraction out into decimal form.

$$\frac{1}{200} \text{ s} = 0.005 \text{ s}$$

If we shift the decimal point *three* places to the right to obtain the figure 5.0, we must multiply by 0.001 or 10^{-3} to restore the original magnitude. From Table 1-3,

$$\frac{1}{200} \text{ s} = 5.0 \times 10^{-3} \text{ s} = \mathbf{5.0 \text{ ms}} \text{ (milliseconds)}$$

When we make calculations involving quantities expressed in units with prefixes, we must remember that the prefixes represent powers of 10. These powers of 10 *must* be included in the numerical computation for the proper magnitude to be maintained in the answer. We can demonstrate this point with the following example. Using unit prefixes, Equation (1-2) in Example 1-4 becomes

$$X_L = 2\pi \times 455 \text{ kHz (kilohertz)} \times 750 \text{ } \mu\text{H (microhenrys)} \qquad \text{(1-5)}$$

EXAMPLE 1-4A

Given $X_L = 2\pi f L$, $f = 455$ kHz and $L = 750$ μH, calculate X_L.

INCORRECT SOLUTION

If we forget the prefixes and just carry out the arithmetic in Equation (1-5), we get

$$X_L = 2 \times \pi \times 455 \times 750 = 2\,144\,000$$

Comparing with the answer to Example 1-4, we note that the answer is *not* in ohms. (To be numerically correct, we must write the symbol for milliohms with the answer above.)

CORRECT SOLUTION

To make sure that we know that the numerical answer is in base units, we include the unit prefixes in the calculation by pressing the calculator's keys as follows:

$$X_L = 2 \times \boxed{\pi} \times 455 \ \boxed{EE} \ 3 \times 750 \ \boxed{EE} \ \boxed{+/-} \ 6 \ \boxed{=} \ 2144 \ \Omega$$

$$\underset{\text{kilo}}{\uparrow} \qquad \underset{\text{micro}}{\underbrace{}} \qquad \underset{\text{root unit}}{\uparrow}$$

Since it is not good form in SI to leave a numerical value greater than 1000, we now make the conversion

$$2144 \ \Omega = 2.144 \times 10^3 \ \Omega = \textbf{2.144 k}\boldsymbol{\Omega} \textbf{ (kilohms)}$$

EXAMPLE 1-5

The calculator display shows a readout of 7.5×10^{-4} henry. Express this answer in a more suitable dimension.

SOLUTION

Table 1-3 has no unit prefix for 10^{-4}. Hence, we must start by simultaneously shifting the decimal point and changing the power of 10 (exponent) until we reach an exponent which is a + or − multiple of 3 and a numerical quantity (mantissa) between 0.1 and 1000. In this particular example, we can shift in either direction. We can *lower* the exponent from −4 to −6 and compensate by shifting the decimal point two places to the *right*, or we can *raise* the exponent from −4 to −3 and shift the decimal point of the mantissa one place to the *left*. Hence,

$$7.5 \times 10^{-4} \ \text{H} = 750 \times 10^{-6} \ \text{H} = 0.75 \times 10^{-3} \ \text{H}$$

from which

$$7.5 \times 10^{-4} \ \text{H} = \textbf{750 } \boldsymbol{\mu}\textbf{H or 0.75 mH}$$

Engineering calculators have an $\boxed{\text{Eng}}$ key which automatically converts a scientific notation answer to display an exponent which is a + or − multiple of 3 and a mantissa between 0.1 and 1000. Other scientific calculators have $\boxed{\text{EE}\uparrow}$ and $\boxed{\text{EE}\downarrow}$ keys so that we can shift both the decimal point and the exponent simultaneously in the proper direction one place at a time. In this example we can press the $\boxed{\text{EE}\downarrow}$ key *twice* to lower the exponent from −4 to −6, or we can press the $\boxed{\text{EE}\uparrow}$ key once to raise the exponent from −4 to −3.

SELF-CHECK

1-6:1. Express the following measurements:

$$f = 5 \ 100 \ \text{Hz},$$
$$C = 0.000 \ 000 \ 000 \ 27 \ \text{F},$$

$$R = 1\ 800\ 000\ \Omega$$
$$\text{and } L = 0.159\ \text{H}$$

in engineering notation.

1-6:2. Express the measurements in Problem 1-6:1 using the proper SI prefix.

1-7 CONVERSION OF UNITS

When only a single conversion from a root unit to a prefixed unit is involved, as in the last step of Example 1-4A, we can usually convert the calculator readout mentally from root units to more appropriate multiple or submultiple units. But some chain calculations may require several conversions before the final numerical answer is displayed. To assist us in determining the correct unit for the final answer, we can resort to an orderly procedure based on **dimensional analysis.** With this technique, we write the units into an algebraic equation along with the numerical quantities. We cancel out the same units in the numerator and denominator and then collect the remaining units to establish the proper units (or *dimension*) of the answer.

EXAMPLE 1-6
Convert $\frac{1}{4}$ hour to seconds.

SOLUTION
We require two conversion factors—hours to minutes and minutes to seconds. We can write the first as

$$60 \text{ minutes} = 1 \text{ hour}$$

We can rearrange the form of this conversion factor by dividing both sides of the equation by 1 hour:

$$\frac{60 \text{ min}}{1 \text{ h}} = \frac{1 \cancel{h}}{1 \cancel{h}} = 1$$

Thus when we express a conversion factor as a ratio, we obtain a dimensionless number equal to unity. We can, therefore, multiply any quantity by 60 min/1 h, or its reciprocal, without altering the value of the original quantity. Hence,

$$\frac{1}{4}\text{ h} = \frac{1}{4} \cancel{h} \times \frac{60 \cancel{\text{min}}}{1 \cancel{h}} \times \frac{60 \text{ s}}{1 \cancel{\text{min}}} = \mathbf{900}\ \boldsymbol{s}$$

We can also use the dimensional analysis technique to convert from one *system* of units into another. Since electrical engineering in North America has already adopted the International System of Units (SI), all calculations in this book will be made in SI units. But for some time to come, we shall encounter quantities in the now obsolescent foot-pound system of units. To make the necessary conversion into SI units for examples in this book, we can probably get by with the two conversion factors of Table 1-4. The

TABLE 1-4

CONVERSION FACTORS
1 inch = 2.54 centimeters
1 pound = 0.4536 kilogram

conversion from inches to centimeters is exact. The conversion from pounds to kilograms is an approximation accurate to four digits.

EXAMPLE 1-7

An electric conductor is 25 ft long. Express its length in SI units.

SOLUTION

$$25 \text{ ft} = 25 \text{ ft} \times \frac{12 \text{ in}}{1 \text{ ft}} \times \frac{2.54 \text{ cm}}{1 \text{ in}} \times \frac{1 \text{ m}}{10^2 \text{ cm}} = \textbf{7.62 m}$$

SELF-CHECK

1-7:1. Given that 1.0 mile = 1.6 km, express the speed 55 mi/h in m/s (meters per second).
1-7:2. Given that 1.0 mile = 1.6 km and 3.785 liters = 1 gallon, express 10.0 km/l in mi/gal.

1-8 PERSONAL COMPUTERS

At the beginning of the 1970s, educational computers were generally limited to costly "mainframe" computers located in college computing centers. A few students in advanced electrical/electronics technology courses had access to such terminals, but computers were not usually available to students in introductory electric circuits courses. By the beginning of the 1980s, rapid advances in VLSI[†] technology made it possible to place thousands of electronic circuits on a silicon chip no larger than a postage stamp. Mass production of **microprocessor** chips, in turn, made it possible to manufacture **micro-computers** with considerable memory and data processing capacity in a package the size of a portable typewriter and at a cost of only a few hundred dollars.

The relatively low cost of the current generation of microcomputers makes it possible for many students to own their own **personal computer.** For the same reason, electrical/electronics technology departments can now provide all students with access to personal computers to assist in electric circuit computations. Since many college students have learned to use such computers in the public school system, we must now consider what place personal computers have in the study of basic electric circuit theory.

The main difference between an electronic calculator and a personal computer is that the computer has a built-in *main memory* that can store many thousands of bits of data and processing instructions. Since each bit of data goes to a specific *address* in the

[†]Very-large-scale integration.

computer's *random-access memory* (RAM), the *control unit* of the computer can locate and process any particular data in microseconds. Most calculators have only a limited memory for temporarily storing intermediate steps in a long chain calculation. Scientific calculators have considerable mathematical data built into a *read-only memory* (ROM). Such data are retained even when the calculator is switched off. They include such data as a numerical value for the constant π, trigonometry and exponential tables, and instructions to make the calculator carry out the appropriate computation when function keys such as ⊞ , ⊡ , and x^2 are pressed.

When we use an electronic calculator, the actual sequence of numerical computations takes place directly as we press each mathematical function key in the proper order, and we see the result of each calculation almost instantly on the display. This direct involvement in the calculation chain is desirable in studying basic electric circuits, since it requires us to think our way through the steps in setting up each numerical example.

Personal computers, on the other hand, use their main memories to store temporarily a *program* for carrying out a series of computations automatically in a predetermined sequence. If the calculation is fairly simple, we usually type the program directly into the computer's input keyboard. For a complex computation or for a special display format, we *load* a prepared program into the computer's main memory from a *software* package consisting of a *permanent memory* file recorded on a magnetic tape cassette or flexible magnetic diskette. The advantage of the personal computer in electric circuit computation is that once a program has been prepared properly and checked, errors are limited to *inputting* incorrect data in response to the computer's requests. The disadvantage is that our attention is focused on the computer's requests for data input, rather than on the electric circuit principles being illustrated by the numerical example. Hence, while we are learning basic electric circuit principles, we should stick to hand-held calculators for simple numerical examples and save the personal computer for more tedious, repetitive calculations.

When a computer is switched off, it loses all the data temporarily stored in its main memory. When it is next switched on, the computer has to reload its main memory with sufficient data and instructions to recognize what we enter via the keyboard and how we want the data to be processed and displayed. For the calculations we encounter in an electric circuits course, personal computers are loaded with one of the variations of BASIC[†] computer language. In spite of the many similarities in the increasing selection of personal computers, each model has its own peculiarities and variations in program format. Hence, it is not feasible for an electric circuits textbook to provide specific computer programming instruction. To duplicate the computer examples on subsequent pages on your computer, you need to be quite familiar with BASIC programming for your particular model. You may also have to modify some of the *keywords* and program formats in the examples.

The computer examples in this book are intended to illustrate the differences between calculator and computer routines for solving various electric circuit examples. In this way, we not only gain some familiarity with personal computers, but also develop some appreciation of where the computer can be more effective than an electronic calculator, and vice versa. We start by considering how we might use a personal computer to solve Example 1-4.

[†]Beginner's All-Purpose Symbolic Instruction Code.

EXAMPLE 1-4B

Given $X_L = 2\pi fL$, $f = 455$ kHz, and $L = 750$ μH, calculate X_L.

SOLUTION 1

With BASIC loaded into the computer, we can use the computer as a simple electronic calculator by omitting the *line numbers* that appear at the beginning of each *statement* on the display. We type into the keyboard

```
PRINT 2*3.1416*455E3*750E-6
```

This statement covers all the numerical data and mathematical operators required to solve the example. We now press the ENTER key and the computer promptly displays **2144.142.**

Each statement starts with a *keyword* which the computer recognizes from a selected BASIC vocabulary. Since some computer keyboards do not have an ☒ key (like calculators), BASIC uses an asterisk * to represent multiplication. For the same reason, BASIC uses a solidus slash / to represent division. Note that BASIC does not provide a numerical value for the constant π.

SOLUTION 2

The *direct-mode* operation of the computer in Solution 1 is an inefficient use of its capabilities. To use the computer's memory, we need a *program* to tell the computer what to do, step by step. The minimum program requires three steps (plus a statement to indicate to the computer that the program is complete). The three steps are (1) input of data, (2) processing the data, and (3) output of results. This minimum program is displayed as follows:

```
10 INPUT F,L
20 LET X=2*3.1416*F*L
30 PRINT X
40 END
```

To start this program, we type RUN and press the ENTER key. The computer displays a ? asking us to enter values for F and L. Once we type and enter the given values, the computer will promptly print the answer **2144.142.**

SOLUTION 3

In practice, we would not write such a "bare-bones" program as Solution 2. With nothing but ? showing on the display, we may not remember what we are supposed to input. A more typical program might take the following format:

```
10 REM CALCULATING INDUCTIVE REACTANCE
20 INPUT "ENTER F,L ";F,L
30 LET X=2*3.1416*F*L
40 PRINT "INDUCTIVE REACTANCE= ";X;" OHMS"
50 END
```

When we RUN this program, it starts off by displaying the identifying caption for the program—CALCULATING INDUCTIVE REACTANCE. The next line displays ENTER F,L followed by a ? symbol. The operator responds by typing in the variables in the order requested, 455E3, 750E-6. The computer will display the answer for X, but it is now part of the statement in quotation marks in line 40:

```
INDUCTIVE REACTANCE=2144.142 OHMS
```

SELF-CHECK

1-8:1. Referring to the equations and values in self-check problem 1-5:1, write a BASIC program that will calculate and print the value of Z.

1-8:2. Referring to the equations and values in self-check problem 1-5:2, write a BASIC program that will calculate and print the value for f.

PROBLEMS

The first two parts of Problem numbers identify the chapter and section illustrated by the problem.

1-2-1. Express the following statements as mathematical equations using standard quantity and unit symbols.
(a) The copper conductor between the lamp and the battery in Figure 1-1 is 0.6 meters long.
(b) The battery in Figure 1-1 has a mass of 0.45 kilograms.

1-2-2. Write the following statements in equation form.
(a) The velocity (v) of a moving object is the distance it travels in unit time.
(b) The kinetic energy (W) of a moving object is one half the product of its mass and the square of its velocity.

1-3-3. Perform the following calculations using an electronic calculator.
(a) $0.005 \times 47\,000 - 958 =$
(b) $0.005(47\,000 - 958) =$
(c) $54\,000 - 75 \times 634 =$
(d) $0.0005^4 =$
(e) $\dfrac{423\,000 + 17\,400}{560 - 0.04} =$
(f) $\sqrt{125^2 + 4 \times 5600} =$

1-4-4. Assuming that all numerical quantities represent measured quantities, express the solutions to the following computations with the appropriate number of significant digits.
(a) $2.34 + 110.7 + 4.569 =$
(b) $180 - 51.74 + 9.3 =$
(c) $456 \times 0.0050 =$
(d) $1225 \div 33 =$
(e) $59.4 + 17 \times 6.28 =$
(f) $\dfrac{59.4 - 17}{6.28} =$

1-5-5. Express each of the following quantities in scientific notation.
(a) 455   (b) 59 000

(c) 10 000   (d) 0.0765

(e) 0.000 37   (f) 0.547

1-5-6. Express each of the following scientific-notation quantities in conventional form.

    (a) 1.2×10^6 (b) 2.37×10^2

    (c) 6.28×10 (d) 5.45×10^{-3}

    (e) 7.7×10^{-6} (f) 8.432×10^{-1}

1-5-7. The distance between two towns is 25 km (kilometers). Express this in meters, using scientific notation.

1-5-8. A coulomb is an electrical unit of quantity representing

$$6\,240\,000\,000\,000\,000\,000 \text{ electrons}$$

Express this in scientific notation.

1-5-9. The mass of an electron is

$$0.000\,000\,000\,000\,000\,000\,000\,000\,000\,899\,9 \text{ g (grams)}$$

Express this in scientific notation.

1-5-10. The effective diameter of an electron is

$$0.000\,000\,000\,000\,2 \text{ cm (centimeters)}$$

Express this in scientific notation.

1-5-11. Perform the following calculations using an electronic calculator.

    (a) $4.2 \times 10^4 + 3.7 \times 10^3 =$ (d) $5.916 \times 10^{-4} \div 6.47 \times 10^5 =$

    (b) $7.76 \times 10^{-3} - 8.43 \times 10^{-4} =$ (e) $\sqrt{2.46 \times 10^5} =$

    (c) $4.32 \times 10^2 \times 5.4 \times 10^{-3} =$

1-5-12. Repeat the calculations of Problem 1-3-3 using the EE key of an electronic calculator after converting all numerical values to scientific notation.

1-6-13. Express each of the following quantities in the preferred engineering notation.

    (a) 47 000 ohms   (b) 45 000 000 hertz

    (c) 1500 watts   (d) 0.0505 second

    (e) 0.0005 volt   (f) 0.000 000 000 39 farad

1-6-14. Convert the following quantities into base units, first in scientific notation and then in conventional notation.

    (a) 3.3 megohms   (b) 35 millimeters

    (c) 0.746 kilowatt   (d) 56 microvolts

    (e) 56 kilometers   (f) 56 kilograms

1-6-15. Express 1650 g in kilograms.

1-6-16. Express 850 mm (millimeters) in meters.

1-6-17. Express 1/60 s in milliseconds.

1-6-18. Express 0.000 067 km in millimeters.

1-6-19. Express 2.2×10^{-2} kg in grams.

1-6-20. Express 0.04 m^2 (square meters) in square centimeters.

1-6-21. Express the cross-sectional area of an electric conductor whose diameter is 4.0 cm in square millimeters.

1-6-22. Express 4.08×10^7 cm^3 (cubic centimeters) in cubic meters.

1-7-23. Solve Example 1-4A using dimensional-analysis techniques.

1-7-24. Derive the conversion factor for miles into kilometers.

1-7-25. Derive the conversion factor for square feet into square meters.

1-7-26. Derive the conversion factor for ounces into grams.

1-7-27. Express a velocity of 50 mi/h (miles per hour) in meters per second.

1-7-28. Express a velocity of 50 ft/s in kilometers per hour.

1-7-29. The specifications for an import automobile give its weight as 1.2 metric tons (megagrams). State its weight in pounds.

1-7-30. The No. 14 AWG (American wire gauge) conductors used in house wiring have a diameter of 64.08 mils. (A mil is one-thousandth of an inch.) What is the cross-sectional area of a No. 14 AWG conductor in square millimeters?

REVIEW QUESTIONS

1-1. What is meant by the term **energy?**

1-2. What is the significance of the **law of conservation of energy?**

1-3. Trace all the energy conversions as we use electricity from a coal-fired generating station to operate a home air conditioner.

1-4. A flashlight is one of the simplest complete electric circuits. It consists of a battery, copper conductors, a switch, and a small lamp. Using standard graphic symbols, draw a schematic diagram for a flashlight.

1-5. Using the flashlight as an example, explain the **law of conservation of energy.**

1-6. Describe a device for converting heat energy into mechanical energy.

1-7. What do the letters **MKS** represent?

1-8. What advantage has a system of units, most of which can be derived from a few basic units?

1-9. What is meant by **scientific notation** in expressing a numerical quantity? What is the advantage of expressing numerical quantities in this manner?

1-10. How does the scientific method of expressing numerical quantities differ from using SI unit prefixes?

1-11. Why is it important that we write down the proper units when we record a calculator solution to a numerical example?

1-12. In the MKS system of units, the wavelength of light was expressed in **millimicrons.** What would be the preferred terminology for the equivalent SI unit?

1-13. What is the meaning of the term **dimensional analysis?**

1-14. How would you apply dimensional analysis when you are doing a chain calculation with your electronic calculator?

1-15. One butcher shop advertises sirloin steak at $3.59 a pound; another sells sirloin steak at $7.59 a kilogram. Assuming equal quality, which is the better price?

1-16. In SI, automobile gasoline consumption is expressed in liters per hundred kilometers. How would you convert miles per gallon to liters per 100 km?

*1-17. Write a BASIC program to accomplish the conversion in Question 1-16.

*1-18. Write a BASIC program to convert Fahrenheit temperature readings into degrees Celsius.

*For students who have access to a personal computer.

2

CURRENT AND VOLTAGE

2-1 THE NATURE OF ELECTRICITY

To understand the behavior of electric circuits, we must know something of the nature of the electrical property of matter. We are probably more familiar with a different property of matter known as its **mass.** This mass property of matter leads to the branch of physical science called **mechanics** and is the basis for mechanical energy in both static and kinetic forms. Although we may not know just why it should exist, we are certainly aware of the force of attraction between two masses. It is this **gravitational force** between the mass of the earth and the mass of objects at the surface of the earth which is responsible for what we call the **weight** of these objects.

To push open a door, we must apply sufficient mechanical force to overcome the mechanical force tending to keep the door closed. This opposing force might be the tension of the spring in the door-closer, or it might be just the friction of the door hinges. The force we apply with our hand in this example is a *contact* force, since it does not act on the door until our hand touches the door. Gravitational force, on the other hand, is not a contact force. Gravitational force can act at a considerable distance, as for example the gravitational force between the mass of the moon and the mass of the earth, which governs the orbit of the moon around the earth. A force, such as gravitational force, which can act at a considerable distance without contact is called a **field force.**

If we carry a heavy parcel against the force of gravity up several flights of stairs, we soon become aware that we are expending a considerable amount of energy. This conversion of the chemical energy stored in the tissue of our body to mechanical energy, which, in turn, moves a mass against the force of gravity is called **work.**

Work is the accomplishment of motion against the action of a force which tends to oppose the motion.

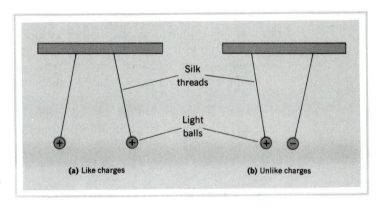

FIGURE 2-1
Like Charges Repel, Unlike Charges Attract.

(a) Like charges   (b) Unlike charges

Because the energy needed to perform this work was present before we climbed the stairs,

Energy is the capacity to do work.

One further basic concept which we must note before we investigate the electrical property of matter is the principle of conservation of energy, which states that

In any closed system, the total energy remains constant.

Energy can be transformed from one form into another, but the total energy after the transformation must exactly equal the total energy going into the transformation. Consequently, **energy** and **work** are numerically the same and we may use the same unit of measurement for both.

In searching for the electrical property of matter which will permit useful work to be accomplished in the basic electric circuit of Figure 1-1, we must look for an electric force and for motion of a particle under the influence of this force. Atoms of matter in their normal state are electrically neutral. Consequently, we do not observe the effect of electric force as readily as we observe the effect of gravitational force.

Some 2500 years ago, Thales of Miletus was one of the first to discover a means of disturbing the normal electrical balance of matter. He noted that when he rubbed a piece of amber,[†] it acquired an ability to attract light pieces of straw and dust. However, it was not until the eighteenth century that serious experiments were conducted in an effort to learn the nature of this force which attracted small particles of straw and dust.

We can duplicate Thales' experiment by rubbing a glass rod with a silk cloth, or an ebonite rod with a piece of cat's fur. Since these rods can then attract bits of paper against the force of gravity, work is being accomplished and energy is being expended. Rubbing the rods must have provided them with some form of energy which they did not reveal in their normal state. We say that we "placed" an electric **charge** on the rod, using the term *charge* to imply potential energy in much the same sense that we speak of a charge of dynamite placed in a drill hole for a blasting operation.

We can investigate the effect of the force produced by rubbing glass or ebonite rods by means of lightweight balls suspended by silk threads, as shown in Figure 2-1. We discover that we can transfer some of the charge from the glass or ebonite rods to the balls

---

[†]Our word *electron* comes from *elektron*, the Greek word for amber.

by touching them with a charged rod. If we touch a charged glass rod to both balls, we discover that they tend to *repel* one another, as in Figure 2-1(a). However, if we touch a charged glass rod to the left ball and a charged ebonite rod to the right ball, we find that they tend to *attract* one another, as in Figure 2-1(b).

From these observations, we can state that the force between the charged balls is a *field* force, since the balls are not in contact. But the force is *not* gravitational force, since gravitational force is always a force of attraction, never repulsion. Second, the charge on the ebonite rod must be different from that on the glass rod, since in one example we produce a force of repulsion, and in the other a force of attraction. Since both balls in Figure 2-1(a) were charged from the same glass rod, the two balls must possess **like charges.** With no knowledge of the exact nature of these charges, the eighteenth-century scientists could only label the charges in Figure 2-1(b) as **unlike charges,** and state that

Like charges repel, unlike charges attract.

We now distinguish the two different types of electric charge by the terms positive charge and negative charge, a designation first introduced by Benjamin Franklin.

Using a more precise version of the apparatus in Figure 2-1, a French physicist, Charles Augustin Coulomb, was able to show that the force between two electrically charged bodies is directly proportional to the product of the magnitudes of the two charges and inversely proportional to the square of the distance between them. We can state this relationship more clearly by writing a mathematical equation in which we represent the various quantities in the statement above by standard letter symbols. Hence, Coulomb's "law" becomes

$$F = k \frac{Q_1 Q_2}{s^2} \tag{2-1}$$

where F is the force between the two electrically charged bodies, Q_1 and Q_2 are the respective electric charges, s is the distance between the charged bodies, and k is a numerical constant which makes the equation balance with the units used for the other quantities. We shall leave consideration of suitable units until Chapter 11.

The property of matter which allows objects to become electrically charged remained a mystery for almost two centuries until Lord Rutherford's experiments with radiation led him to the now familiar concept of the structure of an atom consisting of a **nucleus** around which **electrons** revolve in a manner similar to that in which the planets of our solar system follow orbits around the sun.

From a study of chemistry, we know that all substances can be formed by chemical combination from about one hundred known **elements,** each possessing its own distinctive properties. More than 2000 years ago, Democritus of Thrace proposed that, if we keep dividing a sample of an element into smaller and smaller pieces, we reach a limit beyond which it is impossible for us to continue to divide without losing the characteristic properties of the element. Democritus called this indivisible sample of an element an **atom.**

Rutherford's model of an atom consists of subatomic particles which are held together by *electric* force. These particles are the same for all elements. The atoms of one element differ from those of another only in terms of the number and arrangement of subatomic particles in the model. If we can determine the nature of these particles and the forces acting on them within an atom, we shall then know something about the electrical property of matter.

2-2 THE ATOM

We shall soon discover that the behavior of electric circuits is quite dependent on the behavior of **electrons,** the subatomic particles which travel in planetary orbits around the **nucleus** of an atom. From the orbital motion of electrons within an atom, we can deduce several characteristics of an electron: (1) it must possess energy; (2) it appears to possess a mass which is a small fraction of the mass of the nucleus around which it travels; and (3) there must be a force of attraction between the electron and the nucleus for it to travel in a curved orbit.

Unlike the force which governs the orbits of the planets of the solar system, the force between electron and nucleus is not a gravitational attraction based on mass, since we can show that electrons *repel* one another. The force of attraction between orbital electrons and the nucleus of an atom is an electric force, similar to that illustrated in Figure 2-1(b). Since atomic theory will not permit subdivision of an electron, it follows that every electron must possess as part of its nature a unit of electric charge. The charge on every electron is identical.

For an electric force of attraction to exist between electrons and nucleus, the nucleus must contain particles which possess an *opposite* or *unlike* charge. These particles are called **protons.** Using Franklin's nomenclature to identify the opposite charges of electrons and protons, we find that electrons possess a *negative* electric charge and protons have a *positive* electric charge. The charge on a proton is equal in magnitude to that of an electron. The nucleus of an atom contains as many protons as there are orbital electrons in the atom in its normal state. Consequently, the net charge of a complete atom in its normal state is zero and the atom is electrically neutral. However, protons have a mass that is 1837 times the apparent mass of an electron.

The electrons of every atom are identical. The atoms of one element differ from those of another in terms of the number of orbital electrons and the number of protons in the nucleus. From experimental evidence, it appears that the lightest element, hydrogen, has one electron, as shown in Figure 2-2(a). The next lightest is helium which has

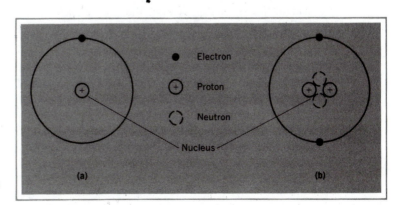

FIGURE 2-2
Atomic Models: (a) Hydrogen Atom; (b) Helium Atom.

two electrons, as shown in Figure 2-2(b). If we arrange all the elements in order of increasing mass, each element has one more electron than the one immediately preceding it in the list. For example, uranium is the ninety-second element in the list and the uranium atom has 92 electrons and 92 protons. To account for the atomic weights of the various elements in this list, we find that all nuclei except hydrogen must possess particles that are slightly heavier than protons. These particles are neither attracted nor repelled by electrons or protons. They are electrically neutral and are called **neutrons.** Recent "atom smashing" research has indicated the existence of several other short-lived particles in the nuclei of heavier atoms. These particles have no particular significance in studying the electrical behavior of atoms.

Since a force of attraction exists between an orbital electron and a nucleus, in order for the electron to maintain a given orbit, it must move at a velocity around the orbit which will develop the centrifugal force required to exactly balance this attraction. And since an electron has a certain apparent mass, the moving electron must possess a kinetic energy proportional to its mass and the square of its average velocity ($W = \frac{1}{2}mv^2$). Therefore, to occupy a certain orbit, an electron must possess a definite amount of energy. Hence, we can discuss electron orbits in terms of the **energy levels** of the electrons in them. To increase the radius of an electron orbit, the electron must acquire additional *potential* energy to move it against the force of attraction between the electron and the nucleus. It follows that the lowest energy level in an atom is the electron orbit closest to the nucleus.

Electrons sometimes behave as if they were electromagnetic waves rather than particles. To behave in this manner, the orbits along which the electrons move must be an integral number of wavelengths. For a given wavelength, only certain radii will permit circumferences that fulfill this condition. Therefore, the electrons of an atom can appear only at certain definite energy levels.[†] The spacing between energy levels is such that, when we catalog the chemical properties of the various elements, it is convenient to group several closely spaced *permissible* energy levels together into **electron shells** and **subshells,** as indicated in Figure 2-3. The gap in energy level between one shell and the next is much greater than the gaps between energy levels within a shell.

The maximum number of electrons which each subshell and shell or energy level can contain is the same for all atoms. The first shell can contain only two electrons, the second eight, the third 18, and so on to a seventh shell for the heaviest atoms.[‡] Since the innermost shell is the lowest energy level, the shells begin to fill up from the nucleus outward as the atomic number of the element increases. Accordingly, no electrons can appear in the second shell until the first shell has its maximum complement of two electrons. And no electrons will appear in the third shell until the second has its full eight electrons. However, after the inner two subshells of the third shell are filled, either one or two electrons will appear in the fourth shell before the outer subshell of the third shell can accept any electrons. Variations of this type become more involved in the fourth, fifth, and sixth shells; nevertheless, a definite pattern is followed.

For our investigation of the electrical property of matter, we need consider only those electrons in the highest occupied energy level of the shell structure of a particular atom. These electrons are called **valence** electrons. By adding a small amount of energy

[†]Niels Bohr proposed this concept of the atom in 1913.

[‡]The maximum number of electrons in a shell is $2n^2$, where n is the shell number, counting from the nucleus.

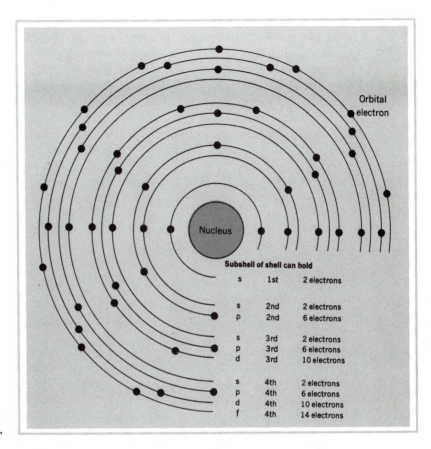

FIGURE 2-3
Possible Location of Electrons in the First Four Energy Levels of an Atom.

	Orbital electron

Subshell of shell can hold

s	1st	2 electrons
s	2nd	2 electrons
p	2nd	6 electrons
s	3rd	2 electrons
p	3rd	6 electrons
d	3rd	10 electrons
s	4th	2 electrons
p	4th	6 electrons
d	4th	10 electrons
f	4th	14 electrons

to the valence electrons of an atom, it is possible to increase the radius of their orbits so that they can escape from the attraction of the nucleus. Whenever valence electrons are thus removed from an atom, the atom now contains more protons than electrons. Consequently, it is no longer electrically neutral but displays a net *positive* charge.

It is also possible to *add* electrons to the valence subshell of an atom. Such atoms display a net *negative* charge. In Section 2-3, we shall discover how the necessary energy to redistribute valence electrons can be provided by chemical reaction. We can also supply the necessary energy in some cases by friction. Hence, the left-hand ball in Figure 2-1(b) is given a positive charge by removing some electrons from its atoms. The right-hand ball is given a negative charge by adding electrons to its normally neutral atoms.

2-3 COMBINATION OF ATOMS

The number of valence electrons possessed by an atom of a particular element governs the manner in which atoms will combine chemically to form stable **molecules.** An atom is in its most stable state when the number of electrons associated with each nucleus in a molecule exactly fills a subshell in the orbital structure. Elements such as helium with its two orbital electrons, and neon with 10 electrons, have no tendency to combine chemically with other elements since they possess completely filled shells. Helium and neon are therefore **inert** gases. However, there are many elements whose atoms have either a few electrons short of a complete subshell or a few electrons in excess of a complete subshell. These atoms tend to combine into chemical compounds in a manner that will permit all atoms in a molecule of a compound to have completely filled subshells. There are three types of atomic **bonding** that can achieve this requirement.

Sodium chloride (common salt) is an example of one form of atomic bonding. A sodium atom has 11 electrons. The first two shells of this atom are filled and there is one electron in the third shell. Chlorine atoms have 17 electrons, one short of a complete subshell. On combining, the sodium atom is quite eager to lend its valence electron to the chlorine atom so that both atoms may have completely filled subshells. Since the sodium atom has lost an electron, it is no longer electrically neutral. An atom that has either lost or gained electrons does not have the same chemical properties as the neutral atom and is called an **ion.** When the sodium atom loses an electron, the positive charge of its nucleus predominates, thus forming a *positive* sodium ion. Similarly, the chlorine has become a *negative* chlorine ion. The attraction between the unlike charges of the sodium and chlorine ions binds them together to form a stable molecule of sodium chloride. This type of bonding is called **ionic bonding.**

Hydrogen, with only one electron, is a very active element. If it cannot acquire an electron to provide two electrons for its first shell, it would just as soon give up the one electron it does possess. As a result, many ionic compounds contain positive hydrogen ions. But to form a stable molecule of pure hydrogen gas, two hydrogen atoms can combine in such a manner that they *share* each other's one electron, as suggested by Figure 2-4. Consequently, each hydrogen atom in the molecule appears to possess two electrons (a completely filled first shell); yet, the molecule as a whole is electrically neutral. This electron sharing is called **covalent bonding.**

Atoms of a group of elements called **metals** are held together by a third form of bonding. The metals with which we are concerned each have a single valence electron. For example, copper atoms have 29 electrons in orbit around each nucleus. After the first three shells are filled (2 + 8 + 18 = 28), there is one lone electron in the fourth shell. Since the copper atom would be more stable without this valence electron, only a very weak attraction holds it in orbit around the nucleus (unlike the 28 tightly bound electrons which fill the first three shells). As a result, the valence electron often finds itself midway between two nuclei, as shown in Figure 2-5. In this position of balanced attraction, the valence electron is then *free* to leave its parent nucleus and to move into the fourth shell of the neighboring atom.

The net effect of the valence electrons of a metal being free to move from one atom to another is that a seemingly solid piece of copper resembles the steel skeleton of a skyscraper with the wind passing through it. The copper atoms donate their valence electrons to form a vast cloud of **free electrons,** thus becoming stable positive copper ions with their 28 tightly bound orbital electrons. Since the positive metal ions repel one another, they will take up fixed geometric positions within the electron cloud, thus giving the metal object its shape. The atoms are held together by the attraction between the positive metal ions and the common cloud of free electrons. This is called **metallic bonding.** Knowing the mass of a single copper atom, we can estimate that there are approximately 85 000 million, million, million (8.5×10^{22}) free electrons in a cubic centimeter of copper.

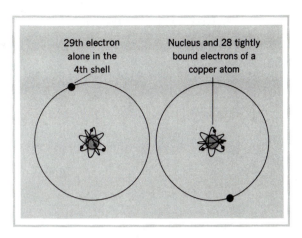

FIGURE 2-4
Atomic Model of a Hydrogen Molecule.

FIGURE 2-5
Atomic Model of Two Adjacent Copper Atoms.

The manner in which various materials behave in electric circuits depends on which of the three types of atomic bonding holds the atoms of each molecule together. We shall investigate some of these distinctions more thoroughly in Chapter 3. However, we can now specify the nature of the particles which flow through the copper conductors of the simple electric circuit of Figure 1-1. Just as a wind will make air flow through the steel skeleton of a building, we can force **free electrons** to move through the lattice of positive metal ions in the seemingly solid copper conductors. This flow of electrically charged particles under the influence of an electric force constitutes the necessary requirement for the transmission of energy in an electrical form. This flow of free electrons in an electric conductor represents an **electric current.**

2-4 ELECTRIC CURRENT

In Section 2-3, we noted that atoms of copper arrange themselves into a crystalline structure so that the positive copper ions consisting of a nucleus and 28 tightly bound electrons are located in a pattern which will allow the twenty-ninth, or valence electron to move freely from atom to atom. The lattice pattern for metals such as copper is in the form of a cube with a metal ion in each of the eight corners of the cube and also in the center of each of the six faces of the cube. Figure 2-6 shows a cross section of such a lattice structure with the outline of one crystal unit sketched in.

Although there is a free electron for each copper ion, Figure 2-7 shows only one of these free electrons so that we may trace its motion through the lattice structure. In a section of a copper conductor lying on a table, this free electron moves from atom to atom, its direction at any time depending on random bits of energy it picks up in its travels from repulsion by other electrons, attraction by copper ions, or from the thermal energy which causes the ions to vibrate in their fixed lattice positions. These random motions average out over a period of time so that there is no *net* motion in any one direction in Figure 2-7(a). Figure 2-7(b) shows the same conductor connected in an electric circuit which is conveying energy. The battery in the basic electric circuit of Figure 1-1 provides

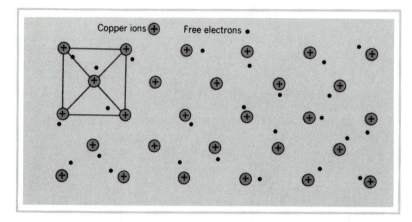

FIGURE 2-6
Cross Section of a Copper Conductor.

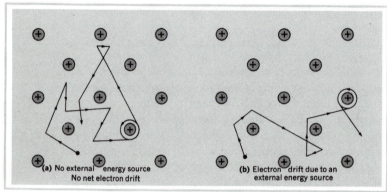

(a) No external energy source
No net electron drift

(b) Electron drift due to an external energy source

FIGURE 2-7
Random Motion of a Free Electron in a Copper Conductor.

an external source of energy for the free electrons in the conductor. The energy picked up by the electrons from this additional source superimposes a net **electron drift** to the right on the random motion of the free electrons.

To produce this electron drift, the battery in Figure 2-8 adds surplus electrons to the left-hand end of the conductor and, at the same time, removes an equivalent number of electrons from the right-hand end of the conductor. At the expense of chemical energy the battery is able to maintain a surplus of electrons (hence, a net negative charge) at its negative terminal, and a deficiency of electrons (net positive charge) at its positive terminal. As we noted in Section 2-1, an electric force of repulsion exists between like electric charges, and an electric force of attraction exists between unlike charges. Since the free electrons in the conductor possess elemental negative charges, they are repelled by the negative charge at the negative terminal of the battery and are attracted by the positive charge at the positive terminal of the battery. Since these electrons are free to move from atom to atom in the lattice of copper ions, the result is a net electron drift from left to right in the conductor in Figure 2-8. As we shall discover in Section 2-8, chemical action within the battery will replenish the supply of surplus electrons at the negative terminal as fast as the free electrons in the conductor drift to the positive terminal of the battery.

To determine the *net* movement of free electrons in an electric conductor, we can consider an imaginary plane cutting across the conductor at right angles to its length, as shown in Figure 2-9. If we were able to observe the free electrons in their random

FIGURE 2-8
Motion of Free Electrons Due to
Electric Energy Supplied by a Battery.

(a) No net electron drift

(b) Net electron drift to the right

FIGURE 2-9
Cumulative Effect of Motion of
Billions of Free Electrons.

wandering, we would discover billions of electrons crossing the plane in both directions in a given time interval. Figure 2-9(a) represents the conductor lying on the table with no external source of energy applied to it. In this case, the number of electrons crossing the plane in one direction in a given period of time equals the number crossing in the opposite direction in the same time interval. There is no net electron drift, no energy is conveyed along the length of the conductor. Hence, by definition, there is no electric current. In Figure 2-9(b), an external energy source causes more electrons to cross the imaginary plane from left to right than from right to left. Consequently, there is now a net electron drift to the right along the length of the conductor. It is this *net* drift of charge carriers which constitutes the **electric current** which we must now be able to express in numerical terms.

> *Electric current is the **net** flow of electric charge carriers past a certain point in an electric circuit in a given period of time.*

Since electrons can be neither created nor destroyed, if there is a flow of charge carriers from the energy source to the load in one electric conductor in the basic electric circuit of Figure 1-1, we must provide a second conductor so that an equivalent number of electrons can circulate from the load back to the energy source.[†] Consequently, electric lamp and appliance cords have two current-carrying conductors.

2-5 THE COULOMB

Since each electron possesses the same elemental quantity of electric charge, we may express the net electron drift across the imaginary plane in Figure 2-9 either in terms of the number of electrons or in terms of the total electric charge possessed by this number of electrons. The first step in expressing electric current in numerical terms is to be able to express the net **quantity** of electric charge crossing the imaginary plane in Figure 2-9.

The elemental quantity of electric charge carried by a single electron is too small for practical purposes. A practical unit for expressing quantity of electric charge must represent the charge carried by many billions of electrons.

The coulomb is the SI unit of quantity of electric charge.[‡]

Physicists have determined that

A coulomb represents the quantity of electric charge carried by 6.24 × 10¹⁸ electrons.

In technical language, we use algebraic equations to state definitions and data in concise and precise form. Therefore, we must have standard *letter symbols* to represent all variable terms in these equations and *unit symbols* for the units of measurement.

The letter symbol for quantity of electric charge is Q.
The unit symbol for coulomb is C.

> **EXAMPLE 2-1**
> If there are 2.40×10^{19} free electrons in a certain piece of copper wire, how many coulombs of electric charge does this represent?
>
> **SOLUTION**
> $$Q = \frac{2.40 \times 10^{19}\text{e}}{6.24 \times 10^{18}\text{e/C}} = \textbf{3.85 C}$$
>
> To enter this equation into a calculator, it becomes
> $$Q = 2.4 \boxed{\text{EE}} \ 19 \div 6.24 \boxed{\text{EE}} \ 18 = \textbf{3.846 C}$$

SELF-CHECK

2-5:1. A charged capacitor contains 2.5 μC of stored electrical charge. How many electrons does this charge represent?

[†]Check the meaning of *circuit* in a dictionary.
[‡]Named in honor of Charles Augustin Coulomb.

2-5:2. A uranium atom has 92 electrons. Determine the charge, in coulombs, represented by the electrons contained in 100 000 uranium atoms.

2-6 THE AMPERE

From our general definition of electric current in Section 2-4, we require units for *two* variables: (1) quantity of electric charge passing a certain point in an electric circuit (coulombs) and (2) in a given period of time. Since the SI unit of time is the *second,* it is quite in order for us to express the magnitude of an electric current in **coulombs per second.**

The International System of Units prefers not to use compound units such as coulombs per second as basic units. As a result, a special unit of electric current has been established to perpetuate the name of one of the pioneer electrical physicists, André Marie Ampère. Thus,

The ampere is the SI unit of electric current.
One ampere is the rate of flow of electric charge (electric current) when one coulomb of electric charge carriers passes a certain point in an electric circuit in one second.

Amperes and **coulombs per second** are synonymous, and in using the ampere as a unit of electric current, we must keep reminding ourselves that current is *rate of charge flow* in coulombs per second.

Any unit which involves *time* is called a *rate* unit. However, there are several types of rate units. *Distance* per unit time is called **velocity**. *Volume* or *quantity* per unit time is called **rate of flow,** or simply **flow.** For example, natural gas *flow* in a pipeline is measured in cubic meters per second. This flow is not the same as the speed or *velocity* at which the gas molecules are moving along the pipe. Similarly, in an electric circuit, we must be careful not to confuse the *rate of flow* of electric charge with the *velocity* of the charge carriers. The ampere is a rate unit in terms of *quantity* per second, not *distance* per second. A traffic engineer measures the rate of flow of cars on a highway by using a traffic counter, which tells him how many cars pass a certain point on the highway in a given period of time. The velocity of the individual cars (in kilometers per hour) may vary considerably and bears no fixed relationship to the number of cars using the highway per day. Similarly, when the electric current in a typical conductor is one ampere, the net *velocity* of electron drift along the length of the conductor is only about one-tenth of a millimeter per second.

The letter symbol for electric current is I.[†]
The unit symbol for ampere is **A.**

We can express the equality between amperes and coulombs per second in equation form as

$$I = \frac{Q}{t} \tag{2-2}$$

[†]*I* is from the French, *Intensité* of electron flow. The letter symbol *C* is reserved to represent capacitance.

where I is current in amperes, Q is quantity of electric charge in coulombs, and t is time in seconds.

EXAMPLE 2-2

What is the current in an electric circuit when 75 C of electric charge pass a certain point in the circuit in half a minute?

SOLUTION

$$I = \frac{Q}{t} = \frac{75 \text{ C}}{30 \text{ s}} = 2.5 \text{ C/s} = \textbf{2.5 A}$$

$$\underbrace{\qquad\qquad}$$

| I | II | III | IV |

As we become more familiar with our calculators, there is a tendency to simply enter the data and come up with a numerical answer all in one step. However, in these early stages, a systematic solution will greatly assist in understanding the behavior of electric circuits. Therefore, we shall take time to identify the sequence of steps involved in solving Example 2-2.

Step I. Note that the problem states two pieces of information and asks for one. We express this information in equation form. *Write down* the symbol for the unknown quantity on the left of the *equals* sign and the symbols for the given data (in their proper relationship) on the right of the *equals* sign.

Step II. Substitute the given data into the equation making sure that powers of ten or unit prefixes are included to preserve the proper magnitude.

Step III. Perform the numerical computation.

Step IV. The magnitude of the calculator display will be in *root* units. Express the answer in appropriate units, using unit prefixes, if necessary, to conform to SI recommended format.

EXAMPLE 2-3

How long will it take 4.0 mC of electric charge to pass through a fuse in an electric circuit if the current is 50 A?

SOLUTION
Since

$$I = \frac{Q}{t} \qquad \therefore t = \frac{Q}{I}$$

$$t = \frac{4 \text{ mC}}{50 \text{ C/s}} = 4 \boxed{\text{EE}} \boxed{+/-} 3 \div 50 \boxed{=} 8 \times 10^{-5} \text{ s} = 80 \times 10^{-6} \text{ s} = \textbf{80 } \boldsymbol{\mu}\textbf{s}$$

SELF-CHECK

2-6:1. What average current is present in a toaster if 380×10^{19} electrons flow through the toaster in 1 minute?

2-6:2. It takes 8 hours for a battery charger to charge a car battery. During that time there was an average electric current of 25 A in the battery. Find the charge that passed through the battery as it was charged. Express your answer (a) in coulombs, and (b) in electrons.

2-7 POTENTIAL DIFFERENCE

In Section 2-4, we noted that there can be no electric current in the conductors of our basic electric circuit until we add a device, such as a battery, which can impart energy to the free electrons in the conductors in such a manner that they flow along the length of the conductor. We say that the battery is the source of an **electron-moving force** or **electromotive force,** usually abbreviated to **emf** (ee-em-eff).

Since we are more familiar with the effects of gravitational force than we are with the effects of electric force, we can consider first of all the operation of the hydroelectric generating station shown in simplified cross section in Figure 2-10. Water from above a waterfall is diverted so that it flows through a "waterwheel" or turbine and is then discharged into the river below the waterfall. In falling 100 meters, the water loses some

FIGURE 2-10
Simplified Cross Section of a Hydroelectric Generating Station.

of its potential energy. In accordance with the law of conservation of energy, the potential energy lost by the water is converted into mechanical energy by the turbine and is then converted into electric energy by the electric generator or **dynamo.**

Since objects at the surface of the earth are 6400 kilometers away from the center of gravity of the earth, the difference in gravitational force acting on a cubic meter of water above and below the generating station shown in Figure 2-10 is negligible. But there is an appreciable difference in the potential energy of a cubic meter of water above and below the station, as we would discover if we had to carry or pump a cubic meter of water from the river below the station 100 meters uphill against the force of gravity to the top of the escarpment. The "force" which operates the hydroelectric generating station is not simply gravitational force. Rather, it is the potential energy difference between a unit quantity of water above and below the generating station. The law of conservation of energy requires the potential energy difference, or simply **potential difference,** between a unit quantity of water above and below the generating station to be equal to the energy expended in raising the unit quantity of water the 100 meters against the force of gravity.

We can now return to the electric circuit of Figure 2-8, in which we connected an electric conductor between the terminals of a battery. This circuit is shown again in Figure 2-11. In this case, we have turned the battery on its side so that the terminology we use in the electric circuit matches the terminology of the hydraulic system of Figure 2-10. Since the battery is the source of the equivalent electromotive "force" in Figure 2-11, we must now find out what happens inside the battery.

FIGURE 2-11
Electric Potential Difference.

We noted in Section 2-4 that free electrons in the conductor flow away from the negative terminal of the battery and toward the positive terminal. To maintain the negative and positive charges at the two battery terminals, *inside* the battery an equivalent number of electrons must be removed from the positive terminal and added to the negative terminal. Thus, inside the battery, electrons must move *away* from the positive terminal and *toward* the negative terminal. In other words, electrons must move *against* the electric forces acting on them, just as water moves against gravitational force when it is pumped uphill. To accomplish this, the electrons must acquire potential energy at the expense of the chemical energy stored by the battery. Consequently, an electron arriving at the − terminal is at a higher potential energy level than it was when it left the + terminal of the battery. There is an electric **potential** energy **difference** between free electrons at the negative and positive terminals of the battery, with electrons at the negative terminal being at a higher potential than those at the positive terminal.

Under the influence of gravitational force, the water in the hydroelectric generating system of Figure 2-10 always tends to fall to a lower potential energy level. Similarly, electrons at the negative terminal of a source tend to "fall" to a lower potential energy level. They can do this via the external conductor connected between the battery terminals. In flowing from the − terminal to the + terminal in the external circuit, electrons lose as much potential energy as they gained in being moved from the + terminal to the − terminal inside the battery. This energy "lost" in the external circuit is converted into light, heat, or some other useful form of energy, depending on the nature of the load component in the electric circuit. In traveling the complete circuit around the closed loop of Figure 2-11, electric charges experience a **potential rise** within the battery and an equivalent **potential fall** or **potential drop** in the external circuit.

2-8 GENERATING A POTENTIAL DIFFERENCE

Before we consider a suitable unit for expressing electric potential difference, we shall consider briefly how a battery converts chemical energy into an electromotive force which, in turn, generates a potential difference between the battery's terminals. Electromotive force is a property of the battery in the electric circuit of Figure 2-11 which distinguishes it from the external load. Emf is a measure of the amount of energy per unit of electric charge which the battery *converts* from chemical to electrical form in the process of *separating electric charges* in order to create a potential difference between its terminals. Electromotive *force* is, therefore, not a true force which we would measure in newtons. Emf is measured in joules per coulomb. Consequently, although we sometimes refer to a battery as a *source of emf*, the term *electromotive force* is disappearing from electrical terminology as we pay more attention to the **potential difference** generated by the battery.

As the term implies, a **battery** originally meant a *group* of chemical cells, but common usage allows us to refer to the simple **voltaic cell** of Figure 2-12 as a single-cell battery. When hydrogen with its one valence electron combines with chlorine, which has 17 orbital electrons, to form the ionic compound hydrochloric acid, the chlorine borrows the one electron from the hydrogen to complete a full subshell in its atomic structure. (Check this by referring to Figure 2-3.) In doing so, the hydrogen atom loses an electron, as indicated by the + sign in Figure 2-12, and the chlorine acquires a surplus electron.

The chemical makeup of hydrogen chloride and zinc chloride is such that energy must be expended to transfer a chlorine atom from a zinc chloride molecule to a hydrogen chloride molecule. Therefore, energy must be released when the chlorine atom moves from a hydrogen chloride molecule to a zinc chloride molecule. This energy is transferred to the surplus electron of the negative chloride ion in the acid as the chloride ion moves toward the zinc rod to combine with atoms of zinc to form zinc chloride. During this chemical process, the surplus electron that the chlorine borrowed from the hydrogen is

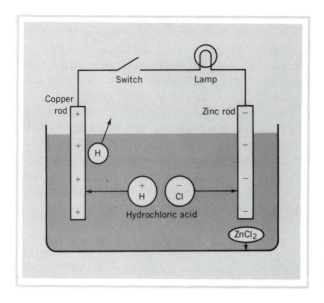

FIGURE 2-12
Producing an Electric Potential Difference with a Simple Chemical Cell.

deposited on the zinc rod, thus building up a surplus of electrons or a negative charge on the zinc rod. As each chlorine ion combines with the zinc, the hydrogen ion which lost the electron to the chlorine ion can now go in search of a source of free electrons to regain its normal state as an atom of hydrogen. The copper rod represents such a source, and as the hydrogen ions rob the copper of free electrons, a positive charge builds up on the copper rod. The number of electrons taken from the copper rod by hydrogen ions is equal to the number of electrons deposited on the zinc rod by chlorine ions.

Thus the electromotive-force property of the battery disturbs the normal ion balance in the hydrochloric acid by *separating* the electrically charged ions. Negative charges are forced to move to the zinc electrode while positive charges go to the copper rod. Hence, we may think of the battery (or any electric energy source) as an electric **charge-separating device.** Or, if we wish to think in terms of our hydroelectric analogy, we may think of an electric energy source (source of emf) as a pump building up a potential difference between its terminals by seemingly pumping electrons from the positive to negative terminal against the electric force opposing such motion.

If we leave the switch open, the chemical action can proceed until sufficient potential difference is established between the copper and the zinc that the negatively charged chlorine ions do not have sufficient chemical attraction to reach the zinc against this electric force. The formation of zinc chloride will then cease. When we close the switch there will be a flow of free electrons through the conductors and the lamp as the electrons in the conductors are repelled by the negative charge on the zinc rod and attracted by the positive charge on the copper rod. This flow is attempting to return the potential difference in the circuit to zero. But as soon as the potential difference drops the least bit, the chemical action recommences, depositing electrons on the zinc rod and removing them from the copper rod. The result will be a steady flow of free electrons (electric current) as the tendency for the electric load (lamp) to lower the potential difference is offset by the action of the cell to raise the potential difference to a level determined by the two particular metals used in the cell. We call this steady flow of electric charge in one direction around the loop a **direct current** (dc).

2-9 THE VOLT

The electric potential difference (PD) between any two points in an electric circuit is the rise (or fall) in potential energy involved in moving a unit quantity of electric charge from one point to the other.

In Section 2-1, we noted that, as a consequence of the law of conservation of energy, work and energy are numerically equal. And we defined work as the accomplishment of motion against the action of a force which tends to oppose the motion.

The letter symbol for work and energy is W.

From the definition of work

$$W = Fs \qquad\qquad (2\text{-}3)$$

where F is force in newtons and s is distance in meters.

We could express work and energy in *newton meters*. But the International System of Units prefers to avoid the compound unit. Hence,

The joule is the SI unit of work and energy.[†]
The unit symbol for joule is **J**.

From the definitions above, we can express electric potential difference in terms of **joules per coulomb.** Again, the International System of Units provides a special *derived* unit for joules per coulomb.

The volt is the SI unit of potential difference.[‡]

From the definition of potential difference,

One volt is the potential difference between two points in an electric circuit when the energy involved in moving one coulomb of electric charge from one point to the other is one joule.
The letter symbol for potential difference is E or **V**.[§]
The unit symbol for volt is V.

We may express this relationship in equation form as

$$E \,(\text{or } V) = \frac{W}{Q} \qquad\qquad (2\text{-}4)$$

where $E\,(or\,V)$ is the potential difference in volts, W is energy in joules, and Q is quantity of electric charge in coulombs.

[†] Named in honor of the English physicist James Joule.
[‡] Named in honor of the Italian physicist Alessandro Volta.
[§] We shall consider the distinction between E and V in the next section.

EXAMPLE 2-4

During the chemical action in the wet cell of Figure 2-12, 50 C of electrons is effectively transferred from the copper to the zinc by the release of 55 J of energy. What is the potential difference between the two rods?

SOLUTION

$$E = \frac{W}{Q} = \frac{55 \text{ J}}{50 \text{ C}} = \textbf{1.1 V}$$

EXAMPLE 2-5

A current of 0.3 A flowing through the filament of a TV picture tube releases 9.45 J of heat energy in 5 s. What is the PD across the filament of the tube?

SOLUTION
Since $I = Q/t$,

$$\therefore Q = It = 0.3 \text{ A} \times 5 \text{ s} = 1.5 \text{ C}$$

and

$$V = \frac{W}{Q} = \frac{9.45 \text{ J}}{1.5 \text{ C}} = \textbf{6.3 V}$$

SELF-CHECK

2-9:1 A battery supplies 36 joules of energy to every 4 coulombs of charge that flow through a smoke detector. What is the potential difference across the battery?

2-9:2 How much energy is supplied each minute to an electric heater that draws 12 A when supplied by a 120-V supply?

2-10 EMF, POTENTIAL DIFFERENCE, AND VOLTAGE

At the beginning of Section 2-7, we used the term **electromotive force** to represent the concept that a generating device must have some property which will *force* electric charge carriers to flow in a closed electric circuit. Emf is, however, not simply the electric force (in newtons) that acts on the electrically charged particles in the circuit to cause a net drift in one direction. By definition, the electromotive force of a generating device is numerically equal to the energy *converted* per unit quantity of electric charge moved from one terminal to the other inside the device. Hence, emf can be expressed in joules per coulomb or volts. But emf and potential rise are numerically equal only when the electric circuit is open ($I = 0$). When there is any current in the electric circuit, inefficiencies in the energy-conversion process mean that the potential rise between the terminals of the generating device will be numerically smaller than the internal emf.

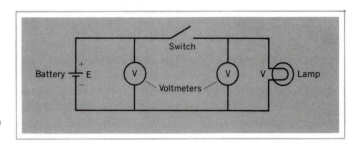

FIGURE 2-13
Illustrating the Distinction between Source Voltage and Voltage Drop in a Basic Electric Circuit.

Electromotive force is a means to an end. The energy conversion is, in turn, responsible for generating a potential difference between the terminals of the generating device. It is this potential difference (often abbreviated to PD) which causes an electric current when we close the switch in the electric circuit. In the electric circuit of Figure 2-13, the voltmeter cannot get at the emf of the battery directly. It can only measure the potential difference between the battery terminals. We can *indirectly* determine the magnitude of the emf if the switch is open and if the voltmeter draws no current from the battery.

In studying electric circuits, it has often been customary to assume that the generating device is perfectly efficient—in which case the potential difference between its terminals and the internal emf would be numerically equal. Such an assumption has allowed us to become careless in our terminology by using the terms *emf* and *potential rise* interchangeably. In this book, we shall follow the more modern practice of avoiding the term emf when we really mean the potential rise (in volts) between the terminals of a generating device.

Many years ago, someone coined the term **voltage** as a simpler expression than *potential difference*. Since "voltage" basically means something that is measured in volts, it could be used in place of potential difference, potential rise, potential drop (or even emf). Over the years, increasing usage has led to voltage being defined as being the same as potential difference. Hence, **voltage** is now the preferred term in much of the technical literature.

In a way, it is unfortunate that the term *potential difference* or PD is not more popular. It is a very descriptive term in that its wording reminds us that *two* particular points in an electric circuit must be stated before we can state the potential *difference*. In the electric circuits of Figures 2-11 and 2-12, we noted that there are two kinds of potential difference. There is the potential *rise* that takes place in the battery and the potential *fall* or *drop* that takes place in the electric conductors and the lamp. We can maintain the distinction between rise and fall in potential and still adopt the more popular terminology by distinguishing between the **source voltage** or **applied voltage** appearing between the terminals of a source as a result of its energy-converting ability and the **voltage drop** that occurs when electric current flows through the external circuit.

With the popularity of the more general term *voltage,* much of the distinction between source voltage or applied voltage and a voltage drop has disappeared from technical literature. Some writers use the same letter symbol (either *E* or *V*) for both applied voltage and voltage drop. Our objective in this book is to *understand* why electric circuits must behave as they do. Hence, we shall maintain a distinction between source voltage and voltage drop by stating that

The letter symbol for source voltage or applied voltage is E.
The letter symbol for voltage drop is V.

That there is indeed a difference between a source voltage and a voltage drop is illustrated in Figure 2-13. When the switch is closed, both voltmeters show the same reading as the current through the lamp creates a voltage drop across the lamp equal to the applied voltage or source voltage between the terminals of the battery. When the switch is open, the voltmeter across the battery terminals still registers the source voltage resulting from the chemical action within the battery. But the voltmeter across the lamp terminals reads zero. A voltage drop can appear across the lamp only when electrons are flowing through it in an attempt to return the potential difference in the circuit to zero.

It is also worth noting at this point the prepositions which we can use with the terms current and voltage. Remembering that current is the rate of flow of electrons, we speak of current *in* or *through* a conductor or other circuit component. To talk about "current *across* the lamp" would be meaningless. Since we are not accustomed to using the term *potential difference,* we must remind ourselves that there is no such thing as a "voltage *at* a certain point." The applied voltage from a **voltage source** and the voltage drop of a load must be measured *between* two points, or *from* one point *with respect to* another, or *across* a certain circuit element. Talking about "voltage *through* a component" would also be meaningless.

We can also invent the term **amperage,** which means something that is measured in amperes. That something is, of course, electric current. The term *amperage* is sometimes used by electricians in referring to the electric current normally carried by an electrical device. The *National Electrical Code*® also uses the term **ampacity** to state the maximum allowable current for a given size of electric conductor in amperes.

SELF-CHECK

2-10:1. Suppose that the battery in Figure 2-13 is a 12-V battery. Determine the voltage drop across the lamp if (a) the switch is open, (b) the switch is closed.

2-11 CONVENTIONAL CURRENT DIRECTION

The most common form of electric conductor is the metallic conductor described in Section 2-4. The charge carriers in metallic conductors are free electrons which flow from the negative terminal of the voltage source toward the positive terminal. However, as we shall discover in Chapter 3, there are several other forms of electrical conduction. Although electric current always consists of a flow of electrically charged particles, these particles are not necessarily free electrons. For example, inside the battery in Figure 2-12, the charge carriers are *not* free electrons. When the switch is closed, *two* types of charge carriers move simultaneously in opposite directions inside the battery. Negatively charged chlorine ions move toward the negative terminal of the battery and positive hydrogen ions move toward the positive terminal.

However, when we are working with pencil-and-paper circuit diagrams, it would be very helpful if we could agree on a standard **conventional direction** for electric current. This is essential if we are to be able to calculate the *algebraic* sum of the currents at a junction point in an electric circuit. For such a chore, the *physical* nature of the actual *flow* of the charge carriers is not important.

Since most electric circuit work deals with metallic conductors, it would have been a very happy arrangement to select the direction of electron flow in such conductors as the mathematically conventional current direction. But experimenters in the days of Benjamin Franklin and Michael Faraday had no knowledge of the atomic structure of matter. They selected a conventional direction for electric current on the basis of observable effects. The clearly observable result of electric current through Faraday's electrolytic cell (described in Chapter 3) suggested that electric current consists of a flow of positively charged particles. As we have noted, positive charge carriers flow in the *opposite* direction to negative charge carriers such as free electrons and negative ions.

Therefore, all the laws and rules for electric circuits were based on a direction *from the positive terminal of the source, through the external circuit, and back to the negative terminal of the source.* As long as we keep in mind the purpose of a mathematical convention for current direction, the fact that the *physical* direction of electron flow is just opposite to the *mathematical* convention due to bad luck should not present any problem. In this text, we associate the term **current** with conventional direction and mark our circuit diagrams accordingly. If we find occasion to refer to the physical process of electrical conduction, we refer to specific charge carriers by using such terms as **electron flow.**

SELF-CHECK

2-11:1. Suppose the switch in Figure 2-13 is closed. (a) What is the direction of the conventional current in the lamp? (b) In which direction do the electrons flow in the lamp?

PROBLEMS

2-6-1. What is the current through a switch in an electric circuit if it takes 6 s for 30 C of electric charge to pass through it?

2-6-2. What is the current through a transistor if it takes 1 min for 1 C of charge carriers to arrive at the collector?

2-6-3. If the net drift of electrons across the imaginary cross section in Figure 2-9(b) is 10^{15} electrons per second, what is rate of electric charge flow in amperes?

2-6-4. Express an electric current of 4.0 mC/min in appropriate SI units.

2-6-5. How many coulombs of electric charge pass through the lamp in Figure 2-12 in 1 min if the current is a steady 300 mA?

2-6-6. How many coulombs of electric charge are added to each of the terminals of the battery by chemical action in Figure 2-12 in $\frac{1}{60}$ s if the current drain is a steady $\frac{1}{2}$ A?

2-6-7. How long will it take for 8 C of electric charge to flow through the switch in Figure 2-12 if the current is 250 mA?

2-6-8. How long will it take for 10^{20} chlorine ions to combine with the zinc if the current through the lamp in Figure 2-12 is a steady 300 mA?

2-9-9. If it takes 5 J of chemical energy to move 20 C of electric charge between the positive and negative terminals of a battery, what is the PD between its terminals?

2-9-10. What PD must be developed across the terminals of a lamp in order for a flow of 0.05 C from one terminal to the other to release 6 J of energy?

2-9-11. How much electric charge moves from one terminal of a load in an electric circuit to the other terminal if 50 J of energy is released at a potential drop of 120 V?

2-9-12. What quantity of electrons will move from the positive terminal to the negative terminal (effectively) of a battery whose terminal voltage is 1.5 V during the conversion of $\frac{1}{4}$ J of chemical energy?

2-9-13. What energy is required to move 5 mC of electric charge through a potential rise of 60 V?

2-9-14. How much electric energy is converted into light and heat energy by an electric lamp when there is a potential drop of 3.0 V as 600 mC of charge carriers move from one terminal to the other?

2-9-15. How much energy is involved in maintaining a 0.5-A current through a battery for 0.5 min if the PD between the battery terminals is 1.1 V?

2-9-16. How long will a steady 250-mA current have to flow through a flashlight lamp with a 3-V drop between its terminals for 6 J of energy to be transferred to the lamp?

2-9-17. What is the applied voltage of a battery in a transistor radio if 13 J of chemical energy is used up in 1 min as the battery maintains a steady current of 24 mA while the radio is operating?

2-9-18. An automobile storage battery is designed to maintain a constant terminal voltage of 12V as 3.5×10^6 J of chemical energy are converted into electric energy. How long can this battery maintain a constant current of 20 A in expending the stored chemical energy?

2-9-19. What current must be flowing through a load in an electric circuit if energy is being delivered to the load at the rate of 18 J/s with a 120-V drop across its terminals?

2-9-20. Assuming perfect efficiency, at what rate is chemical energy being converted in a battery whose voltage rise is 3 V if a lamp connected to the battery draws a steady 150-mA current?

REVIEW QUESTIONS

2-1. The SI unit of energy is the **joule.** The joule is a derived unit equal to one newton meter. Why is it possible to express energy in terms of force and distance?

2-2. How can we demonstrate that electric force is a *field* force rather than a *contact* force?

2-3. What effect would be observed if we touched a charged *ebonite* rod to both balls in Figure 2-1?

2-4. Compare Coulomb's law, which states the factors governing the magnitude of an electric force, with Sir Isaac Newton's law of universal gravitation.

2-5. What is the distinction between the **valence** electrons of an atom and the remainder of the orbital electrons?

2-6. What is the distinction among an **atom,** a **positive ion,** and a **negative ion?**

2-7. How does a **molecule** differ from an **atom?**

2-8. What is the significance of the **shell** concept of atomic structure in determining whether an element is likely to be a good electric conductor?

2-9. Argon has an atomic number of 18 (18 orbital electrons). With reference to the shell structure of an atom, discuss the probable chemical activity of argon.

2-10. Silver has an atomic number of 47. Show how the shell structure of an atom supports the observation that silver is an excellent electric conductor.

2-11. What is the significance of the term **free electron?**

2-12. What is the relationship between free electrons and a metal's ability to act as an electric conductor?

2-13. Distinguish between the behavior of valence electrons in **ionic bonding** and **covalent bonding.**

2-14. From our present knowledge of the structure of atoms, explain why rubbing a glass rod with a silk cloth places a positive charge on the glass rod.

2-15. Although the current in the electric circuit of Figure 2-12 is zero when the switch is open, free electrons in the conductors are not stationary. Describe their motion.

2-16. How is the random motion of free electrons shown in Figure 2-7(a) accounted for?

2-17. What change in electron motion occurs in the circuit of Figure 2-12 when the switch is closed?

2-18. Figure 2-9(b) indicates electrons simultaneously moving in opposite directions in a current-carrying conductor. How is this possible?

2-19. Distinguish between electric **current** and the effective **velocity** at which electrons move along the length of an electric conductor.

2-20. Why is the term **drift** appropriate for describing electric current in a metallic conductor?

2-21. What is the magnitude (in coulombs) of the negative charge of a single electron?

2-22. Although not an approved SI unit, the **ampere-second** is a unit of electrical measurement. In the measurement of what electrical property would it be used?

2-23. Explain the forces acting on a free electron at the center of the conductor shown in Figure 2-8.

2-24. The water in Figure 2-10 loses potential energy as it flows through the turbine of the generating station. How is the energy replenished to maintain the operation of the station over the years?

2-25. In the hydroelectric generating system of Figure 2-10, a potential difference is produced by moving a unit quantity of water against a gravitational force. In an electric circuit, what is the equivalent force which makes a potential difference between two terminals possible?

2-26. Both gravitational force and electric force are expressed in newtons. Why is electromotive "force" not a true force?

2-27. Explain how the copper rod in the simple battery in Figure 2-12 becomes positively charged.

2-28. Inside the battery in Figure 2-12, positively charged hydrogen ions are acting as electric charge carriers. Insofar as these positive charge carriers are concerned, which terminal of the battery is considered to be at the higher potential? Why?

2-29. When the switch in the electric circuit in Figure 2-12 is closed, light energy is produced by the lamp. Where does this energy originally come from?

2-30. Although charge carriers flow from the load back to the source in one of the conductors of an electric circuit, energy is not transferred from the load back to the source. Explain this in terms of potential rise and potential fall.

2-31. Why does chemical action cease in a battery when the switch in the electric circuit is opened?

2-32. Discuss, using numerical examples, how the particular arrangement of letter symbols in the equation $V = W/Q$ satisfies the definition of the **volt.**

2-33. Why is the letter symbol E used in Example 2-4 and the letter symbol V in Example 2-5?

2-34. If we connect a voltmeter across the *open* switch terminals in the circuit of Figure 2-13, it will show the same "voltage" as the voltmeter connected across the battery. Explain.

3

CONDUCTORS AND INSULATORS, SEMICONDUCTORS, AND VOLTAGE SOURCES

3-1 CONDUCTORS

In Section 2-4, we discovered that electric current consists of a net drift or *flow* of electrically charged particles. To be suitable as an electric conductor, a material must possess a large number of such charge carriers per unit volume. It also follows that the greater the number of charge carriers per unit volume, the better that material will be as an electric conductor. As we noted in Section 2-4, metals possess a very large number of charge carriers per unit volume in the form of free electrons. All metals may be used as electric conductors, although some metals possess more free electrons per unit volume than others.

Copper is the most popular electric conductor material. It is reasonably inexpensive, can readily be drawn into wire, and rates second only to silver (which is much more costly) in its abundance of free electrons. We have estimated that a cubic centimeter of copper contains approximately 8.5×10^{22} free electrons. Silver has about 5% more free electrons per unit volume than copper. Aluminum is used commercially as an electric conductor material to some extent because, although it has only 60% as many free electrons per unit volume as copper, it is much lighter and less expensive. We can compensate for aluminum's decreased conductivity by increasing the diameter of aluminum wires. Some of the metal alloys we use in electric heater elements have less than

1% of the number of free electrons that copper has. This still represents something like 6×10^{20} free electrons per cubic centimeter. Such alloys are, therefore, reasonably good electric conductors.

The most familiar form of metallic electric conductor is the copper wire, usually with an insulating sheath. The diameter of a copper conductor depends on the magnitude of the electric current it is required to conduct. SI provides for standard diameters for electric conductors with the SI wire gauge number being 10 times the conductor diameter in millimeters. However, in North America, the standard diameters of conductors are still specified in **mils** (thousandths of an inch) using the American wire gauge table in Appendix 3. To maintain flexibility, the cross-sectional area of larger conductors is broken up into several strands twisted together to form a *stranded* conductor.

Even though copper wires are very good conductors of electric current, not all of the electric energy generated by the voltage source in the basic electric circuit of Figure 1-1 is available for conversion to light (and heat) energy by the load. Some of the energy is lost in the form of heat as free electrons alternately speed up and slow down on their journey through the crystal lattice of metallic conductors (Figure 2-7). In simple, low-current (electronic) circuits, this energy loss may not be significant. But in large power distribution systems, the energy loss in conductors can become a serious problem. Consequently, researchers continue to search for an ideal electric conductor with negligible energy loss.

We call such a conductor a **superconductor**. Until recently, the only means of producing superconductivity has been to cool metallic conductors to almost absolute zero temperature in liquid helium. This technique has been worth the high cost for such devices as the electromagnets of nuclear particle accelerators where extremely large currents are required. At the present time, researchers are achieving some success in synthesizing special ceramic crystal structures which permit the flow of electric current with negligible opposition at temperatures as high as 100 kelvin ($-173°C$). These materials become superconductors when cooled with relatively inexpensive liquid nitrogen. The ultimate goal is to discover a practical material that will superconduct at room temperature. In addition to reducing power system losses, such superconductors would make possible high-speed magnetic levitation trains and greatly increase the speed and capacity of microcomputers.

3-2 INSULATORS

Fortunately, the air surrounding the conductors in the basic electric circuit of Figure 1-1 has relatively few electric charge carriers per unit volume. If air were a reasonably good conductor of electricity, we would not be able to contain electric current within the desired circuit. Also, electric current would continue to flow when we *open* the switch to the position shown in Figure 1-1. Dry air consists mainly of oxygen molecules and nitrogen molecules in which the orbital electrons are all firmly bound to their respective nuclei. Consequently, it is quite difficult for the charge carriers (free electrons) in the metal conductors to cross the boundary and to move from atom to atom through the air surrounding the conductors. Therefore, we rate dry air as a nonconductor of electricity or an electric **insulator.**

In some electric circuits where it is unlikely that anything other than air will come into contact with the conductors, no other insulation is required. In the electric circuit illustrated in Figure 1-1, we can safely use bare copper conductors. Similarly, the conductors of the high-tension transmission lines crossing the countryside suspended from steel towers can be bare metal. However, they must be suspended from the towers by porcelain insulators to prevent leakage of current via the metal of the towers.

When there is a possibility that the conductors of an electric circuit may touch one another (for example, the two conductors of a lamp cord), we must use conductors that are surrounded by a suitable solid insulating material. In ionic-bonded molecules, each atom achieves a filled subshell in its orbital electron structure by one atom lending its valence electron(s) to the other. As a result, all the valence electrons are bound to the negative ions. Hence, ionic solids are (theoretically) electric insulators since none of the valence electrons are free to act as energy carriers. In practice, we find that it is fairly easy to break the ionic bond between the atoms of an ionic solid by adding energy to the molecule. Consequently, we seldom use ionic compounds in manufacturing practical insulating materials.

Polyethylene molecules are formed by the covalent bonding of carbon and hydrogen atoms, as illustrated in Figure 3-1. Carbon has six electrons, two in the first shell and four in the second shell. Four more electrons are needed to completely fill the second shell. In the polyethylene molecule, each carbon atom shares all of its second-shell electrons with neighboring carbon and hydrogen atoms, as shown. One electron is shared with the carbon atom on its left, one with the hydrogen atom above it, one with the carbon atom on its right, and one with the hydrogen atom below it. This particular form of covalent bonding creates a long, filament-like molecule in which every atom now has a completely filled outermost shell. The binding force in such a molecule is so great that it is very difficult for an electron to break away from the pattern. Thus, covalent-bonded

FIGURE 3-1
Structure of a Polyethylene Molecule.

materials of this type, called **polymers,** are excellent electric insulators. Most modern insulating materials such as rubber, plastics, varnish, enamel, silk, cotton, paper, oil, mica, and asbestos are polymers.

If we could manufacture a 100% pure polymer, it would possess no free electrons and would, therefore, be an ideal insulator. But practical insulating materials contain some impurities which do have a few free valence electrons that are not included in the pattern of covalent bonds. Oxidation of polymer molecules makes additional free electrons available for conduction. (Rubber deteriorates rapidly when exposed to ozone, which is created from oxygen by an electric spark.) Moisture absorbed by an insulator introduces ions which, being in a liquid state, can migrate slowly through the insulator. Consequently, polystyrene, which we consider a very good electric insulator, has over 6×10^{10} free electrons per cubic centimeter. Although this seems like a large number, compared with copper it is extremely small. Therefore, for practical purposes, the leakage of the charge carriers through a layer of polystyrene or other polymer is negligible.

Figure 3-2 shows a voltage source connected between the two conductors of a lamp cord. As we noted in Chapter 2, electric force is a *field* force. Hence, the potential difference between the two conductors gives rise to an electric force which acts on the electrons in the valence bonds of the insulating material between the conductors. As the potential difference of the source is increased, a critical level of stress is reached at which billions of valence bonds are broken, resulting in a *rupture* or *breakdown* of the insulator. The potential difference required to rupture a particular insulator will depend on the spacing between the two conductors and the strength of the valence bonds of the insulating material. Some insulating materials can withstand a much greater electric stress for a given thickness than others.[†] Because of the smaller number of atoms per unit volume, gases break down more readily than solids. Consequently, air rates rather poorly in terms of **dielectric strength,** but it is cheap enough so that we can compensate for this by using a thicker sheath of air around electric conductors such as those used in overhead transmission lines.

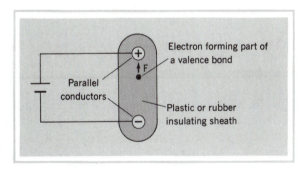

FIGURE 3-2
Electric Stress on Valence Bonds in an Insulator.

3-3 ELECTROLYTIC CONDUCTION

In developing our concept of the **coulomb** and the **ampere,** we considered only free electrons as electric charge carriers. Although by far the most significant type of charge carrier, free electrons are not the only electrically charged particles which can act as

[†] See Table 11-1 on page 290.

charge carriers in an electric circuit. In the chemical battery in Figure 2-12, we found that electric current consists of positive and negative ions moving simultaneously in opposite directions. Michael Faraday made a fairly thorough study of the observable effects of **electrolytic conduction** during the 1830s.

Figure 3-3 shows an **electrolytic cell** or electroplating bath consisting of a silver (Ag) **electrode** and a steel (Fe) **electrode** in the form of a spoon dipped into a solution of silver nitrate ($AgNO_3$). This current-carrying solution is called an **electrolyte.** The silver electrode is connected to the positive terminal of the battery and the spoon is connected to the negative terminal of the source. Faraday reasoned that because he could measure an electric current in the copper conductors connecting the source to the electrodes, electric current must be flowing through the silver nitrate solution to complete the electric circuit. We call the electrode through which free electrons from the metallic conductors enter the electrolyte the **cathode.*** The electrode through which free electrons leave the cell is called the **anode.**

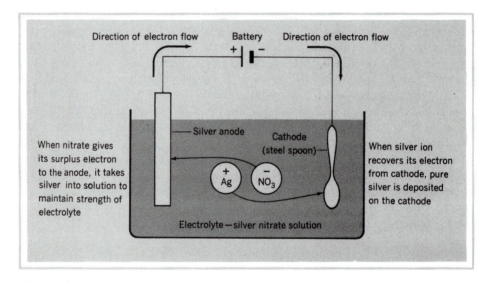

FIGURE 3-3
Simple Electrolytic Cell.

Faraday also noticed that as electric current flowed through the cell, silver disappeared from the anode and the same amount of silver was added to the cathode. With no knowledge of the electron (circa 1835), Faraday was able to state his first law of electrolysis from his observations.

The mass of a substance liberated by the anode and deposited on the cathode varies directly with the quantity (Q) of electricity passing through the cell.

The basis for our present-day concept of the nature of the charge carriers in an electrolyte was suggested in 1887 by Svante August Arrhenius. As a result, we can

---

*Faraday is credited with originating the terminology, including electrode, electrolyte, and electrolysis, which is now universally used with reference to electrolytic conduction of electric current.

explain both the apparent transportation of the silver from the anode to the cathode and the process by which electric current can pass through the silver nitrate solution. As shown in Figure 3-3, silver nitrate is an ionic compound in which the silver lends its valence electron to the nitrate radical to form a molecule consisting of a positive silver ion and a negative nitrate ion. Arrhenius suggested that, when ionic compounds such as acids, bases, and salts are dissolved in water, the ions tend to **dissociate;** that is, the silver ions and nitrate ions can roam around independently in the solution.

If we connect a voltage source to the electrodes, we create a potential difference between them. And since unlike charges have a force of attraction between them, the dissociated positive silver ions are attracted to the negatively charged cathode. As the silver ion touches the metallic cathode and regains its missing electron, it is deposited out of solution on the cathode as pure silver. At the same time an equal number of negative nitrate ions are attracted to the positively charged anode. As the NO_3^- gives up its surplus electron to the positively charged anode, it becomes chemically active and takes an atom of silver from the anode into solution with it, thus maintaining the strength of the electrolyte. No *chemical* energy is used up in this cell, since identical amounts of silver go into solution at the anode and are deposited out at the cathode. The work done in transporting the silver from the anode to the cathode is done at the expense of electric energy taken from the source.

We can now make an important observation which was completely hidden from Faraday. At the same time that electric current in the copper conductors connecting the source to the electrolytic cell consists of a unidirectional flow of free electrons, **two different types of charge carriers—positive silver ions and negative nitrate ions—are flowing simultaneously in opposite directions through the electrolyte.** But the observable effect some 100 years ago was that, when electric current passed through the electrolytic cell of Figure 3-3, the anode lost weight and the cathode gained weight. Hence, it was assumed that the direction of electric current within the cell was from anode to cathode to convey the silver with it. As we noted in Section 2-11, this became the **conventional** direction for electric current.

3-4 CONDUCTION IN VACUUM AND GASES

If we connect a voltage source to a pair of electrodes in a vacuum, as in Figure 3-4(a), no electric current will flow since there are no charge carriers in a vacuum. The free electrons in the electrodes do not possess sufficient energy at room temperature to escape from the surface of the metal. However, if we supply additional energy to the cathode, usually by heating it to red heat with an electric heater, as in Figure 3-4(b), some of the free electrons in the cathode attain sufficient velocity to escape from the surface of the cathode. Once **emitted** from the cathode, these electrons are attracted by the positively charged anode (or plate) and electric current flows around the circuit. Hence, under proper conditions, the charge carriers in a vacuum are electrons, just as in metallic conductors.

Suppose we reverse the terminals of the battery in Figure 3-4(b) so that the anode of the vacuum tube is now connected to the negative terminal of the voltage source and the cathode is connected to the positive terminal. The voltage source will attempt to make electrons flow in the reverse direction through the tube. Since the anode is not heated, it cannot emit electrons as does the cathode. Consequently, just as in Figure 3-4(a), no

(a) No current flow   (b) Electrons emitted by heated cathode

FIGURE 3-4
Electric Current in a Vacuum.

electric current can flow through the tube, even though the cathode is still red hot. Thermionic vacuum tubes, therefore, allow electric current only in one direction.

If we now introduce a small amount of a gas such as neon or mercury vapor into our vacuum tube, as is the case in neon signs and fluorescent lamps, still another type of electric conduction takes place. The electrons emitted by the cathode pick up kinetic energy as they are attracted toward the anode. As these high-speed electrons collide with the gas atoms between the cathode and the anode, they possess enough energy to knock some of the electrons of the gas atoms out of orbit, leaving the gas as positively charged ions. There will now be an increase in the number of electrons arriving at the anode. There will also be a simultaneous flow of positive gas ions to the cathode to recover their missing electrons.

3-5 SEMICONDUCTORS

Column IV of the periodic table of elements includes carbon (six electrons), silicon (14 electrons), and germanium (32 electrons). These elements are grouped because each has four electrons which can act as valence electrons, and each is short four electrons to fill a subshell or energy level. (Check with Figure 2-3.) These three elements tend to form a crystal structure by a covalent-bonding pattern. Pure crystals are rare in nature, but we can "grow" pure or **intrinsic** crystals of germanium and silicon from the purified molten metal. Although the crystal is a three-dimensional structure, we can represent the covalent bonds by a two-dimensional sketch, as in Figure 3-5. Each silicon (or carbon or germanium) atom shares one of its four valence electrons with four adjacent atoms.

Theoretically, we might expect these intrinsic crystals to be insulators. Actually, there is sufficient heat energy in the crystal to cause millions of valence electrons to break away from their valence bonds. As fast as these electrons give up their new-found energy to reform covalent bonds, they are replaced by more free electrons from freshly broken bonds. Hence, at room temperature, there are always sufficient charge carriers in these crystals that we have to rate them as rather poor insulators. Nevertheless, by carefully introducing certain impurities into the intrinsic crystal, we can make use of the equally poor electrical *conductivity* of germanium and silicon crystals in developing

FIGURE 3-5
Covalent Bonds in a Crystal of Pure Silicon.

Si = Silicon nucleus + first 10 electrons

● = Valence electron

electronic devices with rather special electrical characteristics. We call such materials **semiconductors**.

In metals such as copper and aluminum, there is a free electron for every atom. But in semiconductors, the number of natural charge carriers can vary considerably under the influence of various forms of energy. We can also control the number and type of additional charge carriers by controlling the impurities introduced into a pure semiconductor crystal.

In growing a silicon crystal, suppose we *dope* the pure silicon by adding one part in 10 million of a *pentavalent* element such as arsenic from column V of the periodic table. Arsenic atoms take up positions in the lattice structure of the crystal, as shown in Figure 3-6. But only four of the five valence electrons can enter into valence bonds. The fifth electron is only loosely held by the attraction of the arsenic nucleus. Very little energy is required to break these surplus electrons *free* of the attraction of the arsenic nucleus. Adding a pentavalent impurity to a semiconductor thus makes a controlled number of electrons available as electric charge carriers. Hence, we call the arsenic atoms electron-**donor** atoms. And we call the doped semiconductor an **n-type** semiconductor, since negatively charged electrons are available as charge carriers.

Suppose we dope the pure silicon with a small amount of a *trivalent* element such as indium from column III of the periodic table. As indium atoms take up positions in the lattice structure of the semiconductor crystal, one valence bond cannot be completed, since indium has only three valence electrons. As shown in Figure 3-7, there is a **hole** in the covalent bonding pattern where an electron is missing. As a result, very little energy is required for an electron to break a valence bond to move into this electron-**acceptor** void, as shown in Figure 3-8. If we were not quick enough to see this electron

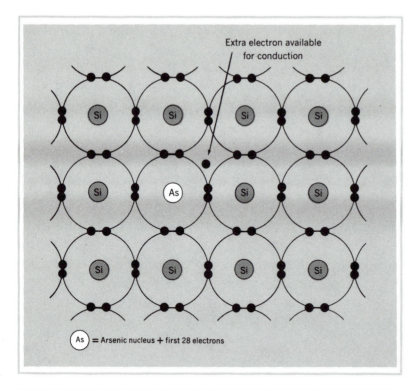

FIGURE 3-6
Donor Atom in an *n*-Type
Semiconductor Crystal.

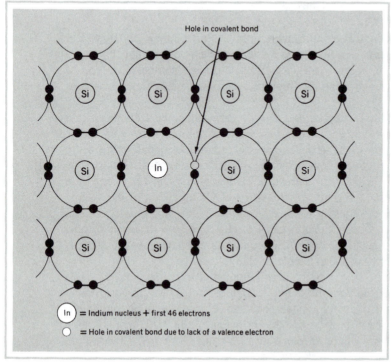

FIGURE 3-7
Acceptor Atom in a *p*-Type
Semiconductor Crystal.

FIGURE 3-8
Hole Flow in a *p*-Type Semiconductor.

move, the observable difference between Figure 3-7 and Figure 3-8 is that the hole has apparently moved to a new location. Another nearby valence bond would then break and the hole would move once more.

It takes a while to become used to the concept of a moving hole where an electron should be but is not. Suppose we place a string of 12 colored lights on a Christmas tree only to discover that we have only 11 lamps. We move lamps from one socket to another as we try to leave the empty socket in the most inconspicuous place. What attracts our attention with each move is not the lamps which have been moved; rather, it is the empty socket, which appears to move every time we switch a lamp to another socket.

In dealing with semiconductors with trivalent impurities, we find it more convenient to consider the **hole flow** than to keep track of electrons jumping from one valence bond to another. Holes move in the opposite direction to electrons, and since a hole is the absence of a negative electron, the electric charge where a hole is located is no longer electrically neutral. Hence, holes are considered to have a *positive* charge equal in magnitude to the negative charge of an electron. Semiconductors doped with acceptor impurities are, therefore, called ***p*-type** semiconductors. These semiconductors allow for electric conduction via positive holes.

3-6 *pn*-JUNCTIONS

We can illustrate the justification for the cost of developing two types of semiconductor crystals having different types of charge carriers by examining an electron device consisting of a layer of a *p*-type semiconductor diffused onto a layer of *n*-type semiconductor, as shown in Figure 3-9. In considering the mechanics of electrical conduction in semiconductor crystals, we have overlooked the equal number of electrons and holes created as

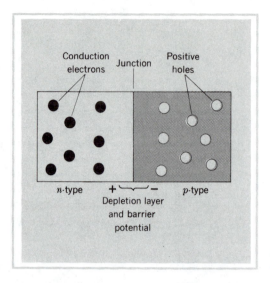

Conduction electrons

Junction

Positive holes

n-type

Depletion layer and barrier potential

p-type

FIGURE 3-9
Cross Section of a *pn*-Junction.

a result of covalent bonds broken by the heat energy possessed by the crystals. The quantity of these **minority carriers** which give rise to **leakage current** in semiconductors is negligible in good semiconductors. The **majority carriers** are those specifically created as described in Section 3-5. In the sketch in Figure 3-9, only the majority carriers are shown.

When *p*-type and *n*-type semiconductors are diffused together, as shown in Figure 3-9, some of the conduction electrons near the junction in the *n*-type semiconductor cross into the *p*-type semiconductor to fill the holes in the valence bonds near the junction. This leaves a **depletion layer** on both sides of the junction where the concentration of charge carriers is greatly reduced. The flow of electrons from left to right across the junction also creates a small potential difference between the two types of semiconductors, with the *p*-type semiconductor now becoming slightly more negative than the *n*-type, which has lost some electrons. Just as the chemical action in the battery in Figure 2-12 ceases when the potential difference builds up to a certain magnitude, electrons will cease flowing across the junction in search of holes in valence bonds when this depletion-layer potential or **barrier potential** reaches a certain magnitude.[†]

We can increase the potential difference between the two semiconductors by connecting them to an external voltage source. If the polarity of the external voltage source *increases* the natural contact potential, as shown in Figure 3-10(a), the electrons in the *n*-type material are attracted *away from* the junction. Similarly, positive holes in the *p*-type semiconductor are attracted away from the junction. As a result, electric current in the circuit is very small. This polarity of external source voltage **reverse biases** the *pn*-junction.

If, on the other hand, we apply the external source voltage with the polarity shown in Figure 3-10(b), the charge carriers in both types of semiconductor are driven *toward* the junction where the holes and electrons recombine into valence bonds. Replacement charge carriers are introduced into the semiconductors by the voltage source and a considerable electric current is maintained. This polarity of external source voltage

[†] $V_B = 0.3$ V for a germanium *pn*-junction and 0.7 V for a silicon *pn*-junction.

54

PART I   THE BASIC ELECTRIC CIRCUIT

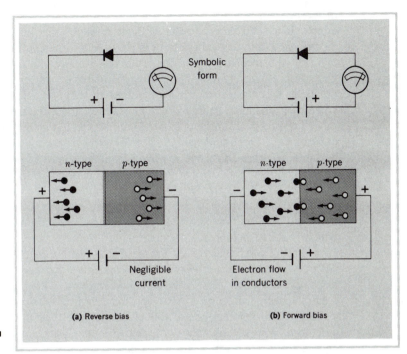

FIGURE 3-10
Rectifier Action of a
pn-Junction.

Symbolic form

n-type p-type

Negligible current

(a) Reverse bias

n-type p-type

Electron flow in conductors

(b) Forward bias

forward biases the *pn*-junction. Hence, like the vacuum **diode** (two-element device) of Section 3-4, the *pn*-junction allows significant electric current only in one direction.

3-7 BATTERIES

For an electric circuit to convey energy from one location to another, we require a device that can generate an electric pressure or potential rise which, in turn, will force free electrons to flow around the circuit. To accomplish this, a practical **voltage source** (such as the simple chemical battery of Figure 2-12) seemingly pumps electrons from one of its terminals to the other. Since this involves moving electrons within the voltage source away from the positive terminal (which attracts electrons) and toward the negative terminal (which repels electrons), some form of energy must be expended. Therefore, as the current in an electric circuit conveys energy to a load, energy must be expended in the voltage source in order to maintain a constant potential rise between its terminals. Actually then, a practical voltage source is an energy *convertor*, which converts some readily available form of energy into electric energy in the form of charge carriers flowing under pressure around the circuit. Similarly, a load in an electric circuit is a device that can *convert* electric energy into heat, light, mechanical, or other useful forms of energy.

We have already encountered one simple voltage source in the form of the simple wet-cell battery of Section 2-8. As long as the switch is closed in Figure 2-12, energy is taken from a chemical form and conveyed by the electric current to the lamp where it is converted into light energy and heat energy. While this energy transfer is taking place, the combining of chlorine ions with the zinc dilutes the acid and eats away the zinc rod. In time then, this battery can no longer maintain a potential difference. We have noted that when we open the switch, chemical action will cease when the potential difference builds up to a level at which the chlorine ion is no longer able to reach the zinc atoms. If we raise the potential difference above this level by connecting some other electric generator to the battery, it is possible to separate the chlorine ions from the zinc chloride, thus **recharging** the battery at the expense of electric energy from the external generator. However, recharging this type of battery in this manner is not economically practical. It is cheaper to replace the used up acid and zinc rod. Cells that are discarded when their chemical energy is exhausted are called **primary** cells.

The practical limitations of the simple wet-cell battery restrict its use to laboratory experiments. The liquid nature of the **electrolyte** means that the battery is not readily portable. Another disadvantage results from the release of hydrogen gas at the copper rod as the hydrogen ions recover their missing electrons from the copper. After the switch in Figure 2-12 has been closed for a few minutes, the copper is blanketed with hydrogen bubbles. Since hydrogen gas is an electric insulator (see Figure 2-4), this layer of hydrogen bubbles prevents hydrogen ions from getting to the copper rod. This action, which is called **polarization,** appreciably limits the current in the circuit.

The **dry-cell** battery of Figure 3-11 overcomes these two disadvantages. The electrolyte must still be *wet* so that the positive and negative ions can migrate to the respective electrodes. But in the so-called *dry* battery the electrolyte either is mixed with a suitable binder to form a paste or saturates an absorbent spacer. Polarization is prevented by surrounding the positive electrode with a chemical **depolarizer** that supplies oxygen to convert any free hydrogen ions into water. As shown in Figure 3-11, the depolarizer occupies most of the space in this type of dry cell.

Most practical dry cells consist of a negative electrode of a pure metal (usually zinc), a positive electrode of a metallic oxide, and an electrolyte separating the two electrodes. In its simplest form, the electro-chemical reaction removes negative oxygen

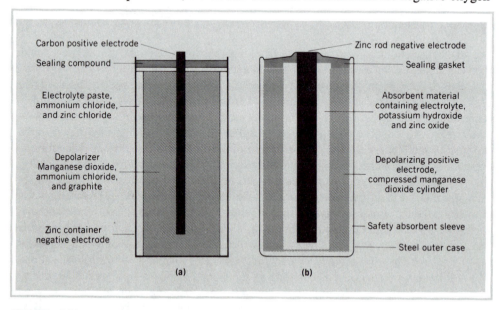

FIGURE 3-11
Simplified Cross Sections: (a) Zinc-Carbon Dry Cell; and (b) Manganese-Alkaline Dry Cell.

ions from the positive electrode via the electrolyte to oxidize the negative electrode. The ion exchange builds up a deficiency of electrons at the positive electrode and a surplus at the negative electrode, thereby creating a potential difference between the electrodes. Chemical action ceases when this PD reaches a critical level. There are many combinations that will generate a potential difference, but only a few can be adapted to viable production methods and costs. Considerable research is still under way to improve existing batteries and to develop new cells with a high ratio of energy output to size and cost.

The **zinc-carbon dry-cell** construction illustrated in Figure 3-11(a) is a familiar version of a voltaic cell patented in 1868 by Georges Leclanché. The zinc container is also the negative electrode. The positive electrode is a carbon rod in contact with the manganese dioxide depolarizer. The electrolyte paste consists mainly of ammonium chloride with some zinc chloride to improve its electrical conductivity. Although the chemical reaction is not quite as simple as that of the metal-acid wet cell of Figure 2-12, the net results are the same. Chemical reaction between the zinc and the ammonium chloride electrolyte adds surplus electrons from the chlorine ions to the negative electrode and releases positive hydrogen ions which, in turn, react with the manganese dioxide to regain their missing electrons as they form water. With the loss of oxygen, the positive electrode thus builds up a deficiency of electrons. Chemical action ceases when the potential difference between the electrodes reaches 1.5 volts (as compared to 1.1 volts for the zinc-copper wet cell).

An improvement on the zinc-carbon battery is the **manganese-alkaline cell,** shown in Figure 3-11(b). As the name indicates, the electrolyte is an alkaline solution of potassium hydroxide and zinc oxide. The negative electrode is still zinc, but in a highly purified form to minimize self-discharge through local action. The positive electrode is made by compressing pure manganese dioxide into cylinders. As a result, the positive electrode provides its own depolarizing action. As shown in Figure 3-11(b), the manganese-alkaline cell is built "inside-out" from the zinc-carbon battery. This provides the outer electrode with maximum surface area to take care of polarization. The complete cell is sealed in a leakproof steel case. The manganese-alkaline cell also generates a voltage of 1.5 volts. Due to the compact, "inside-out" nature of the self-depolarizing positive electrode, the manganese-alkaline battery can maintain its terminal voltage at a given current drain for more than twice as long as a zinc-carbon battery of the same physical size. This increased operating life more than offsets the higher cost associated with the more precise construction of manganese-alkaline batteries.

A third type of dry battery also has a zinc negative electrode and a potassium hydroxide electrolyte. The positive electrode of the **mercury cell** is a self-depolarizing cylinder of mercuric oxide. In addition to the familiar cylindrical form, the mercury cell adapts well to the miniature button form used in hearing aids and electric watches. Such a cell consists of small wafers of amalgamated zinc and mercuric oxide separated by a wafer of absorbent material saturated with potassium hydroxide electrolyte solution. Again the higher cost of the mercury dry battery is offset by its greater output. The nominal terminal voltage (1.3 V) of the mercury cell is slightly lower than that of the zinc-carbon and manganese-alkaline batteries.

Replacing the mercury oxide positive electrode with silver oxide provides a dry cell with improved shelf life and **energy density** (ratio of electric energy output to weight). In addition, the terminal voltage is increased to 1.5 volts. A more recent "high-tech" design uses lithium as the negative electrode, sulphur dioxide as the positive electrode, and a nonaqueous inorganic solvent (thionyl chloride) as the electrolyte. The lithium cell produces a terminal voltage of almost 3 volts with an energy density about three times as great as an equivalent silver oxide cell.

Since voltaic cells are based on electrolytic conduction, it is quite proper to identify the two electrodes as anode and cathode. Indeed, manufacturers' technical data usually use these terms. We have not used these terms in discussing batteries to avoid confusion as to the *polarity* of the voltage between the electrodes. In Section 3-5, we defined *anode* and *cathode* in terms of the electrodes through which electrons either leave or enter the cell—not in terms of the polarity of the potential difference between the electrodes. From this definition (see Glossary), the anode of an electroplating bath (Figure 3-3) or a vacuum tube (Figure 3-4) is connected to the *positive* terminal of the voltage source, thus making the anode positive with respect to the cathode. But in voltaic cells, by definition, the zinc electrode in Figure 3-11 is the anode, since electrons leave the *negative* electrode to flow through the external circuit.

One particular wet cell worthy of note at this point is the lead-acid storage battery. The chemical action is such that no gas is liberated as it discharges its chemical energy. Absence of polarization, along with the ability to convert chemical energy into electric energy very rapidly, allows this battery to maintain an appreciable potential difference between its terminals even when the current through the circuit approaches 100 amperes. The high rate of energy conversion possible with the **lead-acid** battery makes it particularly useful for industrial applications. In this case, recharging the battery by converting surplus electric energy back into energy stored in chemical form is economically feasible. A cell which is recharged in this manner is called a **secondary cell.** The ability of the lead-acid battery to store energy in chemical form accounts for its common classification as a **storage battery.** The electric system of an automobile is a good example of the charge and discharge cycle of the lead-acid storage battery. When we start the car, energy stored in chemical form is converted to electric energy to operate the starter motor. Then, when the car is being driven down the highway, surplus electric energy from the generator (obtained at the expense of mechanical energy from the engine, which in turn came from the chemical energy of the fuel, and so on) reverses the chemical action in the battery to store up energy for the next time the car has to be started.

Another popular rechargeable battery is the **nickel-cadmium** cell. The electrochemical action of the original "ni-cad" battery was developed around 1900 in Sweden to provide a rugged, high-output wet-cell battery for industrial applications. Modern manufacturing methods allow the electrodes to be in the form of thin metallic strips separated by porous spacers saturated with potassium hydroxide (KOH) solution as the electrolyte. This thin "sandwich" is rolled into a cylinder "jelly-roll" fashion to fit into the standard "AA," "C," and "D" dry-cell containers. This compact format makes nickel-cadmium batteries particularly attractive for such portable devices as tape recorders, cordless electric drills, electrical toys, and so on, which discharge primary dry cells quite rapidly. The nickel-cadmium cell produces a terminal voltage of only 1.25 volts— slightly lower than equivalent dry cells. The electrolyte of the nickel-cadmium cell does not change chemically as the cell discharges. Hence, we can readily reverse the chemical reaction by placing a discharged cell in a battery charger that applies a slightly higher potential difference to the electrodes to force current to flow in the opposite direction through the cell.

3-8 OTHER VOLTAGE SOURCES

Batteries are a convenient and not too costly voltage source for such portable applications as flashlights, hearing aids, and portable radios, where the amount of energy converted is relatively small. Batteries are also used for emergency power sources for telephone exchanges, and for supplying energy to the starter motor in an automobile. However, batteries are quite impractical when it comes to supplying the huge amounts of electric energy required by the industries and residences of a modern city.[†]

Most of the world's electric energy is generated at the expense of mechanical energy by rotating machines called **dynamos** (or, through common usage, simply **generators**). In some parts of the world, this mechanical energy is obtained from falling water, as discussed in Section 2-7 when we considered the operation of the hydroelectric generating station shown in Figure 2-10. Much of the required mechanical energy comes from the conversion of the chemical energy of coal and oil into heat energy and then the conversion of this heat energy into mechanical energy by the use of some form of heat engine such as a steam or gas turbine or a diesel engine. The mechanical output of the heat engine, in turn, drives the rotating shaft of a suitable dynamo. In the more recent generating stations, nuclear energy supplies the heat energy to operate the turbines. Since there has been some public resistance in recent years to building more nuclear generating stations, several alternative systems are being investigated.

Until recently, we have not had an efficient means of converting large amounts of chemical energy directly into electric energy. Such a device would be very desirable since it would greatly reduce the initial cost and the maintenance problems associated with rotating machinery. Recent developments indicate that **fuel cells** could become quite important as practical voltage sources in the not too distant future. Consolidated Edison is now operating a demonstration 4.8-megawatt fuel cell generating station in New York City. It is expected that present fuel cell technology can at least match the overall efficiency of coal-fired plants.

The chemical reaction accompanying the conversion of energy in a battery includes either one or both of the electrodes. Consequently, every so often we must restore the electrodes by either replacing the battery or electrically recharging it. But the chemical action in a fuel cell does not affect the electrodes; only the chemical fuel is consumed. Therefore, a fuel cell can maintain a terminal voltage as long as we supply a suitable fuel whose chemical energy is converted directly into electric energy by the cell.

The basic fuel cell uses the energy released as hydrogen and oxygen combine to form water to separate electric charges in the form of a + charge at one electrode and a − charge at the other. Water is an ionic compound in which two hydrogen atoms each lend their single valence electron to an oxygen atom. To create a fuel cell, this combination must take place in two steps. First, the hydrogen and oxygen must be ionized at *different* electrodes in the fuel cell, and then the ions combine to form water. Figure 3-12 is a simplified representation of how this may be achieved. Hydrogen and oxygen (the oxygen may come from air) are fed into separate chambers, each with a catalytic

---

[†] Some electric utility companies are experimenting with huge "storage batteries" in the form of large tanks. The purpose of such devices is to store surplus electric energy generated in off-peak hours (during the night). High-efficiency semiconductor rectifiers and converters allow this stored energy to be fed back into the system to supplement generating capacity during peak-load periods.

FIGURE 3-12
Simplified Cross Section of a Fuel Cell.

electrode to encourage ionization. The chambers are separated by an electrolytic barrier which permits the flow of ions but which will not allow the passage of gas molecules. Consequently, for the hydrogen and oxygen to combine, hydrogen must give its valence electrons to the electrode in the left chamber and the oxygen must acquire electrons from the electrode in the right chamber.

By keeping the oxygen and the hydrogen "fuel" separated during the ionizing process, the fuel cell develops an electric potential difference between its two electrodes. When the switch is closed, this potential difference maintains an electric current through the load. Like the battery, when the switch is open, the potential difference builds up to a certain magnitude where it exceeds the chemical attraction for hydrogen and oxygen ions to combine. Ionization then ceases, and, since gas molecules cannot pass through the electrolytic barrier, consumption of the fuel gas also ceases. Many thousands of individual fuel cells are connected in a series-parallel arrangement to generate sufficient voltage and current for a practical generating unit.

Our increased understanding of semiconductor behavior has permitted development of **photovoltaic cells** that generate a potential difference from light energy striking them. A typical **solar cell** consists of a very thin layer of trivalent boron atoms diffused into the surface of a slab on n-type silicon. The p-type layer thus formed is so thin that light energy passes through it to the n-type base where it supplies the energy needed to drive electrons through the depletion layer to fill the holes in the thin p-type layer, thus creating a potential difference between the two layers. Banks of these cells convert sufficient energy to operate the radio receivers and transmitters of communications satellites for an almost unlimited period of time.

REVIEW QUESTIONS

3-1. What is the charge (coulombs) of the available charge carriers in a cubic centimeter of copper?

3-2. Both copper and sodium chloride (common table salt) are crystalline solids; yet, copper is a good conductor of electricity and NaCl is an insulator. What distinction in the bonding process accounts for this difference?

3-3. Valence bonds in an insulator are broken by heat energy possessed by the insulator. They can also be broken by applying a large electric potential difference across a layer of insulating material. What is the observable difference in the behavior of the material under these two conditions?

3-4. Explain how the silver anode in the electrolytic cell shown in Figure 3-3 obtains the positive charge required to attract nitrate ions.

3-5. Why could electric current not pass through the solution of an ionic compound if dissociation of the ions did not occur?

3-6. What is the difference between the process of electric conduction in a liquid and in a metallic conductor?

3-7. As in incandescent lamps, the heater of a vacuum tube gradually becomes thinner due to vaporization of the metal and finally burns out. What effect does the burning out of the heater have on the operation of a vacuum-tube rectifier?

3-8. At room temperature, there are as many electrons in a *pure* semiconductor crystal as there are positive holes. Explain the source of these charge carriers.

3-9. Explain the mechanism of hole flow in a semiconductor.

3-10. When electric current flows in an intrinsic semiconductor or in an insulator, equal numbers of electrons and positive holes flow in opposite directions. How does this compare with electric charge flow in an electrolyte?

3-11. In a crystal of *n*-type germanium at room temperature, electrons are the **majority** carriers and positive holes are the **minority** carriers. Why are there more electrons than holes?

3-12. The magnitude of the leakage current when a *pn*-junction is **reverse biased** depends on the concentration of *minority* carriers. Explain this with reference to Figure 3-10(a).

3-13. As a mercury cell is discharged, the zinc electrode is oxidized into zinc oxide by combination with negative oxygen ions from the electrolyte. The positive potassium ions reduce the mercuric oxide electrode to mercury, thus maintaining the concentration of the solution. Explain, with a sketch, how this chemical action generates a voltage between the terminals of the cell.

3-14. Describe the various energy conversions that take place from the time some of the chemical energy of an automobile storage battery is used to start the engine of the car until this energy is restored to the battery by the car's electric generator.

3-15 Why do we not define an **anode** as the positive electrode of an electrolytic cell?

3-16. What is the significance of the term **energy density** in specifications for primary cells?

3-17. Why do we not *reverse* the polarity when we use an external voltage source to recharge a secondary cell?

3-18. How does the chemical action of a fuel cell differ from that of a dry cell?

4

RESISTANCE

In the basic electric circuit diagram of Figure 4-1, we have used the generalized form of the graphic symbol to represent a voltage source (Table 1-1). Although the symbol is the same as that for a single-cell battery, we shall use it in the following pages to represent any direct-current voltage source when we are concerned only with the magnitude of the potential rise produced and are not particularly interested in the details of the manner in which the voltage is generated. All that we usually "see" of the voltage source provided by an electrical utility company, for example, are the two terminals of the outlet into which we plug a lamp, toaster, etc. We can show that these two terminals do represent a voltage source by connecting a voltmeter *across* them (see Figure 2-13). The magnitude of the applied voltage is a property of the voltage source and is essentially independent of the electric circuit connected to the source.

We now want to find out just what governs the numerical magnitude of the electric current that will flow when we close the switch in the basic electric circuit of Figure 4-1.

4-1 OHM'S LAW
OF CONSTANT PROPORTIONALITY

Over a hundred years ago, Georg Simon Ohm discovered that every time he closed the switch in a circuit such as that of Figure 4-1, the current became the same constant value. He also discovered that, providing the temperature of the conductor did not change, doubling the applied voltage doubled the current and tripling the applied voltage tripled the current. In other words,

For a given circuit, the ratio of the applied voltage to the current is a constant.

FIGURE 4-1
Schematic Representation of a Basic Electric Circuit.

This statement became known as **Ohm's law of constant proportionality.** We can express Ohm's law in equation form as

$$\frac{E}{I} = k \tag{4-1}$$

where E is the applied voltage in volts, I is the resulting current in amperes, and k is a numerical constant.

Carrying the discovery a step further, Ohm found that changing the conductors or the load to some other size resulted in a different magnitude of electric current, hence a different value for the numerical constant. If the current is one ampere when the applied voltage is 10 volts, the constant becomes 10/1, or simply 10; and if the current is 0.5 ampere when the applied voltage is 10 volts, the constant becomes 10/0.5, or 20. From these results, Ohm concluded that this constant E/I ratio for a given circuit is, therefore, a property of that circuit. Since, for a given applied voltage, the current must *decrease* if the numerical value of the constant *increases,* we can think of this constant as representing the *opposition* of the circuit to flow of charge carriers. This property should then be given a name which suggests *opposition,* such as **resistance.** Therefore,

> *Resistance is the opposition of an electric circuit to electric current through that circuit.*
> *The letter symbol for resistance is R.*

We can substitute the symbol R for the constant k in Equation (4-1). There is another change that we should make in Equation (4-1). As we noted in Figure 2-13, if we open the switch in the circuit of Figure 4-1, the current becomes zero. According to Equation (4-1), the resistance of the circuit is now infinitely high. This is quite true when we remember that the switch is part of the complete circuit. The resistance of an open switch should be infinitely high. But we are usually more interested in determining the resistance of the *load.* As we noted in Section 2-10, when the switch is open and the current is zero, there is also zero voltage drop across the load. Hence, for Equation (4-1) to represent the resistance of the load only, we must think in terms of the voltage drop across the load, rather than of the voltage applied to the total circuit.

Although it would be quite proper to express resistance in terms of **volts per ampere,** it was decided to honor Ohm's original discovery by stating that

> *The ohm is the SI unit of electric resistance.*
> *The unit symbol for ohm is the Greek letter Ω (omega).*

Since Equation (4-1) has now become

$$R = \frac{V}{I}$$

(4-2)

where R is the resistance in ohms, V is the voltage drop across the resistance in volts, and I is the current through the resistance in amperes, the size of the ohm is automatically established. Therefore,

An electric circuit has a resistance of one ohm when a current of one ampere through the circuit causes a voltage drop across the circuit of one volt.

Few people nowadays recall the original statement of Ohm's law of constant proportionality which allowed us to define resistance and to establish a unit of resistance. As a result, Ohm's law is usually stated simply by the equation

$$R = \frac{V}{I}$$

(4-2)

EXAMPLE 4-1

What is the resistance of a lamp if a current of 150 mA flows through the lamp when a voltage of 6.0 V is applied to its terminals?

SOLUTION

$$R = \frac{V}{I} = \frac{6.0 \text{ V}}{150 \text{ mA}} = \textbf{40 } \boldsymbol{\Omega}$$

EXAMPLE 4-2

What value of resistance is required to limit the current through the resistance to 20 μA if the voltage drop across the resistance is to be 480 mV?

SOLUTION

$$R = \frac{V}{I} = \frac{480 \text{ mV}}{20 \text{ } \mu\text{A}} = 24\,000 \text{ } \Omega = \textbf{24 k}\boldsymbol{\Omega}$$

$(R = 480 \boxed{\text{EE}} \boxed{+/-} \; 3 \div 20 \boxed{\text{EE}} \boxed{+/-} \; 6 \boxed{=} 24\,000 = 24 \times 10^3)$

SELF-CHECK

4-1:1. What is the resistance of a toaster that allows 8.0 A of electric current to flow when the voltage across the toaster is 120 V?

4-1:2. Determine the current in a portable radio which uses a 9.0 V battery and has a resistance of 600 Ω.

4-2 THE NATURE OF RESISTANCE

In Section 2-4, we noted that free electrons in metallic conductors move around in a random manner among the metal ions in the crystal structure of the conductor, alternately giving up or adding to their kinetic energy. Most of this energy exchange is due to the vibration of the metal ions in their lattice positions as a consequence of the heat energy possessed by the conductor. When we apply a voltage to the conductor, the random motion continues, but superimposed on this random motion is a net drift along the length of the conductor, as shown in Figure 2-7(b). Since an electron represents a certain mass, as it accelerates along the length of the conductor, potential energy provided by the source is converted into kinetic energy associated with the velocity of the electron.

Before the accelerating free electron gets very far in its journey along the length of the conductor, it collides with one of the metal atoms. This collision results in a considerable reduction in the speed of the electron. As a result, some of the kinetic energy gained by the electron is transferred to the atom with which it collided in the form of heat energy (just as vigorous hand clapping results in a transfer of mechanical energy to heat energy). After the collision, the free electron is again urged by the applied voltage to accelerate and acquire more kinetic energy. Again it will collide with one of the metal atoms and give up some of its kinetic energy in the form of heat. Therefore, the progress of the electrons which constitute the charge carriers in a metallic conductor is characterized by alternate acceleration (during which interval potential energy from the source is transferred to the electron in kinetic form) and sudden deceleration (at which time some of the kinetic energy of the electron is transferred to the atoms of the conductor material in the form of heat).

These many collisions, which constitute an *opposition to electric current,* account for the property of electric conductors known as **resistance.** The consequence of resistance in an electric circuit is the transfer of energy taken from the source into heat energy whenever we force charge carriers to flow in conductors possessing this property.

Resistance is a desirable property in such circuit components as lamps and electric stove elements in which we are interested in producing heat energy. In circuit elements such as the conductors connecting the load to the source in Figure 4-1, resistance is an undesirable property. Not only is some of the energy in the system wasted in the form of heat in the conductors, but if heat is developed at a greater rate than the conductors can dissipate it or pass it on to the surrounding air, the temperature of the conductors can rise to the point where they become a fire hazard.

We often insert resistance into an electric circuit as a means of limiting current. For example, if we wish to dim the lamp in the circuit of Figure 4-2, we can connect a circuit element possessing appreciable resistance in *series* with the lamp and the battery. Now

FIGURE 4-2
Using a Resistor to Limit Current.

the free electrons have to flow through both the lamp and the added resistance in completing their journey from the negative terminal of the battery to the positive terminal. Since this additional resistance produces heat when current flows through it, rather than connecting the lamp to the battery with conductors possessing higher resistance, we prefer to lump the extra resistance into a single electric component termed a **resistor,** which has been designed to dissipate the heat adequately without damage to either itself or nearby objects.

4-3 FACTORS GOVERNING THE RESISTANCE OF METALLIC CONDUCTORS

In stating his law of constant proportionality, Ohm had to add the provision that the temperature must be kept constant. Therefore, temperature must have some effect on the resistance of an electric conductor. We can readily check this in the laboratory by measuring the V/I ratio (resistance) of a lamp operating at the rated value of applied voltage (white hot) and then measuring the V/I ratio when the applied voltage is reduced to 10% of the rated value and the filament gives off practically no light. We shall investigate the exact manner in which temperature affects the resistance of a metallic conductor a little later in this chapter. For the moment, we assume temperature to remain constant at normal room temperature (20° C).

If we cut two lengths of wire from the same reel, one being twice the length of the other, in order for electrons to make the complete trip through the longer wire, they will have twice as many opportunities to collide with atoms of the conductor material. Consequently, the opposition of the longer wire to electric current is twice as great as that of the shorter wire. Therefore,

The resistance of a metallic electric conductor is directly proportional to its length.

If we select a wire size which has double the cross-sectional area of another wire size and cut equal lengths of each, the wire with the larger cross section would have the same cross-sectional area as two pieces of the wire with the smaller cross section connected in *parallel*. Since each of the two smaller diameter wires in parallel will pass the same current when connected to a single voltage source, the two parallel-connected wires pass twice the current of one wire. Since $R = V/I$, double current with no change in the applied voltage represents half as much resistance. Since doubling the cross-sectional area of a single conductor is the same as using two of the smaller diameter wires in parallel, doubling the cross-sectional area cuts the resistance in half. Therefore,

The resistance of a metallic electric conductor is inversely proportional to its cross-sectional area.

We noted in Section 3-1 that some materials possess more free electrons per unit volume than others and that the greater the number of free electrons per unit volume, the better conductor that material becomes. The greater the number of charge carriers in a conductor of certain dimensions, the smaller the percentage of the total energy that is carried by any one electron. Hence, the amount of energy that is converted into heat by collision with atoms of conductor material is proportionately reduced. Thus, a silver wire has a lower resistance than a copper wire with the same dimensions, and the copper wire has a lower resistance than an aluminum wire with the same dimensions. Therefore,

The resistance of a metallic electric conductor is dependent on the type of conductor material.

SELF-CHECK

4-3:1. Conductor "A" has a length of 100 meters and a diameter of 2.0 mm. Conductor "B" has a length of 50 meters and a diameter of 3.0 mm. Find the resistance of "B" if both conductors are made of the same material and measurements show that "A" has a resistance of 2.0 Ω.

4-4 RESISTIVITY

Since we know that the resistance of a conductor is directly proportional to its length, if we are given the resistance of unit length of wire, we can readily calculate the resistance of any length of wire of that particular material having the same diameter. Similarly, since we know that the resistance of a conductor is inversely proportional to its cross-sectional area, if we are given the resistance of a length of wire with unit cross-sectional area, we can calculate the resistance of a similar length of wire of the same material with any cross-sectional area. Combining both of these statements, we find that if we know the resistance of a given conductor, we can calculate the resistance for *any* dimensions of a conductor of the same material at the same temperature.

> EXAMPLE 4-3
> An electric conductor 1 m long with a cross-sectional area of 1 mm^2 has a resistance of 0.017 Ω. What is the resistance of 50 m of wire of the same material with a cross-sectional area of 0.25 mm^2?
>
> SOLUTION
> From the relationships stated above, we can write the equation
>
> $$\frac{R_2}{R_1} = \frac{l_2}{l_1} \times \frac{A_1}{A_2} \qquad (4\text{-}3)$$
>
> $$\therefore R_2 = 0.017 \ \Omega \times \frac{50 \ \text{m}}{1 \ \text{m}} \times \frac{1 \ \text{mm}^2}{0.25 \ \text{mm}^2} = \mathbf{3.4 \ \Omega}$$

As we become more familiar with our calculators, we should resist the tendency to enter data directly from the statement of the example into the calculator. We should always follow the steps outlined in Example 2-2 (as we have done above), so that we have a better chance of spotting an error in the way we made the calculation. To enter the exponent for *square* millimeters in this example, we should note that the prefix *milli* is also squared. (One square meter = 10^6 square millimeters.) However, in this particular example, we might also notice that mm^2 in the denominator of the equation above cancels out mm^2 in the numerator. Hence, we do not need to enter the exponents if we so wish.

As Example 4-3 suggests, a convenient way of showing the effect of the type of material on the resistance of a given conductor is to state the resistance of a standard conductor with unit length and unit cross-sectional area made from that material.

The resistance of a unit length and cross-section conductor of a certain material is called the resistivity of that material. [†]

The letter symbol for resistivity is the Greek letter ρ (rho).

Since the SI base unit for both length and cross section is the meter, the SI standard conductor must take the form of a cube with each of its edges 1 meter in length. The resistivity of a certain electric conducting material will therefore be the resistance between opposite faces of the 1-meter cube. To determine the SI unit of resistivity, we can apply dimensional analysis techniques. Since $R = V/I$, ohms = volts ÷ amperes. But voltage is measured between the ends or across the *length* of a conductor, and current is measured through the *area* of the cross section. Hence, from the definition of resistivity,

$$\rho = \text{volts per unit length} \div \text{amperes per unit cross-sectional area}$$

$$= \frac{\text{volts}}{\text{meter}} \times \frac{\text{meter}^2}{\text{amperes}} = \frac{\text{volts}}{\text{amperes}} \times \text{meters}$$

$$= \textbf{ohm meters}$$

Therefore,

The ohm meter is the SI unit of resistivity.
The unit symbol for ohm meter is $\mathbf{\Omega \cdot m}$.

Since temperature has an effect on the resistance of electric conductors, values of resistivity for various conducting materials hold only for a certain specified temperature, usually 20 degrees Celsius.[‡] Table 4-1 gives the resistivity of the more common metallic conductor materials at 20°C.

In Equation (4-3), if we set $l_1 = 1$ m and $A_1 = 1$ m², then R_1 becomes ρ and has a value as given in Table 4-1. We can now rewrite Equation (4-3) to calculate the resistance of any electric conductor at normal room temperature:

$$R = \rho \times \frac{l}{A} \tag{4-4}$$

where R is the resistance of the conductor in ohms, l is the length of the conductor in meters, A is the cross-sectional area in square meters, and ρ is the resistivity of the conductor material in ohm meters at 20°C.

EXAMPLE 4-4

What is the resistance (at 20°C) of 200 m of an aluminum electric conductor whose cross-sectional area is 4 mm²?

[†] Sometimes called **specific resistance.**

[‡] The Celsius temperature scale, which places the freezing point of water at 0°C and the boiling point of water at +100°C, is used with SI units even though °C is not a true SI unit. (The SI base unit of temperature is the *kelvin*.)

TABLE 4-1

RESISTIVITY OF SOME COMMON ELECTRIC CONDUCTOR MATERIALS

Conductor Material	Resistivity (ohm meters at 20°C)
Silver	1.64×10^{-8}
Copper (annealed)	1.72×10^{-8}
Aluminum	2.83×10^{-8}
Tungsten	5.5×10^{-8}
Nickel	7.8×10^{-8}
Iron (pure)	12.0×10^{-8}
Constantan	49×10^{-8}
Nichrome II	110×10^{-8}

SOLUTION

Using the value of ρ for aluminum from Table 4-1,

$$R = \rho \times \frac{l}{A} = 2.83 \times 10^{-8} \; \Omega \cdot m \times \frac{200 \; m}{4 \times 10^{-6} \; m^2} = \mathbf{1.415 \; \Omega}$$

In entering the cross-sectional area of the conductor in the equation above, the conversion from milli squared to an exponent is shown in order to point out how the unit *meters* cancels out in Equation (4-4).

Most electric conductors have a circular cross section. If we know the diameter of the conductor, we must then express its cross-sectional area in square meters.

EXAMPLE 4-5

What is the resistance at normal room temperature of 60 m of copper wire having a diameter of 0.64 mm?

SOLUTION

$$A = \pi r^2 = \frac{\pi}{4} d^2 = \frac{\pi}{4} \times 0.64^2 \; mm^2 = 3.217 \times 10^{-7} \; m^2$$

$$R = \rho \times \frac{l}{A} = 1.72 \times 10^{-8} \; \Omega \cdot m \times \frac{60 \; m}{3.217 \times 10^{-7} \; m^2} = \mathbf{3.2 \; \Omega}$$

SELF-CHECK

4-4:1. The resistance of the heating wire in a toaster is 15 Ω and the diameter of the wire is 0.8 mm. What length (in meters) of wire is necessary to produce this resistance if the wire is composed of:
(a) Tungsten (b) Nichrome II?

4-4:2. A spool contains a 100 m length of aluminum wire which has a resistance of 0.10 Ω. If the diameters are the same, what length of silver wire would have the same resistance?

4-5 EFFECT OF TEMPERATURE ON RESISTANCE

We have already noted that temperature will have some effect on the resistance of an electric conductor, but so far we have not had to consider it since we have assumed that the temperature remained constant at 20°C. By checking the effect of temperature on the V/I ratio (resistance) of a conductor as previously suggested, we find that for most conducting materials, the resistance increases linearly with an increase in temperature over normal temperature ranges. In Figure 4-3, if R_1 is the resistance of a certain conductor at temperature T_1, then R_2 is its resistance at temperature T_2. The straight line CF represents the manner in which the resistance varies for temperatures between T_1 and T_2. The *slope* of the line CF is dependent on the type of material. Some alloys have been developed (e.g., **constantan**) for which the line CF is almost horizontal. Temperature has very little effect on the resistance of such materials. There are a few materials (e.g., **carbon** and other semiconductors) which have a *negative* temperature effect; that is, the resistance *decreases* as the temperature *increases*.

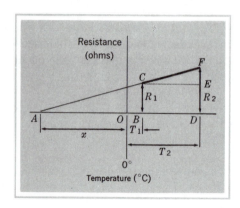

FIGURE   4-3
Effect of Temperature on Resistance.

Since CF in Figure 4-3 is a straight line, we can project it to the left to cut the baseline (zero ohms) at point A, thus producing two similar triangles ABC and ADF. Since these triangles are similar, we can show that

$$\frac{DF}{BC} = \frac{AD}{AB} = \frac{AO + OD}{AO + OB}$$

Substituting the electrical properties that these geometric dimensions represent gives

$$\frac{R_2}{R_1} = \frac{x + T_2}{x + T_1} \qquad (4\text{-}5)$$

where R_1 is the resistance of the conductor at temperature T_1, R_2 is the resistance of the

TABLE 4-2

VALUES OF x FOR SOME COMMON ELECTRIC CONDUCTOR MATERIALS (SEE FIGURE 4-3)	
Conductor Material	$x(°C)$
Silver	243
Copper	234.5
Aluminum	236
Tungsten	202
Nickel	147
Iron	180
Nichrome II	6250
Constantan (55% Cu, 45% Ni)	125 000

conductor at temperature T_2, T_1 and T_2 are the temperatures of the conductor in degrees Celsius, and x is a constant which depends on the type of material used in the conductor.

Table 4-2 gives the numerical value for x for some common electric conductor materials. The value of x is a constant for any given material. From the value of x for such materials as Nichrome II and constantan, it is apparent that x does not represent the actual temperature at which resistance becomes zero ohms. The resistance of metallic conductor materials approaches zero at absolute zero temperature, $-273°C$. The thin line AC in Figure 4-3 is a geometric projection of the linear portion of the resistance-temperature characteristic of various metallic conductor materials so that we can solve for the effect of temperature on resistance by the linear Equation (4-5).

EXAMPLE 4-6
A copper conductor has a resistance of $12\,\Omega$ at 20°C. What is its resistance at 100°C?

SOLUTION

$$\frac{R_2}{R_1} = \frac{x + T_2}{x + T_1} \quad \text{or} \quad R_2 = R_1 \frac{x + T_2}{x + T_1}$$

$$\therefore R_2 = 12\,\Omega \times \frac{(234.5 + 100)}{(234.5 + 20)} = \frac{12 \times 334.5}{254.5} = \mathbf{15.77\,\Omega}$$

We use parentheses when entering the data in the equation above to remind us to use the *parenthesis* keys of our calculator. Parentheses are used to ensure that the mathematical operations are performed by the calculator in the proper sequence. The chain calculation becomes

$$R_2 = 12 \;\boxed{\times}\; \boxed{(}\; 234.5 \;\boxed{+}\; 100 \;\boxed{)}\; \boxed{\div}\; \boxed{(}\; 234.5 \;\boxed{+}\; 20 \;\boxed{)}\; \boxed{=}\; 15.77\,\Omega$$

Note that the units °C cancel out in the equation above. The second last step is included for those solving the example by longhand. These figures will also appear on the calculator display as the chain calculation proceeds.

EXAMPLE 4-7

A precision resistor made of constantan wire has a resistance of 10 000 Ω at 20°C. What is its resistance when its temperature rises 20°C?

SOLUTION

$$R_2 = R_1 \frac{x + T_2}{x + T_1} = 10\,000 \times \frac{(125\,000 + 40)}{(125\,000 + 20)}$$

$$R_2 = \mathbf{10\,001.6\ \Omega}$$

In addition to being able to calculate the effect that temperature has on resistance, we can use the change in resistance of a certain material to calculate the temperature in locations where it is difficult to place a thermometer.

EXAMPLE 4-8

The copper winding of an electric motor which has been standing for several hours in a room at 20°C has a resistance of 0.20 Ω. When the motor has been in use for some hours the resistance of the winding is found to be 0.22 Ω. Calculate the temperature rise of the winding.

SOLUTION

$$\frac{R_2}{R_1} = \frac{x + T_2}{x + T_1}$$

$$\frac{0.22}{0.20} = \frac{234.5 + T_2}{234.5 + 20}$$

To enter the data into a calculator, we rearrange this equation to place T_2 by itself on the left-hand side of the equals sign.

$$T_2 = \frac{0.22}{0.20} \times (234.5 + 20) - 234.5 = 45.45°C$$

$$\therefore \text{temperature rise} = T_2 - T_1 = \mathbf{25.45°C}$$

If we now use Equation (4-4) to substitute for R_1 in Equation (4-5), we obtain

$$R = \rho \frac{l}{A} \left(\frac{x + T}{x + 20} \right) \tag{4-6}$$

With the values of ρ and x given in Tables 4-1 and 4-2, we can now determine the resistance of a given metallic conductor at any temperature.

EXAMPLE 4-9

What is the resistance of 300 m of copper wire with a cross-sectional area of 1.5 mm^2 at 40°C?

SOLUTION

From Table 4-1,

$$\rho = 1.72 \times 10^{-8} \ \Omega \cdot m$$

From Table 4-2,

$$x = 234.5°C$$

$$R = \rho \frac{l}{A} \left(\frac{x + T}{x + 20} \right)$$

$$= 1.72 \times 10^{-8} \ \Omega \cdot m \times \frac{300 \ m}{1.5 \ mm^2} \times \frac{(234.5 + 40)°C}{(234.5 + 20)°C}$$

$$= 3.44 \times \frac{274.5}{254.5} = \mathbf{3.71 \ \Omega}$$

The calculator chain is

$$R = 1.72 \ \boxed{EE} \ \boxed{+/-} \ 8 \times 300 \div 1.5 \ \boxed{EE} \ \boxed{+/-} \ 6 \times (234.5 + 40)$$
$$\div (234.5 + 20) = 3.71 \ \Omega$$

SELF-CHECK

4-5:1. The filament in a light bulb is made of thin tungsten wire. At room temperature (20°C) the resistance of the filament is 15 Ω. What is the filament's resistance when it is white-hot (3600°C)?

4-5:2. The resistance of the Nichrome II heating wire of a toaster is 15 Ω when the wire is red-hot (1500°C). What is this wire's resistance at room temperature?

4-6 TEMPERATURE COEFFICIENT OF RESISTANCE

Although the method we have just derived is quite adequate for calculating the effect of temperature on resistance, the constant x is based on a geometric projection, rather than on a property of the conductor material. So let us return to geometric construction on Figure 4-3 to derive an alternative method of showing the effect of temperature on the resistance of an electric conductor. The line CE is drawn parallel to the baseline. Since resistivity (ρ) is usually stated for a temperature of 20°C, we shall let T_1 represent 20°C.

Since triangle ABC is similar to triangle CEF, we can show that

$$\frac{FE}{BC} = \frac{CE}{AB} \quad \text{or} \quad \frac{FE}{R_1} = \frac{\Delta T}{x + 20}$$

where ΔT is the difference in temperature (degrees Celsius) between T_2 and 20°C.

Therefore,

$$FE = R_1 \left(\frac{\Delta T}{x + 20} \right)$$

Since CE and BD are parallel,

$$R_2 = R_1 + FE$$

$$\therefore R_2 = R_1 + R_1 \left(\frac{\Delta T}{x + 20} \right) = R_1 \left(1 + \frac{\Delta T}{x + 20} \right)$$

If we now call $1/(x + 20)$ the **temperature coefficient of resistance** (at 20°C) and represent it by the Greek letter α (alpha),

$$R_2 = R_1(1 + \alpha \Delta T) \qquad (4\text{-}7)$$

where R_2 is the resistance of a conductor at any specified temperature, R_1 is its resistance at 20°C, α is the temperature coefficient of resistance at 20°C [$\alpha = 1/(x + 20)$], and ΔT is the difference (degrees Celsius) between the specified temperature and 20°C.

Since R_1 represents the resistance of a given conductor at 20°C,

$$R_1 = \rho \frac{l}{A} \qquad (4\text{-}4)$$

Therefore, the general equation for the resistance of a metallic conductor, including the effect of temperature, becomes

$$R = \rho \frac{l}{A}(1 + \alpha \Delta T) \qquad (4\text{-}8)$$

where R is the resistance of the conductor in ohms, ρ is the resistivity of the material in ohm meters at 20°C, l is the length in meters, A is the cross-sectional area in square meters, α is the temperature coefficient of resistance of the material (ohmic change per degree per ohm at 20°C), and ΔT is the difference between the operating temperature and 20°C.

Since the temperature coefficient $(\alpha_{20°}) = 1/(x + 20)$ and x is dependent on the type of material, the temperature coefficient of resistance represents the manner in which each ohm of resistance of that material changes for each degree change in temperature from the reference temperature of 20°C. We can compute a table of temperature coefficients from the relationship we have already stated between α and x (Table 4-3).

EXAMPLE 4-9A
What is the resistance of 300 m of copper wire with a cross-sectional area of 1.5 mm² at 40°C?

SOLUTION
From Table 4-1,

$$\rho = 1.72 \times 10^{-8} \, \Omega \cdot m$$

TABLE 4-3

TEMPERATURE COEFFICIENT OF RESISTANCE OF SOME COMMON ELECTRIC CONDUCTOR MATERIALS AT 20°C	
Conductor Material	$\alpha_{20°}$
Silver	0.0038
Copper	0.003 93
Aluminum	0.0039
Tungsten	0.0045
Nickel	0.006
Iron	0.0055
Nichrome II	0.000 16
Constantan	0.000 008
Carbon	−0.0005

From Table 4-3,

$$\alpha = 0.003\,93$$

$$R = \rho \frac{l}{A}(1 + \alpha \Delta T)$$

$$= 1.72 \times 10^{-8}\,\Omega \cdot m \times \frac{300\ m}{1.5\ mm^2} \times (0.003\,93 \times 20 + 1)$$

$$= 3.44 \times 1.079 = \textbf{3.71 } \boldsymbol{\Omega}$$

[Note that in entering the data into the equation, we changed $(1 + \alpha \Delta T)$ to $(\alpha \Delta T + 1)$ so that we could accomplish the calculation on a calculator having only a single level of parenthesis.]

EXAMPLE 4-10
An electric heater made from Nichrome II wire has a resistance of 16 Ω at 1500°C. What is its resistance at normal room temperature?

SOLUTION
In this example, R_2 in Equation (4-7) is given and R_1 is unknown. Hence,

$$R_1 = \frac{R_2}{(1 + \alpha \Delta T)} = \frac{16}{(0.000\,16 \times 1480 + 1)}$$

$$= \frac{16}{1.237} = \textbf{12.9 } \boldsymbol{\Omega}$$

If the required temperature is below 20°C, ΔT will be a *negative* quantity and

Equation (4-8) becomes effectively:

$$R = \rho \frac{l}{A}(1 - \alpha \Delta T)$$

Solving Equation (4-8) on a calculator involves a fairly long chain of operations, as we discovered in Example 4-9A. If we are only required to solve a few examples to become familiar with Equation (4-8), the calculator is the simplest method. But if we are given an assignment to determine the resistance of a longer list of electric conductors, it may be worthwhile preparing a personal computer program. Once we have written the program in Example 4-11(a), we can RUN it as many times as we like by inserting values for the variables, as in Example 4-11(b).

EXAMPLE 4-11
(a) Write a BASIC computer program to calculate the resistance of any circular-cross-section metallic conductor.
(b) Use this program to determine the resistance of 75 m of aluminum wire with a diameter of 2 mm at a temperature of 36°C.

SOLUTION
(a) Since there are five independent variables in Equation (4-8), we can spread the INPUT statements out into a separate line for each variable. A typical program might consist of

```
100 REM RESISTANCE OF A METALLIC CONDUCTOR
110 INPUT "ENTER RESISTIVITY FOR TYPE OF METAL ";P
120 INPUT "ENTER LENGTH OF CONDUCTOR IN METERS ";L
130 INPUT "ENTER DIAMETER OF CONDUCTOR IN MILLIMETERS ";D
140 INPUT "ENTER TEMP COEFF FOR TYPE OF METAL ";A
150 INPUT "ENTER TEMP OF CONDUCTOR IN DEGREES C ";T
```

Before we write the equation for resistance into the processing statement of the program, we must modify Equation (4-8) to take into account the way the program requests data on conductor *diameter* and temperature. Whereas Equation (4-8) requires cross-sectional area in square meters, our program allows us to enter the conductor's diameter in millimeters. To do this,

$$\text{cross-sectional area} = \frac{\pi D^2}{4} \times 10^{-6} \text{ m}^2$$

$$= 7.854E-7*D^2$$

(^ is the BASIC symbol for an exponent.) Also,

$$\Delta T = T - 20$$

Although BASIC can handle two levels of parentheses, it may be easier if we have to troubleshoot or debug our program to solve Equation (4-8) by splitting it into two functions. Hence, our computer program might continue as

```
160 LET X = 1+A*(T-20)
170 LET R = P*L/(7.854E-7*D^2)*X
180 PRINT "RESISTANCE OF CONDUCTOR = ";R;" OHMS"
190 END
```

(b)

```
RUN
ENTER RESISTIVITY FOR TYPE OF METAL ? 2.83E-8
ENTER LENGTH OF CONDUCTOR IN METERS ? 75
ENTER DIAMETER OF CONDUCTOR IN MILLIMETERS ? 2
ENTER TEMP COEFF FOR TYPE OF METAL ? 0.0039
ENTER TEMP OF CONDUCTOR IN DEGREES C ? 36
RESISTANCE OF CONDUCTOR = 0.717769 OHMS
```

SELF-CHECK

4-6:1. Determine the resistance at 100°C of 1 km of aluminum wire which has a cross-sectional area of 0.5 mm^2.

4-6:2. At what temperature will a piece of iron wire have a resistance of 150 Ω if the wire has a resistance of 50 Ω at room temperature?

4-7 LINEAR RESISTORS

Most electric conductor materials behave in the manner discovered by Georg Ohm in establishing his law of constant proportionality; that is, doubling the applied voltage causes the current to double, tripling the applied voltage triples the current, and so on. We can show this relationship graphically, as in Figure 4-4. The straight-line graph represents a constant value of resistance; the smaller the resistance, the steeper the slope of the linear graph. A resistor that maintains this constant V/I ratio is known as a **linear** resistor. As current through a resistor is increased, electric energy is converted into heat at an increased rate, resulting in a rise in the temperature of the resistor. As indicated in Table 4-3, this increase in temperature causes a slight *increase* in the resistance of most conductor materials. But the percentage change in resistance over the usual temperature ranges encountered in practice is small enough in the case of the common electric conductor materials that the deviation of their resistance graphs from the straight lines of Figure 4-4 is hardly perceptible. Therefore, copper and aluminum are classed as linear resistors. Special alloys such as *constantan* with a temperature coefficient of almost zero have been developed for the production of resistance wire.

As the name suggests, we construct **wire-wound resistors** by winding a wire of one of the metal alloys such as constantan on a hollow porcelain tube and then sealing the

FIGURE 4-4
Volt-Ampere Graphs of Linear Resistors.

wire in position with a porcelain coating. Since temperature has an effect on the resistance of metallic conductors, and since the temperature of a resistor must rise when we increase the current passing through it, the alloy chosen for constructing wire-wound resistors must have almost zero resistance change over normal temperature ranges. The larger the physical dimensions of a resistor, the more readily it can dissipate heat to the surrounding air. Hence, the power rating of a resistor is proportional to its physical size.

Resistors used in electronic devices usually have a resistance of thousands of ohms (kilohms) and pass currents of only a few milliamperes. If, as is usually the case, the conversion of electric energy into heat is at the rate of less than one or two joules per second, a resistance element consisting of a mixture of finely ground carbon and an insulating composition pressed into a tiny rod less than a centimeter long is much more economical than wire-wound resistance elements. The resistance of **carbon-composition resistors** is governed by the ratio of carbon to insulating composition in the resistance element. Electrical connection to the resistance element is made through tinned copper "pigtail" leads imbedded in the ends of the element. The resistance element is sealed in a plastic jacket for protection.

Carbon-composition resistors do not have the linear resistance/temperature characteristic of metal wire resistors shown in Table 4-3. The resistance of a carbon-composition resistor increases if the temperature varies appreciably in *either* direction from normal room temperature, as shown by the expanded scale in Figure 4-5. Again, the physical dimensions of the resistor govern the rate at which it can dissipate heat. If the resistor is operated within its normal rating limits, the temperature variation is small enough for us to think of carbon-composition resistors as linear resistors.

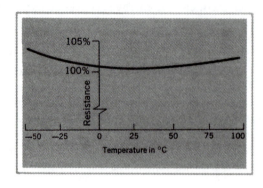

FIGURE 4-5
Resistance-Temperature Characteristic of Carbon Composition Resistors.

Resistors used with the electrical measuring instruments of Chapter 10 must be accurate to within 1% of the specified resistance if the measuring devices are to maintain a similar accuracy. The older form of **precision resistor** consists of a long, thin, insulated wire of an alloy with a very small temperature coefficient. In manufacturing this wire, care is taken to maintain a uniform cross section so that the resistance is a linear function of the length of wire used. The selected length of insulated wire is then spooled onto a small bobbin and connected to terminals on the end of the bobbin. A more recent manufacturing technique consists of depositing a thin film of the appropriate metal on a tiny ceramic or glass cylinder, after which leads are attached. The metal film may be etched if necessary to adjust the resistance to the specified value. Finally, the metal film precision resistor is given a protective coating of an insulating epoxy. Modern production control of both metal and carbon deposition processes has also allowed **thin-film resistors** to compete with carbon-composition resistors for general electronic applications. Thin-film resistors generate less noise due to random electron motion than do carbon-composition resistors.

The resistance element of a **variable resistor** is usually spread out along a 300° arc of a circle. Each end of the resistance element is connected to a fixed terminal. A third terminal is electrically connected to a wiper contact which is moved along the length of the resistance element by a rotatable shaft. Small variable resistors for electronic circuits usually have carbon resistance elements. The resistance elements of larger industrial variable resistors are wire-wound.

4-8 TOLERANCE AND RESISTOR COLOR CODE

In Section 4-7, we noted that the resistance of precision resistors must be within 1% of the specified value. We refer to the permissible error in resistance as the **tolerance** of a resistor. We express tolerance as a percentage of the nominal value. Most electronics applications do not require the precision of measuring devices. Hence, commercial resistors are available with tolerances of 5%, 10%, and 20%.

EXAMPLE 4-12

What are the permissible limits in the resistance of a 68-kilohm resistor with a 10% tolerance?

SOLUTION

Minimum resistance $= 68 \text{ k}\Omega \times 90\% = 61 \text{ k}\Omega$
Maximum resistance $= 68 \text{ k}\Omega \times 110\% = 75 \text{ k}\Omega$

Most carbon-composition and thin-film resistors are too small to print the resistance values on them clearly. Hence, this information is provided by a series of colored bands using a standard **resistor color code.** Note that the bands in Table 4-4 are near one end of the resistor so that there is no mistaking the order in which they are read—from the end toward the center. The resistance value consists of two significant digits multiplied by a power of ten (the number of zeros after the two digits). A tolerance band indicates either $\pm 5\%$ or $\pm 10\%$ tolerance. If this band is missing, the resistance value is within $\pm 20\%$ of the stated value and the resistor costs less.

TABLE 4-4

RESISTOR COLOR CODE

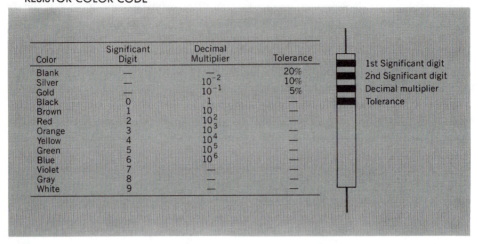

Color	Significant Digit	Decimal Multiplier	Tolerance
Blank	—	—	20%
Silver	—	10^{-2}	10%
Gold	—	10^{-1}	5%
Black	0	1	—
Brown	1	10	—
Red	2	10^2	—
Orange	3	10^3	—
Yellow	4	10^4	—
Green	5	10^5	—
Blue	6	10^6	—
Violet	7	—	—
Gray	8	—	—
White	9	—	—

1st Significant digit
2nd Significant digit
Decimal multiplier
Tolerance

EXAMPLE 4-13

Yellow, violet, yellow $= 47 \times 10^4 \ \Omega = 470 \ \text{k}\Omega \ \pm 20\%$

Green, blue, black, silver $= 56 \times 1 \ \Omega = 56 \ \Omega \quad \pm 10\%$

Red, yellow, gold, gold $= 24 \times 0.1 \ \Omega = 2.4 \ \Omega \quad \pm 5\%$

Not all possible combinations of colors will be found in practice, since resistors are manufactured according to the preferred values given in Table 4-5. Only the two significant digits are shown in the table. Any permissible decimal multiplier can be used with these selected digits.

TABLE 4-5

PREFERRED RESISTANCE VALUES

Tolerance			
5% 10% 20%	5% only	5% and 10%	5% only
Significant Digits			
10	11	12	13
15	16	18	20
22	24	27	30
33	36	39	43
47	51	56	62
68	75	82	91

SELF-CHECK

4-8:1. Listed below are the color bands on a sample of resistors. Identify the resistors.

    (a) brown, black, green, silver

    (b) orange, orange, orange

    (c) blue, red, black, gold

4-8:2. Suppose a resistor has the following color bands: orange, white, red, gold. When measured, the resistor was found to have a resistance of 3650 Ω. Is this resistor within its stated tolerance?

4-9 NONLINEAR RESISTORS

The tungsten filament of an incandescent lamp and the heater of a cathode-ray tube are examples of resistors where the temperature range from room temperature to the operating temperature is such that there is a considerable change in resistance. A 60-watt lamp operating from a 120-volt source has a white hot resistance of 240 ohms to limit the current through it to $I = V/R = 120/240 = 0.5$ ampere. But when the lamp is turned off, its cold resistance is only about 18 ohms. Since its resistance is not reasonably constant over its normal temperature range, we can think of an incandescent lamp as a **nonlinear resistor.**

In the case of the incandescent lamp and tube heater, this nonlinear characteristic is a disadvantage. With a cold resistance of 18 ohms, the *inrush* current at the instant when the lamp is turned on (before the filament has had a chance to heat up) is $I = V/R = 120/18 = 6.6$ amperes. Fortunately, the mass of the lamp filament is small enough that it can reach white heat in less than a millisecond. Therefore, the current surge is of very short duration, as shown in Figure 4-6. Nevertheless, electric switches used with incandescent lamps have to be designed to withstand the inrush current. If the heating elements of electric stoves had the same temperature coefficient as tungsten, the surge would be more serious since stove elements take several seconds to reach red heat. Consequently, electric stove elements are manufactured from an alloy such as Nichrome II, which has a very small temperature coefficient.

A nonlinear resistance characteristic is not always a disadvantage. In the circuit of Figure 4-2, maybe we can arrange for the series resistor to have a large *negative* temperature coefficient so that its resistance is quite high at the instant the switch is closed. As the lamp warms up, its resistance increases. But, as the negative-temperature-coefficient resistor heats up, its resistance *decreases*. The series combination of these two

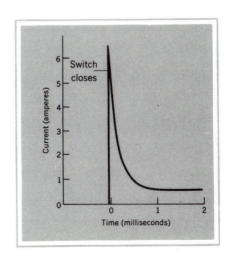

FIGURE 4-6
Inrush Current of an Incandescent Lamp.

nonlinear resistors tends to provide a more uniform current when the switch is first closed, rather than the surge of Figure 4-6. Such negative-temperature-coefficient resistors are available under various trade names for use in limiting the initial surge which occurs when electronic devices are first switched on. These nonlinear resistors have a resistance of over 100 ohms at room temperature, but with a current of approximately 1 ampere through them, their resistance drops to less than 1 ohm after 10 to 15 seconds.

We again look to the mechanics of electrical conduction in semiconductors to account for this remarkable change in resistance. Certain metallic oxides form semiconductor crystals which are mixed with a ceramic binder to form resistors in which the breaking and reforming of covalent bonds make only a limited number of electric charge carriers available at normal room temperature. When electric current flows through these resistors, their temperature rises. With the increase in temperature, many more covalent bonds are broken, many more charge carriers are created, and, consequently, the resistance drops to a fraction of its value at room temperature.

In addition to their application in inrush surge-limiting resistors, tiny beads of these metallic oxide semiconductors can be calibrated in terms of resistance versus temperature to become sensitive temperature-measuring devices. These tiny nonlinear resistors are called **thermistors.** A decrease in temperature of less than 20°C will more than double their resistance.

Semiconductor conduction also accounts for the nonlinear resistance characteristic of a device called a **varistor.** Varistors depend on the nonlinear resistance characteristic of zinc oxide or silicon carbide crystals, which are formed into wafers with a clay binder. Zinc oxide varistors are used to protect low-voltage power supplies from voltage surges. Silicon carbide varistors (commonly known by the trade name **thyrite**) are designed to protect higher voltage systems—lightning arrestors on power transmission lines.

Actually, temperature has little effect on the resistance of a varistor. The rapid increase in the number of charge carriers depends on the potential difference across the varistor wafer. When the applied voltage is increased beyond a certain magnitude, there is a marked *increase* in the breaking of valence bonds and a consequent sharp *decrease* in resistance. As Figure 4-7 indicates, the decrease in resistance is such that many times the normal current can flow through the varistor without appreciably increasing the voltage drop across it. Therefore, connecting a varistor in parallel with an electrical

FIGURE 4-7
Typical Varistor Characteristic.

PART I   THE BASIC ELECTRIC CIRCUIT

device provides that portion of the system with a means of coping with a sudden surge without significantly affecting the voltage across (and the current through) the protected device.

A more recent nonlinear resistor is the **photoresistor.** Light falling on certain semiconductor materials breaks down valence bonds and thus creates additional charge carriers. Unlike the *photovoltaic* cell with its *pn*-junction, the photoresistor consists of a thin zigzag strip of cadmium sulphide behind a window only a few millimeters in diameter. However, its resistance can range from hundreds of kilohms in the dark to less than 100 ohms in bright daylight. Photoresistors have all but replaced photovoltaic cells in photographic exposure meters (auto-exposure cameras) and simple light-operated switches for turning lights on at dusk and off at dawn.

Semiconductor technology has led to the development of three quite different types of nonlinear resistors:

Thermistor: resistance decreases as *temperature* increases.

Varistor: resistance decreases as *applied voltage* increases.

Photoresistor: resistance decreases as *incident light* increases.

To identify these nonlinear resistors in circuit diagrams, we modify the basic resistor symbol as shown in Table 4-6.

TABLE 4-6

GRAPHIC SYMBOLS FOR NON-LINEAR RESISTORS	
Nonlinear Resistor	Graphic Symbol
General	
Thermistor	t°
Varistor OR	V
Varistor	
Photoresistor	

4-10 VOLT-AMPERE CHARACTERISTICS

Ohm's law of constant proportionality [Equation (4-1)] applies only to linear resistors, where the V/I ratio is a reasonably constant value expressed in ohms. However, in Section 4-9 we have used the term **resistance** with nonlinear resistors as well. The difference is that the resistance is no longer a constant, but is a variable depending on such factors as temperature, applied voltage, and incident light. By retaining the concept of resistance, the basic relationship derived from Ohm's law still holds true:

$$R = \frac{V}{I} \tag{4-2}$$

Since $R = V/I$, it is often convenient to show variations in resistance graphically by plotting this volts-per-ampere ratio for the resistor in question. In Figure 4-4, we discovered that the volt-ampere graph for a linear resistor is a straight line whose slope is inversely proportional to the resistance. Figure 4-7 shows a volt-ampere graph for a nonlinear varistor.

In drawing a volt-ampere characteristic, it is customary to plot voltage drop as an independent variable on the x-axis. Current then becomes the dependent variable on the y-axis, as shown in Figure 4-8. Consequently, a low resistance, which by definition has a low volts-per-ampere ratio, appears as a very steep graph. Similarly, a high resistance with its high volts-per-ampere ratio is represented by a graph with a very shallow slope.

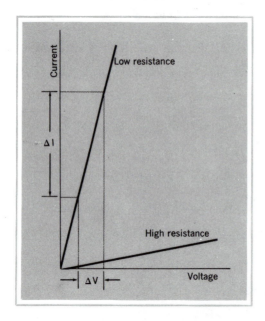

FIGURE 4-8
Volt-Ampere Characteristics of Low and High Linear Resistances.

Both graphs in Figure 4-8 represent linear resistors. For a linear resistor, we can read the resistance from the graph by simply taking the ratio of voltage to current at any point on the graph. All points on a linear characteristic have the same ratio. If the volt-ampere characteristic is not perfectly linear, we often consider the *dynamic* value of resistance in terms of the *change* in voltage drop versus the *change* in current associated with this change in voltage drop, as indicated in Figure 4-8. Hence, for a given ΔV, the ΔI is much greater for a small resistance than it is for a high resistance.

Since linear resistors conform to Ohm's law of constant proportionality within normal operating limits, we seldom bother to plot their straight-line volt-ampere characteristics. We can simply print the resistance in ohms on these resistors either in numerical or color code form. However, the most convenient way of discussing the behavior of a nonlinear resistor such as the varistor of Figure 4-7 is to draw a graph of its particular volt-ampere characteristic.

Both vacuum tubes and transistors behave as nonlinear resistances. Hence, any study of electronic circuits will frequently resort to volt-ampere characteristics. The graphs in Figure 4-9 compare the volt-ampere characteristics of the vacuum-tube diode and the semiconductor diode described in Chapter 3. Since we want to show what happens when we *reverse* the polarity of the applied voltage, we draw the negative x and y axes as well as the usual positive axes. The volt-ampere characteristics show us that both diodes permit electric current when they are forward biased but that the internal resistance of the vacuum-tube diode is much higher than that of the semiconductor diode. When we reverse the polarity of the applied voltage, no current flows through a vacuum-tube diode, whereas the semiconductor diode does permit a small reverse current. Under these conditions, the internal resistance of the semiconductor is very high and the internal resistance of the vacuum-tube diode is infinitely high.

FIGURE   4-9
Volt-Ampere Characteristics of Semi-conductor and Vacuum-Tube Diodes.

4-11 OHM'S LAW APPLIED

At the beginning of the chapter, we set out to determine the factors that govern the current in an electric circuit. By transposing Equation (4-2), which defines Ohm's law, we get $I = V/R$. If we are considering the resistance of the total circuit connected between the terminals of the voltage source, the total voltage drop V will be equal to the applied voltage E of the source (as we noted in Figure 2-13). Hence, the magnitude of the current is dependent on the applied voltage and the total resistance of the circuit. In the electric circuits we shall be considering in this book, applied voltage and resistance are *independent* variables. This means that the potential difference of the batteries of Chapter 3 depends on the two particular metals used and is essentially *independent* of the resistance and the current. Similarly, the resistance of an electric circuit depends on such controllable physical factors as length, diameter, type of material, and temperature. Current, on the other hand, is the *dependent* variable. Whenever the switch is closed in the circuits we have discussed, the current *must* automatically take on a value which will satisfy the relationship

$$I = \frac{E}{R}$$

(4-9)

where I is the current that must flow in the circuit (in amperes), E is the applied voltage (in volts), and R is the total resistance of the circuit (in ohms).

In the circuit of Figure 2-13, we noted that a voltage drop can appear across a resistor *only* when current flows through it. Transposing Ohm's law, Equation (4-2) once more gives

$$V = IR$$

(4-10)

where I is the current through a resistor (in amperes), R is the resistance of the resistor (in ohms), and V is the resulting voltage drop across the resistor (in volts). Because of this relationship, it is quite common to hear *voltage drop* across a resistance referred to as an **IR drop.**

EXAMPLE 4-14

Compute the circuit current when a 4.7-kΩ resistor is connected across a 9-V source.

SOLUTION

$$I = \frac{E}{R} = \frac{9 \text{ V}}{4.7 \text{ k}\Omega} = 1.9 \times 10^{-3} \text{ A} = \textbf{1.9 mA}$$

EXAMPLE 4-15

Calculate the voltage drop that will be developed across a 560-Ω resistor by a 15-mA current through the resistor.

SOLUTION

$$V = IR = 15 \text{ mA} \times 560 \text{ }\Omega = \textbf{8.4 V}$$

SELF-CHECK

4-11:1. When used properly, a light bulb with a resistance of 240 Ω allows 0.5 A of electric current to flow. What is the proper voltage drop across the light bulb?

4-11:2. An electric heater draws 5.0 A when used with a 240 V supply. Assuming that the resistance of the heater remains constant, what current will the heater draw when used with a 200 V supply?

PROBLEMS

4-1-1. What is the resistance of an electric circuit which draws a 2.5-A current from a 120-V source?

4-1-2. What resistance is required to limit the current in a circuit to 2 A when the applied voltage is 30 V?

4-1-3. What is the resistance of an electric stove element when a 5-A current through it causes a 110-V voltage drop across it?

4-1-4. If a current of 0.15 A through the resistor in Figure 4-2 causes a 1.0-V drop across it, what is its resistance?

4-1-5. What current will flow through a 20-Ω resistor connected across a 100-V source?

4-1-6. What current through a 250-Ω resistor will cause a voltage drop of 75 V across the resistor?

4-1-7. A 12-Ω heating element has a voltage drop of 98 V across it. What current is flowing through the heater?

4-1-8. What current will flow in a circuit whose total resistance is 17 Ω when it is connected to a 117-V source?

4-1-9. What voltage must be applied to a 15-Ω resistor to make it pass a 3-A current?

4-1-10. What is the voltage drop across a 125-Ω resistor when a 0.2-A current flows through it?

4-1-11. What voltage drop is produced by a 1.6-A current flowing through a 56-Ω resistor?

4-1-12. What applied voltage is required to cause current at the rate of 0.35 A in a circuit where the total resistance is 124 Ω?

4-1-13. A 15-μA current through a resistor in a radio receiver causes a 30-mV drop across it. What is its resistance?

4-1-14. What is the resistance of an ammeter whose full-scale reading is 10 A with a 30-mV drop across the meter?

4-1-15. What value of resistor is required in order to produce a 7-V drop across the resistor when the current through the resistor is 12 mA?

4-1-16. What value of resistor will draw a 32-mA current when connected to a 2.5-kV source?

4-1-17. A fuse in the power supply of a transistor amplifier has a resistance of 0.02 Ω. What current through the fuse will develop a 500-μV drop across it?

4-1-18. What current will flow when a 50-μV source is connected to a 5-MΩ resistor?

4-1-19. What current will flow through a 2.2-kΩ resistor connected across a 480-mV source?

4-1-20. What current must flow through a 6.8-kΩ voltage-dropping resistor to produce an 80-V drop across the resistor?

4-1-21. What voltage must be applied to a 220-kΩ bleeder resistance in a piece of electronic equipment if the bleeder current is to be 20 μA?

4-1-22. What IR drop is produced by a 6-mA current passing through a 15-kΩ load resistor?

4-1-23. What voltage drop appears across the 2.2-kΩ collector load resistor of a transistor amplifier when the collector current is 1.8 mA?

4-1-24. The operating coil of a relay, which is designed to energize with a current of 24 mA through the coil, has a resistance of 3.7 kΩ. What is the minimum applied voltage which will energize the relay?

NOTE: Normal room temperature (20°C) is assumed in Problems 4-4-25 to 4-4-40.

4-4-25. An electric conductor with a cross-sectional area of 10 mm² has a resistance of 1.72 Ω/km of length. What is the resistance of 200 m of wire of the same material with a cross-sectional area of 4 mm²?

4-4-26. A conductor with a cross-sectional area of 2.5 mm² and a length of 50 m has a resistance of 1.42 Ω. What length of wire of the same material with a cross-sectional area of 0.5 mm² will have a resistance of 5.0 Ω?

4-4-27. An electric conductor with a diameter of 2.0 mm and a length of 80 m has a resistance of 0.44 Ω. What is the resistance of a conductor of the same material having a length of 50 m and a diameter of 0.8 mm?

4-4-28. A conductor 40 m in length and having a diameter of 1.2 mm has a resistance of 17.3 Ω. What diameter of wire of the same material will have a resistance of 1.0 Ω per meter of length?

4-4-29. Using the data of Table 4-1, calculate the resistance of a copper wire 200 m in length with a cross-sectional area of 4 mm².

4-4-30. What is the resistance of a copper bar with a rectangular cross section of 2 × 4 cm and a length of 1.5 m?

4-4-31. Calculate the resistance of 160 m of aluminum wire having a diameter of 1.6 mm.

4-4-32. What is the resistance of a tungsten filament 12 cm in length and 0.1 mm in diameter?

4-4-33. What length of silver wire 0.5 mm in diameter has a resistance of 0.04 Ω?

4-4-34. What diameter of copper wire has a resistance of 1 Ω/m of length at room temperature?

4-4-35. A relay coil requires 144 m of copper wire. If its resistance is to be 50 Ω, what diameter of wire is required?

4-4-36. A coil of constantan wire 0.4 mm in diameter, wound in a single layer on a cylindrical form so that the mean diameter of each turn is 2 cm, has a resistance of 24 Ω. How many turns of wire are there in this wire-wound resistor?

4-4-37. A carbon rod 0.5 cm in diameter and 10 cm in length has a resistance of 0.153 Ω. What is the resistivity of carbon at 20°C in ohm meters?

4-4-38. An electric conductor 40 m in length and 1.8 mm in diameter has a resistance of 17.3 Ω. What material is used for this conductor?

4-4-39. A stranded cable consists of eight copper conductors, each having a diameter of 1.0 mm. Twisting the wires to form the cable requires the length of each conductor to be 3% longer than the length of the cable in its final form. What is the resistance of 1 km of this cable?

4-4-40. What diameter of solid copper conductor will have the same resistance per kilometer as a stranded cable consisting of six aluminum conductors each having a diameter of 1.3 mm? Allow an increase of 3% in the length of the individual aluminum conductors for twisting in forming the cable.

4-5-41. A length of copper telephone line has a resistance of 24 Ω at 20°C. What is its resistance on a hot summer day when its temperature rises to 36°C?

4-5-42. What is the resistance of an aluminum conductor at −20°C if its resistance at +30°C is 1.25 Ω?

4-5-43. What is the resistance at 60°C of 25 m of copper wire having a diameter of 0.5 mm?

4-5-44. What length of Nichrome II wire having a diameter of 0.64 mm has a resistance of 48 Ω at 200°C?

4-6-45. If the resistance-temperature graph for brass is extended in a straight line until the resistance becomes zero, the corresponding temperature would be −480°C. What is the temperature coefficient of brass at 20°C?

4-6-46. The temperature coefficient of platinum at 20°C is 0.003 per degree per ohm. Find the temperature at which the resistance of platinum would become zero if the resistance-temperature graph were extended as a straight line.

4-6-47. An incandescent lamp draws a 1.0-A current from a 110-V source to raise the temperature of its tungsten filament to 2800°C. What is its resistance at normal room temperature?

4-6-48. By how many degrees must the temperature of a nickel rod be changed to increase its resistance to 105% of its resistance at normal room temperature?

4-6-49. A certain conductor has a resistance of 10 Ω at 20°C and 11.35 Ω at 50°C. If you consider only those materials listed in Tables 4-2 and 4-3, of what material is this conductor made?

4-6-50. An electric motor is to be operated at a distance of 250 m from a 120-V source. The current drawn by the motor is 8.0 A. This current raises the temperature of the copper conductors feeding the motor to 40°C. What is the minimum diameter of wire that can be used without the voltage drop in the conductors exceeding 10% of the applied voltage?

4-8-51. What are the resistance and tolerance of resistors color coded as follows?
(a) Three orange bands.
(b) Blue, gray, red.
(c) Brown, black, red, silver.
(d) Orange, white, silver, gold.
(e) Violet, green, gold, gold.
(f) Gray, red, black, silver.

4-8-52. Determine the color code for the following resistors.
(a) 680 kΩ, ±20%.
(b) 22 kΩ, ±20%.
(c) 39 Ω ±10%.
(d) 100 Ω, ±10%.
(e) 0.2 Ω, ± 5%.
(f) 91 Ω, ± 5%.

4-8-53. Calculate maximum and minimum permissible resistances for the three resistors in Example 4-13.

4-8-54. Calculate maximum and minimum permissible resistances for the six resistors in Problem 4-8-52.

4-10-55. Using a voltage scale of 0 to 20 V, and a current scale of 0 to 2 A, draw an accurate V/I graph for a 16-Ω linear resistor.

4-10-56. Superimpose on the graph of Problem 4-10-55 an accurate V/I graph for a nonlinear resistor having a characteristic such that $I = 5 \times 10^{-3} V^2$.

REVIEW QUESTIONS

4-1. Why can Ohm's law be described in terms of **constant proportionability?**

4-2. Why does the V/I ratio of an electric circuit indicate its ability to oppose electric current rather than its ability to permit current?

4-3. If there were no such unit as the **ohm,** how could you express the resistance of an electric circuit?

4-4. If resistance is the *opposition* to electric current, energy must be expended in forcing free electrons to move *against* this opposition. Where does this energy come from? Where does it go?

4-5. What relationship exists between the resistance of a given conductor and the number of free electrons in the conductor?

4-6. An electric fuse consists of a small strip of metal with a low melting temperature. The current in the protected electric circuit flows through this strip of metal. Which will have the greater resistance, a 10-A fuse or a 20-A fuse? Explain.

4-7. The *National Electrical Code* ® limits the current through No. 14 AWG[†] house wiring to 15A. What is the reason for doing so? Is it possible for a current of greater than 15 A to flow through No. 14 wire?

4-8. A resistor made of Nichrome wire wound on a ceramic form has the same resistance and is passing the same current as the filament of an incandescent lamp, yet the temperature of the lamp filament is many times that of the Nichrome resistor. Explain.

[†] American wire gauge; see Appendix 3.

4-9. Why does shortening the length of a given conductor decrease its resistance?

4-10. Why is resistance inversely proportional to the *square* of the diameter of a conductor?

4-11. Why does the term **resistivity** apply to the *material* of an electric conductor, rather than to a particular conductor?

4-12. Given an accurate resistance-measuring device, how would you go about determining the resistivity of a sample of an unknown alloy?

4-13. Draw a graph of the type shown in Figure 4-3 for carbon, using the data given in Table 4-3. What would the temperature x be for carbon?

4-14. In checking the temperature rise of the copper field winding of an electric motor by measuring its resistance, before the motor was started, we found that its resistance (at room temperature) was 50 Ω. After 30 min of operation, its resistance was 53 Ω. A half-hour later its resistance was 54.3 Ω, and after a further 40 min had elapsed, the field coil resistance was 54.7 Ω. It took 3 h in all for the resistance to settle at a steady value of 55 Ω. Plot a graph of temperature rise against time for this heat run. Explain the reasons for the shape of the graph.

4-15. Find the temperature coefficient of resistance for copper at 0°C. From our derivation of temperature coefficient, discuss why the temperature coefficient should be different when 0°C is specified rather than 20°C.

4-16. Suggest a reason why thin-film resistors generate less noise voltage in sensitive electronic circuits than do carbon-composition resistors.

4-17. Table 4-5 shows only six combinations of significant digits for manufacturing resistors with a ±20% tolerance, as compared with 24 combinations for ±5% tolerance resistors. Why is such a limited selection of nominal resistance values satisfactory for ±20% tolerance resistors?

4-18. The volt-ampere characteristic of a carbon-composition resistor is a straight line up to the point where the current reaches the maximum specified value for the resistor. If we increase the current beyond that limit, the V/I graph starts to curve slightly. In which direction will it bend? Why?

4-19. A nonlinear resistor has a characteristic such that $I = 5 \times 10^{-3} V^2$ (see Problem 4-10-55). Would this resistor be a thermistor or a varistor? Explain your selection.

4-20. In the equation $I = E/R$, why do we refer to current as a *dependent variable* in a simple electric circuit?

*4-21. Write a BASIC program which will permit you to enter the nominal value of any color-coded resistor and its tolerance, and print out the maximum and minimum permissible resistance values for that resistor.

*4-22. Write a program from which you can plot the volt-ampere graph for Problem 4-10-55 for values of V from 0 to 20 V in 1-V steps. (If you are familiar with BASIC routines for plotting graphs, write a program to display the resulting V/I graph on fully labeled coordinates.)

5

WORK
AND POWER

5-1 ENERGY AND WORK

In Chapter 2, we defined **work** in mechanical terms:

> *Work is the accomplishment of motion against the action of a force which tends to oppose the motion.*

From this definition, the SI unit for work, the **joule,** is defined as the work accomplished in moving (or the energy required to move) an object through a distance of one meter against an opposing force of one newton.

We also discovered that energy must be expended when we move an object against an opposing force, as in the example of carrying a heavy parcel up several flights of stairs against the opposition of the force of gravity. And since energy can neither be created nor destroyed, energy and work are numerically equal. In the electric circuit shown in Figure 5-1, mechanical energy supplied to the shaft of the dynamo is *converted* into electric energy which is transmitted by a pair of electric conductors to the heater unit where the electric energy is, in turn, *converted* into heat energy. From the sequence of events in this example, it follows that **work is performed whenever energy is converted from one form into another.**

Since the law of conservation of energy must apply to the circuit of Figure 5-1, and since we have already defined the unit of energy and work for a mechanical system, we can also express both electric energy and work and heat energy and work in joules. Since the energy conversions illustrated in Figure 5-1 are typical in practical electric circuits,

FIGURE 5-1
Energy Conversions in a Simple Electric Circuit.

the advantage of expressing mechanical, electric, and heat energy all in joules is obvious. Hence, we can now state that

The joule is the SI unit of electric energy and work.

From Equation (2-4),

One joule of electric energy is required to raise one coulomb of electric charge through a potential difference of one volt.

To illustrate the equivalence among electric, mechanical, and heat energy, we shall consider the following examples. Then, for the remainder of this book, we shall make our energy and work calculations in the *electrical* part of the system.

EXAMPLE 5-1

How much electric energy must be supplied to an electric motor to raise an elevator having a mass of 4 megagrams[†] a distance of 40 m? Assume that 10% of the electric energy supplied to the motor is lost in the form of heat.

SOLUTION

To raise the elevator car, it is necessary to move it against the force of gravity acting on it. From Newton's second principle, the **weight** of an object is its mass times gravitational acceleration.

$$\therefore F_g = mg = \text{kg} \cdot \text{m/s}^2 = \text{newtons}$$

$$= 4 \text{ Mg} \times 9.8 \text{ m/s}^2 = 4 \times 10^3 \text{ kg} \times 9.8 \text{ m/s}^2 = 39\,200 \text{ N}[‡]$$

$$\therefore \text{ mechanical work, } W = F_g s = 39\,200 \text{ N} \times 40 \text{ m} = 1\,568\,000 \text{ J}$$

[†] In the International System of Units, large masses are expressed in *megagrams*. However, SI recognizes that, for some time to come, the old MKS term *tonne* (metric ton) will appear in commercial usage. One tonne = one megagram.

[‡] In entering the exponent for the unit of mass into the calculator in this example, we encounter the one anomaly in SI base units. The base unit of mass is the *kilo*gram.

Although the electric energy going toward raising the elevator is also 1 568 000 J, this represents only 90% of the electric energy supplied to the motor. Therefore, the total electric energy required is

$$W_T = \frac{1\,568\,000 \text{ J}}{0.9} = \mathbf{1\,742\,000 \text{ J}}$$

EXAMPLE 5-2
A teakettle contains 1.6 kg of water at normal room temperature (20°C). If 20% of the electric energy supplied to the stove element under the kettle is lost by radiation and in heating the kettle itself, how much electric energy must be supplied to the stove element to bring the water to the boiling point? (The heat energy required to raise 1 kg of water through 1°C is 4185 J.)

SOLUTION

Heat energy, $W = 1.6 \text{ kg} \times 4185 \text{ J/(kg} \times °C) \times (100 - 20)°C = 535\,680 \text{ J}$

The electric energy going toward heating the water is also 535 680 J. However, the total electric energy supplied to the stove element must be

$$W_T = \frac{535\,680 \text{ J}}{0.8} = \mathbf{670\,000 \text{ J}}$$

SELF-CHECK

5-1:1. How much electric energy is supplied to a car's starter motor if 180 C of charge move through the motor while the potential difference across the motor is 7.5 V?

5-2 POWER

By using a suitable train of gears that would multiply mechanical force by dividing down its speed by the same factor, we could use a small electric motor to raise the elevator in Example 5-1. But in doing so, it would take about five minutes for the elevator to move through the 40 meters. This might be satisfactory for a freight elevator but a passenger elevator would have to be supplied with a much larger motor so that the required amount of mechanical work could be accomplished in a much shorter time. Therefore, in energy-conversion systems, the *time rate* of doing work is just as important as the amount of work done and is given the name **power.**

Power is the rate of doing work.
The letter symbol for power is P.

Since the **joule** is the basic unit of work and the **second** is the basic unit of time, it would be quite satisfactory to express power in **joules per second.** Expressing power in joules per second automatically calls our attention to the fact that power is the *rate* of doing work. But it was decided to honor James Watt for his work in converting heat energy into mechanical energy with his steam engine by calling the unit of electric power the **watt.** Therefore,

The watt is the SI unit of electric power.
The unit symbol for watt is **W.**

The magnitude of the watt is automatically established in terms of joules and seconds:

One watt is the rate of doing work when one joule of work is done in one second.

Expressing this in equation form, we have

$$P = \frac{W}{t} \tag{5-1}$$

where P is power in watts, W is work in joules, and t is time in seconds. Since the term **watt** does not suggest the idea of *rate* as does the term **joules per second,** we must keep reminding ourselves when calculating power that **watts** are exactly the same as **joules per second** and that **power** is the *rate* of doing work.

EXAMPLE 5-3
At what rate must electric energy be supplied to the electric motor in Example 5-1 if the elevator is raised through a distance of 40 m in 4 min?

SOLUTION

$$P = \frac{W}{t} = \frac{1\,742\,000 \text{ J}}{(4 \times 60) \text{ s}} = 7258 \text{ W} \quad \text{or} \quad \textbf{7.26 kW}$$

Equation (5-1) is the basic equation which defines power and establishes the magnitude of the watt. But it does not provide the most convenient method of determining power in electric circuits. By means of the following algebraic substitution, we can develop a more useful relationship for calculating electric power:

From Equation (2-4) defining potential difference,

$$V = \frac{W}{Q} \qquad \therefore W = QV$$

and from Equation (2-2) defining current,

$$I = \frac{Q}{t} \qquad \therefore t = \frac{Q}{I}$$

substituting for W and t in Equation (5-1) gives

$$P = \frac{W}{t} = \frac{QV}{Q/I} = QV \times \frac{I}{Q} = V \times I$$

$$\therefore P = VI \tag{5-2}$$

where *P is power in watts, V is voltage drop in volts, and I is current in amperes.* Also, since $V = IR$ [from Ohm's law, Equation (4-2)], substituting in Equation (5-2) gives

$$P = IR \times I$$

$$\therefore P = I^2R \tag{5-3}$$

where *P is power in watts, I is current through the resistance in amperes, and R is resistance in ohms.* Since $I = V/R$ [from Ohm's law, Equation (4-2)], substituting in Equation (5-2) gives

$$P = \frac{V^2}{R} \tag{5-4}$$

where *P is power in watts, V is voltage drop across the resistance in volts, and R is resistance in ohms.*

EXAMPLE 5-4
A lamp draws a 2-A current when connected to a 120-V source. What is the power rating of the lamp?

SOLUTION

$$P = EI = 120 \text{ V} \times 2 \text{ A} = \textbf{240 W}$$

EXAMPLE 5-5
A 10-kΩ resistor is connected into a circuit where the current through it is 50 mA. What is the minimum safe power rating for a resistor to be used in this circuit?

SOLUTION

$$P = I^2R = (50 \text{ mA})^2 \times 10 \text{ k}\Omega = \textbf{25 W}$$

EXAMPLE 5-6
What is the highest voltage that can be applied to a 5-kΩ 2-W resistor without exceeding its heat-dissipating capability?

SOLUTION
Since

$$P = \frac{V^2}{R}, \quad V^2 = PR \quad \text{and} \quad V = \sqrt{PR}$$

$$\therefore E = V = \sqrt{2 \text{ W} \times 5 \text{ k}\Omega} = \textbf{100 V}$$

SELF-CHECK

5-2:1. An electric iron is connected to a 120-V source. The iron has a resistance of 12 Ω. What is the power rating of the iron?

5-2:2. What is the resistance of a 1600-W electric heater that draws a 15-A current?

5-3 EFFICIENCY

Examples 5-1 and 5-2 suggest that in converting energy from one form into another, some of the input energy is converted to a form of energy that is not useful in performing the work which the equipment was set up to do. This wasted energy is usually in the form of heat. Some of this heat comes from mechanical friction, some of it from current in an electric resistance, and some of it is lost by radiation, as in Example 5-2. Not only does this wasted energy cost money to supply, but the equipment has to be designed to dissipate the wasted heat energy safely.

The ability of an energy-converting device (such as the electric motor of Example 5-1 and the electric stove element of Example 5-2) to convert as much of the input energy as possible into *useful* work is specified in terms of its **efficiency.**

Efficiency is the ratio of useful output energy to total input energy.
The letter symbol for efficiency is the Greek letter η (eta).

$$\eta = \frac{W_{out}}{W_{in}} \tag{5-5}$$

We usually express efficiency as a percentage. For example, the efficiency of the system in Example 5-1 is 90%.

EXAMPLE 5-7

What is the efficiency of an electric hoist if 60 000 J of electric energy has to be supplied to the motor in order to raise a 300-kg mass through 18 m?

SOLUTION

$$\text{Mechanical work, } W = 300 \text{ kg} \times 9.8 \text{ m/s}^2 \times 18 \text{ m} = 52\,920 \text{ J}$$

$$\text{Efficiency, } \eta = \frac{52\,920 \text{ J}}{60\,000 \text{ J}} \times 100\% = \mathbf{88.2\%}$$

From Equation (5-1) defining power, it follows that $W = Pt$. If we substitute this form for W_{out} and W_{in} in Equation (5-5),

$$\text{Efficiency, } \eta = \frac{P_{out} \times t}{P_{in} \times t}$$

from which

$$\eta = \frac{P_{out}}{P_{in}} \qquad\qquad (5\text{-}6)$$

In applying Equation (5-6) to electric motors, P_{out} is *mechanical* power output, and P_{in} is *electric* power input. In the International System of Units, the **watt** applies to both electric and mechanical power. Unfortunately, until conversion to metric units in North America is well established, we shall find that mechanical power output of motors is expressed in terms of a unit invented by James Watt—the **horsepower.** For a few years, therefore, we require a conversion factor to convert mechanical power expressed in *horsepower* into SI units—*watts:*

$$1 \text{ hp} = 746 \text{ W} \qquad\qquad (5\text{-}7)$$

EXAMPLE 5-8

What electric power input must be provided to an electric motor which develops mechanical energy at the rate of 24 hp with an 85% efficiency?

SOLUTION

Mechanical power output, $P_{out} = 24 \times 746 = 17\,904$ W   or   17.9 kW

$$P_{in} = \frac{P_{out}}{\eta} = \frac{17.9 \text{ kW}}{0.85} = \textbf{21 kW}$$

We must note that the **horsepower** is an *English* unit of *mechanical* power. It is not an approved SI unit. Therefore, the horsepower must not be used to express *electric* power.

SELF-CHECK

5-3:1. An automobile starter motor is rated at 1.0 hp. When in use the motor draws 140 A from a 7.5-V source.
(a) What is the electric power input to the starter motor?
(b) What is the efficiency of the starter motor?
5-3:2. Find the power output of an electric motor that draws 10 A from a 240-V source and is 80% efficient.

5-4 THE KILOWATTHOUR

In the examples in this chapter, we purposely left the answers for *work* in base units to show that a relatively small amount of work results in a large numerical answer when we express the answer in joules. Therefore, the base unit of electric energy, the **joule,** is considered to be too small a unit for practical purposes such as computing a monthly electric power bill. The proper SI unit for such purposes is the **megajoule (M J).** However, for commercial purposes, another practical unit, the **kilowatthour,** is used.

Although the International System of Units prefers the term **megajoule,** it does permit the use of the **kilowatthour** for practical purposes for the time being.

Since

$$P = \frac{W}{t} \qquad \therefore W = Pt$$

If P is power in **watts** and t is time in **seconds,** then W must be work in **watt-seconds** or **joules.**

But if P is power in **kilowatts** and t is time in **hours,** then W must be work in **kilowatthours.**

The kilowatthour is the practical unit of electric work or energy.
The unit symbol for kilowatthours is **kWh.**

The kilowatthour is defined by the equation

$$W = Pt \tag{5-8}$$

where W is work or energy in kilowatthours, P is power in kilowatts, and t is time in hours. Since 1 kW = 1000 W and 1 h = 3600 s,

$$1 \text{ kWh} = 1000 \text{ W} \times 3600 \text{ s} = 3.6 \times 10^6 \text{ J} = 3.6 \text{ MJ} \tag{5-9}$$

Note: Kilowatthours = kilowatts *times* hours (not kilowatts/hour).

EXAMPLE 5-9

At 7¢ per kilowatthour, how much will it cost to leave a 60-W lamp burning for 5 days?

SOLUTION

$$W = Pt = 60 \text{ W} \times (24 \times 5) \text{ h} = 7200 \text{ Wh} = 7.2 \text{ kWh}$$

$$\text{Cost} = 7.2 \text{ kWh} \times 7¢/\text{kWh} = \mathbf{50.4¢}$$

The amount of electric energy consumed by a residence or business is measured by a **kilowatthour meter.** It consists of an aluminum disk which rotates on an induction-motor principle at a speed which is directly proportional to the applied voltage and the current drawn by the consumer ($P = EI$). The disk drives a series of 10-to-1 reduction gears connected to dials. Hence the dial reading corresponds to speed times accumulated time, $P \times t$ = kilowatthours. Since there are some costs involved in providing an electric utility service regardless of the energy used, most utilities charge for electric energy on a sliding scale that decreases as more energy is used during the billing period.

EXAMPLE 5-10

A typical electric utility company rate for residential service is

First 100 kWh:	9.9¢/kWh
Next 400 kWh:	5.3¢/kWh
All additional kWh:	3.6¢/kWh

If the present meter reading is 38 770 kWh and the previous reading was 37 530 kWh, what is the bill for the electric energy used in the period between readings?

SOLUTION

Amount of energy used = 38 770 − 37 530 = 1240 kWh

Cost of first 100 kWh = 100 × 9.9¢ = $ 9.90

Cost of next 400 kWh = 400 × 5.3¢ = $21.20

Cost of last 740 kWh = 740 × 3.6¢ = $26.64

Total bill **$57.74**

SELF-CHECK

5-4:1. How many hours will it take for a 12-V battery to supply 1.0 kWh of energy to a lamp that draws 5.0 A?

5-4:2. An air conditioner draws 30 A from a 230-V power line. The air conditioner runs for an average of 9.0 hours each day and the electric company charges $0.07 per kWh of energy usage.
(a) What is the power supplied to the air conditioner?
(b) What is the cost per day to operate the air conditioner?

5-5 INTERRELATIONSHIP OF BASIC ELECTRICAL UNITS

If we gather together all the equations which define the basic electrical units which we have discussed so far, we find that if we know three such circuit parameters as voltage, current, and elapsed time, we can determine any of the other parameters such as resistance, work, and power. This is achieved by a process of algebraic substitution in the defining equations as we did to develop a practical equation for power. Table 5-1 shows some of the more useful derived equations. The student should derive each of the equations shown in this table from the basic defining equations as an aid to understanding and remembering the interrelationship among the basic electrical units. Note that not all possible combinations are shown. Only those which are useful in solving and understanding electric circuits have been included. We may substitute the symbol E for V in this table when our attention is directed toward applied voltage rather than voltage drop in a particular circuit.

TABLE 5-1

INTERRELATIONSHIP OF BASIC ELECTRICAL UNITS

Function	Defining Equation	Unit and Unit Symbol	Definition of Unit	Useful Derived Equations	Derivation
Voltage	$V = W/Q$	Volt V	Joules per coulomb	$V = IR$ $V = P/I$ $V = \sqrt{PR}$	Transpose equation (4-2) Transpose equation (5-2) Transpose equation (5-4)
Current	$I = Q/t$	Ampere A	Coulombs per second	$I = V/R$ $I = P/V$ $I = \sqrt{P/R}$	Transpose equation (4-2) Transpose equation (5-2) Transpose equation (5-3)
Resistance	$R = V/I$	Ohm Ω	Volts per ampere	$R = V^2/P$ $R = P/I^2$	Transpose equation (5-4) Transpose equation (5-3)
Power	$P = W/t$	Watt W	Joules per second	$P = VI$ $P = I^2R$ $P = V^2/R$	Equation (5-2) Equation (5-3) Equation (5-4)
Work and energy	$W = Fs$	Joule J	Newton meters	$W = Pt$	Transpose equation (5-1)

Note: E is interchangeable with V in this table.

PROBLEMS

5-1-1. How many joules of energy are required to carry a 20-kg parcel up four flights of stairs through a vertical distance of 11 m?

5-1-2. How many joules of energy are required to pump 2000 liters[†] of water to the surface from a well which is 5 m deep?

5-1-3. How many joules of energy are required to raise the temperature of 4 liters of water by 35°C?

5-1-4. How many joules of heat energy must be developed by an electric stove element to heat 1 liter of water from 20°C to 85°C, assuming that the container and radiation account for one-third of the heat energy produced by the stove element?

5-2-5. What is the power rating of a toaster which draws a 5-A current from a 120-V source?

5-2-6. What is the power rating of a soldering iron which has a 110-V drop between its terminals when the current through it is 1.2 A?

5-2-7. At what rate is electric energy converted into heat in the heater of a cathode-ray tube that is rated at 6.3 V 0.6 A?

5-2-8. A 4-mA current through a resistor causes a voltage drop across the resistor of 120 V. What is the minimum "wattage" rating this resistor must have?

5-2-9. At what rate must a 75-Ω resistor be able to dissipate heat if the current through it is 2.5 A?

5-2-10. At what rate must a resistor convert electric energy into heat if 6000 J of heat is produced in 10 min?

[†] The liter is the common metric unit of volume. One liter is equal to 10^3 cubic centimeters. Hence a liter of water has a mass of 1 kilogram.

5-2-11. What power rating must a 100-Ω resistor have to pass a current of 20 mA without overheating?

5-2-12. What "wattage" rating must a 10-kΩ resistor have to pass a current of 250 mA safely?

5-2-13. What "wattage" rating would you select for a 560-Ω bias resistor which has to pass a current of 36 mA?

5-2-14. At what rate is electric energy being converted into heat in a 1.5-Ω motor-starting resistor when the starting current through it is 60 A?

5-2-15. What is the maximum current a 33-kΩ, 1-W, carbon-composition resistor can handle without exceeding its power rating?

5-2-16. What is the maximum voltage drop that should be allowed to appear across a 6.8 kΩ, $\frac{1}{2}$-W metallic film resistor?

5-2-17. What is the maximum current that a 20-kΩ 10-W resistor can handle without overheating?

5-2-18. What is the maximum voltage that can be safely applied to a 20-kΩ 10-W resistor?

5-2-19. What voltage drop will there be across a 1-kW stove element whose resistance when hot is 40 Ω?

5-2-20. What voltage drop will there be across a 75-Ω 3-W Christmas tree lamp?

5-2-21. A certain voltmeter has a resistance of 150 kΩ. While it is measuring a voltage source, it draws energy from the source at the rate of 96 mW. What is the PD of the source?

5-2-22. If the 600-mA heater of a television picture tube must produce heat at the rate of 3 W, what is the voltage drop across the heater?

5-2-23. The armature winding of an electric motor has a resistance of 0.2 Ω. At what rate is electric energy being used up by this particular motor loss when the armature current is 40 A?

5-2-24. How many joules of heat are produced in 1 h in the winding of a motor which has an I^2R loss of 320 W?

5-3-25. What current will an electric motor draw from a 250-V source if it is developing mechanical energy at the rate of 5 hp at 90% efficiency?

5-3-26. What mechanical work can be done by a 75% efficient electric motor drawing 2 A from a 120-V source for 8 h?

5-3-27. If 2.5 × 10⁶ J of electric energy are supplied to the motor operating an elevator which raises a total mass of 5 tonnes through 45 m, what is the efficiency of the system?

5-3-28. With the friction of the pipes and pump bearings, the overall efficiency of a water pump is 86%. What mechanical energy is expended in raising 1600 liters of water 5 m to the surface?

5-3-29. What is the efficiency of an electric motor that develops a 2.0-hp mechanical output when the electric power input is 1.8 kW?

5-3-30. What horsepower electric motor is required to operate a pump which can pump water at the rate of 1600 liters per hour through a vertical distance of 5 m if the overall efficiency of the pumping system is 86%?

5-3-31. How long will it take to raise the temperature of 1 liter of water by 40°C if energy is imparted to the water at the rate of 1 kW?

5-3-32. If 60% of the electric energy input to an electric stove element is effective in heating the water in the kettle, what power rating must the stove element have if the temperature of a liter of water is to be raised from 20°C to 85°C in 4 min?

5-4-33. How much electric energy is used by a 750-W heater in 24 h?

5-4-34. If the overall efficiency of a radio transmitter is 48%, how much electric energy is required to produce a power output of 50 kW from 7:00 A.M. to midnight?

5-4-35. How much will it cost to operate a 4000-Ω electric clock from a 110-V power line for one year if electric energy costs 7¢/kWh?

5-4-36. How much electric energy did Mr. Jones use in May? Assume that:
(a) Electric energy costs 6¢/kWh.
(b) The meter reading at the end of May was 74 267 kWh.
(c) He receives a bill from the power company every two months.
(d) The meter reading at the end of February was 73 067 kWh.
(e) His bill for March and April was $48.00.

5-5-37. A fuse element whose resistance is 0.02 Ω is designed to "blow" when its electric power dissipation exceeds 5 W. What is the current rating of the fuse?

5-5-38. A power supply for a transistor amplifier develops an output of 25 V at 2.4 A. If the overall efficiency of the power supply is 80%, what input current will it draw from a 120-V source?

5-5-39. What is the minimum value of resistance that can be placed across the terminals of a 120-V source if the power drawn from the source is not to exceed 0.48 kW?

5-5-40. An electric heater produces heat at the rate of 50 000 J/min when the current through it is 10 A. What is its resistance?

5-5-41. The current drawn by a room air conditioner from a 230-V source is measured as 12.0 A. If the overall efficiency of the unit is 80%, how many joules of heat can the air conditioner remove from the room in 1 hour?

5-5-42. In an automobile electric system, the following devices are simultaneously drawing energy from the storage battery whose terminal voltage is 6.0 V:
(a) two 6-W taillight lamps;
(b) two parking lamps with a hot resistance of 12 Ω each;
(c) a radio drawing a 6.0-A current; and
(d) the heater fan motor which develops $\frac{1}{20}$ hp at 80% efficiency. What is the total rate at which the battery is supplying electric energy?

REVIEW QUESTIONS

5-1. What is the distinction between **work** and **energy**?

5-2. Why is the joule not commonly used in calculating electric energy?

5-3. What is the significance of the unit **watt second** found in some textbooks?

5-4. Why is the amount of energy represented by a joule quite small?

5-5. Why is the joule selected as the basic unit for electric energy even though it is not used in practice?

5-6. Why would a 5-hp motor with a 90% efficiency be physically smaller than a 5-hp motor with a 60% efficiency?

5-7. Why is it possible to express efficiency in terms of the ratio of output **power** to input **power** even though efficiency is defined as the ratio of output **energy** to input **energy**?

5-8. Why must electric kettles be equipped with thermostatic switches to open the circuit when the kettle boils dry?

5-9. Some small heaters consist of an electric heating coil and a small fan. It is noted that the heating element is brighter when the fan is turned off. Explain.

5-10. Is the heater coil in Question 5-9 drawing more current when the fan is turned off? Explain.

5-11. How would you apply the observations of Question 5-9 to the miniaturization of an aircraft radio?

5-12. Two 1000-Ω resistors are wound from Nichrome wire on ceramic tubes. One is rated at 5 W and the other at 50 W. Compare the two resistors for conductor size, overall dimensions, and temperature at rated dissipation.

5-13. A 35-W soldering iron is used to solder miniature radio components, whereas a 150-W iron is needed to solder a lead to the radio chassis. Explain. What would be the effect of using the 150-W iron on the miniature components?

5-14. SI defines the volt as the difference in electric potential between two points of a conductor carrying a constant current of 1 A when the power dissipated between these points is equal to 1 W. Show that this definition is consistent with the one used in Chapter 2.

5-15. Why is it customary to find larger electric motors used on passenger elevators than freight elevators although the freight elevators are the heavier?

5-16. What is wrong with the wording of the question: "How much power is consumed by a toaster drawing 3 A from a 110-V source?"

5-17. What is wrong with the wording of the question: "How many joules are there in a horsepower?"

5-18. What is wrong with the wording of the question: "What is the efficiency of an electric motor which has a power input of 5 hp and a power output of 4 hp?"

REVIEW PROBLEMS

Interrelationship of basic electrical units.

NOTE: The purpose of these short drill problems is to help the student to recognize the relationships that must exist among the basic electrical units. Each of these problems should be solved by the method outlined with Example 2-2. These problems should be attempted without constant reference to Table 5-1.

5-5-43. What voltage drop is produced across a 42-Ω resistor by a 2.4-A current through it?

5-5-44. At what rate is electric energy converted into heat in the resistor in Problem 5-5-43?

5-5-45. How long will it take the resistor in Problem 5-5-43 to convert 2 kWh of energy?

5-5-46. What voltage must be applied to a 34-Ω resistor to make it dissipate energy at the rate of 180 W?

5-5-47. What is the power rating of a 72-Ω heater which passes a 6-A current?

5-5-48. How long will it take the heater in Problem 5-5-47 to consume 9400 J of electric energy?

5-5-49. How many coulombs of electrons pass through the heater in Problem 5-5-48?

5-5-50. What current is flowing in a 17-Ω resistor while it is dissipating energy at the rate of 520 W?

5-5-51. What is the resistance of a load which draws 1.4 A from a 24-V source?

5-5-52. What current will a 40-W lamp draw from a 117-V source?

5-5-53. What resistance will dissipate energy at the rate of 50 W when connected to a 110-V source?

5-5-54. What voltage must be applied to a 555-W heater to make it pass a 12-A current?

5-5-55. What current will a 45-Ω resistor pass while it is producing heat at the rate of 400 W?

5-5-56. What voltage must be applied to a 27-Ω resistor to make it convert $1\frac{1}{2}$ kWh in 7 h?

5-5-57. What is the efficiency of a motor with a 1-kW input and a 1-hp output?

5-5-58. What voltage drop is created by a 6.8-mA current flowing through a 27-kΩ load resistor?

5-5-59. What voltage is developed across a 300-Ω resistance by a power input to the resistance of 50 μW?

5-5-60. What value of resistance is required to obtain a voltage drop of 7 V when the current is 18 mA?

5-5-61. What power rating must the resistor in Problem 5-5-60 have?

5-5-62. What current will flow when a 20-μV signal is applied to a 75-Ω load?

5-5-63. What power is fed to the load in Problem 5-5-62?

5-5-64. What current will a 4-kW load draw from a 220-V source?

5-5-65. If an automobile starter motor develops $1\frac{1}{4}$ hp at 80% efficiency, what current does it draw from a 12-V battery?

5-5-66. How much work can a fully charged 12-V storage battery accomplish if it is rated at 80 ampere hours?

5-5-67. What is the resistance of a voltmeter which reads 120 V when it has a 40-μA current passing through it?

5-5-68. How long will it take for 2.8 C of electric charge to pass through the meter in Problem 5-5-67?

5-5-69. What is the resistance of the copper bus bars feeding an aluminum refining cell if a 4000-A current through them causes a 620-mV drop across them?

5-5-70. How much electric energy is lost as heat in the bus bars of Problem 5-5-69 in 8 h?

5-5-71. What is the resistance of an ammeter shunt if it is designed to have a 5-mV drop across it with a 9.99-A current through it?

5-5-72. What must be the resistance of a 15-A fuse if heat must be developed at the rate of 4.3 W to melt it?

PART II RESISTANCE NETWORKS

PRACTICAL ELECTRIC CIRCUITS are never as simple as the basic electric circuit of Figure 1-1. Nevertheless, the basic relationships among current, voltage, resistance, work, and power summarized in Table 5-1 must hold true for all electric circuits, no matter how complex they may be.

In Part 1, we developed concepts of the physical nature of electric current, potential difference or voltage, resistance, work, and power. Such concepts are part of a qualitative consideration of the principles of electricity. We also learned how to assign numerical values to units of current, voltage, resistance, work, and power so that we can now examine the behavior of practical electric circuits from a quantitative point of view.

Part 2 groups together in four chapters all the principles and techniques we require for the analysis of any resistance network from a simple series circuit to a complicated array including more than one voltage source. The topics are developed step by step in the most popular sequence. Chapter 7 logically follows Chapter 6. But Chapters 8 and 9 are self-sufficient packages which can be reversed in order, or postponed to permit further study of the principles of circuit operation in the first term. Chapter 10 is a supplementary chapter which is intended to show how electrical measurement is an integral part of electric circuit analysis techniques.

6

SERIES AND PARALLEL CIRCUITS

6-1 RESISTORS IN SERIES

In the last two chapters, we have been treating electric current as a numerical quantity expressed in amperes. At this point, we should remind ourselves that electric current is defined as the *rate of flow* of electric charge. In metallic conductors, the flow of electric charge consists of free electrons moving through the circuit as a result of an electric potential difference generated by a voltage source. Keeping this in mind as we examine the simple series circuit of Figure 6-1, a second part of Ohm's law becomes apparent:

The current must be the same in all parts of a simple series circuit.[†]

Common sense tells us that there cannot be more electrons entering R_1 per second than leave R_1 in a second. Nor can more electrons enter R_1 in a second than leave R_2 in the same period of time. Similarly, the same number of electrons must pass through the source per second as pass through any other part of the series circuit in the same time interval. Since we express *electrons per second* in *amperes,* the same current flows through all components of the simple series circuit of Figure 6-1. Because the truth of this statement is quite evident once we understand what constitutes an electric current,

[†] According to the dictionary, *same* means *identical in all respects*. If R_1 and R_2 are equal in the circuit of Figure 6-1, there will be *equal but different* voltage drops across them, not the *same* voltage drop.

FIGURE 6-1
Simple Series Circuit.

Ohm seldom receives credit nowadays for having been the first person to state this as a basic law of electric circuits.

Sometimes series-connected components are described in terms of the manner in which they are physically connected. In the series circuit of Figure 6-1, R_1 and R_2 are connected in series because no other component or branch is connected to the junction of R_1 and R_2. Also from Figure 6-1, we can say that E, R_1, R_2, and R_3 are all in series because there is only *one* path for electrons to follow through them. All characteristics of series circuits are based on the current being common to all components. Hence, we prefer to define a series circuit in terms of this common current:

Two or more electric components are considered to be in series if the same (common) current flows through all these components.

Since the current in a series circuit is common to all components, it is not necessary to use any subscript with the letter symbol I for the circuit of Figure 6-1. Here I represents the current through R_1, R_2, and R_3, and also the current through the source and the wire connecting the components together.

In Chapter 4, we decided that current in simple electric circuits is a *dependent* variable. It depends on what voltage is applied to the circuit and the total resistance of the circuit. Therefore, in solving the simple series circuit of Figure 6-1, we must be able to determine the total resistance of the circuit. Suppose that resistor R_1 is constructed of 2 meters of Nichrome wire, resistor R_2 is made from 1 meter of the same wire, and R_3 consists of 3 meters of the same resistance wire. Therefore, for an electron to travel around the circuit from one terminal of the generator to the other, it must travel through $3 + 1 + 2 = 6$ meters of Nichrome wire. Since the resistance of an electric conductor is directly proportional to its length, and since the total length of resistance wire used in this example is the sum of the individual lengths, it follows that the total resistance of this circuit will be the sum of the individual resistances. Therefore,

The total resistance of a series circuit is

$$R_T = R_1 + R_2 + R_3 + etc. \qquad (6\text{-}1)$$

Once we determine the total resistance of a series circuit, we can then solve for the common current on which all the characteristics of a series circuit are based by Ohm's law: $I = E/R_T$. Actually, the current must take on a magnitude such that it will produce a total voltage drop V_T equal to the applied voltage E.

EXAMPLE 6-1

What current will flow in a series circuit consisting of 20-Ω, 10-Ω, and 30-Ω resistors connected to a 45-V source?

SOLUTION

$$R_T = R_1 + R_2 + R_3 = 20 + 10 + 30 = 60 \ \Omega$$

$$I = \frac{E}{R_T} = \frac{45 \ V}{60 \ \Omega} = \textbf{0.75 A}$$

As far as the source is concerned, the current drawn from it will be exactly the same for a single 60-ohm resistor connected to its terminals as for the 20-ohm, 10-ohm, and 30-ohm resistors in series. Therefore, if it will aid in analyzing circuit performance, we may think of the total resistance as being replaced by a single *equivalent* resistance R_{eq}. And the circuit of Figure 6-2 becomes the **equivalent circuit** of the original circuit of Figure 6-1. The solution we have used in Example 6-1 consists of reducing the original series circuit to a simple equivalent circuit and then solving for current by Ohm's law. As we start to analyze the behavior of more elaborate electric circuits in later chapters, we often resort to circuit simplification by replacing two or more components in a circuit diagram by a single equivalent circuit component.

FIGURE 6-2
Equivalent Circuit of Figure 6-1.

EXAMPLE 6-2

In the series circuit of Figure 6-1, R_1 and R_3 are 10 kΩ and 22 kΩ, respectively. If the potential difference produced by the source is 16 V, what resistance is required for R_2 to limit the current drawn from the source to 340 μA?

SOLUTION

$$R_T = \frac{E}{I} = \frac{16 \ V}{340 \ \mu A} = 47 \ k\Omega$$

$$R_2 = R_T - (R_1 + R_2) = 47 \ k\Omega - (10 \ k\Omega + 22 \ k\Omega) = \textbf{15 k}\boldsymbol{\Omega}$$

SELF-CHECK

6-1:1. Referring to Figure 6-1, suppose that $E = 15$ V and that $R_1 = 1.0$ kΩ, $R_2 = 2.2$ kΩ and $R_3 = 1.8$ kΩ.

(a) Find the total resistance of the series circuit.

(b) Find the current drawn from the source.

(c) Which resistor has the highest current in it?

6-1:2. A circuit contains a 330-Ω resistor in series with a 470-Ω resistor. The current in the circuit is 50 mA. Determine the voltage of the source.

6-2 POLARITY OF VOLTAGE DROP IN A SERIES CIRCUIT

In Figure 6-3, we have redrawn the simple series circuit of Figure 6-1 and added a voltmeter and an ammeter. Moving the positions of R_1 and R_3 in the circuit diagram, so that R_1, R_2, and R_3 are all in a vertical row, in no way alters the electrical behavior of the circuit, because we have not changed the sequence of the components through which the common current must flow. Since the ammeter is to show the magnitude of the common current, it must be connected in *series* with the circuit so that the current being measured will flow *through* it. Since the voltmeter shows the potential difference *between* two points in a circuit, it is connected *between* points B and C in this example. In other words, the voltmeter is connected *across* or *in parallel with* R_2. The ideal voltmeter draws negligible current; consequently, it does not alter the series circuit behavior for present discussions.

FIGURE 6-3
Polarity of Voltage Drops in a Series Circuit.

The analog direct-current meters we shall encounter in Chapter 10 either mark their terminals with + and − signs, as in Figure 6-3, or at least mark one terminal with a + sign. If we reversed the voltmeter or ammeter leads in a practical circuit, the meters would read off-scale to the left of the zero scale marking. To make sure that we connect direct-current meters properly, we must become quite familiar with the **polarity** of applied voltages and voltage drops in dc circuits.

In Section 2-10, we noted that applied voltage E represents a potential rise, and voltage drop V represents a potential drop or fall. In the circuit of Figure 6-3, electrons experience a potential rise *inside* the voltage source as they are moved from the + terminal to the − terminal *against* the electric force which tends to make them flow in the opposite direction. In the *external* circuit, electrons flow *with* the electric field force from the − terminal of the source back to the + terminal. As electrons flow around this external circuit, they gradually lose the potential energy they gained in their trip through the source. Hence, we can say that the electrons at point *D* are at a higher potential than those at point *C*, and electrons at point *B* are at a higher potential than those at point *A*. Similarly, electrons at point *B* are at a lower potential than those at point *C*, and so on.

In Chapter 3 we discovered that there were quite a few different types of electric charge carriers in various electrical devices. We can divide them into two groups. All *negative* charge carriers will behave like the free electrons in the metallic conductors of Figure 6-3. All *positive* charge carriers flow in the *opposite* direction. Hence, positive charge carriers will receive a rise in potential as they move from the − terminal to the + terminal inside the source and will gradually lose the potential energy thus gained as they then flow through the external circuit. It follows that the + charge carriers have a higher potential at point *A* than at point *B*, and so on.

We can avoid possible confusion by concentrating on the **polarity of the voltage drop** rather than on potential rise and fall. As we have noted, electrons are moved from the + to the − terminal inside the voltage source. We call a voltage source an **active** circuit element since it generates electric energy (at the expense of some other form of energy). In **passive** circuit elements (which consume electric energy) electrons flow from the − to the + terminal, the direction in which they are urged to move by the potential difference in the circuit. Note that an ammeter is also a passive circuit element. Hence, we mark its terminals in the same manner. Electrons enter the ammeter through its − terminal and leave through its + terminal.

Given the polarity of the voltage source in the circuit of Figure 6-3, electrons will flow in a counterclockwise direction around the circuit. We can, therefore, proceed to mark the polarities of all voltage drops with − and + signs. The end of the resistor at which electrons enter must be marked − and the end of the resistor from which electrons leave must be marked + so that the potential difference will have the proper polarity to move electrons *with* the resulting electric field force. In a series circuit, it follows that the + end of the voltage drop across R_3, is connected to the − end of the voltage drop across R_2, and so on. We can now talk in terms of point *C* being positive with respect to point *D*, and point *B* being even more positive with respect to point *D*. We also say that point *C* is negative with respect to point *A*, and so on.

In our pencil-and-paper circuit diagrams, the nature of the actual charge carriers is not particularly important. But when we come to parallel circuits where current divides into two or more branches, it is important that we be able to keep track of current direction. As noted in Section 2-11, to avoid confusion which would result from observing the direction of flow of the actual charge carriers, the current practice in electrical engineering is to adopt the direction of flow of positive charge carriers as the **conventional direction** for electric current. Although both conventional current direction and electron flow are marked in Figure 6-3, only conventional direction will be marked in the remainder of the text (unless it is necessary to show the physical behavior of actual charge carriers).

When we mark polarities on circuit diagrams using conventional current direction, positive charge carriers must "flow" from + to − through any passive circuit element.

As shown in Figure 6-3, we end up with exactly the *same* polarities as when we considered electron flow. Similarly, if we encounter a "black-box" circuit element with unknown contents in network diagrams in later chapters, any circuit element in which conventional current direction is from the + terminal to the − terminal *inside* the black box must behave as a *passive* device.

SELF-CHECK

6-2:1. Referring to the circuit in Figure 6-2,
      (a) What is the direction of the (conventional) circuit current?
      (b) What is the direction of the electron flow in the circuit?
      (c) What is the polarity of the voltage drop across R_{eq}?

6-3 DOUBLE-SUBSCRIPT NOTATION

In the circuit diagram of Figure 6-3, the symbol V_1 identifies the voltage drop across R_1. The **polarity** of V_1 is obvious from Figure 6-3. Terminal A is positive with respect to terminal B.

Double-subscript notation provides us with a technique for including polarity information on the voltage drop across R_1. By identifying this voltage drop as V_{AB}, we are talking about the potential at point A *with respect to* point B. **It is standard procedure that the *second* subscript will designate the reference** (with respect to) **point or *node*.** Hence in Figure 6-3, V_{AB} is a positive quantity, whereas V_{BA} is a negative quantity and $V_{AB} = -V_{BA}$.

In dealing with three-phase circuits in Chapter 26, we shall find that double-subscript notation is essential in keeping track of the polarity of voltages and the directions of currents. In dealing with the dc circuits of the next few chapters, double-subscript notation is optional and we use it only when it is essential for clarity. In most examples, single-subscript notation is adequate.

6-4 KIRCHHOFF'S VOLTAGE LAW

Since the rules of algebra allow us to multiply both sides of an equation by the same quantity without changing the equality of the statement, let us multiply both sides of Equation (6-1) by I.

$$IR_T = IR_1 + IR_2 + IR_3 + \text{etc.}$$

Since $\qquad\qquad I = \dfrac{V_T}{R_T} \qquad \therefore IR_T = E \text{ (the applied voltage)}$

and $\qquad\qquad IR_1 = V_1 \text{ (the voltage drop across resistor } R_1)$

$$\therefore E = V_1 + V_2 + V_3 + \text{etc.} \tag{6-2}$$

Therefore,

In a series circuit, the sum of all the voltage drops across the individual resistances must equal the applied voltage.

Gustav Kirchhoff discovered that this applied to any complete electric circuit, whether a simple series circuit as in Figure 6-1 or one loop of an elaborate network. This principle is stated as

Kirchhoff's voltage law: In any complete electric circuit, the algebraic sum of the applied voltages must equal the algebraic sum of the voltage drops.

Kirchhoff's voltage law provides us with an alternate solution to Example 6-1. In this case, the numerical work is essentially the same, but only one step is involved, as compared with the first method in which the first step was to replace the three separate resistances with a single equivalent resistance.

EXAMPLE 6-1A

What current will flow in a series circuit consisting of 20-Ω, 10-Ω, and 30-Ω resistors connected to a 45-V source?

SOLUTION

$$E = IR_1 + IR_2 + IR_3$$

$$45 = 20I + 10I + 30I = 60I$$

$$\therefore I = \frac{45 \text{ V}}{60 \text{ }\Omega} = 0.75 \text{ A}$$

SELF-CHECK

6-4:1. A series circuit contains a 24-V source and two resistors $R_1 = 2.4$ kΩ and $R_2 = 5.6$ kΩ.
   (a) Find the circuit current.
   (b) Find the voltage drop across each resistor.
   (c) Use Kirchhoff's Voltage Law to check your answer to part (b).
6-4:2. A circuit contains a 330-Ω resistor in series with a 470-Ω resistor. The current in the circuit is 50 mA.
   (a) Determine the voltage drop across each resistor.
   (b) Use Kirchhoff's Voltage Law to determine the voltage of the source.

6-5 CHARACTERISTICS OF SERIES CIRCUITS

From Ohm's law, $I = V_1/R_1$, and since the current is the *same* through all components of a series circuit,

$$I = \frac{V_1}{R_1} = \frac{V_2}{R_2} = \frac{V_3}{R_3} = \frac{V_T}{R_T}$$

If we transpose any pair of these equal terms, say $V_1/R_1 = V_2/R_2$, we get $V_1/V_2 = R_1/R_2$. From this, it follows that

In a series circuit, the ratio between any two voltage drops is the same as the ratio of the two resistances across which these voltage drops occur.

We can use this knowledge in solving for some specific information about a circuit without having to complete a detailed solution of the whole circuit.

EXAMPLE 6-3

A 20-kΩ resistor and a 15-kΩ resistor are connected in series to a 140-V source. What is the voltage drop across the 15-kΩ resistor?

Note: There is less chance of making an error in the solution of an electric circuit if, as the first step in any solution, we draw a fully labeled schematic diagram to work from (Figure 6-4).

LONG SOLUTION

$$R_T = R_1 + R_2 = 20 \text{ k}\Omega + 15 \text{ k}\Omega = 35 \text{ k}\Omega$$

$$I = \frac{E}{R_T} = \frac{140 \text{ V}}{35 \text{ k}\Omega} = 4 \times 10^{-3} \text{ A} = 4 \text{ mA}$$

$$V_2 = IR_2 = 4 \text{ mA} \times 15 \text{ k}\Omega = \textbf{60 V}$$

SHORT SOLUTION

$$\frac{V_2}{E} = \frac{R_2}{R_T} \qquad \therefore V_2 = E\frac{R_2}{R_T} = 140 \text{ V} \times \frac{15 \text{ k}\Omega}{35 \text{ k}\Omega} = \textbf{60 V}$$

The calculator chain for the short solution is noticeably simpler than that for the long solution, which is

$$140 \div (20 \text{ EE } 3 + 15 \text{ EE } 3) \times 15 \text{ EE } 3 = 60.$$

FIGURE 6-4
Circuit Diagram for Example 6-3.

Another characteristic of a series circuit that is based on the *same* current flowing through all components is that any change to *one* component of a series circuit will have an effect on the current through *all* the components. Therefore, we must connect in *series* with a load, such *control* components as switches which turn the current on and off, fuses

FIGURE 6-5
Control Components in Series with a Load.

which open the circuit if the current becomes excessive, and variable resistors (rheostats) which control the magnitude of the current, as in Figure 6-5.

To be able to distinguish readily between series and parallel circuit characteristics, we must be thoroughly familiar with the characteristics of series circuits which we have just developed. We can summarize these characteristics as follows:

1. The *current* is the same in all parts of a series circuit.
2. The *total resistance* is the sum of all the individual resistances.

$$R_T = R_1 + R_2 + R_3 + \text{etc.}$$

3. The *applied voltage* is equal to the sums of all the individual voltage drops.

$$E = V_1 + V_2 + V_3 + \text{etc.}$$

4. The *ratio* between voltages is the same as the resistance ratio.
5. Any change to *any* component of a series will affect the current through *all* components.

SELF-CHECK

6-5:1. Refer to Figure 6-5 and suppose that the source voltage is 22 V and the resistances are:

Fuse: $R_F = 0.33\ \Omega$
Rheostat: $R_R = 2.0\ \Omega$
Lamp: $R_L = 5.0\ \Omega$

(a) With the switch closed, find the current in and voltage across each component.
(b) With the switch open, find the current in and voltage across each component.

6-6 INTERNAL RESISTANCE

In Section 2-10 we noted that the terminal voltage generated by a source is numerically equal to the emf produced by the energy-converting action of the source only under open-circuit ($I = 0$) conditions. The emf of a given source depends on the nature of the energy-converting action and is independent of the circuit current. However, as shown in the graph of Figure 6-6, the generated potential difference that appears between the two terminals of a practical voltage source *decreases* as the external load is adjusted to

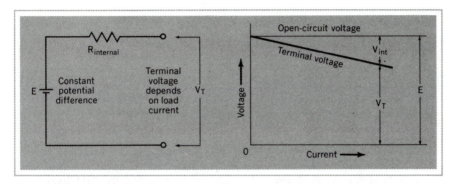

FIGURE 6-6
Schematic and Graphic Representation of a Practical Voltage Source.

draw more current from the source. An example of this change in terminal voltage is the way automobile headlights dim when the starter motor is drawing current from the car battery.

In many cases, the change in terminal voltage with an increase in circuit current is so slight that we can ignore it. But in other cases, particularly in electronic circuits, the decrease in the terminal voltage of a source as the current through it increases must be taken into account. A convenient way of representing this effect is to assume that the *practical* voltage source consists of (1) an *ideal* source which generates a constant potential difference at any current and (2) a resistance connected in series with this ideal source and the actual terminals of the practical source as shown in the schematic representation of Figure 6-6. We call this resistance the **internal resistance** of the source. In addition to causing the terminal voltage to decrease as the current increases, this internal resistance concept can be used to calculate the loss of electric energy through conversion into heat within the practical generator.

If we can measure the voltage between the terminals of the practical source without drawing any current from the source, there will be no voltage drop across the internal resistance (since $V_{int} = IR_{int}$ and I = zero). Therefore, the terminal voltage under these conditions is equal to the constant potential difference generated by the ideal source portion of the practical generator. From this method for determining the potential difference of the ideal voltage source, we call this potential difference the **open-circuit voltage** of the source. In equations we shall represent this open-circuit voltage generated by the ideal portion of the practical source by the letter symbol E. The actual terminal voltage of the source becomes V_T.

The following example illustrates how we can determine the internal resistance of a practical voltage source.

EXAMPLE 6-4
When a voltmeter which draws negligible current is the only circuit connected to the terminals of a battery, it reads 6.0 V. When a 5-Ω resistor is connected to the battery terminals, the voltmeter reads 5.0 V. What is the internal resistance of the battery?

SOLUTION
Since the open-circuit voltage is 6.0 V, the potential difference of the ideal source

FIGURE 6-7
Circuit Diagram for Example 6-4.

(Figure 6-7) is 6.0 V. And since the voltage drop across the 5-Ω load is 5.0 V, according to Kirchhoff's voltage law, the voltage drop across the internal resistance is

$$E = V_{int} + V_{load}$$

$$\therefore V_{int} = E - V_{load} = 6.0 - 5.0 = 1.0 \text{ V}$$

From Ohm's law, the current in the circuit with the 5-Ω resistor connected is

$$I = \frac{V_L}{R_L} = \frac{5.0 \text{ V}}{5 \text{ Ω}} = 1 \text{ A}$$

$$\therefore R_{int} = \frac{V_{int}}{I} = \frac{1.0 \text{ V}}{1 \text{ A}} = 1 \text{ Ω}$$

ALTERNATE SOLUTION

$$\frac{R_{int}}{R_L} = \frac{V_{int}}{V_L}$$

$$\therefore R_{int} = R_L \frac{V_{int}}{V_L} = 5 \text{ Ω} \times \frac{1 \text{ V}}{5 \text{ V}} = 1 \text{ Ω}$$

From Example 6-4, we note that

1. The *terminal voltage* V_T of a practical voltage source is equal to the open-circuit voltage of the source only if there is no voltage drop across its internal resistance.
2. The terminal voltage V_T is equal to the total voltage drop across the load ($V_T = V_L$). Hence, we may say that V_T also represents *total* load voltage drop.
3. From Kirchhoff's voltage law, we may write an equation for the terminal voltage of generators as

$$V_T = E - IR_{int} \tag{6-3}$$

where V_T is the terminal voltage of a practical voltage source, E is the open-circuit voltage of the source, I is the current drawn from the source, and R_{int} is its internal resistance.

Equation (6-3) suggests an interesting possibility—suppose that we can arrange to have the emf of the voltage source *increase* as the current drawn from the source *increases*. If we arrange that the increase in generated voltage (E) just matches the increase in voltage drop (IR_{int}), the terminal voltage (V_T) will remain constant over a wide range of load currents. We call such an arrangement **voltage regulation.** Since the generated emf of a battery depends on the chemical reaction in the cell, we cannot apply voltage regulation to battery voltage sources. However, the control room operators in a utility generating station can certainly increase the voltage generated by rotating machines as the current demand increases. In electronic power supplies, we can readily accomplish control of the source voltage by "feedback" signals monitoring the output terminal voltage. Such a voltage source then becomes a **controlled source.** We shall examine controlled sources in more detail in a later chapter.

As we noted at the beginning of this section, the effect of internal resistance for many practical voltage sources is small enough that we can say that the terminal voltage of the source is essentially the same as its open-circuit voltage. This is generally the case when we are using heavy-duty sources, such as a storage battery or the local electric company supply, which are designed to supply a load current of many amperes. But if the voltage source is an electronic device which has a high internal resistance, we must take care that we are not misled by this circuit parameter, which usually is not shown in the circuit diagram. In solving the problems in this text, if no specific value of internal resistance is stated, we shall consider the internal resistance of the source to be zero, in which case $V_T = E$.

SELF-CHECK

6-6:1. The open circuit voltage of a battery is 13.2 V. When a current of 25 A is drawn from the source the terminal voltage drops to 9.45 V. What is the internal resistance of the battery?

6-6:2. A battery has an open circuit voltage of 26 V and an internal resistance of 0.05 Ω. Determine the current that, when drawn from the source, will cause the terminal voltage to be 24 V.

6-7 MAXIMUM POWER TRANSFER

One effect of internal resistance in a voltage source is a decrease in generator terminal voltage as the current drawn from the source is increased. To determine what other effects internal resistance may have on electric circuit behavior, we can substitute various values of load resistance in the circuit of Figure 6-8. For every value of load resistance, the internal resistance and applied voltage remain constant ($R_{int} = 6\ \Omega$ and $E = 120$ V). We need to perform the following calculations for a range of values of load resistance and to display the results in table format.

$$I = \frac{E}{R_L + R_{int}} = \frac{120}{R_L + 6}$$

$$V_L = IR_L$$

FIGURE 6-8
Circuit Diagram for Table 6-1.

$$P_L = I^2 R_L$$

$$P_{int} = I^2 R_{int} = 6\,I^2$$

$$\eta = \frac{P_L}{P_L + P_{int}} \times 100\% = \frac{100 P_L}{P_L + P_{int}}$$

We have now found an example where a computer can outperform a pocket calculator. To produce the data for Table 6-1 on a calculator, we have to perform over 100 separate calculations. We can program the computer to perform the five calculations listed above, entering E and R_{int} as numerical constants and R_L as the only independent variable. Instead of entering a single value for R_L, we can use the computer's "loop" ability to repeat all calculations automatically with R_L increased by a predetermined

TABLE 6-1

R LOAD OHMS	CURRENT AMPERES	LOAD VOLTAGE	LOAD POWER	INTERNAL POWER	PERCENT EFFICIENCY
0	20.0	0	0	2400	0
2	15.0	30	450	1350	25
4	12.0	48	576	864	40
6	10.0	60	600	600	50
8	8.6	69	588	441	57
10	7.5	75	563	338	63
12	6.7	80	533	267	67
14	6.0	84	504	216	70
16	5.5	87	476	179	73
18	5.0	90	450	150	75
20	4.6	92	426	128	77
22	4.3	94	404	110	79
24	4.0	96	384	96	80
26	3.8	98	366	84	81
28	3.5	99	349	75	82
30	3.3	100	333	67	83
32	3.2	101	319	60	84
34	3.0	102	306	54	85
36	2.9	103	294	49	86
38	2.7	104	283	45	87
40	2.6	104	272	41	87

amount over a predetermined range—for example, from 0 to 40 in 2-ohm steps. (We selected 21 increments for R_L since a typical video screen displays 24 lines with 80 characters per line.) We start the program by instructing the computer how to label and space the columns and how to round off the calculations. The latter chore would be simpler if we had some idea what the final table should look like; hence, our first attempt may produce a ragged table. A typical program might be

```
100 REM   MAXIMUM POWER TRANSFER THEOREM
110 F1$ = "\      \      \      \      \      \      \      \         \      \      \      \    "
120 F2$ = " \   \      \      \      \      \      \      \      \      \      \      \    "
130 F3$ = " ##         ##.#         ###      ###         ####         ## "
140 PRINT USING F1$; "R LOAD";"CURRENT";"LOAD";"LOAD";"INTERNAL";"PERCENT"
150 PRINT USING F2$; "OHMS";"AMPERES";"VOLTAGE";"POWER";"POWER";"EFFICIENCY"
160 FOR R = 0 TO 40 STEP 2
170 LET I = 120/(R + 6)
180 LET VL = I*R
190 LET PL = I^2*R
200 LET PI = 6*I^2
210 LET EFF = 100*PL/(PL + PI)
220 PRINT USING F3$; R;I;VL;PL;PI;EFF
230 NEXT R
240 END
```

When we enter RUN, the computer should then print Table 6-1.

To assist us in analyzing the significance of the data displayed in Table 6-1, we can plot graphs of some of the columns of dependent figures against the stepped values of R_L. The personal computer is quite capable of generating such graphs from the data it has already calculated in the program above. But to guard against losing sight of the purpose of this exercise to the novelty of the computer, we have plotted the graphs of Figure 6-9 from the data of Table 6-1 without resorting to computer graphics. Note in Table 6-1 that

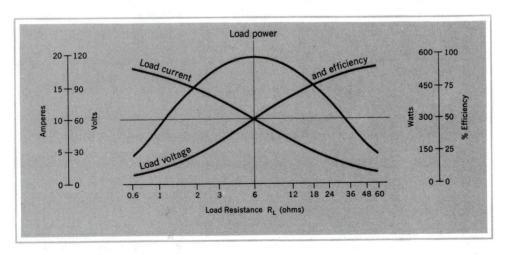

FIGURE 6-9
Graph Showing Requirements for Maximum Power Transfer.

the graph for load power rises to a maximum when R_L reaches 6 ohms and then tapers off gradually as R_L increases further. If we use a linear x-axis scale for R_L, we will produce a lopsided graph (as the computer will). To produce the symmetrical graphs of Figure 6-9, we have used a *logarithmic** scale for the x-axis so that the spacing between 2 and 6 ohms is the same as the spacing between 6 and 18 ohms.

On the y-axis, we first plot load current (column 2) and load voltage (column 3). As we increase the load resistance from 0 ohms (short circuit) to 40 ohms, the current decreases from its short-circuit value of 20 A to 2.6 A when R_L = 40 ohms. Voltage across the load behaves in the opposite manner—rising from 0 volts at short circuit to 104 volts when R_L = 40 ohms. Note that the percentage efficiency graph (from column 6) is the same as the load voltage graph if we use suitable scales.

When we superimpose the graph of load power (from column 4), we immediately note that

1. *Maximum power output (into the load) occurs when*

$$R_L = R_{int} \qquad (6\text{-}4)$$

Also, when R_L is selected for maximum load power,

$$I = 0.5\,I_{sc} \qquad \text{(short-circuit current)} \qquad (6\text{-}5)$$

$$V_L = 0.5\,E \qquad \text{(open-circuit voltage)} \qquad (6\text{-}6)$$

$$\eta = 50\% \qquad \text{(one-half maximum efficiency)} \qquad (6\text{-}7)$$

Note that maximum power output does *not* coincide with maximum efficiency. When a load resistance is selected for maximum power output, there is an equal power dissipation inside the source.

2. If we want to increase the efficiency, a load resistance of from two to three times the internal resistance of the generator results in appreciable reduction in wasted power (as heat in the generator) for only a small reduction in power output.

3. A load resistance less than the internal resistance of the generator not only results in a reduction in power output but also causes a very high dissipation within the generator. In practice, this condition of operation is termed **overload** and must be avoided.

4. If we are interested more in *voltage* output than power output (as in transistor voltage amplifiers), the load resistance should be high in comparison to the internal resistance of the source.

$$R_L \cong 5 \times R_{int} \qquad (6\text{-}8)$$

From Figure 6-9 we note that the *slope* of the load power graph becomes zero at

---

*The x-axis of *semi-log* graph paper is laid out so that each *linear* unit of distance along the x-axis represents a *ten-fold* increase in the value of the independent variable. Hence, the linear horizontal distance of x-axis scale markings is based on the common logarithm of the value of the independent variable. Note that the spacing between 0.6 and 6 ohms in Figure 6-9 is identical to the spacing between 6 and 60 ohms. The y-axis of *semi-log* graph paper is marked with the usual linear scale.

maximum power. Hence, if we are familiar with elementary differential calculus, we can prove the maximum power transfer theorem of Equation (6-4) with a single calculation, as shown in Appendix 2-1.

SELF-CHECK

6-7:1. A battery with an internal resistance of 2.0 Ω and an open circuit voltage of 20 V is supplying power to a lamp which draws maximum power from the battery. Determine the current in the lamp.

6-7:2. A generator has an open-circuit voltage of 120 V and a short-circuit current of 60 A.
   (a) What load resistance will result in the maximum power output?
   (b) What load resistance will result in an efficiency of 90%?

6-8 RESISTORS IN PARALLEL

In dealing with parallel- (or shunt) connected resistors, we find a set of characteristics which, in many respects, are similar but opposite to those of series circuits.[†] Figure 6-10 shows two different ways of drawing a circuit diagram for a simple parallel circuit. Since we usually prefer to draw interconnecting conductors as either horizontal or vertical lines, the diagram of Figure 6-10(a) is the customary configuration. However, to illustrate the definition of a parallel circuit, we can redraw the circuit diagram as in Figure 6-10(b).

(a) Customary configuration.   (b) Alternate configuration.

FIGURE 6-10
Simple Parallel Circuit.

We can define a parallel circuit in terms of the physical connection of the components. In Figure 6-10, E, R_1, R_2, and R_3 are all in parallel because they are all connected between the *same two* **junction points** or **nodes** (A and B). Since each of the resistors is connected directly across the generator terminals, $V_1 = V_2 = V_3 = E$. And since this voltage is common to all components in parallel, we may omit the subscript. In keeping with our definition of a series circuit, we prefer to define a parallel circuit in terms of this

---

[†]We encounter this "similar-but-opposite" characteristic of series and parallel electric circuits quite often. It is sometimes referred to as the principle of **duality.**

common voltage:

Two or more electric components are considered to be in parallel if the same (common) voltage appears across all these components.[†]

In solving the series circuit, we determined the total resistance in order to find the current in the circuit. In the **total current method** of solving parallel circuits, the steps are reversed. If we are given the open-circuit voltage of the source (assuming negligible internal resistance) and the values of each resistance in the circuit of Figure 6-10, we can solve for the current in each branch as an individual circuit by using Ohm's law. If we think of the current in each branch in terms of electrons per second flowing through the branch, it is apparent that the generator current must be the sum of the branch currents.[‡] Therefore,

In a simple parallel circuit, the total current is the sum of all the branch currents.

$$I_T = I_1 + I_2 + I_3 + \text{etc.} \tag{6-9}$$

EXAMPLE 6-5
With reference to Figure 6-10, R_1 is 40 Ω, R_2 is 30 Ω, R_3 is 20 Ω, and E is 120 V. What single resistance would draw the same current from the source?

SOLUTION

$$I_1 = \frac{V_1}{R_1} = \frac{120 \text{ V}}{40 \text{ }\Omega} = 3 \text{ A}$$

$$I_2 = \frac{V_2}{R_2} = \frac{120 \text{ V}}{30 \text{ }\Omega} = 4 \text{ A}$$

$$I_3 = \frac{V_3}{R_3} = \frac{120 \text{ V}}{20 \text{ }\Omega} = 6 \text{ A}$$

$$I_T = I_1 + I_2 + I_3 = 3 + 4 + 6 = 13 \text{ A}$$

$$\therefore R_{eq} = \frac{E}{I_T} = \frac{120 \text{ V}}{13 \text{ A}} = \mathbf{9.23 \text{ }\Omega}$$

From Equation (6-1), we note that the *total* resistance of a series circuit is *greater* than the resistance of any individual resistor. But the answer in Example 6-5 is *less* than the resistance of any individual resistor. Hence, in dealing with parallel circuits, we do not speak of total resistance. Instead, we speak of the **equivalent resistance** (R_{eq}) of two or more resistors in parallel.

[†] Once again we must note that *same* does *not* mean *equal but different.*

[‡] In working with circuits in which there is more than one path for current, it will help us to keep track of the various branch currents if we mark current direction on the schematic diagram with arrows. Note that we prefer to use the *conventional* current direction for this purpose.

SELF-CHECK

6-8:1. Suppose that in Figure 6-10, the source voltage is 12 V and the resistor values are: $R_1 = 1.2$ kΩ, $R_2 = 1.8$ kΩ and $R_3 = 3.6$ kΩ.
   (a) Determine the branch currents: I_1, I_2, and I_3.
   (b) Determine the equivalent resistance of the parallel circuit.
6-8:2. A parallel combination of 330-Ω and 470-Ω resistors is supplied by a 24-V battery.
   (a) Find the equivalent resistance.
   (b) Determine the current drawn from the battery.

6-9 KIRCHHOFF'S CURRENT LAW

Kirchhoff applied the total-current principle of Equation (6-9) to any electric circuit in the form of

Kirchhoff's current law: At any junction point or node in an electric circuit, the algebraic sum of the currents entering the point must equal the algebraic sum of the currents leaving the point.

EXAMPLE 6-6
What current is flowing in the R_2 branch of the circuit of Figure 6-11?

SOLUTION

$$I_1 = \frac{V}{R_1} = \frac{50 \text{ V}}{10 \text{ }\Omega} = 5 \text{ A}$$

The current flowing into node X is I_T and the currents flowing away from node X are I_1 and I_2. Therefore,

$$I_T = I_1 + I_2 \quad \text{or} \quad I_2 = I_T - I_1$$

$$\therefore I_2 = 12 - 5 = \textbf{7 A}$$

FIGURE 6-11
Circuit Diagram for Example 6-6.

SELF-CHECK

6-9:1. Suppose a circuit has two parallel branches and a 20-V source. Measurement shows that the first branch draws 60 mA and the second draws 20 mA. Determine the current in the source.

6-9:2. A parallel combination of 330-Ω and 470-Ω resistors is supplied by a 24-V battery.
(a) Determine the magnitude of each branch current.
(b) Use Kirchhoff's Current Law to determine the battery current.

6-10 CONDUCTANCE

As the results in Examples 6-5 and 6-6 indicate, the total current must always be greater than the current through any branch of a parallel circuit. Therefore, the equivalent resistance must always be *less* than the smallest of the branch resistances. If we continue this line of thought, the *more* resistors we connect in parallel, the *smaller* the equivalent resistance becomes. We can appreciate the behavior of parallel circuits more readily if we turn our attention away from resistance and restate the above remark thus: The more resistors we connect in parallel, the more readily the circuit can pass current since there are more parallel branches for current to flow through. Therefore, since resistance is a measure of the ability of a circuit to *oppose* electric current,

> *Conductance is a measure of the ability of an electric circuit to pass current.*
> *The letter symbol for conductance is G.*
> *The siemens is the SI unit of conductance.* [†]
> *The unit symbol for siemens is S.*

Since conductance is simply the reciprocal point of view of resistance, we can define the size of the siemens by the equation

$$G = \frac{1}{R} \qquad (6\text{-}10)$$

where G is the conductance of a circuit in siemens and R is the resistance of the same circuit in ohms.

When we are given the voltage across a group of parallel branches, the total-current method described in the preceding section is a convenient method of solving parallel circuits. Even if this voltage is not known, we can assume a suitable voltage in order to solve for the equivalent resistance. But, in this case, the concept of conductance offers a more straightforward approach to the solution of parallel circuits. If we divide both sides of Equation (6-9) by E (or V, since they are the same for simple parallel circuits),

---

[†] Before the IEEE adopted the SI name **siemens** for the unit of conductance, the common name for the unit of conductance was the **mho** (ohm spelled backward). The unit symbol was (logically) an *inverted* Greek letter "omega." The siemens was named in honor of a British engineer, Sir William Siemens.

we get

$$\frac{I_T}{E} = \frac{I_1}{V} + \frac{I_2}{V} + \frac{I_3}{V} + \text{etc.}$$

But $R = V/I$, and since $G = 1/R$, $G = I/V$

$$\therefore G_T = G_1 + G_2 + G_3 + \text{etc.} \tag{6-11}$$

Therefore,

> *In parallel circuits, the total conductance is equal to the sum of the conductances of all the individual branches.*

Having determined the total conductance, the equivalent resistance is simply $R_{eq} = 1/G_T$.

EXAMPLE 6-5A

With reference to Figure 6-10, R_1 is 40 Ω, R_2 is 30 Ω, and R_3 is 20 Ω. What single resistance would draw the same current from the source?

SOLUTION

$$G_T = G_1 + G_2 + G_3 = \frac{1}{40} + \frac{1}{30} + \frac{1}{20} \text{ S}$$

$$= 2.5 \times 10^{-2} + 3.33 \times 10^{-2} + 5 \times 10^{-2} = 0.1083 \text{ S}$$

$$\therefore R_{eq} = \frac{1}{G_T} = \frac{1}{0.1083 \text{ S}} = \mathbf{9.23 \ \Omega}$$

Using the reciprocal $1/x$ key, the calculator chain for G_T becomes 40 $1/x$ + 30 $1/x$ + 20 $1/x$ = 0.1083 S. Pressing the reciprocal key once more converts G_T into R_{eq}.

When only *two* resistors in parallel are involved, we can reduce Equation (6-11) to a convenient form for determining the equivalent resistance directly.

$$G_T = G_1 + G_2 = \frac{1}{R_1} + \frac{1}{R_2} = \frac{R_1 + R_2}{R_1 R_2}$$

$$\therefore R_{eq} = \frac{R_1 R_2}{R_1 + R_2} \tag{6-12}$$

Therefore,

> *For two resistors in parallel, the equivalent resistance equals their product over their sum.*

This does *not* apply to more than two resistors, as we shall discover by trying to simplify Equation (6-11) with three or more terms.

To deal with the situation where we want to know what resistance R_2 we must connect in parallel with a given resistance R_1 to obtain a certain equivalent resistance, we can rearrange Equation (6-12) as

$$R_2 = \frac{R_1 R_{eq}}{R_1 - R_{eq}} \tag{6-13}$$

To determine the "wattage" rating needed for resistor R_2, we want to express its current I_2 as a fraction of the total current through the parallel combination of R_1 and R_2. Since $V_2 = V_1$, $I_2 R_2 = I_1 R_1$. And from Kirchhoff's current law, $I_2 R_2 = (I_T - I_2) R_1$ from which,

$$I_2 = I_T \left(\frac{R_1}{R_1 + R_2} \right) \tag{6-14}$$

Another special example of Equation (6-11) is the case of *N equal* resistors in parallel. A moment's reflection will show us that

$$R_{eq} = \frac{R}{N} \tag{6-15}$$

where R is the resistance of each of the parallel resistors, and N is the number of resistors connected in parallel.

EXAMPLE 6-7
What is the equivalent resistance of a 1-kΩ and a 4-kΩ resistor in parallel?

SOLUTION

$$R_{eq} = \frac{R_1 R_2}{R_1 + R_2} = \frac{1 \text{ k}\Omega \times 4 \text{ k}\Omega}{(1 \text{ k}\Omega + 4 \text{ k}\Omega)} = \frac{4\,000\,000}{5000} = \mathbf{800\ \Omega}$$

(For verification, solve this example using the calculator procedure of Example 6-5A.)

SELF-CHECK

6-10:1. Find the conductance of each branch and the total conductance of the circuit in Figure 6-11.

6-10:2. A parallel combination of a 330-Ω and a 470-Ω resistor is supplied by a 24-V battery. (a) Find the conductance of each branch. (b) Use the conductances to find the branch currents.

6-11 CONDUCTIVITY

We defined resistivity as a property of a conducting *material*, being the resistance of a unit length and cross-section conductor of that material. Similarly, then,

The conductivity of a material is the conductance of a unit length and cross-section conductor of that material.

The letter symbol for conductivity is the Greek letter σ (sigma).

Since conductance is the reciprocal of resistance, conductivity is the reciprocal of resistivity. Therefore,

$$\sigma = \frac{1}{\rho} \qquad (6\text{-}16)$$

Conductivity is expressed in **siemens per meter.**

6-12 CHARACTERISTICS OF PARALLEL CIRCUITS

From Ohm's law, $V = IR$, and since $R = 1/G$, $V = I/G$, and since the voltage is the *same* across all components of a parallel circuit,

$$V = \frac{I_1}{G_1} = \frac{I_2}{G_2} = \frac{I_3}{G_3} = \frac{I_T}{G_T} \qquad (6\text{-}17)$$

If we take any pair of these equal terms, say $I_1/G_1 = I_2/G_2$ and transpose it we get $I_1/I_2 = G_1/G_2 = R_2/R_1$. From this, it follows that

In a parallel circuit, the ratio between any two branch currents is the same as the ratio of their conductances or the inverse of their resistance ratio.

EXAMPLE 6-8

The total current drawn by a 12.5-kΩ resistor (R_1) and a 50-kΩ resistor (R_2) in parallel is 15 mA. What is the current through the 50-kΩ resistor?

SOLUTION 1

From the circuit diagram of Figure 6-12,

$$I_1 + I_2 = 15 \text{ mA}$$

$$\frac{I_1}{I_2} = \frac{R_2}{R_1} = \frac{50 \text{ k}\Omega}{12.5 \text{ k}\Omega} = 4$$

$$\therefore I_1 = 4I_2$$

Substituting this value of I_1 in the first equation gives

$$4I_2 + I_2 = 15 \text{ mA}$$

$$\therefore I_2 = \frac{15 \text{ mA}}{5} = \mathbf{3 \text{ mA}}$$

FIGURE 6-12
Circuit Diagram for Example 6-8.

SOLUTION 2

$$R_{eq} = \frac{R_1 R_2}{R_1 + R_2} = \frac{12.5 \text{ k}\Omega \times 50 \text{ k}\Omega}{(12.5 \text{ k}\Omega + 50 \text{ k}\Omega)} = 10 \text{ k}\Omega$$

$$V = I_T R_{eq} = 15 \text{ mA} \times 10 \text{ k}\Omega = 150 \text{ V}$$

$$\therefore I_2 = \frac{V}{R_2} = \frac{150 \text{ V}}{50 \text{ k}\Omega} = \textbf{3 mA}$$

Note: There is more than one way of solving most electric circuit problems. Although it is wise to try to understand the various methods, we should select the method by which we can visualize what each step represents. It is much better to achieve accurate results by a longer method than to make an error in a shortcut through not appreciating the significance of each step of the solution.

Another characteristic of parallel circuits is based on the internal resistance of the source being negligible. If such is the case, altering the resistance of one branch will not affect the voltage across and thus the current through the other branches. Therefore, changes in one branch of a parallel circuit have negligible effect on the other branches. In house wiring, each lighting circuit is connected in parallel with the others to the 117-volt source so that switching one circuit on or off does not affect the operation of the other circuits, as shown in Figure 6-13.

FIGURE 6-13
Parallel Connection of Loads to a Branch Circuit in House Wiring.

We may summarize the characteristics of parallel circuits as follows:

1. The *voltage* is the same across all components in a parallel circuit.
2. The *total conductance* is the sum of all the individual branch conductances.

$$G_T = G_1 + G_2 + \text{etc.}$$

3. The total *current* is the sum of all the individual branch currents.

$$I_T = I_1 + I_2 + I_3 + \text{etc.}$$

4. The *ratio* between branch currents is the *same* as the conductance ratio and the *inverse* of the resistance ratio.
5. Each branch is essentially independent of any changes in the other branches, providing the voltage across the parallel circuit is kept constant.

SELF-CHECK

6-12:1. Consider the circuit shown in Figure 6-14. Determine the current in each component and the voltage across each component (a) with the switch closed and (b) with the switch open.

FIGURE   6-14

PROBLEMS

Draw a fully labeled schematic diagram for each problem.

6-1-1. Given $R_1 = 5\ \Omega$, $R_2 = 10\ \Omega$, and $R_3 = 15\ \Omega$, what is the total resistance when they are connected in series?

6-1-2. What is the total resistance of a series circuit consisting of $R_1 = 47\ k\Omega$, $R_2 = 330\ k\Omega$, and $R_3 = 2.2\ k\Omega$?

6-1-3. What is the total resistance of a network consisting of six 680-Ω resistors in series?

6-1-4. What current will a circuit consisting of a 330-kΩ resistor and a 1.5-kΩ resistor in series draw from a 24-V source?

6-1-5. What resistance must be connected in series with a 470-Ω resistor to bring the total resistance to 800 Ω?

6-1-6. What resistance must be connected in series with a 22-kΩ resistor to limit the current to 1.5 mA when the series combination is connected to a 120-V source?

6-1-7. A 6-V car radio draws a current of 6.5 A. What resistance must be connected in series with it to operate in a car with a 12-V battery? What "wattage" resistor is required?

6-1-8. An electric stove element is rated at 300 W when connected to a 110-V source. Assuming negligible change in resistance for any change in temperature, determine the total rate of energy conversion when two of these elements are connected in series.

6-1-9. What resistance must be connected in series with a 100-Ω resistor for the 100-Ω resistor to dissipate heat at the rate of 30 W when the combination is connected to a 120-V source?

6-1-10. What resistance must be connected in series with a 100-Ω resistor for the unknown resistance to dissipate heat at the rate of 30 W when the combination is connected to a 120-V source?

6-4-11. If the series circuit of Problem 6-1-1 is connected to a 120-V source, what is the voltage drop across each resistor?

6-4-12. A 390-Ω resistor is in series with an 1.8-kΩ resistor. If the voltage drop across the 1.8-kΩ resistor is 48 V, what is the voltage drop across the 390-Ω resistor? What is the source voltage?

6-4-13. Three resistors are connected in series to a 120-V generator. The first has a resistance of 50 Ω, the second passes a current of 0.5 A, and the third has a voltage drop across it of 50 V. Calculate the resistances of the second and third resistors.

6-4-14. Three resistors are connected in series to a 120-V source. The voltage drop across R_1 and R_2 together is 80 V, and the voltage drop across R_2 and R_3 together is 90 V. If the total resistance is 8 kΩ, what is the resistance of each of the three resistors?

6-4-15. A 10-kΩ 20-W resistor, an 80-kΩ 100-W resistor, and a 20-kΩ 200-W resistor are connected in series. What is the maximum voltage that can be applied to the network without exceeding the power rating of any resistor?

6-4-16. A Christmas tree light set for use with a 110-V source consists of eight 6-W lamps in series. What is the hot resistance of each lamp?

6-5-17. A 22-Ω resistor in series with a 68-Ω resistor are connected to a 12-V source. What is the voltage drop across each resistor?

6-5-18. Given $R_1 = 40$ Ω, $R_2 = 60$ Ω, and $R_3 = 100$ Ω connected in series. If the voltage drop across R_2 is 12 V, what is the voltage drop across R_1? What is the source voltage?

6-6-19. A storage battery has an open-circuit voltage of 6 V and an internal resistance of 0.05 Ω.
(a) What is the terminal voltage of the battery when a 0.2-Ω load is connected to it?
(b) What power will be dissipated by a 0.5-Ω load?

6-6-20. The high-voltage power supply of a television receiver produces an open-circuit voltage of 15 kV. Its internal resistance is 2 MΩ.
(a) What is the voltage at the high voltage anode of the picture tube when the anode current is 600 μA?
(b) What will the anode voltage be if the anode current is increased by 50%?
(c) What is the short-circuit current of this power supply?

6-6-21. When a 180-Ω load is connected to a voltage source, the terminal voltage of the source decreases by 10%. What is the internal resistance of the source?

6-6-22. When the current drawn from a voltage source is increased from 50 mA to 80 mA, the terminal voltage drops from 30 V to 24 V. What is the internal resistance of the source? What is its open-circuit terminal voltage?

6-6-23. A generator has a terminal voltage of 110 V when a 5.5-Ω load is connected to its terminals. The terminal voltage becomes 105 V when the load is 3.5 Ω. What is the internal resistance of the generator?

6-6-24. A 200-Ω resistor dissipates heat at the rate of 8 W when connected to a certain generator. If a 300-Ω resistor is connected in series with the 200-Ω resistor, the dissipation of the 200-Ω resistor becomes 2 W. What is the open-circuit voltage of the generator?

6-7-25. In the circuit of Problem 6-6-19,
(a) What is the efficiency of the system when a 0.15-Ω load is connected to the battery?
(b) What is the maximum power that any load can draw from the battery?

6-7-26. A phonograph pickup develops an open-circuit voltage of 50 mV and has an internal resistance of 1.2-kΩ.
(a) What voltage will it develop across a 4-kΩ load resistance?
(b) What value of load resistor must be used for a 75% efficiency?
(c) What is the maximum power output of the pickup?

6-7-27. A generator whose internal resistance is 1 Ω feeds a load at the end of a line, each wire of which has a resistance of 2 Ω. The generator voltage is adjusted to produce 120 V across the load at full load. If the total dissipation of the two wires is not to exceed 10% of the total generated power, what is the maximum power that can be delivered to the load?

6-7-28. An audio amplifier has an internal resistance at its output terminals of 4 Ω. The loudspeaker is connected at the end of a line, each wire of which has a resistance of 2 Ω. What resistance load must the loudspeaker present to obtain maximum power in the loudspeaker?

6-7-29. A 12-V source with an internal resistance of 2 Ω, a 0 to 10-Ω variable resistor, and a fixed 8-Ω load are connected in series. What setting of the variable resistor will produce maximum power in the fixed 8-Ω load?

6-7-30. The generator of a lighting plant driven by a gasoline engine has an open-circuit voltage of 32 V and an internal resistance of 0.2 Ω. The generator is designed for a constant duty power output of 300 W.
(a) What value of load resistor will draw energy from the generator at the rate of 300 W?
(b) What is the efficiency of the system under this condition?

6-8-31. Given $R_1 = 5\ \Omega$, $R_2 = 10\ \Omega$, and $R_3 = 15\ \Omega$, what is the total current when they are connected in parallel to a 120-V source?

6-8-32. What is the equivalent resistance of a parallel circuit consisting of $R_1 = 47$ kΩ, $R_2 = 330$ kΩ, and $R_3 = 2.2$ kΩ?

6-8-33. What is the equivalent resistance of a circuit consisting of six 680-Ω resistors in parallel?

6-8-34. Three lamps operating in parallel on a 110-V circuit are rated at 40 W, 60 W, and 100 W, respectively. What is the equivalent hot resistance of this load?

6-9-35. A 400-Ω resistor and an unknown resistor in parallel draw a total current of 120 mA from a 32-V source. What is the resistance of the unknown resistor?

6-9-36. A 10-kΩ resistor and a resistor R are connected in parallel across a voltage source. The voltage drop across the 10-kΩ resistor is 25 V and the dissipation of resistor R is 1/8th W. What is the magnitude of the current entering the junction of the two resistors?

6-9-37. Three resistors in parallel pass a total current of 0.6 A. The first resistor has a resistance of 400 Ω, the second passes a current of 60 mA, and the third has a voltage drop across it of 150 V. Calculate the resistances of the second and third resistors.

6-9-38. Three resistors connected in parallel draw energy from a voltage source at a total rate of 840 W. R_1 draws a current of 5 A, the voltage drop across R_2 is 60 V, and the dissipation of R_3 is 360 W. Find the resistance of each of the three resistors.

6-9-39. If the three resistors in Problem 6-8-32 are each rated at $\frac{1}{2}$ W, what is the maximum total current that the network can handle without overheating any resistor?

6-10-40. A circuit element having a conductance of 150 μS is connected in parallel with a branch having a conductance of 750 mS. What is the equivalent resistance of the circuit?

6-10-41. What is the total conductance when four parallel branches have conductances of 400 mS, 560 mS, 700 mS, and 340 mS, respectively? What is the equivalent resistance of the network?

6-10-42. What is the total conductance when $R_1 = 22$ kΩ, $R_2 = 33$ kΩ, and $R_3 = 47$ kΩ are connected in parallel? What is the equivalent resistance of the network?

6-10-43. What resistance must be placed in parallel with a 15-kΩ resistor to reduce the equivalent resistance to 10 kΩ?

6-10-44. The equivalent resistance of three resistors in parallel is 2.5 kΩ. If R_1 is 10 kΩ and R_2 is 20 kΩ, what is the resistance of R_3?

6-12-45. The total current passed by a 10-kΩ resistor, a 15-kΩ resistor, and a 20-kΩ resistor in parallel is 20 mA. What is the current through each branch?

6-12-46. Three resistors connected in parallel have an equivalent resistance of 2.5 kΩ. R_1 has a resistance of 15 kΩ, R_2 has a voltage drop of 25 V, and R_3 dissipates electric energy at the rate of 25 mW. Determine the resistances of R_2 and R_3.

REVIEW QUESTIONS

6-1. How would you justify the statement that the current is the same in all parts of a simple series circuit?

6-2. How would you justify the statement that the total resistance of a series circuit is the sum of all the individual resistances?

6-3. In the electric circuit of Figure 6-15, one of the "black boxes" represents a voltage source and the other a resistor. The voltmeter and ammeter read correctly when their terminals are connected with the polarity markers as shown. Which "black box" is the source?

6-4. Kirchhoff's voltage law may be stated thus: "The voltage between any two points in an electric network is the same via any path between those two points." Prove this statement numerically by considering the two ends of R_2 in Figure 6-1, using the values given in Example 6-1.

6-5. Prove that in a series circuit, the ratio between any two power dissipations is the same as the ratio between the two resistances.

6-6. A lamp, rated at 120 V, 100 W and one rated at 120 V, 25 W are connected in series to a 120-V source. Which one will glow more brightly? Explain.

6-7. Explain the visible effect when the 25-W lamp in Question 6-6 is (a) open-circuit and (b) short-circuit.

6-8. What disadvantage does the "series string" type of Christmas tree lamp possess as compared to the 110-V parallel type of Christmas tree lamp?

6-9. How would you go about determining the internal resistance of a given flashlight battery?

FIGURE 6-15
Polarities in an Electric Circuit.

6-10. Explain why the kitchen light becomes a bit dimmer when a toaster is turned on.

6-11. When a storage battery is discharging, its terminal voltage is less than its open-circuit voltage; but when it is being charged, its terminal voltage is greater than its open-circuit voltage. Explain.

6-12. Using the data of Table 6-1, draw a graph of load power vs. load resistance using linear graph paper. Explain the shape of the graph.

6-13. Using the data of Table 6-1, draw a graph of load power vs. load current. Express the condition for maximum power output in relation to short-circuit and open-circuit current.

6-14. There are two numerically correct answers to Problem 6-7-30(a). But we select only one as the electrically correct answer. Why?

6-15. Generator B develops three times the open-circuit voltage of generator A but has three times as great an internal resistance. Each generator is developing the same power in its load. Which has the greater efficiency?

6-16. A generator is operated in such a manner that its terminal voltage is kept constant by increasing its generated potential difference as the load current is increased. Does this mean that the efficiency is 100%? Explain.

6-17. In operating electronic equipment, it is customary to make the load resistance equal to the internal resistance. Why would we do this for electronic apparatus but not in the case of a power-generating station?

6-18. Why must the equivalent resistance of a parallel circuit always be less than that of the smallest of the branch resistances?

6-19. Show that the statement that the total current in a parallel circuit is the sum of all the branch currents confirms Kirchhoff's current law.

6-20. Why is it preferable to think in terms of **conductance** in working with parallel circuits rather than in terms of resistance?

6-21. If the two lamps of Question 6-6 are connected in parallel, explain the visible effect when the 25-W lamp is (a) open-circuit and (b) short-circuit.

6-22. An electric stove element has a resistance of 50 Ω with a center-tap connection. Draw circuit diagrams to show three means of connecting it to a 110-V source to obtain three different rates of conversion of electric energy to heat.

6-23. Most homes have a lighting fixture that can be switched on or off from two different locations. This is accomplished with two single-pole, double-throw switches (which electricians call "three-way switches"). Draw a circuit diagram.

6-24. Figure 6-13 is not a simple parallel circuit because the fuse is in *series* with the remainder of the circuit. Why must it be located in the position shown?

6-25. Two 50-kΩ resistors are connected in series across a voltage source. If a voltmeter which has a resistance of 100 kΩ is connected first across one resistor and then the other and finally across both, the sum of the first two readings does not equal the third reading. Is this third reading greater or less than the sum of the other two? Explain.

*6-26. Adapt the program shown in Section 6-7 for your computer and then program the computer to draw a graph of load power vs. load resistance.

*6-27. Write a BASIC program for a voltage source E with an internal resistance R connected to a variable load resistor (as in Figure 6-8). The program should ask the operator to input the values of E and R and load power. It should then print out the two possible values of load resistance that meet these requirements. (See Problem 6-1-10.)

*6-28. Write a BASIC program which will print out the equivalent resistance of any four resistances in parallel. How would you use this program for only three resistors in parallel?

7

SERIES-PARALLEL CIRCUITS

7-1 SERIES-PARALLEL RESISTORS

In Chapter 6, we considered only simple series and simple parallel circuits. In practice, electric networks are seldom this simple. If we consider the circuit of Figure 7-1 as a *whole*, it does not comply with the characteristics of either a series or a parallel circuit. But if we consider only R_2 and R_3, since they are connected between the same two junction points in the circuit, they must both have the *same* voltage drop. Therefore, R_2 and R_3 *do* fill the definition of a parallel circuit. If we are given the values for R_2 and R_3, we can use the rules of parallel circuits to solve for a single equivalent resistance. As far as the generator is concerned, the simplified circuit of Figure 7-2 is the same as the original circuit of Figure 7-1. In Figure 7-2, R_1 is in series with the equivalent resistance of R_2 and R_3 in parallel. Therefore, we can solve the circuit of Figure 7-2 by the rules of series circuits.

Another arrangement of series-parallel resistors is shown in Figure 7-3(a). In this case, R_2 and R_3 have the same current through them and, therefore, are in series. We may replace them, if we wish, by an equivalent resistor, as in Figure 7-3(b), which is equal to $R_2 + R_3$. We can now solve the simplified circuit of Figure 7-3(b) by the rules for simple parallel circuits.

As these examples show, we may define a series-parallel circuit as one in which some portions of the circuit have the characteristics of simple series circuits and other portions have the characteristics of simple parallel circuits. Whenever two or more components of an electric network are in parallel, all the characteristics of parallel

FIGURE 7-1
Simple Series-Parallel Circuit.

FIGURE 7-2
Equivalent Circuit of Figure 7-1.

FIGURE 7-3
(a) Parallel-Series Circuit; (b) Simplified Version.

circuits must apply to these components. And, whenever two or more components are in series, all the characteristics of series circuits must apply.

7-2 EQUIVALENT-CIRCUIT METHOD

Figures 7-2 and 7-3(b) suggest a method whereby we may solve some of the simpler series-parallel networks, that is, by substituting the equivalent resistance for various portions of the circuit until the original circuit is reduced to a simple series or parallel circuit.

EXAMPLE 7-1
Complete Table 7-1 with reference to the circuit diagram of Figure 7-1.

TABLE 7-1

Component	Resistance	Voltage	Current	Power
R_1	12 Ω			
R_2	10 Ω			
R_3	40 Ω			
Totals		100 V		

FIGURE 7-4
Circuit Diagram for Example 7-1.

SOLUTION

Step I. Draw a fully labeled schematic diagram for this particular circuit.

Step II. Visual inspection of Figure 7-4 shows that R_2 and R_3 are in parallel.

$$\therefore R_{eq} = \frac{R_2 R_3}{R_2 + R_3} = \frac{10 \times 40}{(10 + 40)} = 8 \ \Omega$$

With reference to Figure 7-4, the total resistance becomes

$$R_T = R_1 + R_{eq} = 12 + 8 = \textbf{20} \ \boldsymbol{\Omega}$$

Enter this answer in the appropriate blank of Table 7-1.

Step III. From Ohm's law,

$$I_T = \frac{E}{R_T} = \frac{100 \ V}{20 \ \Omega} = \textbf{5 A}$$

Step IV. Since R_1 is directly in series with the source,

$$I_1 = I_T = \textbf{5 A}$$

Step V. From Ohm's law,

$$V_1 = I_1 R_1 = 5 \ A \times 12 \ \Omega = \textbf{60 V}$$

Step VI. From Kirchhoff's voltage law,

$$V_{eq} = E - V_1 = 100 - 60 = 40 \ V$$

Returning now to the original circuit,

$$V_2 = V_3 = V_{eq} = \textbf{40 V}$$

Step VII. From Ohm's law,

$$I_2 = \frac{V_2}{R_2} = \frac{40 \text{ V}}{10 \text{ } \Omega} = \textbf{4 A}$$

$$I_3 = \frac{V_3}{R_3} = \frac{40 \text{ V}}{40 \text{ } \Omega} = \textbf{1 A}$$

As a check on our calculations, we can note that according to Kirchhoff's current law, $I_1 = I_2 + I_3$ for this circuit, which does check with our calculations.

Step VIII. Since $P = VI$,

$$P_1 = V_1 I_1 = \quad 60 \text{ V} \times 5 \text{ A} = \textbf{300 W}$$

$$P_2 = V_2 I_2 = \quad 40 \text{ V} \times 4 \text{ A} = \textbf{160 W}$$

$$P_3 = V_3 I_3 = \quad 40 \text{ V} + 1 \text{ A} = \textbf{40 W}$$

$$P_\text{T} = EI_\text{T} = 100 \text{ V} \times 5 \text{ A} = \textbf{500 W}$$

Again as a check, P_T should equal $P_1 + P_2 + P_3$, which does check with our calculations.

EXAMPLE 7-2

What is the resistance at the input terminals of the **ladder** network in Figure 7-5 (a) with the output terminals open-circuit and (b) with a 600-Ω load resistance connected to the output terminals?

FIGURE 7-5
Circuit Diagram for Example 7-2.

SOLUTION

(a) *Step I.* With the output terminals open-circuit, no current flows in the two right-hand 100-Ω resistors. Thus they are effectively out of the circuit, and it is equivalent to Figure 7-6(a).

Step II. The two 200-Ω resistors and the right-hand 800-Ω resistor have the *same* current through them. Hence they must have the characteristics of a *series* circuit.

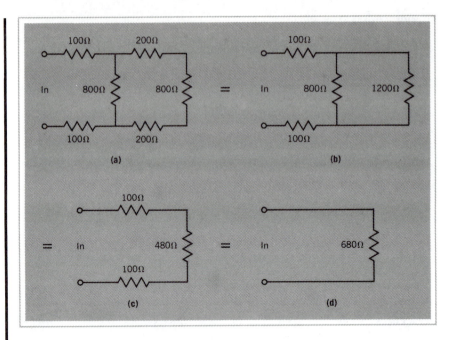

FIGURE 7-6
Equivalent Circuits for Example 7-2(a).

$$\therefore R_{eq} = 200 + 800 + 200 = 1200 \ \Omega$$

This gives the equivalent circuit of Figure 7-6(b).

Step III. Since the 800-Ω and 1200-Ω resistors in the equivalent circuit of Figure 7-6(b) are connected between the same junction points, they are in *parallel*.

$$\therefore R_{eq} = \frac{800 \times 1200}{(800 + 1200)} = 480 \ \Omega$$

Step IV. The equivalent circuit of Figure 7-6(c) is a simple series circuit.

$$\therefore R_{in} = 100 + 480 + 100 = \textbf{680 } \mathbf{\Omega}$$

(b) *Step I.* When a 600-Ω load resistor is connected to the output terminals, it forms a series circuit with the two right-hand 100-Ω resistors. Hence,

$$R_{eq} = 100 + 600 + 100 = 800 \ \Omega$$

Step II. We can replace the two 800-Ω resistors in parallel in the equivalent circuit of Figure 7-7(a) with a single 400-Ω resistance, as in Figure 7-7(b).

Step III. The two 200-Ω resistors and the 400-Ω resistor of Figure 7-7(b) are in series and may be replaced by the single 800-Ω equivalent resistance in Figure 7-7(c).

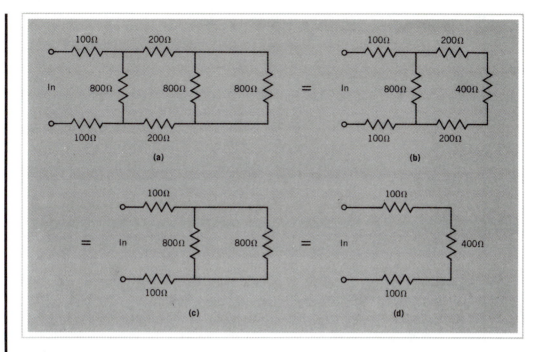

FIGURE 7-7
Equivalent Circuits for Example 7-2(b).

Step IV. Once again, two 800-Ω resistors in parallel are equivalent to the single 400-Ω resistance of Figure 7-7(d).

Step V. Finally, the simple series circuit of Figure 7-7(d) has a total resistance of

$$R_{in} = 100 + 400 + 100 = \mathbf{600\ \Omega}$$

SELF-CHECK

7-2:1. For the parallel-series circuit shown in Figure 7-3(a) suppose the values are as follows:

source voltage, $E = 36$ V,

$R_1 = 1.8$ kΩ, $R_2 = 2.4$ kΩ and $R_3 = 1.2$ kΩ.

Complete a table for this circuit similar to Table 7-1.

7-3 KIRCHHOFF'S-LAWS METHOD

Kirchhoff's laws provide us with an alternate method of solving series-parallel circuits, leaving them in their original form and not reducing them to a simple series or parallel circuit by substituting equivalent resistances.

KIRCHHOFF'S-LAW SOLUTION TO EXAMPLE 7-1

Referring to Figure 7-4, from Kirchhoff's current law,

$$I_1 = I_2 + I_3$$

But from Ohm's law, $I_1 = V_1/R_1$, etc.

$$\therefore \frac{V_1}{R_1} = \frac{V_2}{R_2} + \frac{V_3}{R_3} \quad \text{and} \quad \frac{V_1}{12} = \frac{V_2}{10} + \frac{V_3}{40}$$

Since R_2 and R_3 are in parallel, $V_2 = V_3$, and from Kirchhoff's voltage law,

$$V_2 = V_3 = E - V_1$$

Substitution in the preceding equation gives

$$\frac{V_1}{12} = \frac{100 - V_1}{10} + \frac{100 - V_1}{40}$$

Multiplying through by the least common denominator (120) to clear the fractions

$$10V_1 = 1200 - 12V_1 + 300 - 3V_1$$

Collecting the terms gives

$$25V_1 = 1500$$

from which

$$V_1 = 60 \text{ V}$$

$$\therefore V_2 = V_3 = E - V_1 = 100 - 60 = 40 \text{ V}$$

We may now determine the various currents by Ohm's law: $I_1 = V_1/R_1$, etc.

The information given in Example 7-1 was such that we had a choice of method in its solution. However, with the data given for Example 7-3, we cannot readily solve for the equivalent resistance, which would enable us to reduce the circuit to a simple series circuit. In this case, we can again use Kirchhoff's laws to help us to set up algebraic equations from which we can obtain the required data.

EXAMPLE 7-3

A resistor passing a 20-mA current is in parallel with a 5-kΩ resistor. This combination is in series with another 5-kΩ resistor, the whole network being connected to a 500-V source. What is the resistance of the resistor which is passing the 20-mA current?

SOLUTION

With reference to the circuit diagram of Figure 7-8, from Kirchhoff's voltage law,

FIGURE 7-8
Circuit Diagram for Example 7-3.

$$V_1 + V_2 = E$$

and, since V = IR,

$$I_1 R_1 + I_2 R_2 = E$$

From Kirchhoff's current law,

$$I_2 = I_1 - 0.02$$
$$\therefore R_1 I_1 + R_2(I_1 - 0.02) = E$$

Substituting given values for R_1, R_2, and E,

$$5000I_1 + 5000(I_1 - 0.02) = 500$$

From which $I_1 = 60$ mA. Therefore,

$$I_2 = I_1 - 0.02 = 0.06 - 0.02 = 0.04 \text{ A} \quad \text{or} \quad 40 \text{ mA}$$

Since R_2 and R_3 are in parallel,

$$\therefore V_2 = V_3$$

But
$$V_2 = I_2 R_2 = 40 \text{ mA} \times 5 \text{ k}\Omega = 200 \text{ V}$$

and then
$$R_3 = \frac{V_3}{I_3} = \frac{200 \text{ V}}{20 \text{ mA}} = \textbf{10 k}\boldsymbol{\Omega}$$

It is also possible to start the solution by substituting the appropriate V/R in the Kirchhoff's current-law equation, $I_1 = I_2 + 0.02$.

SELF-CHECK

7-3:1. Referring to the circuit shown in Figure 7-3(a), suppose that $R_1 = 5.6$ kΩ and $R_2 = 1.0$ kΩ and that the following currents are known: 5.0 mA in R_1 and 10.0 mA in R_2.

Determine the following:

(a) the current in R_3,
(b) the current in the source,
(c) the source voltage,
(d) the voltage across R_3 and
(e) the resistance of R_3.

7-4 VOLTAGE-DIVIDER PRINCIPLE

In the series circuit of Figure 7-9, Kirchhoff's voltage law states that $E = V_1 + V_2 + V_3$. In other words, the total applied voltage is *divided* among the three resistors. Whereas we could measure only one voltage across the terminals of the source, we now have *six* possible combinations of terminals A, B, C, and D, across which we can connect a voltmeter, as shown in Figure 7-9. The series combination of R_1, R_2, and R_3, therefore, becomes a **voltage divider.**

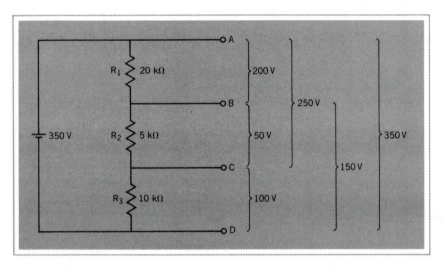

FIGURE 7-9
Voltage-Divider Principle.

We can solve for the six possible voltages in Figure 7-9 by using the long solution given in Example 6-3; that is, calculate R_T, find I from Ohm's law, and then calculate the voltage drop across each resistance from $V_1 = IR_1$, and so on. However, since our attention is directed to the *division* of voltage, we prefer the short solution, using the principle developed in Section 6-5. This is often referred to as the

Voltage-divider principle: In a series circuit, the ratio between any two voltage drops is the same as the ratio of the two resistances across which these voltage drops occur.

Given any resistor in a voltage divider,

$$\frac{V_x}{E} = \frac{R_x}{R_T}$$

from which

$$V_x = E\frac{R_x}{R_T} \qquad (7\text{-}1)$$

Hence, Equation (7-1) becomes an algebraic statement of the voltage-divider principle.

EXAMPLE 7-4

What is the voltage between terminals B and D in the circuit of Figure 7-9?

SOLUTION

$$V_{BD} = E\frac{R_2 + R_3}{R_1 + R_2 + R_3}$$

$$= 350 \text{ V} \times \frac{(5 \text{ k}\Omega + 10 \text{ k}\Omega)}{(20 \text{ k}\Omega + 5 \text{ k}\Omega + 10 \text{ k}\Omega)}$$

$$= 350 \text{ V} \times \frac{15 \text{ k}\Omega}{35 \text{ k}\Omega} = \mathbf{150 \text{ V}}$$

We often require a continuously variable terminal voltage, rather than the fixed voltages provided by the simple voltage divider shown in Figure 7-9. This can be obtained by a voltage divider in which a sliding contact moves along either a wire-wound or a carbon-composition resistance element to form the **potentiometer** circuit shown in Figure 7-10.

FIGURE 7-10
Voltage Divider or Potentiometer.

SELF-CHECK

7-4:1. A series circuit contains a 330-Ω resistor, a 470-Ω resistor, and a 24-V battery. Use the voltage-divider principle to find the voltage across the 330-Ω resistor.

7-4:2. Referring to Figure 7-10, suppose that $E = 60$ V and $R_T = 1.0$ kΩ. At what R_x should the potentiometer be set so that V_x is 25 V?

7-5 VOLTAGE DIVIDERS

The voltage-divider principle is widely used in electronic circuits where, for the sake of economy, the one voltage source must supply all the various voltages required by a piece of equipment. The series dropping resistor of Figure 7-11 provides the simplest method of obtaining the required voltage drop across a certain circuit element.

FIGURE 7-11
Series Dropping Resistor.

EXAMPLE 7-5
A portion of an electronic circuit requires an operating voltage of 15 V at a current drain of 20 mA. If the supply terminal voltage is 25 V, what value of series dropping resistor is required?

SOLUTION
From Ohm's law, we can represent this load circuit by a resistor, as in Figure 7-11.

$$R_L = \frac{V_L}{I_L} = \frac{15 \text{ V}}{20 \text{ mA}} = 750 \text{ } \Omega$$

From Kirchhoff's voltage law, the voltage drop across the series dropping resistor must be

$$V_D = E - V_L = 25 - 15 = 10 \text{ V}$$

Since this is a simple series circuit,

$$I_D = I_L = 20 \text{ mA}$$

$$\therefore \text{ dropping resistance} = \frac{V_D}{I_D} = \frac{10 \text{ V}}{20 \text{ mA}} = \mathbf{500 \text{ } \Omega}$$

To complete the design, we must know the minimum power rating that the chosen resistor must possess.

$$P = VI = 10 \text{ V} \times 20 \text{ mA} = \mathbf{0.2 \text{ W}}$$

We would probably select a $\frac{1}{2}$-W resistor in practice so that it would operate well below rated temperature, thus improving the reliability of the equipment.

The advantage of the simple series dropping resistor is that the current drain on the power supply is no greater than the current required by the circuit element in question. But this circuit has the disadvantage that any change in load resistance will cause appreciable change in the current through the series dropping resistor and, therefore, in the voltage drop across it. This, in turn, will allow appreciable change in the voltage supplied to the load.

In practical electronic circuits where R_L in Figure 7-11 is a transistor, it is quite possible for R_L to be infinitely high (when the transistor is biased to cutoff). Under such circumstances, the voltage drop across the load must rise to the full applied voltage. Even though the transistor is not drawing current, this high terminal voltage may be sufficiently high to cause voltage breakdown of the device. To prevent this breakdown from happening, in some voltage-divider designs we include a **bleeder** resistor[†] in parallel with the load, as in Figure 7-12. When the load resistance rises to infinity, the current through the bleeder maintains some current through the series dropping resistor. Consequently, there is still appreciable voltage drop across the series dropping resistor. Hence, the terminal voltage will not rise to full applied voltage.

FIGURE 7-12
Simple Voltage Divider.

In the design of power supplies for electronic equipment, a bleeder current of from 10 to 25% of the total current drawn from the source will provide sufficient protection against excessive terminal voltage under no-load conditions. With an increase in the bleeder current in relation to the normal load current, variations in load current will have even less effect on the load voltage. The improved **voltage regulation** thus obtained is achieved at the expense of extra current drain from the source and extra heat produced in the voltage divider resistors. In designing voltage dividers for loads consisting of a single transistor, the voltage regulation is a more significant consideration than an extra few milliamperes of bleeder current. We define voltage regulation as the percentage change in output voltage from no load to full load with respect to the full load voltage. Hence,

$$\text{Voltage regulation} = \frac{V_{NL} - V_{FL}}{V_{FL}} \times 100\% \qquad (7\text{-}2)$$

[†]So called because bleeder resistors may also serve to discharge filter capacitors in power supply units.

EXAMPLE 7-6

Allowing a bleeder current of 50 mA, design a voltage divider to supply 15 V at 20 mA from a 25-V source. Calculate the open-circuit terminal voltage and voltage regulation of the voltage divider.

SOLUTION

Since the bleeder resistor is in parallel with the load (Figure 7-12), the voltage drop across it is 15 V. Therefore,

$$\text{bleeder resistance} = \frac{V_L}{I_B} = \frac{15\text{ V}}{50\text{ mA}} = \textbf{300 }\boldsymbol{\Omega}$$

$$\text{power rating of bleeder} = V_L I_B = 15\text{ V} \times 50\text{ mA} = \textbf{0.75 W}$$

Voltage drop across series dropping resistor,

$$V_D = E - V_L = 25 - 15 = 10\text{ V}$$

and

$$I_D = I_B + I_L = 50\text{ mA} + 20\text{ mA} = 70\text{ mA}$$

$$\therefore \text{ series dropping resistor} = \frac{V_D}{I_D} = \frac{10\text{ V}}{70\text{ mA}} = \textbf{143 }\boldsymbol{\Omega}$$

$$\text{power rating} = V_D I_D = 10\text{ V} \times 70\text{ mA} = \textbf{0.7 W}$$

We would select at least 1-W resistors for this voltage divider.

Under open-circuit conditions, we can use the voltage-divider principle of Equation (7-1):

$$V_T = V_B = 25 \times \frac{300}{143 + 300} = \textbf{16.9 V}$$

$$\text{Voltage regulation} = \frac{16.9\text{ V} - 15\text{ V}}{15\text{ V}} \times 100\% = \textbf{12.7\%}$$

(Compared with the 67% voltage regulation in Problem 7-5.)

We recall that it is meaningless to talk about the voltage "at a point" in an electric circuit unless we add "with respect to some other point." Particularly in electronic circuits where we obtain several different voltages from one power supply, it would be very convenient to agree on one particular point in the circuit as a reference point for all voltage measurements. We usually connect this reference point electrically to the metal chassis or frame of the unit (which we represent by the graphic symbol shown in Figure 7-14). In many cases, we **ground** the chassis or frame electrically (to a water pipe or to a metal rod driven into the ground) so that there is zero potential difference between the chassis and the earth. We indicate that the reference point in the power supply voltage divider of Figure 7-13 is grounded by using the graphic symbol for **ground**. We may now say that the voltage at point *A* is +250 V with respect to ground.

FIGURE 7-13
Voltage Divider with Multiple Output Voltages.

EXAMPLE 7-7
Design a voltage divider for the specifications shown in Figure 7-13.

SOLUTION
    Step I. It will be easier to trace the various currents if we draw in resistors to represent the various loads given in the specifications. We should also draw in current-direction tracing arrows for each branch. Conventional current direction is shown in Figure 7-13. None of the 50-mA current drawn from terminal *A* flows through the voltage divider resistors. Hence, this current does not enter into our calculations.

    Step II. Starting with the bleeder resistor, calculate the current through each resistor of the voltage divider and mark the circuit diagram of Figure 7-13 accordingly.
    The only current through R_B is the 10-mA bleeder current. R_2 passes bleeder current + the 10-mA current drawn by the 100-V load.

$$\therefore I_2 = 10 + 10 = 20 \text{ mA}$$

Finally, R_1 passes I_2 + the 25-mA current drawn by the 150-V load.

$$\therefore I_1 = 20 + 25 = 45 \text{ mA}$$

    Step III. Calculate the resistance and power rating for each of the three resistors of the voltage divider.

$$R_B = \frac{V_B}{I_B} = \frac{100 \text{ V}}{10 \text{ mA}} = \textbf{10 k}\boldsymbol{\Omega}$$

and
$$P_B = V_B I_B = 100 \text{ V} \times 10 \text{ mA} = \textbf{1 W}$$

The voltage drop across R_2 is the difference between the potentials at points B and C (with respect to ground)

$$\therefore V_2 = 150 - 100 = 50 \text{ V}$$

Hence,

$$R_2 = \frac{V_2}{I_2} = \frac{50 \text{ V}}{20 \text{ mA}} = \mathbf{2.5 \text{ k}\Omega}$$

and

$$P_2 = V_2 I_2 = 50 \text{ V} \times 20 \text{ mA} = \mathbf{1 \text{ W}}$$

(For a reliability margin, we would select 2-W resistors or even 5-W resistors for R_2 and R_B.)

Finally,

$$V_1 = 250 - 150 = 100 \text{ V}$$

Hence,

$$R_1 = \frac{V_1}{I_1} = \frac{100 \text{ V}}{45 \text{ mA}} = \mathbf{2.22 \text{ k}\Omega}$$

and

$$P_1 = V_1 I_1 = 100 \text{ V} \times 45 \text{ mA} = \mathbf{4.5 \text{ W}}$$

(Select a 10-W resistor.)

Transistor power supplies may require either the $+$ or the $-$ terminal to be grounded, depending on the type of transistor used. Or, more likely, a transistor circuit will require *both* $+$ and $-$ voltages with respect to the common chassis connection. We can obtain both positive and negative voltages (with respect to chassis) from the one power supply by connecting the appropriate tap on a voltage divider to chassis, as shown in Figure 7-14.

FIGURE 7-14
Voltage Divider for a Transistor Amplifier Power Supply.

EXAMPLE 7-8
The power supply for a transistor amplifier develops a full-load terminal voltage of 24 V dc. Design a voltage divider which will provide $+$ and -12-V outputs (with respect to chassis) when the current drain on the $+12$-V source is 400 mA and the drain on the -12-V source is 200 mA. The total current drain on the power supply is 0.5 A.

SOLUTION

Since R_1 is in parallel with Load 1, $V_1 = 12$ V. From Kirchhoff's current law applied to junction A,

$$I_1 = 500 \text{ mA} - 400 \text{ mA} = 100 \text{ mA}$$

$$R_1 = \frac{V_1}{I_1} = \frac{12 \text{ V}}{100 \text{ mA}} = \mathbf{120 \ \Omega}$$

and

$$P_1 = V_1 I_1 = 12 \text{ V} \times 100 \text{ mA} = \mathbf{1.2 \ W}$$

Similarly,

$$V_2 = 12 \text{ V} \quad \text{and} \quad I_2 = 500 \text{ mA} - 200 \text{ mA} = 300 \text{ mA}$$

$$\therefore R_2 = \frac{12 \text{ V}}{300 \text{ mA}} = \mathbf{40 \ \Omega} \quad \text{and} \quad P_2 = 12 \text{ V} \times 300 \text{ mA} = \mathbf{3.6 \ W}$$

SELF-CHECK

7-5:1. Suppose that you are to design a circuit similar to the one shown in Figure 7-12. Your circuit is to draw 24 mA from a 75-V source and supply 6 mA to a 3.3-kΩ load, R_2. Find the proper resistance and power rating for
(a) the bleeder resistor and
(b) the series dropping resistor.

7-6 CURRENT-DIVIDER PRINCIPLE

In Example 7-6, the total current drawn from the voltage source *divides* between the load and the bleeder resistor. To assist us in the solution of problems in which we are concerned with the effects of current division between two parallel branches, we can apply the parallel-resistance principle developed in Section 6-12. This parallel circuit *dual* of the voltage-divider principle for series circuits is often called the

Current-divider principle: In a parallel circuit, the ratio between any two branch currents is the same as the ratio of the two conductances through which these currents flow.

To translate this principle into a useful algebraic equation, since V is common for two resistors in parallel,

$$\frac{I_1}{G_1} = \frac{I_2}{G_2} = \frac{I_T}{G_T} = \frac{I_x}{G_x} \tag{6-17}$$

from which

$$I_x = I_T \frac{G_x}{G_T} \tag{7-3}$$

Since we are usually provided with data on the *resistance* of the parallel branches, rather than their conductances, we can substitute $G_x = 1/R_x$ and $G_T = 1/R_{eq}$ in Equation (7-3) giving

$$I_x = I_T \frac{R_{eq}}{R_x} \tag{7-4}$$

Note that the current ratio is the *inverse* of the *resistance* ratio. Hence, we may restate the current-divider principle:

> **In a parallel circuit, the ratio between any two branch currents is the inverse of their resistance ratio.**

For two resistors in parallel,

$$R_{eq} = \frac{R_1 R_2}{R_1 + R_2} \tag{6-12}$$

$$\therefore I_1 = I_T \frac{R_1 R_2}{R_1(R_1 + R_2)}$$

and

$$I_1 = I_T \frac{R_2}{R_1 + R_2} \tag{7-5}$$

Similarly,

$$I_2 = I_T \frac{R_1}{R_1 + R_2} \tag{7-6}$$

Note in Equations (7-5) and (7-6) that the numerator resistance has the *opposite* subscript to the unknown branch current.

EXAMPLE 7-9

Calculate the bleeder current in the voltage divider designed in Example 7-6 if the load resistance increases to 3 kΩ. See Figure 7-15.

SOLUTION

The equivalent resistance of the bleeder resistor and the load resistance in parallel is

$$R_{eq} = \frac{300\ \Omega \times 3\ k\Omega}{(300\ \Omega + 3\ k\Omega)} = 273\ \Omega$$

FIGURE 7-15
Circuit Diagram for Example 7-9.

$$R_T = R_s + R_{eq} = 143 + 273 = 416 \ \Omega$$

$$I_T = \frac{E}{R_T} = \frac{25 \ V}{416 \ \Omega} = 60.1 \ mA$$

$$I_B = I_T \frac{R_L}{R_B + R_L} = 60.1 \ mA \times \frac{3 \ k\Omega}{3.3 \ k\Omega} = \textbf{54.6 mA}$$

EXAMPLE 7-10

An ammeter has an internal resistance of 50 Ω and reads full scale when the current through it is 1.0 mA. What resistance of shunt must be placed in parallel with the meter so that it will read full scale with a circuit current of 1.0 A? See Figure 7-16.

FIGURE 7-16
Circuit Diagram for Example 7-10.

SOLUTION
From Kirchhoff's current law,

$$I_{sh} = I_T - I_M = 1000 \ mA - 1 \ mA = 999 \ mA$$

According to the current-divider principle,

$$\frac{R_{sh}}{R_M} = \frac{I_M}{I_{sh}}$$

$$\therefore R_{sh} = R_M \times \frac{I_M}{I_{sh}} = 50 \times \frac{1}{999} = \textbf{0.050\,05} \ \Omega$$

In Example 7-2, we calculated the input resistance of a typical **ladder** network, but we did not complete the circuit analysis by finding out how the output voltage and current compares with the input voltage and current. Now that we understand the current-divider principle, we can complete the task without resorting to a series of tedious Ohm's-law calculations.

EXAMPLE 7-11

With an input voltage of 204 mV applied to the ladder network of Figure 7-17 (and also Figure 7-6), what is the output voltage (a) with the output terminals open-circuit and (b) with a 600-Ω load resistance connected to the output terminals?

SOLUTION
(a) From the solution of Example 7-2(a),

$$R_{in} = 680 \ \Omega$$

FIGURE 7-17
Circuit Diagram for Example 7-11.

Then
$$I_{\text{in}} = \frac{E}{R_{\text{in}}} = \frac{204 \text{ mV}}{680 \ \Omega} = 300 \ \mu A$$

This current splits into two branches at node A. From the current-divider Equation (7-6) and Figure 7-6(b),

$$I_2 = 300 \ \mu A \frac{800 \ \Omega}{800 \ \Omega + 1200 \ \Omega} = 120 \ \mu A$$

Since the output terminals are open-circuit,

$$I_L = 0 \quad \text{and} \quad I_3 = I_2$$

Also, there will be zero voltage drop across the two right-hand 100-Ω resistors.

$$\therefore V_L = V_{\text{CD}} = I_2 R_{\text{CD}} = 120 \ \mu A \times 800 \ \Omega = \mathbf{96 \ mV}$$

(b) From the solution of Example 7-2(b),

$$R_{\text{in}} = 600 \ \Omega$$

Then
$$I_{\text{in}} = \frac{E}{R_{\text{in}}} = \frac{204 \text{ mV}}{600 \ \Omega} = 340 \ \mu A$$

This current splits into two branches at node A. From Figure 7-7(c), we note that each branch between nodes A and B has the same resistance.

$$\therefore I_2 = 0.5 \times 340 \ \mu A = 170 \ \mu A$$

At node C, I_2 splits into two branches. From Figure 7-7(a),

$$I_L = 0.5 \times 170 \ \mu A = 85 \ \mu A$$
$$\therefore V_L = I_L R_L = 85 \ \mu A \times 600 \ \Omega = \mathbf{51 \ mV}$$

SELF-CHECK

7-6:1. Suppose a circuit has two parallel resistors whose values are $R_1 = 1.1$ kΩ and $R_2 = 3.3$ kΩ. The total current drawn by the parallel combination is 60 mA. Use the current-divider principle to find I_1 and I_2.

7-6:2. In a 2-branch parallel circuit, "branch 1" draws 15 mA and "branch 2" draws 25 mA. If "branch 2" contains a 1.0 kΩ resistor, determine the resistance of "branch 1."

PROBLEMS

7-2-1. Calculate the equivalent resistance of each of the resistance networks in Figure 7-18.

7-2-2. What is the resistance measured from terminal R to ground in each of the resistance networks of Figure 7-19? (Note the calculator solution for Example 6-5A.)

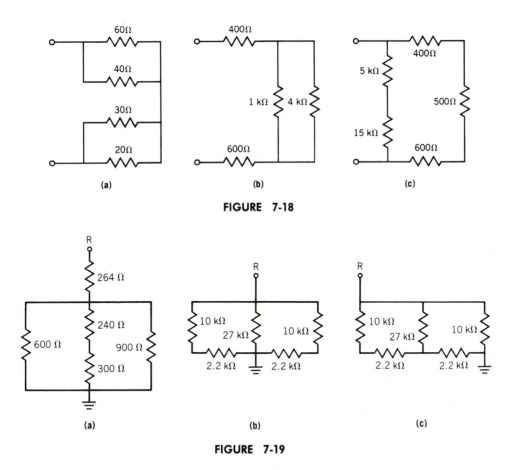

FIGURE 7-18

FIGURE 7-19

7-2-3. What current is drawn from the voltage source in each of the circuits in Figure 7-20?

7-2-4. Determine the source current in each of the circuits in Figure 7-21.

FIGURE 7-20

FIGURE 7-21

7-2-5. By connecting two leads to various terminals of the resistance network in Figure 7-22, six different values of equivalent resistance can be obtained. Calculate all six possibilities.

7-2-6. Calculate R_{AB}, R_{BC}, and R_{CA} in each of the networks in Figure 7-23.

FIGURE 7-22   **FIGURE 7-23**

7-2-7. Determine the equivalent resistance of the circuit in Figure 7-24
(a) With the switch open.
(b) With the switch closed.

7-2-8. What is the voltage drop across the switch in the circuit of Figure 7-24 when the switch is open?

7-2-9. What is the input resistance of the ladder network in Figure 7-25
(a) With a 75-Ω load resistor connected as shown?
(b) With the load resistor disconnected?

7-2-10. What voltage must be applied to the input terminals of the ladder network in Figure 7-25 to produce a 300-μW dissipation in the 75-Ω load resistor?

FIGURE 7-24

FIGURE 7-25

For Problems 7-2-11 to 7-3-16, draw fully labeled diagrams similar to Figure 7-1 and prepare tables similar to Table 7-1.

7-2-11. $E = 24$ V; $R_1 = 4\ \Omega$; $R_2 = 6\ \Omega$; $R_3 = 12\ \Omega$.

7-2-12. $P_T = 300$ W; $R_1 = 8\ \Omega$; $R_2 = 5\ \Omega$; $R_3 = 20\ \Omega$.

7-3-13. $E = 300$ V; $R_1 = 8\ \Omega$; $I_2 = 1$ A; $V_3 = 150$ V.

7-3-14. $E = 100$ V; $P_T = 75$ W; $R_2 = 100\ \Omega$; $I_3 = 0.5$ A.

7-3-15. $E = 180$ V; $R_1 = 5\ \Omega$; $R_2 = 15\ \Omega$; $I_3 = 4$ A.

7-3-16. $E = 250$ V; $R_1 + R_2 = 10$ kΩ; $V_2 = 100$ V; $I_3 = 5$ mA.

7-3-17. A 12-V generator has an internal resistance of 0.05 Ω. Two loads are connected in parallel to its terminals, one drawing a 12-A current and the other dissipating energy at the rate of 200 W. What is the terminal voltage of the generator with this load?

7-3-18. When the automatic gain control voltage of a radio receiver is measured with an electronic voltmeter whose input resistance is 10 MΩ, it reads 4.5 V. When an ordinary voltmeter whose input resistance is 10 kΩ is also connected across the AGC voltage source, both meters read 0.40 V. What is the internal resistance of the AGC voltage source?

7-3-19. Determine the resistances of R_1 and R_2 in the circuit of Figure 7-26 (a) when $R_L = 400\ \Omega$, $I_L = 50$ mA and (b) when $R_L = 200\ \Omega$, $I_L = 75$ mA.

FIGURE 7-26

7-3-20. Figure 7-27 shows the focus control for a cathode-ray tube. In the one extreme position, the focus anode voltage is to be 600 V, at which setting the current drawn by the focus anode is 8 μA. At the other extreme, the focus anode voltage is to be 1200 V for an anode current of 12 μA. Calculate the resistance of R_1 and the focus control.

FIGURE 7-27

7-5-21. What bleeder resistor is required to complete a voltage divider to give 300 V at 40 mA from a 500-V source if the series dropping resistor is 4 kΩ?

7-5-22. Design a voltage divider to supply 100 V at 20 mA from a 250-V source if the total drain on the source is to be 100 mA.

7-5-23. What is the percent voltage regulation of the voltage divider in Problem 7-5-21?

7-5-24. What is the voltage regulation of the voltage divider in Problem 7-5-22?

7-5-25. Design a voltage divider to deliver 100 V at 1 mA from a 240-V source so that a 100% increase in load current causes only a 5% reduction in load voltage.

7-5-26. Design a voltage divider to supply 10 V at 10 mA and 25 V at 25 mA from a 40-V dc supply with a bleeder current (I_3 in Figure 7-28) of 100 mA.

7-5-27. Assuming that load 2 in Figure 7-14 acts like a linear resistance, determine the voltage distribution (with respect to chassis) in Example 7-8 if load 1 becomes open-circuit.

FIGURE 7-28

FIGURE 7-29

7-5-28. A voltage divider consists of a 10-kΩ resistor with an adjustable tap. The tap is set so that it feeds 80 V at 5 mA to a load when the voltage divider is connected across a 200-V source. What is the resistance of the bleeder portion of the voltage divider?

7-5-29. A transistorized hi-fi amplifier system requires the following dc supplies: −25 V at 2.0 A, −12 V at 120 mA, and +12 V at 30 mA. All voltages are measured with respect to ground. Design a voltage divider which will permit this amplifier to be operated from a single dc supply which develops a terminal voltage of 40 V with a total current drain of 2.5 A.

7-5-30 The voltage-divider circuit in Figure 7-29 provides the transistor with the proper currents for operation as an amplifier. If $V_{BE} = 0.6$ V and $V_E = 1.1$ V, calculate V_C.

7-6-31. Calculate the current through each resistor in the circuit of Figure 7-30(a).

7-6-32. What value of resistor must be connected across the terminals of the constant-current source shown in Figure 7-30(b) to reduce the terminal voltage to 50 V?

7-6-33. In the circuit of Figure 7-31, what value of resistor R will draw a current equal to 25% of the source current?

(a)   (b)

FIGURE 7-30

FIGURE 7-31

7-6-34. In the circuit of Figure 7-31, what value of resistor R is required for the source current to be 200 mA?

REVIEW QUESTIONS

7-1. With respect to series-parallel resistance networks, what is meant by an "equivalent resistance"?

7-2. What characteristics must an "equivalent resistance" possess?

7-3. What circuit rules allow us to replace R_2 and R_3 in Figure 7-32 with an equivalent resistance?

7-4. What advantage does solution of series-parallel resistance networks by writing Kirchhoff's-law equations have as compared with the "equivalent resistance" method?

7-5. Write one Kirchhoff's current-law and two Kirchhoff's voltage-law equations for the network of Figure 7-32.

7-6. What circuit condition must exist for the **voltage-divider principle** to be applied in network solutions?

7-7. The four-terminal potentiometer network of Figure 7-33 (sometimes called an "L" attenuator) is the volume-control circuit of an audio amplifier. Explain its operation.

FIGURE 7-32

FIGURE 7-33
Volume Control.

7-8. What is the purpose of a **bleeder** resistor in a practical voltage-divider circuit?

7-9. Calculate the load voltage in Examples 7-5 and 7-6 if the load current decreases to 10 mA. Explain the difference in the two values.

7-10. Why does a decrease in the bleeder resistance of a given voltage divider improve the **voltage regulation** of the output of the voltage divider?

7-11. What factor governs the voltage regulation of a voltage source?

7-12. One voltage source has a 10% voltage regulation and the other has a 90% voltage regulation. Which source is preferable?

7-13. In Chapter 2, we noted that voltage is measured *between* two points in a circuit; yet in some electronic circuits the data might state that "The voltage at Point A is +12 V." Explain.

7-14. A dc power supply for an electronic amplifier has only two output terminals. Explain how it is possible to obtain simultaneously both positive and negative potentials with respect to the chassis of the amplifier.

7-15. What circuit conditions must exist for the **current-divider principle** to be applied in network solutions?

7-16. The current-divider principle is sometimes called the parallel-circuit *dual* of the voltage-divider principle. Explain the duality relationship.

7-17. Explain why the current-divider principle states that the ratio of the currents in two parallel branches is the *inverse* of their resistance ratio.

*7-18. Write a BASIC program which will ask the operator to enter values for E, R_1, R_2, and R_3 with respect to the series-parallel circuit of Figure 7-1. After these values have been entered, the program should then have the computer display the calculations in a table similar to Table 7-1.

*7-19. For a voltage divider similar to that shown in Figure 7-13, prepare a BASIC program which can be stored in permanent memory (cassette or diskette) which will ask the operator to input values for

   1. Source voltage
   2. Bleeder current
   3. Output current from terminals A and D
   4. Voltage and current for terminals B and D
   5. Voltage and current for terminals C and D

The computer should then display in table format the resistance and power ratings of R_1, R_2, and R_B.

*7-20. Write a BASIC program which will print out the required value of shunt resistor to convert a basic 1-mA meter movement with an internal resistance of 50 Ω into an ammeter for any desired magnitude of full-scale current reading simply by inputing the required full-scale reading.

8
RESISTANCE NETWORKS

8-1 NETWORK EQUATIONS FROM KIRCHHOFF'S LAWS

At first glance, the **Wheatstone bridge** shown in Figure 8-1 may appear to be a simple series-parallel circuit. But no two resistors meet the requirements for either series or parallel components. Hence, we cannot use the *equivalent resistance* techniques of Chapter 7 to analyze a bridge circuit. There are two general methods we can use to solve such networks. The more recent method uses various *network theorems* to simplify the original network so that we can once more use equivalent-circuit techniques to analyze the complete circuit. This method is particularly suited to pocket calculators.

The second method is more traditional and is based on Kirchhoff's voltage and current laws, which we encountered in Sections 6-4 and 6-9. As fundamental *laws* of electric circuit behavior, Kirchhoff's laws must hold true, not only in the simple series and parallel circuits of the preceding chapters, but also in the most complex electric circuits we may encounter. Once we understand the significance of Kirchhoff's laws, we can write a series of equations (the algebraic statement of Kirchhoff's laws) which will enable us to analyze the behavior of the more sophisticated electric and electronic circuits we shall consider in later chapters and subsequent courses.

To apply the Kirchhoff's-law equations method, we need to learn a few basic rules which will permit us to write an orderly set of simultaneous equations. This structured approach makes this method particularly suited to the development of circuit-analysis software packages for computers. Hence, the traditional circuit analysis methods of this chapter are experiencing a resurgence in popularity.

The disadvantage of these **mesh** and **nodal** analysis techniques in an introductory

FIGURE 8-1
Wheatstone Bridge.

electric circuits course is that we can become more involved with plugging numbers into computer programs than with learning to understand the principles of electric circuit behavior. The equivalent-circuit and network-theorem methods of subsequent chapters accomplish the same results with less emphasis on mathematical formats.

An understanding of the principle of **source conversion** will allow us to simplify the final formats we use in writing Kirchhoff's-law equations for various resistance networks. Hence, we shall digress a bit into equivalent-circuit techniques for the next three sections of this chapter.

8-2 CONSTANT-VOLTAGE SOURCES

We have already discovered that most practical sources generate a terminal voltage which falls as we draw current from the source. We explained this effect in Section 6-6 by assuming that a practical voltage source consists of a constant-voltage source producing a constant E = open-circuit terminal V with an internal resistance in series with this constant-voltage source. Although we did not use the particular terminology at the time, we discovered an **equivalent circuit** for a practical source. This type of equivalent circuit is called a **constant-voltage source.** From a circuit-analysis point of view, the load in the circuit of Figure 8-2(a) "sees" the source as a constant-voltage source which includes an internal series resistance, as shown in the equivalent circuit of Figure 8-2(b).

In replacing portions of a pencil-and-paper circuit diagram containing a practical voltage source with its equivalent constant-voltage source, we follow a mental procedure

FIGURE 8-2
Equivalent Circuit for a
Practical Voltage Source.

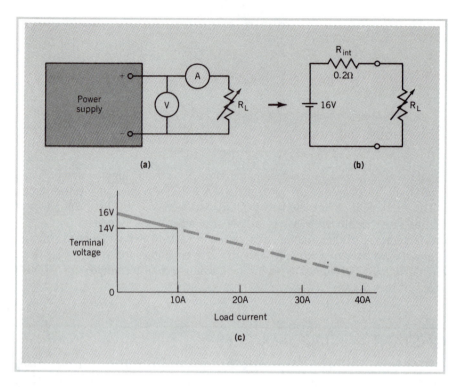

FIGURE 8-3
Equivalent Circuit for a DC
Power Supply.

which duplicates a laboratory technique for finding out what is sealed into a "black box" that has only two terminals exposed. For example, let us consider a power supply which provides 16 volts dc to operate a transistor amplifier. As shown in Figure 8-3(a), this power supply is essentially a "black box" which derives its energy from a 120-V 60-Hz power line and delivers a dc output voltage from a pair of terminals. But as far as the load R_L is concerned, it can "see" only a source from which direct current may be drawn. Details of what happens inside the "black box" are of no significance to the electric circuit connected to its output terminals.

Given a voltmeter that draws negligible current, we can measure the **open-circuit terminal voltage** by disconnecting the load and connecting the voltmeter across the power supply output terminals. As we reduce R_L from an infinitely high value, current drawn from the power supply increases and the terminal voltage falls, as shown by the graph in Figure 8-3(c). In our laboratory experiment, we can reduce the load resistance to a point where the current drawn from the power supply is high enough to "blow" the protective fuse. In our pencil-and-paper circuit analysis, we are not hampered by such limitations, and we can proceed along the dashed portion of the graph until $R_L = 0$ ohms and we read the **short-circuit current.** With no knowledge of the details of the internal operation of this particular power supply, all that we can state from the data shown in Figure 8-3(c) is that the power supply behaves as a constant-voltage source with an internal resistance of $R_{int} = \Delta V/\Delta I = \frac{2}{10} = 0.2$ ohm, as shown in Figure 8-3(b). Extending the graph of Figure 8-3(c) in a straight line until it cuts the x-axis, the short-circuit current is 80 amperes. Once we know the open-circuit voltage and the short-circuit current, the internal resistance becomes

$$R_{\text{int}} = \frac{\text{open-circuit terminal voltage}}{\text{short-circuit terminal current}} \qquad (8\text{-}1)$$

Hence, R_{int} in the constant-voltage source of Figure 8-3(b) is $V_{\text{oc}}/I_{\text{sc}} = 16\text{ V}/80\text{ A} = 0.2\ \Omega$.

8-3 CONSTANT-CURRENT SOURCES

As we have discovered on many occasions, there are always two ways of looking at the same property in electric circuits. For example, *resistance* is the ability of a circuit to *oppose* electric current. Looking at the same property from the opposite point of view, *conductance* is the ability of a circuit to *permit* electric current. From these definitions, it follows that conductance is the reciprocal of resistance, and $G = 1/R$. We select the point of view which is more convenient for the task at hand. In dealing with series circuits, we prefer to think and work in terms of resistance. But, in dealing with parallel circuits, it is often more convenient to switch our point of view and to think and work in terms of conductance.

Perhaps there is an opposite point of view of what the sealed-box power supply of Figures 8-3(a) and 8-4(a) might contain that will produce the same results in an external load. Our voltmeter/ammeter technique tells us that the sealed box has an open-circuit terminal voltage of 16 volts and a short-circuit terminal current of 80 amperes. Theoretically, the box could contain a special generator with a completely variable terminal voltage but which always produces a constant current equal to the short-circuit terminal current. For this constant current to flow when the terminals of the box are open-circuit, its internal resistance must now be shown in *parallel* with the terminals of the constant-current generator, as shown in Figure 8-4(b). To produce the required open-circuit terminal voltage, it follows that the value of R_{int} must be

$$R_{\text{int}} = \frac{\text{open-circuit terminal voltage}}{\text{short-circuit terminal current}} \qquad (8\text{-}1)$$

Note that this is the same equation we used to find the internal resistance of the constant-

(a)   **(b)**

FIGURE 8-4
Constant-Current Source.

voltage source in Figure 8-3(b). The direction of the arrow in the symbol for a constant-current source shows the conventional current direction required to maintain the same polarity of voltage across R_L in Figure 8-4(b) as in Figure 8-4(a).

Suppose that we set R_L in the circuits shown in Figures 8-3(b) and 8-4(b) at 0.6 ohm. With the constant-voltage source of Figure 8-3(b), the voltage-divider principle tells us that

$$V_L = \frac{R_L}{R_{int} + R_L} \times V_{oc} = \frac{0.6}{0.2 + 0.6} \times 16 \text{ V} = 12 \text{ V}$$

and

$$I_L = \frac{V_L}{R_L} = \frac{12 \text{ V}}{0.6 \text{ } \Omega} = 20 \text{ A}$$

With the constant-current source of Figure 8-4(b), we switch to the current-divider principle.

$$I_L = \frac{R_{int}}{R_{int} + R_L} \times I_{sc} = \frac{0.2}{0.2 + 0.6} \times 80 \text{ A} = 20 \text{ A}$$

and

$$V_L = I_L R_L = 20 \text{ A} \times 0.6 \text{ } \Omega = 12 \text{ V}$$

For any value of load resistance we may choose, we find that the constant-voltage source of Figure 8-3(b) and the constant-current source of Figure 8-4(b) are exact equivalents, as "seen" by R_L. The constant-current equivalent source of Figure 8-4(b) is credited to an American engineer, Edward L. Norton, and is the **dual** of the constant-voltage source of Figure 8-3(b).

In most practical electric circuits, we are used to the concept of *opening* the circuit (disconnecting the load) to switch the circuit *off*. When we do so, we expect the current through the source, as well as the current in the load, to become zero. The power dissipation in the internal resistance of the source would also be zero under open-circuit conditions. This is the situation with a constant-voltage source. Hence, the equivalent circuit of Figure 8-3(b) represents the "real-world" situation in electric circuit diagrams.

It is difficult to imagine a current source existing in a practical electrical system. If we disconnect R_L in the circuit of Figure 8-4(b), the current through R_{int} must *increase* to the full constant-current value and the internal dissipation of the source becomes $V_{oc} \times I_{sc} = 1280 \text{ W}$. Theoretically, to switch the circuit of Figure 8-4(b) *off*, we *short-circuit* R_L so that the terminal voltage and the dissipation become zero. The constant-current equivalent source of Figure 8-4(b) is primarily a pencil-and-paper device to assist in network analysis.

Although we will not encounter current sources in electric power system networks, we do find devices which function as if they were current sources in low-power electronic circuits. For example, it is easier to analyze transistor circuits if we show a current source in their equivalent circuit. Also, it is possible to use electronic feedback circuits to construct a dc power supply which will maintain a constant load current with a terminal voltage which varies as the load resistance connected to it varies. Such constant-current power supplies are not as common as voltage-regulated power supplies which maintain a constant terminal *voltage* as the load resistances connected to them vary.

SELF-CHECK

8-3:1. An automobile battery has an open-circuit voltage of 13.2 V and a short-circuit current of 150 A.

(a) Determine the R_{int} of the battery.

(b) Draw the constant-voltage and the constant-current equivalent sources for this battery.

8-4 SOURCE CONVERSION

In developing either the constant-voltage source of Figure 8-3(b) or the constant-current source of Figure 8-4(b), we used the *same* "black-box" technique to determine R_{int}.

$$R_{int} = \frac{E_{oc}}{I_{sc}} \qquad (8\text{-}1)$$

Hence, if we are given a constant-voltage source, as in Figure 8-5(a), R_x in the equivalent constant-current source of Figure 8-5(b) has the *same* value but appears in *parallel* with the ideal current source. Similarly, if we are given the constant-current source of Figure 8-5(c), R_x in the equivalent constant-voltage source of Figure 8-5(d) has the *same* value but appears in *series* with the ideal voltage source.

FIGURE 8-5
Source Conversion.

To be able to replace a constant-voltage source in a network with its constant-current equivalent, there must be some resistance R_x in *series* with an ideal voltage source. R_x may be an internal resistance of the source, or it may be one of the network resistors in series with the voltage source. Similarly, there must be some form of resistance in parallel with an ideal current source before we can replace it with its constant-voltage equivalent.

Given the constant-voltage source of Figure 8-5(a), from Equation (8-1), I_x for the equivalent constant-current source of Figure 8-5(b) is the short-circuit current

$$I_x = \frac{E_{oc}}{R_x} \qquad (8\text{-}2)$$

And given the constant-current source of Figure 8-5(c), E_x for the equivalent constant-voltage source of Figure 8-5(d) is the open-circuit voltage

$$E_x = I_{sc}R_x \qquad (8\text{-}3)$$

EXAMPLE 8-1
(a) A voltage source develops an open-circuit voltage of 120 V and has an internal resistance of 6 Ω. Determine the equivalent constant-current source.
(b) A current source develops a constant current of 40 A with an internal resistance of 0.8 Ω. Determine the equivalent constant-voltage source.
(c) Verify the conversion in part (a) by connecting a 12-Ω load to both sources.

SOLUTION
(a) R_x for the constant-current source is the same as for the voltage source. $R_x = \mathbf{6\ \Omega.}$

$$I_x = \frac{E_{oc}}{R_x} = \frac{120\ \text{V}}{6\ \Omega} = \mathbf{20\ A}$$

(b) R_x for the constant-voltage equivalent source is the same as for the current source. $R_x = \mathbf{0.8\ \Omega.}$

$$E_x = I_{sc}R_x = 40\ \text{A} \times 0.8\ \Omega = \mathbf{32\ V}$$

(c) When we connect a 12-Ω load to the voltage source of Figure 8-5(a),

$$I_L = \frac{E_x}{R_x + R_L} = \frac{120\ \text{V}}{(6 + 12)\Omega} = \mathbf{6.67\ A}$$

and $\qquad V_L = 6.67\ \text{A} \times 12\Omega = \mathbf{80\ V}$

When we connect a 12-Ω load to the current source of Figure 8-5(b),

$$V_L = I_x \times \frac{R_x \times R_L}{R_x + R_L}$$

$$= 20\ \text{A} \times \frac{6 \times 12}{6 + 12} = \mathbf{80\ V}$$

and $\qquad I_L = \frac{V_L}{R_L} = \frac{80\ \text{V}}{12\ \Omega} = \mathbf{6.67\ A}$

Once we are familiar with the concept, we can convert voltage sources to equivalent current sources, and vice versa, for network analysis purposes with no more effort than a bit of Ohm's-law mental arithmetic. Sometimes we may encounter resistance networks containing both voltage and current sources, particularly networks containing transistor equivalent circuits. Such networks are usually easier to solve if we use source conversion so that the network contains all voltage sources or all current sources.

SELF-CHECK

8-4:1. A voltage source has an open-circuit voltage of 24 V. When connected to a 10-Ω load the terminal voltage of the voltage source drops to 20 V.
   (a) Find the internal resistance of the voltage source.
   (b) Convert the voltage source to an equivalent constant-current source.

8-5 KIRCHHOFF'S VOLTAGE-LAW EQUATIONS: LOOP PROCEDURE

As we discovered in Chapter 6, we can write two distinct types of equations for the same circuit: one set based on Kirchhoff's voltage law, and the other based on Kirchhoff's current law. Network equations based on Kirchhoff's voltage law are called **loop** or **mesh** equations. Network equations based on Kirchhoff's current law are called **nodal** equations. Each method has its advantages and disadvantages. To avoid confusion, we shall consider the two methods separately.

There are several ways of stating Kirchhoff's voltage law. In Section 6-4, we used the form.

In any complete electric circuit, the algebraic sum of the source voltages must equal the algebraic sum of the voltage drops.

This statement leads to equations in the form

$$E = V_1 + V_2 + V_3 \qquad (8\text{-}4)$$

Some texts prefer to state the same law thus,

In any closed loop the algebraic sum of all voltage rises and voltage drops must equal zero.

This statement leads to an equation in which we must carefully record the sign of voltage drops and voltage rises.

$$E - V_1 - V_2 - V_3 = 0 \qquad (8\text{-}5)$$

To enable us to set up our equations in a form that is suited to solution by determinants, we rearrange our original statement and say that

In any complete electric circuit (closed loop), the algebraic sum of the voltage drops must equal the algebraic sum of the source voltages.

This leads to equations in the form

$$V_1 + V_2 + V_3 = E \tag{8-6}$$

Perhaps the best way to develop our loop procedure is to refer to a specific example. We can start with an example which we can readily check by the equivalent-resistance techniques of Chapter 7 by removing the center resistor R_5 in the bridge circuit of Figure 8-1.

EXAMPLE 8-2

Determine the magnitude and polarity of the potential difference between the open-circuit terminals A and B in the bridge network of Figure 8-6.

FIGURE 8-6
Bridge Network for Example 8-2.

SOLUTION

Step I.   The first step is to draw a circuit diagram large enough for us to mark on tracing loops, polarities of voltages, and any other data we need in solving the network. We may change the *physical* location of components so long as we do not change their *electrical* connection. (Figure 8-7).

Step II.   We emphasize the polarity of the sources by marking the $+$ and $-$ terminals for two reasons: (1) if there is more than one source, reversing the polarity of one will completely alter all current and voltage relationships in the network; and (2) for our procedure, we draw current **tracing loops** on the circuit diagram, beginning at the $+$ terminal of the source and proceeding along a path through the network back to the $-$ terminal of the source. These tracing loops will help us to write our Kirchhoff's voltage-law equations. We need sufficient tracing loops to include *all* components and to provide as many separate equations as there are unknowns. In this example, we require two tracing loops for the two unknown loop currents I_1 and I_2.

Step III.   In Section 6-2, we considered the polarities of voltage drops and voltage

FIGURE 8-7
Circuit Diagram for Example 8-2.

rises in relation to current direction. Extending what is obvious in the simple circuit of Figure 8-8 to the tracing loops of Figure 8-7,

> ***In any* passive *component (such as a resistor) in an electric network, the tracing direction is from positive to negative through that component.***

FIGURE 8-8
Polarity of Voltage Drop and Source Voltage in Relation to Direction of Current Tracing Loop.

Hence, once we have established the directions of tracing loops I_1 and I_2, we can mark the polarity of the various voltage drops on the circuit diagram of Figure 8-7.

 Step IV. We are now ready to write our Kirchhoff's voltage-law equations. For the I_1 loop,

$$V_{R_{int}} + V_{R_1} + V_{R_3} = E$$

If we think now of I_1 as the current through the R_1, R_3 branch, from Ohm's law,

$$V_{R_1} = I_1 R_1 \quad \text{and} \quad V_{R_3} = I_1 R_3$$

Considering the junction point C in Figure 8-7, we see that the current I_1 *leaves* this point through R_1 and R_3, and the current I_2 *leaves* this point through R_2 and R_4. Hence, from Kirchhoff's current law, the current through R_{int} is $I_1 + I_2$. Thus,

$$V_{R_{int}} = (I_1 + I_2)R_{int}$$

The Kirchhoff's voltage-law equation for the I_1 tracing loop becomes

$$(I_1 + I_2)R_{int} + I_1R_1 + I_1R_3 = E$$

This is the standard form we shall use for writing our Kirchhoff's voltage-law equations, minimizing the number of unknowns by writing voltage drop as *IR* drops.

Step V. The Kirchhoff's voltage-law equation for the I_1 tracing loop becomes

$$10(I_1 + I_2) + 20I_1 + 30I_1 = 80 \text{ V}$$

Collecting terms, $\qquad\qquad\qquad 60I_1 + 10I_2 = 80 \qquad\qquad\qquad (1)$

Similarly, for the I_2 tracing loop,

$$10(I_1 + I_2) + 15I_2 + 10I_2 = 80 \text{ V}$$

from which $\qquad\qquad\qquad 10I_1 + 35I_2 = 80 \qquad\qquad\qquad (2)$

We then solve Equations (1) and (2) simultaneously:

Elimination method:

$$
\begin{array}{rl}
6 \times (2) \text{ gives} & 60I_1 + 210I_2 = 480 \\
\textit{subtracting } (1) & 60I_1 + 10I_2 = 80 \\
\hline
 & 200I_2 = 400
\end{array}
$$

from which $\qquad\qquad\qquad\qquad I_2 = 2 \text{ A}$

Substituting this value in Equation (1), we obtain

$$I_1 = 1 \text{ A}$$

Determinant method:[†]

$$60I_1 + 10I_2 = 80 \qquad\qquad\qquad (1)$$

$$10I_1 + 35I_2 = 80 \qquad\qquad\qquad (2)$$

[†] The traditional algebraic procedure for solving simultaneous equations is the *Gaussian elimination* method in which equations with two or more unknowns are subtracted until a single equation with only one unknown is left. No fixed procedure is written down for this method since a certain amount of individual judgment is required in selecting the unknowns to be eliminated.

However, as more and more computation is turned over to computers, it is desirable to sacrifice flexibility for a standard procedure which we can use to solve *any* array of simultaneous equations. The procedure currently favored is the *determinant* method. For those whose background in mathematics does not include matrix algebra, a brief description of the determinant method is included in Appendix 1.

$$I_1 = \frac{\begin{vmatrix} 80 & 10 \\ 80 & 35 \end{vmatrix}}{\begin{vmatrix} 60 & 10 \\ 10 & 35 \end{vmatrix}} = \frac{2800 - 800}{2100 - 100} = \frac{2000}{2000} = 1 \text{ A}$$

$$I_2 = \frac{\begin{vmatrix} 60 & 80 \\ 10 & 80 \end{vmatrix}}{\begin{vmatrix} 60 & 10 \\ 10 & 35 \end{vmatrix}} = \frac{4800 - 800}{2000} = \frac{4000}{2000} = 2 \text{ A}$$

Step VI. We can now find the voltage drops across all components of the network. In this example, we need know only two of them.

$$V_{R_3} = I_1 R_3 = 1 \text{ A} \times 30 \text{ }\Omega = \mathbf{30 \text{ V}}$$

and $$V_{R_4} = I_2 R_4 = 2 \text{ A} \times 10 \text{ }\Omega = \mathbf{20 \text{ V}}$$

To determine the potential difference between terminals A and B, we recall that a *voltage drop* in a resistance is just another way of expressing a *potential difference* between the two terminals of the resistance. Hence, terminal A in Figure 8-7 is 30 V more positive than junction D (the junction point of R_3 and R_4). Likewise, terminal B is 20 V more positive than junction D. Therefore,

Terminal A *is 10 V more positive than terminal* B.

At this point, let's consider how a personal computer might assist us in solving general resistance network examples. As specialized circuit-analysis software packages become more numerous, we can solve all sorts of examples without learning anything about loop equations or determinants or network theorems. These techniques are buried within the computer program. A typical circuit-analysis program asks us first of all to describe the network layout by identifying the type of circuit elements (resistor, voltage source, etc.) and the two specific nodes between which the element is connected. We then input numerical values for these elements and the computer proceeds to print out the complete solution of the network. These software packages are designed primarily to assist people in industry in optimizing a circuit design by rapidly printing out the results of many variations of circuit configuration and values. Since they do not assist in learning the principles of network analysis, we shall not pursue this type of computer program.

However, if we intend to solve quite a few examples involving simultaneous equations, it could be worthwhile setting up programs to solve second- and third-order determinants. We do this by following the convention for coefficients and the Cramer's rule formula developed in Appendix 1. The advantages of having such programs stored in permanent memory is that we can concentrate on setting up the Kirchhoff's-law equations from the network data without having to worry about arithmetic errors and the time required for individual calculations with a pocket calculator. A typical program that we could use with Example 8-1 might be

```
100 REM   SECOND ORDER DETERMINANTS
110 REM   SET UP TWO EQUATIONS IN THE FORM
120 PRINT  "AX + BY = K"
130 REM   A - SIGN IN THE EQUATION BELONGS TO THE
          COEFFICIENT
140 PRINT "ENTER COEFFICIENTS FOR EQUATION 1"
150 INPUT  A1,B1,K1
160 PRINT "ENTER COEFFICIENTS FOR EQUATION 2"
170 INPUT  A2,B2,K2
180 LET D=A1*B2-A2*B1
190 LET X=(K1*B2-K2*B1)/D
200 LET Y=(A1*K2-A2*K1)/D
210 PRINT "X = ";X
220 PRINT "Y = ";Y
230 END
```

For Example 8-2, this program produces the following result:

```
RUN
AX + BY = K
ENTER COEFFICIENTS FOR EQUATION 1
? 60, 10, 80
ENTER COEFFICIENTS FOR EQUATION 2
? 10, 35, 80
X = 1
Y = 2
```

If we add a meter between terminals A and B of the bridge circuit of Figure 8-6, the network becomes a Wheatstone bridge which is used for very accurate resistance measurements. To include the meter current in our Kirchhoff's-law equations, we now require one more tracing loop—I_3 in Figure 8-9. We have chosen a tracing loop for I_3 passing through R_1, R_M (the resistance of the meter), and R_4. We might have chosen a

FIGURE 8-9
Solving Wheatstone Bridge by Loop Procedure.

tracing loop for I_3 passing through R_2, R_M, and R_3. Since I_3 would then pass in the *opposite* direction through R_M, it follows that it would turn out to have the *opposite* sign in our solution. If a tracing loop current turns out to have a negative sign, we simply treat the current as a negative quantity when we add the loop currents *algebraically*.

Using the I_3 loop shown in Figure 8-9, the Kirchhoff's voltage-law equations for the three tracing loops are

$$\text{From loop 1,} \quad R_1(I_1 + I_3) + R_3I_1 = E$$

$$\text{From loop 2,} \quad R_2I_2 + R_4(I_2 + I_3) = E$$

$$\text{From loop 3,} \quad R_1(I_1 + I_3) + R_MI_3 + R_4(I_2 + I_3) = E$$

Substituting the given values of resistance and source voltage, we can solve these three simultaneous equations for the three loop currents from which

Total current is	$I_1 + I_2 + I_3$
Meter current is	I_3
Current through R_1 is	$I_1 + I_3$
Current through R_2 is	I_2
Current through R_3 is	I_1
Current through R_4 is	$I_2 + I_3$

EXAMPLE 8-3
If a galvanometer whose resistance is 50 Ω is connected between terminals A and B of the bridge circuit of Figure 8-7, what is the meter current?

SOLUTION
For the I_1 loop,

$$10(I_1 + I_2 + I_3) + 20(I_1 + I_3) + 30I_1 = 80 \text{ V}$$

for the I_2 loop,

$$10(I_1 + I_2 + I_3) + 15I_2 + 10(I_2 + I_3) = 80 \text{ V}$$

and for the I_3 loop,

$$10(I_1 + I_2 + I_3) + 20(I_1 + I_3) + 50I_3 + 10(I_2 + I_3) = 80 \text{ V}$$

Collecting terms,

$$60I_1 + 10I_2 + 30I_3 = 80 \tag{1}$$

$$10I_1 + 35I_2 + 20I_3 = 80 \tag{2}$$

$$30I_1 + 20I_2 + 90I_3 = 80 \tag{3}$$

Since we are asked only to determine the meter current, we need only solve for I_3. Thus, using the determinant method,

$$I_3 = \frac{\begin{vmatrix} 60 & 10 & 80 \\ 10 & 35 & 80 \\ 30 & 20 & 80 \end{vmatrix}}{\begin{vmatrix} 60 & 10 & 30 \\ 10 & 35 & 20 \\ 30 & 20 & 90 \end{vmatrix}}$$

$$= \frac{168\,000 + 24\,000 + 16\,000 - 84\,000 - 96\,000 - 8000}{189\,000 + 6000 + 6000 - 31\,500 - 24\,000 - 9000}$$

$$= \frac{20\,000}{136\,500} = \mathbf{0.15\ A}$$

SELF-CHECK

8-5:1. For the circuit shown in Figure 8-10:
  (a) Write Kirchhoff's voltage-law equations for I_1 and I_2.
  (b) Write the determinants that will lead to the solutions for I_1 and I_2.
  (c) Solve the determinants for I_1 and I_2.

FIGURE 8-10

8-6 NETWORKS WITH MORE THAN ONE VOLTAGE SOURCE

Even an electrical system as simple as a flashlight usually has more than one voltage source. To provide the lamp with a potential difference higher than that of a single battery, we connect two or more batteries in *series aiding*. Since both voltage sources in Figure 8-11(a) attempt to make current flow in the *same* direction through any load, the total effective source voltage becomes the *sum* of the individual source voltages. In pencil-and-paper electric circuits, we can replace the individual series-connected voltage

FIGURE 8-11
Voltage Sources: (a) Series Aiding;
(b) Series Opposing.

sources with a single equivalent source. [The automobile storage battery in Figure 8-12(a) consists of six cells in series aiding.]

It is quite possible for us to load one of the flashlight (or radio or camera) batteries wrong-way-round. With such a *series opposing* connection [shown in Figure 8-11(b)] each source tries to make current flow in opposite directions in any load. As a result, the net terminal voltage for the equivalent source is the *difference* between the individual source voltages. We avoid series opposing connection of voltage sources in practical electric circuits for two reasons: (1) the equivalent source voltage is actually *reduced;* and (2), since the higher voltage source determines the current direction, current will flow the wrong way (charging) through the lower voltage source when we connect a load to the source in Figure 8-11(b).

Figure 8-12(a) shows a practical electrical system in which both a storage battery and a generator driven by an automobile engine provide current to a common load circuit. The terminals of these two voltage sources (each with its own internal resistance) are connected in *parallel*. If the auto engine stops, the generator emf must fall to zero and a heavy current—limited only by the two internal resistances in series—flows from the battery through the generator. In a practical automobile electrical system, we must arrange for the generator to be automatically disconnected from the rest of the system when the engine is stopped. Parallel connection of voltage sources requires special precautions.

FIGURE 8-12
Examples of Networks with More than One Voltage Source.

The innocent-looking network of Figure 8-12(b) is the coupling circuit for a transistor "flip-flop" circuit. Since dozens of stages of an electronic system are fed from a single power supply, we usually avoid cluttering circuit diagrams with long interconnecting leads by marking the dc supply voltages for individual stages, as shown in Figure 8-12(b). We can redraw this network in the form shown in Figure 8-16. It now becomes apparent that we have another example of a network containing two separate voltage sources. Since both networks in Figure 8-12 include more than one voltage source, we cannot solve them by simple Ohm's-law calculations.

8-7 LOOP EQUATIONS IN MULTISOURCE NETWORKS

The procedure we developed in Section 8-5 for writing Kirchhoff's voltage-law equations for closed tracing loops applies equally well to multisource networks. Again, we can illustrate the procedure and note possible points of difficulty by referring to specific examples. As our first example, we shall use the automobile electrical system of Figure 8-12(a).

EXAMPLE 8-4

An automobile generator with an internal resistance of 0.2 Ω develops an open-circuit voltage of 16.0 V. The storage battery has an internal resistance of 0.1 Ω and an open-circuit voltage of 12.8 V. Both sources are connected in parallel to a 1.0-Ω load. Determine the generator current, battery current, and load current.

SOLUTION 1

If we wish, we can redraw the circuit diagram of Figure 8-12(a) as shown in Figure 8-13 to make it easier to draw our tracing loops.

Writing a Kirchhoff's voltage-law equation for the generator loop I_G in Figure 8-13,

$$V_{RG} + V_L = E_G$$

But from Ohm's law, any voltage drop across a resistance is always equal to the

FIGURE 8-13
Circuit Diagram for Example 8-4.

product of the resistance and the current through it. Therefore,

$$V_{RG} = 0.2I_G$$

Since the diagram of Figure 8-13 shows both the I_G and the I_B loops going in the *same* direction through the load, from Kirchhoff's current law, the load current is

$$I_L = I_G + I_B \quad \text{and} \quad V_L = 1.0(I_G + I_B)$$

Hence, the Kirchhoff's voltage-law equation for the generator loop becomes

$$0.2I_G + 1.0(I_G + I_B) = E_G$$

or
$$1.2I_G + 1.0I_B = 16.0 \text{ V} \tag{1}$$

Similarly, the Kirchhoff's voltage-law equation for the battery loop becomes

$$0.1I_B + 1.0(I_G + I_B) = E_B$$

or
$$1.0I_G + 1.1I_B = 12.8 \text{ V} \tag{2}$$

Solving Equations (1) and (2) simultaneously,

Elimination method:

$$5 \times (1) \text{ gives} \quad 6I_G + 5.0I_B = 80$$
$$6 \times (2) \text{ gives} \quad \underline{6I_G + 6.6I_B = 76.8}$$

subtracting,
$$-1.6I_B = 3.2$$

from which
$$I_B = -2 \text{ A}$$

The negative answer for I_B simply tells us that the battery is not supplying current to the load, as we had supposed in setting up the direction for the tracing loop for I_B. However, the numerical answer is quite correct. In this example, the storage battery is *charging* at the rate of 2 A. To avoid having to change our tracing directions in Figure 8-13, we can carry on with the original tracing loops as long as we treat I_B as a negative quantity. Substituting this value of I_B in Equation (1),

$$1.2I_G + (1.0) \times (-2) = 16$$

or
$$1.2I_G - 2 = 16$$

from which
$$I_G = 15 \text{ A}$$

and
$$I_L = I_G + I_B = 15 + (-2) = 13 \text{ A}$$

Determinant method:

$$1.2I_G + 1.0I_B = 16.0$$
$$1.0I_G + 1.1I_B = 12.8$$

$$I_G = \frac{\begin{vmatrix} 16.0 & 1.0 \\ 12.8 & 1.1 \end{vmatrix}}{\begin{vmatrix} 1.2 & 1.0 \\ 1.0 & 1.1 \end{vmatrix}} = \frac{17.6 - 12.8}{1.32 - 1.0} = \frac{4.8}{0.32} = \textbf{15 A}$$

$$I_B = \frac{\begin{vmatrix} 1.2 & 16.0 \\ 1.0 & 12.8 \end{vmatrix}}{\begin{vmatrix} 1.2 & 1.0 \\ 1.0 & 1.1 \end{vmatrix}} = \frac{15.36 - 16.0}{0.32} = \frac{-0.64}{0.32} = \textbf{-2 A}$$

$$I_L = I_G + I_B = 15 - 2 = \textbf{13 A}$$

SOLUTION 2

The tracing loops we selected for Solution 1 are the normal choice from a circuit-operation point of view. However, other choices all lead to the same final solution. To illustrate the loop technique further, we can return to the circuit layout given in the original circuit diagram of Figure 8-12(a) and reproduced in Figure 8-14.

FIGURE 8-14
Circuit Diagram for Example 8-4, Second Solution.

We select one tracing loop I_G from the + terminal of the generator through the internal resistance of the generator and the battery and through the battery back to the − terminal of the generator. The I_L tracing loop follows the same path as the I_B loop in Solution 1. But I_L is *not the same current* as I_B. In this case, the total load current is I_L, since this is the only tracing loop through R_L. The battery current in this solution is the discharge current I_L minus the charge current I_G.

Note that the voltage drop across R_B will be $R_B \times$ the *difference* of the two loop currents. In writing loop equations, the tracing loop we use is always the *positive* current direction; any current we encounter going *against* the tracing direction is negative. Hence, Kirchhoff's voltage-law equation for the I_L loop becomes

$$0.1(I_L - I_G) + 1I_L = 12.8 \text{ V}$$

from which
$$-0.1I_G + 1.1I_L = 12.8 \qquad (1)$$

In writing the equation for the I_G loop, we note that I_G enters the $+$ terminal of the battery, just as it enters the $+$ terminal for voltage drops across resistors. Thus, E_B can appear as a positive term on the left-hand side of the equation. If we wish, we can think of E_B as a voltage source *bucking* the generator voltage and put E_B on the right-hand side of the equation as a *negative* quantity. The loop equation for the I_G loop becomes

$$0.2I_G + 0.1(I_G - I_L) + 12.8 = 16.0 \text{ V}$$

from which
$$0.3I_G - 0.1I_L = 3.2 \qquad (2)$$

Once the loop equations are written, the rest of the solution is merely algebraic solution of simultaneous equations

$$-0.1I_G + 1.1I_L = 12.8 \qquad (1)$$
$$0.3I_G - 0.1I_L = 3.2 \qquad (2)$$

Generator current, I_G

$$I_G = \frac{\begin{vmatrix} 12.8 & 1.1 \\ 3.2 & -0.1 \end{vmatrix}}{\begin{vmatrix} -0.1 & 1.1 \\ 0.3 & -0.1 \end{vmatrix}} = \frac{-1.28 - 3.52}{+0.01 - 0.33} = \frac{-4.80}{-0.32} = \textbf{15 A}$$

Load current, I_L

$$I_L = \frac{\begin{vmatrix} -0.1 & 12.8 \\ 0.3 & 3.2 \end{vmatrix}}{D} = \frac{-0.32 - 3.84}{-0.32} = \frac{-4.16}{-0.32} = \textbf{13 A}$$

Battery current $= I_L - I_G = 13 - 15 = \textbf{-2 A}$
Again we find that the battery is charging, rather than discharging.

For a second example, we can consider the Edison three-wire distribution system shown in Figure 8-15. Originally designed to reduce I^2R power loss in transmission lines, it provides us with a good example for our network analysis techniques.[†]

[†]Because residential loads are many miles from generating stations, there is a costly loss of energy in the interconnecting transmission lines. We can reduce this loss if we can reduce the current required to feed a given load. Since $P = VI$, one method is to increase system voltage. For example, a 120-volt 1200-watt toaster draws a current of 10 amperes, but a 240-volt 1200-watt toaster requires a current of only five amperes. For safety reasons, it is not desirable to raise the operating voltage of household appliances too much. The three-wire distribution system reduces transmission line losses by arranging to have loop currents at least partially *cancel* in the *neutral* lead. If we arrange for equal values of R_L in Figure 8-15, the load is *balanced* and the neutral current is zero. The *unbalanced* load shown does increase power loss in the transmission lines, but not as much as operating both loads in parallel from a two-wire system with the same potential difference between lines.

FIGURE 8-15
Three-Wire Distribution System.

EXAMPLE 8-5

(a) Calculate the total power loss in the three 1-Ω conductors in the three-wire distribution system shown in Figure 8-15.

(b) Calculate the total power loss if the two loads were fed in parallel from a single 110-V source through a pair of 1-Ω conductors.

SOLUTION

(a) The voltage drop across the resistance of the neutral conductor is 1 Ω × the *difference* of the two loop currents. As in the second solution for Example 8-4, tracing loop direction is the *positive* quantity and any current going *against* the tracing direction has a *negative* sign. Therefore, the Kirchhoff's voltage-law equation for the I_1 loop becomes

$$1I_1 + 10I_1 + 1(I_1 - I_2) = E_1$$

or
$$12I_1 - I_2 = 110 \tag{1}$$

For the I_2 loop,

$$1(I_2 - I_1) + 15I_2 + 1I_2 = E_2$$

or
$$-I_1 + 17I_2 = 110 \tag{2}$$

$$I_1 = \frac{\begin{vmatrix} 110 & -1 \\ 110 & 17 \end{vmatrix}}{\begin{vmatrix} 12 & -1 \\ -1 & 17 \end{vmatrix}} = \frac{1870 - (-110)}{204 - (+1)} = \frac{1980}{203} = 9.75 \text{ A}$$

$$I_2 = \frac{\begin{vmatrix} 12 & 110 \\ -1 & 110 \end{vmatrix}}{D} \frac{1320 - (-110)}{203} = \frac{1430}{203} = 7.04 \text{ A}$$

Power in line 1, $P_1 = I_1^2 R_1 = 9.75^2 \times 1 = 95 \text{ W}$

Power in line 2, $P_2 = I_2^2 R_2 = 7.04^2 \times 1 = 50 \text{ W}$

Neutral current, $I_N = I_1 - I_2 = 9.75 - 7.04 = 2.71 \text{ A}$

Neutral power, $P_N = I_N^2 R_N = 2.71^2 \times 1 = 7 \text{ W}$

Total power loss in conductors,

$$P_T = P_1 + P_2 + P_N = 95 + 50 + 7 = \textbf{152 W}$$

(b) If the two loads are connected in parallel, the equivalent load resistance is

$$R_{eq} = \frac{10 \times 15}{10 + 15} = 6 \ \Omega$$

Adding the resistance of two conductors, $R_T = 8 \ \Omega$. Hence, the circuit current becomes

$$I = \frac{E}{R_T} = \frac{110 \text{ V}}{8 \ \Omega} = 13.75 \text{ A}$$

Power in *each* line $= I^2 R = 13.75^2 \times 1 = 189 \text{ W}$

Therefore, total power loss in conductors is

$$P_T = 2 \times 189 \text{ W} = \textbf{378 W}$$

For our third example, we shall tackle the transistor coupling circuit shown in Figure 8-12(b). Even though the original circuit diagram does not show the actual voltage sources, we draw them in our working diagram of Figure 8-16 to enable us to draw complete tracing loops.

FIGURE 8-16
Circuit Diagram for Example 8-6.

EXAMPLE 8-6

If R_B in the circuit shown in Figures 8-12(b) and 8-16 is 1 kΩ, determine the magnitude and polarity of the voltage drop across R_B.

SOLUTION

We require three tracing loops to include all components in this circuit. Several combinations of loops are possible. Those selected in Figure 8-16 are the "natural" choice. Writing the three Kirchhoff's voltage-law equations from our loop procedure is quite straightforward. Since volts = milliamperes × kilohms, we can write the equation for the I_1 loop as

$$68(I_1 - I_2) + 22I_1 = 150 \text{ V} \tag{1}$$

(where I_1 and I_2 are in milliamperes). From the I_2 loop,

$$33(I_2 + I_3) + 56I_2 + 68(I_2 - I_1) = 100 \text{ V} \tag{2}$$

and from the I_3 loop,

$$33(I_3 + I_2) + 1I_3 = 100 \text{ V} \tag{3}$$

The tedious part of the solution is the numerical computation in solving third order simultaneous equations. However, we need only solve for I_3 to provide the requested information in this example. Collecting terms,

$$90I_1 - 68I_2 \qquad = 150 \tag{1}$$

$$-68I_1 + 157I_2 + 33I_3 = 100 \tag{2}$$

$$33I_2 + 34I_3 = 100 \tag{3}$$

$$I_3 = \frac{\begin{vmatrix} 90 & -68 & 150 \\ -68 & 157 & 100 \\ 0 & 33 & 100 \end{vmatrix}}{\begin{vmatrix} 90 & -68 & 0 \\ -68 & 157 & 33 \\ 0 & 33 & 34 \end{vmatrix}}$$

$$= \frac{1\,413\,000 + 0 - 336\,600 - 0 - 297\,000 - 462\,400}{480\,420 + 0 + 0 - 0 - 98\,010 - 157\,216}$$

$$= \frac{317\,000}{225\,194} = 1.4 \text{ mA}$$

Hence, the voltage drop across R_B is

$$V_B = 1.4 \text{ mA} \times 1 \text{ kΩ} = \textbf{1.4 V}$$

Since I_3 turned out to be a positive quantity, the top end of R_B is positive with respect to chassis.

SELF-CHECK

8-7:1. For the circuit shown in Figure 8-17:
(a) Write Kirchhoff's voltage-law equations for the loops traced by I_1 and I_2.
(b) Use the determinant method to solve for I_1 and I_2.
(c) Find the true current in the 30-kΩ resistor.

FIGURE 8-17

8-8 MESH EQUATIONS

Now that we have a "feel" for writing equations which satisfy Kirchhoff's voltage law, we can turn our attention to a format procedure which is designed to simplify and speed up the task of writing the set of simultaneous equations we need to solve various resistance networks. In writing **mesh equations,**[†] the emphasis is on following a set of rules, rather than on the mental inspection of the network which we used with the loop procedure.

The price we pay for a short cut format is that mesh equations cannot handle some of the networks that we can solve with the loop procedure. We define a **mesh** as a closed loop within a network such that there are no other circuit elements within the loop. Therefore, the circuit diagram cannot contain any conductors crossing without there being an electrical junction at the point of crossing. In other words, the network diagram must be strictly two-dimensional or **planar.** Second, the mesh format requires all sources to be *voltage* sources. If there are any *current* sources, we must convert them to equivalent voltage sources before we use the mesh format. Third, we cannot arrange our current loops so that we only have to solve a single equation, as in Example 8-3.

After writing equations for voltage drops around the current loops in the loop procedure, we then rewrite the Kirchhoff's voltage-law equations to collect the current terms. In the mesh format, we go directly to this second step. To illustrate the mesh format, we can refer to the network shown in Figure 8-18. We have laid out our circuit

---

[†] Mesh equations were developed over a century ago by James Clerk Maxwell, the British physicist who worked out the electromagnetic theory of light and predicted the existence of radio waves before they were discovered by Heinrich Hertz in 1888.

FIGURE 8-18
Mesh Network for Example 8-7.

diagram so that the network is clearly planar and the four meshes have no internal electrical components. The next step is to draw a current loop for each mesh. All mesh currents must be in the same direction—either clockwise or counterclockwise. The more popular choice is clockwise. The result is that, for any particular mesh current (e.g., I_A in Figure 8-18), all other currents through common circuit elements (I_B and I_C for mesh A) are in the *opposite* direction. This provides a standard format for every mesh equation. The mesh equation for mesh A has the form

$$R_T I_A - R_{AB} I_B - R_{AC} I_C = \Sigma E \qquad (8\text{-}7)$$

Since I_A passes through all resistors around mesh A, we insert the total resistance around mesh A for R_T. R_{AB} is the value of the resistor which is common to mesh A and mesh B—similarly for R_{AC}. ΣE is the algebraic sum of the voltage sources around the mesh. Where I_A passes from $-$ to $+$ *inside* a voltage source, this is its normal positive direction and that E is a positive quantity. Similarly, if the direction of I_A is from $+$ to $-$ inside a voltage source, that E would have a negative sign.

EXAMPLE 8-7
Find the magnitudes and polarities of the voltage drops across the 30-Ω, 50-Ω, and 60-Ω resistors in the network shown in Figure 8-18.

SOLUTION
Using the mesh format Equation (8-7) to write a Kirchhoff's voltage-law equation for each of the four meshes in Figure 8-18,

$$(10 + 30 + 50)I_A - 30I_B - 50I_C = 20 \text{ V}$$
$$(30 + 20 + 40 + 60)I_B - 30I_A - 60I_D = 0 \text{ V}$$
$$(80 + 70 + 50)I_C - 50I_A = -30 \text{ V}$$
$$(90 + 60)I_D - 60I_B = (30 - 40) \text{ V}$$

Collecting the terms,

$$90I_A - 30I_B - 50I_C = 20 \tag{1}$$

$$-30I_A + 150I_B - 60I_D = 0 \tag{2}$$

$$-50I_A + 200I_C = -30 \tag{3}$$

$$-60I_B + 150I_D = -10 \tag{4}$$

Unfortunately, the rules for solving fourth-order determinants are not as simple as those for second- and third-order determinants. We can get around this problem by eliminating one of the unknowns. From Equation (4),

$$I_D = \frac{60I_B - 10}{150}$$

Substituting in Equation (2),

$$-30I_A + 150I_B - (24I_B - 4) = 0$$

This leaves us with three simultaneous equations,

$$90I_A - 30I_B - 50I_C = 20 \tag{1}$$

$$-30I_A + 126I_B = -4 \tag{5}$$

$$-50I_A + 200I_C = -30 \tag{3}$$

$$I_A = \frac{\begin{vmatrix} 20 & -30 & -50 \\ -4 & 126 & 0 \\ -30 & 0 & 200 \end{vmatrix}}{\begin{vmatrix} 90 & -30 & -50 \\ -30 & 126 & 0 \\ -50 & 0 & 200 \end{vmatrix}} = \frac{291\,000}{1\,773\,000} = 164.13 \text{ mA}$$

$$I_B = \frac{\begin{vmatrix} 90 & 20 & -50 \\ -30 & -4 & 0 \\ -50 & -30 & 200 \end{vmatrix}}{D} = \frac{13\,000}{1\,773\,000} = 7.33 \text{ mA}$$

$$I_C = \frac{\begin{vmatrix} 90 & -30 & 20 \\ -30 & 126 & -4 \\ -50 & 0 & -30 \end{vmatrix}}{D} = \frac{-193\,200}{1\,773\,000} = -108.97 \text{ mA}$$

$$I_D = \frac{60 \times 7.33 \times 10^{-3} - 10}{150} = -63.73 \text{ mA}$$

$$I_{30} = I_A - I_B = 164.13 \text{ mA} - 7.33 \text{ mA} = 156.8 \text{ mA}$$

$$\therefore V_{30} = 30 \ \Omega \times 156.8 \text{ mA} = \textbf{4.7 V} \quad \text{positive at the top}$$

$$I_{50} = I_A - I_C = 164.13 \text{ mA} + 108.97 \text{ mA} = 273.1 \text{ mA}$$

$$\therefore V_{50} = 50 \ \Omega \times 273.1 \text{ mA} = \textbf{13.7 V} \quad \text{positive on the right}$$

$$I_{60} = I_B - I_D = 7.33 \text{ mA} + 63.73 \text{ mA} = 71.06 \text{ mA}$$

$$\therefore V_{60} = 60 \ \Omega \times 71.06 \text{ mA} = \textbf{4.3 V} \quad \text{positive on the right}$$

As an exercise in writing mesh equations, we can now repeat Examples 8-2 to 8-6 using the mesh format. Obviously, we should end up with the same numerical results.

EXAMPLE 8-2A
In the circuit diagram of Figure 8-7, there are two meshes. Mesh I_1 is the same as loop I_1. The second mesh is the closed loop containing R_3, R_1, R_2, and R_4 and no voltage source. The two mesh equations are

$$(10 + 20 + 30)I_1 - (20 + 30)I_2 = 80 \text{ V}$$

$$(30 + 20 + 15 + 10)I_2 - (20 + 30)I_1 = 0 \text{ V}$$

$$I_{R_3} = I_1 - I_2 \quad \text{and} \quad I_{R_4} = I_2$$

The remainder of the solution is the same as in Example 8-2.

EXAMPLE 8-3A
In the circuit diagram of Figure 8-9, there are three meshes. Mesh I_1 is the same as loop I_1. The second mesh is the closed loop containing R_1, R_2, and R_M. The third mesh contains R_3, R_M, and R_4. After collecting terms, the three mesh equations become

$$60I_1 - 20I_2 - 30I_3 = 80 \tag{1}$$

$$-20I_1 + 85I_2 - 50I_3 = 0 \tag{2}$$

$$-30I_1 - 50I_2 + 90I_3 = 0 \tag{3}$$

We need to solve for both I_2 and I_3, since $I_M = I_2 - I_3$.

EXAMPLE 8-4A
In the circuit diagram of Figure 8-13, we can write mesh-current equations by simply reversing the direction of I_B. The second solution for Example 8-4 using the circuit configuration of Figure 8-14 already satisfies the rules for the mesh format and will yield the same simultaneous equations when the terms are collected.

EXAMPLE 8-5A
The circuit layout of Figure 8-15 also satisfies the mesh format rules.

EXAMPLE 8-6A

Before we write the mesh equations for the circuit shown in Figure 8-16, all mesh currents must be in the same direction. Reversing I_3 so that all mesh currents are counterclockwise,

$$(22 + 68) \text{ k}\Omega \times I_1 - 68 \text{ k}\Omega \times I_2 = 150 \text{ V}$$

$$(33 + 56 + 68) \text{ k}\Omega \times I_2 - 68 \text{ k}\Omega \times I_1 - 33 \text{ k}\Omega \times I_3 = 100 \text{ V}$$

$$(1 + 33) \text{ k}\Omega \times I_3 - 33 \text{ k}\Omega \times I_2 = -100 \text{ V}$$

Collecting terms,

$$90 \text{ k}\Omega \times I_1 - 68 \text{ k}\Omega \times I_2 \qquad = \quad 150 \qquad (1)$$

$$-68 \text{ k}\Omega \times I_1 + 157 \text{ k}\Omega \times I_2 - 33 \text{ k}\Omega \times I_3 = \quad 100 \qquad (2)$$

$$-33 \text{ k}\Omega \times I_2 + 34 \text{ k}\Omega \times I_3 = -100 \qquad (3)$$

To solve for V_B, we need to solve only for I_3, which turns out to be -1.4 mA, since we reversed the direction of I_3.

SELF-CHECK

8-8:1. For the circuit shown in Figure 8-19:
(a) Use the standard format method to write the mesh equations for mesh A, mesh B, and mesh C.
(b) Use the determinant method to solve for I_A, I_B, and I_C.

FIGURE 8-19

8-9 KIRCHHOFF'S CURRENT-LAW EQUATIONS

The loop-current and mesh-equation methods of solving resistance networks are based on writing a sufficient number of Kirchhoff's voltage-law equations to include all circuit elements. The number of simultaneous equations required depends on the number of loops or meshes in the network. To solve the voltage-law equations, we substitute an IR drop for each voltage drop and solve for the various loop or mesh currents.

FIGURE 8-20
Identifying Nodes in a Network.

Kirchhoff's current law allows us to write a different set of equations for currents entering and leaving junction points or **nodes** in an electric circuit. Before we can write such current-law equations, we must be able to identify the nodes in a given network. In the network shown in Figure 8-20, we do not consider the junction of E and R_1, and the junction point of R_3 and R_6 as nodes, since the current-law equation $I_1 = I_1$ is redundant. Also, as indicated by the alternate circuit layout of Figure 8-20(b), we consider all connections to the common grounded conductor in Figure 8-20(a) as a single node. Hence, there is a total of three nodes in the network of Figure 8-20.

Some texts prefer to state Kirchhoff's current law as a mathematical statement that the *algebraic* sum of currents at a node must equal zero. The difficulty in translating this statement into an equation is that we have to remember which currents are positive and which are negative. In setting up a nodal procedure, we prefer to state Kirchhoff's current law as

At any node in an electric circuit, the sum of the currents leaving the node equals the sum of the currents entering the node.

For the three nodes in the network of Figure 8-20, the above definition leads to the following equations.

$$\text{Node A:} \quad I_4 + I_2 = I_1 \tag{1}$$

$$\text{Node B:} \quad I_5 + I_3 = I_2 \tag{2}$$

$$\text{Reference node:} \quad I_1 = I_4 + I_5 + I_3 \tag{3}$$

Note that if we add Equations (1) and (2), we obtain Equation (3). Hence, the Kirchhoff's current-law equation for the third node is redundant in setting up the simultaneous equations. The number of Kirchhoff's current-law equations we need to solve any network is always one less than the total number of nodes. We label the redundant node as the **reference node.** We usually select as reference node one that is common to as many branch currents as possible. The *grounded* part of a network is a good choice for reference node, as shown in Figure 8-20.

The two equations for nodes A and B have five unknown currents. However, there are only *two* node voltages V_A and V_B with respect to the reference node in the network shown in Figure 8-20. To reduce the unknowns in the current-law equations to two, we substitute V/R ratios $(I = V/R)$ for each current in such a manner that we are solving for V_A and V_B. Once we know the node voltages, we can find any other unknown in the network by Ohm's law. Hence, in the network shown in Figure 8-20, $V_4 = V_A$ and $V_5 = V_B$.

For node A:

$$I_4 = \frac{V_A}{R_4}$$

$$I_2 = \frac{V_2}{R_2} = \frac{V_A - V_B}{R_2}$$

and

$$I_1 = \frac{V_1}{R_1} = \frac{E - V_A}{R_1}$$

Substituting these expressions in Equation (1) gives

$$\frac{V_A}{R_4} + \frac{V_A - V_B}{R_2} = \frac{E - V_A}{R_1} \tag{8-8}$$

For node B:

$$I_5 = \frac{V_B}{R_5}$$

$$I_3 = \frac{V_B}{R_3 + R_6}$$

and

$$I_2 = \frac{V_A - V_B}{R_2}$$

$$\therefore \frac{V_B}{R_5} + \frac{V_B}{R_3 + R_6} = \frac{V_A - V_B}{R_2} \tag{8-9}$$

Given numerical values for E and all resistors, the only unknowns are V_A and V_B.

EXAMPLE 8-4B

An automobile generator with an internal resistance of 0.2 Ω develops an open-circuit voltage of 16.0 V. The storage battery has an internal resistance of 0.1 Ω and an open-circuit voltage of 12.8 V. Both sources are connected in parallel to a 1-Ω load. Determine the load current.

FIGURE 8-21
Circuit Diagram for Example 8-4B.

SOLUTION

Although there are two nodes in Figure 8-21, except for a reversal of all signs, the same Kirchhoff's current-law equation applies to both nodes. Hence, we start the nodal solution by writing *one* Kirchhoff's current-law equation,

$$I_L = I_G + I_B$$

Note that we again assume that the battery is discharging. In Figure 8-21,

$$I_L = \frac{V_L}{R_L}, \quad I_G = \frac{E_G - V_L}{R_G} \quad \text{and} \quad I_B = \frac{E_B - V_L}{R_B}$$

Thus, we substitute V/R ratios for the currents in the Kirchhoff's current-law equation to obtain

$$\frac{V_L}{1} = \frac{16.0 - V_L}{0.2} + \frac{12.8 - V_L}{0.1}$$

Clearing the fractions by multiplying the equation by 0.1,

$$0.1\, V_L = 8.0 - 0.5\, V_L + 12.8 - V_L$$

from which

$$V_L = \frac{20.9}{1.6} = 13 \text{ V}$$

and

$$I_L = \frac{V_L}{R_L} = \frac{13 \text{ V}}{1 \ \Omega} = \mathbf{13 \ A}$$

Since we only required one nodal equation in the solution of Example 8-4 B, we avoided the simultaneous equations of Example 8-4, which required two tracing loops. However, loop (and mesh) equations are free from algebraic fractions, since we substitute a *product*—the *IR* drop—for each voltage drop. In the nodal equation of Example 8-4 B, we substitute a *V/R ratio* for each current term, leaving us with an equation in fractional form. We can avoid the fraction produced by the expression $I = V/R$ if we switch our point of view from resistance to conductance, which is the reciprocal of resistance (Section 6-10). Now the equivalent expression becomes $I = VG$.

SELF-CHECK

8-9:1. For the circuit shown in Figure 8-22:
  (a) Write Kirchhoff's current-law equations at Node A, Node B, and the Reference Node.
  (b) Substitute the V/R ratios for the resistor currents in the Node A and Node B equations.
  (c) Solve these two node equations for the node voltages V_A and V_B.

FIGURE 8-22

8-10 NODAL ANALYSIS

Nodal analysis is a circuit-analysis *format* which combines Kirchhoff's current-law equations with the source conversions we developed in Section 8-4. Converting all voltage sources to equivalent constant-current sources allows us to standardize the way we write the Kirchhoff's current-law equations. For the nodal analysis format, we consider source currents to flow *into* a node. If the arrows in a circuit diagram show that a source current actually flows out of a particular node, then that current is a *negative* quantity as far as that node is concerned. Similarly, we consider all resistor currents to flow *out of* a node. We are thus restating Kirchhoff's current law in the form

> *At any independent node, the algebraic sum of the resistor currents leaving the node equals the algebraic sum of the source currents entering the node.*

This definition leads to Kirchhoff's current-law equations having the format

$$I_{R_1} + I_{R_2} + \cdots = I_{S_1} + I_{S_2} + \cdots \tag{8-10}$$

The next step is to replace resistor currents on the left-hand side of Equation (8-10) with VG products. The voltage term depends on the two nodes between which the particular resistor is connected. When we write the nodal equation for node V_1 in the circuit of Figure 8-23, the current through R_1 becomes $V_1 G_1$. But the voltage across R_3

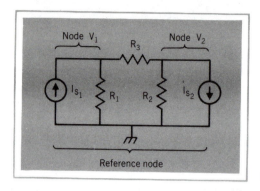

FIGURE 8-23
Circuit Diagram for Writing Nodal Equations.

is the algebraic difference between node voltages V_1 and V_2. To assure the correct sign for the current through R_3, we always subtract the adjacent node voltage (V_2) from the voltage "at" the node for which we are writing the equation (V_1). Hence the Kirchhoff's current-law equation for node V_1 becomes

$$V_1 G_1 + (V_1 - V_2)G_3 = I_{S_1}$$

Collecting the voltage terms gives

$$(G_1 + G_3)V_1 - G_3 V_2 = I_{S_1} \tag{8-11}$$

Equation (8-11) is the standard format for writing a nodal equation for every independent node in the network. The conductance and current terms become the coefficients for determinant matrices.

We can write Equation (8-11) directly by noting that the positive voltage term on the left-hand side of the equation is the unknown voltage for the node in question multiplied by the sum of the conductances connected to that node. From this positive term, we must subtract a term for the node voltage at every adjacent node to which a resistor is connected from the node in question. This term consists of the adjacent node voltage multiplied by the conductance between the two nodes. Following these rules, we can write the nodal equation for node V_2 as[†]

$$(G_2 + G_3)V_2 - G_3 V_1 = -I_{S_2}$$

Nodal analysis is particularly useful in analyzing networks where a common portion of the network is fed from several sources in parallel. Our automobile generator example is just such a network.

EXAMPLE 8-4C

An automobile generator with an internal resistance of 0.2Ω develops an open-circuit voltage of 16.0 V. The storage battery has an internal resistance of 0.1Ω and an open-circuit voltage of 12.8 V. Both sources are connected in parallel to a 1-Ω load. Determine the load current.

[†] Note the **duality** between the nodal format equation and Equation (8-7) for mesh format equations.

FIGURE 8-24
Circuit Diagram for Example 8-4C.

SOLUTION

Step I. Now that we are familiar with source conversion, we can skip the original circuit layout of Figure 8-21 and draw the equivalent circuit of Figure 8-24 with all sources converted to constant-current sources. The short-circuit current for the battery is

$$I_B = \frac{E_B}{R_B} = \frac{12.8 \text{ V}}{0.1 \text{ } \Omega} = 128 \text{ A}$$

Similarly, the short-circuit current of the generator is

$$I_G = \frac{E_G}{R_G} = \frac{16.0 \text{ V}}{0.2 \text{ } \Omega} = 80 \text{ A}$$

Hence, we can fill in the required data on the circuit diagram of Figure 8-24 with nothing more than a bit of mental arithmetic. Note that the generator and battery currents in this *equivalent* circuit are not the same as the real currents in the original circuit.

Step II. To determine the conductance of each branch, we simply enter the value of the resistor into our calculator and press the reciprocal key. Hence,

$$G_G = \frac{1}{R_G} = 0.2 \text{ } \Omega \text{ } \boxed{1/x} = 5 \text{ S} \qquad G_B = \frac{1}{R_B} = 0.1 \text{ } \Omega \text{ } \boxed{1/x} = 10 \text{ S}$$

and

$$G_L = \frac{1}{R_L} = 1 \text{ S}$$

Step III. Although it might appear at first glance that there are six junction points in Figure 8-24, there are only *two* significant nodes, because the *same* voltage appears across all parallel branches. For our nodal analysis, we can redraw the circuit in the form shown in Figure 8-25. If we label the lower node as the reference node, there is only one other independent node. Consequently, we require only one Kirchhoff's current-law equation for the circuit of Figure 8-25. Using Equation (8-10), at the single independent node,

$$I_5 + I_1 + I_{10} = I_G + I_B$$

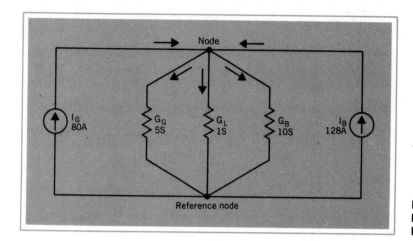

FIGURE 8-25
Equivalent Circuit for
Example 8-4C.

Or, we can go directly to the format of Equation (8-11) and substitute coefficients,

$$(G_5 + G_1 + G_{10})V = I_G + I_B$$

from which

$$V = \frac{80 + 128}{5 + 1 + 10} = \frac{208 \text{ A}}{16 \text{ S}} = 13 \text{ V}$$

and

$$I_L = VG_L = 13 \text{ V} \times 1 \text{ S} = \mathbf{13 \text{ A}}$$

Nodal analysis allowed us to solve Example 8-4C with not much more than mental arithmetic. This good fortune was due to a single independent node in the network. Hence, we did not need the simultaneous equations and determinants of the loop and mesh solutions. Unfortunately, many networks have more than one independent node. The transistor coupling circuit of Figures 8-12(b) and 8-16, and again in Figure 8-26, is such an example.

FIGURE 8-26
Circuit Diagram for Example 8-6B.

EXAMPLE 8-6B
If R_B in the circuit shown in Figure 8-26 is 1 kΩ, determine the magnitude and polarity of the voltage drop across R_B.

SOLUTION

Step I. In the circuit diagram of Figure 8-26, we can consider the 22-kΩ resistor to be the internal resistance of a 150-V constant-voltage source connected across the 68-kΩ resistor. Similarly, we can think of the 33-kΩ resistor as the internal resistance of a 100-V constant-voltage source connected across R_B. Therefore, when we convert the voltage sources to constant-current sources, we obtain the circuit of Figure 8-27 where

$$I_1 = \frac{150\ V}{22\ k\Omega} = 6.8\ mA \quad \text{and} \quad I_2 = \frac{100\ V}{33\ k\Omega} = 3.03\ mA$$

FIGURE 8-27
Equivalent Circuit for Example 8-6B.

We must take care in marking the directions of I_1 and I_2 to be consistent with the polarities of the voltage sources they replace.

Step II. Converting each of the five resistances into its equivalent conductance, we obtain the values shown (in microsiemens) in Figure 8-27.

Step III. The whole bottom conductor (chassis) in the circuit of Figure 8-27 becomes the reference node. This leaves two independent nodes—node 1 and node 2 in Figure 8-27. Each of these nodes has one source current "entering" the node and three resistor currents "leaving" the node.[†] To solve this network, we require a nodal equation for each of the two independent nodes. Using the format of Equation (8-11), we note that the source current I_1 will be a negative quantity, since it *leaves*, rather than enters, node 1.[†]

[†] At node 1, all currents turn out to be negative quantities since the source current actually *leaves* the node and resistor currents *enter* the node. If we solve for V_1, node 1 turns out to be negative with respect to chassis.

For node 1,

$$(45.4 + 14.7 + 17.9)V_1 - 17.9V_2 = -6800$$

For node 2,

$$(17.9 + 30.3 + 1000)V_2 - 17.9V_1 = 3030^\dagger$$

Collecting terms in matrix form,

$$78V_1 - 17.9V_2 = -6800 \tag{1}$$

$$-17.9V_1 + 1048.2V_2 = 3030 \tag{2}$$

Step IV. To find the voltage drop across R_B, we need solve only for V_2.

$$V_2 = \frac{\begin{vmatrix} 78 & -6800 \\ -17.9 & 3030 \end{vmatrix}}{\begin{vmatrix} 78 & -17.9 \\ -17.9 & 1048 \end{vmatrix}} = \frac{237\,340 - 121\,720}{81\,744 - 320} = \frac{115\,620}{81\,424} = \mathbf{1.42\ V}$$

Hence, the voltage drop across R_B is 1.4 V and since it turned out to be a positive quantity, node 2 is positive with respect to chassis.

Although we had to solve simultaneous equations in Example 8-6B, they were only second-order equations, rather than the third-order simultaneous equations we encountered in the loop and mesh solutions.

In the Wheatstone bridge circuit of Figure 8-28(a), there are four nodes. Labeling one as a reference node, we would still need three nodal equations to solve the Wheatstone bridge by nodal analysis. Since this is no improvement on the three simultaneous equations required in the loop or mesh procedures, we would probably not select nodal analysis in this case. If we use the loop or mesh format to solve the circuit in Figure 8-28(b), we require *four* simultaneous equations. However, as shown in Figure 8-28(b), we would require only *one* nodal equation to solve the whole network by the nodal analysis procedure just outlined.

SELF-CHECK

8-10:1. For the circuit shown in Figure 8-29:
      (a) Convert Source A to a constant-current source.

---

† Since microamperes = volts × microsiemens, we can avoid repetitive use of the exponent key of our calculator by expressing current in microamperes and conductance in microsiemens in this equation. We then enter the digits without the customary exponent for the prefix *micro.*

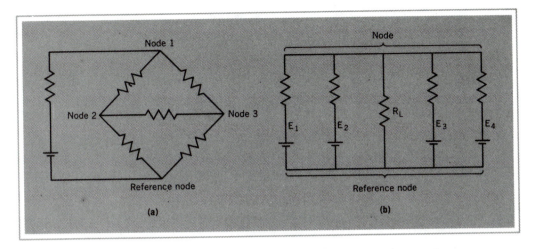

FIGURE 8-28
(a) Wheatstone Bridge; (b) Four Parallel Sources Supplying
a Common Load.

FIGURE 8-29

(b) Use the standard format to write the nodal equations for Node 1 and Node 2.

(c) Solve for V_1 and V_2.

8-11 THE SUPERPOSITION THEOREM

A characteristic of the Kirchhoff's-laws methods for solving resistance networks is that we have to solve a set of simultaneous linear equations. As long as we can solve networks with only two tracing loops or nodes, the numerical computation is not particularly onerous. But when we need three or more simultaneous equations, we start looking for alternative solutions. We can avoid simultaneous equations in multisource networks by applying a fundamental principle of linear networks known as the **superposition theorem.** This theorem is useful only for networks containing more than one source and for determining the current through or voltage drop across *one* branch of these networks. However, in many cases, once we know the current through one branch of a network, we can solve the remainder of the network quite readily. We shall not worry about the proof of the superposition theorem, which may be stated thus:

The current that flows in any branch of a network of resistors resulting from the simultaneous application of a number of voltage sources distributed in any manner throughout the network is the algebraic sum of the component currents in that branch that would be caused by each source acting independently in turn while the others are replaced in the network by their respective internal resistances.

In both the loop-current and nodal methods of solving resistance networks, the currents appearing in the simultaneous equations are not necessarily *actual* currents that we could measure at various points in the network. To obtain *actual* branch currents, we have to add the appropriate loop or mesh currents algebraically. Similarly, currents in a network with some of the sources inactive are not real currents. Component currents calculated by the superposition theorem are only a means of calculating the actual current in a certain branch and do not represent currents that we can measure in a network. In using the superposition theorem, we must take care to use the component currents only for the purpose stated in the theorem.

EXAMPLE 8-4D

An automobile generator with an internal resistance of 0.2 Ω develops an open-circuit voltage of 16.0 V. The storage battery has an internal resistance of 0.1 Ω and an open-circuit voltage of 12.8 V. Both sources are connected in parallel to a 1.0-Ω load. Determine the load current.

SOLUTION

Because there are *two* sources, there will be *two* component currents through the load resistance. For the purpose of solving this example by the superposition theorem, we can redraw the circuit of Figures 8-12(a) and 8-13 as the two equivalent circuits of Figure 8-30.

FIGURE 8-30
Equivalent Circuit for Example 8-4D.

Solving the first equivalent circuit of Figure 8-30(a) by the series-parallel resistance technique,

$$R_T = R_G + \frac{R_L \times R_B}{R_L + R_B} = 0.2 + \frac{1 \times 0.1}{1 + 0.1} = 0.29 \ \Omega$$

$$I_T = \frac{E_G}{R_T} = \frac{16.0 \ V}{0.29 \ \Omega} = 55 \ A$$

From the current divider principle, the first component of the load current is

$$I_{L_1} = I_T \times \frac{R_B}{R_L + R_B} = 55 \times \frac{0.1}{1.1} = 5 \ A$$

Similarly, for the circuit of Figure 8-30(b),

$$R_T = R_B + \frac{R_G \times R_L}{R_G + R_L} = 0.1 + \frac{0.2 \times 1}{0.2 + 1} = 0.267 \ \Omega$$

$$I_T = \frac{E_B}{R_T} = \frac{12.8 \ V}{0.267 \ \Omega} = 48 \ A$$

The second component of the load current is

$$I_{L_2} = I_T \times \frac{R_G}{R_G + R_L} = 48 \times \frac{0.2}{1.2} = 8 \ A$$

Since the direction for the component load currents is the *same* in Figures 8-30(a) and (b), the *actual* load current is the *sum* of the component currents.

$$\therefore I_L = I_{L_1} + I_{L_2} = 5 + 8 = \mathbf{13 \ A}$$

EXAMPLE 8-6C
If R_B in the circuit shown in Figure 8-12(b) is 1 kΩ, determine the magnitude and polarity of the voltage drop across R_B.

SOLUTION
Again there are two sources, hence the superposition theorem will yield two component currents. In the equivalent circuit of Figure 8-31(a), the 100-V source has been replaced by its zero-ohm internal resistance so that only the 150-V source contributes to the current through R_B. Again we can solve the circuit as a series-parallel network. However, in this case, we solve for R_T in two steps. First we replace the 56-kΩ and 33-kΩ resistors and R_B with their equivalent resistance, R_{eq} in Figure 8-31(b). For the equivalent circuits of Figures 8-31(a) and (b),

$$R_{eq} = 56 \ k\Omega + \frac{33 \ k\Omega \times 1 \ k\Omega}{33 \ k\Omega + 1 \ k\Omega} = 57 \ k\Omega$$

FIGURE 8-31
Equivalent Circuit for Example 8-6C.

$$R_T = 22\,k\Omega + \frac{68\,k\Omega \times 57\,k\Omega}{68\,k\Omega + 57\,k\Omega} = 53\,k\Omega$$

$$I_T = \frac{E}{R_T} = \frac{150\ V}{53\ k\Omega} = 2.83\ mA$$

We must now use the current-divider principle twice to find the component current through R_B. First, in the equivalent circuit of Figure 8-31(b), the current through R_{eq} is

$$I_{eq} = 2.83\ mA \times \frac{68\ k\Omega}{68\ k\Omega + 57\ k\Omega} = 1.54\ mA$$

In the circuit of Figure 8-31(a),

$$I_{B_1} = 1.54\ mA \times \frac{33\ k\Omega}{33\ k\Omega + 1\ k\Omega} = 1.49\ mA$$

Similarly, when the 100-V source operates alone, as in Figure 8-31(c), we start by replacing the 56-kΩ, 22-kΩ, and 68-kΩ resistors with an equivalent resistance. Solving for the second component current,

$$R_{eq} = 56\ k\Omega + \frac{22\ k\Omega \times 68\ k\Omega}{22\ k\Omega + 68\ k\Omega} = 72.6\ k\Omega$$

$$R_T = 33 \text{ k}\Omega + \frac{72.6 \text{ k}\Omega \times 1 \text{ k}\Omega}{72.6 \text{ k}\Omega + 1 \text{ k}\Omega} = 34 \text{ k}\Omega$$

$$I_T = \frac{E}{R_T} = \frac{100 \text{ V}}{34 \text{ k}\Omega} = 2.94 \text{ mA}$$

In Figure 8-31(d), we need only apply the current-divider principle once to find the second component of the current through R_B.

$$I_{B_2} = 2.94 \text{ mA} \times \frac{72.6 \text{ k}\Omega}{72.6 \text{ k}\Omega + 1 \text{ k}\Omega} = 2.90 \text{ mA}$$

We note from Figures 8-31(a) and (c) that the two component currents flow in *opposite* directions through R_B. Hence, the real current through R_B is

$$I_B = 2.90 - 1.49 = 1.41 \text{ mA}$$

and
$$V_B = 1.4 \text{ mA} \times 1 \text{ k}\Omega = \textbf{1.4 V}$$

Because the component current I_{B_2} is greater than I_{B_1}, the real current has the same direction through R_B as I_{B_2} in Figure 8-31(c). Thus, the top end of R_B is 1.4 V positive with respect to chassis.

Although we avoided simultaneous equations in Example 8-6C, we probably had to do just as much numerical computation to arrive at a solution as we did in Example 8-6, which we solved by loop equations.

SELF-CHECK

8-11:1. For the circuit shown in Figure 8-32:
  (a) Use the superposition theorem to redraw the circuit as two equivalent circuits.
  (b) Solve for the currents in each of the equivalent-circuit pairs.
  (c) Superpose these currents to find the true current in the 3.0-kΩ resistor.

FIGURE 8-32

PROBLEMS

NOTE: Problems in which the section number is replaced by a ? may be solved by more than one method. For practice in circuit analysis techniques, solve problems using one method and check the solution with a different method.

8-2-1. As the current drawn from a dc power supply is increased from 600 mA to 800 mA, the terminal voltage decreases from 24 V to 23 V. Determine the equivalent constant voltage source.

8-3-2. A voltage source has an open-circuit terminal voltage of 120 V and an internal resistance of 0.6 Ω. Determine the equivalent constant-current source.

8-3-3. Convert the voltage sources of Figure 8-33 into equivalent constant-current sources.

8-3-4. Replace the current sources shown in Figure 8-34 with equivalent constant-voltage sources.

8-?-5. Determine the magnitude and polarity of the voltage drop across the 50-Ω resistor in the network shown in Figure 8-35(a).

(a) (b) (c)

FIGURE 8-33

(a) (b) (c)

FIGURE 8-34

(a) (b)

FIGURE 8-35

8-?-6. Find the current drawn from each of the sources in the network shown in Figure 8-35(b).

8-?-7. As the car slows down, the open-circuit terminal voltage of the generator in Example 8-4 drops to 14.0 V. What is the battery current?

8-?-8. (a) What must the open-circuit voltage of the generator be in Example 8-4 for the battery current to be zero?
(b) What open-circuit voltage must the generator develop in Example 8-4 if the battery is to charge at a steady 10 A?

8-?-9. What will the neutral current be in the circuit of Example 8-5 if a second 15-Ω load is connected in parallel with the existing 15-Ω load?

8-?-10. Calculate the total power loss in the three 1-Ω conductors in the three-wire distribution system shown in Figure 8-15 if the polarity of source E_2 is reversed.

8-?-11. Two batteries are connected in parallel to feed a 12-Ω load. Battery A has an open-circuit voltage of 6.3 V and an internal resistance of 1.5 Ω. Battery B has an open-circuit voltage of 6.0 V and an internal resistance of 2.1 Ω. Find the current drain from each battery.

8-?-12. If the 12-Ω load in Problem 8-11 is replaced by a battery charger with an open-circuit voltage of 7.2 V and an internal resistance of 0.2 Ω, find the charging current of each battery.

8-?-13. What is the current through each voltage source in the circuit of Figure 8-36?

8-?-14. What is the magnitude and direction of the current through the 2.2-kΩ resistor in the circuit of Figure 8-36?

8-?-15. Determine the current through each of the five resistors in the circuit of Figure 8-37(a).

FIGURE 8-36

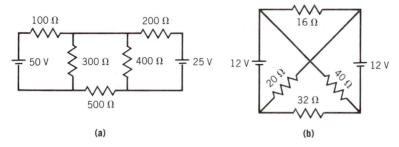

(a)   (b)

FIGURE 8-37

8-?-16. In the network shown in Figure 8-37(b), there is *no* electrical connection where the leads of the 20-Ω and 40-Ω resistors cross. Nevertheless, redrawing the network shows that it is a *planar* network. Determine the voltage drop across the 40-Ω resistor.

8-?-17. Determine the current through the 400-Ω resistor in the network of Figure 8-38(a).

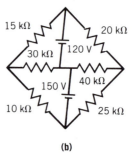

(a)

(b)

FIGURE 8-38

8-?-18. Replace the 6-V source in the network of Figure 8-38(a) with an ammeter having a resistance of 0 ohms, and place the 6-V source in series with the 400-Ω resistor. What is the ammeter reading? Compare this with the current through the 400-Ω resistor in Problem 8-17. This is an example of the **reciprocity theorem**, which states that in any linear network with a single voltage source, when a voltage source in branch *A* is moved to branch *B*, then the current that was in branch *B* now appears in branch *A*.

8-8-19. Use loop or mesh equations to determine all eight branch currents in the network of Figure 8-38(b).

8-11-20. Use the superposition theorem to determine the voltage drop across the 40-kΩ resistor in Figure 8-38(b).

8-?-21. Find the current drawn from each of the sources in the network shown in Figure 8-35(b) when a 15-kΩ resistor is connected across the top of the network (from the + terminal of the 100-V source to the + terminal of the 200-V source).

8-?-22. The 56-kΩ resistor in the circuit of Figure 8-12(b) is replaced with a variable resistor which is adjusted until the voltage across R_B is exactly zero. What is the resistance of the variable resistor under these circumstances?

8-?-23. In the balance-detecting circuit of Figure 8-39, the voltmeter has a resistance of 100 kΩ. What is its reading?

8-?-24. The current through the unknown resistor in the circuit shown in Figure 8-40 is 0.5 A. What is its resistance?

FIGURE 8-39   **FIGURE 8-40**

8-?-25. What is the current through a 500-Ω load connected to the output terminals of the bridged-T attenuator of Figure 8-41 when the input voltage is 2 V?

8-?-26. If 360 V dc is applied to the input terminals of the lattice network of Figure 8-42, determine the current through a 48-Ω resistor connected to the output terminals.

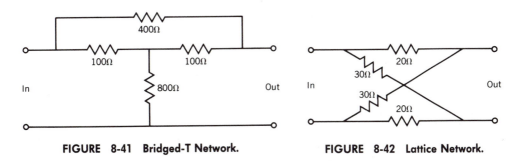

FIGURE 8-41 Bridged-T Network. FIGURE 8-42 Lattice Network.

8-10-27. Determine the voltage drop across each of the three resistors in the network shown in Figure 8-43(a).

8-10-28. Determine the voltage drop across the 300-Ω resistor in the network shown in Figure 8-43(b).

8-?-29. In the network shown in Figure 8-44:
(a) What is the voltage drop across the 2.2-kΩ resistor?
(b) What is the voltage drop across the 4.7-kΩ resistor?

8-11-30. In the network shown in Figure 8-44, determine the voltage drop across the 10-kΩ resistor by the superposition theorem.

(a) (b)

FIGURE 8-43

FIGURE 8-44

(a)   (b)

FIGURE   8-45

8-?-31. In the network shown in Figure 8-45(a), determine the current drawn from each of the voltage sources.

8-?-32. Determine the voltage drop across the 50-Ω resistor in the network shown in Figure 8-45(b).

8-?-33. A pulse generator circuit acts as a switch in series with a 10-kΩ resistor, as shown in Figure 8-46. Calculate the output voltage of the interstage coupling network across a 470-kΩ resistor connected from the output terminal to ground
   (a) With the switch open.
   (b) With the switch closed.

8-?-34. The circuit shown in Figure 8-46 is modified to form the circuit shown in Figure 8-47. Calculate the voltage drop across a 470-kΩ resistor connected from the output terminal to ground
   (a) With the switch open.
   (b) With the switch closed.

FIGURE   8-46

FIGURE   8-47

8-?-35. A trolleybus is halfway along a section of trolley wire which is fed at one end from a 600-V source and at the other end from a 596-V source. The internal resistances of the sources are negligible, but each trolley wire in the section has a resistance of 1.0 Ω. This trolleybus is the only one in the section at the moment and it draws a 50-A current from the trolley wires. What is the voltage drop across the trolleybus?

REVIEW QUESTIONS

8-1. Define a **constant-voltage source.**

8-2. Why is the terminal voltage of a constant-voltage source not necessarily constant as the load resistance is varied?

8-3. Describe a laboratory procedure for determining the internal resistance of an automobile storage battery.

8-4. What is the significance of the term **short-circuit current?**

8-5. How would you determine the short-circuit current of an automobile battery?

8-6. What is the significance of the term **open-circuit voltage?**

8-7. Define a **constant-current source.**

8-8. Why is the current drawn from a constant-current source not necessarily constant as the load resistance is varied?

8-9. What happens to the terminal voltage of a constant-current source when a load resistor is connected to it? Explain.

8-10. Compare equivalent constant-voltage and constant-current sources in terms of their open-circuit voltage and short-circuit current.

8-11. Describe the *two* methods for determining the internal resistance of constant-voltage and constant-current sources.

8-12. Why can we not replace the source shown in Figure 8-1 with an equivalent constant-current source?

8-13. Show that Equations (8-5) and (8-6) are numerically equal and, therefore, represent two different ways of stating Kirchhoff's voltage law.

8-14. Explain the procedure for determining the polarity of voltage drops around a current tracing loop, using Figure 8-7 as an example.

8-15. How does this procedure apply when determining the voltage drop across R in the circuit of Figure 8-40?

8-16. If, in solving Example 8-3, we had chosen the I_3 tracing loop (Figure 8-9) from the + terminal of the source through R_2, R_M, and R_3 back to the − terminal of the source, the solution for I_3 would have been −0.15 A. What is the significance of a negative result in solving loop equations?

8-17. Is it likely that I_1 or I_2 in Figure 8-9 would ever work out to be negative quantities? Explain.

8-18. Given the following data for the Wheatstone bridge of Figure 8-9, $R_1 = 40$ Ω, $R_2 = 10$ Ω, $R_3 = R_4 = 20$ Ω; what path would you select for the I_3 tracing loop in seeking to avoid a negative solution for I_3?

8-19. How do you write loop equations for the circuit of Figure 8-12(b) when no complete loops appear in the circuit diagram?

8-20. Loop equations are based on Kirchhoff's voltage law but the unknown quantities in the loop equations are *currents*. Explain the presence of current terms in voltage-law equations.

8-21. In the second solution to Example 8-4, how do we know that the battery current should be $I_L - I_G$ rather than $I_G - I_L$?

8-22. In distributing loads on a three-wire residential service, why is it desirable to have the two values of R_L in Figure 8-15 as nearly equal as possible?

8-23. Although the circuit diagram of Figure 8-37(b) shows conductors crossing with *no* electrical connection, the circuit is actually a *planar* circuit. Explain.

8-24. Why are we more likely to encounter negative current terms with the mesh format for writing Kirchhoff's voltage-law equations than with the loop procedure?

8-25. Many of the currents calculated with either the loop or mesh formats in a pencil-and-paper network cannot be measured in the real network. Explain.

8-26. How do we determine whether the source voltage in a mesh equation is a positive or negative quantity?

8-27. How do we determine the polarity of the voltage drop across a resistor in a network where the actual current is made up of two different pencil-and-paper mesh currents?

8-28. In writing loop equations, we insert branch currents in the basic Kirchhoff's voltage-law equation before we collect terms in the form of loop currents. With the mesh format, we skip the in-between step and write the term directly in terms of mesh currents. What features of the mesh format permit us to take this short cut?

8-29. Node-voltage equations are based on Kirchhoff's *current* law, but the unknown quantities in the equations are *voltages*. Explain the presence of voltage terms in current-law equations.

8-30. When we write node-voltage equations, we require one less equation than the total number of nodes. Explain.

8-31. What do we mean when we refer to "the voltage *at* node X"?

8-32. What factors would lead us to select nodal analysis in preference to loop or mesh analysis in solving a given network?

8-33. Under what circumstances are loop equations preferable to nodal analysis?

8-34. What advantage does the superposition theorem have over Kirchhoff's-law equations in network solutions?

8-35. What factors tend to limit the application of the superposition theorem in network analysis?

8-36. Are we likely to encounter negative solutions for component currents when applying the superposition theorem to a network problem? Explain.

8-37. Is it possible that we shall have to *subtract* component currents when finding the real branch current in a network by the superposition theorem? Explain.

8-38. To use mesh equations, we convert any current sources to equivalent constant-voltage sources. To use nodal analysis, we convert any voltage sources to equivalent constant-current sources. Do we have to make such conversions when using the superposition theorem?

8-39. A certain circuit-analysis computer program solves networks in terms of voltages between up to seven independent nodes and a reference node (node 0). The program asks the operator for the network configuration in terms of the names of the components connected between specific nodes. It assumes that the "TO" node is positive with respect to the "FROM" node and that component values are entered in basic units. The computer display informs us that its memory already contains the following network:

```
REF     NAME        FROM    TO    VALUE
1       BATTERY      0       1      24
2       BATTERY      0       4      20
3       RESISTOR     2       1      50
4       RESISTOR     0       2     100
5       RESISTOR     3       2     200
6       RESISTOR     0       3     100
7       RESISTOR     3       4     300
8       RESISTOR     4       1     500
```

Draw and label the circuit diagram for this network.

*8-40. Prepare a computer program that you can store for future use in solving third-order determinants.

*8-41. Write a BASIC program that will print out the magnitude and direction of the current through the milliammeter in the Wheatstone bridge circuit of Figure 8-9 when the operator enters values for source voltage and the five resistances (including R_M).

9

EQUIVALENT-CIRCUIT THEOREMS

The advantages of the classical Kirchhoff's-law methods for solving resistance networks in Chapter 8 are (1) we do not have to tamper with the original configuration of the network, and (2) we can solve almost any network ranging from simple to very elaborate. The disadvantages are the large number of tedious computations and the many possibilities for writing an equation incorrectly or for making a numerical error. We learn these methods because they form the basis for computer circuit-analysis software packages. The universal application of Kirchhoff's laws allows the software to handle many different network configurations. And (after the original program debugging) computer programs can handle a very large number of computations rapidly and without error. Only the final solution is displayed by the computer.

In this chapter we shall investigate more contemporary methods used to simplify network calculations. Since these techniques involve substituting equivalent circuits for portions of the original network, they are not particularly adaptable to computer software packages. However, since they do involve thinking about how electric circuits behave, they are particularly suited to an introductory electric circuits course where our objective is to *understand* just how electric circuits must behave. With equivalent-circuit techniques, there are many ways of solving a particular network. Hence, an electronic calculator is better suited to most of the examples in this chapter than a personal computer.

9-1 EQUIVALENT CIRCUITS

One of the most useful circuit-analysis techniques is to simplify a circuit diagram by replacing several components with a simpler configuration which performs identically as far as the remainder of the circuit is concerned. We encountered this **equivalent-circuit** technique in Section 7-1, when we replaced series and parallel resistors with a single equivalent resistance, as shown in Figure 9-1. As far as the source is concerned, it can "see" no difference among the three circuits of Figure 9-1.

FIGURE 9-1
Equivalent Circuit for a Series-Parallel Circuit.

Figure 9-2(a) shows the complete circuit diagram for one stage of a transistor audio amplifier. Actually, this diagram contains two distinct circuits. A transistor requires certain dc potentials at its base, emitter, and collector to perform its function as an amplifying device. As far as direct current is concerned, the three capacitors are open circuits. Hence, the dc equivalent circuit becomes simply a voltage-divider system providing the necessary dc potentials, as shown in Figure 9-2(b).

Since the terminal voltage of the 16-volt dc power supply cannot vary at an audio frequency, the −16-volt dc terminal in the circuit of Figure 9-2(a) appears to be grounded as far as the audio signal is concerned. If the three capacitors have been selected properly, at a midrange audio frequency of 1 kilohertz they appear as short circuits by comparison with the resistances in the circuit. Consequently, for midrange audio frequencies, we can simplify the transistor amplifier circuit to the ac equivalent circuit shown in Figure 9-2(c).

The next step in the analysis of the performance of this circuit is to replace the transistor by an equivalent circuit composed of basic electrical properties. This is a chore which we must leave for a separate study of electronic circuits.[†]

9-2 THÉVENIN'S THEOREM

In Section 8-2, we adapted a laboratory "black-box" strategy to enable us to replace any practical voltage source in a network with a simple constant-voltage source (or constant-current source) which is an exact equivalent as far as the network is concerned. In this section, we shall expand this principle into one of the most useful of the circuit simplification theorems.

[†] We shall examine the equivalent circuit for a transistor briefly in Section 25-4.

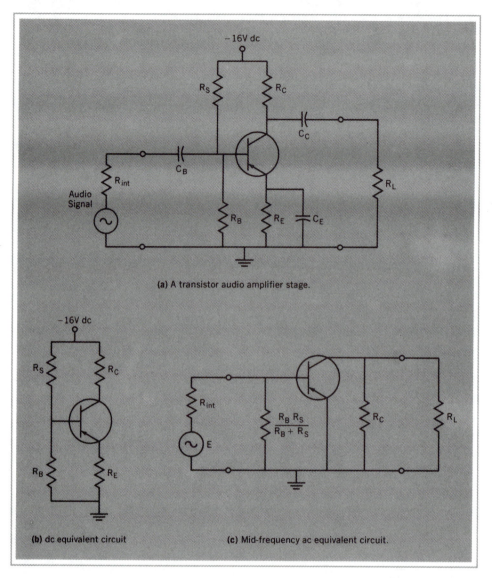

(a) A transistor audio amplifier stage.

(b) dc equivalent circuit

(c) Mid-frequency ac equivalent circuit.

FIGURE 9-2
Equivalent Circuits for a Transistor Audio Amplifier.

Suppose that the electric circuit of Figure 9-3(a) is sealed into a "black box" (represented by the shaded area) so that only the two terminals *A* and *B* are exposed. We can attempt to establish the nature of the contents of the box by first connecting a sensitive voltmeter across terminals *A* and *B*. The voltmeter will tell us that the box contains a voltage source, and, since the voltmeter draws negligible current, it shows the **open-circuit terminal voltage** of the source. As far as we can tell without opening the box, it contains an 80-volt source, as shown in the equivalent constant-voltage source of Figure 9-3(b).

FIGURE 9-3
(a) Original Voltage Source;
(b) Thévenin-Equivalent Source.

Since we do not have to worry about blowing a fuse in our pencil-and-paper analysis, we can short-circuit terminals *A* and *B* through an ammeter having zero resistance. For the black box of Figure 9-3(a), we find that the short-circuit current is 2 milliamperes. Consequently, without any knowledge of the exact contents, the box appears to contain an 80-volt source which will permit a 2-milliampere short-circuit current. Hence, the source appears to have an internal resistance of

$$R_x = \frac{E_{oc}}{I_{sc}} = \frac{80 \text{ V}}{2 \text{ mA}} = 40 \text{ k}\Omega$$

as shown in Figure 9-3(b). We can check the equivalency by opening the box and solving the network it actually contains in Figure 9-3(a).

No matter how elaborate the network inside the sealed box may be, as long as it contains one or more voltage sources, all that we can determine about it without opening the box is that it is equivalent to a simple constant-voltage source with a single internal resistance in series with it. Any load connected to terminals *A* and *B* of the equivalent circuit of Figure 9-3(b) will draw exactly the same current with exactly the same voltage drop as it would if it were connected to terminals *A* and *B* of the original network of Figure 9-3(a).

Rather than leaving this demonstration as an interesting fundamental principle of electric circuits, Léon Charles Thévenin stated the principle in the form of a theorem that has become one of the most useful methods of simplifying electrical networks to assist in their pencil-and-paper analysis. We may state Thévenin's theorem as follows:

Any two-terminal network of fixed resistances and voltage sources may be replaced by a single voltage source having an equivalent voltage equal to the open-circuit voltage at the terminals of the original network and having an internal resistance equal to the resistance looking back into the network from the two terminals with all the voltage sources replaced by their internal resistances.

We can apply Thévenin's theorem to any of the resistance networks we have solved so far by treating one branch of the network as a *load* and the remainder of the network as a *two-terminal network containing one or more voltage sources*. Having decided which branch of the original network we are going to treat as a load, we *remove* it from the original network and place it in a Thévenin-equivalent circuit and apply the rules set forth in the theorem to determine the remainder of the equivalent circuit. Note that Thévenin uses an "ohmmeter" approach to determine the internal resistance "looking back into" the open-circuit terminals of the source network, rather than the short-circuit current method we used with our black-box technique. Either line of thought leads to the same answer for R_x. To illustrate the Thévenin method, we shall follow his instructions to determine the Thévenin-equivalent circuit for Figure 9-3(a).

EXAMPLE 9-1
What current will a 10-kΩ resistor draw when it is connected to a 100-V source through the "T" network shown in Figure 9-3(a)?

SOLUTION
Step I. Rather than connect the 10-kΩ resistor to the original circuit of Figure 9-3(a), we connect it to a Thévenin-equivalent source consisting of a constant-voltage source E_x with a series internal resistance R_x, as in Figure 9-4(b).

FIGURE 9-4
Equivalent Circuit for Step II of Example 9-1.

According to Thévenin's theorem, E_x of the equivalent circuit will be the open-circuit voltage between terminals A and B of Figure 9-3(a). Since an open circuit draws no current, there will be no voltage drop across the 20-kΩ resistor. Hence, the open-circuit terminal voltage will be the same as the voltage drop across the 100-kΩ resistor. Because terminal A is open-circuit, the 25-kΩ and the 100-kΩ resistors form a simple series circuit. Thus, the voltage-divider principle of Section 7-4 and a bit of mental arithmetic will provide us with the voltage drop across the 100-kΩ resistor, which is the open-circuit voltage between terminals A and B. This voltage, in turn, is E_x in the Thévenin-equivalent circuit.

$$E_x = \frac{100}{25 + 100} \times 100 \text{ V} = 80 \text{ V}$$

Step II. The next step in following the instructions stated in Thévenin's theorem is to replace the actual source in Figure 9-3(a) by its own internal resistance, which,

214

in this case, is zero ohms. After we do this, the original circuit of Figure 9-3(a) becomes the series-parallel resistance network of Figure 9-4(a). Using techniques shown in Figure 9-1, the internal resistance "looking back into" terminals A and B is

$$R_x = 20 \text{ k}\Omega + \frac{25 \text{ k}\Omega \times 100 \text{ k}\Omega}{25 \text{ k}\Omega + 100 \text{ k}\Omega}$$

$$= 20 \text{ k}\Omega + 20 \text{ k}\Omega = 40 \text{ k}\Omega$$

Step III. To complete the solution, we simply solve for the current through the 10-kΩ resistor in the Thévenin-equivalent circuit of Figure 9-4(b).

$$I = \frac{E_x}{R_T} = \frac{80 \text{ V}}{40 \text{ k}\Omega + 10\text{k}\Omega} = \textbf{1.6 mA}$$

To illustrate further the procedure of Thévenin's theorem and, at the same time, to assess its usefulness, we can consider Example 7-3 which we solved originally by using Kirchhoff's-law equations.[†]

EXAMPLE 9-2
A resistor passing a 20-mA current is in parallel with a 5-kΩ resistor. This combination is in series with another 5-kΩ resistor, the whole network being connected to a 500-V source. What is the resistance of the resistor which is passing the 20-mA current?

SOLUTION
Step I. Having first drawn a circuit diagram for the circuit described in the example, we select the unknown resistance passing the 20-mA current as the load and *remove* it from the original circuit, as in Figure 9-5(a), and place it in the Thévenin-equivalent circuit of Figure 9-5(b).

FIGURE 9-5
Circuit Diagram for Step I of Example 9-2.

[†] See page 141.

Using the voltage-divider principle again, we can now state the open-circuit terminal voltage of the remaining circuit in Figure 9-5(a) by simple mental arithmetic. Hence, E_x for the Thévenin-equivalent circuit of Figure 9-5(b) is

$$E_x = \frac{5}{5 + 5} \times 500 \text{ V} = 250 \text{ V}$$

Step II. If we short out the original voltage source (and replace it with its internal resistance of zero ohms), the circuit of Figure 9-5(a) becomes the parallel-resistance network of Figure 9-6(a). This network is simple enough that we can again solve for the internal resistance "looking back into" terminals A and B by mental arithmetic. Thus,

$$R_x = \frac{5 \text{ k}\Omega \times 5 \text{ k}\Omega}{5 \text{ k}\Omega + 5 \text{ k}\Omega} = 2.5 \text{ k}\Omega$$

Step III. We can now solve the Thévenin-equivalent circuit of Figure 9-6(b) quite readily by Ohm's law.

$$R_T = \frac{E_x}{I} = \frac{250 \text{ V}}{20 \text{ mA}} = 12.5 \text{ k}\Omega$$

$$\therefore R_L = R_T - R_x = 12.5 \text{ k}\Omega - 2.5 \text{ k}\Omega = \textbf{10 k}\Omega$$

FIGURE 9-6
Equivalent Circuit for Step II of Example 9-2.

Thévenin's theorem allows us to solve a Wheatstone bridge network without resorting to the third-order simultaneous equations we required with the Kirchhoff's-laws solution of Chapter 8.

EXAMPLE 9-3
In the Wheatstone bridge circuit of Figure 8-1, R_1 and R_4 have a resistance of 300 Ω each. R_2 and R_3 each have a resistance of 150 Ω. R_5 is a galvanometer whose resistance is 50 Ω, and the source voltage is 100 V. What is the current in R_5?

FIGURE 9-7
Solving Wheatstone Bridge by Thévenin's Theorem.

SOLUTION
   Step I. Since we are asked to find the current through only *one* branch of the Wheatstone bridge, Thévenin's theorem allows us to *remove* R_5 (the meter) from the original circuit, as in Figure 9-7(a), to the Thévenin-equivalent circuit, as in Figure 9-7(b).

   Step II. In the remaining circuit of Figure 9-7(a), from the voltage-divider principle,

$$V_3 = 100 \text{ V} \times \frac{150}{450} = 33.33 \text{ V}$$

and

$$V_4 = 100 \text{ V} \times \frac{300}{450} = 66.67 \text{ V}$$

From Kirchhoff's voltage law,

$$V_{AB} = V_3 - V_4 = 33.33 - 66.67 = -33.34 \text{ V}$$

(with terminal *A* *negative* with respect to terminal *B*). The open-circuit voltage V_{AB} becomes E_x of the Thévenin-equivalent circuit of Figure 9-7(b).

   Step III. Assuming that the source has negligible internal resistance, we can short-circuit the generator symbol in Figure 9-7(a) to determine the internal resistance between terminals *A* and *B*. This becomes R_x for the Thévenin-equivalent circuit of Figure 9-7(b). With the generator short-circuited, R_1 is in parallel with R_3, and R_2 is in parallel with R_4. These two parallel circuits are then in series as far as terminals *A* and *B* are concerned. Therefore,

$$R_x = \frac{R_1 R_3}{R_1 + R_3} + \frac{R_2 R_4}{R_2 + R_4}$$

$$= \frac{300 \times 150}{300 + 150} + \frac{150 \times 300}{150 + 300} = 200 \ \Omega$$

Step IV. In the Thévenin-equivalent circuit, the meter current then becomes

$$I_M = \frac{E_x}{R_x + R_M} = \frac{33.34 \text{ V}}{250 \text{ } \Omega} = \mathbf{133 \text{ mA}}$$

We can determine the current through any other one branch of the bridge circuit in exactly the same manner.

Thévenin's theorem applies to networks having more than one voltage source. Its usefulness depends on whether we can determine E_x without resorting to simultaneous equations. With only two sources, this is usually not a problem, as we can show with the network of Figure 9-8, which is similar to the automobile generator example of Chapter 8.

FIGURE 9-8
Network with Two Voltage Sources.

EXAMPLE 9-4
Determine the current in the 4.8-Ω resistor in the network shown in Figure 9-8.

SOLUTION
Step I. Remove the 4.8-Ω resistor from the original circuit of Figure 9-8 and place it in the Thévenin-equivalent circuit of Figure 9-9(b).

Step II. According to Kirchhoff's voltage law, the potential difference between

FIGURE 9-9
Circuit Diagram for Example 9-4.

points C and D in Figure 9-9(a) must be the *same* via both paths. One path is the 8-V source. From inspection of the other path, it follows that there must be a 2-V drop across the 6-Ω and 4-Ω resistors in series. From the voltage-divider principle, these voltage drops become 1.2 V and 0.8 V, respectively, as shown in Figure 9-9(a).

When we trace the potential difference between points A and B by either path, the open-circuit voltage E_x becomes 7.2 V.

Step III. Replacing the sources by their internal resistances (zero ohms) results in the 6-Ω and 4-Ω resistors appearing in *parallel* between points A and B. Looking back into the source, we then find that

$$R_x = \frac{6 \times 4}{6 + 4} = 2.4 \ \Omega$$

Step IV. From the Thévenin-equivalent circuit of Figure 9-9(b),

$$I_L = \frac{E_x}{R_x + R_L} = \frac{7.2 \ \text{V}}{(2.4 + 4.8) \ \Omega} = 1 \ \text{A}$$

SELF-CHECK

9-2:1. Referring to the circuit shown in Example 7-9:
   (a) Determine the Thévenin-equivalent circuit as "seen" by the load resistor, R_L.
   (b) Use the Thévenin-equivalent circuit to solve for the current in R_L.
9-2:2. Referring to the circuit shown in Figure 9-10:
   (a) Determine the Thévenin-equivalent circuit as "seen" by the load, R_L.
   (b) Use the Thévenin-equivalent circuit to solve for the voltage drop across the load, R_L.

FIGURE 9-10

9-3 NORTON'S THEOREM

Figure 9-3 shows how we can reduce a network containing one or more voltage sources into a simple Thévenin-equivalent constant-voltage source. In Chapter 8, we found that we can readily convert a constant-voltage source into an exact equivalent constant-current source. Therefore, it follows that we should be able to reduce any network containing one or more voltage sources directly to a simple equivalent constant-current source, as shown in Figure 9-11. Norton's theorem combines Thévenin's theorem with constant-current source conversion into a single procedure. We may state Norton's theorem as follows:

Any two-terminal network of fixed resistances and voltage sources may be replaced by a single **constant-current** *source whose current is equal to the current drawn by a short circuit across the terminals of the original network and having in* **parallel** *with the constant-current source a resistance equal to the resistance looking back into the network from the two terminals.*

(a)

(b)

FIGURE 9-11
(a) Original Voltage Source;
(b) Equivalent Constant-Current Source.

EXAMPLE 9-5
Determine the Norton-equivalent constant-current source for the network of Figure 9-11(a).

SOLUTION 1
Step I. Following the instructions of Norton's theorem, we place a short-circuit across terminals A and B in Figure 9-11(a). The total resistance "seen" by the 100-V source (in kilohms) is

$$R_T = 25 + \frac{100 \times 20}{100 + 20} = 41.67 \text{ k}\Omega$$

The current drawn from the 100-V source is

$$I_T = \frac{E}{R_T} = \frac{100 \text{ V}}{41.67 \text{ k}\Omega} = 2.4 \text{ mA}$$

From the current-divider principle,

$$I_x = I_{sc} = \frac{100}{100 + 20} \times 2.4 \text{ mA} = \textbf{2.0 mA}$$

Step II. Using the "ohmmeter" method of determining internal resistance, as stated in Norton's theorem, we replace the 100-V source with 0 Ω internal resistance.

Then the resistance seen looking back into terminals A and B (in kilohms) is

$$R_x = 20 + \frac{25 \times 100}{25 + 100} = \mathbf{40\ k\Omega}$$

I_x and R_x are the required values for the Norton-equivalent constant-current source of Figure 9-11(b).

SOLUTION 2

Step I. We can avoid the current-divider calculations of Solution 1 by first finding the Thévenin-equivalent source of Figure 9-12(a). From Figure 9-11(a),

$$E_x = E_{oc} = \frac{100}{100 + 25} \times 100\ V = 80\ V$$

FIGURE 9-12
(a) Thévenin-Equivalent Constant-Voltage Source; (b) Norton-Equivalent Constant-Current Source.

Step II. We find R_x by the same "ohmmeter" technique as in Solution 1. $R_x = \mathbf{40\ k\Omega}$.

Step III. Using source conversion to obtain the Norton-equivalent constant-current source of Figure 9-12(b), with reference to the Thévenin-equivalent source of Figure 9-12(a),

$$I_x = I_{sc} = \frac{E_x}{R_x} = \frac{80\ V}{40\ k\Omega} = \mathbf{2\ mA}$$

EXAMPLE 9-4A

Determine the current in the 4.8-Ω resistor in the circuit shown in Figure 9-8.

SOLUTION

Step I. Remove the 4.8-Ω resistor from the original circuit and place it in the Norton-equivalent circuit of Figure 9-13(b).

(a)   (b)

FIGURE 9-13
Circuit Diagram for Step II of Example 9-4A.

Step II.   We replace the 4.8-Ω resistor in the original circuit by a short circuit, as shown in Figure 9-13(a). Drawing arrows for current directions shows that the short-circuit current is the *sum* (Kirchhoff's current law) of the currents resulting from placing the 6-Ω resistor directly across the 6-V source and placing the 4-Ω resistor directly across the 8-V source. Therefore, the short-circuit current, I_x in Figure 9-13(b) is

$$I_x = \frac{E_1}{R_1} + \frac{E_2}{R_2} = \frac{6}{6} + \frac{8}{4} = 3 \text{ A}$$

Step III.   Replacing the sources by their internal resistance (zero ohms) gives the circuit of Figure 9-14(a). Looking back into the source, we find that R_x becomes the parallel equivalent of the 6-Ω and 4-Ω resistors, Hence,

$$R_x = \frac{6 \times 4}{6 + 4} = 2.4 \text{ Ω}$$

(a)   (b)

FIGURE 9-14
Circuit Diagram for Step III of Example 9-4A.

Step IV. Using the current-divider principle in the Norton-equivalent circuit of Figure 9-14(b),

$$I_L = \frac{2.4}{2.4 + 4.8} \times 3 \text{ A} = 1 \text{ A}$$

SELF-CHECK

9-3:1. Referring to the circuit shown in Figure 9-10:
    (a) Determine the Norton-equivalent circuit "seen" by the load, R_L.
    (b) Use the Norton-equivalent circuit to solve for the current in the load, R_L.

9-4 THÉVENIN'S THEOREM IN MULTISOURCE NETWORKS (MILLMAN'S THEOREM)

The general statement of Thévenin's theorem indicates that a two-terminal network containing *any number* of voltage sources can be replaced by a single voltage source with a single series resistance. If the original network contains only a single voltage source, finding the Thévenin-equivalent source is quite simple. Often we can handle *two* sources with little difficulty when, as shown in Figure 9-15, we can reduce the original network to a single loop by removing a resistor to create the two terminals.

FIGURE 9-15
Circuit Diagram for Example 9-6.

EXAMPLE 9-6
An automobile generator with an internal resistance of 0.2 Ω develops an open-circuit voltage of 16.0 V. The storage battery has an internal resistance of 0.1 Ω and an open-circuit voltage of 12.8 V. Both sources are connected in parallel to a 1-Ω load. Determine the load current.

SOLUTION
    Step I. Removing the load resistance from the original circuit leaves the circuit

shown in Figure 9-15(a), which we can solve by one simple Kirchhoff's voltage-law equation. Because the two sources are in opposition as far as current around the single closed loop of Figure 9-15(a) is concerned, the current direction is established by the source with the greater terminal voltage. Therefore,

$$E_G - E_B = IR_G + IR_B$$

$$16.0 - 12.8 = 0.2I + 0.1I$$

from which

$$I = 10.67 \text{ A}$$

and

$$V_{RG} = IR_G = 10.67 \times 0.2 = 2.13 \text{ V}$$

$$V_{RB} = IR_B = 10.67 \times 0.1 = 1.07 \text{ V}$$

If we inspect Figure 9-15(a) and note the polarities of the voltage drops, we find that the total potential difference between terminals A and B is either

$$E_x = E_G - V_{RG} = 16.0 - 2.13 = 13.87 \text{ V}$$

or

$$E_x = E_B + V_{RB} = 12.8 + 1.07 = 13.87 \text{ V}$$

Step II. If we short-circuit the generator and battery symbols in Figure 9-15(a) so that only their internal resistances remain in the circuit, we find that R_G and R_B are connected in parallel between terminals A and B. Consequently, following the procedure set out in Thévenin's theorem,

$$R_x = \frac{R_G \times R_B}{R_G + R_B} = \frac{0.2 \times 0.1}{0.2 + 0.1} = 0.067 \text{ }\Omega$$

Step III. We can now solve the Thévenin-equivalent circuit of Figure 9-15(b) by Ohm's law.

$$R_T = R_x + R_L = 0.067 + 1 = 1.067 \text{ }\Omega$$

$$I_L = \frac{E_x}{R_T} = \frac{13.87 \text{ V}}{1.067 \text{ }\Omega} = \textbf{13 A}$$

If there are *three* or more voltage sources in parallel feeding a common load, as in Figure 9-16(a), we cannot find the Thévenin-equivalent voltage E_x directly without resorting to simultaneous equations. However, we can avoid such equations by converting *each* voltage source in the original network to its Norton-equivalent constant-current source, as in Figure 9-16(b). It is then quite simple to reduce the three parallel constant-current sources to a single Norton-equivalent source, as in Figure 9-16(c). Finally, we convert back to the required Thévenin-equivalent source of Figure 9-16(d).

EXAMPLE 9-7
Find the Thévenin-equivalent source for the three sources in parallel shown in Figure 9-16(a).

FIGURE 9-16
Equivalent Circuits for Example 9-7.

SOLUTION

Step I. As for the Thévenin's-theorem procedure, the load is removed from the original circuit of Figure 9-16(a) and connected to the final Thévenin-equivalent circuit of Figure 9-16(d).

Step II. We now follow a procedure which is similar to the one we used in nodal analysis, the difference being that we have removed the load in this case. Replace each of the sources by its Norton-equivalent constant-current source. Again we substitute the internal *conductance* for internal resistance in the resulting parallel circuit. We also use microamperes and microsiemens as units (microamperes = volts × microsiemens).

$$I_1 = \frac{E_1}{R_1} = \frac{150 \text{ V}}{100 \text{ k}\Omega} = 1500 \ \mu\text{A} \quad \text{and} \quad G_1 = \frac{1}{R_1} = 10 \ \mu\text{S}$$

Similarly, $$I_2 = \frac{75 \text{ V}}{250 \text{ k}\Omega} = 300 \ \mu\text{A} \quad \text{and} \quad G_2 = 4 \ \mu\text{S}$$

and $$I_3 = \frac{300 \text{ V}}{500 \text{ k}\Omega} = 600 \mu\text{A} \quad \text{and} \quad G_3 = 2 \ \mu\text{S}$$

In entering these data on Figure 9-16(b), we must make sure that the current direction for the constant-current source is consistent with the polarity of its equivalent constant-voltage source. Note the reversal of direction of I_2 with respect to I_1 and I_3.

Step III. Determining the Norton-equivalent source for Figure 9-16(b) is quite simple. It will have a single constant-current source equal to the *algebraic* sum of the three current sources of Figure 9-16(b). Hence, the Norton-equivalent constant-current source in Figure 9-16(c) is

$$I_x = I_1 + I_2 + I_3 = 1500 + (-300) + 600 = 1800 \ \mu A$$

The internal conductance in Figure 9-16(c) will be the sum of the three internal conductances in Figure 9-16(b).

$$G_x = 10 + 4 + 2 = 16 \ \mu S$$

Step IV. Finally, we convert the constant-current source of Figure 9-16(c) into the equivalent constant-voltage source of Figure 9-16(d).

$$R_x = \frac{1}{G_x} = \frac{1}{16 \ \mu S} = \textbf{62.5 k}\boldsymbol{\Omega}$$

and $\qquad\qquad E_x = I_x R_x = 1800 \ \mu A \times 62.5 \ k\Omega = \textbf{112.5 V}$

EXAMPLE 9-8

If R_B in the circuit shown in Figure 9-17 is 1 kΩ, determine the magnitude and polarity of the voltage drop across R_B.

FIGURE 9-17
Circuit Diagram for Example 9-8.

SOLUTION

Step I. Removing R_B from the original circuit leaves the circuit shown in Figure 9-18(a). We can replace everything to the left of the dashed line in Figure 9-18(a) by its Thévenin-equivalent source, as shown in Figure 9-18(b), where

$$E_x = \frac{68 \ k\Omega}{22 \ k\Omega + 68 \ k\Omega} \times 150 \ V = 113.3 \ V$$

and $\qquad\qquad R_x = \frac{22 \ k\Omega \times 68 \ k\Omega}{22 \ k\Omega + 68 \ k\Omega} + 56 \ k\Omega = 72.6 \ k\Omega$

Step II. We now set out to find the single Thévenin-equivalent source to replace the two parallel sources of Figure 9-18(b), using either of the techniques just de-

FIGURE 9-18
Equivalent Circuits for Example 9-8.

scribed. Converting each voltage source to its equivalent constant-current source,

$$I_1 = \frac{E_1}{R_1} = \frac{113 \text{ V}}{72.6 \text{ k}\Omega} = 1561 \ \mu\text{A} \quad \text{and} \quad G_1 = 13.8 \ \mu\text{S}$$

Similarly, $\quad I_2 = \frac{100 \text{ V}}{33 \text{ k}\Omega} = 3030 \ \mu\text{A} \quad \text{and} \quad G_2 = 30.3 \ \mu\text{S}$

Step III. We can replace the circuit of Figure 9-18(c) with a single Norton-equivalent source. Noting the current directions,

$$I_x = I_1 + I_2 = -1561 + 3030 = 1469 \ \mu\text{A}$$

and $\qquad\qquad G_x = G_1 + G_2 = 13.8 + 30.3 = 44.1 \ \mu\text{S}$

Step IV. If we wish, we may solve for the answer directly from the Norton-equivalent circuit of Figure 9-18(d). Or we can carry on to the final Thévenin-equivalent circuit of Figure 9-18(e) where

$$R_x = \frac{1}{G_x} = \frac{1}{44.1 \ \mu\text{S}} = 22.7 \text{ k}\Omega$$

and $\qquad\qquad E_x = I_x R_x = 1469 \mu\text{A} \times 22.7 \text{ k}\Omega = 33.3 \text{ V}$

Step V. From the voltage-divider principle,

$$V_B = \frac{1\ k\Omega}{22.7\ k\Omega + 1\ k\Omega} \times 33.3\ V = \mathbf{1.4\ V}$$

The procedure we used in Examples 9-7 and 9-8 is sometimes referred to as **Millman's theorem.** Instead of going through Example 9-7 in four steps, we can summarize the procedure into two equations. From Figure 9-16,

$$R_x = \frac{1}{G_x} \quad \text{and} \quad G_x = G_1 + G_2 + G_3$$

Therefore, in the Thévenin-equivalent source for a group of parallel voltage sources,

$$R_x = \frac{1}{G_1 + G_2 + G_3 + \cdots + G_N} \tag{9-1}$$

Also, $\quad E_x = I_x R_x$

But $\quad I_x = E_1 G_1 + E_2 G_2 + E_3 G_3$

$$\therefore E_x = \frac{E_1 G_1 + E_2 G_2 + E_3 G_3 + \cdots + E_N G_N}{G_1 + G_2 + G_3 + \cdots + G_N} \tag{9-2}$$

Equations (9-1) and (9-2) represent the formal statement of Millman's theorem. Rather than work from memorized formulas, we should continue to use the separate steps outlined in Examples 9-7 and 9-8.

SELF-CHECK

9-4:1. For the circuit in Figure 9-19:
(a) Find the Thévenin-equivalent source for the network supplying the load, R_L.
(b) Find the voltage drop across the load, R_L.

FIGURE 9-19

9-5 DEPENDENT SOURCES

The emf generated by the voltage sources we have encountered so far has been determined entirely by the internal energy-conversion process of the source. The open-circuit terminal voltage is, therefore, *independent of what happens elsewhere in a network containing the source*—hence the name *constant-voltage source* or **independent source.** Even though we know that the terminal voltage of a practical voltage source decreases as load current increases, we account for this effect in network diagrams by including an internal resistance in series with the ideal constant-voltage source. From the above definition, the constant-current source of Section 8-3 is also an independent source. Most of the sources we encounter in practical electric circuits are independent sources.

By using electronic devices such as transistors, we can create an effect in an electric circuit which amounts to a voltage or current source whose magnitude is *dependent on* or *controlled by a voltage or current appearing in some other branch of the network.* We can represent such devices in our network diagrams by **dependent** or **controlled sources.** To distinguish dependent sources from independent sources in circuit diagrams, the most popular symbol at present is a diamond-shaped symbol with + and − polarity indicators for a voltage source and a current arrow for a current source. As shown in Table 9-1, there are four possible dependent sources.

Figure 9-20 shows a simple network containing a current-controlled voltage source in series with the more familiar 120-volt independent voltage source and its internal resistance. This dependent source is designed so that its terminal voltage is[†]

$$V = 5 \times I_1$$

[†] To be technically rigorous, we assign the unit *ohm* to the numerical constant $k = 5$ in this equation to satisfy the units for Ohm's law.

TABLE 9-1

DEPENDENT SOURCES		
Type of Source	Magnitude	Symbol
Voltage-controlled voltage source	$V = kV_1$	
Current-controlled voltage source	$V = kI_1$	
Current-controlled current source	$I = kI_1$	
Voltage-controlled current sources	$I = kV_1$	

By substituting various values for I_1 (as we change R_L), we soon discover that the dependent source adds a voltage which exactly offsets the IR drop across R_{int}. Consequently, V_L remains constant at 120 volts for any setting of R_L. We have created a voltage-regulated power supply.

We think of the CCVS in Figure 9-20 as a *source* because the relationship between the polarity of the voltage between its terminals and the current through it is that of a *source*. However, dependent sources differ from real energy sources. If we disconnect R_L in the circuit of Figure 9-20 so that $I_1 = 0$, the voltage across the CCVS disappears—unlike independent voltage sources. Hence, we use the symbol V rather than E for dependent sources. Although Kirchhoff's laws and network theorems hold true for networks containing dependent sources, we shall find some situations where we cannot use some of the methods to solve the network.

FIGURE 9-20
Network with a Dependent Source.

At first glance, it would seem that we could write two mesh equations based on Kirchhoff's current law to solve the network shown in Figure 9-21. But the rules for mesh equations require us to convert all current sources to equivalent voltage sources. We cannot convert the VCCS in Figure 9-21 to an equivalent VCVS because there is no resistor which we can use as R_x in parallel with the original current source. We can, however, "Thévenize" the portion of the network to the left of V_1 *without affecting V_1* to create the single loop of Figure 9-22.

FIGURE 9-21
Circuit Diagram for Example 9-9.

FIGURE 9-22
Equivalent Circuit for Solution 1 for Example 9-9.

EXAMPLE 9-9
Determine V_L in the network shown in Figure 9-21.

SOLUTION 1

In the equivalent circuit of Figure 9-22, from the VCCS,

$$I = 0.01 \, V_1$$

and from Ohm's law,

$$V_1 = 4 - 160 \, I$$

$$\therefore V_1 = 4 - 160 \times 0.01 \, V_1 = \frac{4}{2.6} = 1.54 \text{ V}$$

$$I = 0.01 \, V_1 = 15.4 \text{ mA}$$

and
$$V_L = IR_L = 15.4 \text{ mA} \times 500 \, \Omega = 7.7 \text{ V}$$

We cannot use superposition theorem for this example since removing the 5-V source will drop V_1 to zero. Hence, I will also be zero. But we can convert the independent voltage source to its equivalent constant-current source and use nodal analysis.

SOLUTION 2

In the equivalent circuit of Figure 9-23, the single nodal equation is

$$\left(\frac{1}{200} + \frac{1}{800} \right) V_1 = 25 \text{ mA} - 10^{-2} V_1$$

FIGURE 9-23
Equivalent Circuit for Solution 2 for Example 9-9.

This reduces to
$$V_1 = \frac{25 \text{ mA}}{16.25 \text{ mS}} = 1.54 \text{ V}$$

Then
$$I_L = 10^{-2} V_1 = 0.01 \times 1.54 = 15.4 \text{ mA}$$

and
$$V_L = I_L R_L = 15.4 \text{ mA} \times 500 \, \Omega = 7.7 \text{ V}$$

In the circuit shown in Figure 9-24, we have replaced the VCCS of Figure 9-21 with a current-controlled voltage source. The superposition theorem still will not work. And

FIGURE 9-24
Circuit Diagram for Example 9-10.

we cannot replace the left-hand part of the network with its Thévenin-equivalent source because we will lose the I_1 which governs the voltage produced by the CCVS. However, we can use both mesh and nodal analysis techniques.

EXAMPLE 9-10
Determine V_L in the network shown in Figure 9-24.

SOLUTION 1
From Kirchhoff's voltage law,

$$(200 + 800)I_A - 800I_B = 5$$

$$(500 + 800)I_B - 800I_A = 100I_1 = 100(I_A - I_B)$$

Collecting the terms,

$$1000I_A - 800I_B = 5$$

$$900I_A - 1400I_B = 0$$

$$I_B = \frac{\begin{vmatrix} 1000 & 5 \\ 900 & 0 \end{vmatrix}}{\begin{vmatrix} 1000 & -800 \\ 900 & -1400 \end{vmatrix}} = \frac{-4500}{-680\,000} = 6.62 \text{ mA}$$

$$V_L = I_B R_L = 6.62 \text{ mA} \times 500 \text{ }\Omega = \mathbf{3.31 \text{ V}}$$

SOLUTION 2
Converting both voltage sources in the original network of Figure 9-24 into their Norton equivalents for the circuit of Figure 9-25;

Independent source:

$$I_A = \frac{5 \text{ V}}{200 \text{ }\Omega} = 25 \text{ mA}$$

FIGURE 9-25
Equivalent Circuit for Solution 2 for Example 9-10.

Dependent source:

$$I_B = \frac{V_B}{500}$$

But

$$V_B = 100I_1 \qquad \therefore I_B = 0.2I_1$$

and

$$I_1 = \frac{V_N}{800} \qquad \therefore I_B = \frac{V_N}{4000}$$

For the single independent node, the nodal equation is

$$\left(\frac{1}{200} + \frac{1}{800} + \frac{1}{500}\right)V_N = 25 \text{ mA} - \frac{V_N}{4000}$$

Bringing the term for the dependent source current to the left-hand side of the equation and using the calculator reciprocal key to clear the fractions,

$$V_N = 25 \text{ mA} \times 117.65 \ \Omega = 2.94 \text{ V}$$

Note that V_N is *not* the same as V_L in the original network of Figure 9-24. Returning this value of V_N to the original network,

$$I_1 = \frac{V_N}{800} = \frac{2.94 \text{ V}}{800 \ \Omega} = 3.676 \text{ mA}$$

and

$$V_B = 100I_1 = 0.3676 \text{ V}$$

Hence,

$$V_L = V_N + V_B = 2.94 + 0.37 = \textbf{3.31 V}$$

In the solutions for Examples 9-9 and 9-10, we have not tried to include a dependent source in any Thévenin transformation. We can include dependent sources as long as we are careful how we determine R_x. If the controlling variable is within the portion of the network to be included in the Thévenin-equivalent source, we cannot use the shortcut method stated in Thévenin's theorem for finding R_x by "looking back" into the open-circuit terminals. We must use the short-circuit current of the "black-box" method of Section 8-2, where we found that

$$R_x = \frac{\text{open-circuit terminal voltage}}{\text{short-circuit terminal current}} \bullet \qquad (8\text{-}1)$$

EXAMPLE 9-11
Removing the 500-Ω R_L from the network shown in Figure 9-24, determine the Thévenin-equivalent source which can replace the remainder of the network.

SOLUTION
For the open-circuit condition shown in Figure 9-26(a),

$$I_1 = \frac{5 \text{ V}}{(200 + 800) \; \Omega} = 5 \text{ mA}$$

FIGURE 9-26
Dependent Source in a Thévenin Transformation.

The dependent source voltage is

$$V = 100I_1 = 0.5 \text{ V}$$

$$\therefore V_{oc} = E_x = 5 \text{ mA} \times 800 \; \Omega + 0.5 \text{ V} = \mathbf{4.5 \text{ V}}$$

To find the short-circuit current, we can write two mesh equations for the circuit of Figure 9-26(b).

$$(200 + 800)I_A - 800I_B = 5$$

and

$$800I_B - 800I_A = 100I_1 = 100(I_A - I_B)$$

Collecting the terms,

$$1000I_A - 800I_B = 5 \qquad (1)$$

$$900I_A - 900I_B = 0 \qquad (2)$$

$$I_{sc} = I_B = \frac{\begin{vmatrix} 1000 & 5 \\ 900 & 0 \end{vmatrix}}{\begin{vmatrix} 1000 & -800 \\ 900 & -900 \end{vmatrix}} = \frac{-4500}{-180\,000} = 25 \text{ mA}$$

$$R_x = \frac{E_{oc}}{I_{sc}} = \frac{4.5 \text{ V}}{25 \text{ mA}} = \mathbf{180\ \Omega}$$

To check E_x and R_x, from Figure 9-26(c),

$$V_L = \frac{500}{180 + 500} \times 4.5 \text{ V} = \mathbf{3.31 \text{ V}}$$

Note that R_x is *not* the same as the resistance "seen" looking back into the open-circuit terminals of Figure 9-26(a).

SELF-CHECK

9-5:1. Consider the 2-source network in Figure 9-27.
   (a) Use the standard format method to write the nodal equations which will allow the determination of V_1 and V_2.
   (b) Solve for V_1 and V_2.
   (c) Determine the current in R_L.

FIGURE 9-27

9-6 DELTA-WYE TRANSFORMATION

Another useful type of circuit transformation which replaces a given circuit with an equivalent circuit applies to the three-terminal resistance network of Figure 9-28. If we can state the conditions under which the "wye" circuit of Figure 9-28(a) is equivalent to the "delta" circuit of Figure 9-28(b), we can substitute the wye for the delta, and vice versa, in circuit simplification.

FIGURE 9-28
(a) Y-Network; (b) Δ-Network.

If we ignore terminal C for the moment, the resistance between terminals A and B must be the same for both networks. For the wye circuit of Figure 9-28(a), the circuit between A and B is a simple series circuit giving

$$R_{AB} = R_A + R_B$$

For the delta circuit of Figure 9-28(b), there are two branches in parallel between terminals A and B, giving

$$R_{AB} = \frac{R_Z(R_X + R_Y)}{R_Z + (R_X + R_Y)}$$

$$\therefore R_A + R_B = \frac{R_X R_Z + R_Y R_Z}{R_X + R_Y + R_Z} \tag{1}$$

Similarly,
$$R_B + R_C = \frac{R_X R_Y + R_X R_Z}{R_X + R_Y + R_Z} \tag{2}$$

and
$$R_C + R_A = \frac{R_Y R_Z + R_X R_Y}{R_X + R_Y + R_Z} \tag{3}$$

Subtracting (2) from (1) gives

$$R_A - R_C = \frac{R_Y R_Z - R_X R_Y}{R_X + R_Y + R_Z} \tag{4}$$

Adding (3) and (4) gives

$$2R_A = \frac{2R_Y R_Z}{R_X + R_Y + R_Z}$$

from which
$$R_A = \frac{R_Y R_Z}{R_X + R_Y + R_Z} \tag{9-3}$$

Likewise,
$$R_B = \frac{R_X R_Z}{R_X + R_Y + R_Z} \tag{9-4}$$

and
$$R_C = \frac{R_X R_Y}{R_X + R_Y + R_Z} \qquad (9\text{-}5)$$

These are the **delta to wye transformation equations.** In using them, we must be careful to associate the numerical values of the original delta circuit with their proper counterparts in the transformation equations. Note that there is a simple pattern to the three equations.

$$R_A = \frac{\text{product of the two } \Delta \text{ arms connected to terminal } A}{\text{sum of all three } \Delta \text{ arms}}$$

In using these equations with a calculator, we first calculate the sum $R_X + R_Y + R_Z =$ and store this number in the calculator's memory. Then the calculation for R_A becomes simply $R_Y \times R_Z \div$ Memory recall.

We can also reduce Equations (1), (2), and (3) algebraically to solve for the equivalent delta circuit from a given wye circuit. The **wye to delta transformation equations** are usually stated as

$$R_X = \frac{R_A R_B + R_B R_C + R_C R_A}{R_A} \qquad (9\text{-}6)$$

$$R_Y = \frac{R_A R_B + R_B R_C + R_C R_A}{R_B} \qquad (9\text{-}7)$$

$$R_Z = \frac{R_A R_B + R_B R_C + R_C R_A}{R_C} \qquad (9\text{-}8)$$

Again the three equations have a pattern.

$$R_X = \frac{\text{sum of products of each } pair \text{ of } Y \text{ arms}}{\text{opposite } Y \text{ arm}}$$

In this case, we calculate the *numerator* and store this number in the calculator's memory to be divided by the appropriate denominator in using the wye to delta transformation equations.

Since Equations (9-6), (9-7), and (9-8) are more elaborate than Equations (9-3), (9-4), and (9-5), suppose we look for an alternative format. Since the delta circuit has *parallel* branches, we can *invert* Equation (9-6) and insert $1/G$ for each resistance.

$$G_X = \frac{\dfrac{1}{G_A}}{\dfrac{1}{G_A G_B} + \dfrac{1}{G_B G_C} + \dfrac{1}{G_C G_A}} = \frac{\dfrac{1}{G_A}}{\dfrac{G_A + G_B + G_C}{G_A G_B G_C}}$$

From which
$$G_X = \frac{G_B G_C}{G_A + G_B + G_C} \qquad (9\text{-}9)$$

We can establish similar equations for G_Y and G_Z. These equations are the **duals** of Equations (9-3), (9-4), and (9-5).

Inverting Equation (9-9) gives

$$R_X = R_B R_C(G_A + G_B + G_C) \tag{9-10}$$

Similarly,

$$R_Y = R_A R_C(G_A + G_B + G_C) \tag{9-11}$$

and

$$R_Z = R_A R_B(G_A + G_B + G_C) \tag{9-12}$$

This *dual* form of the wye to delta transformation equations is particularly useful when dealing with complex impedances and admittances in ac circuits.

EXAMPLE 9-3A

In the Wheatstone bridge circuit of Figure 9-29(a), R_1 and R_4 have a resistance of 300 Ω each. R_2 and R_3 each have a resistance of 150 Ω. The source voltage is 100 V. What is the current through the galvanometer if its resistance is 50 Ω?

FIGURE 9-29
Solving Wheatstone Bridge by Delta-Wye Transformation.

SOLUTION

Step I.   Replace R_1, R_2, and R_M of the original bridge circuit of Figure 9-29(a) (which form a delta) with their equivalent wye circuit, as in Figure 9-29(b).

$$R_A = \frac{300 \times 150}{300 + 150 + 150} = 90 \ \Omega$$

$$R_B = \frac{300 \times 50}{500} = 30 \ \Omega$$

$$R_C = \frac{50 \times 150}{500} = 15 \ \Omega$$

Step II.   The load on the source is now a simple series-parallel circuit with an equivalent resistance of

$$R_{eq} = R_A + \frac{(R_B + R_3) \times (R_C + R_4)}{(R_B + R_3 + R_C + R_4)}$$

$$= 90 + \frac{180 \times 315}{495} = 204.55 \ \Omega$$

$$\therefore I_T = \frac{E}{R_{eq}} = \frac{100 \ V}{204.55 \ \Omega} = 489 \ mA$$

Step III. From the current-divider principle,

$$I_3 = 489 \ mA \times \frac{315}{495} = 311 \ mA$$

and
$$I_4 = 489 \ mA \times \frac{180}{495} = 178 \ mA$$

Since R_3 and R_4 were not altered by the transformation, I_3 and I_4 are the *same* as for the original circuit. We can now return to the original circuit, and in Figure 9-29(a),

$$V_3 = I_3 R_3 = 311 \ mA \times 150 \ \Omega = 46.65 \ V$$

and
$$V_4 = I_4 R_4 = 178 \ mA \times 300 \ \Omega = 53.4 \ V$$

Step IV. From Kirchhoff's voltage law,

$$V_M = V_4 - V_3 = 53.4 - 46.65 = 6.75 \ V$$

(with the right-hand terminal positive with respect to the left-hand terminal)

and
$$I_M = \frac{6.75 \ V}{50 \ \Omega} = \mathbf{135 \ mA}$$

SELF-CHECK

9-6:1. Determine the resistance values that will make the circuits shown in Figure 9-30 equivalent.

FIGURE 9-30

9-7 SUBSTITUTION THEOREM

In Chapter 7 we discovered that we could *substitute* a single equivalent resistor for a branch in a network consisting of a combination of series and/or parallel resistors, as shown in Figure 9-31(a) and (b). By selecting (calculating) the appropriate value for this substitute resistor, the voltage drop across and current through the branch is identical with the original network. Hence, the rest of the network "sees" no difference with the substitution.

FIGURE   9-31
Substitution Theorem.

The substitution principle is such a fundamental property of electric circuits that we can expand it into the **substitution theorem.**

Any branch in a linear network can be replaced by any combination of circuit elements which will produce the **same** *voltage across and current through that branch.*

Figure 9-31(c), (d), and (e) shows several possible combinations which produce the *same* voltage and current as the circuit of Figure 9-31(a) and, therefore, will not affect the rest of the network. We should note that:

1. Since we have to know the voltage across and the current through the selected branch to make this type of substitution, this theorem is not useful for *solving* network examples. It is sometimes used by circuit designers to check or optimize a circuit design.

2. When we replace resistors by sources, current must flow the "wrong way" through voltage sources and the voltage drop must be the "wrong polarity" across current sources so that they act like the resistors they replace.

3. We cannot replace branches containing discharging sources by resistors alone (which cannot supply energy). In theory, we can call a voltage source a *negative resistance*.

As an exercise, check that the substitutions shown in Figure 9-31 do satisfy the substitution theorem.

9-8 RECIPROCITY THEOREM

We can also expand another principle based on a fundamental property of electric circuits into the **reciprocity theorem.**

In any linear network containing only a single voltage source, if that source located in branch A causes a current I in branch B, then moving that voltage source (but not its internal resistance) to branch B will cause a current I in branch A equal to the original current in branch B.

Although it applies to any network configuration, the reciprocity theorem is useful primarily in showing the bilateral property of the four-terminal networks we shall encounter in Chapter 25. There is also a version of the reciprocity theorem for a *current* source in branch A causing a *voltage drop* across branch B. Unlike the substitution theorem, the currents and voltages in the other branches of the network are *not* unchanged when we move the source from one branch to another. Check by determining the three currents within the four-terminal T-attenuator in both circuits shown in Figure 9-32.

FIGURE 9-32
Reciprocity Theorem.

PROBLEMS

9-2-1. Determine the Thévenin-equivalent constant-voltage source for the two-terminal network of Figure 9-33.

9-2-2. Determine the Thévenin-equivalent source for the network of Figure 9-34.

FIGURE 9-33
T-Network.

FIGURE 9-34
π-Network

FIGURE 9-35

9-2-3. Use Thévenin's theorem to determine the current through the 500-Ω resistor in the circuit of Figure 9-35.

9-2-4. Determine the current in the 100-Ω resistor in the circuit of Figure 9-35.

9-2-5. What resistance must be connected to the source shown in Figure 9-33 to draw a 200-mA current through the unknown resistor? Use Thévenin's theorem.

9-2-6. Use Thévenin's theorem to determine the current through a 5-kΩ load connected to the source shown in Figure 9-34.

9-2-7. What is the maximum power that can be developed in a load connected to the source shown in Figure 9-33?

9-2-8. What value of load resistance will dissipate energy at the rate of 100 mW when connected to the source shown in Figure 9-34? (two answers)

9-2-9. Solve for the current drawn from the 20-V source in Figure 9-36 by replacing the portion of the circuit to the left of the dashed line with its Thévenin-equivalent source.

9-2-10. Find the voltage drop across the 25-Ω resistor in Figure 9-37 by replacing the current source with an equivalent constant-voltage source.

FIGURE 9-36

FIGURE 9-37

9-2-11. Determine the Thévenin-equivalent source as "seen" by the 10-kΩ resistor in the network shown in Figure 9-38.

9-2-12. Determine the Thévenin-equivalent source as "seen" by the 33-kΩ resistor in the network shown in Figure 9-38.

9-2-13. The switch in the pulse generator circuit of Figure 9-39 alternately closes for 1 ms and opens for 4 ms. Draw the open-circuit voltage waveform (with respect to chassis) and determine its amplitude accurately.

FIGURE 9-38   **FIGURE 9-39**

9-2-14. Given the following data for the Wheatstone bridge of Figure 9-29(a), use Thévenin's theorem to determine the meter current. $E = 200$ V, $R_1 = 10$ kΩ, $R_2 = 2.5$ kΩ, $R_3 = 5$ kΩ, $R_4 = 20$ kΩ, and $R_M = 25$ kΩ.

9-2-15. Use Thévenin's theorem to determine the current through a 500-Ω load connected to the output terminals of the bridged-T attenuator of Figure 9-40 when the input voltage is 2 V.

9-2-16. If the voltage applied to the four-terminal lattice network of Figure 9-41 is 360 V, what value of load resistor must be connected to the output terminals to draw a load current of 1 A? Use Thévenin's theorem.

9-3-17. Determine the Norton-equivalent constant-current source for the two-terminal network of Figure 9-33.

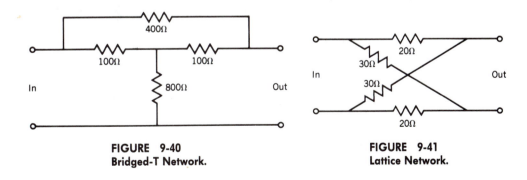

FIGURE 9-40
Bridged-T Network.

FIGURE 9-41
Lattice Network.

9-3-18. Determine the Norton-equivalent source for the network of Figure 9-34.

9-3-19. What current will a 500-Ω load draw from the source shown in Figure 9-33? Use Norton's theorem.

9-3-20. Use Norton's theorem to determine the value of load resistance required to draw a 40-μA current from the source shown in Figure 9-34.

9-3-21. Determine the current through the center 50-Ω resistor in the network of Figure 9-36 by replacing each voltage source with its Norton-equivalent source. (nodal analysis)

9-3-22. Find the voltage drop across the 25-Ω resistor in Figure 9-37 by replacing the voltage source with an equivalent constant-current source.

9-3-23. Find the current through the 100-Ω resistor in the circuit of Figure 9-37 by leaving the sources in their original form and using the superposition theorem.

9-3-24. Determine the Norton-equivalent source as "seen" by the 10-kΩ resistor in the network shown in Figure 9-38.

9-4-25. Determine the current through the 60-Ω resistor in the network of Figure 9-42.

9-4-26. Determine the voltage drop across the 5-kΩ resistor in the network of Figure 9-43.

9-4-27. Three voltage sources in a network ($E_1 = 40$ V, $R_1 = 2.2$ kΩ; $E_2 = 24$ V, $R_2 = 3.3$ kΩ; and $E_3 = -24$ V, $R_3 = 4.7$ kΩ) all feed a common component in parallel. Determine the Millman-equivalent voltage source.

9-4-28. A battery charger has an open-circuit voltage of 16.0 V and an internal resistance of 0.1 Ω. It is charging four storage batteries in parallel with open-circuit voltages and internal resistances as follows: $V_1 = 12.8$ V, $R_1 = 0.2$ Ω; $V_2 = 12.3$ V, $R_2 = 0.3$ Ω; $V_3 = 13.5$ V, $R_3 = 0.25$ Ω; and $V_4 = 12.6$ V, $R_4 = 0.15$ Ω. What is the total battery charger current?

9-5-29. The output equivalent circuit of a typical transistor amplifier is shown in Figure 9-44. Calculate the output voltage if the input signal current $I_1 = 24$ μA, the forward current ratio $h_{fe} = 40$, the output admittance $h_{oe} = 5 \times 10^{-5}$ S, and the load resistance $R_L = 22$ kΩ.

FIGURE 9-42

FIGURE 9-43

FIGURE 9-44
Transistor Amplifier Equivalent Output Circuit.

9-5-30. Determine the Thévenin-equivalent source for the transistor output circuit of Problem 9-5-29.

9-5-31. Determine V_L in the circuit of Figure 9-45(a).

(a)   (b)

FIGURE 9-45

9-5-32. Determine V_L in the circuit of Figure 9-45(b).

9-5-33. Find the Thévenin-equivalent voltage source for the network of Figure 9-46(a).

9-5-34. Find the Thévenin-equivalent voltage source for the network of Figure 9-46(b).

9-5-35. Determine V_L in the amplifier equivalent circuit of Figure 9-47.

(a)   (b)

FIGURE 9-46

FIGURE 9-47

9-5-36. A 2000-Ω feedback resistor is connected between the junction of the 200-Ω and 800-Ω resistors and the top of the VCCS in Figure 9-47. Determine V_L.

9-6-37. Given the π-network of Figure 9-48(a), use a delta-wye transformation to determine the equivalent T-network.

9-6-38. Given the T-network of Figure 9-48(b), use a wye-delta transformation to determine the equivalent π-network.

9-6-39. Given the following data for the Wheatstone bridge of Figure 9-29(a), use a delta-wye transformation to determine the meter current. $E = 100$ V, $R_1 = 10$ Ω, $R_2 = 20$ Ω, $R_3 = 30$ Ω, $R_4 = 40$ Ω, and the meter resistance $R_M = 50$ Ω.

9-6-40. Use a delta-wye transformation to determine the current through a 300-Ω load connected to the output terminals of the bridged-T attenuator of Figure 9-40 when the input voltage is 4 V.

(a) (b)

FIGURE 9-48
(a) π-Network; (b) T-Network.

9-6-41. Use a delta-wye transformation to determine what voltage must be applied to the input terminals of the lattice network of Figure 9-41 in order to draw a 2-A current from the source when a 20-Ω load is connected to the output terminals of the network.

9-8-42. A voltage source with zero internal resistance and an open-circuit voltage of 20 V is connected to the input terminals of the T-network of Figure 9-48(b). What is the reading on a milliammeter with zero resistance connected across the output terminals? What is the milliammeter reading when the voltage source is connected to the output terminals and the milliammeter is connected to the input terminals?

REVIEW QUESTIONS

9-1. What is meant by a **Thévenin-equivalent source?**

9-2. What is the difference between the "black-box" method and the Thévenin's-theorem method of determining the internal resistance of a Thévenin-equivalent source?

9-3. Show that the Thévenin-equivalent resistance, as found by the short-circuit method in the "black-box" experiment, is the same as that found by the procedure set forth in the statement of Thévenin's theorem.

9-4. Draw a Thévenin-equivalent circuit for the source shown in Figure 9-49 and express the equivalent voltage E_x and the equivalent internal resistance R_x in terms of the parameters of the original circuit.

9-5. If a load R_L is connected to the original circuit of Figure 9-49, derive a single equation for I_L.

9-6. In a Thévenin-equivalent circuit, $I_L = E_x/(R_x + R_L)$. Show that this is equal to the equation for I_L in Question 9-5, thereby verifying Thévenin's theorem.

9-7. What is meant by a **Norton-equivalent source?**

9-8. What happens to the terminal voltage of a Norton-equivalent source when a load resistor is connected to its terminals? Explain.

FIGURE 9-49

9-9. Attempt a solution for Problem 9-2-3 by Norton's theorem. Note that in this problem it is easier to find the Norton-equivalent source by first finding the Thévenin-equivalent source and then converting to the equivalent constant-current source.

9-10. Explain the polarity of the voltage drops across R_G and R_B in Figure 9-15(a).

9-11. Show how you would use Millman's theorem to obtain the Thévenin-equivalent circuit of Figure 9-15(b).

9-12. The sources feeding the network shown in Figure 9-38 have negligible internal resistance. How do we go about replacing them with equivalent constant-current sources in applying Millman's theorem to the network?

9-13. The numerical constant for the current-controlled voltage source in Figure 9-20 is assigned the unit *ohm*. What unit is assigned to the numerical constant for
(a) A voltage-controlled voltage source?
(b) A voltage-controlled current source?

9-14. What restrictions are there on using the superposition theorem in a network containing dependent sources?

9-15. What restrictions are there on using Thévenin's and Norton's theorems in a network containing dependent sources?

9-16. What value of R_x is "seen" looking back into the terminals of the dependent voltage source of Figure 9-50? (*Hint:* Connect a fixed voltage source to the terminals and calculate the terminal voltage and current—the voltmeter-ammeter method of determining R_x.)

9-17. Derive the wye-delta transformation Equation (9-6).

9-18. Figure 9-29(b) shows a method for solving a Wheatstone bridge using a delta-wye transformation. Draw a diagram showing how we can solve the same bridge circuit using a wye-delta transformation.

9-19. Under what circumstances can we substitute a resistor for a voltage source in an electrical network?

9-20. The mathematical process of addition in electronic analog computers is performed by the circuit shown in Figure 9-51. By selecting several numerical combinations for E_1 and E_2, show that the open-circuit E_{out} is always directly proportional to $E_1 + E_2$. (All voltages are expressed with respect to ground.)

FIGURE 9-50   FIGURE 9-51

*9-21. Write a computer program to print out E_x and R_x for the Thévenin-equivalent source when given up to four constant-voltage sources in parallel. Use the Millman's theorem equations (9-1) and (9-2).

*9-22. Revise the computer program of Question 8-41 to solve for the current in the balance-indicating meter of a Wheatstone bridge by using Thévenin's theorem.

10

ELECTRICAL MEASUREMENT

We have already encountered electric circuit diagrams (Figure 8-3) showing how we connect a voltmeter in *parallel* with part of the circuit to measure the voltage *between* two points, and how we connect an ammeter in *series* with a branch in the circuit to measure the current *through* that branch. So far, we have assumed that the ideal voltmeter has an infinitely high resistance, so that it draws no extra current from the circuit. Similarly, the ideal ammeter has zero resistance, so that there is no additional voltage drop across it. In this chapter, we shall consider how practical electrical measuring instruments go about obtaining data on what is happening in electric circuits, and how practical meters might disturb normal circuit parameters. We shall not attempt to cover details of construction, characteristics, and applications of this equipment. Such topics can be covered more effectively in a concurrent course on electrical measurements.

10-1 MECHANICAL TORQUE FROM ELECTRIC CURRENT

As we shall investigate in more detail in Chapter 13, when electric current flows in a conductor, it produces a magnetic field around the conductor. To assist us in visualizing the nature of a magnetic field, Michael Faraday suggested that we think of a magnetic field as being made up of **magnetic lines of force.** Figure 10-1(a) shows a cross section of an electric conductor in which the conventional current direction is away from us into the page (as indicated by the feathered end of an arrow). From this perspective, the

FIGURE 10-1
Force Acting on a Current-Carrying Conductor in a Magnetic Field.

electric current produces magnetic lines of force which appear as concentric circles. Figure 10-1(b) shows a permanent magnet with its magnetic lines of force crossing the air gap between the N and S poles of the magnet.

If we now place the current-carrying conductor between the poles of the U-shaped permanent magnet of Figure 10-1(b), we cannot simply draw the magnetic field of the conductor superimposed on that of the permanent magnet. One of the basic rules of the concept of lines of force is that they cannot intersect. Therefore, the two component magnetic fields must combine to form a single composite magnetic field. Below the conductor in Figure 10-1(c), the direction of the lines of force created by the current in the conductor is *opposite* to the direction of the permanent magnet field. This results in a cancelling effect that reduces the flux density below the conductor. Above the conductor, the direction of both component fields is the *same,* thus increasing the flux density. We can think of the effect of the current in the conductor as causing the magnetic lines of force of the permanent magnet in the vicinity of the current-carrying conductor to *detour* above the conductor.

Since magnetic lines of force tend to become as short as possible and also tend to repel one another, the lines of force of the composite magnetic field of Figure 10-1(c) will attempt to straighten out and to space themselves uniformly in an effort to regain the flux pattern of Figure 10-1(b). To do so, the magnetic field must force the current-carrying conductor to move *downward*. The magnitude of this force is directly dependent on the strength of the two component magnetic fields. The strength of the permanent magnet field is represented by its flux density B. The strength of the magnetic field around the conductor is proportional to the current I in the conductor and also the length l of the conductor within the magnetic field of the permanent magnet. Since the international standard ampere is defined in terms of the force between parallel current-carrying electric conductors (Section 13-3),

$$F = BIl \tag{10-1}$$

where F is the force acting on the conductor in newtons, B is the flux density of the permanent magnet field in teslas, I is the current in the conductor in amperes, and l is the length in meters of the conductor perpendicular to the magnetic field.

Since the flux density produced by a permanent magnet remains constant with a given magnetic circuit, both B and l in Equation (10-1) are constant. Therefore, the force acting on the conductor is directly proportional to the current in it. If we can devise a means of indicating this mechanical force, we then have a device for measuring the current in the conductor. Figure 10-2 shows a spring scale, like those used for weighing fish, attached to the conductor. According to Hooke's law, the extension of the spring is directly proportional to the mechanical force applied to it. This force, in turn, is directly proportional to the electric current in the conductor.

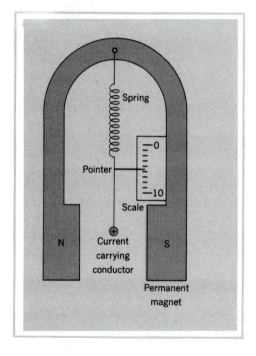

FIGURE 10-2
Simple Current-Measuring Device.

A more practical form for a current-measuring instrument is based on two conductors (the two sides of a loop) supported in the field of a permanent magnet by a pivot X. If the current direction is away from us in conductor B, it must be toward us in the other side of the loop, conductor A. This current causes the composite magnetic field to be distorted as shown in Figure 10-3. The mechanical force acting downward on conductor B and upward on conductor A results in a turning force or **torque** that attempts to rotate the loop in a clockwise direction about its pivot. This torque is counterbalanced by the

FIGURE 10-3
Developing a Mechanical Torque from Electric Current.

countertorque produced by a flat spiral spring fastened from the pivot to some stationary point. Since the countertorque of the hairspring is directly proportional to the angle through which the pivot is rotated (Hooke's law), the angle through which the loop rotates is directly proportional to the current through it. Since torque equals force times the length of the lever arm, the torque produced by each side of the loop will be

$$T = Fr$$

where T is the torque in newton meters, F is the force acting on the conductors in newtons [Equation (10-1)], and r is the radius of the loop in meters.

The total torque contributed by *both* sides of the loop will be

$$T = 2Fr = Fd$$

where d is the diameter of the loop in meters.

If the loop consists of more than one turn of wire, each side of the loop can be thought of as a stranded conductor, with each strand carrying a current equal to the current in the coil. Therefore, the effective current in the loop will be

$$I_{eff} = NI$$

where N is the number of turns of wire in the pivoted coil.

We can modify Equation (10-1) to apply to a loop in a magnetic field.

$$\therefore T = BIlNd \qquad \qquad (10\text{-}2)$$

EXAMPLE 10-1

What torque is developed by a 1-mA current flowing in a 120-turn coil wound on a 1-cm × 1-cm square form? The flux density in the air gap is 0.1 T.

SOLUTION

$$T = BIlNd = 0.1 \text{ T} \times 1 \text{ mA} \times 1 \text{ cm} \times 120 \times 1 \text{ cm}$$

$$= 1.2 \times 10^{-6} \text{ N} \cdot \text{m}$$

Equation (10-2) is the basic **electric-motor** equation. But because we are using a countertorque spring to prevent the loop from rotating freely, we can adapt the electric-motor principle to the **moving-coil** meter movement shown in Figure 10-4. This movement is sometimes referred to as a D'Arsonval movement, after the man who first patented a current-measuring device based on the moving-coil principle. Because Edward Weston developed the moving-coil instrument from the delicate laboratory apparatus of D'Arsonval to the reasonably rugged panel type of meter in use today, the moving-coil movement is also known as the Weston movement.

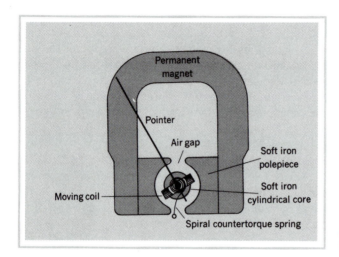

FIGURE 10-4
Moving-Coil Movement.

Since B, l, N, and d are constant for a given meter movement, the torque is directly proportional to the current in the moving coil; and, since the countertorque of the spiral spring is directly proportional to its angular extension, the angular displacement of the pointer is directly proportional to the current in the moving coil. If we adjust the spiral spring so that the coil occupies the position shown in Figure 10-3 when the current is zero, the movement is called a **galvanometer.** A galvanometer can indicate the flow of direct current in either direction since a clockwise torque will move the coil (and the pointer) in one direction from center, and a counterclockwise torque will move it in the other direction. We usually use a current meter or **ammeter** in a circuit in which the current direction does not change. By readjusting the stationary end of the spiral spring, we can set the moving coil and pointer in the position shown in Figure 10-4. Current through the coil in the proper direction will cause the pointer to be deflected across the scale, but reversed meter leads will cause the pointer to read off scale to the left, coming up against a mechanical stop.

SELF-CHECK

10-1:1. Suppose that the magnet in Figure 10-2 is 5.0 cm long and its flux density is 10^{-2} T. Determine the magnitude of the current in the conductor if the magnetic force on the conductor is 4.0×10^{-3} N.

10-1:2. A coil is wound on a 1.4-cm square form and will be immersed in a magnetic field of 250-mT flux density. How many turns of wire are necessary to produce a torque of 220×10^{-6} N $\cdot$ m when the coil is carrying a 50-mA current?

10-2 THE AMMETER

By using magnetic circuits that produce a high flux density in the air gap and delicate countertorque springs, we can build commercial movements which will read full scale with a current of 20 microamperes or less through the moving coil. The same spiral springs that provide the countertorque are used to connect the moving coil to the meter terminals. By using less turns of a larger wire size on the moving coil and a stronger spiral spring, moving-coil movements can be built which require currents in the order of several milliamperes to obtain full-scale deflection. If the physical size of the meter is limited to about eight centimeters in diameter, the maximum wire size which can be conveniently used for the moving coil limits the meter to about 30 milliamperes for full-scale deflection. To use the moving-coil instrument as an ammeter for larger currents, we must apply our knowledge of resistors in parallel.

For the following example, we shall use a moving-coil movement which reads full scale when one milliampere flows through its moving coil. A one-milliampere movement is a reasonable compromise between the more sensitive 50-microampere movements used in electronic-circuit measurements and the more rugged 5-milliampere movements used for measurements in electric power circuits. The moving coil of the 1-milliampere movement in question has a resistance of 27 ohms. It is usually customary to include a **calibrating resistor** in series with the moving coil to bring the total resistance of the movement up to some convenient round number, as shown in Figure 10-5. It is common practice to select this calibrating resistor so that the full-scale current through the movement and its calibrating resistor produces a 50-mV drop between the meter terminals. For the movement in question,

$$R_{\mathrm{T}} = \frac{V}{I} = \frac{50 \text{ mV}}{1 \text{ mA}} = 50 \text{ } \Omega$$

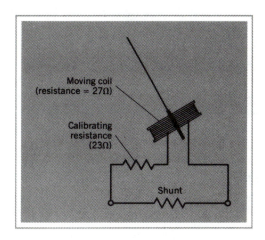

FIGURE 10-5
Moving-Coil Ammeter with Shunt Resistor.

and, therefore, the resistance of the calibrating resistor will be

$$R_C = 50 - 27 = 23 \ \Omega$$

The calibrating or **swamping** resistor is included in the case of the meter and is made from precision resistance wire having a small negative temperature coefficient to offset the positive temperature coefficient of the moving coil.

To use the one-milliampere movement to measure a current in the order of 0.6 ampere, as is the case in the circuit of Figure 10-6, we must arrange the meter circuit so that the movement and its **shunt** resistor form two parallel branches with 999 times as much current flowing through the shunt as through the meter movement. Since the shunt and the movement are in parallel, the *same* voltage drop appears across both and their current ratio is inversely proportional to their resistance ratio. Therefore,

$$V_{sh} = V_M$$

from which

$$I_{sh} \times R_{sh} = I_M \times R_M$$

and

$$R_{sh} = \frac{1}{999} \times R_M = \frac{50}{999} = 0.050\,05 \ \Omega$$

FIGURE   10-6
Measuring Electric Current.

If the current flowing through the lamp in Figure 10-6 is sufficient to make the movement read full scale, the current in the moving-coil branch will be 1 milliampere. Therefore the current through the shunt resistor will be 999 milliamperes and the current through the lamp is 1 + 999 mA = 1 ampere. As long as this movement and shunt are used together, we can interpret the meter scale as being calibrated for 1 ampere full scale.

If the lamp current happens to be 0.6 ampere, this total current will split so that one part (0.6 mA) will flow through the moving coil and 999 parts (0.599 4 A) will flow through the shunt. Therefore, we can again interpret the 0.6 scale marking of the basic 1-milliampere movement as indicating 0.6 ampere. For currents up to about 25 amperes, the shunt is physically small enough to be placed inside the meter case. For larger currents, external shunt resistors are required.

Connecting an ammeter into an electric circuit will have some slight effect on the circuit since we are adding a little extra resistance to the circuit. The total resistance of the meter and its shunt in Figure 10-6 is 0.05 ohm. The hot resistance of the lamp is in the order of 200 ohms. Therefore, the error in circuit current due to adding the ammeter to the circuit is only 0.05 part in 200, or 0.025%. In practice, the main problem with making current measurements is that we must place the ammeter in *series* with the circuit whose current is to be measured. Unless the meter is to be left in the circuit permanently, this involves opening the circuit in order to include the meter and then reconnecting the circuit when we remove the meter. Because of its low resistance (0.05 Ω for the 1-ampere meter), we must guard against connecting an ammeter *across* a source.

When an ammeter is to be permanently connected in a circuit, a single range is quite satisfactory. But for a test ammeter, it would be desirable to have several ranges so that an unknown current can be read on a range which provides a reading somewhere around two-thirds full scale where the meter calibration is most accurate. We can accomplish this by using several shunts, as shown in Figure 10-7(a). The switch must be of the *make-before-break* type so that the meter cannot be damaged by being in the circuit without a shunt as we change the range.

FIGURE 10-7
(a) Multirange Shunt; (b) Ayrton Shunt.

The Ayrton shunt of Figure 10-7(b) eliminates the possibility of the meter being in circuit without a shunt. This advantage is gained at the price of a slightly higher meter resistance. The Ayrton shunt also provides us with an excellent opportunity to apply our resistance network theory to a practical circuit.

EXAMPLE 10-2
Using a 1-mA movement with a total resistance of 50 Ω in the circuit shown in Figure 10-7(b), determine R_A, R_B, and R_C so that when the common + terminal and terminal 1 are used, the meter is a 1-A meter; when the common + terminal and terminal 5 are used, the meter is a 5-A meter; and when the common + terminal and terminal 10 are used, the meter is a 10-A meter.

SOLUTION 1
On the 1-A range, $R_A + R_B + R_C$ forms the shunt, and it must take 999 parts of the

total current when the meter movement takes 1 part. Therefore, as in the example of the simple shunt,

$$R_A + R_B + R_C = \frac{1}{999} \times 50 = 0.050\,05 \ \Omega \tag{1}$$

On the 5-A range, $R_A + R_B$ forms the shunt and R_C is in series with the moving coil. For the 5-A range, 4999 parts of the total current must flow through the shunt when the meter movement passes 1 part. Therefore,

$$R_A + R_B = \frac{50 + R_C}{4999} \tag{2}$$

On the 10-A range, R_A is the shunt and R_B and R_C are in series with the meter. For the 10-A range, R_C takes 9999 parts of the total current, while 1 part flows through the meter and R_B and R_C. Therefore,

$$R_A = \frac{50 + R_B + R_C}{9999} \tag{3}$$

The solution now becomes an algebra problem involving solution of the simultaneous equations (1), (2), and (3) for the values of R_A, R_B, and R_C.

$$\begin{array}{rl}
4999 \times (1) \text{ gives} & 4999R_A + 4999R_B + 4999R_C = 250.2 \\
(2) \text{ gives} & 4999R_A + 4999R_B - R_C = 50 \\
\hline
\text{Subtracting,} & 5000R_C = 200.2
\end{array}$$

$$R_C = \frac{200.2}{5000} = \mathbf{0.040\,04 \ \Omega}$$

$$\begin{array}{rl}
9999 \times (1) \text{ gives} & 9999R_A + 9999R_B + 9999R_C = 500.45 \\
(3) \text{ gives} & 9999R_A - R_B - R_C = 50 \\
\hline
\text{Subtracting,} & 10\,000R_B + 10\,000R_C = 450.45
\end{array}$$

Substituting for R_C, $10\,000R_B + 400.4 = 450.45,$

$$R_B = \frac{50.05}{10\,000} = \mathbf{0.005\,005 \ \Omega}$$

Substituting for R_B and R_C in Equation (1),

$$R_A + 0.005\,005 + 0.040\,04 = 0.050\,05$$

$$\therefore R_A = 0.050\,05 - 0.045\,045 = \mathbf{0.005\,005 \ \Omega}$$

SOLUTION 2

We can avoid the simultaneous equations of Solution 1 by equating the voltage drops across the two parallel branches. On the 1-A range,

$$I_{sh}(R_A + R_B + R_C) = I_M R_M$$

from which $\qquad R_A + R_B + R_C = \dfrac{0.001}{0.999} \times 50 = 0.050\,05 \; \Omega$

as in Solution 1. On the 5-A range,

$$\bar{R}_{sh} = R_A + R_B = (R_A + R_B + R_C) - R_C = 0.050\,05 - R_C$$

and the resistance of the meter branch is $50 + R_C$. Hence

$$I_{sh}(0.050\,05 - R_C) = I_M(50 + R_C)$$

Solving for R_C,

$$R_C = \frac{I_{sh}(R_A + R_B + R_C) - I_M R_M}{I_{sh} + I_M}$$

$I_M R_M$ remains the same for all ranges $= 0.001 \times 50 = 0.05$ V. Note that the denominator becomes the full-scale current for the range. Hence

$$R_C = \frac{4.999 \times 0.050\,05 - 0.05}{5.000} = \mathbf{0.040\,04 \; \Omega}$$

Similarly,

$$R_B + R_C = \frac{9.999 \times 0.050\,05 - 0.05}{10.000} = 0.045\,045$$

from which $\qquad R_B = 0.045\,045 - 0.040\,04 = \mathbf{0.005\,005 \; \Omega}$

and $\qquad R_A = 0.050\,05 - 0.045\,045 = \mathbf{0.005\,005 \; \Omega}$

SELF-CHECK

10-2:1. An ammeter is to be constructed from a 10-mA moving coil movement that has a resistance of 50 Ω and which is connected in series with a 20-Ω calibrating resistor. What value of shunt resistor will give this ammeter a full-scale range of (a) 100 mA, (b) 10 A?

10-2:2. A 1.0-A ammeter is to be constructed so that the voltage drop across the ammeter is 30 mV when a current produces full-scale deflection. The meter movement has a resistance of 100 Ω and the calibrating resistor is 50 Ω. Determine the resistance of the shunt resistor.

10-3 THE VOLTMETER

The same moving-coil movement forms the basis of a practical **voltmeter.** As we have already noted, if our 1-milliampere movement has a total resistance of 50 ohms, a voltage drop of $V = IR = 0.001 \times 50 = 0.05$ V will appear across it when it is reading

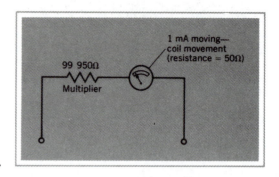

FIGURE 10-8
Simple Voltmeter.

full scale. If we now connect a 99 950-ohm resistor in series with the meter (as in Figure 10-8) to bring the total resistance up to 100 kilohms, the voltage we must apply to produce a 1-milliampere current through the meter (and full-scale deflection) is $V = IR = 0.001 \times 100\,000 = 100$ volts. If we apply a 50-volt potential difference to the voltmeter, the current through it will be $I = V/R = 50/100\,000$ ampere $= 0.5$ mA and the meter will read half scale. Therefore, we can provide this particular voltmeter with a linear scale marked from 0 to 100 volts. The series resistor which is required to convert the basic moving-coil movement into a voltmeter is called a **multiplier resistor.**

To measure open-circuit voltage, a perfect voltmeter should draw no current. Since current must flow in a moving-coil movement to obtain a reading, a 50-microampere basic movement will make a much more accurate voltmeter for use in high-resistance circuits than the 1-milliampere movement. The resistance of a standard 50-microampere movement and its calibrating resistor is 1000 ohms. Therefore, to obtain a 100-volt full-scale range, the total resistance must be $R_T = 100/0.000\,05$ ohms $= 2$ megohms; and the multiplier resistance must be $2\,000\,000 - 1000 = 1\,999\,000$ ohms. To obtain a 500-volt scale with a 50-microampere movement, the total resistance must be ten megohms, and to obtain a 1-volt full-scale range, $R_T = 1/0.000\,05$ ohms $= 20$ kilohms.

No matter what full-scale voltage we select for the 50-microampere basic movement, the total resistance required is 20 kilohms for *each* volt of the full-scale reading. Therefore, when we use a 50-microampere movement as a voltmeter, it has a **sensitivity of 20,000 ohms per volt.** We determine voltmeter sensitivity by dividing the full-scale current into 1 volt. We find the total resistance of a meter movement and the multiplier resistor for a given range by multiplying the full-scale voltage by the voltmeter sensitivity. Therefore, when we use a 1-milliampere movement as a voltmeter, its sensitivity is $1/0.001 = 1000$ ohms per volt, and the total resistance required for a 500-V range will be 500×1000 ohms $= 500$ kilohms.

Multirange-voltmeter circuits are quite straightforward, as shown in Figure 10-9.

EXAMPLE 10-3
A 1-mA full-scale moving-coil movement with a resistance of 50 Ω is to be made into a voltmeter with scales of 5 V, 50 V, 150 V, and 500 V. Calculate R_A, R_B, R_C, and R_D in the circuit of Figure 10-9.

SOLUTION
Voltmeter sensitivity $= 1/0.001 = 1000$ ohms per volt. Therefore, the total resis-

FIGURE 10-9
Multirange Voltmeter.

tance for the 5-V range = 5 kΩ, and

$$R_A = 5000 - 50 = \textbf{4950 } \boldsymbol{\Omega}$$

The total resistance for the 50-V range = 50 kΩ, and

$$R_B = 50\,000 - 5000 = \textbf{45 k}\boldsymbol{\Omega}$$

The total resistance for the 150-V range = 150 kΩ, and

$$R_C = 150 - 50 = \textbf{100 k}\boldsymbol{\Omega}$$

The total resistance for the 500-V range = 500 kΩ, and

$$R_D = 500 - 150 = \textbf{350 k}\boldsymbol{\Omega}$$

The voltmeter is a very useful instrument for checking the operating conditions in an electric circuit. With suitable test leads, we can clip a voltmeter *across* any part of an electric circuit without having to disconnect or disturb the circuit being checked, as is the case when we test with an ammeter.

SELF-CHECK

10-3:1. A voltmeter is to be constructed from a 200-μA movement that has a moving-coil resistance of 100 Ω. (a) What is the sensitivity of this voltmeter? What value of multiplier resistor will give this voltmeter a full-scale reading of (b) 100 mV, (c) 10 V?

10-4 VOLTMETER LOADING EFFECT

Since an ammeter is connected in *series* with a circuit in which the current is to be measured, an ideal ammeter should have zero ohms internal resistance. If the internal resistance of an ammeter is comparable to the resistance of the circuit under test, adding the ammeter to the circuit will increase the total resistance of the circuit appreciably.

Hence, the current indicated by the ammeter will be less than the circuit current when the ammeter is shorted out. However, as we noted in Section 10-2, in most applications the ammeter resistance is sufficiently small in comparison with the circuit resistance that the loading effect of the ammeter is negligible.

Since a voltmeter is connected in *parallel* with the portion of the circuit in which the voltage drop is to be measured, an ideal voltmeter should draw zero current and should, therefore, have an infinitely high internal resistance. Unfortunately, the moving-coil type of voltmeter must draw some current to obtain a pointer deflection. The smaller the current required for full-scale deflection, the more sensitive the voltmeter and the smaller the loading effect it will have on the circuit being tested. As noted in Section 10-3, voltmeter sensitivity is expressed in ohms per volt of full-scale voltage, rather than in terms of full-scale current drain. We can show the effect of voltmeter loading in practical circuits by the following numerical example.

EXAMPLE 10-4
(a) What is the actual voltage drop across the 200-kΩ resistor in Figure 10-10 with no meter in the circuit?
(b) What will a 20 000-ohm-per-volt meter with 150 V full scale read when connected across the 200-kΩ resistor?
(c) What will a 1000-ohm-per-volt meter with 150 V full scale read when connected across the 200-kΩ resistor?

FIGURE 10-10
Effect of Voltmeter Sensitivity.

SOLUTION
(a) From the voltage-divider principle,

$$V = \frac{200 \text{ k}\Omega}{300 \text{ k}\Omega} \times 200 \text{ V} = \mathbf{133.3 \text{ V}}$$

(b) A 20 000-ohm-per-volt meter with a 150-V range has a total resistance of 20 000 × 150 Ω = 3 MΩ. This 3-MΩ resistance is in parallel with the 200-kΩ resistor, thus reducing the resistance of the parallel combination to

$$R = \frac{200 \times 3000}{200 + 3000} = 187.5 \text{ k}\Omega$$

and
$$V = \frac{187.5 \text{ k}\Omega}{287.5 \text{ k}\Omega} \times 200 \text{ V} = \mathbf{130.4 \text{ V}}$$

(c) A 1000-ohm-per-volt meter on a 150-V range has a total resistance of $1000 \times 150 \ \Omega = 150 \text{ k}\Omega$. Therefore, the 200-k$\Omega$ resistor and the voltmeter in parallel have an equivalent resistance of

$$R = \frac{200 \times 150}{200 + 150} = 85.7 \text{ k}\Omega$$

and
$$V = \frac{85.7 \text{ k}\Omega}{185.7 \text{ k}\Omega} \times 200 \text{ V} = \mathbf{92.8 \text{ V}}$$

When measuring voltage in electric power circuits where fairly large currents are flowing, an extra five mA does not alter the circuit conditions appreciably. Thus, the more rugged and less sensitive voltmeters are quite satisfactory. However, in electronic circuits where the normal circuit current is very small, an insensitive voltmeter can alter circuit conditions considerably. Even the 20 000-ohm-per-volt meter used in part (b) of Example 10-4 introduced an error in excess of 2%. Consequently, the simple moving-coil type of voltmeter is not suitable for voltage measurement in high-resistance electronic circuits.

We can still use the simple moving-coil milliammeter to measure voltage by measuring the *drain* current of a field-effect transistor. Over the operating range, the drain current of the FET in the simplified circuit of Figure 10-11 is inversely proportional to the voltage applied between the *gate* and *source* terminals of the FET with the polarity shown. Hence, we can calibrate the milliammeter scale in terms of dc input voltage. As long as the gate of the FET is negative with respect to the source terminal, the gate/source *pn*-junction is *reverse*-biased and the gate current is negligible. Hence, the internal resistance of the **electronic voltmeter** is determined by the value of the gate return resistor in Figure 10-11, which can be several megohms.

FIGURE 10-11
Electronic Voltmeter Principle.

SELF-CHECK

10-4:1. Suppose that you have a simple series circuit that has a dc voltage source supplying a 50-kΩ load. The source has an internal resistance of 10 kΩ and an open-circuit voltage of 12 V. You are going to connect a voltmeter, which has a 50-V full-scale range and a sensitivity of 1000 Ω/V, across the load to monitor the voltage drop.

    (a) What is the "true" voltage drop across the load when the voltmeter is not connected to the circuit?

    (b) What is the reading on the voltmeter when it is connected across the load?

10-4:2. Repeat Self-Check Problem 10-4:1 using a voltmeter with sensitivity of 1 MΩ/V.

10-5 THE POTENTIOMETER

Although the electronic voltmeter is quite satisfactory for most test purposes in high-resistance electronic circuits, its accuracy is not great enough for measuring the open-circuit voltage of sources with small terminal voltages and high internal resistances, such as the cell used to measure the pH of a chemical solution. We can still use a moving-coil movement for this purpose by making use of an accurately known reference voltage and a **potentiometer.**

    When the galvanometer switch in Figure 10-12 is in the center open-circuit position, no current is drawn from the sliding tap B of the potentiometer. Hence, the current through R_{AB} is the *same* as the current through R_{BC}. Consequently,

$$\frac{V_{BC}}{R_{BC}} = \frac{V_{AC}}{R_{AC}}$$

**FIGURE 10-12
Potentiometer Method of
Measuring an Unknown
Source Voltage.**

But both V_{AC} and R_{AC} are fixed. Therefore, the open-circuit output potential difference V_{BC} of the potentiometer is directly proportional to the resistance R_{BC}. We can construct an experimental potentiometer by fastening a length of uniform-resistance wire to a one-meter ruler (see Figure 10-24). The open-circuit output voltage would then be directly proportional to the *length* of wire representing R_{BC}. Commercial potentiometers are usually precision wire-wound resistors with the variable sliding contact arranged so that the ratio of R_{BC} to R_{AC} exactly equals the percentage of the full-scale rotation of the calibrated dial. Although we do not know the absolute magnitude of V_{BC} accurately (since the terminal voltage of the dry cell is not precise), we can *compare* two different values of V_{BC} quite accurately by the ratio of the two potentiometer pointer readings.

To determine the magnitude of the unknown source voltage in Figure 10-12, we carefully adjust the potentiometer until there is no trace of galvanometer deflection when we throw the galvanometer switch to the upper (E_x) position. If there is no meter deflection, there must be no current through the moving coil and, therefore, no potential difference across the meter movement. This, in turn, means that $V_{BC} = E_x$. And, because the meter current is zero, we are measuring the *open-circuit* voltage of the unknown source, regardless of its internal resistance.

Since we do not know the absolute magnitude of V_{BC}, we must compare it with a known reference voltage. The Weston standard cell is a precision laboratory standard intended for use only with potentiometers. It cannot be used to operate a lamp and still serve as a standard. With proper treatment, the Weston cell develops an open-circuit voltage of 1.0183 volts. Having recorded the potentiometer scale reading when $V_{BC} = E_x$, we adjust the potentiometer for zero meter deflection when we throw the switch to the lower (E_s) position. We can now calculate E_x, knowing that the ratio of the two open-circuit voltages equals the ratio of the two potentiometer readings.

EXAMPLE 10-5

The potentiometer circuit of Figure 10-12 gives zero meter deflection in the E_x position for a potentiometer scale reading of 478. Balance is obtained in the E_s position when the potentiometer reading is 662. What is the open-circuit voltage of the unknown source?

SOLUTION

$$\frac{E_x}{E_s} = \frac{478}{662}$$

$$\therefore E_x = 1.0183 \times \frac{478}{662} = \mathbf{0.735\ V}$$

10-6 RESISTANCE MEASUREMENT

We can determine the resistance of the unknown resistor in Figure 10-13 by applying Ohm's law to the readings obtained from the voltmeter and ammeter. This system of

FIGURE 10-13
Measuring Resistance with a
Voltmeter and an Ammeter.

resistance measurement has the disadvantage of having to connect the unknown resistance into a special circuit. However, we can use this method in the electrical laboratory as a means of determining very small resistances such as the armature resistance of a motor, particularly if we want the resistance to be measured with its normal current through it.

With the voltmeter connected as shown in Figure 10-13, the ammeter indicates the sum of the currents through R_x and the voltmeter. Unless the voltmeter current is an insignificant fraction of the current through R_x, the V/I ratio taken from the meter readings (which is the equivalent resistance of the two parallel branches) will be lower than the actual value of R_x. We can check for voltmeter loading by watching the ammeter reading as we disconnect the voltmeter. If there is any noticeable decrease in the current, we then connect the voltmeter on the source side of the ammeter. The ammeter now reads only the current through R_x. But the voltmeter now reads the sum of the voltage drops across the ammeter and R_x. However, if R_x is large enough to show a noticeable voltmeter loading effect, the voltage drop across the ammeter will be insignificant in comparison with V_x and we can neglect ammeter loading effect. If there is no noticeable change in current when we disconnect the voltmeter from its original position across R_x, we reconnect it in its original position, as shown in Figure 10-13.

For quick checks of circuit resistance, we need a simpler apparatus than that of Figure 10-13. We can once more use the basic moving-coil movement to construct an **ohmmeter.** The simple ohmmeter shown in Figure 10-14(a) consists of a 1-milliampere movement, a 4.5-volt battery, and sufficient resistance to permit a current of 1 milliampere when we short-circuit the ohmmeter terminals. A portion of the total resistance is adjustable so that we can calibrate the meter to read exactly full scale when we connect the two test leads together. In this example, the total resistance (including the meter movement) must be

$$R_T = \frac{E}{I} = \frac{4.5 \text{ V}}{1 \text{ mA}} = 4.5 \text{ k}\Omega$$

FIGURE 10-14
Simple Ohmmeter.

EXAMPLE 10-6

In preparing the scale for the ohmmeter of Figure 10-14, determine what resistance values must be marked at full scale, center scale, one-quarter of full scale, and one-tenth of full scale.

SOLUTION

As we have already noted, the total internal resistance of the ohmmeter is adjusted so that the meter reads exactly full scale when the test leads are short-circuited. This represents connecting the test leads to a zero-ohm resistor. Therefore, the full-scale point on the scale must be marked **zero ohms.**

For the meter to read half scale, we must connect the test leads to a resistor which has the same value as the total ohmmeter resistance so that the total resistance in the series loop is now doubled, thus reducing the current to 0.5 mA. Therefore, the center scale point must be marked **4.5 kΩ.**

$$I = \frac{E}{R_M + R_x} = \frac{4.5 \text{ V}}{4.5 \text{ k}\Omega + 4.5 \text{ k}\Omega} = 0.5 \text{ mA}$$

For one-quarter of full scale, the current must be 0.25 mA. Therefore, the total resistance in the loop must be

$$R_T = \frac{4.5 \text{ V}}{0.25 \text{ mA}} = 18 \text{ k}\Omega$$

and

$$R_x = 18 \text{ k}\Omega - 4.5 \text{ k}\Omega = \textbf{13.5 k}\boldsymbol{\Omega}$$

Finally, the resistance value for the one-tenth of full-scale mark will be

$$R_x = \frac{4.5 \text{ V}}{0.1 \text{ mA}} - 4500 \text{ }\Omega = \textbf{40.5 k}\boldsymbol{\Omega}$$

As shown in Figure 10-14(b), the scale of an ohmmeter is nonlinear as far as the spacing of the ohms values is concerned. But since a few ohms one way or the other make a greater *percentage* difference to a low resistance than to a high resistance, this nonlinear scale is not a disadvantage. The simple ohmmeter of Figure 10-14 is limited in its usefulness in that it is not accurate for very low or very high resistances, since these appear at the ends of the scale. For use in electronic circuit testing where we need a high-range ohmmeter, we would start with a 50-microampere movement rather than a 1-milliampere movement. If we use a 50-microampere movement in Figure 10-14(a), the total internal resistance of the ohmmeter will be

$$R_T = \frac{E}{I} = \frac{4.5 \text{ V}}{50 \text{ }\mu\text{A}} = 90 \text{ k}\Omega$$

and, therefore, the center-scale reading of the meter will also be 90 kilohms.

FIGURE 10-15
Low-Range Ohmmeter.

From these examples, we note that the centerscale resistance mark of the ohmmeter is inversely proportional to the full-scale current of the meter. Therefore, we can convert the basic ohmmeter into a low-range ohmmeter by placing a shunt across the moving coil, as shown in Figure 10-15.

EXAMPLE 10-7
Design an ohmmeter using a 1-mA movement with an internal resistance of 50 Ω and a 4.5-V battery to read 45 Ω center scale rather than 4.5 kΩ.

SOLUTION
The total internal resistance of the ohmmeter must be 45 Ω for the meter to read full scale with only the 45 Ω of the meter in circuit and half scale when the total loop resistance is

$$R_x + R_M = 45 + 45 \ \Omega$$

Therefore, full-scale current must be

$$I = \frac{E}{R} = \frac{4.5 \ \text{V}}{45 \ \Omega} = 0.1 \ \text{A}$$

Since the meter movement passes 1 mA at full scale, the shunt current must be 99 mA and the shunt resistance

$$R_{\text{sh}} = \frac{1}{99} \times 50 = \mathbf{0.505 \ \Omega}$$

The total resistance of the meter movement and shunt in parallel is

$$R = \frac{50 \times 0.505}{50 + 0.505} = 0.500 \; \Omega$$

Therefore, the total resistance of the series resistor and the "ohms adjust" rheostat will be

$$R_s = 45 - 0.5 = \mathbf{44.5 \; \Omega}$$

As a test instrument, the ohmmeter has the same advantage as the voltmeter in that we can connect it *across* a portion of a circuit without disconnecting any wiring. However, since the ohmmeter is calibrated on the basis of its own voltage source (the 4.5-volt battery), we must make sure that no other voltage source is acting in the circuit while we measure its resistance.

When we wish to measure a very high resistance such as the leakage between the two conductors of a cable, as in Figure 10-16, we can apply our knowledge of electric circuit theory to use a voltmeter and a separate voltage source to form a high-resistance ohmmeter. This system has the added advantage that we can measure resistance under actual applied-voltage conditions and thus check for voltage breakdown of an insulator.

FIGURE 10-16
Using a Voltmeter as a
High-Resistance Ohmmeter.

EXAMPLE 10-8

The voltmeter in Figure 10-16 is a 20 000-ohm-per-volt meter with a 150-V scale. When the switch is closed, it reads 120 V and when the switch is open it reads 10 V. What is the leakage resistance of the cable insulation?

SOLUTION

The resistance of the voltmeter is

$$R_M = 20\,000 \; \Omega/V \times 150 \; V = 3 \; M\Omega$$

The voltage drop across the leakage resistance of the cable is $V = 120 - 10 = 110$ V. And since the resistance of the voltmeter and the leakage resistance of the cable form

a simple series circuit,

$$\frac{R_x}{R_M} = \frac{V_x}{V_M}$$

$$\therefore R_x = 3 \text{ M}\Omega \times \frac{110 \text{ V}}{10 \text{ V}} = 33 \text{ M}\Omega$$

For precision resistance measurements, we can use various forms of the Wheatstone bridge shown in Figure 10-17. If we adjust the circuit of Figure 10-17 in such a way that there is absolutely no deflection of the galvanometer when we close the switch, the voltage drops across R_x and R_A must be exactly the same, so that no potential difference appears across the galvanometer to cause current in its moving coil. With no current through the meter movement, $I_x = I_y$. Hence,

$$V_x = I_x R_x \quad \text{and} \quad I_x = \frac{E}{R_x + R_y}$$

FIGURE 10-17
Measuring Resistance with a Wheatstone Bridge.

Therefore,

$$V_x = \frac{E R_x}{R_x + R_y}$$

Similarly,

$$V_A = \frac{E R_A}{R_A + R_B}$$

Therefore, for perfect balance,

$$\frac{E R_x}{R_x + R_y} = \frac{E R_A}{R_A + R_B}$$

Cancelling out E and cross-multiplying gives

$$R_x R_A + R_x R_B = R_x R_A + R_A R_y$$

Canceling out $R_x R_A$,

$$R_x R_B = R_y R_A$$

or

$$R_x = \frac{R_A R_y}{R_B}$$

(10-3)

For the bridge to be balanced, the product of one pair of opposite arms must equal the product of the other pair of opposite arms of the bridge.

Since E does not appear in Equation (10-3), the magnitude of the source voltage used with a bridge circuit has no effect on the accuracy of the measurement. The source merely causes a deflection of the galvanometer pointer if the bridge is not properly balanced. With precision resistors for R_A, R_B, and R_y, one or more of which must be adjustable so that we can obtain a balance, we can then use Equation (10-3) to determine the value of R_x to a high degree of accuracy.

EXAMPLE 10-9

The Wheatstone bridge circuit of Figure 10-17 is balanced when $R_A = 1.0\ \Omega$, $R_B = 50\ \Omega$, and $R_y = 17\ \Omega$. What is the resistance of R_x?

SOLUTION

$$R_x = \frac{R_A R_y}{R_B} = \frac{1.0 \times 17}{50} = \textbf{0.34}\ \boldsymbol{\Omega}$$

SELF-CHECK

10-6:1. You are building a simple ohmmeter, similar to that shown in Figure 10-14, using a 5.0-V battery. The center-scale resistance value is to be 100 kΩ.
(a) What is the internal resistance necessary for your ohmmeter?
(b) What is the full-scale current rating on the movement you will need?
10-6:2. A Wheatstone bridge has fixed resistances of $R_A = 10.0$ kΩ and $R_B = 200\ \Omega$. When the resistance of a 3.3 k$\Omega \pm 10\%$ (nominal) resistor is measured, the bridge balances when $R_y = 65\ \Omega$. Is the resistor within tolerance?

10-7 MOVING-COIL MOVEMENT IN AC CIRCUITS

Since the magnetic field produced by the *permanent* magnet of the moving-coil movement must maintain a fixed direction, it follows that current flowing one way through the moving coil develops a clockwise torque, and current flowing in the other direction develops a counterclockwise torque. With the spiral spring adjusted so that the zero-current position of the pointer is center-scale (a galvanometer), the movement can indicate current in either direction. If we place the instrument in a circuit where the current builds up first in one direction, drops back to zero, and then builds up in the other direction, the torque will change from instant to instant, as shown in Figure 10-18(a). A negative value of torque on this graph simply indicates that the torque is attempting to rotate the pointer in the opposite direction to that produced by a positive value of torque.

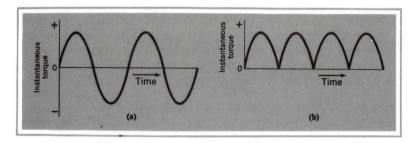

FIGURE 10-18
Instantaneous Torque Developed in a Moving-Coil Movement: (a) by an
Alternating Current; (b) by a Rectified Alternating Current.

If this variation in current is quite slow (once a second), the pointer will be able to follow
the changing torque by swinging to the right and then to the left of zero. Since the pointer
is never stationary, it is difficult to obtain a reading having any significance.

If the current direction alternates many times a second, the mechanical inertia of
the moving-coil assembly is too great for the pointer to follow the instantaneous variations in torque (and in current through the moving coil). Therefore, the pointer must take
up a position dictated by the *average* value of the changing torque. The average value
of the sine wave in Figure 10-18(a) is zero. The pointer must then remain at the zero mark
on the scale, both for the center-zero galvanometer and for those movements with spiral
springs adjusted to place the zero mark at the left-hand end of the scale. Because of their
permanent magnets, moving-coil movements are basically *direct-current* instruments.

The advantage of using batteries as the voltage sources in the resistance networks
we have encountered in preceding chapters is that we do not have to consider the *time*
element in our calculations. The *direct* currents in the various branches of dc networks
have constant magnitudes and directions. In practical electrical systems, however, most
voltage sources generate an *alternating* emf which reverses its polarity periodically at
rates ranging from several times a second (power-line frequencies) to many millions of
times a second (radio frequencies).[†] Hence, we require measuring devices which do not
have the limitations of the basic moving-coil movement.

In Section 3-6, we discovered that semiconductor *pn*-junctions permit current to
flow freely in one direction but have a high resistance to current in the opposite direction.
We can use a **bridge rectifier** composed of four such *pn*-junctions to reverse the leads
of a moving-coil meter movement automatically for each half-cycle of alternating current, as shown in Figure 10-19. The arrow of the graphic symbol for a semiconductor
rectifier indicates the direction of minimum resistance for conventional current. The
current through the meter movement, and consequently the instantaneous torque, now
has the form shown in Figure 10-18(b). As a result, the pointer deflection will be
proportional to the *average* value of the alternating current.[‡]

To use the rectifier-type meter as an ammeter, we require an ammeter shunt resistor
across the input terminals of the bridge rectifier in Figure 10-19. Since the moving coil

[†] See Section 16-2 for a detailed description of an alternating voltage.

[‡] As defined in Section 16-10. If the meter is to be used in sine-wave ac circuits, the fixed 1.1 : 1 ratio
between rms and average values of a sine wave (form factor) allows us to calibrate the meter scale to read
rms values.

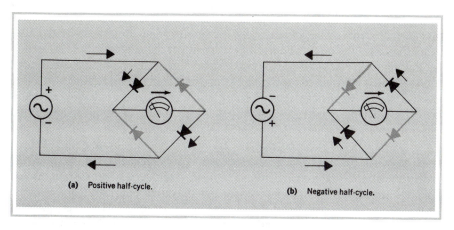

(a) Positive half-cycle.

(b) Negative half-cycle.

FIGURE 10-19
Moving-Coil Movement with a Bridge Rectifier.

is not a pure resistance, the split in current between the shunt and the meter depends on the frequency of the alternating current. Consequently, a rectifier-type ammeter is accurate only for a specific frequency at which it is calibrated. Hence, moving-coil movements are seldom used in ac ammeters.

Since the multiplier resistance of a voltmeter is in *series* with the moving coil, slight variations in coil impedance are not a problem. Therefore, the rectifier and moving-coil movement ac voltmeter is widely used in **multimeters** in which a single meter movement is used for both dc and ac voltage measurements.

10-8 MOVING-IRON MOVEMENT

A more economical movement for use primarily in alternating-current circuits is the **moving-iron** movement, shown in Figure 10-20. It consists of a stationary solenoid with

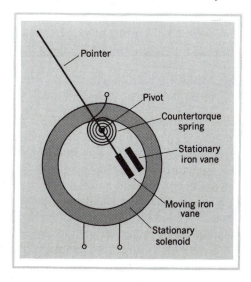

Pointer

Pivot

Countertorque spring

Stationary iron vane

Moving iron vane

Stationary solenoid

FIGURE 10-20
Moving-Iron Movement.

two soft-iron vanes within the coil. One vane is stationary, but the other is fastened to the pivot which carries the pointer and the counter-torque spring. Since the soft-iron vanes are located at the point where the flux density of the magnetic field created by the current through the solenoid is maximum, they become temporary magnets. Since the S pole of one vane is adjacent to the S pole of the other (the two N poles are also adjacent), the pivoted vane is *repelled* by the stationary vane. This repulsion is proportional to the current in the solenoid. Figure 10-20 shows only one of several possible shapes in which the soft-iron vanes may appear. The scale of this moving-iron instrument is not as linear as that of the moving-coil instrument; it is cramped at the low-current end of the scale. Scale linearity can be improved by the use of specially shaped soft-iron vanes.

We can calibrate the moving-iron movement in terms of *direct* current through the solenoid and sometimes use it as an inexpensive direct-current instrument where we do not require the greater accuracy of the moving-coil instrument. However, the moving-iron instrument was developed primarily for measuring *alternating* currents at power-line frequencies. When the direction of the current in the solenoid is reversed, the N and S poles of *both* vanes reverse, and the moving vane is still *repelled* by the stationary vane. Therefore, when alternating current flows in the solenoid, the instantaneous torque developed by the moving-iron vane is in the form of pulses which are always in the same direction, as shown in Figure 10-18(b). Because the mechanical inertia of the whole pointer and moving-vane assembly is too great to allow it to follow the instantaneous variations in torque, the pointer will indicate the *average* value of the torque. When calibrating the scale of a moving-iron instrument for alternating current, we must allow for hysteresis losses in the iron vanes. Therefore, the calibration of an ac moving-iron instrument is accurate only for power-line frequencies, not for direct current or high-frequency alternating currents.

Since the stationary solenoid of the moving-iron movement can be wound with a few turns of heavy wire or many turns of fine wire in order to produce the required magnetic field for full-scale deflection, the range of full-scale current readings available is much greater than for moving-coil instruments. The basic ammeter movement for ac power circuits has a five-ampere full-scale reading. Shunts similar to those used with moving-coil movements are used for larger currents. Although it is not possible to produce moving-iron movements that are as sensitive as moving-coil instruments, we can use one with a solenoid consisting of many turns of fine wire along with suitable series multiplier resistors to form a voltmeter for use in ac circuits.

10-9 ELECTRODYNAMOMETER MOVEMENT

As we noted in the preceding section, the fixed direction of the magnetic field of the permanent magnet is responsible for limiting the moving-coil movement to dc (or rectified ac) circuits. If we replace the two poles of the *permanent* magnet with an *electromagnet* consisting of two stationary coils, as shown in Figure 10-21, we have an **electrodynamometer** movement in which we can reverse the direction of the stationary magnetic field at will. Since it contains no iron, we can use the electrodynamometer as the basic movement for either dc or ac instruments. We can use it to measure *alternating* current by connecting the stationary and moving coils in series. Whenever the current in the moving coil reverses its direction, the magnetic field produced by the stationary coil

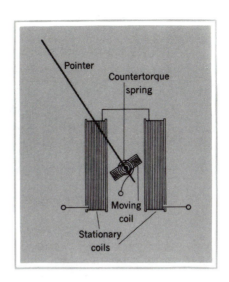

FIGURE 10-21
Electrodynamometer Movement.

also reverses. Therefore, the resulting torque tends to move the pointer in a clockwise direction, regardless of the direction of the current in the coils. However, for voltmeter and ammeter applications, the electrodynamometer movement is too costly to compete with either the moving-coil movement for dc measurement or the moving-iron movement for ac measurement.

In developing Equation (10-2) for the torque of a moving coil, we noted that the force acting on a conductor is proportional to the *product* of the stationary magnetic field strength and the current in the moving coil. Therefore, the scale reading of an electrodynamometer movement is proportional to the product of the stationary-coil and moving-coil currents. If the wire on the stationary coils is heavy enough to be connected in series with a load, as in Figure 10-22, the stationary-coil current is the same as the load current. If the moving coil with a multiplier resistor in series with it is connected across the load, the moving-coil current is directly proportional to the load voltage. Therefore, the scale reading of the electrodynamometer movement is proportional to the product of the load current and voltage. Since $P = IV$, we have a direct-reading **wattmeter.** If we connect a wattmeter to a load in an ac circuit, both load current and load voltage reverse their direction the same number of times per second, thus producing a net positive torque. Therefore, we can use the electrodynamometer movement as a wattmeter in either dc or ac circuits. Although we can calculate the power in a load by taking separate ammeter and voltmeter readings, a direct-reading wattmeter is a very useful instrument, particularly in ac circuits.

If we reverse the leads to *either* the current coil *or* the voltage coil when we connect a wattmeter into a circuit, a *negative* torque, which moves the pointer *down scale,* will be developed. To assist us in connecting a wattmeter properly, it is customary in practice to identify one end of each coil with ± marks, as shown in Figure 10-22. Standard connection requires that we connect the identified terminal of the current coil to the generator and the identified terminal of the voltage coil to the line containing the current coil, as shown in Figure 10-22.

FIGURE 10-22
Electrodynamometer Movement as a Wattmeter.

10-10 DIGITAL METERS

We now call watches with hands *analog* watches to distinguish them from *digital* watches. Similarly, the meter movements we have considered in this chapter are *analog* instruments. The term *analog* pertains to data in the form of some *continuously* variable quantity such as the angular position of watch hands or the angular displacement of a meter pointer across the scale of the meter. A *digital* device, on the other hand, displays data as discrete numerals with no in-between positions.

For many years, the most common piece of electrical test equipment has been the **multimeter** or VOM (volt-ohm-milliammeter) in which a multiposition switch combines the dc milliammeter ranges of Figure 10-17(b), the dc and ac voltmeter ranges of Figures 10-9 and 10-19, and the ohmmeter ranges of Figures 10-14 and 10-15. The multimeter's single moving-coil movement requires a multiplicity of scales which are not always easy to read accurately. As improved production techniques reduce the size and cost of integrated-circuit components, digital multimeters are rapidly gaining in popularity. They have the advantages of a high input resistance like the electronic voltmeter of Figure 10-11, greater accuracy than moving-coil meters, and nonambiguous readout of the numerical value of the measured quantity.

Digital meters usually operate by *counting* a number of pulses from an accurate oscillator circuit. These pulses are fed to the digital counter through a gate circuit which switches the flow of pulses on and then off after a precise time interval. This time interval is determined by an analog to digital converter circuit that develops a gating pulse which is directly proportional in width (time duration) to the magnitude of the electrical quantity being measured. The number of oscillator pulses getting by the gate, and therefore the numerals displayed on the digital readout, are calibrated to represent the voltage, current, or resistance being measured to well within 1% accuracy.

PROBLEMS

10-1-1. What force in newtons acts on a conductor which is perpendicular to a magnetic field whose flux density is 0.8 T if 15 cm of the conductor are in the magnetic field and the current in the conductor is 5 A?

10-1-2. The countertorque spring of a certain moving coil instrument develops a torque of 10^{-5} newton meter when the pointer is at full scale. The flux density in the air gap is 5×10^{-2} T, and there are 80 turns of wire wound on a form whose dimensions are 2 cm parallel to the pivot and 1.5 cm at right angles to the pivot. What current must flow through the coil for half-scale deflection?

10-2-3. The full-scale voltage drop across a 5-mA movement is 50 mV. Calculate the resistance of the shunt required if the movement is to be used as a 1-A full-scale ammeter.

10-2-4. If the shunt calculated in Problem 10-2-3 is used with a 1-mA movement having a total resistance of 50 Ω, what current would its full-scale reading represent?

10-2-5. Calculate an Ayrton shunt to provide 100-mA, 1-A, and 10-A ranges with a 50-μA movement whose total resistance is 1000 ohms.

10-2-6. Calculate an Ayrton shunt for 30-mA, 100-mA, 300-mA, and 1-A ranges with a 1-mA movement whose total resistance is 50 ohms.

10-3-7. Calculate the multiplier resistors required to form a voltmeter with 10-V, 50-V, 100-V, and 500-V ranges, using the meter movement of Problem 10-2-5.

10-3-8. The resistance of the movement for a 200 ohm-per-volt voltmeter is 10 ohms. Calculate the multiplier resistors required for 30-V, 150-V, and 1500-V ranges.

10-4-9. The terminal voltage of a 12-V source whose internal resistance is 100 kΩ is checked by a 20 000-ohm-per-volt voltmeter on its 50-V range. What is the voltmeter reading?

10-4-10. The bias voltage of a FET is applied to its gate terminal through a 470-kΩ resistor. If a 1000-ohm-per-volt meter on its 10-V range measures the voltage at the gate terminal (with respect to ground) as 0.8 V, what will it be when the voltmeter is removed if there is negligible gate current?

10-4-11. Resistor R_2 in the voltmeter shown in Figure 10-23 is open-circuit, thus its value cannot be determined. However, R_1 and R_3 can be measured as 97 kΩ and 300 kΩ, respectively. What value of resistance for R_2 is required to repair the voltmeter?

FIGURE 10-23
Multirange Voltmeter.

10-4-12. Using the data from Problem 10-4-11, calculate the meter resistance, full-scale current, and voltmeter sensitivity for the voltmeter shown in Figure 10-23.

10-5-13. In the potentiometer circuit of Figure 10-12, the Weston standard-cell voltage equals the potentiometer output voltage when R_{BC} is 71.0% of full-scale resistance. To obtain zero galvanometer deflection with the switch in the E_x position, we must set the potentiometer at 34.5% of full-scale resistance. What is the open-circuit voltage of the unknown source?

10-5-14. We want to use the potentiometer in Figure 10-12 to produce an open-circuit output voltage V_{BC} of 1.0 V. We first calibrate the potentiometer with a Weston standard cell, which requires a potentiometer setting of 654 to produce a balance. What potentiometer setting is required for the desired 1.0-V output?

10-6-15. What is the center-scale reading of a basic ohmmeter constructed with a 50-μA movement and a $1\frac{1}{2}$-V battery?

10-6-16. What battery voltage would be required to use the same scale on a 1-mA movement?

10-6-17. What resistance must be connected to the terminals of the ohmmeter in Problem 10-6-15 for the pointer to read 20% of full-scale deflection?

10-6-18. Design an ohmmeter with 1000 Ω appearing at center scale, using a $1\frac{1}{2}$-V battery and a 50-μA movement having a resistance of 1000 ohms.

10-6-19. The insulation resistance of a solenoid is checked by connecting one terminal of a 1-kV source to the solenoid and the other terminal of the source through a 20 000-ohm-per-volt voltmeter having a 1500-V full-scale reading to the iron core of the solenoid. If the voltmeter reads 360 V, what is the insulation resistance?

10-6-20. What would the meter read in Problem 10-6-19 if the leakage resistance between the solenoid winding and the core dropped to 0.5 MΩ?

10-6-21. In the Wheatstone bridge circuit of Figure 10-17, balance is attained when $R_A = 5$ kΩ, $R_B = 1000$ Ω, and $R_y = 42$ kΩ. What is the resistance of R_x?

10-6-22. What value must R_y be set at to obtain a balance in the Wheatstone bridge of Figure 10-17 if $R_A = 1000$ Ω, $R_B = 1200$ Ω, and $R_x = 600$ Ω?

10-6-23. When the Wheatstone bridge in Figure 10-17 is balanced, V_A is 2.4 V and I_B is 0.24 A. If the total current drawn from the battery is 300 mA, what is the resistance of R_x?

10-6-24. Figure 10-24 is a variation of the Wheatstone bridge known as the **slide-wire bridge.** The slide wire is a length of uniform resistance wire 1 m long. The standard resistor $R_s = 24$ Ω. What is the value of R_x if the bridge is balanced when the slider is 37 cm from the end that is connected to the standard resistor?

10-6-25. Figure 10-25 is a variation of the Wheatstone bridge known as the **Varley loop.** It is used for locating grounds on long transmission lines. To accomplish this, the far ends of two identical conductors are connected together. Assume negligible resistance between the ground on the switch and the fault ground on the transmission line. With the switch in the *line* position, the bridge balances with $R_A = 100$ Ω, $R_B = 1000$ Ω, and $R_C = 200$ Ω. When the switch is thrown to the *ground* position, R_A and R_B are unchanged but R_C is changed to 150 Ω to obtain balance. How far is the ground from the end of the cable at which the apparatus is located?

FIGURE 10-24
Slide-Wire Bridge.

FIGURE 10-25
Varley Loop.

REVIEW QUESTIONS

10-1. Draw a sketch showing the composite field when an electric conductor with a conventional current direction coming out of the page is placed in an air gap in a magnetic circuit in which the lines of force pass from left to right across the air gap.

10-2. Explain why the composite magnetic field in Question 10-1 has the shape you have shown.

10-3. Explain why the composite magnetic field in Figure 10-3 has the pattern shown.

10-4. Why is the force acting on a coil of wire in a magnetic field directly proportional to the number of turns in the coil?

10-5. Why is the flat spiral spring essential in all the meter movements described in this chapter?

10-6. What is the relationship between Hooke's law and scale markings on a moving-coil movement?

10-7. What would be the effect on the operation of a moving-coil movement if the soft-iron cylinder (shown in Figure 10-4) were left out during the assembly of the instrument?

10-8. What is a **galvanometer?**

10-9. Why is a galvanometer used in the circuit of Figure 10-17 rather than a microammeter?

10-10. What is the purpose of the **calibrating resistor** used with moving-coil movements?

10-11. Suggest an advantage in arranging that moving-coil movements have a 50-mV drop across their terminals when full-scale current flows through them. (Refer to Problem 10-2-4.)

10-12. Show that

$$R_S = R_M \left(\frac{I_M}{I_T - I_M} \right)$$

where R_S is the resistance of an ammeter shunt, R_M is the resistance of the moving-coil movement and its calibrating resistor, I_M is the current in the moving coil required for full-scale deflection, and I_T is the desired full-scale ammeter reading.

10-13. Figure 10-26 shows the right and wrong ways of connecting an ammeter with an external shunt into a circuit. What would be the effect of connecting the circuit incorrectly as in Figure 10-26(a)?

10-14. Why is the insertion resistance of a multirange ammeter equipped with an Ayrton shunt greater than that of an ammeter which uses a separate shunt for each range?

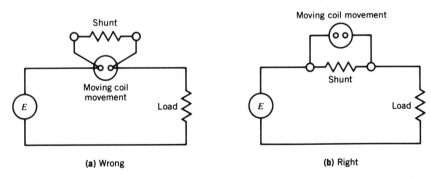

(a) Wrong　　　　　　　　　　　**(b) Right**

FIGURE 10-26
Connecting an Ammeter with an External Shunt into a Circuit.

10-15. The basic moving-coil movement is a *current*-indicating device. How is it possible to calibrate it as a voltmeter?

10-16. What is the significance of **voltmeter sensitivity?**

10-17. Why is a 20 000-ohm-per-volt meter more suitable than a 1000-ohm-per-volt meter for checking electronic circuitry?

10-18. Why is a 200-ohm-per-volt meter satisfactory for use with electrical machinery?

10-19. Why is voltmeter loading effect more of a problem in practical measurements than ammeter insertion resistance?

10-20. Why is the loading effect of an electronic voltmeter considerably less than that of a 20 000-ohm-per-volt meter?

10-21. With a precision potentiometer, we can readily measure an unknown voltage to a three-figure accuracy. Does the galvanometer scale have this degree of accuracy? Explain.

10-22. With the switch of the potentiometer circuit in Figure 10-12 in the E_x position, when we rotate the potentiometer dial so that R_{BC} increases, the meter deflection decreases but does not reach zero. Explain the condition which would give this effect.

10-23. What effect will a 2% decrease in the dry-cell voltage in the potentiometer circuit of Figure 10-12 have on the accuracy of the instrument? Explain.

10-24. Discuss the possible errors that might occur when using the voltmeter/ammeter method of determining the resistance of a 50-kΩ resistor. (Draw a circuit diagram.)

10-25. What precaution must be observed when checking the resistance of an electric circuit with an ohmmeter?

10-26. Lay out a scale (see Figure 10-14) for a simple ohmmeter, using a 50-μA movement and a 1.5-V battery.

10-27. Lay out a scale for the low-resistance ohmmeter shown in Figure 10-27.

FIGURE 10-27
Low-Resistance Ohmmeter.

10-28. What would be the effect of connecting an ammeter *across* a load?

10-29. What would be the effect of connecting a voltmeter in *series* with a load?

10-30. What circuit information could be obtained from the voltmeter reading in Question 10-29?

10-31. Which of the following movements would you select for the Wheatstone bridge circuit of Figure 10-17: a 50-μA movement or a 5-mA movement? Explain your selection.

10-32. What effect does the accuracy of the galvanometer scale calibration have on the accuracy of the resistance measurements made with a Wheatstone bridge?

10-33. Why does a moving-coil movement read zero in an ac circuit?

10-34. Explain why a bridge rectifier "automatically reverses the leads of the meter movement for each half-cycle of alternating current."

10-35. A typical multirange volt-ohm-milliammeter (VOM) has the following ranges:
Volts dc: 2.5 V, 10 V, 50 V, 250 V, 1000 V
Volts ac: 2.5 V, 10 V, 50 V, 250 V, 1000 V
Ohms: $R \times 1$, $R \times 10$, $R \times 1000$, $R \times 10\,000$
Milliamperes dc: 100 μA, 1 mA, 10 mA, 50 mA, 250 mA, 1 A, 10 A
Why is there no provision for measuring alternating current?

10-36. Why is the pointer of a moving-iron movement able to indicate a steady reading in an ac circuit?

10-37. Why is it possible to use the same electrodynamometer-type wattmeter in both dc and ac circuits?

*10-38. Design a personal computer program which will print out the three resistor values in the Ayrton shunt of Figure 10-7(b) upon entering meter current for full-scale deflection, meter resistance, and the three desired current ranges.

*10-39. Having acquired a 50-μA moving-coil meter movement in good condition, you plan to construct a basic ohmmeter (similar to Figure 10-4) using a 3-V battery and suitable series resistance. The linear scale of the meter is laid out in 2-μA divisions from 0 to 50 μA. Develop a personal computer program to prepare a table of resistance values for each mark on the original scale in preparation for plotting a new ohmmeter scale for the meter.

PART III CAPACITANCE AND INDUCTANCE

THERE ARE THREE BASIC PROPERTIES that an electric circuit may possess. In all the electric circuits we have considered so far, we have encountered only one property—the tendency of the circuit to oppose electric current through it. This property is called **resistance**. No matter how complex an electric circuit may be, there are only two additional properties that it may possess. A second property tends to oppose any *change* in the electric current through it. This property is called **inductance**. The third property tends to oppose any *change* in the potential difference or voltage between two points in an electric circuit. This property is called **capacitance**.

In Part 3, we shall investigate the nature of these two additional properties and discover their behavior in direct-current circuits. Since inductance depends on the magnetic fields associated with an electric current, we shall also review the principles of magnetism and magnetic circuits.

11

CAPACITANCE

11-1 STATIC ELECTRICITY

Since capacitance is associated with potential difference, we must first review some of the basic concepts concerning potential difference that we developed in Chapter 2. We noted that atoms and molecules in their normal state are electrically neutral. To create an electric potential difference, we must first *separate* some valence electrons from one group of neutral atoms—thus leaving them with a deficiency of electrons or a net positive electric charge—and then add these surplus electrons to a second group of atoms thus providing them with a net negative electric charge. To *maintain* a continuous electric current in a dc circuit, we employ electric "charge separators," such as batteries, which maintain a potential difference between their terminals by continuous expenditure of chemical energy.

But electric charge separation by friction between electric insulating materials was discovered many centuries before the earliest electric batteries were invented. Electric charges that are separated by friction between insulating materials are known as **static electricity**. The study of the behavior of these charges is called **electrostatics**.

Although static electricity is often no more than an annoyance encountered as a result of the electric charge separation that takes place when shoe leather is rubbed against a carpet (particularly when the air is dry), we shall find that the behavior of capacitance in electric circuits is based on the principles of electrostatics demonstrated by Charles Coulomb some 200 years ago.

11-2 THE NATURE OF AN ELECTRIC FIELD

In Section 2-1, we noted that Coulomb, using a delicate torsion balance, was able to measure the force that exists between two electrically charged pith balls. The result of these measurements led to

Coulomb's law of electrostatic force: The electric force between two point electric charges is directly proportional to the product of the two charges and inversely proportional to the square of the spacing between the charges.

We can now add units to Equation (2-1), giving

$$F = k\frac{Q_1 Q_2}{s^2} \tag{11-1}$$

For F to represent electric force in newtons, Q_1 and Q_2 the two charges in coulombs, and s the spacing between the charges in meters, the constant of proportionality k has a numerical value of 8.99×10^9.[†]

We have already observed that electric force requires no physical contact between the two electric charges. Hence, we describe electric force as a *field* force. We can extend this field concept by examining what happens when a tiny positive charge (such as a hydrogen ion which consists of one proton) is placed in the vicinity of a fixed positive charge. We now know that the repulsion of the *like* charges will cause the proton to move radially away from the fixed positive charge. We say that the tiny positive charge is acted upon by the **electrostatic** (or, simply **electric**) **field** of the fixed positive charge. We can show this electric field diagrammatically by drawing the paths along which the tiny + charge will move away from the fixed charge, as shown in Figure 11-1(a). It follows that the electric field surrounding a fixed negative charge will *attract* the tiny + charge, resulting in the diagrammatic representation of Figure 11-1(b). We can say that

An electric field is that region in which an electric charge is acted upon by an electric force.

FIGURE 11-1
Electric Fields Surrounding Point Electric Charges.

[†] Strictly speaking, the constant k also has to have a unit such that other units in Equation (11-1) cancel out leaving only newtons. Using the dimensional analysis techniques of Section 1-7, the unit for k becomes $N \cdot m^2/C^2$.

In studying electric circuits, we are not particularly concerned with the electric fields surrounding isolated electric charges in space. Since we are concerned with electric potential difference between two points in an electric circuit, we are more interested in the electric field between two unlike electric charges. We can create such a field by connecting two parallel metallic plates to the two terminals of a voltage source, as shown in Figure 11-2. The left-hand plate becomes positively charged, since the + terminal of the battery removes some electrons from it; and the right-hand plate becomes negatively charged as the − terminal of the battery forces a surplus of electrons into it. If we now place a tiny positively charged particle between the plates, it will move toward the right-hand plate. If the particle has a negative charge, it will move toward the left-hand plate. Again, we can draw lines to show the paths followed by the tiny charged particle. These lines represent the direction of action of the electric force on the charged particle. Hence, we can think of an electric field as consisting of invisible electric lines of force, where

An electric line of force represents the path along which a weightless electrically charged particle will move from one electrically charged plate to the other.

FIGURE 11-2
Electric Field Between Parallel Plates.

Since a positively charged particle can travel only toward the right-hand plate and a negatively charged particle can travel only toward the left-hand plate, we can consider that electric lines of force possess *direction*. To agree with the conventional direction for electric current, we assume that

The direction of an electric line of force is the direction in which a positively charged particle will travel between the two plates.

From the sketches in Figures 11-1, 11-2, and 11-3, we note that

Electric lines of force always begin and end on charged bodies and always meet a charged surface at right angles to that surface.

Although we are concerned mainly with the electric field between parallel plates, two other electric field patterns are of interest to us—the electric field between parallel

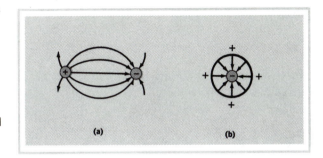

FIGURE 11-3
Electric Field between (a) Parallel Conductors, and (b) Concentric Conductors.

(a)   (b)

conductors, as shown in Figure 11-3(a), and the electric field between concentric conductors, as shown in Figure 11-3(b).

As we have noted, a positively charged particle in the electric field between the two parallel plates will be repelled by the + plate and attracted by the − plate. Because we developed the concept of an electric field to show the force that will be exerted on charged particles, we must be able to express the **strength** or **intensity*** of the electric field numerically.

The letter symbol for electric field strength is the Greek letter E (capital epsilon).

To distinguish the symbol for electric field strength from the symbol for applied voltage, we can use **boldface** type. Boldface type is used to call attention to *vector* quantities which possess *direction* as well as magnitude. No special unit is required for electric field strength since we can express it in terms of the charge on the particle and the force acting on the particle.

An electric field has a strength of one unit at a certain point if a force of one newton is exerted on an electric charge of one coulomb located at that point.

We may express this in equation form as

$$\mathbf{E} = \frac{\mathbf{F}}{Q} \tag{11-2}$$

where $\mathbf{E}$ is the electric field strength in **newtons per coulomb,** *$\mathbf{F}$ is the force acting on the charge in newtons, and Q is the magnitude of the charge in coulombs.* Since we show electric field strength as a vector quantity on the left-hand side of Equation (11-2), we should also show the vector nature of $\mathbf{F}$ on the right-hand side.

In the special case of the electric field between parallel plates, the electric field is quite uniform between the plates. Therefore, a constant force is exerted on a charged particle as it moves from one plate to the other. (As the repulsion of one plate becomes weaker with an increase in distance, the attraction of the other plate increases as the distance decreases.) As the charge moves the complete distance from one plate to the

---

*Electrical engineering prefers the term **electric field strength**, rather than **electric field intensity** sometimes used in physics.

other, the work done is

$$W = Fs \qquad \text{(2-3)}$$

Since we established the magnitude of the volt in terms of the work done in moving a charge through a potential difference,

$$W = QV$$
$$\therefore Fs = QV \qquad \text{(2-4)}$$

Substituting for F from Equation (11-2),

$$EQs = QV$$

from which

$$E = \frac{V}{s} \quad \text{(parallel plates)} \qquad \text{(11-3)}$$

where E is the electric field strength in **newtons per coulomb** *or* **volts per meter,** *V is the potential difference between the plates in volts, and s is the spacing between the plates in meters.*

This expression for electric field strength leads to an interesting characteristic of electric fields. According to Kirchhoff's voltage law, when we apply 400 volts to the two parallel plates of Figure 11-4, there must be a 400-volt potential difference between them. Since the electric lines of force begin at the + plate and end at the − plate, this 400-volt potential difference appears across each line of force. We can imagine an "ideal" voltmeter which draws zero current with one lead connected to the bottom plate. Since Figure 11-4 shows this bottom plate connected to ground, we may now refer all our voltage readings to ground without having to say "with respect to" for every point at which we we locate the probe connected to the other voltmeter terminal.

FIGURE 11-4
Showing the Voltage Gradient along an Electric Line of Force.

When we touch the voltmeter probe to the top plate, the meter will read 400 volts. As we move the voltmeter probe one quarter of the way down from the top plate, the meter reads 300 volts; when the probe is halfway between the plates, the meter reading

PART III   CAPACITANCE AND INDUCTANCE

is 200 volts, and so on. Therefore, in the case of parallel plates, the total potential difference is distributed uniformly along the length of the electric line of force as if it were a resistance voltage divider. From this characteristic, we can speak of the **voltage gradient** in volts per meter. Because of the uniform electric field between parallel plates, in this case the voltage gradient is numerically equal to the electric field intensity. Therefore, we can express E in volts per meter.

Since every electric line of force has the total potential difference distributed along its length, we can mark a point on each line of force having the same potential difference with respect to ground. Joining these points produces what is known as an **equipotential surface** for the electric field. Equipotential-surface "maps" are very useful in applying the theory of electrostatics to the design of cathode-ray tubes. Equipotential surfaces for parallel plates and parallel conductors are shown by the shaded lines in Figure 11-5.

FIGURE 11-5
Equipotential Surfaces in the Electric Field between (a) Parallel Plates, and (b) Parallel Conductors.

(a)   (b)

If we think of an electric line of force joining every surplus electron in the negatively charged plate to every atom which is deficient an electron in the positively charged plate, the number of electric lines of force is directly proportional to the electric charge. The total number of electric lines of force in an electric field is called the **electric flux**.

The letter symbol for electric flux is the Greek letter ψ (psi).

Since the electric flux is directly proportional to the charge, we need no special unit for electric flux. We simply state that

$$\psi \equiv Q \tag{11-4}$$

where ψ is the electric flux in MKS units, and Q is the charge on each plate in coulombs.

The letter symbol for electric flux density is D.

Since the distribution of electric lines of force is quite uniform for the special case of two parallel plates, the electric flux density is also uniform and becomes simply

$$D = \frac{\psi}{A} \quad \text{(parallel plates)} \tag{11-5}$$

SELF-CHECK

11-2:1. Determine the force present on an electron that is in an electric field of strength 10^6 N/C.

11-2:2. Two charged metal plates are placed parallel to each other 1.0 mm apart. The electric field strength between the plates is 50 000 N/C. What is the potential difference between the plates?

11-3 ELECTROSTATIC INDUCTION

Figure 11-6 shows an electric conductor in the electric field between two parallel plates. The conductor is shorter than the distance between the plates, so there is no electrical connection between it and the plates. Because there are countless free electrons drifting around in a conducting material, some of these electrons will be attracted by the + plate and repelled by the − plate. As a result, a surplus of electrons (negative charge) will appear at the end of the conductor closer to the + plate and a deficiency of electrons (positive charge) at the end nearer the − plate. Since there is no electrical connection between the conductor and the plates, these two equal and opposite charges that appear when the conductor is placed in the electric field are said to be **electrostatically induced** charges.

Copper rod

FIGURE 11-6
Electrostatically Induced Charges.

Because a net movement of electrons in one direction constitutes an electric current, there is a slight surge of current in the conductor while the charges take up their positions. But since the conductor is insulated from the plates, there can be no continuous current. And since $V = IR$, when the **displacement** current ceases, there can be no voltage drop between the ends of the conductor. This means that there must be a redistribution of the electric field between the two parallel plates as shown in Figure 11-6.

11-4 DIELECTRICS

As a final step in this brief study of electrostatics, we shall place a piece of insulating material between the charged plates of Figure 11-7. Insulating materials exposed to electric fields are known as **dielectrics**. Although the atomic structure of an insulator is

FIGURE 11-7
Effect of an Electric Field on an Atom of a Dielectric.

such that it is very difficult for electrons to move from one atom to another, the electrons orbiting around each nucleus will be attracted by the + plate and repelled by the − plate. As a result, the orbits of the electrons in each atom of the dielectric will be displaced as shown for the single atom in Figure 11-7. The net effect is that, although electrons cannot move from atom to atom in an insulator, each individual atom behaves like the conductor in Figure 11-6. The center of the positive charge of each atom (the nucleus) no longer coincides with the center of the negative charge (the center of the electron orbit). The effect of the positive charge of the atom being closer to one side of the atom and the negative charge being closer to the other is called **polarization** of the atom.

The extent to which the orbits of the electrons in the atoms of the dielectric are distorted (and the extent to which the atoms become polarized) depends on the strength of the electric field. As we increase the strength of the electric field by increasing the potential difference between the two parallel plates, we reach a value at which so much force is exerted on the orbital electrons that they are torn free of their orbits, causing a breakdown of the dielectric. The dielectric then becomes a conductor and the two plates are short-circuited. Some dielectrics are able to withstand a much greater electric field strength than others. The field strength required to break down a dielectric is called its **dielectric strength**. The dielectric strength of some materials can vary considerably as a result of variations in manufacturing. The values of dielectric strength listed in Table 11-1 are typical average values for some of the common dielectric materials.

After exposing a dielectric to a high electric field strength in Figure 11-7, we remove the source and momentarily connect the parallel plates together to allow their charges to become neutralized. The stress on the electron orbits of the dielectric is, therefore, removed and they can return to their normal positions. When we remove the connection between the two plates after the initial discharge, we find that with some dielectrics a small potential difference still appears between the two plates. This indicates that the electron orbits in the dielectric did not instantly return to their original positions. Therefore, even though the plates were connected together, the residual electron orbit displacement in the dielectric held an induced charge in the plates in the same manner that charges were induced into the conductor in Figure 11-6. This effect is called **dielectric absorption**.

TABLE 11-1

Dielectric	Average Dielectric Strength (kilovolts per millimeter)
Air	3
Barium-strontium titanate (ceramic)	3
Porcelain (ceramic)	8
Transformer oil (organic liquid)	16
Bakelite (plastic)	16
Paper	20
Rubber	28
Teflon (plastic)	60
Glass	120
Mica	200

SELF-CHECK

11-4:1. Two charged metal plates are placed parallel to each other 1.0 cm apart. What is the maximum potential difference that can be placed across these plates if the space between the plates is filled with (a) air or (b) teflon?

11-5 CAPACITORS

So far, we have considered only what goes on in the insulated space between a pair of charged parallel plates. While this is important in a study of such devices as cathode-ray tubes, at the moment we are interested in the behavior of electric circuits. Therefore, it is time for us to consider what effect including a component consisting of a pair of parallel conducting plates with a dielectric between them has on the rest of the circuit. Such a component is called a **capacitor**.[†]

A capacitor may be constructed in the parallel-plate form by supporting two insulated metal plates with air between them, or by coating the two sides of a ceramic disk with metal, as in Figure 11-8(a). Some capacitors consist of interleaved parallel plates with either air or mica dielectric, as in Figure 11-8(b). Others consist of two long strips of aluminum foil rolled up with two strips of dielectric, as in Figure 11-8(c). Originally, waxed paper was used as the dielectric for such tubular capacitors. Modern tubular capacitors use thin films of a plastic such as polyethylene. Capacitors constructed by the methods shown in Figure 11-8 range in physical size from about the size of a match head in miniaturized electronic circuitry to industrial capacitors as large as a 10-liter gasoline can. Their capacitance is, however, limited to less than 1 microfarad.

An ingenious means of obtaining a reasonably large capacitance in a fairly compact space employs the electrolysis principle we investigated in Chapter 3. We can replace the plastic film dielectric in the construction of Figure 11-8(c) with a thin gauze soaked in

[†]Sometimes referred to by the now obsolete term **condenser**.

FIGURE 11-8
Construction of Capacitors.

an aluminum hydroxide solution. Like any electrolyte, the aluminum hydroxide solution becomes a conductor by dissociating into positive and negative ions. When first assembled, such an **electrolytic capacitor** is an electrical short circuit. To make it into a capacitor, it is necessary to **form** a dielectric by passing a direct current through the electrolyte between the two aluminum foils. The negative hydroxide ions are attracted to the positive foil where they chemically react with the aluminum to form a very thin layer of aluminum oxide which becomes the dielectric of the capacitor.

The forming current soon diminishes to a very small value and we now have a capacitor where one plate is the positive aluminum foil, the other is the electrolyte in contact with the negative foil, and the dielectric is the thin oxide layer. Electrolytic capacitors are used where capacitances in the range 1 to 100 microfarads are required.

However, there are some limitations in the circuit application of electrolytic capacitors. They must always be used in such a way that the polarity of the potential difference between their terminals is the same as the polarity of the forming voltage. Reversing the polarity of the operating voltage reverses the electrolytic action and tends to remove the oxide film and turn the capacitor into an electrolytic conductor in the process. Hence, the leads of electrolytic capacitors are clearly marked with $(+)$ and $(-)$ polarity indicators or color-coded wires. Electrolytic capacitors are not generally suited to alternating-current circuits.[‡]

In order to obtain a high capacitance, the manufacturer keeps the oxide film of. electrolytic capacitors as thin as possible. Hence the breakdown voltage of electrolytic capacitors is often quite low and is designated on the capacitor as the **working voltage** that cannot be exceeded without danger of dielectric breakdown. And, as we would expect, this type of dielectric is not a perfect insulator. There is always some leakage current through an electrolytic capacitor.

The development of semiconductor devices has introduced another interesting type of capacitor which is particularly useful in manufacturing microminiature circuit modules. As shown in Figure 3-10(a), a *reverse-biased pn*-junction acts as a nonconductor. Under these conditions, we have the basic requirements for a capacitor—two conducting bodies with a nonconducting depletion layer between them. As long as we maintain a reverse bias, a *pn*-junction can be used as a capacitor. We can even change the effective thickness of the dielectric by changing the magnitude of the reverse bias voltage. The resulting "voltage-variable capacitor," often called a **varactor**, is the heart of most automatic frequency control circuits in radio and television receivers. However, if the

[‡] It is possible to use two electrolytic capacitors connected in series opposing as a starting capacitor for ac motors.

polarity of the potential difference between the terminals of a semiconductor capacitor ever reverses, it immediately becomes a short circuit.

In considering the manner in which a capacitor behaves as part of an electric circuit, we represent a capacitor by a standard graphic symbol which somewhat resembles the parallel plates of a physical capacitor. Table 1-1 shows two current styles for the capacitor symbol. The parallel-plate form is preferred internationally; but for the past 25 years, North American practice has moved toward the second style with one *curved* plate, as shown in Figure 11-9. We shall adopt this symbol for the remainder of this book.

11-6 CAPACITANCE

When we first connected a pair of parallel plates to a voltage source in Figure 11-2, the charges could not appear on the plates *instantly*.[†] Electrons had to *flow* from the battery onto the right-hand plate to give it a negative charge, and a similar number of electrons had to flow into the battery from the left-hand plate to give it a positive charge. Since the current is the same in all parts of a series circuit, until we close the switch in the circuit of Figure 11-9, there can be no charging of either plate of the capacitor, even though the bottom plate is permanently connected to the − terminal of the battery. If neither plate possesses any charge, there will be no electric field between them. And since $E = V/s$, there will be no potential difference between plates.

FIGURE 11-9
Charging a Capacitor.

When we close the switch in Figure 11-9, the **galvanometer** will register a sudden flow of electrons flowing counterclockwise around the circuit. Since electrons cannot flow right through the insulated area between the plates, a negative charge builds up on the bottom plate and a positive charge builds up on the top plate. As the charges build up, the strength of the electric field increases and a potential difference builds up between the plates. According to Kirchhoff's voltage law, the potential difference between the two plates must stop building up when it is equal to the potential difference of the source. The galvanometer must, therefore, read zero after the initial charging surge is completed.

[†] Because of the attraction between unlike charges, the charges are concentrated at the surface of each plate facing the other plate. Therefore, we speak of the charge *on* the plate rather than the charge *in* the plate.

If, after the charging surge in Figure 11-9 is over, we suddenly double the applied voltage, there must be another similar surge as the capacitor plates charge sufficiently to raise the potential difference between them to equal the new voltage. We find that there is a fixed ratio between the potential difference between any given pair of insulated conducting plates and the charge required to establish this potential difference. Therefore, for any given pair of insulated conductors,

$$\frac{Q}{V} = \text{a constant} \tag{11-6}$$

If we now open the switch in Figure 11-9, there is no longer any electric circuit connection between the two plates. And thus there is no way for the surplus electrons in the bottom plate to get to the atoms in the top plate (which are deficient some electrons) in order to neutralize the charges. As a result, the pair of parallel plates (a capacitor) *stores* electric charge when a potential difference is applied between the plates. Since Q/V is a constant, as long as charges remain on the plates, a potential difference exists between them. Therefore, we can say that

A capacitor is an electric circuit component constructed for the purpose of storing an electric charge with a potential difference between its terminals.

To determine just how much charge a capacitor can store when we connect it to a given voltage source, we must determine the factors that govern the numerical value of the constant in Equation (11-6). This numerical constant, which represents the ability of a capacitor to store an electric charge, is called its **capacitance.**[†]

The letter symbol for capacitance is C.
The farad is the SI unit of capacitance.[‡]
The unit symbol for farad is F.
An electric circuit has a capacitance of one farad when a charge of one coulomb is required to raise the potential difference by one volt.

$$\therefore C = \frac{Q}{V} \tag{11-7}$$

where C is the capacitance in farads, Q is the charge in coulombs, and V is the potential difference in volts.

To satisfy the specifications for the SI units, the basic unit of capacitance had to be defined in terms of *one* coulomb and *one* volt. But the farad represents a much greater capacitance than appears in practical circuits. In practice, the largest unit size used to express capacitance is the **microfarad** (1 μF = 10^{-6} farad). In many cases, even the microfarad is too large and we use the **picofarad** (1 pF = 10^{-12} farad). As North America becomes more accustomed to SI unit prefixes, we also encounter the **nanofarad**

[†] The ability of a capacitor to store a charge at a certain potential difference is sometimes referred to as its **capacity.** However, in present-day usage, the term capacitance (which rhymes with resistance and inductance) has replaced the older form—capacity.

[‡] Named in honor of Michael Faraday.

($1 \text{ nF} = 10^{-9}$ farad). However, when substituting for C in equations, we must remember that these equations were set up in terms of the *basic* unit of capacitance, the *farad*.

Note that in defining the farad we spoke in general terms of the capacitance of a *circuit*. This was done purposely so that we would not be suggesting that capacitance appears in a circuit only when we connect a capacitor into the circuit. We find exactly the same conditions existing in the circuit of Figure 11-10 when we connect a voltage source to a pair of conductors which form an open-circuit transmission line. Capacitance, therefore, is an electric circuit property possessed by *any* pair of conductors that are insulated from each other. A capacitor is merely a device for *lumping* this property into a small physical space.

FIGURE 11-10
Capacitance between Parallel Conductors.

One of the characteristics of capacitance is its ability to store electric charge. But we would like to have a definition for capacitance that is in keeping with our definition for resistance. We have already noted that a capacitor cannot charge instantly. It takes time, even if only a fraction of a second, for the charge to build up. Consequently the potential difference across a capacitor cannot appear at the instant the switch is closed. It takes *time* for the potential difference across a capacitor to rise. And when we open the switch, the potential difference across the capacitor does not disappear. It remains until we supply a conducting path for electrons to flow from the negatively charged plate to the positive plate. Therefore, since the time element is involved in any change in the potential difference between the plates of a capacitor, we may think of the capacitance of an electric circuit as opposing any change in the potential difference across it. (Such is not the case for resistance.) Therefore, we can say that

Capacitance is that property of an electric circuit that opposes any change in the voltage across that circuit.

Resistance is that property of an electric circuit that opposes electric current through that circuit.

SELF-CHECK

11-6:1. A capacitor has a capacitance of $0.01 \ \mu\text{F}$. What is the potential difference across the capacitor when the device is charged with:
(a) 2 nC of electric charge.
(b) 2 μC of electric charge.

11-6:2. What amount of electric charge is stored in a 47 μF electrolytic capacitor charged to 30 V?

11-7 FACTORS GOVERNING CAPACITANCE

In setting up equations for electric field strength and electric flux, we considered only the case of the uniform electric field between two parallel plates. Therefore, we shall develop only those factors governing the capacitance of parallel plates.[†] We can consider all the capacitors shown in Figure 11-8 as parallel-plate capacitors. Equation (11-7), $C = Q/V$, is an *electrical* expression defining the magnitude of the farad. In order to determine the *physical* factors governing the capacitance of a circuit, we can solve for those factors governing the magnitude of the electric field, since the electric flux is directly proportional to the charges on the plates [Equation (11-4)] and the electric field strength is directly proportional to the potential difference between the plates [Equation (11-3)].

If we double the area of the plates, we have room for twice as many electric lines of force with the same voltage across them. Hence, the charge that a capacitor can hold at a given potential difference is doubled, and since $C = Q/V$, the capacitance is doubled. Therefore,

The capacitance of parallel plates is directly proportional to their area.

To assist us in visualizing the effect of the spacing between the plates on their capacitance, we can refer back to Figure 11-6. When we placed an insulated conductor between the two plates, the positively charged plate attracted the electrons in the conductor toward the end closer to the positively charged plate. The negatively charged plate repelled some electrons from the end of the conductor closer to the negatively charged plate. The two parallel plates of a capacitor have a similar effect on each other. The positively charged plate helps to attract a greater number of surplus electrons from the − terminal of a voltage source onto the negatively charged plate of the capacitor. The negatively charged plate, in turn, helps to drive a greater number of electrons out of the positively charged plate into the + terminal of the source. As we bring the plates closer together, these forces of attraction and repulsion increase and the capacitor is able to store a greater charge when connected to a given voltage source. Therefore,

The capacitance of parallel plates is, inversely proportional to their spacing.

Note the similarity between the dimensional factors governing the conductance of an electric conductor and those governing the capacitance of a capacitor.

$$G \propto \frac{A}{l} \qquad \text{(from)} \quad (4\text{-}4)$$

$$C \propto \frac{A}{s} \qquad\qquad (11\text{-}8)$$

[†]Formulas for the capacitance of parallel conductors, concentric conductors, etc., are found in electrical and electronics handbooks.

where C is capacitance (farads), A is area of the parallel plates (square meters), and s is the spacing between the plates (meters).

We also noted that the conductance (and resistance) of a metallic conductor depends on the type of conductor material. Similarly, if we slide a sheet of glass or mica between the plates of the capacitor in Figure 11-9 after the charging current has ceased, the galvanometer would show an additional surge of charging current. Since the charge Q increased but the potential difference V did not change, replacing the air dielectric with either glass or mica as a dielectric has increased the capacitance of the capacitor. Therefore,

The capacitance of parallel plates is dependent on the type of dielectric between the plates.

To express the effect of different types of conductor materials, we defined conductivity of a material as the conductance between parallel faces of a one meter cube of the material. Similarly, we can express the "capacitivity" of different dielectric materials by measuring the capacitance between two parallel plates 1 square meter in area and spaced 1 meter apart. Since capacitance is *directly* proportional to the charge it can store at a given potential difference [Equation (11-7)], capacitance is directly proportional to the ability of the dielectric to *permit* the setting up of electric lines of force. Hence,

The capacitance between opposite faces of a unit length and cross section of a dielectric material is called the permittivity of the dielectric.
The letter symbol for permittivity is the Greek letter ϵ (epsilon).

Just as
$$R = \rho \frac{l}{A} \tag{4-4}$$

and
$$G = \sigma \frac{A}{l}$$

it follows from the definition of permittivity that

$$C = \epsilon \frac{A}{s} \qquad \text{(parallel plates)} \tag{11-9}$$

where C is the capacitance in farads, ϵ is the absolute permittivity in MKS units, A is the area of each plate in square meters, and s is the spacing between the plates in meters.

To determine the unit of permittivity, we transpose Equation (11-9):

$$\epsilon = C\frac{s}{A} = \text{farads} \times \frac{\text{meters}}{\text{meters}^2} = \textbf{farads/meter}$$

Substituting for C in the equation above from the defining Equation (11-7) for capacitance,

$$\epsilon = \frac{Q}{V} \times \frac{s}{A} = \frac{Q}{A} \times \frac{s}{V}$$

But from Equations (11-4) and (11-5), for parallel plates,

$$\frac{Q}{A} = D \qquad (11\text{-}10)$$

and from Equation (11-3), for parallel plates,

$$\frac{V}{s} = E \qquad (11\text{-}3)$$

By making these substitutions, we can also express permittivity in terms of the electric flux density and the strength of the electric field set up in the dielectric between the plates of the capacitor.

$$\epsilon = \frac{D}{E} \qquad (11\text{-}11)$$

where ϵ is the absolute permittivity of a given dielectric in farads per meter, D is the electric flux density in coulombs per square meter, and E is the electric field strength in volts per meter.

It is not absolutely essential to have a dielectric material between the plates of a capacitor in order to set up an electric field. If we place two parallel plates in a vacuum, an electric field will be set up between them when we apply a voltage to the plates. The ratio of electric flux density to electric field strength will be

$$\frac{D}{E} = 8.85 \times 10^{-12} = \epsilon_v \qquad (11\text{-}12)$$

where ϵ_v is called the permittivity of free space in farads per meter.

When we do place a dielectric between the plates of a capacitor, the electron orbits of the atoms are forced off-center, as shown in Figure 11-7. As a result, the negative charge of each atom of the dielectric is a little closer to the positive plate and the positive charge of each atom of the dielectric is a little closer to the negative plate. This has the same effect on the capacitance as moving the parallel plates a little closer together. Therefore, the absolute permittivity of most solid and liquid dielectrics is considerably greater than that of free space. Because of their low density, the absolute permittivity of air and other gases is very close to that of free space. If we substitute Equation (11-12) in Equation (11-9), the capacitance of a parallel-plate *air dielectric* capacitor becomes

$$C = \frac{8.85\,A}{10^{12}\,s} \qquad (11\text{-}13)$$

where C is the capacitance in farads, A is the area of each plate in square meters, and s is the spacing between the plates in meters.

SELF-CHECK

11-7:1. Two charged metal plates are placed parallel to each other with 0.1 mm of air between them. Each plate has an area of 100 cm^2 and the pair is charged with 20 nC of electric charge.
(a) Find the capacitance of the capacitor.
(b) What is the potential difference across the plates?
11-7:2. How does the capacitance of the capacitor in the previous problem change if the plates are doubled in area and moved twice as far apart?

11-8 DIELECTRIC CONSTANT

Rather than state the absolute permittivity of liquid and solid dielectrics, it has become the custom to prepare tables stating the ratio between the absolute permittivity of a given dielectric and the absolute permittivity of free space. This ratio is called the **relative permittivity** or the **dielectric constant** of that material.

$$k = \frac{\epsilon_r}{\epsilon_v} \tag{11-14}$$

where k is the dielectric constant, ϵ_r is the absolute permittivity of a given dielectric, and ϵ_v is the absolute permittivity of free space.

Therefore, the equation for the capacitance of *any* parallel-plate capacitor becomes

$$C = \frac{8.85kA}{10^{12}s} \qquad \text{(parallel plates)} \tag{11-15}$$

where C is the capacitance in farads, k is the dielectric constant, A is the area of each plate in square meters, and s is the spacing between plates in meters.

The dielectric constant of a given type of material can vary considerably as a result of variation in manufacturing. Table 11-2 gives typical average values of dielectric constant for those dielectric materials listed in Table 11-1. On comparing Tables 11-1 and 11-2, we note that there is no correlation between dielectric strength and dielectric constant. In manufacturing a capacitor for use in high-voltage circuits, often we are not able to choose a dielectric with a high dielectric constant because its dielectric strength is too low. Distilled water has a dielectric constant of 80, but its dielectric strength is so low that it is not practical to build water dielectric capacitors.

Some ceramics such as barium-strontium titanate have molecules which **polarize** very easily when exposed to an electric field. As a result, they have dielectric constants which are extremely high in comparison with the average dielectric materials. Development of these ceramics during the past thirty years has made it possible to manufacture capacitors which are much more compact for a given capacitance and voltage rating than the older mica and waxed-paper dielectric capacitors. (See Figure 11-8.) These ceramics are called **ferroelectric** dielectrics. Oddly enough, the term **ferroelectric** does not imply any iron content in such materials. It results from a comparison with a similar behavior

TABLE 11-2

Dielectric	Average Dielectric Constant
Air	1.0006
Barium-strontium titanate (ceramic)	7500
Porcelain (ceramic)	6
Transformer oil	4
Bakelite (plastic)	7
Paper	2.5
Rubber	3
Teflon (plastic)	2
Glass	6
Mica	5

of ferrous compounds in *magnetic* circuits. Just as iron permits a much greater magnetic flux density for a given magnetizing force than nonferrous materials, a ferroelectric dielectric permits a much greater electric flux density for a given electric field strength than most dielectrics [Equation 11-11]. The dielectric *constant* of ferroelectric materials varies slightly with changes in electric field intensity.

EXAMPLE 11-1

A capacitor consists of a disk of barium-strontium titanate 1 cm in diameter and 0.2 mm thick, silver-plated on each side. Using the average values of dielectric strength and dielectric constant given in Tables 11-1 and 11-2 calculate (a) its capacitance and (b) the maximum voltage that should be applied to it.

SOLUTION

(a)
$$C = \frac{8.85kA}{10^{12}s} = \frac{8.85 \times 7500 \times \pi \times (0.5 \text{ cm})^2}{10^{12} \times 0.2 \text{ mm}}$$

$$= 2.6 \times 10^{-8} \text{ F} = \textbf{0.026 } \boldsymbol{\mu}\textbf{F}$$

(b)
$$V_m = 3 \text{ kV/mm} \times 0.2 \text{ mm} = \textbf{600 V}$$

SELF-CHECK

11-8:1. Two parallel-plate capacitors are identical except that the first capacitor has an air dielectric and the second has a rubber dielectric. Compare the capacitances of these capacitors.

11-8:2. A capacitor consists of two metal plates, each with an area of 10^{-2} m^2, placed parallel to each other with 0.1 mm of dielectric material between them. What voltage is necessary to put 0.2 μC of electric charge on the capacitor if the dielectric material is:
(a) air,
(b) glass,
(c) barium-strontium titanate?

FIGURE 11-11
Capacitors: (a) in Parallel; (b) in Series.

11-9 CAPACITORS IN PARALLEL

As even the graphic symbols in Figure 11-11(a) suggest, connecting capacitors in parallel is like increasing the area of the plates of a single capacitor. Therefore, the total capacitance is greater than that of either one. Since both capacitors have to charge from the same battery, the total charge drawn from the battery is

$$Q_T = Q_1 + Q_2 \tag{11-16}$$

Since Figure 11-11(a) is a simple parallel circuit,

$$E = V_1 = V_2$$

$$\therefore \frac{Q_T}{E} = \frac{Q_1}{V_1} + \frac{Q_2}{V_2}$$

and

$$C_T = C_1 + C_2$$

Therefore,

> **When we connect capacitors in parallel, the total capacitance is the sum of all the individual capacitances.**

$$C_T = C_1 + C_2 + C_3 + \text{etc.} \tag{11-17}$$

EXAMPLE 11-2

In the circuit of Figure 11-11(a), C_1 and C_2 have capacitances of 0.01 μF and 0.05 μF, respectively. What single capacitance can be used to replace C_1 and C_2?

SOLUTION

$$C_T = C_1 + C_2 = 0.01 \ \mu\text{F} + 0.05 \ \mu\text{F} = \mathbf{0.06 \ \mu F}$$

SELF-CHECK

11-9:1. Given three capacitors with the following capacitances:
1.0 μF, 2.2 μF and 3.3 μF, what total capacitance results by connecting these devices in parallel?

PART III   CAPACITANCE AND INDUCTANCE

11-9:2. Suppose the parallel combination of capacitors from Self-Check Problem 11-9:1 is connected to a 10-V battery. Determine the voltage across each capacitor and the charge in each capacitor. Find the total charge stored in this parallel combination.

11-10 CAPACITORS IN SERIES

When we connect capacitors in series, the charging action will be similar to that shown in Figure 11-6. The bottom plate of C_1 in Figure 11-11(b) and the top plate of C_2 will be charged by electrostatic induction. And, as the graphic symbols suggest, the effect is much the same as increasing the spacing between the plates of a single capacitor. The total capacitance is less than that of either one, but the combination is capable of withstanding a higher total potential difference than either one by itself. Since Figure 11-11(b) is a simple series circuit,

$$E = V_1 + V_2$$

Since the current is the same in all parts of a simple series circuit,

$$Q_T = Q_1 = Q_2$$

$$\therefore \frac{E}{Q_T} = \frac{V_1}{Q_1} + \frac{V_2}{Q_2}$$

or

$$\frac{1}{C_{eq}} = \frac{1}{C_1} + \frac{1}{C_2}$$

Therefore,

The equivalent capacitance of series capacitors is

$$C_{eq} = \frac{1}{1/C_1 + 1/C_2 + 1/C_3 + \text{etc.}} \tag{11-18}$$

┌ *EXAMPLE 11-3*

In the circuit of Figure 11-11(b), C_1 and C_2 have capacitances of 0.01 μF and 0.05 μF, respectively. What single capacitance can be used to replace C_1 and C_2?

SOLUTION

$$C_{eq} = \frac{1}{(1/0.01 \ \mu\text{F}) + (1/0.05 \ \mu\text{F})} = 8.33 \times 10^{-3} \ \mu\text{F}$$

$$= 0.008\,33 \ \mu\text{F}$$

Equation (11-18) is easy to use with the reciprocal key of a calculator by first calculating the denominator and then reading the reciprocal. We can also reduce the calculator operations by omitting the exponent for *micro* in entering the capacitance. The display result will automatically be in microfarads. Hence, the calculator chain becomes

$$.01 \boxed{1/x} + 0.5 \boxed{1/x} = \boxed{1/x}$$

With only two capacitors in *series,* we can use the same simplification that we used for two resistors in *parallel.* Equation (11-18) then becomes

$$C_{eq} = \frac{C_1 C_2}{C_1 + C_2} = \frac{0.01 \times 0.05}{(0.01 + 0.05)} = 8.33 \times 10^{-3} \ \mu F = \mathbf{0.008\,33 \ \mu F}$$

To avoid compound fractions such as Equation (11-18), we sometimes use a technique we developed when we considered parallel resistors in Section 6-9. Since connecting *more* resistors in parallel *reduces* the equivalent resistance, we consider the reciprocal property—the *conductance* of the various branches. Since connecting *more* capacitors in series *reduces* the equivalent capacitance, we can again switch our thinking to the reciprocal property. Because capacitance is proportional to the ability of a dielectric to *permit* the setting up of electric lines of force in the dielectric [Equations (11-9) and (11-11)],

> *Elastance is the opposition to the setting up of electric lines of force in an electric insulator or dielectric.*
> *The letter symbol for elastance is S.*

From the definition,

$$S = \frac{1}{C} \tag{11-19}$$

where S is elastance in reciprocal farads,[†] and C is capacitance in farads. When we substitute in Equation (11-18), for capacitors in series,

$$S_T = S_1 + S_2 + S_3 + \text{etc.}$$

Since
$$Q_1 = C_1 V_1 \quad \text{and} \quad Q_2 = C_2 V_2,$$

and, since $Q_1 = Q_2$ in a series circuit,

$$C_1 V_1 = C_2 V_2$$

from which

$$\frac{V_1}{V_2} = \frac{C_2}{C_1} \tag{11-21}$$

[†]The International System of units does not provide a special name for the reciprocal of the farad. In some textbooks, the unit of elastance is called the **daraf** (*farad* spelled backwards).

When capacitors are connected in series, the ratio between any two potential differences is the *inverse* ratio of their capacitances.

EXAMPLE 11-4

A 0.01-μF capacitor and a 0.04-μF capacitor are connected first in parallel and then in series to a 500-V source. (In each case, the capacitors are discharged before the connections are made.)

(a) What is the total capacitance in each case?

(b) What is the total charge in each case?

(c) What is the charge on each capacitor and the potential difference across each capacitor in each case?

SOLUTION

(a) Parallel: $\quad C_T = C_1 + C_2 = 0.01 + 0.04 = \mathbf{0.05\ \mu F}$

  Series: $\quad C_{eq} = \dfrac{C_1 C_2}{C_1 + C_2} = \dfrac{0.01 \times 0.04}{0.01 + 0.04} = \mathbf{0.008\ \mu F}$

(b) Parallel: $\quad Q_T = C_T V_T = 0.05 \times 500 = \mathbf{25\ \mu C}$

  Series: $\quad Q_T = C_{eq} V_T = 0.008 \times 500 = \mathbf{4\ \mu C}$

(c) Parallel: $\quad V_1 = V_2 = E = \mathbf{500\ V}$

  $\quad Q_1 = C_1 V_1 = 0.01 \times 500 = \mathbf{5\ \mu C}$

  $\quad Q_2 = C_2 V_2 = 0.04 \times 500 = \mathbf{20\ \mu C}$

  Series: $\quad Q_1 = Q_2 = Q_T = \mathbf{4\ \mu C}$

  $\quad V_1 = \dfrac{Q_1}{C_1} = \dfrac{4}{0.01} = \mathbf{400\ V}$

  $\quad V_2 = \dfrac{Q_2}{C_2} = \dfrac{4}{0.04} = \mathbf{100\ V}$

SELF-CHECK

11-10:1. Given three capacitors with the following capacitances: 1.0 μF, 2.2 μF and 3.3 μF, what equivalent capacitance will result from connecting these devices in series?

11-10:2. Suppose the series combination of capacitors from Self-Check Problem 11-10:1 is connected to a 10-V battery. Determine the voltage across each capacitor and the charge in each capacitor. Find the total charge stored in this series combination.

PROBLEMS

11-2-1. What force of attraction (newtons) will act on a single electron located 0.5 cm from a point charge of +5 μC?

11-2-2. Two electric charges, one of which is four times as large as the other, develop a mutual force of repulsion of 16 newtons when they are located 12 cm apart. Determine the magnitude of each charge.

11-2-3. Two parallel plates 1 cm apart are connected to a 500-V source. What force will be exerted on a free electron between the plates?

11-2-4. If 2×10^{10} electrons are removed from one parallel plate and added to the other, what is the total electric flux between the plates?

11-6-5. If moving 2×10^{10} electrons from one parallel plate to the other produces a potential difference between the plates of 180 V, what is the capacitance of the plates?

11-6-6. How many electrons must be removed from one plate of a 270-pF capacitor and added to the other to raise the voltage between the plates to 420 V?

11-7-7. A neutralizing capacitor in a radio transmitter consists of two aluminum disks, each 10 cm in diameter and $\frac{1}{2}$ cm apart with air between them. What is its capacitance?

11-7-8. If the electric flux density between the parallel plates of the capacitor in Problem 11-7-7 is $1.0 \ \mu C/m^2$, what is the voltage between the plates?

11-7-9. An interleaved parallel-plate capacitor consists of 10 sheets of mica 1 cm long $\times$ 0.5 cm wide and 0.025 mm thick. Eleven sheets of aluminum foil are pressed firmly between the mica sheets and connected as shown in Figure 11-8(b). What is the capacitance of this capacitor?

11-7-10. Two sheets of aluminum foil 2.5 cm wide and 1.0 m long and two sheets of 0.1 mm thick waxed paper 3.0 cm wide and 1.0 m long are rolled as shown in Figure 11-8(c) to form a tubular capacitor. The waxed paper has a dielectric strength of 6 kV/mm and a dielectric constant of 3. What is the capacitance and voltage rating of this capacitor?

11-9-11. A 10-μF and a 40-μF capacitor are connected in parallel to a 400-V source.
  (a) What is the total capacitance?
  (b) What is the magnitude of the charge stored by each capacitor?

11-9-12. The 10-μF capacitor in the circuit of Figure 11-12 is completely discharged. Draw a graph showing how the voltage across the 10-μF capacitor changes with time as the switch is operated for five complete cycles. Label the exact voltage after each cycle of switch operation. (Assume no leakage.)

FIGURE 11-12

11-10-13. A 10-μF, a 20-μF, and a 40-μF capacitor are connected in series to a 400-V source.
  (a) What is the equivalent capacitance?
  (b) What is the magnitude of the charge stored by each capacitor?
  (c) What is the voltage across each capacitor?

11-10-14. After charging, the three capacitors in Problem 11-10-13 are disconnected from the source and from each other. They are then connected in parallel, with + plate connected to + plate. What is the voltage across each capacitor?

11-10-15. A sheet of glass 0.3 cm thick is placed between the plates of the neutralizing capacitor of Problem 11-7-7. What is its capacitance? (*Hint:* Capacitors in series.)

11-10-16. If the potential difference between the plates of the capacitor in Problem 11-10-15 is 1 kV, what is the voltage gradient in the air and in the glass?

11-10-17. Find the equivalent capacitance and the voltage across each of the capacitors in the circuit of Figure 11-13.

11-10-18. Find the equivalent capacitance and the voltage across each of the capacitors in the circuit of Figure 11-14.

11-10-19. Given three capacitors with capacitances of 0.01 μF, 0.02 μF, and 0.05 μF, respectively, and a rated working voltage of 400 V for each capacitor, what is the highest voltage that can safely be applied to the series combination of the three capacitors?

11-10-20. What is the total elastance of a combination of the three capacitors of Problem 11-10-19 connected (a) in series and (b) in parallel?

FIGURE 11-13   FIGURE 11-14

REVIEW QUESTIONS

11-1. Why must energy be expended by an electric "charge separator," which removes electrons from neutral atoms?

11-2. How does charge separation in an electric battery differ from charge separation by rubbing a glass rod with a silk cloth?

11-3. Coulomb's law of electrostatic force states that electric force is proportional to the product of two electric charges; yet, Figure 11-1 shows an electric field for a *single* isolated electric charge. Explain the purpose of sketches such as those in Figure 11-1.

11-4. Why does a potential difference exist between two electric conductors possessing unlike charges?

11-5. Why are we more concerned with the electric field between unlike charges in electric circuits than we are with the repulsion between like charges which we can demonstrate in electrostatics?

11-6. What is the significance of the arrows on the sketch of electric lines of force between two parallel plates?

11-7. What characteristic of an electric field is referred to by the term **electric field strength**?

11-8. What is meant by the **potential gradient** of an electric field?

11-9. How are **electric field strength** and **potential gradient** related in the case of charged parallel plates?

11-10. What factors govern the **flux density** of the electric field between charged parallel plates?

11-11. Draw a sketch showing the equipotential surfaces in the electric field between two charged concentric conductors.

11-12. Draw a sketch showing the equipotential surfaces in the electric field of Figure 11-6.

11-13. Why must the + and − charges at the two ends of the insulated conductor in Figure 11-6 be equal?

11-14. Define a **displacement current.**

11-15. What is meant by a **polarized** atom?

11-16. What type of atoms can become polarized?

11-17. Why is the strength of a dielectric expressed in terms of the **voltage gradient** of an electric field rather than the total potential difference?

11-18. Define **dielectric absorption.**

11-19. Why must the ratio between the charge on a given pair of insulated conductors and the potential difference between them be a constant?

11-20. Define the term **capacitance** in terms of Q and V.

11-21. Define the term **capacitor** in terms of Q and V.

11-22. Why can we define capacitance as that property of an electric circuit that opposes any change in voltage across that circuit?

11-23. Why does the spacing between two parallel plates have an effect on their capacitance?

11-24. Why does the voltage applied to two parallel plates not have an effect on their capacitance?

11-25. Why is it possible to have capacitance between two parallel plates when the space between them is evacuated?

11-26. Why does the presence of an insulating material between the parallel plates increase their capacitance as compared with that of the same plates in a vacuum?

11-27. Since we are concerned with electric circuits, we defined the **farad** in terms of the charge on the conducting *plates* and the voltage between the *plates*. Define the **farad** in terms of flux density and intensity of the electric *field* in the dielectric between the plates.

11-28. Distinguish between the **absolute permittivity** and the **relative permittivity** of a given dielectric.

11-29. Why is it possible to define the **dielectric constant** of a given insulating material as the ratio between the capacitance of a pair of parallel plates with that material as a dielectric and the same parallel plates with air as a dielectric?

11-30. When two capacitors are connected in series, one plate of each has no electrical connection to any voltage source. How is it possible for these capacitors to become charged?

11-31. Why must capacitors in series all take the same charge?

11-32. A 0.1-μF 400-V capacitor and a 0.05-μF 200-V capacitor are connected in series. What is the maximum voltage that can be safely applied to this combination?

11-33. Some of the older electrolytic capacitors packaged in aluminum cans have been known to explode when connected across a dc voltage source such that the insulated terminal is negative with respect to the can. Explain what happens.

11-34. Explain why a motor-starting capacitor consisting of two electrolytic capacitors connected in series opposing can be used safely in an ac circuit.

* 11-35. Design a personal computer program which will request the following input data: area of a capacitor plate specified in square centimeters, thickness of the dielectric specified in millimeters, and dielectric constant. After these data have been entered, the computer should display the capacitance of a simple parallel-plate capacitor specified in picofarads.

* 11-36. Write a personal computer program which can print out the equivalent capacitance for the circuits of *either* Figure 11-13 *or* Figure 11-14 by entering data on the manner in which the three capacitors are connected, as well as their capacitances.

12

CAPACITANCE IN DC CIRCUITS

Now that we know something about capacitance as a basic property of electric circuits, we can examine the behavior of practical electric circuits when capacitance is present intentionally in the form of capacitors, or as a byproduct of other circuit configurations, such as a reverse-biased *pn*-junction. In either case, we represent circuit capacitance in our schematic diagrams by the standard graphic symbol shown in Figure 12-1.

12-1 CHARGING A CAPACITOR

Strictly speaking, when we speak of *charging* a capacitor, we should think in terms of the *charge* stored by it in *coulombs*. But $Q = CV$, and for a given capacitor, C is a constant based on such physical factors as the area of the plates, the distance between plates, and the type of dielectric. Therefore, the charge is directly proportional to the potential difference between the plates. Since we have voltmeters for reading potential difference directly, it is more convenient to speak of the charging of a capacitor in terms of the potential difference built up between its plates. This point of view is consistent with our definition of capacitance.

At the instant we close the switch in the circuit of Figure 12-1, there can be no potential difference across the capacitor, since it takes *time* for electrons to flow into the bottom plate and out of the top plate in order to build up a potential difference between the plates. But Kirchhoff's voltage law states that there must be a voltage drop in the circuit equal to the applied voltage when the switch is closed. Although no resistance is shown, we shall assume that the resistance of the wiring and the internal resistance of the battery add up to 0.5 ohm. Therefore, at the instant we close the switch, since there is

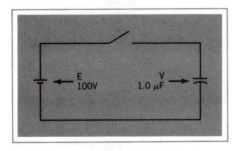

FIGURE 12-1
Charging a Capacitor Without a
Current-Limiting Resistor.

no potential difference across the capacitor, there must be an *IR* drop in the circuit equal to the applied voltage. To produce this voltage drop, the current which must flow the instant we close the switch is

$$\text{initial } I = \frac{E}{R} = \frac{100 \text{ V}}{0.5 \ \Omega} = 200 \text{ A}$$

Since current is the rate of flow of electrons past a certain point in a circuit, a 200-ampere current will charge the plates of the capacitor very rapidly, and the potential difference across the capacitor will take only a few microseconds to rise to 100 volts. When the potential difference across the capacitor equals the applied voltage, current must cease. Therefore, if we charge a capacitor by connecting it directly across a voltage source, there is a very short duration pulse of very high current. This is rather rough on the source so, in practice, we prefer to charge capacitors through a resistance which limits the peak value of the initial current surge.

At this point, we must note an important convention regarding the letter symbols we use to represent electrical properties in equations. The resistance of an electric circuit is dependent on such physical factors as length, cross section, and type of material. Capacitance depends on such physical factors as area of the conducting surfaces, spacing between the surfaces, and the type of dielectric material between the conducting surfaces. The numerical values of these circuit parameters are essentially independent of *time*. Consequently, we represent resistance and capacitance by upper-case letter symbols. In the circuits we have considered in preceding chapters, the source voltage, the resulting current, the voltage drop, and the power have also been independent of any exact instant in time. Accordingly, we have represented all these **steady-state** values with uppercase letter symbols. But, in considering the charging of a capacitor, we note that the magnitude of the charging current and also the potential difference across the capacitor change from instant to instant and depend on how long the switch has been closed. We represent such **instantaneous** values of voltage and current by *lowercase* letter symbols. Note that we have always represented **time** itself by the lowercase *t*.

Lowercase letter symbols represent instantaneous values which are dependent on the exact instant in time for their numerical value.

Uppercase letter symbols represent steady-state values which are not dependent on the exact instant in time for their numerical value.

All the time the switch is closed in the series circuit of Figure 12-2, the sum of the

FIGURE 12-2
Charging a Capacitor through a
Current-Limiting Resistor.

potential difference across the capacitor and the *IR* drop across the resistor must equal the applied voltage. Therefore,

$$E = v_C + iR \qquad (12\text{-}1)$$

Again, at the instant we close the switch, there will be no potential difference across the capacitor. Hence, the initial current must be such that the *IR* drop in the circuit is equal to the applied voltage. Therefore,

$$\text{initial } I = \frac{E}{R} \qquad (12\text{-}2)$$

The initial current in the circuit of Figure 12-2 is only 50 microamperes. This current immediately starts to charge the capacitor. Thus, a potential difference starts to build up across the capacitor. To satisfy Kirchhoff's voltage law, the *IR* drop across the resistor must decrease. Consequently, the instantaneous current (the dependent variable of our electric circuits) must become smaller. As a result, the potential difference across the capacitor must rise more slowly. Therefore, the *rate of rise* of the instantaneous voltage across the capacitor depends on the magnitude of the instantaneous current, which depends on how much *IR* drop must appear across the resistance of the circuit, which, in turn, depends on the instantaneous voltage across the capacitor, which depends on how long the switch has been closed. Before we can sort out these interdependences, we must consider for a moment just how we can express rate of rise or, in more general terms, *rate of change* of voltage numerically.

SELF-CHECK

12-1:1. Referring to the circuit shown in Figure 12-2, what value of resistor would be necessary to ensure that the instantaneous current is never greater than 20 μA?

12-1:2. Suppose you have a circuit similar to the one shown in Figure 12-2 but with the following component values:
$$E = 15 \text{ V}, R = 120 \text{ k}\Omega, \text{ and } C = 0.47 \ \mu\text{F}.$$
(a) Determine the current in the circuit the instant the switch is closed.
(b) What happens to the value of this current as the capacitor charges?

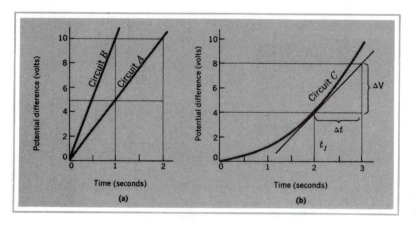

FIGURE 12-3
Graphical Determination of Rate of Change of Voltage.

12-2 RATE OF CHANGE OF VOLTAGE

For the moment, we can assume the existence of some special electric circuits in which the potential difference between two points in these circuits is represented by the graphs of Figure 12-3. When we close the switch in the circuits represented by the graphs of Figure 12-3(a), the potential difference increases as a linear function of time (rather than immediately taking on a steady-state value, as in the case of simple resistance circuits). Since these are straight-line graphs, we can read the rate of change of voltage directly from the graph by noting the number of volts increase in potential difference during 1 second. For circuit A, the voltage rises from 0 to 5 volts in the first second and from 5 to 10 volts during the second second. In either case, the *rate of change* of voltage is 5 volts per second. Similarly, the rate of change of voltage in circuit B is 10 volts per second. Note that the graph for circuit B is twice as steep as the graph for circuit A. Hence, we represent the rate of change of voltage by the *slope* of the graph of voltage as a function of time.

Circuit C represents an electric circuit in which the magnitude of the potential difference between two specified points is a function of the *square* of the elapsed time. Since this graph is not a straight line, the rate of change of voltage is not the same for all time intervals. We must select some specific time t_1 at which we want to know the rate of change of voltage. By drawing a *tangent* to the graph at time t_1 in Figure 12-3(b), we can now read the rate of change of voltage at time t_1 by noting the change in voltage ΔV for any convenient change in time Δt (because the tangent is a straight line). Hence, the rate of change of voltage at time t_1 is $\Delta V / \Delta t$ volts per second. In practice, we adopt the calculus symbol for *rate of change*.[†]

[†] Calculus provides us with a more precise method of expressing *rate of change* of voltage by differentiating the equation for the instantaneous voltage as a function of time. The equation for the instantaneous voltage in circuit A in Figure 12-3(a) is

$$v = 5t$$

By differentiating, we obtain the rate of change of voltage

$$\frac{dv}{dt} = 5 \text{ V/s}$$

*The symbol for rate of change of voltage is **dv/dt** volts/second.*

When we close the switch in the circuit of Figure 12-2, current must flow instantly at its initial value and then become progressively smaller as the voltage across the capacitor rises, as shown by the graph of Figure 12-4. Since resistance is a physical constant, the instantaneous voltage drop across it must always be proportional to the instantaneous current through it (Ohm's law). Thus, the graph of the *IR* drop across the resistor will have the same shape as the instantaneous current graph of Figure 12-4.

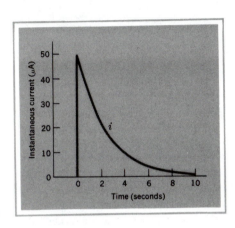

FIGURE 12-4
Instantaneous Charging Current.

Since the initial current in the *CR* circuit of Figure 12-2 is only 50 microamperes, it will take appreciable time for the capacitor to charge, that is, for the voltage across the capacitor to rise to the same value as the applied voltage. Because we know the magnitude of the initial current, we can determine the *initial* rate of change of voltage across the capacitor. Since

$$Q = CV \tag{11-7}$$

and
$$Q = It \tag{2-2}$$

$$CV = It$$

And since
$$\text{initial } I = \frac{E}{R} \tag{12-2}$$

$$CV = \frac{E}{R} \times t$$

Similarly, in circuit *B*

$$v = 10t \quad \text{and} \quad \frac{dv}{dt} = 10 \text{ V/s}$$

Circuit *C* is just as easy to solve.

Since
$$v = t^2, \quad \frac{dv}{dt} = 2t \text{ V/s}$$

FIGURE 12-5
Rise of Potential Difference across a Capacitor Charging through a Resistance.

from which

$$\text{initial } \frac{dv}{dt} = \frac{E}{CR}$$

(12-3)

Since Equation (12-3) contains all the independent variables in the circuit of Figure 12-2, no matter what numerical values we choose, the graph of the actual instantaneous voltage across the capacitor must start to rise with the same slope and at the same instant in time as the dashed line representing the initial rate of change of capacitor voltage in Figure 12-5. The slope of the actual potential difference must then become progressively more gradual until it finally ends up tangent to the dashed line representing the applied voltage.*

SELF-CHECK

12-2:1. Use the current information in the graph of Figure 12-4 to determine the voltage drops across the resistor and the capacitor in Figure 12-2 at the following instants:
(a) 2 seconds after the switch is closed
(b) 10 seconds after the switch is closed.

12-2:2. Suppose you have a capacitor in a circuit, similar to the one shown in Figure 12-2, with the following component values:

$$E = 15 \text{ V}, R = 120 \text{ k}\Omega, \text{ and } C = 0.47 \text{ }\mu\text{F}$$

(a) Determine the slope of the instantaneous voltage versus time graph an instant after the switch is closed.
(b) What happens to the value of this slope as the capacitor charges?

12-3 TIME CONSTANT

If we assume for the moment that the potential difference across the capacitor can rise at a steady rate equal to its initial rate until it equals the applied voltage, we can quite readily derive a convenient expression for the charging time of the capacitor. This time

---

*In mathematics, we call these dashed construction lines **asymptotes**.

interval is called the **time constant** of the *CR* circuit. On the basis of the assumption above, we can define time constant thus:

The time constant of a CR circuit is the time it would take the potential difference across the capacitor to rise to the same value as the applied voltage if it were to continue to rise at its initial rate of change for the whole time interval.
The letter symbol for time constant is the Greek letter τ (tau).

From Figure 12-5, if the rate of change of voltage remains constant, the final potential difference across the capacitor becomes simply

$$V = \tau \times \text{initial} \frac{dv}{dt}$$

But
$$V = E \quad \text{and} \quad \text{initial} \frac{dv}{dt} = \frac{E}{CR} \tag{12-3}$$

from which
$$\tau = CR \text{ seconds} \tag{12-4}*$$

where τ is the length of time in seconds defined as a time constant, C is the capacitance of the capacitor in farads, and R is the resistance through which the capacitor charges in ohms. Since the farad is not a practical size of unit, we can still make Equation (12-4) balance if we express capacitance in microfarads and resistance in megohms.

Note that the time constant of a *CR* circuit is directly proportional to the capacitance. If we double the capacitance, twice the charge must flow in order to raise the potential difference across the capacitor to the applied voltage. But if the resistance is not changed, the initial current remains unchanged, and, therefore, it will take twice as long for the capacitor to charge. If we double the resistance, the initial current is cut in half, and it will again take twice as long for the potential difference of a given capacitor to rise until it equals the applied voltage. Note also that the magnitude of the applied voltage has no effect on the time constant. If we double the applied voltage, the capacitor must store twice the charge, but the initial current will be twice as great. Therefore, it will take exactly the same time for the potential difference of the capacitor to reach the applied voltage.

Because of the fixed relationship between the solid exponential graph of Figure 12-5, representing the actual rise in potential difference across the capacitor, and the dashed lines from which we solved for the time constant, the actual charging time of a *CR* circuit bears a fixed relationship to the time constant. The instantaneous potential difference across a capacitor will always be 63% of the applied voltage one time constant after the switch is closed. For all practical purposes,

The potential difference across the capacitor will reach a steady-state value equal to the applied voltage after a time interval equal to five time constants has elapsed.

⌐ *EXAMPLE 12-1*
(a) What is the initial rate of change of potential difference across the capacitor in Figure 12-2?

$$* CR = \frac{coulombs}{volts} \times \frac{volts}{amperes} = \frac{coulombs}{amperes} = \frac{amperes \times seconds}{amperes} = seconds.$$

(b) How long will it take the capacitor in Figure 12-2 to charge to a potential difference of 100 V?

SOLUTION

(a)
$$\text{Initial } \frac{dv}{dt} = \frac{E}{CR} = \frac{100 \text{ V}}{1 \text{ }\mu\text{F} \times 2 \text{ M}\Omega} = 50 \text{ V/s}$$

(b)
$$\tau = CR = 1 \text{ }\mu\text{F} \times 2 \text{ M}\Omega = 2 \text{ s}$$

Since
$$100 \text{ V} = E = \text{full charge}$$
$$\therefore t = 5\tau = 5 \times 2 \text{ s} = 10 \text{ s}$$

SELF-CHECK

12-3:1. Suppose you have a capacitor in a circuit, similar to the one shown in Figure 12-2, with the following component values:

$$E = 20 \text{ V}, R = 250 \text{ k}\Omega, \text{ and } C = 0.40 \text{ }\mu\text{F}.$$

The switch is closed and the capacitor begins to charge.
(a) How long will it take the capacitor voltage to reach 63% of the 20 V supplied by the source?
(b) How long will it take to fully charge the capacitor to a potential difference of 20 V?

12-3:2. Determine the time necessary to fully charge a 1.0 μF capacitor when the capacitor is in a series circuit that also contains:
(a) a 20-V battery and a 500-kΩ resistor and
(b) a 50-V battery and a 2-MΩ resistor.

12-4 GRAPHICAL SOLUTION FOR INSTANTANEOUS POTENTIAL DIFFERENCE

Because of the fixed relationship between the time constant and the time it actually takes the potential difference of a capacitor to rise to the same value as the applied voltage, we can use the exponential curve of Figure 12-5 to solve graphically for the instantaneous potential difference of any CR circuit at any instant in time. To do this, we carefully replot the exponential curve of Figure 12-5 with a horizontal (time) axis calibrated in terms of time constants rather than seconds and with a vertical (instantaneous voltage) axis marked in percentage of the applied voltage, as shown in Figure 12-6.

EXAMPLE 12-2
(a) What is the instantaneous potential difference across the capacitor in the circuit of Figure 12-2 three seconds after the switch is closed?
(b) How long will it take the potential difference across the capacitor to rise from zero to 55 V?

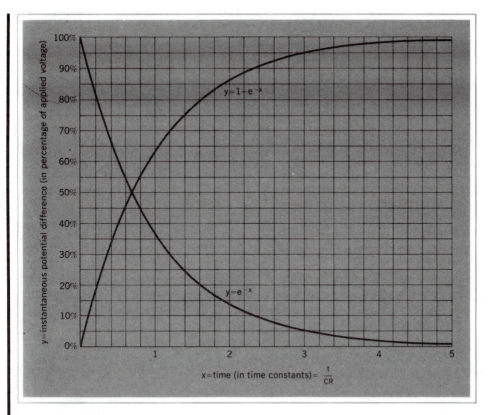

FIGURE 12-6
Universal Exponential Curves for the Graphical Solution of the Charge and
Discharge of Capacitors in Direct-Current Circuits.

SOLUTION

(a)
$$\tau = CR = 1 \ \mu\text{F} \times 2 \ \text{M}\Omega = 2 \ \text{s}$$

$$\therefore 3 \ \text{s represents} \ \frac{3 \ \text{s}}{2 \ \text{s}} = 1.5 \ \tau$$

From Figures 12-6 and 12-7(a), when $t = 1.5\tau$,

$$v_C = 77\% \text{ of E} = 0.77 \times 100 \ \text{V} = \textbf{77 V}$$

(b)
$$55 \ \text{V represents} \ \frac{55 \ \text{V}}{100 \ \text{V}} \times 100\% = 55\% \text{ of } E$$

From Figures 12-6 and 12-7(b), when $v_C = 55\%$ of E,

$$t = 0.8 \text{ time constant} = 0.8 \times 2 \ \text{s} = \textbf{1.6 s}$$

FIGURE 12-7
Using the Graph of Figure 12-6 to
Solve Example 12-2.

SELF-CHECK

12-4:1. Suppose you have a capacitor in a circuit, similar to the one shown in Figure 12-2, with the following component values:

$$E = 20 \text{ V}, \quad R = 250 \text{ k}\Omega, \text{ and } C = 0.40 \text{ }\mu\text{F}.$$

The switch is closed at time $t = 0$ s and the capacitor begins charging. Use the charging curve in Figure 12-6 to answer the following:
(a) At what instant is the voltage across the capacitor equal to 10 V?
(b) How long does it take the capacitor to charge to a potential difference of 19 V?

12-4:2. For the circuit in Self-Check Problem 12-4:1, find the voltages across the capacitor and the resistor 200 ms after the switch is closed.

12-5 DISCHARGING A CAPACITOR

If we assume that the switch in the circuit of Figure 12-8(a) has been in position 1 for a length of time equal to at least five time constants, the potential difference across the capacitor will be equal to the applied voltage, 200 volts. If we now place the switch in the open-circuit position 2, the potential difference across the capacitor will remain at

FIGURE 12-8
Discharging a Capacitor.

200 volts for some considerable time since there is no conducting path between the plates except a bit of leakage through the dielectric. At the instant we throw the switch to position 3, the potential difference across the capacitor will still be 200 volts since it takes *time* for the surplus electrons to flow from the bottom plate to the top plate of the capacitor. To satisfy Kirchhoff's voltage law, the initial current will have to be such that a 200 volt *IR* drop will be developed across the resistance of the conductor connecting the two plates together. Therefore, the initial current is very high and the capacitor discharges very rapidly.

If we wish to reduce the magnitude of the initial discharge current, we can change the circuit to that shown in Figure 12-8(b) so that, in switch position 3, a resistance of 100 kilohms is connected across the charged capacitor. The initial current is now limited to

$$\text{initial } I = \frac{V_{\text{init}}}{R} = \frac{200 \text{ V}}{100 \text{ k}\Omega} = 2 \text{ mA} \tag{12-5}$$

Since the charge stored in the capacitor is being released at a slower rate $i = dq/dt$, it will take much longer for the capacitor to discharge fully.

The potential difference across the capacitor is at its maximum value just as the switch is thrown to position 3. Therefore, the initial current is the maximum value that the instantaneous discharge current can have and, as a result, the voltage across the capacitor starts to fall fairly rapidly. But, as the charge diminishes, the voltage across the capacitor must also decrease and, according to Kirchhoff's voltage law, the *IR* drop must have the same value as the potential difference of the capacitor. Therefore, the current must decrease and as a result, the capacitor will discharge more slowly. This interdependence among the potential difference, the *IR* drop, the current, and the rate of decrease of the potential difference results in the exponential discharge of Figure 12-9(a). The current in the 100-kilohm resistor flows in the opposite direction when the capacitor of Figure 12-8(b) is discharging with the switch in position 3 to that when the capacitor is charging in switch position 1. Consequently, the polarity of the *IR* drop across the resistance is reversed when the capacitor discharges. We indicate this reversal in the graph of the instantaneous current and *IR* drop in Figure 12-9(b) by using the *negative*

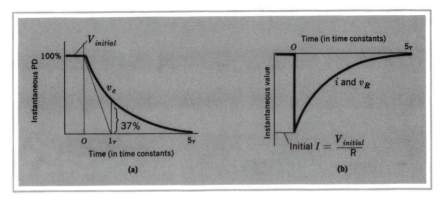

FIGURE   12-9
Instantaneous Voltage and Current as a Capacitor Discharges.

vertical axis to plot instantaneous current or instantaneous voltage drop across the resistance.

With the same procedure we used in deriving Equation (12-3), the initial rate of change of potential difference across the capacitor as it starts to discharge becomes

$$\frac{dv}{dt} = -\frac{V_{init}}{CR} \qquad (12\text{-}6)$$

The − sign is a result of the discharge current flowing through the resistance in a direction opposite from that followed during the charging of the capacitor. Therefore, the rate of change of voltage is a *negative* quantity, which indicates that the voltage across the capacitor is decreasing rather than increasing. If we define the **discharge time constant as the time it would take for the capacitor to discharge completely** *if* **it were to continue to discharge at the initial rate of fall of potential difference for the whole time interval,** Equation (12-6) reduces to

$$\tau = CR \qquad (12\text{-}4)$$

Therefore, **we determine the discharge time constant in exactly the same manner as the charge time constant.** Again, the exponential discharge curve bears a fixed relationship to the time constant. In a time interval equal to one time constant, the potential difference of the capacitor will have dropped to 37% of its initial value. Or, if we wish, we can say that the capacitor loses 63% of its initial potential difference in a time interval of one time constant. For practical purposes, the time required for the capacitor to completely discharge will again be five time constants. In Figure 12-8(b),

$$\tau = CR = 0.05 \ \mu\text{F} \times 100 \ \text{k}\Omega = 5 \ \text{ms}$$

and total discharge time = 5 × 5 ms = 25 ms.

EXAMPLE 12-3

After the switch has been in position 1 for a considerable period of time,
(a) What is the initial rate of change of potential difference across the capacitor in the circuit of Figure 12-8(b) when the switch is placed in position 3?
(b) What will the potential difference across the capacitor be 7 ms after the switch is placed in position 3?
(c) How many milliseconds will elapse after the switch is placed in position 3 before the instantaneous voltage drop across the 100-kΩ resistor reaches 110 V?

SOLUTION

(a)
$$\text{Initial } \frac{dv}{dt} = -\frac{V_{init}}{CR}$$

$$= -\frac{200 \ \text{V}}{0.05 \ \mu\text{F} \times 100 \ \text{k}\Omega} = -40 \ \text{kV/s}$$

(b)
$$\tau = CR = 0.05 \ \mu F \times 100 \ k\Omega = 5 \ ms$$

$$\therefore 7 \ ms \ represents \ \frac{7 \ ms}{5 \ ms} = 1.4 \ \tau$$

From Figures 12-6 and 12-10(a), when $t = 1.4\tau$,

$$v_C = 25\% \ of \ V_{init} = 0.25 \times 200 \ V = \textbf{50 V}$$

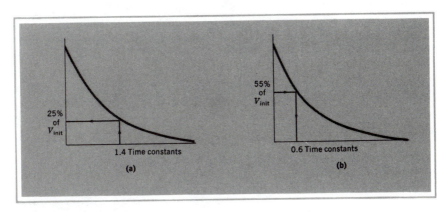

FIGURE 12-10
Using the Graph of Figure 12-6 to Solve Example 12-3.

(c) When the switch is in position 3, the resistor is connected directly across the capacitor. From Kirchhoff's voltage law,

$$v_C + v_R = 0 \quad from \ which \quad v_R = -v_C$$

$$110 \ V \ represents \ \frac{110}{200} = 55\% \ of \ V_{init}$$

From Figures 12-6 and 12-10(b), when $v_C = 55\%$ of V_{init},

$$t = 0.6\tau = 0.6 \times 5 \ ms = \textbf{3 ms}$$

SELF-CHECK

12-5:1. Given a capacitor in a circuit similar to the one shown in Figure 12-8(b), with the following component values:

$$E = 100 \ V, \quad R = 250 \ k\Omega, \quad and \quad C = 0.20 \ \mu F.$$

The switch is originally in position "1" and the capacitor is fully charged. At time $t = 0$ s, the switch is moved to position "3" and the capacitor begins to discharge through the resistor.

(a) What is the initial discharge current in the resistor?

(b) How long does it take (practically) to discharge the capacitor?

12-5:2. For the circuit used in Self-Check Problem 12-5:1, use the discharge graph in Figure 12-6 to find the voltages across the capacitor and the resistor 60 ms after the discharge begins.

12-6 *CR* WAVESHAPING CIRCUITS

Suppose we replace the manually operated switches of Figures 12-2 and 12-8 with a motor-driven switch, as in Figure 12-11(a), which rotates at 500 revolutions a second. Consequently, the moving contact dwells for one millisecond on each stationary contact. The graph for the output voltage appears as shown in Figure 12-11(b), and the circuit of Figure 12-11(a) constitutes a simple **square-wave** generator. Since the polarity of the output voltage does not reverse, we describe it as a **pulsating dc** waveform. Although mechanical "choppers" are sometimes used to generate *intermittent* direct current from a steady dc source, the switching action is usually performed by transistors in the pulse circuits we encounter in radar, television, and digital computers.

FIGURE 12-11
Simple Square-Wave Generator.

In Figure 12-12, we have connected a *CR* network across the output of the square-wave generator. In Figure 12-12(a), the output voltage is v_C; and, in Figure 12-12(b), the output voltage is v_R. If the time constant of the *CR* network is short compared with the duration of each half-cycle of the square wave, the surge of charging and discharging current will be completed each half-cycle before the switch changes position. In this case, it is a fairly simple matter to sketch the output-voltage waveform.

Suppose that the *CR* networks of Figure 12-12 have time constants of 0.1 millisecond. Five time constants will then be 0.5 millisecond, and the charging and discharging current ceases halfway through each half-cycle of the square-wave input voltage. If we follow through the charge and discharge cycle of the capacitor, we find that the output voltages in the circuits of Figure 12-12 have the waveforms shown in the graphs of Figure 12-13.

However, if the time constant of the *CR* network is sufficiently long that charging

FIGURE 12-12
Simple *CR* Waveshaping Networks.

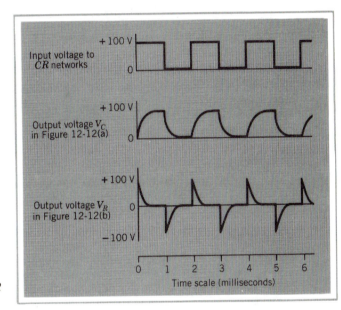

FIGURE 12-13
**Voltage Waveform Graphs for the *CR*
Networks of Figure 12-12.**

or discharging of the capacitor is not completed during a half-cycle of the square-wave input, calculation of the waveform becomes a bit of a chore, as illustrated by Example 12-4.

EXAMPLE 12-4
Starting from position 1, the switch in the circuit of Figure 12-14 dwells for 1 ms in each position. Calculate v_C 3 ms after the switch first moves to position 2.

SOLUTION
Step I. Starting from position 1, the capacitor is completely discharged. In position 2, the charging time constant is

$$\tau = CR = 0.01 \ \mu\text{F} \times 100 \ \text{k}\Omega = 1 \ \text{ms}$$

FIGURE 12-14
Circuit Diagram for Example 12-4.

Since the switch remains in position 2 for 1 ms, the capacitor will charge for a period of time equal to one time constant. Hence, v_C will rise to 63% × 500 V = 315 V.

Step II. As the switch returns to position 1, the discharging time constant becomes

$$\tau = CR = 0.01 \ \mu F \times (100 \ k\Omega + 100 \ k\Omega) = 2 \ ms$$

While the switch is in position 1, the capacitor will discharge for a period of time equal to one-half a time constant.

From Figure 12-6, v_C will drop to 60% of its initial value. Therefore, at the end of the second millisecond,

$$v_C = 0.6 \times 315 \ V = 189 \ V$$

As we start the third millisecond, the capacitor starts to charge again. But, this time, it starts from an initial potential difference of 189 V. We have a choice of two lines of thought in handling the rise in voltage across a partially charged capacitor.

Step IIIA. 189 V represents $\frac{189}{500}$ = 38% of E. On the charge curve of Figure 12-6, if the capacitor had started its charging from zero, a time interval of 0.48 time constant would have elapsed [Figure 12-15(a)] for the voltage across the capacitor to reach 189 V.

Since 1 ms = 1 charging time constant, one millisecond after the switch returns to the *charge* position, the potential difference across the capacitor will be the same as if it had charged from zero for a period of 0.48 + 1.0 = 1.48τ. From Figure 12-6, when $t = 1.48\tau$,

$$v_C = 77\% \text{ of } E = 0.77 \times 500 \ V = \mathbf{385 \ V}$$

Step IIIB. The alternative method is to think in terms of the *additional* potential difference to which the capacitor can charge.

Since $E = 500$ V and the voltage across the capacitor is already 189 V, the maximum *additional* applied voltage is only 500 V − 189 V = 311 V.

In a charging time interval of one time constant, the voltage can rise by 63% of this additional E.

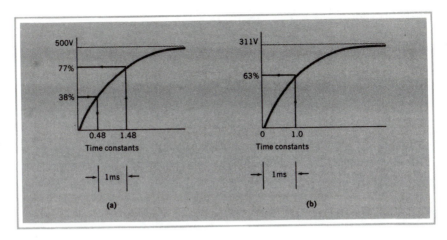

FIGURE 12-15
Graphical Solution of Step III of Example 12-4.

$$\therefore \text{ additional } v_C = 0.63 \times 311 \text{ V} = 196 \text{ V} \quad [\text{Figure 12-15(b)}]$$

Therefore, the final voltage across the capacitor is its *initial* voltage plus its *additional* increase in potential difference.

$$\therefore v_C = 189 \text{ V} + 196 \text{ V} = \mathbf{385 \text{ V}}$$

12-7 CALCULATOR SOLUTION FOR INSTANTANEOUS POTENTIAL DIFFERENCE

When we require greater accuracy than the universal exponential graph of Figure 12-6 provides, we may solve for instantaneous voltage across a capacitor by using the equations from which the curves of Figure 12-6 were plotted. Since a purely algebraic solution does not provide an opportunity to visualize what is going on in the circuit, it is well worthwhile accompanying any calculator solution with a freehand graphic solution to check any serious error in the more accurate algebraic solution. We start with the Kirchhoff's voltage-law equation for the basic circuit of Figure 12-2.

$$E = iR + v_C \tag{12-1}$$

Because the instantaneous current is the instantaneous rate of charging,

$$i = \frac{dq}{dt}$$

And since the instantaneous charge is the product of the capacitance and the instantaneous PD across it, $q = Cv_C$. Therefore,

$$i = C \frac{dv_C}{dt} \tag{12-7}$$

Substituting for i in Equation (12-1),

$$E = CR \frac{dv_C}{dt} + v_C \tag{12-8}$$

Since Equation (12-8) is a differential equation, to solve for the instantaneous potential difference v_C, we must resort to some routine calculus. The solution of Equation (12-8) is developed in Appendix 2-2. Hence, **the instantaneous voltage across a capacitor** *charging* **from 0 volts is**

$$v_C = E(1 - e^{-x}) \tag{12-9}$$

where $e = 2.718$ (the base of natural logarithms) *and* $x = t/CR$.

Since
$$\tau = CR \text{ seconds}, \quad x = \frac{t}{\tau}$$

We can again think of x as representing elapsed time measured in **time constants.**

Rather than repeat the solution of the differential Equation (12-1) to find the instantaneous charging current i_C, we can simply substitute the expression [Equation (12-9)] we obtained for the instantaneous PD across the capacitor directly into Equation (12-1). Therefore,

$$E = iR + E(1 - e^{-x})$$

from which
$$i_C = \frac{E}{R} e^{-x} \tag{12-10}$$

When the switch in the circuit of Figure 12-8(b) is in position 3, there is no voltage source in the discharge loop. Hence, the applied voltage E is zero. Consequently, the IR drop across the resistance and the potential difference across the capacitor must be numerically equal at all instants in time in order to satisfy Kirchhoff's voltage law. Therefore,

$$0 = iR + v_C = CR \frac{dv_C}{dt} + v_C$$

Solving for v_C, **the instantaneous voltage across a** *discharging* **capacitor is**

$$v_C = V_0 e^{-x} \tag{12-11}^\dagger$$

[†] See Appendix 2-2 for solution.

where V_0 is the initial voltage across the capacitor, e = 2.718 (the exponential function key of a calculator), and $x = t/CR$.

Equations (12-9) and (12-11) are the algebraic expressions from which the two curves of Figure 12-6 used in the graphical solution of CR examples were prepared.

EXAMPLE 12-2A

(a) What is the instantaneous voltage across the capacitor in the circuit of Figure 12-2 three seconds after the switch is closed?

(b) How long will it take the voltage across the capacitor to rise from 0 V to 55 V?

SOLUTION

(a)
$$x = \frac{t}{CR} = \frac{3\text{ s}}{1\ \mu\text{F} \times 2\text{ M}\Omega} = 1.5$$

$$v_C = E(1 - e^{-x}) = 100 \times (1 - e^{-1.5}) = \textbf{77.7 V}$$

The calculator chain for this computation becomes

$$100\ \boxed{\times}\ (1\ \boxed{-}\ 1.5\ \boxed{+/-}\ \boxed{e^x}\ )\ \boxed{=}$$

enter value of x \qquad converts to $-x$ \qquad solves for e^{-x}

When you press the $\boxed{e^x}$ key, the calculator displays the value of e^{-x} as 0.22313.

(b)
$$55\text{ V} = 100\text{ V}(1 - e^{-x})$$

from which $e^{-x} = 0.45$.

Solving for $-x$ with a calculator, we use the $\boxed{\textbf{ln}}$ key.

Hence,
$$-x = 0.45\ \boxed{\textbf{ln}}\ = -0.7985$$

from which
$$x = 0.8 = \frac{t}{CR} \qquad (\text{Press}\ \boxed{+/-}\ \text{key})$$

Hence,
$$t = 0.8 \times 2\text{ s} = \textbf{1.6 s}$$

EXAMPLE 12-4A

Starting from position 1, the switch in the circuit of Figure 12-14 dwells for 1 ms in each position. Calculate v_C 3 ms after the switch first moves to position 2.

SOLUTION

Step I. When the switch first moves to position 2, the capacitor will charge from 0 V toward 500 V. At any instant,

$$v_C = E(1 - e^{-x}) = 500(1 - e^{-x})$$

After 1 ms,

$$x = \frac{t}{CR} = \frac{1 \text{ ms}}{0.01 \text{ } \mu\text{F} \times 100 \text{ k}\Omega} = 1$$

(Hence, x represents a time interval of one time constant.)

$$\therefore v_C = 500(1 - e^{-1})$$
$$= 500 \times (1 - 0.3679) = 316 \text{ V}.$$

Step II.   As the capacitor discharges in position 1,

$$v_C = V_0 e^{-x} = 316 \times e^{-x}$$

After 1 ms,

$$x = \frac{t}{CR} = \frac{1 \text{ ms}}{0.01 \text{ } \mu\text{F} \times 200 \text{ k}\Omega} = 0.5$$
$$\therefore v_C = 316 \times e^{-0.5}$$
$$= 316 \times 0.6065 = 192 \text{ V}$$

Again, we can solve the additional charge of a partially charged capacitor by either of the lines of thought we used in Steps IIIA and IIIB of the solution for Example 12-4. For the moment, we will use the longer A method.

Step III.   When the switch returns to position 2, v_C starts at 192 V. The length of time it would take the capacitor to charge from 0 V to 192 V is

$$192 = 500(1 - e^{-x})$$

from which

$$e^{-x} = \frac{500 - 192}{500} = 0.616$$

Using the **ln** key of a calculator,

$$x = 0.616 \boxed{\text{ln}} \boxed{+/-} = 0.485$$

Since

$$x = \frac{t}{CR}$$

$$\therefore t = 0.4845 \times 1 \text{ ms} = 0.4845 \text{ ms}$$

One millisecond later, the capacitor will reach a voltage equivalent to having charged for a total time interval of $0.4845 + 1.0 = 1.4845$ ms. Hence,

$$x = \frac{t}{\tau} = \frac{1.4845 \text{ ms}}{1 \text{ ms}} = 1.4845$$

and
$$v_C = 500(1 - e^{-1.4845})$$
$$= 500 \times (1 - 0.2266) = \mathbf{387 \ V}$$

EXAMPLE 12-5

After being open for a considerable period of time, the switch in the circuit shown in Figure 12-16 is closed at $t = 0$ s and is opened again at $t = 1$ s. What is the voltage across the capacitor at $t = 2$ s?

FIGURE 12-16
Thévenin-Equivalent Circuit of a Capacitor Charging Circuit.

SOLUTION

The discharge circuit when the switch is open is quite straightforward. However, to calculate the charge time constant when the switch is closed, we need Thévenin's theorem.

Step I. Considering the capacitor to be the load, we remove it from the original circuit of Figure 12-16(a) and place it in the Thévenin-equivalent circuit of Figure 12-16(b). Once again, mental arithmetic is all that is required to tell us that E_x and R_x for the equivalent circuit of Figure 12-16(b) are 400 V and 160 kΩ, respectively.

Step II. Charging time constant is, therefore,

$$\tau = CR_x = 2 \ \mu\text{F} \times 0.16 \ \text{M}\Omega = 0.32 \ \text{s}$$

$$\therefore 1 \text{ s} = \frac{1.0}{0.32} = 3.125 \ \tau$$

Since the switch was originally open, the capacitor starts to charge with zero volts between its plates. From the algebraic solution of Equation (12-9), after charging for 1 s,

$$v_C = 400(1 - e^{-3.125}) = 400(1 - 0.044) = 382.4 \text{ V}$$

Step III. Discharging time constant is simply

$$\tau = CR_{\text{dis}} = 2 \ \mu\text{F} \times 800 \ \text{k}\Omega = 1.6 \text{ s}$$

$$\therefore 1 \text{ s} = \frac{1.0}{1.6} = 0.625 \ \tau$$

From Equation (12-11), at $t = 2$s,

$$v_C = 382.4 \times e^{-0.625} = 382.4 \times 0.535 = \textbf{205 V}$$

BASIC computer language includes a function which allows us to include e^{-x} in a calculation by entering EXP($-$X). We can solve Examples 12-2A and 12-4A by using a personal computer in the calculator mode, or by writing a simple program which requests input of values for the independent variables E, T, C, and R, and then prints out the solution for

```
V=E-E*EXP(-T/(C*R))
```

A more useful application of the computer's abilities would employ its ability to repeat a series of calculations rapidly with small increments in one of the independent variables (*time* in this case). Hence, we can program a computer to prepare the table from which we can draw the universal exponential graph of Figure 12-6 accurately. Just for fun (because the resulting graph is not accurate enough for graphical solution of examples) we can also program the computer to draw the charging curve for a *CR* circuit on either a video display or a hard-copy printer.

SELF-CHECK

12-7:1. Suppose you have a circuit similar to the one shown in Figure 12-16, with the following component values:

$$E = 100 \text{ V, each } R = 330 \text{ k}\Omega, \text{ and } C = 4.7 \ \mu\text{F}.$$

The capacitor is initially discharged. The switch is closed at time $t = 0$ s and the capacitor begins to charge.
(a) How long does it take to charge the capacitor?
(b) What is the potential difference across the fully charged capacitor?
(c) What is the potential difference across the capacitor at time $t = 1$ s?

12-7:2. Referring to the circuit used in Self-Check Problem 12-7:1, suppose that after the capacitor has become fully charged, the switch is opened again and the capacitor begins to discharge.

(a) How long does it take for the capacitor to discharge?

(b) At what time is the voltage across the discharging capacitor equal to 10 V?

12-8 TRANSIENT RESPONSE

Now that we are familiar with the way a capacitor charges and discharges in a dc circuit, and have algebraic procedures for calculating instantaneous voltages and currents, we can consider a third method of looking at the response of CR networks to opening and closing switches in direct-current circuits. Let's consider the charging of a partially charged capacitor, which we encountered in Step III of Example 12-4. In Step IIIB, we discovered that the final instantaneous capacitor voltage is equal to the initial voltage plus a percentage of the *additional* voltage based on elapsed time measured in time constants. Adapting this statement to Equation (12-9) gives

$$v_C = V_0 + (V_F - V_0)(1 - e^{-x}) \tag{12-12}$$

where V_0 *is the initial capacitor voltage and* V_F *is the steady-state voltage to which the capacitor can charge* (after an interval of at least five time constants). Carrying out the multiplication,

$$v_C = V_0 + V_F - V_0 - V_F e^{-x} + V_0 e^{-x}$$

Collecting the terms gives us a **universal equation for instantaneous capacitor voltage in a dc CR network.**[†]

$$v_C = V_F + (V_0 - V_F)e^{-x} \tag{12-13}$$

where v_C *is the instantaneous voltage across the capacitor,* V_0 *is the initial voltage across the capacitor, and* V_F *is the final steady-state voltage after an interval of at least five time constants has elapsed.*

In the purely resistive network of Figure 12-17(a), when we close the switch at $t = 0$, the output voltage of the two-terminal network instantly drops from 500 V to its new steady-state value of 100 V, as shown by the time graph of Figure 12-17(b). But in the CR network of Figure 12-17(c), the output voltage is the voltage across the capacitor which takes five time constants to discharge from its initial voltage V_0 of 500 V to its final steady-state voltage V_F of 100 V, as shown by the time graph of Figure 12-17(d).

Equation (12-13) represents the instantaneous output voltage as being made up of two components: the **steady-state response** (V_F) of Figures 12-17(b) and 12-18, and the **transient response** ($V_0 - V_F$)e^{-x} shaded in Figure 12-18. This transient response exists only for the short interval from $t = 0$ to $t = 5\tau$. Therefore

A transient is the time interval during which a network's instantaneous voltages and currents adjust from one steady state to another.

[†] We can also derive Equation (12-13) by considering a charged capacitor *discharging* toward a steady-state voltage other than zero.

FIGURE 12-17
Transient Response of a dc *CR* Network.

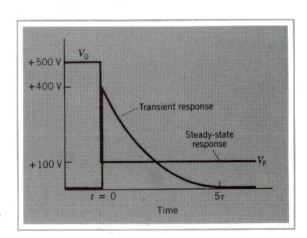

FIGURE 12-18
Separating *v*ₒᵤₜ into Steady-State and Transient Components.

EXAMPLE 12-6
If the switch in the two-terminal network of Figure 12-17(c) has been open for several minutes, what is the instantaneous output voltage 0.5 s after the switch is closed?

SOLUTION
From inspection of the network,

$$V_0 = 500 \text{ V} \quad \text{and} \quad V_F = 100 \text{ V}$$

The *discharge* time constant is governed by the Thévenin-equivalent resistance.

$$\therefore x = \frac{0.5 \text{ s}}{2 \text{ } \mu\text{F} \times 0.16 \text{ M}\Omega} = 1.5625$$

$$v_{\text{out}} = V_F + (V_0 - V_F)e^{-x}$$

$$= 100 + (500 - 100)e^{-1.5625}$$

$$= 100 + 400 \times 0.2096 = \mathbf{184 \text{ V}}$$

In the simple *CR* charging circuit of Figure 12-2, for the capacitor, $V_0 = 0$ and $V_F = 100$ V. Since $v_R = E - v_C$, for the resistor, $V_0 = 100$ V and $V_F = 0$. With this reversal of V_0 and V_F, *as long as we are careful in specifying V_0 and V_F for each steady-state situation, Equation (12-13) also applies to the instantaneous voltage drop across series-charging and shunt-discharging resistors.*

┌ *EXAMPLE 12-2B*
What is the instantaneous voltage across
(a) The capacitor
(b) The series charging resistor in the circuit of Figure 12-2 3 s after the switch is closed?

SOLUTION

$$x = \frac{t}{CR} = \frac{3 \text{ s}}{1 \text{ } \mu\text{F} \times 2 \text{ M}\Omega} = 1.5$$

(a) For the capacitor, $V_0 = 0$ and $V_F = 100$ V

$$v_C = V_F + (V_0 - V_F)e^{-x}$$

$$= 100 \text{ V} + (0 - 100 \text{ V})e^{-1.5}$$

$$= 100 \text{ V} - 100 \text{ V} \times 0.22313 = \mathbf{77.7 \text{ V}}$$

(b) For the resistor, $V_0 = 100$ V and $V_F = 0$

$$v_R = V_F + (V_0 - V_F)e^{-x}$$

$$= 0 + (100 \text{ V} - 0)e^{-1.5}$$

$$= 100 \text{ V} \times 0.22313 = \mathbf{22.3 \text{ V}}$$

Check: $E = v_C + v_R = 77.7 \text{ V} + 22.3 \text{ V} = 100 \text{ V}$

One advantage of the universal transient response point of view is that we can use the single Equation (12-13) in place of both Equations (12-9) and (12-11) and the procedure we used in Step III of Example 12-4A for determining the voltage across a partially charged capacitor.

EXAMPLE 12-4B
Starting from position 1, the switch in the circuit of Figure 12-14 dwells for 1 ms in each position. Calculate v_C 3 ms after the switch first moves to position 2.

SOLUTION

Step I.   When the switch first moves to position 2,

$$V_0 = 0 \quad \text{and} \quad V_F = 500 \text{ V}$$

Just before the switch moves back to position 1,

$$x = \frac{1 \text{ ms}}{0.01 \ \mu\text{F} \times 0.1 \text{ M}\Omega} = 1.0$$

$$\therefore v_C = 500 + (0 - 500)e^{-1.0}$$

$$= 500 - 500 \times 0.3679 = 316 \text{ V}$$

Step II.   When the switch first moves to position 1,

$$V_0 = 316 \text{ V} \quad \text{and} \quad V_F = 0$$

Just before the switch moves back to position 2,

$$x = \frac{1 \text{ ms}}{0.01 \ \mu\text{F} \times 0.2 \text{ M}\Omega} = 0.5$$

$$\therefore v_C = 0 + (316 - 0)e^{-0.5}$$

$$= 0 + 316 \times 0.6065 = 192 \text{ V}$$

Step III.   When the switch moves back to position 2,

$$V_0 = 192 \text{ V} \quad \text{and} \quad V_F = 500 \text{ V}$$

At the end of the 3-ms interval,

$$x = \frac{1 \text{ ms}}{1 \text{ ms}} = 1.0$$

and

$$v_C = 500 + (192 - 500)e^{-1.0}$$

$$= 500 - 308 \times 0.3679 = \mathbf{387 \text{ V}}$$

SELF-CHECK

12-8:1. Suppose that CR network shown in Figure 12-17 (c) has the capacitor changed to 47 μF.
(a) How long will it take, after the switch is closed, for v_{out} to reach its steady-state value?
(b) What is the voltage at v_{out} 10 seconds after the switch is closed?

FIGURE 12-19

12-8:2. Referring to the circuit shown in Figure 12-19, suppose that the switch has been in position "1" long enough so that the capacitor is fully charged. At time $t = 0$ the switch is moved to position "2". Determine the following for this circuit:
   (a) the initial capacitor voltage.
   (c) the time required for the capacitor to reach steady-state.
   (d) the instant that the capacitor voltage passes through zero volts.

12-9 ENERGY STORED BY A CAPACITOR

While a capacitor is charging, a displacement current flows in the circuit. As a result, energy is taken from the voltage source. Some of this energy is converted into heat as the charging current flows through the series current-limiting resistance in the circuit. This accounts for only half of the energy drawn from the source. The other half is stored by the capacitor. Energy is expended as the charging current forces a charge into a capacitor against the mounting opposition of the potential difference building up between the two plates. When we disconnect a charged capacitor from the source, its potential difference remains at a value equal to the original applied voltage for a considerable period of time, provided there is negligible leakage of the charge through the dielectric between the plates.

Since no current is required to maintain the potential difference between the plates of a charged capacitor, we can consider that capacitance *stores* electric energy in a **static** form. We think of this energy as being stored in the electric field which builds up between the plates of the capacitor and which produces a stress on the orbits of electrons bound to the molecules of the dielectric, as shown in Figure 11-7. We might compare the storing of electric energy by a capacitor to the storing of a charge of air under pressure in a steel cylinder. Energy is expended in forcing the air into the cylinder against the mounting back pressure. This stored energy can be recovered as desired by using the compressed air to operate a paint spray, pump up an automobile tire, etc. Similarly, we can recover the electric energy stored by a charged capacitor at will, as we shall discover in Section 12-10.

We can plot the graph of the *rate* at which a capacitor stores energy, since instantaneous power represents this rate, and $p = vi$. By calculating the product of the instantaneous potential difference across the capacitance and the instantaneous charging current for each division along the horizontal (time) axis of the graph of Figure 12-20,

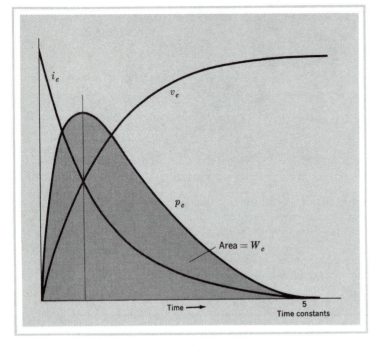

FIGURE 12-20
Graphical Representation of the Energy Stored by a Capacitor.

we can plot a graph of the rate at which the capacitance stores energy from instant to instant. As indicated by the shaded vertical line in Figure 12-20, this instantaneous power input to the capacitor reaches its peak value at the same instant in time that the graphs of instantaneous charging current and potential difference cross, that is, when the instantaneous voltage has risen to one-half of its final value and the instantaneous charging current has dropped to one-half of its initial value. From Figure 12-6, peak power input occurs 0.6 time constant after the capacitor commences to charge.

After a period of time equal to five time constants, the charging current has ceased, and the instantaneous power has, therefore, decreased to zero. The capacitor has now stored all the energy it is capable of storing at the particular potential difference applied to it. Since the total energy is the product of the average power and the time during which the capacitor is receiving energy, we can represent the energy stored by the capacitance by the *area* contained by the instantaneous power graph in Figure 12-20.

Solving for the area under the instantaneous power graph of Figure 12-20 requires basic integral calculus, as outlined in Appendix 2-3. However, we can simplify the task of determining how much energy a capacitor can store by returning to the notion of a capacitor charging at a *constant* rate equal to the initial rate of change of voltage. To achieve this, we connect the capacitor to a *constant-current* source whose terminal voltage can vary but whose current output always remains constant at the initial value of $I = E/R$, as shown in Figure 12-21(a). Since the voltage across the capacitor now rises at a *constant* rate from zero to its final value of $V = E$, as shown in Figure 12-21(b), the *average* voltage during this time interval must be one-half of V. For this linear charging condition,

$$W = Pt \quad \text{becomes} \quad W = V_{av} \times I \times t$$

FIGURE 12-21
Determining the Energy Stored by a Capacitor.

But
$$I \times t = Q \tag{2-2}$$

where Q is the charge stored by the capacitor after a time interval equal to t seconds. As shown in Figure 12-21(b), after t seconds, the voltage across the capacitor will have risen to V volts,

$$\therefore Q = CV \tag{11-7}$$

Since $V_{av} = 0.5\,V$, $W = Pt$ becomes

$$W = 0.5\,V \times C \times V = 0.5\,CV^2 \tag{12-14}$$

where W is the energy stored by the capacitance in joules, C is the capacitance in farads, and V is the potential difference across the capacitance in volts.

EXAMPLE 12-7

How much energy is stored in the 500-pF high-voltage filter capacitor of a television receiver when it is charged to 15 kV?

SOLUTION

$$W = 0.5\,CV^2 = 0.5 \times 500 \text{ pF} \times (15 \text{ kV})^2 = 5.63 \times 10^{-2} \text{ J} = \mathbf{56.3 \ mJ}$$

SELF-CHECK

12-9:1. A 33-μF electrolytic capacitor has a peak working voltage of 50 V. Find the energy stored in the capacitor when the voltage across the capacitor is (a) 50 V and (b) 25 V.

12-9:2. Suppose you have two identical capacitors except that the second is charged to ten times the voltage of the first. Compare the energy stored in these capacitors.

12-10 CHARACTERISTICS OF CAPACITIVE DC CIRCUITS

The ability of a capacitor to store electric charge and, as a consequence, to oppose a rapid change in potential difference between its terminals, makes the capacitor a particularly useful circuit component in practical dc circuits. Its ability to oppose any instant change in the voltage across it makes the capacitor very suitable for filtering or smoothing any fluctuations in the output from a pulsating voltage source such as a rectifier, as shown in Figure 12-22(a). When the instantaneous source voltage drops below the voltage across the capacitor, the capacitor maintains the load voltage by discharging some of its stored energy into the load. The capacitor will then recover this charge the next time that the instantaneous source voltage rises to its peak value, as shown in Figure 12-22(b). This application of a capacitor is similar to the use of a reservoir tank with an air compressor used with a paint spray. Since the pressure of the air in the tank cannot change without the flow of the stored air, the tank maintains a fairly constant pressure at the paint spray in spite of the fluctuations in the pressure developed by the piston action of the compressor. (The compressed air tank analogy provides a fairly accurate means of visualizing the charge and discharge action of capacitance, with the friction of the air hose representing resistance.)

FIGURE 12-22
Filtering Voltage Fluctuations with a Capacitor.

The energy-storage ability of a capacitor is very useful in cases where we wish to obtain a very high current for a short period of time without subjecting the source to a severe current surge. With a circuit patterned after the basic circuit of Figure 12-23, the capacitor charges slowly from the source and then discharges rapidly when we connect the low-resistance load across it. The peak current drawn from the source is limited by the series resistor. This system is used to obtain a one microsecond pulse for radar transmitter magnetrons with a peak power of over a million watts during the pulse. This system is also used in electric spot welding of aluminum where a short-duration surge of very high current is required to make the weld. The same system is used on a smaller scale in making the delicate welds which hold cathode-ray tube elements in place. In this case, the circuit of Figure 12-23 is used, not because of its ability to limit the peak current drawn from the source, but because we are able to control very precisely the amount of energy in each discharge pulse by setting the potential difference to which the capacitor is allowed to charge.

The instantaneous potential difference across the capacitor of a given *CR* combination is an exponential function of elapsed time and will always rise along exactly the same

FIGURE 12-23
Obtaining Short-Duration, High-Current Pulses from a Reservoir Capacitor.

FIGURE 12-24
Using a *CR* Time Constant for Timing Purposes.

curve. Therefore, we can use a series *CR* circuit as the basis for a very accurate timing device for time intervals of from a small fraction of a second to several seconds. In the simple circuit of Figure 12-24, the neon lamp "strikes" (glows) when the potential difference across it reaches 100 volts and extinguishes when the voltage drops below 20 volts. When a neon lamp is not struck, the gas is an insulator and we may consider the lamp to be an open circuit. When the lamp strikes, the gas becomes ionized and becomes an excellent electric conductor with the free electrons knocked out of the gas molecules flowing in one direction and the positively charged gas ions moving in the other direction. The resistance of the neon lamp under these conditions is only around 100 ohms. If the time constant of the series *CR* circuit in Figure 12-24 is several seconds, it will take several seconds for the potential difference across the capacitor to rise to the striking voltage of the neon lamp. The low resistance of the ionized neon will then quickly discharge the capacitor down to the extinguishing voltage of the lamp, whereupon the resistance of the neon lamp again becomes extremely high and the capacitor starts to recharge. The time interval between the short flashes from the neon lamp is, therefore, dependent on the time constant of *C* and *R*.

In Section 12-6, we noted one of the more significant applications of capacitance characteristics to waveshaping circuits. Before we complete this section, we should also note that capacitance can have undesirable characteristics in some instances. So far, we have assumed that there is negligible leakage of charge through the dielectric of capacitors. This is essentially true for practical purposes with such dielectrics as ceramic, mica, and plastic film. However, it is not true for electrolytic capacitors of the type described in Section 11-5. We can represent capacitor leakage in practical circuits by an appropriate high resistance in parallel with the capacitance of the capacitor in a circuit diagram, as shown in Figure 12-25.

Capacitance in electric circuits is not limited to capacitors. As we noted in Figure 11-10, capacitance can exist between any pair of electric conductors which are at different potentials. In many circuit applications, this **stray capacitance** is small enough that its effect on the behavior of the circuit is negligible. But in modern digital computer circuits, where timing pulse voltages must rise and fall in a time interval of a few nanoseconds, the tendency of stray capacitance to oppose such rapid changes in voltage can have a serious effect on circuit performance. In the circuit of Figure 12-26, the capacitance across the *pn*-junctions of the transistor is directly in parallel with the input

FIGURE 12-25
Equivalent Circuit of a "Leaky"
Capacitor.

FIGURE 12-26
Effect of Stray Capacitance.

to the amplifier (even though it would not appear in the actual circuit diagram). Hence, the input signal to the transistor is the potential difference across this stray capacitance. Since this capacitance must charge and discharge through the internal resistance of the pulse generator, if the time constant is more than a small fraction of the pulse duration, there will be appreciable distortion of the desired pulse.

PROBLEMS

NOTE: Solutions need not be restricted to the method indicated by the section number.

12-3-1. When the switch is first closed in the circuit shown in Figure 12-27(a)
    (a) What is the initial current?
    (b) What is the initial voltage across the capacitor?
    (c) What is the initial rate of rise of voltage across the capacitor?
    (d) What is the charging time constant?
    (e) How long will it take the capacitor to charge?

FIGURE 12-27

12-3-2. In the circuit shown in Figure 12-27(b), $v_C = 0$ at $t = 0$.
    (a) What is the initial charging current?
    (b) What is the initial rate of change of voltage across the capacitor?
    (c) What is the charging time constant?
    (d) What is the voltage across the capacitor at $t = \tau$?
    (e) What is the charge on the capacitor at $t = \tau$?

12-4-3. Using the universal exponential curves of Figure 12-6, for the circuit of Figure 12-27(a) determine
    (a) The voltage across the capacitor 5 s after the switch is closed.
    (b) The instantaneous charging current 2 s after the switch is closed.
    (c) The length of time for the voltage across the capacitor to reach 200 V.
    (d) The length of time for the voltage across the capacitor to rise from 50 V to 150 V.

12-4-4. Using the universal exponential curves of Figure 12-6, for the circuit of Figure 12-27(b) determine
    (a) How long it will take the capacitor to charge to 16 V.
    (b) How long it will take the capacitor to charge to 10 V.
    (c) The instantaneous voltage across the capacitor at $t = 0.5$ ms.
    (d) The instantaneous charging current at $t = 0.25$ ms.
    (e) How long it will take the voltage across the capacitor to rise from 4 V to 12 V.

12-5-5. A 40-μF capacitor has been charged to a voltage of 500 V.
    (a) How long will it take to discharge the capacitor through a 470-kΩ resistor?
    (b) What is the maximum value of the discharge current?

12-5-6. The capacitor of Problem 12-5-5 is discharged by shorting its terminals with an electric conductor having a resistance of 0.4 Ω.
    (a) What is the peak discharge current?
    (b) How long will it take to discharge the capacitor?

12-5-7. A 40-μF capacitor, which has been charged to a voltage of 500 V, is discharged through a 470-kΩ resistor. Using the universal exponential curves, determine
    (a) How long it will take the voltage across the capacitor to drop to 50 V.
    (b) The voltage across the capacitor 10 s after it starts to discharge.
    (c) The instantaneous discharge current 20 s after it starts to discharge.
    (d) How long it will take the voltage across the capacitor to drop from 400 V to 100 V.

12-5-8. If the resistor in the circuit of Figure 12-24 is 5 MΩ and the capacitor has a capacitance of 8 μF, calculate the time interval between flashes of the neon lamp. The lamp strikes at 100 V, and its resistance then becomes 100 Ω. It extinguishes when the voltage across it drops to 20 V, and its resistance then becomes infinitely high.

12-6-9. The switch in the circuit of Figure 12-28 is alternately thrown to position 1 for 1 ms and then to position 2 for 9 ms.

FIGURE 12-28

(a) Draw an accurate graph of the voltage between terminals *A* and *C* for an interval of 30 ms.

(b) Superimpose (in another color) on the graph of part (a), a graph of the instantaneous voltage between terminals *B* and *C* during the same time interval. Label the magnitude of the instantaneous voltage at each instant of switch operation.

(c) Draw a separate graph of the voltage drop between terminals *A* and *B* for the same time interval.

12-6-10. Repeat Problem 12-6-9 with the switch remaining 1 ms in each position. Show a time interval of 10 ms.

12-6-11. A pulse generator having negligible internal resistance feeds a 20-kHz pulse waveform, shown in detail in Figure 12-29, into the *CR* waveshaping network shown.

(a) Draw an accurate graph of two complete cycles of v_{out}.

(b) Repeat with the positions of the capacitor and resistor reversed.

FIGURE 12-29

12-6-12. A 10-kΩ resistor is connected in series with the capacitor in the circuit of Figure 12-29. Draw an accurate graph of two complete cycles of v_{out} across the series combination of the capacitor and the 10-kΩ resistor.

12-7-13. Repeat Problem 12-4-3 using algebraic expressions for instantaneous values of voltage and current.

12-7-14. Repeat Problem 12-4-4 using algebraic equations to solve for instantaneous voltage and current.

12-7-15. After a considerable period of time in position 2, the switch in the circuit of Figure 12-30 is thrown to position 1 at $t = 0$.
(a) Calculate i_C and dv_C/dt at $t = 0$.
(b) How long must the switch remain in position 1 for v_C to rise to 100 V?
(c) The switch is thrown to position 2 at $t = 5$ min. Calculate i_C and dv_C/dt as the switch makes contact in position 2.
(d) If the switch returns to position 1 at $t = 5.5$ min, calculate v_C at $t = 10$ min.

12-7-16. After a considerable period of time in position 1, the switch in the circuit of Figure 12-31 is thrown to position 2 for 1 ms and is then returned to position 1.
(a) Calculate v_C 1 ms after the switch returns to position 1.
(b) Draw a graph of the output voltage at terminal A with respect to ground for this 2-ms interval.

FIGURE 12-30   FIGURE 12-31

12-7-17. Commencing from the open position, the switch in the circuit of Figure 12-32 closes for 1 ms and then opens. Calculate the voltage across the capacitor 1 ms after the switch reopens.

12-7-18. Repeat the procedure of Problem 12-7-17 for the circuit of Figure 12-33.

12-8-19. If the switch in the two-terminal network of Figure 12-17(c) has been closed for several minutes, what is the instantaneous output voltage 0.5 s after the switch is opened?

FIGURE 12-32   FIGURE 12-33

12-8-20. When the switch in the circuit of Figure 12-31 is thrown from position 1 to position 2 at $t = 0$ and back to position 1 at $t = 5$ ms, we can think of the instantaneous output voltage at terminal

A (with respect to ground) as consisting of a square pulse plus two transients. Plot separate fully labeled graphs from $t = 0$ to $t = 10$ ms for

(a) The steady-state change in V_0 (as if the capacitor were open-circuit).
(b) The transient component only $[(V_0 - V_F)e^{-x}]$.
(c) The actual instantaneous output voltage.

12-8-21. The switch in the network of Figure 12-34 has been in position 1 long enough for the initial charging transient to have been completed. What is the instantaneous output voltage (with respect to ground) 2 ms and 4 ms after the switch is thrown to position 2?

12-8-22. Assuming that the charging transient is completed before the switch is operated in each case, what is the instantaneous output voltage (with respect to ground) in the network of Figure 12-35
(a) 2 ms after the switch closes?
(b) 2 ms after the switch opens?

FIGURE 12-34   **FIGURE 12-35**

12-9-23. Calculate the energy stored by the capacitors in the circuits of Figure 12-27(a) and (b) when they are fully charged.

12-9-24. Calculate the energy stored by the capacitor in Problem 12-5-5
(a) Before it starts to discharge.
(b) After it has been discharging for one time constant.

12-9-25. A 50-μF capacitor charged to a potential difference of 32 V is discharged through a 20-kΩ resistor. How much energy is dissipated by the resistor as the capacitor is completely discharged? If the capacitor had been discharged through a 680-Ω resistor, how much energy would have been dissipated by the resistor?

12-9-26. In Problem 12-5-8, how much energy is transferred to the neon lamp with each flash?

REVIEW QUESTIONS

12-1. What factors determine the initial charging current of a capacitor?

12-2. What factors determine the rate of change of potential difference between the plates of a capacitor?

12-3. When a discharged capacitor is connected to a voltage source, the current changes instantly from zero to a value of E/R. Why then can the voltage across the capacitor not change instantly?

12-4. Account for the exponential shape of the graph of v_C in Figure 12-5.

12-5. Why must the rate of change of voltage across a capacitor diminish as the charge on its plates increases?

12-6. What is the advantage of using the **time constant** of a CR circuit as a unit of time in determining the instantaneous voltage?

12-7. Why does the initial charge on a capacitor have no effect on the length of time required for the capacitor to discharge through a given resistance?

12-8. Draw an accurately labeled graph of the instantaneous voltage across the capacitor in the circuit of Figure 12-36 plotted against time as the double-pole, double-throw switch is thrown to the opposite position.

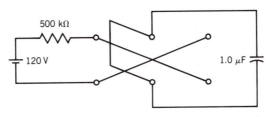

FIGURE 12-36

12-9. Figure 12-6 shows a graph of the algebraic equation $y = 1 - e^{-x}$. Why do we use this equation to solve for v_C rather than the simple exponential equation $y = e^{-x}$?

12-10. In the circuit of Figure 12-32, we use the Thévenin-equivalent resistance in the *charge* time constant, but not in the *discharge* time constant. In the circuit of Figure 12-33, we use the Thévenin-equivalent resistance in the *discharge* time constant, but not in the *charge* time constant. Explain the difference.

12-11. Why are the charge and discharge time constants the same in the circuit of Figure 12-34?

12-12. Using the techniques we employed in producing Figure 12-20, plot a graph of the instantaneous power in a resistor when a charged capacitor is connected across it.

12-13. An air-dielectric capacitor is charged from a voltage source and then disconnected. The plates are then moved twice as far apart. Assuming no leakage of the charge, account for the difference in the energy stored by the capacitor before and after moving the plates.

12-14. Instead of changing the spacing between the plates in Question 12-13, a dielectric with a dielectric constant of 2.0 is placed between the plates after the capacitor is disconnected from the source. Account for any change in the energy stored by the capacitor.

12-15. The circuit of Figure 12-12(b) resembles the CR coupling used between stages of an electronic amplifier. However, the output waveform shown in Figure 12-13(c) is not a faithful replica of the input waveform of Figure 12-13(a). What conditions must exist for the CR coupling circuit to transmit the square wave with minimum distortion?

12-16. Figure 12-37 shows the circuit diagram of a BC-type (battery-capacitor) photoflash unit for standard flash bulbs. Explain its operation and advantages.

12-17. When steady-state voltages are suddenly switched in a dc CR network, the transient current always starts from zero and returns to zero, having either a positive or negative direction, as shown by the graph of Figure 12-38. Does this mean that the current in resistors of the network must be zero except during the transient? Explain.

FIGURE 12-37

FIGURE 12-38
Positive and Negative _CR_ Transients.

12-18. Draw superimposed graphs (in different colors) to show how a graph of v_C is obtained by combining a sudden change in steady-state voltage with a transient having the form of either the positive or negative transient shown in Figure 12-38.

*12-19. Design a computer program to print out a labeled table of x and e^{-x} for values of x ranging from 0 to 5 in steps of 0.25.

*12-20. Design a computer program to display a labeled graph of the function $V = 100*(1 - EXP(-X))$ for values of X ranging from 0 to 5.

13

MAGNETIC CIRCUITS

13-1 MAGNETISM

The third property of an electric circuit, **inductance,** depends on the relationship between electric current in a conductor and the magnetic field around the conductor created by the electric current (just as capacitance depends on the relationship between potential difference and the electric field in a dielectric created by the potential difference). Before we can proceed with an investigation of inductance in electric circuits, we must be able to identify magnetic properties and express their magnitudes in suitable units.

In Chapter 2, we considered two natural phenomena—gravitation and electricity. We found that both phenomena were revealed through a force which can act without physical contact. Hence, we define both electric force and gravitational force as **field forces,** and we measure both forces in newtons. **Magnetism** is a third form of natural phenomenon, the *effects* of which were observed almost two thousand years ago by Chinese mariners, who used a crude form of magnetic compass for navigation. If we bring a magnet close to a compass needle (which itself is a tiny magnet carefully pivoted at its center of gravity), as in Figure 13-1, we find that the compass needle will be deflected so that one end of the needle always points to one end of the magnet. As we move the magnet, the compass needle follows the motion. Once again, the phenomenon is revealed by a force which acts without physical contact. Consequently, magnetic force is a *field* force which also can be measured in newtons.

Although magnetism and electricity may appear to be independent natural phenomena, in 1819, Hans Christian Oersted showed that magnetic fields and electric current are closely associated. Oersted discovered that a magnetic compass needle always takes up a position perpendicular to a current-carrying electric conductor placed near the compass needle. Although we do not as yet know as much about the nature of magnetism as we

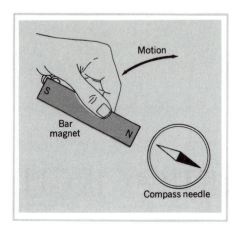

FIGURE 13-1
Detecting the Presence of a
Magnetic Field.

know about the nature of electricity, modern theories of magnetism suggest that all magnetic properties stem from the *motion* of electrons in an atom. (Electric properties stem from the *position* of electrons in an atom.) As we have already discovered, electric *current* also consists of electrons in motion, but, in this case, the motion is detached from the parent atom.

The laws of magnetism were developed when the CGS system of units was the universal standard. The CGS magnetic units honor such pioneers as William Gilbert, Karl Friedrich Gauss, James Clerk Maxwell, and Hans Christian Oersted. There has been considerable reluctance to abandon the CGS units in favor of the presently accepted MKS units, which are part of the International System of Units. There has also been a tendency in North America to use a hybrid set of magnetic units which combined CGS or MKS units with the English unit of linear measurement—the inch. We can avoid all this confusion by using the currently preferred SI units exclusively in this text.

13-2 MAGNETIC FIELDS

In studying the effects of electric force, we thought of the space surrounding an electric charge as constituting an electric **field.** We defined this electric field in terms of the region surrounding an electric charge in which a second elemental electric charge is acted upon by an electric force. Similarly,

> *A magnetic field is that region in which a magnetic material is acted upon by a magnetic force.*

In Chapter 11, we found it convenient to visualize an electric field as consisting of invisible electric lines of force, which are, in fact, the paths along which elemental positive electric charges move as a result of being acted upon by the electric force. Although we have not discovered a magnetic equivalent of an elemental positive electric charge, Michael Faraday suggested that we still think of a magnetic field as being made up of **magnetic lines of force.**

We can outline the magnetic field surrounding a bar magnet by placing it under a piece of paper sprinkled with fine particles of a magnetic material such as iron filings,

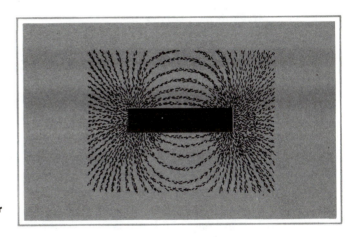

FIGURE 13-2
Pattern Formed by
Sprinkling Iron Filings over
a Bar Magnet.

as in Figure 13-2. We note that the filings tend to gather around two areas, one at each end of the magnet. These areas where the effect of the magnetic field is concentrated are called the **poles** of the magnet. By applying Faraday's magnetic lines of force concept to the magnetic field detected by the iron filings in Figure 13-2, we can sketch the magnetic field of a bar magnet in the form shown in Figure 13-3. If we could invent such a thing as an isolated magnetic pole (a concept used in some theories of magnetism),

A magnetic line of force represents the path along which a theoretical isolated magnetic pole would move from one pole of a magnet to the other.

If we explore the magnetic fields of various shapes of magnets with a small compass needle, as in Figure 13-1, or with iron filings, as in Figure 13-2, we become aware of several important characteristics of magnetic lines of force. Since all the electromagnetic effects which we shall be studying can be traced directly to these characteristics, it is important that we learn them thoroughly at this point.

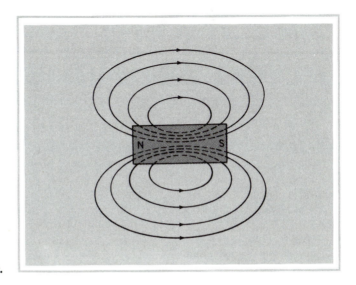

FIGURE 13-3
Diagram of the Magnetic
Field around a Bar Magnet.

Magnetic lines of force possess direction.

Combining Figures 13-3 and 13-1, we can say that the compass needle in Figure 13-1 aligns itself with the magnetic lines of force. If we now bring the opposite pole of the bar magnet near the compass needle in Figure 13-1, we find that the compass needle will again align itself with the magnetic lines of force, but it will now have the *opposite* end of the needle pointing at the pole of the bar magnet. For the compass needle to be able to distinguish one pole of the bar magnet from the other, we must think of the magnetic lines of force as possessing a positive direction. Since magnetic lines of force are hypothetical, this direction will have to be arbitrarily assigned. However, since the earth itself has a magnetic field, we can identify the poles of a magnet more specifically. If we suspend the bar magnet by a thread tied around its center of gravity, it will act like a compass needle and align itself with the earth's magnetic field. One pole of the magnet will always point north and is, therefore, referred to as the **north pole** of the magnet. Similarly, the other pole is called the **south pole.** Having identified the poles of a magnet, we can now state that the positive direction of the magnetic lines of force in the region around a magnet shall be considered as being *from* the north pole of the magnet *toward* the south pole.

Magnetic lines of force always form complete loops.

They do not begin at the north pole of a magnet and end at the south pole but continue between south pole and north pole inside the magnet to form complete closed loops. We can show this by cutting the magnet of Figure 13-4(a) in two. As we pull the two sections apart, we can detect the presence of the magnetic lines of force in the gap between the two halves. New north and south poles are formed on each side of the gap, as shown in Figure 13-4(b). No matter how many times we cut the magnet, each section

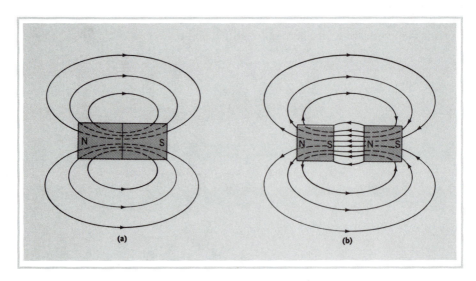

FIGURE 13-4
Showing that Magnetic Lines of Force Form Complete Loops by Continuing from South Pole to North Pole within the Magnet.

has its own north and south poles. We cannot continue the subdivision to a point where we find an isolated magnetic pole.

Magnetic lines of force represent a tension along their length which tends to make them as short as possible.

This effect would also be noticed as we separate the two halves of the magnet in Figure 13-4(b). In pulling the two halves apart, we would be lengthening the lines of force, and we would notice a definite attraction between the two pieces as the magnetic lines of force try to pull them together again in an attempt to shorten the lines of force. The analogy of a magnetic field composed of rubber bands sometimes helps in grasping this idea. Since the two pieces of the one magnet in Figure 13-4(a) became separate magnets in Figure 13-4(b), this tendency for magnetic lines of force to become as short as possible accounts for the familiar statement that **unlike poles of a magnet attract one another.**

Magnetic lines of force repel one another.

We can see this effect by examining the magnetic field around a bar magnet (Figure 13-4). The magnetic lines of force tend to diverge as we move away from the poles rather than to converge or even to remain parallel. As a result of this mutual repulsion,

Magnetic lines of force cannot intersect but must always form individual closed loops.

This mutual repulsion of magnetic lines of force accounts for the effect shown in Figure 13-5 when like magnetic poles approach one another.

Like poles of magnets repel one another.

The combination of these basic characteristics is responsible for the shape of the magnetic field of a magnet. The magnetic lines of force are kept in equilibrium by each

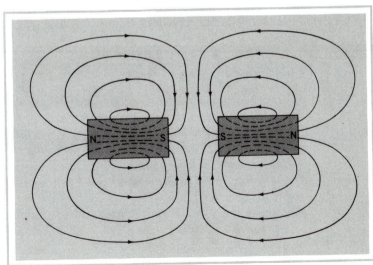

FIGURE 13-5
Repulsion of Like Poles of Magnets.

line tending to shrink its loop length and at the same time being prevented from doing so by the repulsion of the next smaller line-of-force loop. The magnetic field around a uniform magnet will be symmetrical unless there are other magnetic materials within the field. (Compare Figures 13-3 and 13-5.)

If we place a piece of soft iron in the field of a magnet, we find that some of the magnetic lines of force alter their normal paths to include the iron bar in their circuit, as shown in Figure 13-6. This is because iron is a better "conductor" of magnetic lines of force than air. The net result of the change in the pattern of the magnetic field is that the total number of lines of force is increased. Therefore,

Magnetic fields always tend to arrange themselves in such a manner that the maximum number of lines of force is set up.

FIGURE 13-6
Magnetic Lines of Force
Detour to Include Soft-Iron
Bar in Their Circuit.

While it is in the field of the main magnet, the piece of soft iron in Figure 13-6 exhibits the properties of a magnet. It takes on a south pole where the direction of the magnetic lines of force is toward the soft iron, and a north pole where their direction is away from the iron. The piece of soft iron has become a **temporary** magnet by **induction.** It will lose most of its magnetism when we remove it from the field of the main magnet.

We have noted the attraction of unlike poles as magnetic lines of force tend to become as short as possible. The magnetic lines of force which detour through the soft-iron bar in Figure 13-6 can become shorter if the iron object approaches the nearer pole of the main magnet. This accounts for the attraction of unmagnetized ferromagnetic objects such as nails and steel washers to the closer pole of a magnet. We can measure this magnetic force of attraction in newtons. Powerful electromagnets produce a force of attraction sufficient to lift several tonnes (megagrams) of scrap iron.

Although there is no "insulator" for magnetic lines of force, it is possible to shield a certain space such as a sensitive meter movement or a cathode-ray tube from stray magnetic fields by enclosing this space in a cylinder of soft iron. Magnetic lines of force will go out of their usual path to include the soft iron as part of their magnetic circuit. This **magnetic shielding** leaves the space within the cylinder comparatively free of magnetic lines of force, as shown in Figure 13-7.

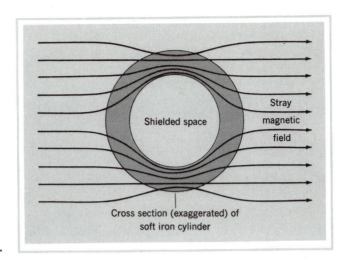

Shielded space

Stray magnetic field

Cross section (exaggerated) of
soft iron cylinder

**FIGURE 13-7
Magnetic Shielding.**

13-3 ELECTROMAGNETISM

So far we have discussed only the magnetic field around a permanent magnet. This field is due to the alignment of the iron atoms in the permanent magnet in such a manner that their orbiting electrons generate an overall magnetic field. The disadvantage of this type of magnet is that we cannot turn the magnetic field on and off at will. However, Oersted discovered that a magnetic field is produced around an electric conductor whenever current flows through it. Since we can control an electric current quite readily, we, therefore, have a means of controlling the production of a magnetic field.

By exploring the magnetic field around a current-carrying conductor with iron filings or small compass needles, as in Figure 13-8, we find that the lines of force form concentric circles in a plane which is at right angles to the axis of the conductor. As the current increases from zero, a magnetic line of force loop of infinitesimal radius forms at the center of the conductor. As the current increases, the radius of the magnetic line of force increases and another loop forms at the center of the conductor. As the result of mutual repulsion of magnetic lines of force, this new loop forces the first line of force to expand still further. As long as the current is *increasing,* a whole series of closed-loop magnetic lines of force *expand* outward from the center of the conductor. When the current becomes steady, the magnetic field becomes stationary. Because of the tendency for magnetic lines of force to shorten like rubber bands, they will be concentrated at the center of the conductor and spaced farther and farther apart as we move outward from the center of the conductor. When the current is turned off, this tension causes each line of force to collapse back toward the center of the conductor.

Although the magnetic field around a straight current-carrying conductor has no north and south poles, the individual lines of force must still have direction. This direction can be checked by the compass needle in Figure 13-8(b). Reversing the direction of the current through the conductor causes all the compass needles to reverse their positions. We can determine the direction of the magnetic field around a straight current-carrying conductor by the right-hand rule illustrated in Figure 13-9.

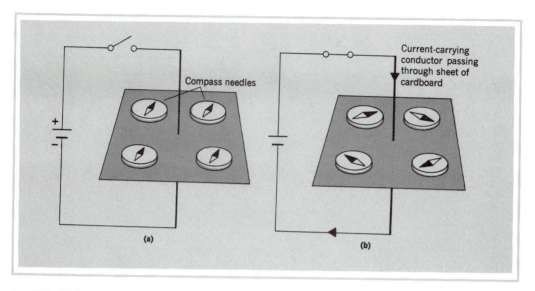

FIGURE 13-8
(a) With No Current, Compass Needles All Indicate Earth's Magnetic Field; (b) When Current Flows, Compass Needles Indicate Circular Magnetic Lines of Force around the Conductor.

FIGURE 13-9
Right-Hand Rule for Determining Direction of Magnetic Lines of Force around a Straight Current-Carrying Conductor.

Grasp the conductor (figuratively speaking) with the right hand so that the thumb points in the conventional current direction; the fingers circling the conductor then point in the direction of the magnetic lines of force around the conductor.

If we form the current-carrying conductor into a loop, as in Figure 13-10, we note that the magnetic lines of force all pass through the center of the loop in the *same* direction. For a given current, the total number of lines of force has not changed; therefore, we have strengthened the magnetic field by concentrating it into a smaller physical area.

FIGURE 13-10
Concentrating the Magnetic Field by Forming the Conductor into a Loop.

We can concentrate the magnetic field of the conductor still further by winding the conductor around a cardboard tube, as in Figure 13-11, to form a **solenoid.** Because the current in the adjacent turns of the solenoid is traveling in the same direction around the circumference of the coil, the solenoid behaves as a single loop of a stranded conductor, each strand carrying a current equal to the actual solenoid current. This will result in a magnetic field around the solenoid having the pattern shown in Figure 13-11. Since this magnetic field pattern is similar to that of the bar magnet of Figure 13-3, we have produced an **electromagnet.**

We can adapt the right-hand rule for the direction of the magnetic lines of force around a straight conductor to apply to a solenoid, as illustrated in Figure 13-12.

Grasp the solenoid with the right hand so that the fingers follow the conventional current direction around the circumference of the solenoid; the thumb then points in the direction of the magnetic lines of force through the center of the solenoid.

Because we can think of the solenoid as an electromagnet, we can think of the end of the solenoid where the direction of the magnetic lines of force is away from the coil (Figure 13-12) as being the north pole of the electromagnet. Therefore, the thumb in this hand rule also points to the north pole end of the electromagnet.

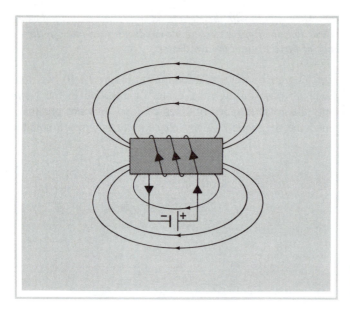

FIGURE 13-11
Magnetic Field Pattern of a
Current-Carrying Solenoid.

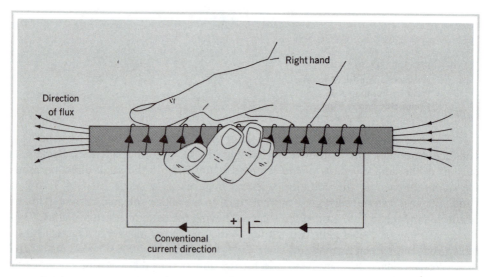

FIGURE 13-12
Right-Hand Rule for Determining Direction of Magnetic Lines of
Force through a Current-Carrying Solenoid.

As shown in Figures 13-4(b) and 13-5, interaction between the magnetic fields of two permanent magnets produces a magnetic force between them. Similarly, a magnetic force exists between two parallel current-carrying electric conductors. In Figure 13-13(a), equal currents flow in opposite directions in the cross-sectional view of two parallel conductors. The tail feathers of an arrow on the left-hand conductor represent a conventional current direction *into* the page, and the point of the arrow on the right-hand conductor represents a conventional current direction *out* of the page (Check the direction

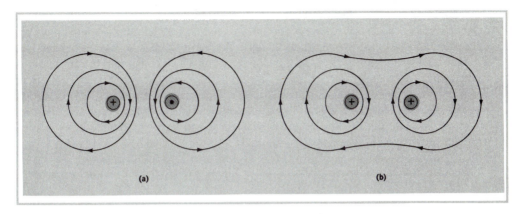

FIGURE 13-13
Magnetic Field around Parallel Current-Carrying Conductors.

of the magnetic lines of force with Figure 13-9.) Between the two conductors in Figure 13-13(a), the magnetic lines of force all have the same direction. Since they tend to repel one another, there is a resultant magnetic force of *repulsion* between the two conductors.

In Figure 13-13(b), where the conventional current direction in both conductors is into the page, the magnetic lines of force between the conductors have *opposite* directions. As a result, the magnetic lines of force which would have come midway between the two conductors can become shorter by combining to form single loops encompassing both conductors. The tendency of these lines of force to become still shorter results in a magnetic force of *attraction* between the two conductors. The magnitude of these magnetic forces depends on the length and spacing of the two conductors and on the magnitude of the electric current in them.

In Section 1-2, we stated that the **ampere** is the fourth basic unit which extends the meter-kilogram-second system of units to include electrical measurement. If we now define the ampere in terms of the *magnetic* force in *newtons* between two *current-carrying* electric conductors, we can include magnetic measurements in the MKSA units, which form the basis of the International System of Units. Hence, the international ampere is the constant current that, if maintained in two straight parallel conductors that are of infinite length and negligible cross section and are separated from each other by a distance of 1 meter in a vacuum, will produce between these conductors a force equal to 2×10^{-7} newton per meter of length.

13-4 MAGNETIC FLUX

In keeping with the designation we used in dealing with electric fields, when we wish to refer collectively to the total *number* of lines of force in a magnetic field, we use the term **magnetic flux.** Therefore, **magnetic flux** and **magnetic lines** (plural) **of force** are synonymous.

The letter symbol for magnetic flux is the Greek letter Φ (phi).

For current to flow in an electric circuit, the circuit must form a closed loop. Since magnetic lines of force must always form closed loops, we can refer to the path we can trace around these loops as a **magnetic circuit.** Although there is no *flow* associated with a magnetic circuit comparable to the flow of electrons which constitutes the current in an electric circuit, we shall discover many similarities between current in an electric circuit and flux in a magnetic circuit. Since electricity and magnetism are closely related natural phenomena, we find that for every magnetic property there is a comparable electrical property. We note these comparisons as we proceed to develop suitable units of measurement for magnetic properties.

Magnetic flux in a magnetic circuit is the counterpart of electric current in an electric circuit.

Because of the close relationships between magnetism and electricity, in establishing a unit of magnetic flux, we should base it on already established electrical units. In Figure 13-11, we noted that when we caused an electric current to flow in a solenoid, we established a magnetic field around it. Michael Faraday replaced the battery illustrated in Figure 13-11 with a sensitive current-indicating **galvanometer,** as shown in Figure 13-14. Faraday then thrust a permanent magnet into the solenoid. As he did so, he noticed a momentary deflection of the meter pointer. Therefore, not only does an electric current produce a magnetic field, but when we cause a magnetic field to *link* an electric circuit, we also cause an energy conversion from the *moving* **magnetic field to** *induce* an electromotive force by moving electric charge carriers from one end of the coil to the other. As we noted in Section 2-10, the induced emf, in turn, produces a potential difference between the ends of the coil which (under open-circuit conditions) is numerically equal to the induced emf. If the electric circuit is a closed loop (as it is through the galvanometer in Figure 13-14), the induced voltage, in turn, causes current to flow in the electric circuit. Faraday discovered that the induced voltage appeared only when he moved the magnet with respect to the electric circuit and that the faster he moved the magnet, the greater the deflection of the meter pointer. We shall pursue the consequences of this **electromagnetic induction** on the behavior of electric circuits in Chapter 14. But for the moment, Faraday's discovery gives us a means of specifying the basic unit of

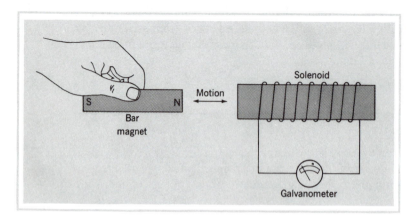

FIGURE 13-14
Faraday's Demonstration of Electromagnetic Induction.

magnetic flux in terms of the emf induced into a *single-turn* loop (similar to Figure 13-10) as we cause the flux *linking* (passing through the center of) the loop to build up from zero.

> **The weber is the SI unit of magnetic flux.** [†]
>
> **If the flux linking a single-turn coil builds up from zero at a rate which will induce an average emf of one volt in the coil, at the end of one second the flux linking the coil will have a magnitude of one weber.**
>
> **The unit symbol for weber is Wb.**

13-5 MAGNETOMOTIVE FORCE

Just as an electric current cannot flow in an electric circuit until we connect it to a voltage source, magnetic flux (magnetic lines of force) cannot be established until a **magneto-motive force** is produced.

> **Magnetomotive force in a magnetic circuit is the counterpart of electromotive force in an electric circuit.**

With the electromagnet of Figure 13-11, magnetic flux appears only when electric current flows in the solenoid. Therefore, magnetomotive force (mmf) must be a direct result of electric current. We can, therefore, establish a unit of mmf on the basis of the electric current in a *single-turn* coil of wire. Hence,

> **The ampere is the SI unit of magnetomotive force.**

Since adding turns of wire to a solenoid produces the same effect as adding extra strands to a single-turn coil, each strand carrying the input current to the coil, the *effective* amperes of magnetomotive force is the product of the coil current and the number of turns in the coil.

> **The letter symbol for magnetomotive force is F_m.** [‡]

$$F_m = NI \tag{13-1}$$

where F_m is the magnetomotive force in amperes (effective), N is the number of turns of wire in the coil, and I is the actual current through the coil in amperes.

So that we do not forget that the total magnetomotive force is the product of actual coil current and the number of turns of wire in the coil, we can say that

> **The ampere-turn is the practical unit of magnetomotive force.**

[†] The SI unit of magnetic flux was named in honor of Wilhelm Eduard Weber, who established many of the fundamental theories of magnetism.

[‡] Although current standards for letter symbols avoid the use of script letters, the script letter $\mathscr{F}$ is sometimes used to distinguish magnetomotive force from the magnetic force of attraction and repulsion.

Equation (13-1) becomes the defining equation for magnetomotive force. We shall use the practical unit—the ampere-turn—for clarity in this text.

SELF-CHECK

13-5:1. Determine the magnetomotive force in a 50-turn coil that is carrying a 2-A electric current.

13-6 RELUCTANCE

For a given electric circuit, Ohm discovered a constant proportionality between the applied voltage and the resulting current. He termed this constant the **resistance** of the electric circuit. Similarly, for a given magnetic circuit (of nonferrous materials), there is a constant proportionality between the magnetomotive force and the resulting magnetic flux. This constant is called the **reluctance** of the magnetic circuit.

Reluctance in a magnetic circuit is the counterpart of resistance in an electric circuit.

From the so-called "Ohm's law for magnetic circuits,"

Reluctance is the opposition of a magnetic circuit to the establishing of magnetic flux. The letter symbol for reluctance is R_m. [†]

From the definition of reluctance,

$$R_m = \frac{F_m}{\Phi}$$

(13-2)

where R_m is reluctance, F_m is magnetomotive force in ampere-turns, and Φ is the magnetic flux in webers.

For present purposes, we can express reluctance in **ampere-turns per weber,** as indicated by Equation (13-2). The SI unit of reluctance is the *reciprocal henry.*

SELF-CHECK

13-6:1. Determine the reluctance of a magnetic circuit if a 50-turn coil carrying a 2-A electric current results in 50 mWb of magnetic flux.
13-6:2. How many turns are necessary on a coil which is to carry a 50-mA current and produce 200 mWb of magnetic flux in a magnetic circuit having a reluctance of 100 H^{-1}?

[†] Although current standards for letter symbols avoid the use of script letters, the script letter $\mathscr{R}$ is sometimes used to distinguish reluctance from resistance.

13-7 PRACTICAL MAGNETIC CIRCUITS

Since magnetic fields are produced around current-carrying electric conductors, it follows that magnetic fields are associated with most electrical apparatus. The operation of much of this apparatus (for example, motors, generators, transformers, loudspeakers, and relays) depends on the magnetic fields associated with it. It is just as important to be able to control the path and strength of the magnetic fields in such devices as it is to control the flow of electric current in them.

Most materials are nonmagnetic and have no more effect on a magnetic field than does free space. **Ferromagnetic** materials (iron) have a unique molecular structure that reduces their magnetic reluctance to a very small fraction of the reluctance of nonmagnetic materials. We can make use of this magnetic property of iron to construct magnetic circuits to contain the magnetic field within the chosen iron path. And, by applying the relationship stated by the so-called "magnetic Ohm's law" of Equation (13-2), we can determine the magnetomotive force required to develop the desired field strength.

In some magnetic circuits, such as the loudspeaker and the meter movement in Figure 13-15, the required magnetomotive force is supplied by permanent magnets. But, in most electric machines, the mmf is produced by current-carrying coils of wire

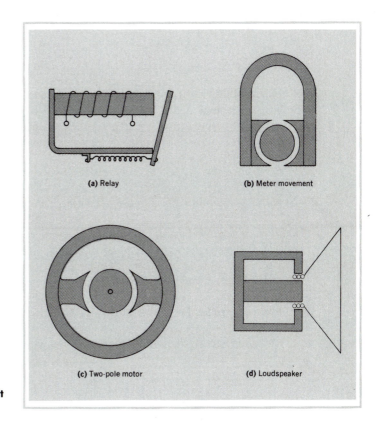

(a) Relay

(b) Meter movement

(c) Two-pole motor

(d) Loudspeaker

FIGURE 13-15
Typical Magnetic Circuit Cross Sections.

(solenoids, field windings, armatures, operating coils, etc.). The shape and the location of the resulting magnetic field depends, in turn, on the reluctance of the magnetic circuit and the available magnetomotive force. We can determine the magnetomotive force we require to produce a certain flux in a given magnetic circuit in much the same manner that we determined the applied voltage (resulting from an electromotive force) needed to produce a certain electric current in an electric circuit possessing a given resistance.

13-8 PERMEABILITY

As shown in Figure 13-6, placing a ferromagnetic material in a magnetic field *increases* the number of magnetic lines of force set up by a given magnetomotive force. Hence, in working with practical magnetic circuits, it will be more convenient to think in terms of their ability to permit the setting up of magnetic flux, rather than in terms of their opposition (reluctance). Consequently,

> **Permeance is a measure of the ability of a magnetic circuit to permit the setting up of magnetic lines of force.**
> **The letter symbol for permeance is P_m.**[†]

From the definition of permeance,

$$P_m = \frac{1}{R_m} \qquad (13\text{-}3)^{‡}$$

From Equation (13-3), we may express permeance in **webers per ampere-turn.** The SI unit of permeance is the *henry.*

In comparing the magnetic properties of various materials, we would then compare the permeance of sections of each material with unit length and cross-sectional area.

> **Permeance per unit length and cross-sectional area is called permeability.**

Note that **permeability** is a figure indicating the ability of a *material* to permit the setting up of magnetic lines of force, whereas **permeance** is a measure of the ability of a given *magnetic circuit* to permit the setting up of magnetic lines of force. Since permeability relates to the magnetic property of materials, it also has a counterpart relating to the property of materials in *electric* fields. In Section 11-6, we defined *permittivity* of a dielectric in terms of its ability to permit the setting up of *electric* lines of force.

> **The letter symbol for permeability is the Greek letter μ (mu).**

Unfortunately, the letter symbol for the permeability of magnetic materials is the

[†] Although current standards for letter symbols avoid the use of script letters, the script letter $\mathcal{P}$ is sometimes used to distinguish permeance from power.

[‡] Note that permeance in a magnetic circuit is the counterpart of conductance in an electric circuit. See Equation (6-10).

same as the prefix for *micro*. From the definition of permeability

$$\mu = P_m \frac{l}{A} = \frac{l}{R_m A}$$ (13-4)

where μ is the permeability, P_m is permeance in webers per ampere-turn, R_m is reluctance in ampere-turns per weber, l is the length of the magnetic circuit in meters, and A is the cross-sectional area in square meters.

For present purposes, we can express permeability in terms of **webers per ampere-turn for a one-meter cube.** In SI, permeability is expressed in *henrys per meter*.

SELF-CHECK

13-8:1. A magnetic circuit is to be created in a metal ring by the use of an electric current in a toroid winding (see Figure 13-17). The ring has a cross-sectional area of 0.25 cm², an average circumference of 15 cm, and a reluctance of 10^7 H^{-1}. Find the permeance of the ring's magnetic circuit.

13-8:2. Find the permeability of the metal ring described in Self-Check Problem 13-8:1.

13-9 FLUX DENSITY

Permeability is a measure of the ability of a *material* to permit the setting up of magnetic lines of force. We set up Equation (13-4) so that permeability could be expressed in terms of permeance of a cube of that material having unit length and cross-sectional area. Therefore, in thinking of permeability, we are not dealing with total flux. Rather, we are interested in the *flux per unit cross-sectional area*.

Flux per unit cross-sectional area is called flux density.
The letter symbol for flux density is B.
The tesla is the SI unit of flux density.[†]
The unit symbol for tesla is T.

$$B = \frac{\Phi}{A}$$ (13-5)

where B is flux density in teslas (or webers per square meter), Φ is the total flux in a magnetic circuit in webers, and A is the cross-sectional area of the magnetic circuit in square meters.

Equation (13-5) automatically defines a **tesla** as being equal to a **weber per square meter** (T = Wb/m²).

[†]Named in honor of the inventor, Nikola Tesla, who immigrated from Hungary to the United States where he was responsible for much of the early design of rotating electric machines.

SELF-CHECK

13-9:1. A magnetic circuit is to be created in a metal ring by the use of an electric current in a toroid winding (see Figure 13-17). The ring has a cross-sectional area of 0.25 cm², an average circumference of 15 cm, and a reluctance of 10^7 H⁻¹. If the electric current produces 2 μWb of magnetic flux in the ring, find the flux density in the metal ring.

13-10 MAGNETIC FIELD STRENGTH

Since permeability is based on a *cube* of the magnetic material, we are not particularly concerned with the total magnetomotive force required for the full length of the magnetic circuit. Rather, we are interested in the magnetomotive force required to create a certain flux density in a *unit length* of the magnetic circuit.

Magnetomotive force per unit length is called magnetizing force.
The letter symbol for magnetizing force is H.

From the definition of magnetizing force,

$$H = \frac{F_m}{l} \qquad \text{(13-6)}$$

where H is the magnetizing force in ampere-turns per meter,[†] F_m is the magnetomotive force in ampere-turns, and l is the length of the magnetic circuit in meters.

Now that we have defined permeability, flux density, and magnetizing force, we can derive a very useful relationship among these three magnetic quantities.

Since $\qquad R_m = \dfrac{F_m}{\Phi} \qquad$ and $\qquad \therefore P_m = \dfrac{\Phi}{F_m}$

substituting in Equation (13-4) gives

$$\mu = \frac{\Phi}{F_m} \times \frac{l}{A} \qquad \text{or} \qquad \mu = \frac{\Phi}{A} \times \frac{l}{F_m}$$

But $\qquad \dfrac{\Phi}{A} = B \qquad$ and $\qquad \dfrac{F_m}{l} = H$

$$\therefore \mu = \frac{B}{H} \qquad \text{(13-7)}$$

where μ is permeability in henrys per meter (or in webers per ampere-turn for a

---

[†]Strictly speaking, in SI, magnetomotive force is expressed simply in effective *amperes*. Hence, magnetizing force is expressed in *amperes per meter*.

one-meter cube), B is flux density in teslas (or in webers per square meter), and H is the magnetizing force in amperes per meter (ampere-turns per meter).

Magnetizing force is only one of several terms that we can express in *amperes per meter*. Because we now define the *ampere* in terms of the force between current-carrying conductors (Section 13-3), we can also express the *strength* or *intensity* of a magnetic field in amperes/meter. We use Equation (13-7) to define **magnetic field strength** or **magnetic field intensity.** A fourth (seldom used) term, **mmf gradient,** is described by Equation (13-6). Because all four terms—magnetic field strength, magnetic field intensity, magnetizing force, and mmf gradient—are numerically equal, electrical engineering now prefers **magnetic field strength** as the general term for H (the magnetic counterpart of electric field strength in a dielectric).

Comparing Equation (13-7) with Equation (11-11),[†] we note that

Magnetic flux density in a magnetic circuit is the counterpart of electric flux density in an electric circuit.

Also,

Magnetic field strength in a magnetic circuit is the counterpart of electric field strength in an electric circuit.

and,

Permeability in a magnetic circuit is the counterpart of permittivity in an electric circuit.

SELF-CHECK

13-10:1. For the toroidal coil described in Self-Check Problem 13-9:1, find the magnetic field strength within the metal ring.

13-11 LINEAR MAGNETIC CIRCUITS

In a linear *electric* circuit, we can determine the resistance (at a certain temperature) by Equation (4-4).

$$R = \rho \frac{l}{A} \qquad (4\text{-}4)$$

For R to remain constant, the resistivity ρ must be independent of the applied voltage and/or the resulting current. Transposing Equation (13-4) gives

$$R_{\mathrm{m}} = \frac{l}{\mu A} \qquad (13\text{-}8)$$

[†] Page 297.

Hence, a linear magnetic circuit must be composed of a material that has a constant permeability. Unfortunately, the reluctance (and the permeability) of a magnetic circuit containing a *ferro*magnetic material depends on the degree of magnetization. Consequently, we can use only *non*magnetic materials (such as brass, plastic, air) in a *linear* magnetic circuit.

We can use Equation (13-7) to determine the permeability of all nonmagnetic materials by measuring the B/H ratio of a magnetic field in free space.

The permeability of free space is

$$\mu_v = 4\pi \times 10^{-7} \text{ H/m} \tag{13-9}$$

If we know the length and cross-sectional area of a given magnetic circuit composed of nonmagnetic materials, we can combine Equation (13-8) with the magnetic "Ohm's law" Equation (13-2) to calculate the total flux produced by a given magnetomotive force.

With the simple solenoid wound on a nonmagnetic core, as shown in Figure 13-16, it is difficult to calculate (or control) the length and cross-sectional area of the *complete* loops of the magnetic lines of force. But by winding the coil completely around a ring-shaped core to form the **toroid**[†] winding shown in Figure 13-17, we confine the magnetic field to a path completely within the coil for either nonmagnetic or ferromagnetic core materials.

FIGURE 13-16
Solenoid Wound on a
Nonmagnetic Core.

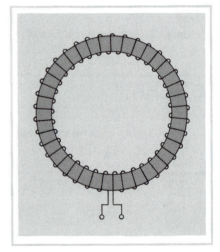

FIGURE 13-17
Construction of a Toroid
Winding.

EXAMPLE 13-1

The brass core in Figure 13-17 has a circular cross section. The inside diameter of the core is 8 cm and the outside diameter is 12 cm. If there are 2000 turns in the winding, what total flux will be produced by a 1-A current in the coil?

[†] Toroid means a donut-shaped surface.

SOLUTION

Step I.   Some care is required in determining the correct values for the length and cross-sectional area of magnetic circuits. From the dimensions given, it follows that the diameter of the cross section of the core is 2 cm.

$$\therefore A = \pi \times (1 \text{ cm})^2 = \pi \times 10^{-4} \text{ m}^2$$

The length of the magnetic circuit will be the *average* path length. In this example, the inside circumference is $\pi \times 8$ cm and the outside circumference is $\pi \times 12$ cm. Hence, the average of these two extreme path lengths is $\pi \times 10$ cm.

Step II. $R_{\mathrm{m}} = \dfrac{l}{\mu A} = \dfrac{\pi \times 10 \text{ cm}}{4\pi \times 10^{-7} \times \pi \times 10^{-4} \text{ m}^2} = 7.958 \times 10^8$ At/Wb

Step III. $F_{\mathrm{m}} = NI = 2000\ t \times 1 \text{ A} = 2000$ At

Step IV. $\Phi = \dfrac{F_{\mathrm{m}}}{R_{\mathrm{m}}} = \dfrac{2000 \text{ At}}{7.958 \times 10^8 \text{ At/Wb}} = \mathbf{2.5 \times 10^{-6}}$ **Wb**

13-12 MAGNETIZATION CURVES

The simple solution for Example 13-1 depends on the permeability of nonmagnetic materials remaining constant. Unfortunately, the permeability of *magnetic* materials depends on the extent to which they have been *magnetized* by the applied magnetic field.

Atoms of ferromagnetic materials possess unique electron orbit patterns that cause each atom to generate a tiny magnetomotive force called a **magnetic moment.** These atoms have the ability to rotate their magnetic axes in groups of some 10^{15} (more or less) adjacent atoms called magnetic **domains** in response to the applied magnetic field. The extent of the magnetization, and thus the permeability of a magnetic core, depends on the flux density of the applied magnetic field. Even though we can prepare a graph, as in Figure 13-18, showing how the permeability of a given sample of iron varies with flux density, we discover an interdependence which becomes evident when we rearrange Equation (13-7).

$$B = \mu H$$

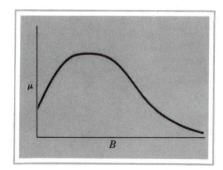

FIGURE   13-18
Manner in Which the Permeability of Cast Steel Varies with Flux Density.

For a given magnetic field strength H, the flux density B depends on the permeability μ, which, in turn, depends on the flux density B. Such a situation leads to a "cut-and-try" solution for flux density.

To avoid this situation, we can eliminate one dependent variable by disregarding permeabililty and plotting an accurate graph of the manner in which the flux density in a certain type of iron varies with the magnetic field strength. Since B and H as units of measurement are both based on a cube with unit dimensions, the **magnetization curve** (or simply **BH curve**) of Figure 13-19 depends only on the *type* of iron and not on the dimensions of the magnetic circuit.

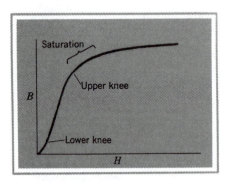

FIGURE   13-19
**BH Curve for a Typical
Cast-Steel Sample.**

Most of the magnetization of ferromagnetic cores occurs for applied magnetizing forces (and resulting magnetic field strengths) between the lower and upper knees of the magnetization curve of Figure 13-19. In this steep region of the BH curve, the magnetic axes of the domains are *switching* to act as closely as possible in the same direction as the applied magnetic field. As we increase H beyond the upper knee, note that there is very little increase in flux density. This is due to almost all domains having finally aligned themselves with the applied magnetic field. We call this condition **saturation** of the magnetic core.

13-13 HYSTERESIS

Although the BH curve of Figure 13-19 represents the manner in which the flux density of an *unmagnetized* sample of iron rises as we increase the value of H, it does *not* represent the manner in which the flux density decreases as we decrease the magnetic field strength. Most magnetic materials have some **retentivity.** When the magnetizing force is returned to zero, the residual magnetism of the iron will produce an appreciable value of flux density. In order to get rid of this residual flux, it is necessary to pass some current through the solenoid in the *opposite* direction. The amount of *negative* magnetic field strength or magnetizing force required to demagnetize a particular sample of iron is called the **coercive force.**

Figure 13-20 shows the complete cycle of magnetization of a piece of iron as we pass an *alternating* current through the solenoid. If we assume that the sample started from an unmagnetized state, as we increase the current from zero to maximum in one

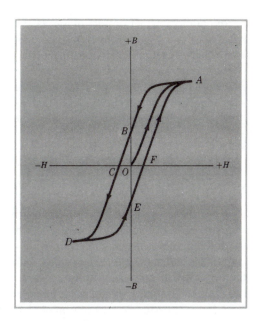

FIGURE 13-20
Typical Hysteresis Loop.

direction through the solenoid (positive value of H), the flux density increases along the customary magnetization curve OA. As we reduce the current to zero, the flux density decreases from saturation to a residual flux density at point B. To bring the flux density in the iron to zero, we must reverse the current through the coil and increase it to a value represented by OC. As we now increase the current in this direction, the iron will again saturate at point D, with the magnetic domains all lined up in the opposite direction to that represented at point A. Again, as we reduce the current to zero, the flux density in the iron will decrease only to point E. Again we must reverse the current direction to bring the flux density to zero at point F. Increasing the current to its maximum value in the positive direction completes the cycle by saturating the iron and returning the flux density to point A in the graph of Figure 13-20.

As shown in Figure 13-20, when H is due to an *alternating* current, the flux density B in a ferromagnetic core tends to *lag* behind the magnetizing force that creates it. This lagging of the magnetization of iron behind the magnetizing force is called **hysteresis.** We call the graph of Figure 13-20 that shows this lag a **hysteresis loop.**

Hysteresis is due to some of the magnetic domains in a ferromagnetic material not wanting to return to their original orientation and having to be forced to do so by a certain amount of reversed magnetizing force (coercive force). Whenever motion is accomplished against an opposing force, an energy transfer must take place. This energy is taken from the source of alternating voltage, which is responsible for the magnetic domains having to change their orientation, and is transferred to the molecules of the ferromagnetic material in the form of heat. The higher the frequency of the alternating current in the solenoid, the more rapidly the magnetic domains have to change their alignment and, therefore, the greater the **hysteresis loss.**

The greater the retentivity of a particular type of iron, the greater the coercive force that is required to demagnetize it. This means an increased opposition by the magnetic domains to reorientation, which results in a greater transfer of energy into heat. Therefore, hysteresis loss is also proportional to the retentivity of the iron. The iron selected for use in the magnetic circuits of transformers which are continually subject to an alternating mmf should have an absolute minimum retentivity if hysteresis loss is to be kept at a low value. Consideration of Figure 13-20 will show us that as the residual flux density of a sample of iron increases, the area within the hysteresis loop increases. Therefore, the area within the hysteresis loop for a given iron sample is a useful indication of its hysteresis loss.

13-14 NONLINEAR MAGNETIC CIRCUITS

In Section 4-9, we discovered that we could not provide data on nonlinear resistors by simply stating a resistance value in ohms, since their resistance is dependent on the current through them and/or the voltage drop across them. To analyze electric circuits containing nonlinear resistors, we are provided with data on their behavior in electric circuits in the form of **volt-ampere graphs** similar to those shown in Figures 4-7 to 4-9. Similarly, to analyze nonlinear magnetic circuits, we require comparable data on the magnetic properties of a particular type of iron or ferrite used in the magnetic circuit. This information is usually published in the form of accurate **magnetization curves** such as the *BH* graphs of Figure 13-21.

  If we know the dimensions and the *type* of iron used in a given magnetic circuit, the data provided by Figure 13-21 allow us to calculate the magnetomotive force required to produce a certain magnetic flux in the core, or vice versa.

⌐ *EXAMPLE 13-2*
The toroid winding of Figure 13-17 is wound on a circular cast-steel core with a *square* cross section. The inside diameter of the ring is 12 cm, and the outside diameter is 15 cm. If there are 2000 turns in the coil, what current must flow through it to develop a total flux in the core of 3×10^{-4} Wb?

SOLUTION 1
  Step I. Determine the dimensions of the magnetic circuit:

  Cross-sectional area: $A = (1.5 \text{ cm})^2 = 2.25 \text{ cm}^2 = 2.25 \times 10^{-4} \text{ m}^2$

  Average path length: $l = \pi \times 13.5 \text{ cm} = 42.4 \text{ cm} = 0.424 \text{ m}$

  Step II. Determine flux density:

$$B = \frac{\Phi}{A} = \frac{3 \times 10^{-4} \text{ Wb}}{2.25 \times 10^{-4} \text{ m}^2} = 1.333 \text{ T}$$

  Step III. Determine permeability: From the magnetization curve for cast steel in Figure 13-21, a magnetizing force H of 1750 At/m is required to produce a flux density B of 1.33 T.

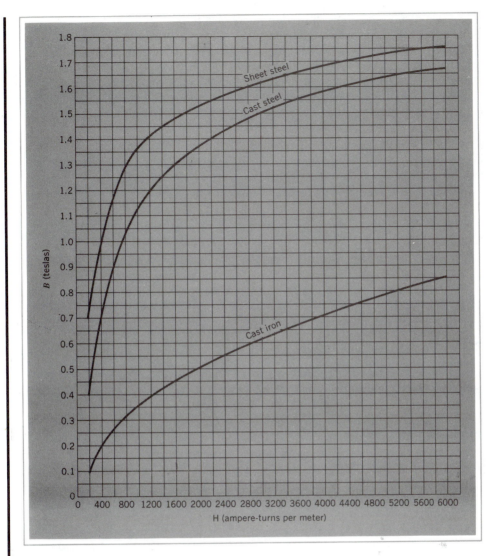

FIGURE 13-21
Typical Magnetization Curves.

$$\therefore \ \mu = \frac{B}{H} = \frac{1.33 \text{ T}}{1750 \text{ At/m}} = 7.6 \times 10^{-4} \text{ H/m}$$

Step IV. $\quad R_m = \dfrac{l}{\mu A} = \dfrac{0.424 \text{ m}}{7.6 \times 10^{-4} \times 2.25 \times 10^{-4} \text{ m}^2} = 2.48 \times 10^6 \text{ At/Wb}$

Step V. $\quad F_m = \Phi R_m = 3 \times 10^{-4} \text{ Wb} \times 2.48 \times 10^6 \text{ At/Wb} = 744 \text{ At}$

Step VI. $\quad \therefore I = \dfrac{F_m}{N} = \dfrac{744 \text{ At}}{2000 \text{ t}} = 0.372 \text{ A} = \mathbf{372 \ mA}$

Actually, the above procedure represents a rather roundabout way of solving magnetic circuits in which reference to *BH* curves is involved. Once we have determined the flux density, the magnetization curve tells us how much magnetizing force *H* we require for each unit length of the magnetic circuit. We can, therefore, solve for the mmf directly without going through Steps III and IV. Hence, we completely bypass both permeability and reluctance in the second method of solving magnetic circuits.

SOLUTION 2

Step I. Dimensions of the magnetic circuit (as in Solution 1):

$$A = 2.25 \times 10^{-4} \text{ m}^2 \quad \text{and} \quad l = 0.424 \text{ m}$$

Step II.
$$B = \frac{\Phi}{A} = \frac{3 \times 10^{-4} \text{ Wb}}{2.25 \times 10^{-4} \text{ m}^2} = 1.333 \text{ T}$$

Step III. From the *BH* curve for cast steel in Figure 13-21, $H = 1750$ At/m

Step IV.
$$F_m = Hl = 1750 \text{ At/m} \times 0.424 \text{ m} = 742 \text{ At}$$

Step V.
$$I = \frac{F_m}{N} = \frac{742 \text{ At}}{2000 \text{ t}} = 0.371 \text{ A} = \textbf{371 mA}$$

In a simple magnetic circuit where *all* the magnetomotive force is devoted to producing flux in the *uniform* cross section of the iron core, it is not difficult to solve for the flux when we are given the magnetomotive force.

EXAMPLE 13-3

A current of 0.5 A is passing through a coil consisting of 960 turns wound on a ring-type cast-steel core whose cross-sectional area is 12 cm² and whose average path length is 30 cm. What is the total flux in the core?

SOLUTION

$$F_m = NI = 960 \text{ t} \times 0.5 \text{ A} = 480 \text{ At}$$

$$H = \frac{F_m}{l} = \frac{480 \text{ At}}{30 \text{ cm}} = 1600 \text{ At/m}$$

From the *BH* curve for cast steel (Figure 13-21),

when
$$H = 1600 \text{ At/m}, B = 1.3 \text{ T}$$

$$\therefore \Phi = BA = 1.3 \text{ T} \times 12 \text{ cm}^2 = \textbf{1.56} \times \textbf{10}^{-3} \textbf{ Wb}$$

SELF-CHECK

13-14:1. Use Figure 13-21 to find the permeability of a cast iron ring in which the magnetic field has a flux density of 0.5 T.

13-14:2. A 1200-turn toroidal coil is wound on a cast iron ring which has the magnetization curve shown in Figure 13-21. The ring has a cross-sectional area of 3 cm^2 and an average circumference of 40 cm. Determine the following for the magnetic field within the ring when the coil is carrying a 2.0-A current:
(a) the field strength
(b) the flux density
(c) the total flux

13-15 LEAKAGE FLUX

In the preceding section, we wound the energizing coil along the full length of the magnetic circuit (toroid fashion) in an attempt to keep as much of the flux within the core as possible. This type of winding is expensive since it has to be wound by threading the wire through the center of the core. It is used only in cases where it is imperative that the number of lines of force traveling through the air outside the core be kept to a minimum. If a magnetic circuit is constructed as shown in Figure 13-22, the coil can be wound on a bobbin and then the magnetic core can be assembled. From our investigation of parallel resistors, we learned that *all* the current does not flow through the path of least resistance. Actually, the current splits up in proportion to the conductance of the parallel branches. When high-permeability iron is used for the full length of the magnetic circuit, the majority of the lines of force will follow the path established by the iron. But, since the air in the center of the core is not a magnetic insulator, and since magnetic lines of force have a tendency to try to become as short as possible, some of the flux will complete its circuit outside the core, as shown in Figure 13-22. These particular lines of force are called the **leakage flux.**

FIGURE 13-22
Leakage Flux in a Magnetic Circuit.

In designing a simple magnetic circuit similar to that of Figure 13-22, we can do several things to keep the leakage flux to a minimum. If we keep the magnetic circuit compact, the length of a magnetic line of force taking the short cut through the air is not much shorter than the path length through the iron. The higher the permeance (the lower the reluctance) of the magnetic circuit, the greater will be the number of lines of force which stay within the magnetic circuit. High permeance is achieved by making sure that the cross-sectional area is great enough that the required flux density does not cause the core to saturate. As suggested by Figure 13-15, many practical magnetic circuits contain air gaps. The presence of the air gap greatly increases the total reluctance of the magnetic circuit. Accordingly, leakage flux will be more pronounced in such circuits.

In the magnetic circuit of Figure 13-22, the total flux through the top section of the magnetic circuit is less than that through the center of the coil, due to the leakage flux bypassing the top section. The effect of the leakage flux, then, is that we must increase the current in the coil slightly in order to obtain as much flux in the top section of the core as there would be if there were no leakage flux. The ratio between the total developed flux and the useful flux is known as the **leakage factor.** Since the leakage flux is only a very small percentage of the total flux, we may quite safely neglect it in making our magnetic circuit calculations in this chapter if a specific value of leakage factor is not given.

13-16 SERIES MAGNETIC CIRCUITS

Practical magnetic circuits are seldom as simple as the iron ring of Figure 13-17. The magnetic circuit of Figure 13-22 consists of two sections, a straight section and a U-shaped section in *series*. The two sections are in series since tracing around the complete loop of a line of force takes us first through one section and then through the other. As we have already noted, the equation for reluctance in magnetic circuits resembles Ohm's law of electric circuits. Therefore, just as the total resistance of a series electric circuit is the sum of the individual resistances, it follows that

> *The total reluctance of a series magnetic circuit is the sum of all the individual reluctances.*

For electric circuits:

$$R_T = R_1 + R_2 + R_3 + \text{etc.} \tag{6-1}$$

For magnetic circuits:

$$R_{mT} = R_{m1} + R_{m2} + R_{m3} + \text{etc.} \tag{13-10}$$

There are no standard graphic symbols for drawing *schematic* diagrams of magnetic circuits. However, since we can compare a series magnetic circuit with a series electric circuit, we might represent the series magnetic circuits of Figures 13-22 and 13-24 by the schematic diagram of Figure 13-23.

Using the relationship for total reluctance in a series magnetic circuit stated by Equation (13-10), we can solve the series magnetic circuit of Example 13-4 by the first method we used to solve Example 13-2. In many series magnetic circuits, the cross-sectional areas of the various sections may not be the same, in which case we must determine the flux density B for each section. Note that part of the magnetic circuit shown in Figure 13-24 is made up of a stack of many layers of insulated sheet-steel **laminations** to minimize **eddy-current** losses.*

---

* The rising and falling magnetic field created by *alternating* current in a solenoid induces a voltage into the electrically conductive iron core. This voltage causes current to flow in the core in a plane perpendicular to the direction of the magnetic lines of force. Breaking a solid core into many thin sheet-steel laminations insulated from each other by thin layers of varnish minimizes the actual eddy current.

FIGURE 13-23
Schematic Representation of a Series Magnetic Circuit.

Reluctance of U-shaped
cast steel section

R_{mc}

F_m

R_{ms} Reluctance of I-shaped
sheet steel section

Φ

$R_{mT} = R_{mc} + R_{ms}$

FIGURE 13-24
Magnetic Circuit for Example 13-4.

4 cm

15 cm

2.5 cm

10 cm

12.5 cm

Cast
steel

2.5 cm

Laminated sheet steel

EXAMPLE 13-4

What magnetomotive force must be developed by a coil wound on the magnetic circuit of Figure 13-24 to develop a total flux of 1.2×10^{-3} Wb? (The varnish on each of the laminations accounts for 10% of the lamination's thickness.)

SOLUTION 1

Step 1. Determine the reluctance of the U-shaped section.

Examination of the specifications shows that all three arms of the *cast*-steel section have the same width (2.5 cm).

$$\therefore A_c = 2.5 \text{ cm} \times 4 \text{ cm} = 10 \text{ cm}^2 = 10^{-3} \text{ m}^2$$

The *average* path is marked out by the dotted line in Figure 13-24. Its length can be obtained by averaging the outside length of the U-shaped section (12.5 + 15 + 12.5 = 40 cm) and the inside length (10 + 10 + 10 = 30 cm).

$$\therefore \, l_c = \frac{1}{2}(40 + 30) = 35 \text{ cm} = 0.35 \text{ m}$$

$$B_c = \frac{\Phi}{A_c} = \frac{1.2 \times 10^{-3} \text{ Wb}}{10^{-3} \text{ m}^2} = 1.2 \text{ T}$$

From the *BH* curves of Figure 13-21, for

$$B_c = 1.2 \text{ T in cast steel}$$
$$H_c = 1220 \text{ At/m}$$

$$\therefore \, \mu_c = \frac{B_c}{H_c} = \frac{1.2 \text{ T}}{1220 \text{ At/m}} = 9.84 \times 10^{-4} \text{ H/m}$$

$$R_{mc} = \frac{l}{\mu_c A_c} = \frac{35 \text{ cm}}{9.84 \times 10^{-4} \text{ H/m} \times 10^{-3} \text{ m}^2} = 3.557 \times 10^5 \text{ At/Wb}$$

Step II. Determine the reluctance of the I-shaped stack of laminations.

$$A_s = 2.5 \text{ cm} \times 4 \text{ cm} \times 90\% = 9 \text{ cm}^2 = 9 \times 10^{-4} \text{ m}^2$$

$$l_s = \tfrac{1}{2}(20 + 10) \text{ cm} = 15 \text{ cm} = 0.15 \text{ m}$$

$$B_s = \frac{\Phi}{A_s} = \frac{1.2 \times 10^{-3} \text{ Wb}}{9 \times 10^{-4} \text{ m}^2} = 1.333 \text{ T}$$

From the *BH* curves of Figure 13-21, for

$$B_s = 1.33 \text{ T in } \textit{sheet} \text{ steel}$$
$$H_s = 900 \text{ At/m}$$

$$\therefore \, \mu_s = \frac{B_s}{H_s} = \frac{1.33 \text{ T}}{900 \text{ At/m}} = 1.48 \times 10^{-3} \text{ H/m}$$

$$R_{ms} = \frac{l_s}{\mu_s A_s} = \frac{15 \text{ cm}}{1.48 \times 10^{-3} \text{ H/m} \times 9 \times 10^{-4} \text{ m}^2} = 1.126 \times 10^5 \text{ At/Wb}$$

Step III. $R_{mT} = R_{mc} + R_{ms} = (3.557 \times 10^5) + (1.126 \times 10^5)$

$$= 4.683 \times 10^5 \text{ At/Wb}$$

and $\quad F_m = \Phi R_{mT} = 1.2 \times 10^{-3} \text{ Wb} \times 4.683 \times 10^5 \text{ At/Wb} = \textbf{562 At}$

Another characteristic of series electric circuits is that the total voltage drop is the sum of all the individual voltage drops. Similarly, in magnetic circuits,

The total magnetomotive force is the sum of the mmf's required for each section of the series magnetic circuit.

For electric circuits:

$$V_T = V_1 + V_2 + V_3 + \text{etc.} \qquad (6\text{-}2)$$

For magnetic circuits:

$$F_{mT} = F_{m1} + F_{m2} + F_{m3} + \text{etc.} \qquad (13\text{-}11)$$

Using Equation (13-6) to substitute for the separate mmfs,

$$F_{mT} = H_1 l_1 + H_2 l_2 + H_3 l_3 + \text{etc.} \qquad (13\text{-}12)^\dagger$$

Applying Equation (13-12) to Example 13-4, we may again completely bypass determination of permeability and reluctance in solving series magnetic circuits. Once we have determined the magnetic field intensity required for each section of the circuit, we can go right ahead and find the magnetomotive force required for that section.

SOLUTION 2
For the U-shaped section,

$$H_c = 1220 \text{ At/m} \qquad \text{(from Solution 1)}$$
$$\therefore F_{mc} = H_c l_c = 1220 \text{ At/m} \times 35 \text{ cm} = 427 \text{ At}$$

For the I-shaped section,

$$H_s = 900 \text{ At/m} \qquad \text{(from Solution 1)}$$
$$\therefore F_{ms} = H_s l_s = 900 \text{ At/m} \times 15 \text{ cm} = 135 \text{ At}$$

and
$$F_{mT} = F_{mc} + F_{ms} = 427 + 135 = \textbf{562 At}$$

13-17 AIR GAPS

In all the magnetic circuits shown in Figure 13-15, the complete loop of any line of force is not wholly within the high permeability path provided by the iron. Every line of force has to cross an air gap to complete its loop. This gap is necessary in the case of motors, meters, and loudspeakers in order to provide space in which electric conductors can move. Even in stationary equipment like the magnetic circuit of Figure 13-24, an air gap may be inserted into the magnetic circuit at points X and Y between the U-shaped and the I-shaped sections. This air gap increases the total reluctance of the circuit, to assist in preventing saturation of the iron portion of the magnetic circuit when a heavy current flows through the coil wound on the core. In such apparatus, the "air" gap is often a sheet

---

† Shortly after Oersted announced his discovery of the relationship between electric current and magnetic fields in 1819, Ampère stated this magnetic circuit equivalent of Kirchhoff's voltage law. We sometimes refer to Equation (13-12) as **Ampère's circuital law.**

of any nonmagnetic material such as fiberboard or brass, so that the length of the air gap can be accurately and rigidly set.

Since all the magnetic lines of force must pass through the air gaps in these circuits, the gaps are in *series* with the remainder of the circuit. Therefore, we can solve magnetic circuits with air gaps just like the series circuit of Example 13-4. Before we start to calculate the flux density in the air gap, we must note how the magnetic lines of force try to spread out as they pass through the air gap, as shown in Figure 13-25. This effect follows from the characteristic of magnetic lines of force in that they tend to repel one another. Since the air outside of the gap is just as capable of passing magnetic lines of force as the air within the gap, the lines of force take advantage of this opportunity to get further apart as they cross the gap. This spreading apart of the magnetic lines of force as they cross an air gap is called **fringing.** Fringing results in the flux density in the air gap being slightly less than that in the adjacent iron sections of the magnetic circuit.

If the length of the air gap is small, the fringing is small, and we can make correction for it in magnetic circuit calculations by assuming that the cross-sectional dimensions of an air gap are greater than the cross-sectional dimensions of the adjacent iron by an amount equal to the length of the air gap.

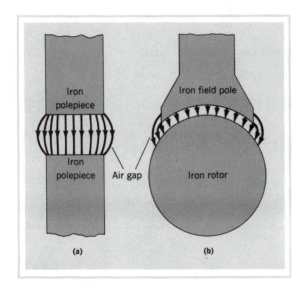

FIGURE 13-25
Fringing of Magnetic Lines of Force in an Air Gap.

EXAMPLE 13-5

What magnetomotive force must be developed by a coil wound on the magnetic circuit of Figure 13-24 to develop a total flux of 1.2×10^{-3} Wb if pieces of fiberboard 0.1 mm thick are inserted into the magnetic circuit at points X and Y? (Allow 10% of the lamination thickness for varnish.)

SOLUTION 1

Since the total flux is the same as in Example 13-4, the reluctance of the U-shaped and I-shaped steel sections will be the same as we have already calculated, 3.557×10^5 and 1.126×10^5 At/Wb, respectively. When we allow for fringing, the cross-sectional area of the air gap is

$$A_A = 2.51 \text{ cm} \times 4.01 \text{ cm} = 10.07 \text{ cm}^2 = 1.007 \times 10^{-3} \text{ m}^2$$

$$R_{mA} = \frac{l_A}{\mu_A A_A} = \frac{0.1 \text{ mm}}{4\pi \times 10^{-7} \text{ H/m} \times 1.007 \times 10^{-3} \text{ m}^2} = 7.9 \times 10^4 \text{ At/Wb}$$

Since each magnetic line of force has to pass through *both* air gaps, they are in *series*. Since both air gaps are identical, each has a reluctance of 7.9×10^4 At/Wb.

$$R_{mT} = (3.557 \times 10^5) + (1.126 \times 10^5) + 2(7.9 \times 10^4)$$

$$= 6.26 \times 10^5 \text{ At/Wb}$$

$$F_m = \Phi R_{mT} = 1.2 \times 10^{-3} \text{ Wb} \times 6.26 \times 10^5 \text{ At/Wb}$$

$$= \textbf{751 At}$$

SOLUTION 2
The mmf required for the iron will be the same as in Example 13-4, 562 At. For each air gap,

$$B_A = \frac{\Phi}{A_A} = \frac{1.2 \times 10^{-3} \text{ Wb}}{1.007 \times 10^{-3} \text{ m}^2} = 1.19 \text{ T}$$

$$H_A = \frac{B_A}{\mu_A} = \frac{1.19 \text{ T}}{4\pi \times 10^{-7} \text{ H/m}} = 9.47 \times 10^5 \text{ At/m}$$

$$F_{mA} = H_A l_A = 9.47 \times 10^5 \text{ At/m} \times 0.1 \text{ mm} = 94.7 \text{ At}$$

$$\therefore F_{mT} = 562 + (2 \times 94.7) = \textbf{751 At}$$

13-18 PARALLEL MAGNETIC CIRCUITS

If we examine the magnetic circuits of Figure 13-15(c) and 13-15(d), we note that, although *all* the magnetic flux passes through some portions of the magnetic circuit, in other portions it splits, with half going one way and half going the other. In electric circuits, this is the condition that exists when two equal resistors are connected in *parallel*. Half the total current flows through each resistor and the *same* voltage drop appears across each branch. Designers have found that supplying more than one possible return path for magnetic lines of force appreciably reduces the leakage flux. Fortunately, in most designs, the parallel paths are limited to two symmetrical paths so that we can assume that the total flux splits equally.

EXAMPLE 13-6
If a 1000-turn coil is placed on the center leg of the shell-type core of Figure 13-26 to form a filter choke, what is the maximum current it can pass without the total flux in the core exceeding 2×10^{-3} Wb? The core is made of sheet-steel laminations stacked to a total thickness of 4 cm. (Allow 5% of the thickness for varnish.) The width of the outside legs of the core is 2.5 cm and that of the center leg is 5 cm. The average path length (see Figure 13-26) is 25 cm.

Average path length

$\mathbb{C}$

FIGURE 13-26
Magnetic Circuit for Example 13-6.

SOLUTION

We can divide the magnetic circuit down the centerline of Figure 13-21 and stack one section on top of the other without changing the total reluctance of the magnetic circuit. The cross-sectional area then becomes

$$A = 2 \times 2.5 \text{ cm} \times 4 \text{ cm} \times 95\% = 19 \text{ cm}^2$$

$$B = \frac{\Phi}{A} = \frac{2 \times 10^{-3} \text{ Wb}}{19 \text{ cm}^2} = 1.05 \text{ T}$$

From the *BH* curve for sheet steel (Figure 13-21), for

$$B = 1.05 \text{ T}$$

$$H = 460 \text{ At/m}$$

$$F_m = Hl = 460 \text{ At/m} \times 25 \text{ cm} = 115 \text{ At}$$

$$I = \frac{F_m}{N} = \frac{115 \text{ At}}{1000 \text{ t}} = \textbf{115 mA}$$

We could arrive at the same answer if we considered only *half* of the magnetic circuit accommodating only *half* of the total flux. The flux density in the iron would still be

$$B = \frac{\Phi}{A} = \frac{1 \times 10^{-3} \text{ Wb}}{9.5 \text{ cm}^2} = 1.05 \text{ T}$$

and again $F_m = 115$ At. This *same* mmf would cause the other 1×10^{-3} Wb to be set up in the other half of the magnetic circuit.

PROBLEMS

13-3-1. Draw the magnetic lines of force around the conductor in Figure 13-27(a) and indicate their direction.

13-3-2. Mark the conventional current direction on the conductor in Figure 13-27(b).

13-3-3. Draw the magnetic field around the coil in Figure 13-28(a) and indicate its direction.

(a) (b)

FIGURE 13-27

(a) (b)

FIGURE 13-28

13-3-4. Mark the conventional current direction on the coil of Figure 13-28(b).

13-6-5. What is the reluctance of a magnetic circuit in which a total flux of 2×10^{-3} Wb is created by a 5-A current flowing in a solenoid consisting of 200 turns of wire?

13-6-6. What current must be passed through a 500-turn solenoid to produce a total flux of 1.2 mWb in a magnetic circuit whose reluctance is 2×10^6 At/Wb?

13-6-7. How many turns of wire are there in a solenoid if a 500-mA current through it establishes a total flux of 8×10^{-4} Wb in a magnetic circuit whose reluctance is 5×10^5 At/Wb?

13-6-8. What total flux is created in a magnetic circuit whose reluctance is 2.5×10^6 reciprocal henrys by a current of 1.2 A flowing in a 500-turn solenoid wound around the magnetic circuit?

13-8-9. The magnetic circuit in Problem 13-6-5 has a uniform cross-sectional area of 5 cm^2 and an average path length of 25 cm. What is the permeability of the core material?

13-8-10. The core of a solenoid consists of a brass cylinder 10 cm in length and 2 cm in diameter. What is its reluctance?

13-8-11. A piece of iron 10 cm in length and having a rectangular cross section of 1×2 cm has a reluctance of 1.25×10^6 At/Wb. What is the permeability of the iron?

13-8-12. The magnet of Figure 13-4 has a rectangular cross section of 1×2 cm. If the spacing between the two sections in Figure 13-4(b) is 1 mm, what is the reluctance of the air gap?

13-9-13. What is the flux density in the magnetic circuit of Problems 13-6-5 and 13-8-9?

13-9-14. Electric current in a solenoid with an inside diameter of 2 cm creates a flux density inside the solenoid of 0.2 T. What is the total flux inside the solenoid?

13-10-15. A uniform magnetic circuit has a path length of 20 cm. What is the magnetizing force when a 500-mA current flows in an 800-turn coil wound on the magnetic circuit?

13-10-16. A magnetic field strength of 2000 At/m produces a flux density of 1.0 T in a certain type of iron. What is its permeability at this flux density?

13-11-17. The core for a low-loss inductor in a crossover filter is made from a piece of flat plastic stock 6 mm thick and 24 mm wide by 15 cm long formed into a closed loop. 1200 turns of wire form a toroid winding on this form. What current must flow in the coil to develop a flux of 10^{-5} Wb in the core?

13-11-18. A 120-mA current flowing in a 1500-turn toroid winding creates a flux density of 8×10^{-4} T in the nonmagnetic core. What is the average path length of the core?

13-12-19. Use the magnetizing curves of Figure 13-21 to determine the flux density created in sheet steel by a magnetizing force of 2800 At/m.

13-12-20. Using the *BH* graphs of Figure 13-21, determine the magnetic field strength required to maintain a constant flux density of 1.5 T in cast steel.

13-14-21. If the core in Problem 13-11-17 were a cast-iron ring with the same dimensions, what must the current be to produce a total flux of 10^{-4} Wb?

13-14-22. If the cast-steel magnetic circuit shown in Figure 13-22 has a uniform cross section of 8 cm² and an average path length of 44 cm, calculate the total magnetic flux through a 120-turn coil in which the current is 150 mA.

13-16-23. The horseshoe-shaped cast-steel electromagnet shown in Figure 13-29 is 3.0 cm in diameter and has an average path length of 30 cm. What current must flow through the windings of the electromagnet to develop a total flux of 1.0 mWb in the sheet-steel bar held by the magnet? The dimensions of the bar are $15 \times 4 \times 2$ cm.

13-16-24. The current through the windings of the horseshoe-shaped electromagnet of Figure 13-29 is increased to 0.5 A. Calculate the total flux through the sheet-steel bar. *Hint:* Start by assuming that the total mmf splits between the two sections of the magnetic circuit in the same proportions as in Problem 13-16-23. If this does not produce equal fluxes in the two sections, alter the ratio until the calculated fluxes are close to being equal.

13-17-25. The magnetic circuit of Figure 13-30 is made from a sheet-steel bar 15 cm long with a 12-mm × 20-mm rectangular cross section. It is bent into a ring, leaving an air gap 0.25 mm long. What current must flow in the 800-turn coil wound on this core to produce a total flux of 3×10^{-4} Wb in the air gap?

FIGURE 13-29

FIGURE 13-30

13-17-26. If a sheet of aluminum 1.5 mm thick is placed between the sheet-steel bar and the electromagnet of Figure 13-29, what current must flow in the coils to develop a total flux of 7.5×10^{-4} Wb in the sheet-steel bar if the leakage factor has now become 1.15?

11.5 cm

2 cm

7.5 cm

2 cm   4 cm   2 cm

2.5 cm

FIGURE 13-31

13-18-27. If the sheet-steel laminations of Figure 13-31 are stacked to a total thickness of 2.5 cm, how many turns of wire must be wound on the center leg for a 100-mA current to develop a total flux of 1.2×10^{-3} Wb in the center leg? Assume no air gap and negligible leakage flux. Allow 5% of each lamination thickness for varnish.

13-18-28. Calculate the number of turns required in Problem 13-18-27 if there is a 0.5-mm plastic spacer between the E-shaped and I-shaped sections of the magnetic circuit.

13-29. The tendency of magnetic lines of force to become as short as possible creates a magnetic force is

$$F = \frac{B^2 A}{2\mu} \qquad (13\text{-}13)$$

where F is the force acting to close the air gap in newtons, and μ is the permeability of air. What force is exerted in the steel ring in Problem 13-17-25 as the magnetic field attempts to close the air gap?

13-30. What force is required to pull the sheet-steel bar away from the electromagnet in Problem 13-16-23?

REVIEW QUESTIONS

13-1. Draw a diagram of the magnetic field around a horseshoe-shaped permanent magnet.

13-2. With reference to the characteristics of magnetic lines of force, show why a nail is attracted to either pole of a horseshoe-shaped permanent magnet.

13-3. Show, with sketches, how it is possible to determine the direction of the current in an electric conductor with the aid of a compass needle.

13-4. Why is the flux density of a magnetic field greater in the center of a current-carrying solenoid than around a straight wire carrying the same current?

13-5. How is the magnitude of the **weber** established?

13-6. What is the significance of the term **reluctance** to a solenoid with a nonmagnetic core?

13-7. What is the significance of the **magnetic moment** of an iron atom?

13-8. What is the distinction between **magnetomotive force** and **magnetizing force?**

13-9. Why is permeability a more useful property than permeance?

13-10. Account for the upper knee on the *BH* graph of ferromagnetic materials.

13-11. Draw a *BH* curve for aluminum.

13-12. Compare permanent and temporary magnets in terms of **coercive force.**

13-13. Describe a **toroid** winding. What is its advantage?

13-14. In free space, there is always a fixed ratio between **magnetic field strength** and **flux density.** What is its numerical value?

13-15. How does the **permeability** of a ferromagnetic material differ from that of a nonmagnetic material?

13-16. What causes **leakage flux** in magnetic circuits? State two means of reducing leakage flux in magnetic circuit design.

13-17. Laminating the iron core of a transformer greatly reduces the eddy-current losses. What effect does the laminating process have on hysteresis losses? Explain.

13-18. Some magnetic circuits consist of powdered iron pressed into the desired form with a ceramic binder. Compare such a core with a solid core of the same type of iron in terms of
   (a) **permeability,**
   (b) **hysteresis,** and
   (c) **eddy current.**

13-19. What characteristics of magnetic circuits allow us to calculate the total magnetomotive force required in a circuit made up of several different sections?

13-20. Many magnetic circuits contain an **air gap** in the form of a sheet of plastic or brass. Why do such materials constitute "air gaps"?

13-21. In an all-iron magnetic circuit, the relationship between the applied magnetomotive force and the resulting total magnetic flux is quite nonlinear, as indicated by the graphs of Figure 13-21. By purposely including an air gap in the magnetic circuit, there is a more linear relationship between applied mmf and total flux. Explain.

13-22. What causes **fringing** of the magnetic lines of force in an air gap?

13-23. If the two parallel paths for magnetic lines of force in the magnetic circuit of Figure 13-26 were not symmetrical, what complication would be introduced into our solution of an example such as Example 13-6?

13-24. Mechanical work is required to pull a soft-iron bar away from a U-shaped permanent magnet. Where does the energy go?

14

INDUCTANCE

In Chapter 11, we discovered that capacitance in an electric circuit depends on electric fields. Similarly, the third property of electric circuits, **inductance,** depends on the interaction of *magnetic fields* with electric circuits.

14-1 ELECTROMAGNETIC INDUCTION

We have already encountered the interaction of magnetic fields with electric circuits in defining the SI unit of magnetic flux—the weber. In Section 13-4, we noted Faraday's discovery of electromagnetic induction as he thrust a permanent magnet into a solenoid (Figure 13-14). Figure 14-1 shows another arrangement for demonstrating Faraday's discovery. As the conductor is moved downward through the magnetic lines of force, the galvanometer pointer swings one way from its center zero. As the conductor is moved upward, the galvanometer pointer swings in the opposite direction. But when there is no motion, there is no pointer deflection.

For current to flow through the galvanometer, there must be some voltage source in the loop consisting of the galvanometer and the conductor in the magnetic field. And, as experiment has shown, this voltage appears only when the conductor is *cutting across* the magnetic lines of force. Moving the conductor parallel to the lines of force will produce no deflection of the galvanometer pointer. In Figure 14-1, the conductor is moved while the magnetic field remains stationary; whereas, in Figure 13-14, the magnetic field is moved while the electric conductor remains stationary. In both cases, there is a cutting across magnetic lines of force by an electric conductor, and, in both cases, a voltage appears in the conductor. The generation of a voltage when an electric conductor cuts across magnetic lines of force is called **electromagnetic induction,** and the resulting voltage is called an **induced** voltage.

FIGURE 14-1
Demonstrating Electromagnetic Induction.

Another method of demonstrating electromagnetic induction is shown in Figure 14-2. In this example, *relative* motion between the magnetic lines of force and the electric conductor is achieved without mechanical motion of either the magnet or the electric conductor. As we reduce the resistance of the rheostat, the current in the left-hand (**primary**) winding increases. This causes an increase in the number of magnetic lines of force appearing in the iron core. Since the right-hand (**secondary**) winding is wound on the same core, the number of lines of force passing through the winding is increasing, and the deflection of the galvanometer pointer indicates that a voltage is being induced into the secondary winding. Since most of the magnetic lines of force are confined to the iron core, it is more usual for us to think of the flux as **linking** the secondary winding rather than as cutting across the turns of the secondary winding.

FIGURE 14-2
Demonstrating Mutual Induction.

When we increase the resistance of the rheostat, the current becomes smaller and the number of lines of force linking the secondary becomes less. At the same time, the pointer of the galvanometer swings in the opposite direction, indicating that the induced voltage has the opposite polarity. But as long as the current in the primary is steady, there is no deflection of the galvanometer pointer. The generation of a voltage in a secondary winding by a *changing* current in a primary winding is called **mutual induction.**

14-2 FARADAY'S LAW

One observation we can note in the demonstration of mutual induction in Figure 14-2 is that the faster the flux in the core builds up or collapses, the greater the deflection of the galvanometer pointer. When we consider the reason for this, we come to the conclusion which is referred to as

Faraday's law: The voltage induced in an electric circuit is proportional to the time rate of change of the flux of magnetic induction linked with the circuit.

We have already used Faraday's law in Section 13-4 to establish the SI unit of magnetic flux—the *weber*. To express *rate of change* of flux numerically, we adopt the convenient calculus derivative form which we discussed on page 310. Hence, from the definition of the weber,[†]

$$\text{voltage per turn (V)} = \frac{d\phi}{dt} \text{ (Wb/s)}$$

If the secondary winding in Figure 14-2 has more than one turn, the *same* changing flux links all turns and thus induces identical voltages in each turn. Since these "voltages per turn" are all in series, the total induced voltage is N times the voltage per turn. Hence, We may express Faraday's law in equation form as

$$e_T = N\frac{d\phi}{dt} \tag{14-1}$$

where e_T is the instantaneous value of the induced voltage between the terminals of the winding in volts, N is the number of turns in the winding, and $d\phi/dt$ is the rate of change of magnetic flux in webers per second.

Combining Equations (13-2) and (13-1),

$$\Phi = \frac{F_m}{R_m} = \frac{NI}{R_m}$$

The number of turns in the primary winding is fixed; and if we keep the flux density below the saturation point of the iron core, the reluctance of the magnetic circuit is reasonably constant. Therefore, the *rate of change of flux* is directly proportional to the *rate of*

[†] Note the use of lowercase ϕ to indicate *changing* flux.

change of current. When we are dealing with electromagnetic induction due to a changing current rather than mechanical motion, we may restate Faraday's law:

The magnitude of the induced voltage is directly proportional to the rate of change of current.

SELF-CHECK

14-2:1. A 50-turn coil is in a magnetic field where the linked flux is changing at 0.1 Wb/s. Find the voltage that is induced in the coil.
14-2:2. A coil with 250 turns is in a magnetic field. The flux linked with the coil changes from 1 mWb to 15 mWb in two seconds. What is the average voltage induced in the coil during those two seconds?

14-3 LENZ'S LAW

We have already noted that the polarity of the induced voltage depends on the direction in which the conductor is moved across the magnetic lines of force as in Figure 14-1, or whether the current in the primary is increasing or decreasing as in Figure 14-2. We can apply a law stated by Heinrich Lenz to determine the exact polarity of the induced voltage. Lenz reasoned that the process of induction must conform to the well-known principle in physics that reaction is equal and opposite to action. Of several ways of stating Lenz's law, we shall use the following form:

Lenz's law: The polarity of the induced voltage must be such that any current resulting from it will develop a flux which tends to oppose any change in the original flux.

When we first close the switch in the mutual induction demonstration of Figure 14-3(a), the primary flux will increase from zero. Since a resistor is connected across the

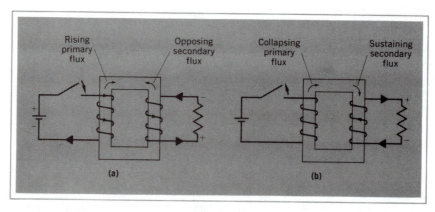

FIGURE 14-3
Determining the Direction of an Induced Voltage by Lenz's Law.

secondary winding to form a complete closed circuit, current will flow in the secondary winding when a voltage is induced in it by the rising primary flux. According to Lenz's law, the flux produced by this secondary current must try to oppose the increase in primary flux. Applying our right-hand rule to the primary winding, we find that the primary flux will have a clockwise direction around the core. To oppose an increase in this clockwise primary flux, the flux produced by the secondary current must have a *counterclockwise* direction. Again applying the right-hand rule, this time to the secondary winding, we can establish the direction in which the secondary current must flow through the resistor, and thus determine the polarity of the induced voltage in terms of the polarity of the voltage drop across the resistor.

When we open the switch, as in Figure 14-3(b), the primary current stops flowing and the primary flux collapses. According to Faraday's law, this reduction in flux will induce a voltage into the secondary, and, according to Lenz's law, the secondary current must try to oppose the collapse of the primary flux. To help sustain the primary flux, which has a clockwise direction in the core, the secondary current must also produce a clockwise flux direction. Applying the right-hand rule, we obtain the polarity shown in Figure 14-3(b). The polarity of the induced voltage when the primary current is decreasing is, therefore, opposite to that induced when the primary current is increasing.

We can also use Lenz's law to determine the direction of the induced voltage in the demonstration of Figure 14-1. The galvanometer completes the circuit so that current will flow in the conductor as we move it through the magnetic field. Current in the conductor will, in turn, produce a magnetic field around the conductor. If we consider the two magnetic fields separately, the field of the stationary magnet has the form shown in Figure 14-4(a). If we assume for the moment that the conventional current direction in the moving conductor is out of the page, the magnetic field around the conductor has the form shown in Figure 14-4(b). Before we place the conductor between the magnetic poles in the sketch and superimpose the two magnetic fields, we must recall that magnetic lines of force can never intersect. Therefore, the two component fields must combine to form a single composite magnetic field pattern, as shown in Figure 14-5.

Stationary permanent
magnet field.

(a)

Magnetic field around
current-carrying
conductor.

(b)

FIGURE 14-4
Component Magnetic Fields for the Electromagnetic Induction Demonstration
of Figure 14-1.

FIGURE 14-5
Using Lenz's Law to Determine the Direction of the Current Produced by an Induced Voltage.

Above the conductor [Figure 14-4(b)], the direction of the magnetic lines of force is *opposite* to that of the permanent magnet field [Figure 14-4(a)]. Consequently, there is a cancelling effect which bends the resultant magnetic field away from the conductor, as shown in Figure 14-5. Below the conductor, the direction of its magnetic field is the *same* as that of the permanent magnet field. As a result, the flux density is increased below the conductor in the composite flux pattern of Figure 14-5. The net effect of a conventional current out of the page in the conductor in Figures 14-1 and 14-5 is to warp the magnetic field produced by the permanent magnet so that the magnetic lines of force in the vicinity of the conductor *detour* below the conductor.

Another characteristic of magnetic lines of force is that they tend to become as short as possible and tend to repel one another. Therefore, the lines of force in the composite magnetic field of Figure 14-5 will attempt to straighten out and to space themselves uniformly, in order to regain the flux pattern of Figure 14-4(a). To accomplish this, the magnetic field will attempt to force the current-carrying conductor to move upward. According to the principle stated in Lenz's law, this force will *oppose* the actual motion of the conductor through the stationary magnetic field. Consequently, when the electric conductor in the demonstration shown in Figure 14-1 moves downward through the stationary magnetic field, the voltage induced into the conductor has a polarity such as to cause the current in the closed loop to have a direction out of the page.

14-4 SELF-INDUCTION

A voltage induced by a *changing* current is not restricted to mutual induction of a voltage in a secondary winding by current changing in a primary winding. When we first close the switch applying a voltage to the simple two-turn coil in Figure 14-6, the increasing flux produced by the rising current in turn *A* will induce a voltage into turn *B*. Similarly, the same rising current in turn *B* will induce a voltage into turn *A*. These voltages are in series and, according to Lenz's law, both will have a polarity such as to oppose the increasing flux and the rising current. From Faraday's law equation (14-1), the total induced voltage again is $N\, d\phi/dt$. The generation of a voltage in an electric circuit by a *changing* current in the *same* circuit is called **self-induction**.

As we discovered in Section 2-8, generation of a potential difference requires a transfer of energy in order to perform the separation of electric charge—placing a surplus

FIGURE 14-6
Self-Induction.

of electrons at one terminal and a deficiency at the other. This energy conversion constitutes an electromotive force in joules per coulomb, since it is responsible for generating the induced voltage. In older terminology, since the induced voltage has a polarity such as to *oppose* the force that creates it, the self-induced voltage is sometimes called a **counter emf,** or **cemf.** Since we can only measure the final induced voltage, we are inclined to forget the steps in its generation. When current is drawn from the voltage source in Figure 14-6, the voltage source is providing energy to build up a magnetic field around the coil. The rising magnetic field induces an emf *in* the turns of the coil. The emf, in turn, through electric charge-separation produces a potential difference *across* the coil. In modern terminology, we call this potential difference an **induced voltage.**

When we open the switch in Figure 14-6, the magnetic field collapses and again induces an emf in turns *A* and *B*. In this case, the polarity of the self-induced voltage is reversed as the coil tries to maintain current flow by having the polarity of its induced voltage such that it adds to the source voltage, thus raising the total voltage in an attempt to maintain current flow in the circuit.

14-5 SELF-INDUCTANCE

From the foregoing discussion, we find that the simple electric circuit of Figure 14-6 has the property of tending to oppose any *change* in current through it (or in the magnetic flux linking it). This property is called its **self-inductance.** Therefore,

Self-inductance is the property of an electric circuit that opposes any change in current in that circuit.

Because it will be quite apparent from the circuit diagram whether the induction action in question is *self*-induction or *mutual*-induction, it is customary to omit the *self* and simply speak of the **inductance** of an electric circuit. And since inductance is a property of electric circuits, we shall require a letter symbol to represent inductance and a unit of measurement.

The letter symbol for inductance is L.
The henry is the SI unit of inductance.[†]
The unit symbol for henry is **H.**

[†] Named in honor of the pioneer American physicist Joseph Henry.

In the International System of Units, since we define the *ampere* in terms of the magnetic force between parallel current-carrying conductors, and the *weber* in terms of the voltage induced by a changing current, magnetic units are completely compatible with electrical units. We can define the unit of inductance either in terms of the change in flux resulting from a given change in current or in terms of the induced voltage resulting from a given change in current. The general definition of the henry is based on flux linkages. Hence,

An electric circuit has an inductance of one henry when a change in current of one ampere produces a change in total flux linkages of one weber.

As we noted in Section 13-5, the total flux linkages of a multiturn coil are N times as great as for a single turn carrying the same current. Hence, the defining equation for the henry becomes

$$L = N\frac{d\phi}{di} \qquad (14\text{-}2)^{\dagger}$$

where L is the inductance of a circuit in henrys, N is the number of turns linked by the magnetic flux, and $d\phi/di$ is the change in flux for a given change in current in webers per ampere.

Rearranging Equation (14-1) gives

$$N = \frac{e_T}{d\phi/dt}$$

and rearranging Equation (14-2) gives

$$N = \frac{L}{d\phi/di}$$

from which

$$\frac{L}{d\phi/di} = \frac{e_T}{d\phi/dt}$$

and

$$L = e_T \times \frac{d\phi}{di} \times \frac{dt}{d\phi} = e_T \times \frac{dt}{di}$$

or

$$L = \frac{e_L}{(di/dt)} \qquad (14\text{-}3)$$

[†] Since we have just defined the henry in terms of webers per ampere, we can now appreciate how we can express magnetic permeance (webers per effective ampere) in henrys, magnetic reluctance (effective amperes per weber) in *reciprocal* henrys, and magnetic permeability in henrys per meter on pages 358 and 361.

where L is the inductance of a circuit in henrys, e_L is the voltage induced into the circuit in volts by current changing at the rate of di/dt amperes per second.

Thus we can also define the magnitude of the henry completely in electrical units without having to determine the magnetic flux.

> *An electric circuit has an inductance of one henry when current changing at the rate of one ampere per second induces a voltage of one volt into that circuit.*

SELF-CHECK

14-5:1. The magnetic field of 50-turn coil has linked flux that changes with current at a rate of 0.1 Wb/A. Determine the inductance of the coil.

14-5:2. A coil with an inductance of 100 mH is carrying an electric current. Determine the voltage induced in the coil if the current in the coil is:
(a) 2 A and steady
(b) 2 A and changing at 0.5 A/s

14-6 FACTORS GOVERNING INDUCTANCE

Even a straight wire possesses inductance. As the current through a conductor increases, magnetic lines of force first appear as tiny loops at the center of the conductor and then expand outward. As they expand, they cut across the copper of the conductor, thus inducing a voltage into the conductor. This induced voltage produced by the changing current is the evidence of the presence of inductance in the circuit. However, the inductance of a straight wire is so small that we can neglect it, except at very high radio frequencies.

If we wish to construct a circuit component in which the inductance effect is very pronounced, we would wind the conductor into a coil of many turns. From Equations (13-1) and (13-2),

$$\Phi = \frac{F_m}{R_m} = \frac{NI}{R_m}$$

Dividing both sides of the equation by t,

$$\frac{\Phi}{t} = \frac{N}{R_m} \times \frac{I}{t}$$

from which

$$\frac{d\phi}{dt} = \frac{N}{R_m} \times \frac{di}{dt}$$

Substituting in Equation (14-2) gives

$$e_T = \frac{N^2}{R_m} \times \frac{di}{dt}$$

And substituting this value for induced voltage into Equation (14-3),

$$L = \frac{N^2}{R_\mathrm{m}}$$ (14-4)

Hence, the inductance of a coil of wire is dependent on the number of turns and the reluctance of the magnetic circuit on which the coil is wound. A circuit component that is constructed for the express purpose of displaying the property of inductance is called an **inductor.** As Equation (14-4) indicates, we can greatly increase the inductance of an inductor by using an iron core to reduce the reluctance of the magnetic circuit. And since doubling the number of turns of wire in the coil of Figure 14-6 not only doubles the flux linking the coil for a given current (thus doubling the induced voltage per turn), but also doubles the number of turns that this flux links, the total induced voltage (and consequently the inductance) is proportional to the *square* of the number of turns.

Rearranging Equation (13-4) provides us with the factors governing the reluctance of the magnetic circuit of an inductor.

$$R_\mathrm{m} = \frac{l}{\mu A}$$ (13-8)

Substituting this value of R_m in Equation (14-4),

$$L = \frac{N^2 \mu A}{l}$$ (14-5)

where L is the inductance of the inductor in henrys, N is the number of turns in the coil, μ is the permeability of the magnetic circuit in henrys per meter, A is the cross-sectional area of the magnetic circuit in square meters, and l is the length of the magnetic circuit in meters (length of the coil for air-core coils).

In applying Equation (14-5) to practical inductors, we must realize that it is based on *all* the flux linking *all* the turns. In the case of iron-core coils and toroids, this equation is reasonably accurate. But, for solenoid-type air-core coils where there is appreciable leakage flux, empirical formulas for inductance have been developed. We find these in electrical and radio handbooks.

EXAMPLE 14-1

2000 turns of wire are wound toroid-fashion on a cast-steel ring whose cross-sectional area is 2.25 cm², whose average path length is 42.4 cm, and whose permeability is 7.6×10^{-4} henrys per meter. (See Example 13-2.) What is the inductance of the toroid?

SOLUTION

$$L = \frac{N^2 \mu A}{l} = \frac{2000^2 \times 7.6 \times 10^{-4}\mathrm{H/m} \times 2.25 \times 10^{-4}\ \mathrm{m}^2}{0.424\ \mathrm{m}}$$

$$\therefore L = \mathbf{1.61\ H}$$

(Note how dimensional units cancel.)

SELF-CHECK

14-6:1. How many turns of wire would have to be wound on the steel ring described in Example 14-1 to produce a coil with an inductance of 500 mH?
14-6:2. Compare the inductances of two air-core coils which have the same cross-sectional area but where the second coil has twice the number of turns and half the length of the first coil.

14-7 INDUCTORS IN SERIES

If we connect two inductors in series, as in Figure 14-7(a), the same current flows through both, and therefore, both will be subject to the same rate of change of current. For the present, we shall assume that the two inductors are physically located so that the magnetic field of one cannot induce a voltage into the other. Since inductance is directly proportional to the induced voltage, such mutual induction would alter the apparent self-inductance of the inductors. A changing current in the circuit of Figure 14-7(a) will induce a voltage of e_1 in L_1 and e_2 in L_2. The total induced voltage in the series circuit is

$$e_T = e_1 + e_2$$

From our second definition of the henry,

$$L_T = \frac{e_T}{di/dt} = \frac{e_1 + e_2}{di/dt} = \frac{e_1}{di/dt} + \frac{e_2}{di/dt} = L_1 + L_2$$

Therefore,

When there is no mutual coupling between inductors in series, the total inductance is the sum of the individual inductances.

FIGURE 14-7
(a) Air-Core Inductors in Series; (b) Iron-Core Inductors in Parallel.

(a)   (b)

$$L_T = L_1 + L_2 + L_3 + \text{etc.} \qquad \text{(14-6)}$$

Note that the equation for inductors in series is similar to the equation for the total resistance of resistors in series.

SELF-CHECK

14-7:1. Three coils with inductances of:

$$L_1 = 100 \text{ mH}, \qquad L_2 = 220 \text{ mH} \qquad \text{and} \qquad L_3 = 330 \text{ mH}$$

are connected in series in a circuit. The current in the circuit is 2 A and is changing at 10 A/s.
(a) Determine the total inductance of the series combination.
(b) Determine the total induced voltage across the series combination.

14-8 INDUCTORS IN PARALLEL

If we now connect two inductors in parallel, as in Figure 14-7(b), each must have the same potential difference between its terminals. Since this voltage is produced by self-induction, the current in L_1 must change at the rate of di_1/dt, and current in L_2 must change at the rate of di_2/dt. Therefore, the total current must change at the rate of $(di_1 + di_2)/dt$.

$$\therefore L_{eq} = \frac{e_L}{\dfrac{di_1 + di_2}{dt}}$$

To simplify this equation, we invert both sides.

$$\frac{1}{L_{eq}} = \frac{\dfrac{di_1 + di_2}{dt}}{e_L} = \frac{\dfrac{di_1}{dt} + \dfrac{di_2}{dt}}{e_L} = \frac{\dfrac{di_1}{dt}}{e_L} + \frac{\dfrac{di_2}{dt}}{e_L}$$

$$\therefore \frac{1}{L_{eq}} = \frac{1}{L_1} + \frac{1}{L_2}$$

Therefore,

The equivalent inductance of inductors in parallel is

$$L_{eq} = \frac{1}{1/L_1 + 1/L_2 + 1/L_3 + \text{etc.}} \qquad \text{(14-7)}$$

For only two inductors in parallel, this reduces to

$$L_{eq} = \frac{L_1 \times L_2}{L_1 + L_2} \qquad (14\text{-}8)$$

Note again that we derived Equations (14-6), (14-7), and (14-8) on the assumption that no mutual induction takes place between the individual inductors. We shall leave these more complex calculations until Section 25-6.

SELF-CHECK

14-8:1. Three coils with inductances of:

$$L_1 = 100 \text{ mH}, \qquad L_2 = 220 \text{ mH} \qquad \text{and} \qquad L_3 = 330 \text{ mH}$$

are connected in a parallel combination. The total current to the parallel combination is 2 A and is changing at 10 A/s.
(a) Determine the equivalent inductance of the parallel combination.
(b) Determine the total induced voltage across the parallel combination.

PROBLEMS

14-5-1. Current changing at the rate of 200 mA/s in an inductor induces in it a voltage of 50 mV. What is its inductance?

14-5-2. What inductance must the collector load of a transistor have if collector current rising at a uniform rate from 2 mA to 7 mA in 50 μs induces in it a voltage of 20 mV?

14-5-3. What voltage will be induced into a 10-H inductor in which the current changes from 10 A to 7 A in 90 ms?

14-5-4. How long will it take the current in a 2-H inductor to rise from zero to 6 A if the voltage induced into the inductor is constant at 3 V?

14-6-5. If we can assume no leakage flux, what is the approximate inductance of an air-core coil of 20 turns with an inside diameter of 2 cm and a length of 2 cm?

14-6-6. How many turns must we wind on an air-core coil 3 cm long and 3 cm in diameter to obtain an inductance of 20 μH?

14-6-7. What is the inductance of the inductor in Problem 13-17-25 of Chapter 13?

14-6-8. What is the inductance of the filter choke in Problem 13-18-27 of Chapter 13?

14-6-9. A 2-H inductor has 1200 turns. How many turns must we add to raise its inductance to 3 H?

14-6-10. A 500 mH inductor has 750 turns. Where must we place a tap to obtain an inductance of 250 mH between one end and the tap? What inductance is available if we use the other end and the tap?

14-7-11. Three inductors whose inductances are 30 mH, 40 mH, and 50 mH, respectively, are located so that there is no mutual induction between them. What is the total inductance when the three coils are connected in series?

14-7-12. Assuming no mutual induction, what inductance must be connected in series with a 360-mH inductor to bring the total inductance up to 0.5 H?

14-8-13. What is the equivalent inductance when the three coils of Problem 14-7-11 are connected in parallel?

14-8-14. What inductance must be placed in parallel with an inductance of 60 μH to reduce the net inductance to 15 μH?

REVIEW QUESTIONS

14-1. What is meant by **electromagnetic induction?**

14-2. What factors govern the magnitude of the emf induced into an electric conductor?

14-3. What factors govern the direction of the emf induced into an electric conductor?

14-4. Mark the conventional current direction in the sketches of Figure 14-8. The arrows show the direction of motion.

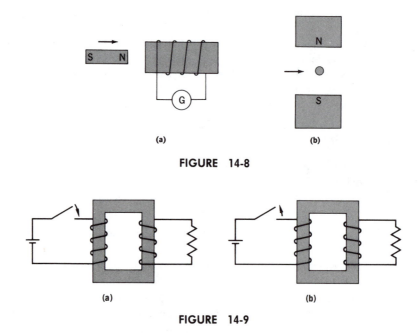

FIGURE 14-8

FIGURE 14-9

14-5. Mark the conventional current direction in both primary and secondary windings in Figure 14-9:
(a) As the switch is first closed.
(b) As the switch is opened.

14-6. What is meant by **self-induction?**

14-7. How is the unit of inductance derived from previously established units?

14-8. Why is the self-inductance of an inductor proportional to the *square* of the number of turns?

14-9. Why does the inductance of an iron-core coil depend on the current through it?

14-10. What effect will the addition of an air gap to the magnetic circuit of a choke coil have on its inductance?

14-11. Why was it necessary to stipulate that there must be no magnetic coupling between the inductors in establishing the equation $L_T = L_1 + L_2 + L_3$ + etc. for inductors in series?

14-12. Why is the total inductance of series inductors equal to the sum of the individual inductances?

15

INDUCTANCE IN DC CIRCUITS

15-1 CURRENT IN AN IDEAL INDUCTOR

To start our investigation of the behavior of inductance in dc circuits, we assume that the inductor of Figure 15-1 has no resistance. Such a situation is hypothetical since, at the present time, there is no conductor material with zero resistance at room temperature. However, it will help us to determine what happens when we close the switch in Figure 15-1 if, for the moment, we do consider the resistance to be zero.

According to Kirchhoff's voltage law, all the time that the switch is closed in an electric circuit, a voltage drop exactly equal to the applied voltage must appear across the external circuit. If this circuit contains only *resistance,* this voltage drop is produced by current flowing through the resistance. Therefore, when we close the switch, the current in a resistance circuit must *instantly* assume a magnitude of $I = V/R = E/R$. But if there is no resistance in the circuit (as we have assumed in Figure 15-1), none of this voltage drop can be made up of *IR* drop.

In Section 14-4, we discovered that such a voltage is the result of an emf induced into the coil by a *changing* current in the coil. In Equation (14-3) we represented the induced voltage by the letter symbol e_L. Using e rather than v indicated a voltage resulting from an induced emf, and the lowercase e indicated that the magnitude of the voltage depended on the *rate of change* of current. However, we do not think of the inductor as a voltage source in Figure 15-1. Since the voltage between the terminals of an inductor exists only when there is a *changing* current in the coil, this induced voltage becomes the voltage drop required by Kirchhoff's voltage law. Thus, we can rewrite Equation (14-3)

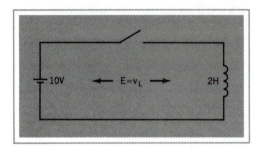

FIGURE 15-1
Simple Electric Circuit Containing Inductance.

as

$$v_L = L \times \frac{di}{dt} \tag{15-1}$$

We can determine the current that must flow when we close the switch in Figure 15-1 by rearranging Equation (15-1) as

$$\frac{di}{dt} = \frac{v_L}{L}$$

Since the applied voltage is constant, the voltage across the inductor must also be constant, which allows us to represent it by the uppercase V_L. And since inductance depends on the number of turns and the reluctance of the magnetic circuit, the inductance of a given air-core coil is also constant. As a result, the V/L ratio is a constant. Hence, the rate of change of current must also be constant, which allows us to drop the d operator, giving

$$\frac{I}{t} = \frac{V}{L}$$

from which
$$I = \frac{V}{L} \times t$$

For the numerical values given in Figure 15-1,

$$I = \frac{10\ V}{2\ H} \times t = 5t\ A$$

Therefore, 0.5 second after we close the switch, the current will be 2.5 amperes; 1 second after we close the switch, it will be 5 amperes; at $t = 2$ s, the current must be 10 amperes; and so on, as shown by the graph of Figure 15-2.

Since the induced voltage in this case must remain constant, the rate of change of current must be constant and, therefore, the graph of Figure 15-2 is a straight line. The *slope* of this graph of current plotted against time represents the *rate of change* of current. If we reduce the inductance of the circuit, the current must rise more rapidly in order to

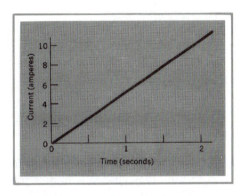

FIGURE 15-2
Rise of Current in a Direct-Current Circuit
Containing Only Inductance.

generate the required voltage. If the inductance is very small (approaching zero), the current must rise almost instantly to an infinitely high value, and we have a typical *short circuit*. As the graph of Figure 15-2 indicates, the presence of appreciable inductance in the circuit acts to oppose this sudden change in current from zero to a very high value by forcing the current to rise at a slower, steady rate governed by the ratio V/L.

15-2 RISE OF CURRENT IN A PRACTICAL INDUCTOR

Because we do not have at our disposal any wire which has zero ohms resistance at room temperature, all practical inductors possess some resistance as well as inductance. Since there is only the one path for charge carriers to flow through, we must consider the practical inductor as consisting of inductance and resistance in *series,* as shown in Figure 15-3.

When we close the switch in the circuit of Figure 15-3, the voltage drop across the resistance of the coil plus the voltage drop across the inductance of the coil (resulting from the cemf generated in it by the rising current) must equal the applied voltage in order to satisfy Kirchhoff's voltage law. Therefore,

$$E = v_R + v_L$$

FIGURE 15-3
Simple Electric Circuit Containing a
Practical Inductor.

From Ohm's law,

$$v_R = iR$$

and from Equation (15-1),

$$v_L = L\frac{di}{dt}$$

Note that we are using lowercase letters for current and voltage because, as we shall discover shortly, they change with time immediately after the switch is closed.

Substituting for v_R and v_L in the Kirchhoff's-law equation,

$$E = iR + L\frac{di}{dt} \tag{15-2}$$

If we can arrange this equation to solve for the instantaneous current in the circuit, we can then determine the voltage drop across the resistance from Ohm's law. Subtracting v_R from E then gives us the inductive voltage drop. However, solving Equation (15-2) for i involves rather more than basic algebra.[†] Therefore, we shall set this equation aside for the moment and see if we can estimate the circuit behavior from our prior investigation of the rise in current in the circuit of Figure 15-1.

The inductance of the practical inductor and the applied voltage in Figure 15-3 are the same as for the theoretical inductor of Figure 15-1. Therefore, the current cannot rise at a more rapid rate than that shown in the graph of Figure 15-2, otherwise the induced voltage of the inductance alone would be greater than the applied voltage. Since there was no resistance in the circuit of Figure 15-1, there was no limit to the value of the instantaneous current. But, in the case of the practical inductor, it is not possible for the current to rise to a value such that the IR drop across the resistance exceeds the applied voltage, otherwise Kirchhoff's law again would not hold true. The greatest value that the instantaneous current can have is such that $v_R = E$. When this occurs, $v_L = 0$, and for the voltage across the inductance to be zero, $di/dt = 0$, and, therefore, the current must be a *steady* value which allows us to use the uppercase letter symbol;

$$\therefore I_m = \frac{E}{R} \tag{15-3}$$

These two limits on the graph of the instantaneous current in a practical inductor are shown by the shaded lines in Figure 15-4.

Inductance opposes any *change* in current; consequently, as indicated by the graph of Figure 15-2, current cannot flow instantly when the switch is closed. It must build up from zero. At the instant we close the switch in Figure 15-3, the current through the resistance is zero and there will be no voltage drop across it. Therefore, at the instant the switch is closed, an inductive voltage equal to the applied voltage must appear across the

[†]See Appendix 2-4.

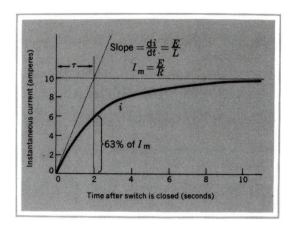

FIGURE 15-4
Rise of Current in a Direct-Current Circuit Containing Inductance and Resistance.

inductance. Since $di/dt = v_L/L$,

$$\text{initial } \frac{di}{dt} = \frac{E}{L} \tag{15-4}$$

This is the same rate of change of current as for Figure 15-1 (since the resistance cannot make its presence known if there is no current through it). Hence, the graph of the instantaneous current in Figure 15-4 starts off with the same slope as in Figure 15-2.

However, since the current is rising in order to induce a voltage in the inductance, the voltage drop across the resistance must be rising. As a result, the induced voltage required of the inductance for $v_R + v_L = E$ must decrease; and consequently, the rate of change of current must become smaller. The slope of the instantaneous current curve in Figure 15-4, therefore, becomes more gradual. Although the current is now rising less rapidly, nevertheless it is still rising, and consequently, the inductive voltage must be decreasing. As v_L becomes smaller, the rate of change of current must become smaller. As a result, the instantaneous current curve in Figure 15-4 continues to become still more gradual. After an appreciable period of time, the current reaches its maximum value, which produces a voltage drop across the resistance equal to the applied voltage. Consequently, the current is no longer changing but has settled to the steady value given by Equation (15-3).

Therefore, when we close the switch in a dc circuit containing inductance and resistance in series, the current cannot instantly take on a value of E/R as would be the case if there were no inductance in the circuit. The effect of the inductance is to force the current to rise to this steady-state value along the solid curve in Figure 15-4. Once the current reaches this steady value, the effect of the inductance in the circuit disappears. Since $v_R = iR$, and R is a constant, the graph of the voltage drop across the resistance of the circuit must have the same shape as the instantaneous-current graph (Figure 15-5). We obtain the graph of the instantaneous voltage across the inductance by subtracting the instantaneous IR drop from the constant applied voltage.

$$v_L = E - v_R$$

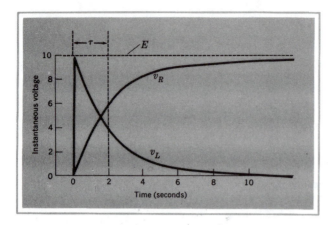

FIGURE 15-5
Instantaneous Voltage across
the Resistance and
Inductance.

SELF-CHECK

15-2:1. Using the graph in Figure 15-4 determine the following values for the series *LR* circuit in Figure 15-3 at a time two seconds after the switch is closed:
(a) the circuit current
(b) the voltage drop across the resistance
(c) the voltage drop across the inductance

15-2:2. Suppose that you have a series *LR* circuit similar to that shown in Figure 15-3 but with the following component values:

$$E = 20 \text{ V}, \quad R = 10 \; \Omega \quad \text{and} \quad L = 500 \text{ mH}.$$

Determine the following for the circuit after the switch is closed:
(a) the steady-state current in the inductor
(b) the current in the inductor the instant the switch is closed
(c) the initial rate of change of current in the inductor

15-3 TIME CONSTANT

When we close the switch in a dc circuit containing inductance it takes *time* for the current to rise to its steady-state value. Later on, we shall calculate this time element directly from the equation of the curve for instantaneous current in Figure 15-4. We can derive a simple and convenient method of indicating this time interval by assuming for the moment that the current can continue to rise at a *constant* rate equal to its initial rate of rise until it reaches the steady-state value, as shown by the shaded lines in Figure 15-4. This time interval is called the **time constant** of the *LR* circuit. On the basis of the above assumption, we can define time constant thus:

The time constant of an LR circuit is the time it would take the current to rise to its steady-state value if *it were to continue to rise at its initial rate of change for the whole time interval.*

The letter symbol for time constant is the Greek letter τ (tau).

If the rate of change of current remains constant (as shown by the straight shaded line in Figure 15-4), the steady-state current would be simply

$$I_m = \tau \times \text{initial } \frac{di}{dt}$$

But

$$I_m = \frac{E}{R} \tag{15-3}$$

and

$$\text{initial } \frac{di}{dt} = \frac{E}{L} \tag{15-4}$$

from which

$$\tau = \frac{L}{R} \tag{15-5}$$

where τ is the length of time in seconds defined as a time constant, L is the inductance of the circuit in henrys, and R is the resistance of the circuit in ohms.

Note that the time constant of an *LR* circuit (practical inductor) is directly proportional to the inductance. If we double the inductance, the initial rate of rise of current will only be half as great, and it will, therefore, take twice as long to reach the steady-state value. Also, the time constant is inversely proportional to the resistance. If we double the resistance, the steady-state current will be only half as much. So, for a given inductance, the current (rising at the same rate) will only take half as long to reach the steady-state value. Note also that the applied voltage has no effect on the time constant. If we double the applied voltage, we double the initial rate of rise of current E/L, but we also double the steady-state current E/R. Therefore, it will take exactly the same time to reach the steady-state value.

However, with resistance in the circuit, the instantaneous current does *not* continue to rise at its initial rate; it rises along an **exponential** curve, as shown by the solid graph in Figure 15-4. Nevertheless, this exponential curve must bear a fixed relationship to the two shaded straight lines in Figure 15-4.* It starts at the same point and with the same slope as the initial rate of change of current line, and its slope gradually changes until it merges with the horizontal steady-state current line. Therefore, no matter what values of *L* and *R* are involved, in the time interval defined as a time constant, the instantaneous current actually rises to 63% of the steady-state value. If we wish, we may redefine time constant as the time it takes the instantaneous current in an *LR* circuit to reach 63% of its steady-state value. Although we might argue that the instantaneous current theoretically never does quite reach the steady-state value, we shall consider that, for practical purposes,

The instantaneous current will reach the steady-state value of E/R after a time interval equal to five time constants has elapsed.

⌐ EXAMPLE 15-1
(a) What is the steady-state current in the circuit of Figure 15-3?
(b) How long does it take the current to reach this value after the switch is closed?

---

* In mathematics, we call these dashed lines **asymptotes**.

SOLUTION

(a)
$$I_{\mathrm{m}} = \frac{E}{R} = \frac{10\ \mathrm{V}}{1\ \Omega} = \mathbf{10\ A}$$

(b)
$$\tau = \frac{L}{R} = \frac{2\ \mathrm{H}}{1\ \Omega} = \mathbf{2\ s}$$

∴ total rise time $= 5\,\tau = 5 \times 2\ \mathrm{s} = \mathbf{10\ s}$

SELF-CHECK

15-3:1. Suppose that you have an *LR* series circuit similar to that shown in Figure 15-3 but with the following component values:

$$E = 20\ \mathrm{V}, \quad R = 10\ \Omega \quad \mathrm{and} \quad L = 500\ \mathrm{mH}.$$

How long will it take the circuit current to reach its steady-state value after the switch is closed?

15-3:2. Compare the times to reach steady-state currents for the following two series *LR* circuits:
(a) Circuit 1 contains $L = 50$ mH, $R = 10\ \Omega$ and $E = 12$ V.
(b) Circuit 2 contains $L = 1$ H, $R = 200\ \Omega$ and $E = 100$ V.

15-4 GRAPHICAL SOLUTION FOR INSTANTANEOUS CURRENT

Sometimes we wish to know what the instantaneous current is at a certain instant in time after the switch is closed or how long it takes the current to reach a certain percentage of its steady-state value. If we draw the instantaneous-current graph of Figure 15-4 accurately enough, we can solve such problems graphically from the exponential curve. We have already noted that, regardless of the numerical values of applied voltage, inductance, and resistance, the initial rate of rise of current will always be E/L; the final steady-state current will always be E/R; and, for all practical purposes, it will take a time equal to $5L/R$ to reach that steady-state value. This allows us to prepare a **universal exponential graph,** as in Figure 15-6, which will apply to *all* numerical values of E, L, and R. We calibrate the vertical (instantaneous current) axis of the graph in percentage of the final current and calibrate the horizontal (time) axis in time constants rather than in seconds.

EXAMPLE 15-2

(a) What is the instantaneous current in the circuit of Figure 15-3 1 s after the switch is closed?
(b) How long will it take the instantaneous current to rise from 0 A to 5 A?

SOLUTION
(a) We have already determined that the time constant of this circuit is

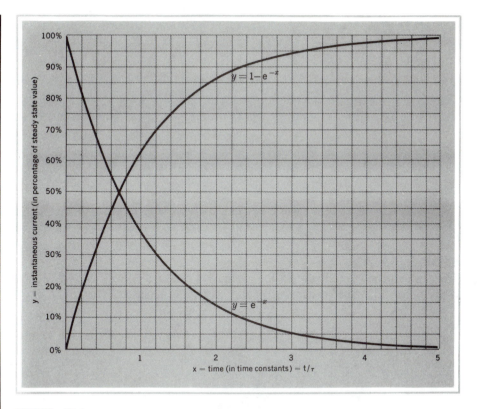

FIGURE 15-6
Universal Exponential Curves for the Graphical Solution of the Rise and Fall of Current in Inductive Direct-Current Circuits.

$$\tau = \frac{L}{R} = \frac{2\,\text{H}}{1\,\Omega} = 2\,\text{s}$$

Therefore, one second represents a time interval of 0.5 time constant on the universal graph of Figure 15-6. From this graph, 0.5 time constant after the switch is closed, as shown in Figure 15-7(a), the instantaneous current is

$$i = 39\% \times \frac{E}{R} = 0.39 \times \frac{10\,\text{V}}{1\,\Omega} = \textbf{3.9 A}$$

(b) Since $I_m = E/R = 10\,\text{V}/1\,\Omega = 10\,\text{A}$,

$$5\,\text{A} = \frac{5}{10} \times I_m = 50\% \text{ of } I_m$$

The universal curve of Figure 15-6 shows that it takes 0.7 time constant [Figure 15-7(b)] for the instantaneous current to reach 50% of I_m. Therefore,

$$t = 0.7\tau = 0.7 \times 2\,\text{s} = \textbf{1.4 s}$$

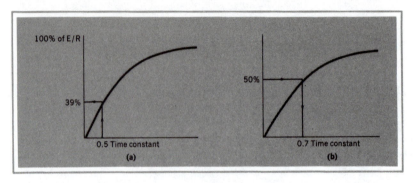

100% of E/R

39%

0.5 Time constant

(a)

50%

0.7 Time constant

(b)

FIGURE 15-7
Using the Universal Exponential Graph to Solve Example 15-2.

EXAMPLE 15-3
In the circuit shown in Figure 15-8, switch S_2 is closed 1 s after switch S_1 is closed. What is the instantaneous current 1 s after switch S_2 is closed?

S_2

S_1

4Ω

1Ω

10 V

2H

FIGURE 15-8
Circuit Diagram for Example 15-3.

SOLUTION
   Step I.   While S_1 is closed and S_2 is open, the total resistance in the circuit is

$$R_T = 4 + 1 = 5\ \Omega$$

Time constant

$$\tau = \frac{L}{R} = \frac{2\ \text{H}}{5\ \Omega} = 0.4\ \text{s}$$

$$\therefore\ 1\ \text{s} = \frac{1}{0.4} \times \tau = 2.5\ \tau$$

From the curve of Fig. 15-6, when $t = 2.5\tau$,

$$i = 92\% \text{ of } \frac{E}{R} = 0.92 \times \frac{10\text{ V}}{5\text{ }\Omega}$$

$$= 1.84\text{ A}$$

There are two lines of thought that we may use to solve the second step which is the 1-s interval after S_2 is closed.

Step IIA. The first method is to consider the instantaneous current curve for the new circuit, which has only 1 Ω of resistance in circuit. The final steady-state current now becomes

$$I_m = \frac{E}{R} = \frac{10\text{ V}}{1\text{ }\Omega} = 10\text{ A}$$

But at the instant S_2 is closed, i is already 1.84 A. Therefore, we must locate this point on the graph and mark off the additional 1 s from there [Figure 15-9(a)].

$$1.84\text{ A} = \frac{1.84}{10} \times I_m = 18.4\% \text{ of } I_m$$

From the curve of Figure 15-6, when $i = 18.4\%$ of I_m,

$$t = 0.2\tau = 0.2\frac{L}{R} = 0.2 \times \frac{2\text{ H}}{1\text{ }\Omega} = 0.4\text{ s}$$

Therefore, the instant in time we are concerned with will be 1 s later, or the 1.4-s mark on the curve of Figure 15-9(a). In terms of time constants,

$$1.4\text{ s} = \frac{1.4}{2}\tau = 0.7\tau$$

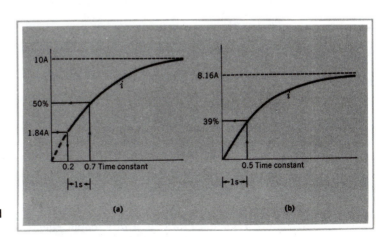

FIGURE 15-9
Graphical Solution of Step II
of Example 15-3.

Again, from Figure 15-6, when $t = 0.7\tau$,

$$i = 50\% \text{ of } I_m = 0.5 \times 10 \text{ A} = \mathbf{5 \text{ A}}$$

Step IIB. The alternative method requires only one reading from the universal exponential graph and is based on thinking in terms of the *additional* rise in current after S_2 is closed. Since we start Step II with 1.84 A already flowing in the inductance, and since $I_m = E/R = 10 \text{ V}/1 \text{ } \Omega = 10 \text{ A}$, the maximum *additional* rise in current can be only $10 - 1.84 = 8.16$ A.

The time constant $= L/R = 2 \text{ H}/1 \text{ } \Omega = 2$ s; therefore, a time interval of 1 s after S_2 closes represents a time interval of $0.5 \text{ } \tau$. From Figure 15-6, in half a time constant, the instantaneous current will rise 39% of the maximum *additional* current [Figure 15-9(b)].

$$\therefore \textit{additional } i = 39\% \text{ of } 8.16 \text{ A} = 3.18 \text{ A}$$

The instantaneous current 1 s after S_2 is closed will be 3.18 A greater than when S_2 is first closed. Therefore,

$$i = 1.84 + 3.18 = \mathbf{5 \text{ A}}$$

SELF-CHECK

15-4:1. Suppose that you have an *LR* series circuit similar to that shown in Figure 15-3 but with the following component values:

$$E = 30 \text{ V}, R = 10 \text{ } \Omega, \text{ and } L = 500 \text{ mH}$$

Use the rising current curve of Figure 15-6 to determine the current in the circuit 80 ms after the switch is closed.

15-4:2. Referring to the circuit used in Self-Check Problem 15-4:1, use Figure 15-6 to determine at what time the circuit current is equal to 0.5 A.

15-5 CALCULATOR SOLUTION FOR INSTANTANEOUS CURRENT

The graphic solution for instantaneous current is sufficiently accurate for most purposes and has the advantage that it helps us to visualize the behavior of inductance in dc circuits. If we are required to determine the instantaneous current with greater accuracy, we must return to an algebraic solution based on Equation (15-2).

$$E = iR + L\left(\frac{di}{dt}\right) \tag{15-2}$$

Since Equation (15-2) is a differential equation, to solve for the instantaneous current i, we must once more resort to the calculus solution of Appendix 2-4, which yields

$$i = \frac{E}{R}(1 - e^{-x}) \qquad (15\text{-}6)$$

where e = 2.718 *(the base of natural logarithms) and* $x = tR/L = t/\tau$.

Since
$$\tau = \frac{L}{R}, \qquad x = \frac{t}{\tau}$$

Therefore, we can think of x as representing elapsed time measured in **time constants.**

EXAMPLE 15-2A
(a) What is the instantaneous current in the circuit of Figure 15-3 1 s after the switch is closed?
(b) How long will it take the instantaneous current in this circuit to rise to 5 A?

SOLUTION
(a)

$$x = t \times \frac{R}{L} = 1\text{s} \times \frac{1\ \Omega}{2\ \text{H}} = 0.5$$

$$i = \frac{E}{R}(1 - e^{-x}) = \frac{10\ \text{V}}{1\ \Omega} \times (1 - e^{-0.5}) = 10 \times (1 - 0.6065) = \mathbf{3.93\ A}$$

The calculator chain for this computation becomes

$$\boxed{10}\ \boxed{\times}\ \boxed{(}\ 1\ \boxed{-}\ 0.5\ \boxed{+/-}\ \boxed{e^x}\ \boxed{)}\ \boxed{=}$$

(b)
$$i = \frac{E}{R}(1 - e^{-x})$$

$$5 = \frac{10\ \text{V}}{1\ \Omega} \times (1 - e^{-x})$$

from which $e^{-x} = 0.5$. $-x = \ln 0.5 = -0.693$

$$\therefore t = x \times \frac{L}{R} = 0.693 \times \frac{2\ \text{H}}{1\ \Omega} = \mathbf{1.39\ s}$$

EXAMPLE 15-3A
In the circuit shown in Figure 15-8, switch S_2 is closed 1 s after switch S_1 is closed. What is the instantaneous current 1 s after switch S_2 is closed?

SOLUTION
   Step I.   One second after S_1 is closed,

$$x = t \times \frac{R}{L} = 1 \text{ s} \times \frac{5 \text{ }\Omega}{2 \text{ H}} = 2.5$$

$$\therefore i = \frac{E}{R}(1 - e^{-x}) = \frac{10 \text{ V}}{5 \text{ }\Omega} \times (1 - e^{-2.5}) = 2 \times (1 - 0.0821) = 1.84 \text{ A}$$

Step IIA. Using the method illustrated by Figure 15-9(a),

$$i = \frac{E}{R}(1 - e^{-x})$$

$$1.84 = \frac{10 \text{ V}}{1 \text{ }\Omega} \times (1 - e^{-x})$$

$$\therefore e^{-x} = 0.816$$

from which

$$x = 0.203 = t \times \frac{R}{L}$$

$$\therefore t = 0.203 \times \frac{2 \text{ H}}{1 \text{ }\Omega} = 0.406 \text{ s}$$

One second later, $\qquad t = 1.406 \text{ s}$

and now $\qquad x = 1.406 \text{ s} \times \frac{1 \text{ }\Omega}{2 \text{ H}} = 0.703$

$$\therefore i = \frac{E}{R}(1 - e^{-x}) = \frac{10 \text{ V}}{1 \text{ }\Omega} \times (1 - e^{-0.703}) = 10 \times (1 - 0.495) = \mathbf{5.05 \text{ A}}$$

Step IIB. Using the method illustrated in Figure 15-9(b),

$$\text{Maximum } additional \text{ current} = 10 - 1.84 = 8.16 \text{ A}$$

For a time interval of 1 s,

$$x = t \times \frac{R}{L} = 1 \text{ s} \times \frac{1 \text{ }\Omega}{2 \text{ H}} = 0.5$$

Actual additional current 1 s after S_2 is closed,

$$i = 8.16(1 - e^{-0.5}) = 8.16(1 - 0.6065) = 3.21 \text{ A}$$

Therefore, the total instantaneous current 1 s after S_2 is closed is

$$i = 1.84 + 3.21 = \mathbf{5.05 \text{ A}}$$

The *LR* network of Figure 15-10(a) includes a 4-ohm **discharge resistor** in parallel

FIGURE 15-10
Thévenin-Equivalent Circuit for Determining Time Constant.

(a)   (b)

with a practical inductor. We shall consider the function of this resistor in Section 15-7. In the meantime, to determine the time constant for the buildup of current in the inductor, we may consider the 4-ohm resistor to be part of the source network, as "seen" by the inductor. This allows us to use the Thévenin-equivalent circuit of Figure 15-10(b) to determine the rise of current in the inductor when the switch is first closed.

EXAMPLE 15-4

Calculate the instantaneous current in the inductor 1 s after the switch in the circuit of Figure 15-10(a) is first closed.

SOLUTION

Step I.   Disconnecting the inductor from the rest of the circuit, the open-circuit voltage across the 4-Ω resistor is

$$E_x = 10 \text{ V} \times \frac{4}{4 + 1} = 8 \text{ V}$$

and

$$R_x = \frac{4 \times 1}{4 + 1} = 0.8 \ \Omega$$

Hence, the time constant is

$$\tau = \frac{L}{R_T} = \frac{2 \text{ H}}{(1 + 0.8)\Omega} = 1.111 \text{ s}$$

Step II.

$$x = \frac{t}{\tau} = \frac{1 \text{ s}}{1.111 \text{ s}} = 0.9$$

and

$$i = \frac{E_x}{R_T}(1 - e^{-x}) = \frac{8 \text{ V}}{1.8 \ \Omega}(1 - e^{-0.9})$$

$$= 4.444 \text{ A}(1 - 0.4066) = \mathbf{2.64 \text{ A}}$$

SELF-CHECK

15-5:1. Figure 15-11 shows a practical inductor with an internal resistance of 0.5 Ω and inductance of 500 mH in a network. Determine the following values for that circuit at the instant 0.10 s after the switch is closed:
(a) the current in the inductor,
(b) the voltage across the inductor.

15-5:2. For the circuit in Figure 15-11, determine at what time the current in the practical coil is equal to 2.0 A.

FIGURE   15-11

15-6 ENERGY STORED BY AN INDUCTOR

If we connect an ideal inductor to a voltage source having no internal resistance, the voltage generated by the inductance must remain constant and equal to the applied voltage; therefore, the current must rise at a constant rate, as in Figure 15-12(b). The source is supplying electric energy to the ideal inductor at the rate of $p = Ei$. But, unlike resistance, inductance cannot convert this energy into heat or light. The energy that the source supplies is stored in the magnetic field, since the rising current forces the magnetic lines of force to expand against their natural characteristic of trying to become as short

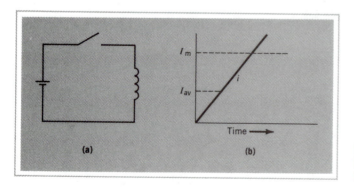

FIGURE   15-12
Determining the Energy Stored by an Inductor.

as possible. We can think of the storing of energy in an inductor as being similar to winding up the rubber band motor of a model airplane. As we apply torque to the propeller with our finger, we twist the rubber band and build up a countertorque in it. Since we are accomplishing motion against this countertorque, we are doing work and thus are storing energy in the rubber band. When we release the propeller, the energy stored in the rubber band is released, and it drives the propeller until the stored energy is all used up.

To determine the energy stored by an inductor, we must pick some final value of current [I_m in Figure 15-12(b)] and find out how much energy is stored as the current rises to this value. In resistance circuits, where the current and voltage do not change with a change in time, we found that $W = Pt = VIt$. Although the voltage remains constant in the circuit of Figure 15-12, the current steadily increases as time elapses. However, since the rate of change of current is constant, the *average* value of the instantaneous current as it rises from zero to I_m in Figure 15-12(b) will be $0.5I_m$. Therefore,

$$W = V \times 0.5I_m \times t$$

But
$$V = L\frac{di}{dt} \qquad (15\text{-}1)$$

and, since the rate of change of current is constant,

$$V = L \times \frac{I_m}{t}$$

Substituting gives
$$W = L \times \frac{I_m}{t} \times 0.5I_m \times t$$

$$\therefore W = 0.5LI_m^2 \qquad (15\text{-}7)$$

In the case of the practical inductor, both the instantaneous voltage and current are changing and their rate of change is not constant. By first plotting accurate graphs of the instantaneous voltage across and current through the inductance on a piece of graph paper, we can then plot a graph of the instantaneous *rate* at which the inductance of the circuit stores energy from the relationship $p = vi$. By calculating instantaneous power for each division along the horizontal axis of the graph, we obtain the instantaneous power graph shown in Figure 15-13. When the current in a practical inductor reaches its steady-state value of $I_m = E/R$, the magnetic field ceases to expand, the voltage generated by the inductance has dropped to zero; hence, the instantaneous power $p = vi$ becomes zero. Therefore, the inductor increases the amount of energy stored in its magnetic field only while the current is building up to its steady-state value. As long as the current remains constant at this steady-state value, a constant amount of energy is stored by the magnetic field.

Although no additional energy is *stored* by the inductance of the practical inductor when I_m is reached, energy will be constantly dissipated by the resistance of the inductor at a steady rate of $P = I_m^2R$. Energy is stored in an inductor in a *dynamic* form, and current must continue to flow for the magnetic field to be maintained. In this respect, energy storage by an inductor differs from energy storage by a capacitor, where the

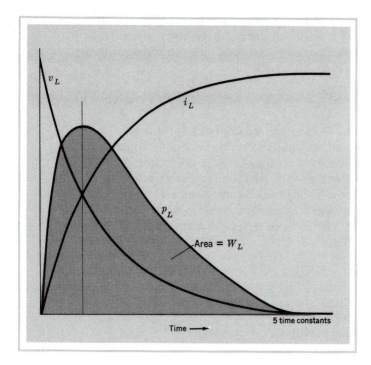

FIGURE 15-13
Graphical Representation of the
Energy Stored by a Practical Inductor.

charging current ceases when the capacitor is charged. The electric field of a capacitor stores energy in a *static* form.

Note the similarity between Figure 15-13 and Figure 12-19. The significant difference is that the voltage curve for one graph becomes the current curve for the other. Since $p = vi$, the instantaneous power graph is the same in both sketches. Once again, the instantaneous rate of storage of energy p reaches its peak value when i_L rises to $0.5 I_m$ and, at the same instant in time, v_L drops to $0.5 E$.

Again, the *area* contained by the instantaneous power graph of Figure 15-13 represents the product of the *average power* and *time,* which, in turn, represents the *energy* stored by the inductance. By resorting to the basic integral calculus solution of Appendix 2-5 to determine this area, the energy stored in the magnetic field of the practical inductor of Figure 15-13 by the time the current reaches its steady-state value is

$$W = 0.5 LI^2 \qquad \qquad \text{(15-8)}$$

where W is the stored energy in joules, L is the inductance in henrys, and I is the steady current in amperes. $(I = E/R.)$

EXAMPLE 15-5
What energy is stored by the magnetic field of an inductor whose inductance is 5 H and whose resistance is 2 Ω when it is connected to a 24-V source?

SOLUTION

$$I = \frac{E}{R} = \frac{24 \text{ V}}{2 \text{ }\Omega} = 12 \text{ A}$$

$$W = 0.5LI^2 = 0.5 \times 5 \text{ H} \times (12 \text{ A})^2 = \mathbf{360 \text{ J}}$$

SELF-CHECK

15-6:1. A practical inductor has a resistance of 0.5 Ω and an inductance of 500 mH. What current must flow in the inductor in order to store 10 J of energy in the device?

15-6:2. What current must flow in a 30-mH coil to store the same amount of energy as that stored in a 100-mH coil carrying a 50-mA current?

15-7 FALL OF CURRENT IN AN INDUCTIVE CIRCUIT

Interrupting the flow of current in an inductive circuit is like removing our finger after winding the rubber band motor of a model airplane; the stored energy is released. We have defined inductance as the circuit property that tends to oppose *any* change in current. If we examine the graphs of instantaneous current in this chapter, we note that it is not possible for the current to change instantly. It takes *time* for the current to change, even if only a few microseconds.

Therefore, at the instant we open the switch in the circuit of Figure 15-14, the current in the inductor must be the same as it was just before the switch was opened. To assist us in determining the effect of interrupting current flow in an inductive direct-current circuit, we shall give a specific value (500 kilohms) to the leakage resistance of the open switch. The steady-state current before the switch is opened is

$$I = \frac{E}{R} = \frac{12 \text{ V}}{2 \text{ }\Omega} = 6 \text{ A}$$

FIGURE 15-14
Interrupting Current in an Inductor.

At the instant we open the switch, the current in the inductor must still be 6 amperes. Since this is a simple series circuit, this six-ampere current must flow through the leakage resistance of the switch, creating a voltage drop between the switch contacts of

$$V = IR = 6 \text{ A} \times 500 \text{ k}\Omega = 3\,000\,000 \text{ V}$$

This extremely high voltage drop is matched by the voltage generated in the turns of the inductor by the rapidly collapsing magnetic field. If we allow such a high voltage to be induced in practice, not only will there be arcing of the switch contacts, but the insulation of the inductor will break down. Therefore, we must design dc circuits so that we never suddenly interrupt current in an inductor.

EXAMPLE 15-6
What is the voltage across the switch contacts in the circuit of Figure 15-15 at the instant that the switch is opened?

FIGURE 15-15
Circuit Diagram for Example 15-6.

SOLUTION
The steady-state current through the inductor is

$$I = \frac{E}{R} = \frac{12 \text{ V}}{2 \text{ }\Omega} = 6 \text{ A}$$

Although a 1-A current passes through the 12-Ω resistor with a conventional direction, as shown in Figure 15-15(a), this 1 A does *not* flow through the inductance of the circuit. When the switch is opened, the inductor will momentarily act as a generator as its magnetic field collapses. Across this generator, there is a total load of 12 Ω and 2 Ω in series. And since the current through the inductor cannot change instantly, a current of 6 A with a conventional direction, as shown in Figure 15-15(b), will flow through the 12-Ω resistor. This will create a voltage drop across the 12-Ω

resistor of

$$v = iR = 6 \text{ A} \times 12 \ \Omega = 72 \text{ V}$$

Between the switch contacts, we have the 72-V drop across the 12-Ω resistor and the 12-V potential difference of the battery. Their polarities are such that they *add*. Therefore, the total voltage across the switch contacts at the instant we open the switch is

$$v = 72 + 12 = \textbf{84 V}$$

According to Kirchhoff's voltage law, the initial induced voltage must equal the voltage drop across the total resistance at the instant we open the switch. Hence,

$$E_{\text{m}} = I_0 R_{\text{T}} \tag{15-9}$$

where E_{m} is the peak voltage induced by the collapsing magnetic field, I_0 is the current through the inductor at the instant the switch is opened (which must be the same as the instant before the switch is opened), and R_{T} is the total resistance of the loop containing the inductor. In Example 15-6,

$$E_{\text{m}} = 6 \text{ A} \times 14 \ \Omega = 84 \text{ V}$$

According to Lenz's law, the direction of the induced voltage will always be such as to oppose any change in current. When the current rises, the direction of the induced voltage tends to cancel out some of the applied voltage. But when the source is removed, the magnetic field around the inductor collapses. The polarity of the voltage induced by a collapsing field is the opposite of that generated by a rising field. This reversed polarity tends to take the place of the original source to maintain current flow in the original direction through the inductor.

The 12-ohm resistor in the circuit of Figure 15-15 is called a **discharge resistor.** As the numerical results of Example 15-6 have shown, the presence of this discharge resistor in the circuit protects both the switch contacts and the insulation of the inductor from an excessively high voltage surge. Some such provision must always be made with inductive direct-current circuits which store appreciable energy, such as the field coils of dc motors and generators. Thyrite resistors are excellent for this purpose.

Since the induced voltage in Example 15-6 is much smaller than when no discharge resistor is used, the magnetic field must be collapsing at a much slower rate and, therefore, it takes longer to discharge the stored energy when a discharge resistor is included in the circuit. Since the magnetic field *is* collapsing and since the magnetic field strength depends on the instantaneous current in the windings, the current must start to decrease as soon as the switch is opened. In determining the rise time for the current in an inductive circuit, we used the term **time constant** based on a constant rate of change of current. Using the same technique, we may define **discharge time constant** as **the time it would take the current to drop to zero if it continued to drop at its initial rate for the whole time interval.** If the current continues to drop at a constant rate equal to

FIGURE 15-16
Fall of Current in an Inductive
Direct-Current Circuit.

its initial rate, from Figure 15-16,

$$\text{initial } \frac{di}{dt} = \frac{I_0}{\tau}$$

But $\qquad$ $\text{initial } \dfrac{di}{dt} = \dfrac{E_m}{L}$ $\qquad$ and $\qquad$ $E_m = I_0 R_T$ $\qquad$ (15-9)

from which $\qquad\qquad\qquad$ $\tau = \dfrac{L}{R_T}$ $\qquad\qquad$ (15-5)

Therefore, **we determine the time constant of an _LR_ circuit in exactly the same manner for both rise and fall of current.**

However, the current cannot fall at a constant rate. If it were to do so, the induced voltage would remain constant. But since the current is decreasing, the voltage drop across the resistance must also decrease. Because Kirchhoff's voltage law requires the induced voltage to equal the voltage drop across the resistance at every instant in time, the induced voltage must also decrease. As the induced voltage decreases the magnetic field must collapse more slowly. Thus, the rate of change of current must become progressively smaller with the passing of time. Since this sequence of events is similar to the rise of current when the switch was first closed, the instantaneous discharge graph (Figure 15-16) has the same basic exponential shape. Figure 15-6 includes an accurate graph for graphical solution of current decay in inductive dc circuits.

⌐ _EXAMPLE 15-7_
The switch in the circuit of Figure 15-15 is closed for 3 s and then opened. What is the instantaneous current in the inductor 0.25 s after the switch is opened?

SOLUTION
   Step I.

$$\text{Charge } \tau = \frac{L}{R} = \frac{4 \text{ H}}{2 \text{ }\Omega} = 2 \text{ s}$$

Expressing 3 s in time constants,

$$3 \text{ s} = \frac{t}{\tau} = \frac{3}{2} = 1.5 \text{ time constants}$$

From the graph for current *rise* in Figure 15-6, when $t = 1.5$ time constants, $i_L = 78\%$ of I_m.

Hence,
$$i_L = 0.78 \times \frac{E}{R} = 0.78 \times \frac{12 \text{ V}}{2 \text{ }\Omega} = 4.7 \text{ A}$$

Step II.

$$\text{Discharge } \tau = \frac{L}{R_T} = \frac{4 \text{ H}}{(12 \times 2)\Omega} = 0.286 \text{ s}$$

Expressing 0.25 s in time constants,

$$0.25 \text{ s} = \frac{t}{\tau} = \frac{0.25}{0.286} = 0.875 \text{ time constant}$$

From the graph for current *decay* in Figure 15-6, when $t = 0.88$ time constant, $i_L = 42\%$ of I_0.

Hence,
$$i_L = 0.42 \times 4.7 \text{ A} = \mathbf{2 \text{ A}}$$

A second method of preventing sparking of the switch or breaker contacts when we open the switch in the circuit of Figure 15-14 is shown in Figure 15-17. When we place a capacitor across the switch contacts, the potential difference across the capacitor is zero while the switch is closed. Since the voltage cannot rise instantly when we open the switch, there is sufficient time to move the switch contacts far enough apart that they do not spark.[†] Some of the energy released by the collapsing magnetic field of the inductor

[†] This principle is used to prevent sparking of the breaker points in an automobile ignition system.

FIGURE 15-17
Supressing a Voltage Surge with a Capacitor.

is transferred to the capacitor as it charges. The capacitor then partially discharges, building up a smaller magnetic field around the inductor in the opposite direction. Several cycles of diminishing energy transfer take place until the stored energy is all dissipated in the resistance of the load. We must leave the transient response of dc circuits containing both inductance and capacitance until we have investigated **resonance** in Chapter 23.

SELF-CHECK

15-7:1. Refer to the circuit shown in Figure 15-11, but now suppose that the switch has been closed for a while and the circuit is in its steady-state condition, when the switch is suddenly opened.
    (a) What is the voltage drop across the inductor the instant before and the instant after the switch is opened?
    (b) How long does it take for the current in the inductor to drop to zero (practically)?
15-7:2. Use the information and the circuit from Self-Check Problem 15-7:1 and the discharge graph in Figure 15-6 to determine the following:
    (a) the voltage across the practical inductor 250 ms after the switch is opened.
    (b) at what time, after the switch is opened, the voltage across the inductor is equal to 1.5 V?

15-8 CALCULATOR SOLUTION FOR INSTANTANEOUS DISCHARGE CURRENT

When we write the Kirchhoff's voltage-law equation for the circuit of Figure 15-15(b), the applied voltage in the circuit is zero. Therefore,

$$0 = iR_T + L\frac{di}{dt}$$

Solving this equation for the instantaneous current,[†]

$$i = I_0 e^{-x} \tag{15-10}$$

where I_0 is the initial current at the instant the switch is opened, e $= 2.718$, and $x = t(R_T/L) = t/\tau$.

> EXAMPLE 15-7A
> The switch in the circuit of Figure 15-15 is closed for 3 s and then opened. What is the instantaneous current in the inductor 0.25 s after the switch is opened?
>
> SOLUTION
>   Step I. When the switch is closed,

[†] See Appendix 2-4.

$$x = t \times \frac{R}{L} = 3 \text{ s} \times \frac{2\ \Omega}{4\ \text{H}} = 1.5 \text{ time constants}$$

$$\therefore i_L = \frac{E}{R}(1 - e^{-x}) = \frac{12\ \text{V}}{2\ \Omega} \times (1 - e^{-1.5})$$

$$= 6\ \text{A} \times (1 - 0.223) = 4.66\ \text{A}$$

Step II. When the switch is opened,

$$x = t \times \frac{R_T}{L} = 0.25 \text{ s} \times \frac{14\ \Omega}{4\ \text{H}} = 0.875 \text{ time constant}$$

$$\therefore i_L = I_0 \times e^{-x} = 4.66\ \text{A} \times e^{-0.875}$$

$$= 4.66 \times 0.417 = \mathbf{1.94\ A}$$

SELF-CHECK

15-8:1. Use the information and the circuit from Self-Check Problem 15-7:1 and calculate the following:
   (a) the voltage across the practical inductor 250 ms after the switch is opened.
   (b) at what time, after the switch is opened, the voltage across the inductor is equal to 1.5V?
   (Note: this is the same problem that you solved graphically in Self-Check Problem 15-7:2)

15-9 TRANSIENT RESPONSE

When we use either mechanical or electronic switches to change source voltage(s) suddenly in a dc circuit, a purely resistive network responds with instant changes in steady-state current. But LR networks respond by *delaying* the change to a new steady-state current. We think of the time interval between the instant of switching ($t = 0$) and the final steady-state current ($t = 5\tau$) as the **transient response** of the LR network.

In the graphical solution for LR transients, we used a separate $(1 - e^{-x})$ curve for rising current and e^{-x} curve for falling current, to help us to understand the behavior of inductance in dc circuits. This led to the two separate equations: Equation (15-6) for rising instantaneous current and Equation (15-10) for falling instantaneous current. However, we can use the transient concept to derive a single equation for any LR transient, as long as we are careful in specifying the initial steady-state current (I_0) just before the switch is operated, and the final steady-state current (I_F) which flows after the transient has ended.

To establish this general case, let's consider the situation in Step IIB of Example 15-3A (Figure 15-8), where closing S_2 attempts to change the current in the circuit from $I_0 = 1.84$A at $t = 1$ s to $I_F = 10$ A. We noted that the instantaneous current during the transient is equal to the initial current plus a percentage of the *additional* current, based on the elapsed time measured in time constants. Expressing this statement in algebraic form,

$$i_L = I_0 + (I_F - I_0)(1 - e^{-x}) \tag{15-11}$$

where I_0 *is the initial current at the instant of switching and* I_F *is the final steady-state current* (after an interval of at least five time constants has elapsed). Carrying out the multiplication,

$$i_L = I_0 + I_F - I_0 - I_F e^{-x} + I_0 e^{-x}$$

Collecting the terms gives us the **universal equation for instantaneous inductor current** in a dc *LR* network.

$$i_L = I_F + (I_0 - I_F)e^{-x} \tag{15-12}$$

where i_L *is the instantaneous inductor current,* I_0 *is the initial inductor current, and* I_F *is the final steady-state inductor current after an interval of at least five time constants has elapsed.*

Equation (15-11) expresses rising instantaneous current as *initial* current plus a percentage of the additional difference between initial and final current. In rearranging the terms, Equation (15-12) now expresses the instantaneous current during an *LR* transient in terms of the *final* steady-state current and an e^{-x} transient caused by the change in steady-state current.

EXAMPLE 15-8
In the *LR* circuit shown in Figure 15-18(a), the switch is closed at $t = 0$ and opened again at $t = 1.0$ s.
(a) What is the instantaneous current at $t = 1.0$ s?

FIGURE 15-18
Transient Response in a DC *LR* Network.

(b) What is the voltage across the switch contacts at $t = 1.1$ s?

(c) Draw a graph of the voltage across the switch from $t = 0$ to $t = 5$ s.

SOLUTION

With the switch open, the steady-state current is

$$I_F = \frac{10 \text{ V}}{(4 + 1)\Omega} = 2 \text{ A}$$

and

$$\tau = \frac{L}{R_T} = \frac{2 \text{ H}}{5 \, \Omega} = 0.4 \text{ s}$$

With the switch closed, the steady-state current is

$$I_F = \frac{10 \text{ V}}{1 \, \Omega} = 10 \text{ A}$$

and

$$\tau = \frac{2 \text{ H}}{1 \, \Omega} = 2.0 \text{ s}$$

(a) When the switch closes at $t = 0$, the initial current through the coil is 2 A. Hence, from $t = 0$ to $t = 1.0$ s,

$$i = I_F + (I_0 - I_F)e^{-t/\tau}$$
$$= 10 + (2 - 10)e^{-1.0/2.0}$$
$$= 10 - 8 \times 0.6065 = \textbf{5.15 A}$$

(b) At $t = 1.1$ s, the switch has been open for 0.1 s.

and

$$i = I_F + (I_0 - I_F)e^{-t/\tau}$$

becomes

$$i = 2 + (5.15 - 2)e^{-0.1/0.4}$$
$$= 2 + 3.15 \times 0.7788 = 4.453 \text{ A}$$

Since the switch is in parallel with the 4-Ω resistor,

$$v_{sw} = 4.453 \text{ A} \times 4 \, \Omega = \textbf{17.8 V}$$

(c) While the switch is open, the 2-A steady-state current through the 4-Ω resistor creates a steady voltage drop of 8 V. When the switch closes, the voltage across the resulting short circuit instantly becomes zero (in spite of a current transient through the switch).

When the switch opens at $t = 1.0$ s, the 5.15-A current creates a voltage drop of 5.15 A $\times$ 4 Ω = 20.6 V across the switch contacts, creating the transient shown in Figure 15-18(b). Since five time constants represent 2 s with the switch open, the transient will have disappeared by $t = 3$ s.

SELF-CHECK

15-9:1. Returning to Example 15-8, now suppose that the switch is again closed at time $t = 5$ s.
   (a) Find the circuit current at time $t = 6$ s.
   (b) Find the circuit current at time $t = 16$ s.
   (c) What is the voltage across the coil at time $t = 9$ s?

15-10 CHARACTERISTICS OF INDUCTIVE DC CIRCUITS

In summarizing the behavior of inductance in dc circuits, we should note that the most important characteristic is that the inductance develops an induced voltage which always tends to oppose any *change* in current. As a result of this characteristic, the current in inductive circuits cannot change instantly. A definite time interval is required for current to either rise or fall between any two instantaneous values. This time interval depends on the inductance and resistance of the circuit.

When we close the switch in an inductive circuit, it takes time for the current to reach its steady-state level. This delay is somewhat of an advantage in that it suppresses a sudden surge in the current drawn from the source. But when current in an inductive circuit is interrupted, an extremely high induced voltage transient is developed which can damage switch contacts and insulation unless we provide suitable circuit design so that current can decay gradually. Switches for dc circuits are designed so that their contacts move far apart as rapidly as possible to prevent arcing.

There are, however, some cases where we can use the very high induced voltage of a collapsing magnetic field to advantage. The ignition coil of an automobile engine is an example. Energy is stored gradually as current from the car battery builds up a magnetic field in the core of the coil. Then the current is suddenly interrupted by the breaker points, and a 10-kilovolt induced voltage, generated by the rapidly collapsing magnetic field, causes a spark between the points of the spark plug.

We can use the tendency of inductance to oppose any change in current in **filtering** or **smoothing** variations in load current that would otherwise occur when the load is fed from a dc source such as a rectifier system which provides a pulsating terminal voltage. As the source voltage in Figure 15-19 rises above its average value, the tendency for the current in the circuit to increase causes the inductance to store additional energy in its expanding magnetic field. Then, when the source voltage drops below average, the stored energy is released to help maintain a constant power input to the load.

FIGURE 15-19
Filter Choke.

In Chapter 12, we compared the energy storage ability of a capacitor with the *static* energy stored in a tank of compressed air. We may think of the energy storage ability of an inductor in terms of the *dynamic* energy stored by the *flywheel* of a piston engine.

PROBLEMS

NOTE: Solutions need not be restricted to the method indicated by the section number.

15-3-1. How long after the switch is closed will it take the current to reach its maximum value in a circuit whose resistance is 8 Ω and whose inductance is 2.4 H?

15-3-2. What is the resistance of a solenoid whose inductance is 4 H if it takes 0.2 s for the current to reach its maximum value?

15-3-3. What voltage must be applied to an 80-μH inductor if the initial rate of change of current is to be 10 000 A/s?

15-3-4. A 120-mH inductor has a resistance of 6 Ω. The initial rate of rise of current is 1000 A/s. What is the final current?

15-3-5. A choke with an inductance of 20 H and a resistance of 16 Ω is connected through a switch to a 24-V source.
(a) What is the initial current when the switch is closed?
(b) What is the initial rate of change of current when the switch is closed?
(c) What is the final steady-state current?
(d) How long does it take the current to reach its maximum value?

15-3-6. An inductor has an inductance of 20 H and a resistance of 500 Ω. In parallel with this coil is a 1000-Ω resistor. This combination is connected across a 120-V dc source whose internal resistance is negligible.
(a) What initial current is drawn from the source when the switch is closed?
(b) What is the initial rate of change of current in the coil?
(c) What is the steady-state current in the coil?
(d) How long does it take the current to reach its steady-state value?

15-4-7. (a) How long after the switch is closed will it take the current in the circuit of Problem 15-3-1 to reach 63% of its maximum value?
(b) How long will it take the current to reach 50% of its maximum value?
(c) If the current reaches a magnitude of 2.0 A in 0.5 s after the switch is closed, what is its maximum value?

15-4-8. (a) How long after the switch is closed will it take the current in the inductor of Problem 15-3-4 to reach a magnitude of 16 A?
(b) What is the instantaneous current 10 ms after the switch is closed?

15-5-9. (a) How long after the switch is closed will it take the instantaneous current in the choke of Problem 15-3-5 to reach a magnitude of 1 A?
(b) What is the instantaneous current 2·s after the switch is closed?

15-5-10. A 20-Ω resistor is connected in series with the choke of Problem 15-3-5, and this resistor is short-circuited 1 s after the switch is closed. What induced voltage is being developed by the inductance 1 s after the 20-Ω resistor is short-circuited?

15-6-11. (a) How much energy is stored by the choke in Problem 15-3-5 when the current reaches its steady-state value?
(b) At what rate is energy being drawn from the source when the current through the choke reaches its maximum value?

15-6-12. (a) How much energy is stored in the magnetic field of the inductor in Problem 15-3-6 when the current has reached its steady-state value?

(b) At what rate is electric energy being converted into heat when the current in the circuit has reached its maximum value?

15-7-13. The field coils of a dc motor have an inductance of 8 H, a resistance of 40 Ω, and are connected across as 120-V source.

(a) What maximum value of discharge resistor must be connected across the field coils if the voltage between the terminals of the coils must never exceed 200 V?

(b) How long will it take to discharge the energy stored in the field coils?

(c) How much energy is dissipated by the discharge resistor after the source is disconnected?

15-7-14. (a) What voltage will appear across the switch contacts when the switch in the circuit of Problem 15-3-6 is opened?

(b) How long does it take the coil to discharge its stored energy?

(c) If the switch is opened 1 s after it was first closed, how long does it take the current in the coil to drop 25 mA?

15-8-15. In the filter network of Figure 15-20(a), the 80-V source and the 40-mH inductor both have negligible internal resistance.

(a) What is the instantaneous output voltage 5 μs after the switch closes?

(b) If the switch remains closed for at least five time constants, how long after the switch opens will it be before the instantaneous output voltage becomes 50 V?

(a)   **(b)**

FIGURE 15-20

15-8-16. If the inductor in the circuit of Figure 15-20(b) has negligible resistance

(a) What is the instantaneous output voltage 5 μs after the switch closes?

(b) If the switch remains closed for at least five time constants, how long after the switch opens will it be before the instantaneous output voltage becomes +15 V (with respect to ground)?

15-9-17. (a) If the switch in the circuit of Figure 15-20(a) opens 10 μs after it closes, what is the instantaneous output voltage (with respect to ground) 10 μs after the switch opens?

(b) If the switch dwells alternately for 50 μs in each position, draw a labelled graph of two complete cycles of the output voltage waveform.

15-9-18. Repeat Problem 15-9-17 for the circuit of Figure 15-20(b).

15-9-19. The switch in the circuit of Figure 15-21 is closed for 2 s and then opened. What is the instantaneous current in the inductor 1 s after the switch is opened?

15-9-20. Solve Problem 15-9-19 by replacing the source with an equivalent constant-current source.

FIGURE 15-21

REVIEW QUESTIONS

15-1. Why is the graph of Figure 15-2 a straight line?

15-2. What is the significance of lowercase letter symbols?

15-3. Why are lowercase letter symbols used in representing the rise of current in inductive circuits?

15-4. Why must the rise of current follow the exponential curve of Figure 15-4?

15-5. What effect does the magnitude of the resistance of an inductor have on:
(a) The initial rate of change of current?
(b) The final steady-state current?
(c) The time it takes the current to reach its steady-state value?

15-6. What effect does the magnitude of the inductance of an inductor have on:
(a) The initial rate of change of current?
(b) The final steady-state current?
(c) The time it takes the current to reach its steady-state value?

15-7. Justify the following description of a time constant: "A time constant is the time it takes the current in an inductive circuit to rise to 63% of its final steady-state value."

15-8. Use dimensional analysis to show that the unit for time constant τ is the *second*.

15-9. Why does the value of the voltage applied to an inductive circuit have no effect on the time it takes the instantaneous current to reach a steady-state value?

15-10. Draw a detailed graph of the actual instantaneous current in Example 15-3.

15-11. Why is there no further increase in the energy stored in an inductor when the current has reached its steady-state value? Where is the energy supplied by the source going while a steady current flows in the circuit?

15-12. What is the effect of attempting to interrupt current in an inductive dc circuit instantly?

15-13. How does a discharge resistor prevent insulation breakdown in an inductor?

15-14. Why can we say that the current through an inductor the instant after the switch is opened is the same as the instant before the switch is opened?

15-15. Why is a **thyrite** type of varistor particularly suitable for use as a discharge resistor?

15-16. Some designers use a semiconductor rectifier which has a very high resistance to current in one direction and a very low resistance to current in the other direction in place of a discharge resistor, as shown in Figure 15-22. What are the advantages of such a circuit?

15-17. What effect does the magnitude of the resistance of the discharge loop of an inductive circuit have on:
(a) The initial current when the switch is opened?

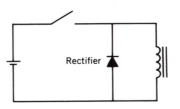

FIGURE 15-22
Magnetic Field Discharge by a
Rectifier.

(b) The initial rate of fall of current?

(c) The time it takes the current to fall to zero?

15-18. The ignition "tune-up" kit for an automobile engine contains a set of breaker points and a "condenser" to connect across the points. Why do these components need replacing from time to time to improve engine performance?

15-19. Derive Equation (15-12) by considering what happens in the circuit of Figure 15-8 when S_2 is opened but S_1 remains closed.

15-20. *LR* networks are sometimes used as waveshaping circuits in electronic pulse circuits. If a 1-kHz symmetrical square wave from a generator with negligible internal resistance is fed into the *LR* waveshaping network of Figure 15-23, draw an accurate graph of the output voltage waveform.

FIGURE 15-23
***LR* Waveshaping Circuit.**

*15-21. Design a computer program identified as "INSTANTANEOUS CURRENT IN AN LR CIRCUIT" which will ask the operator to input values for initial current, final steady-state current, inductance, net resistance, and time interval. The program should then display the value of the instantaneous current in amperes.

*15-22. Design a computer program which will display an accurate, fully labeled graph for part (c) of Example 15-8.

PART IV

ALTERNATING CURRENT

IN ALTERNATING-CURRENT CIRCUITS, the magnitude and direction of the source voltage is continuously changing. To investigate the behavior of various resistance networks in the chapters of Part 2, we restricted voltage sources to such constant-voltage devices as batteries (and solar cells). The advantage of starting with direct-current circuits is that the time factor does not appear in our steady-state calculations—except when we want to know the total amount of energy dissipated in a certain time interval ($W = Pt$).

In certain chapters of Part 3, we encountered the effect of time on the instantaneous voltages and currents as capacitors and inductors charged and discharged their stored energy. However, after a short transient interval following the closing or opening of switches, even CR and LR circuits settle down to a steady-state condition where time is not a factor in a dc system.

In the remaining chapters of Parts 4 and 5, we encounter voltage sources in which the instantaneous generated emf continuously changes and produces an open-circuit terminal voltage whose magnitude and direction depend on the exact instant in time. Before we can work with the many forms of alternating-current networks in Part 5, we must investigate the rules for dealing with alternating currents in the chapters of Part 4.

16

ALTERNATING CURRENT

16-1 A SIMPLE ROTATING GENERATOR

The only practical voltage source we have considered so far is the battery. Because of the nature of the chemical action within it, the battery is able to maintain a constant terminal voltage for a considerable period of time. Although the battery is a convenient voltage source for flashlights and portable radios, it does not offer a suitable means of developing the large amounts of electric energy which we require for both home and industry.

In Chapter 14, we discussed a second method of generating a voltage, based on Faraday's discovery of electromagnetic induction. As an electric conductor is moved in such a manner that it cuts across magnetic lines of force, a voltage is induced in it. In order to maintain a potential difference by this method, we must keep the conductor in motion within a stationary magnetic field. We can achieve this by the arrangement shown in Figure 16-1. The electric conductor is in the form of a loop which can be rotated continuously within the magnetic field in the air gap between a north and a south pole. Electrical connection to the rotating loop is maintained through a pair of **sliprings** and **brushes**.

As the loop rotates, the two sides of the loop must always cut across the magnetic lines of force in opposite directions. As a result, the two induced voltages are connected in series so that the brushes receive the *sum* of the two voltages (just as two flashlight cells may be connected in series to double the applied voltage). If we connect a load resistor to the brushes of this simple generator, the induced voltage will cause current to flow in the closed circuit consisting of the loop and the load. This load current flowing in the rotating loop produces a magnetic field which, according to Lenz's law, reacts with

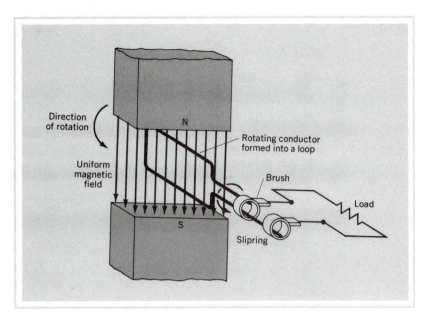

FIGURE 16-1
Simple AC Generator.

Labels in figure: Direction of rotation · N · Rotating conductor formed into a loop · Uniform magnetic field · Brush · Load · S · Slipring

the stationary magnetic field in such a manner as to *oppose* the rotation of the loop. Therefore, in order to maintain rotation, we must rotate the loop against this mechanical force opposing its rotation. Whenever motion is accomplished against an opposing force, mechanical energy is expended. Therefore, the electric energy that is converted into heat energy in the load resistance is generated directly at the expense of the mechanical energy maintaining the rotation of the loop.

16-2 THE NATURE OF THE INDUCED VOLTAGE

Let us consider the sequence of events as the conductor forming either side of the loop in Figure 16-1 rotates through a complete revolution (360°). First of all, it must cut across the magnetic lines of force in one direction and then move back across the same magnetic lines of force in the opposite direction. Since the induced voltage appearing at the brushes of this simple generator must reverse its polarity with each half-revolution, the basic rotating machine generator for converting mechanical energy into electric energy develops an **alternating** voltage.

Since most of our electric energy is generated by rotating machines, it is necessary for us to be able to cope with an *alternating* voltage in electric circuits just as readily as we have been able to determine the relationships among voltage, current, resistance, and power with the *direct* voltage from a battery, which we have been using up to this point. Figure 16-2(a) is a cross-sectional drawing of the simple generator of Figure 16-1. Only one side of the loop is shown in this sketch. The one conductor is shown in twelve positions during one rotation. In order to be able to relate the rotation of the conductor to numerical calculations, it is conventional to consider the normal or *positive* direction of rotation as being counterclockwise, starting from an angular position representing three o'clock.

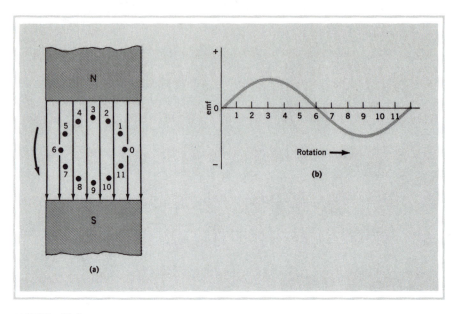

FIGURE 16-2
Nature of Induced Voltage.

At position 0, the motion of the conductor is parallel to the magnetic lines of force. Therefore, at that moment, no lines of force are being cut by the conductor. Hence, the induced voltage at the instant the conductor passes through point 0 is 0 volts. At position 1, the conductor has started to cut across the lines of force. Thus, appreciable voltage is induced into the conductor as it passes through point 1. At position 2, the conductor is cutting across the magnetic lines of force at an even steeper angle. Thus, the *rate* of cutting and, therefore, the voltage induced into the conductor as it passes through point 2 is even greater. At position 3, the conductor is cutting across the magnetic lines of-force at right angles, which results in the greatest possible rate of cutting. Therefore, maximum voltage is induced into the conductor as it passes through point 3.

We can compare the manner in which the rate of cutting increases as the conductor rotates through the first 90° to four men walking along a sidewalk at the same velocity. As they reach a certain point, the first man continues to walk parallel to the street and, therefore, never reaches the other side. So his *rate of crossing the street* is zero. The second man makes a 30° turn and crosses the street at an angle. The third man makes a 60° turn and crosses the street at a steeper angle, while the fourth man makes a 90° turn and goes straight across the street. With all four men turning at the same instant and walking at the same speed, the fourth man has the shortest distance to go and, therefore, gets to the other side first. His rate of crossing the street is maximum even though his absolute velocity is the same as that of the other three. Number three man is the next to reach the other side, and number two man is the third to reach the opposite sidewalk. Their rates of crossing are proportionately less than that of number four man. Since the magnetic lines of force are parallel like the opposite sidewalks on a street, the same situation exists if we consider the actual direction in which the conductor is moving at each position.

From positions 3 to 6 in Figure 16-2, the *rate of cutting* across magnetic lines of force becomes progressively less, so the induced voltage at each position of the conductor gradually decreases back to zero volts as it passes through point 6, where it is again momentarily moving parallel to the lines of force. From positions 6 to 12, the same sequence of events occurs, with the induced voltage rising to a maximum at position 9 and then dropping back to zero at position 12, except that the conductor is now cutting across the magnetic lines of force in the opposite direction. As a result, the polarity of the induced voltage is reversed during the second half of the revolution. The + sign on the graph of Figure 16-2(b) indicates the reference polarity of the terminal voltage. The − sign simply indicates that the polarity of the voltage is reversed with respect to the reference polarity. One complete rotation of the loop constitutes one **cycle.** The time taken to complete one cycle is called the **period** of the waveform.

16-3 THE SINE WAVE

In the preceding section, we observed the manner in which the alternating voltage induced into a conductor rotating in a uniform magnetic field varies continually as the angular position of the conductor changes. We can show the nature of this continual change best by plotting a graph showing the magnitude and polarity of the induced voltage at any position of the loop, as shown in Figure 16-2(b). To appreciate fully the manner in which electric circuits behave when the voltage applied to the system is of the alternating type, we must be able to determine exactly what the induced voltage is at any instant during the rotation of the conductor.

For such an analysis, we replace the cross-sectional sketch of Figure 16-2(a) with a much simpler diagram, as shown in Figure 16-3. The straight line *OX* represents the rotating conductor pivoted at point *O* (for origin). The end that is free to rotate is indicated by an arrowhead. We call this symbolic representation of the rotating conductor a **phasor.** As we have already noted, mathematical convention requires this phasor to rotate in a counterclockwise direction, starting from a position representing three o'clock, referred to as the *reference axis* in Figure 16-3. Instead of showing its position at twelve particular points, as in Figure 16-2(a), the position of the phasor at a certain instant in time is located by stating the angle ϕ through which it has rotated from the reference axis. Also, to keep the sketch as simple as possible, in Figure 16-3, we have

FIGURE 16-3
Phasor Representation of the Rotating Conductor.

.not shown the magnetic lines of force, which are assumed to be uniformly spaced and in a vertical plane, as they were in Figure 16-2(a).

Faraday's law tells us that the voltage induced in a rotating conductor is directly proportional to the *rate* at which it cuts across the magnetic lines of force. If we assume a uniform flux density and a constant angular velocity, the rate of cutting will depend only on the angular position of the loop. At 0° (with respect to the reference axis), the loop moves parallel to the magnetic lines of force; hence, no voltage is induced into the loop at that instant. As the loop swings around toward 90°, the rate of cutting increases, becoming maximum at 90°. To determine the exact value of the induced alternating voltage at any instant, we must be able to determine the exact rate of cutting of magnetic lines of force by the conductor at that instant with respect to the maximum rate of cutting of lines of force.

In Figure 16-4, OX represents the rotating conductor after it has traveled through an angular distance ϕ from the reference axis. Since OX is pivoted at O, the line XY drawn perpendicular to OX represents the direction of motion of the tip of the phasor (conductor) at that particular moment. If the angular velocity of the conductor is constant, the linear velocity of the tip of the phasor must also be constant. We can represent this by drawing $XY = OX = $ a constant for any value of ϕ. We can resolve the motion represented by XY into a vertical component of motion XZ, and a horizontal component XW. Since XZ is parallel to the magnetic lines of force, this component is ineffective in inducing any voltage. Since XW cuts across the magnetic lines of force at right angles, the induced voltage will be directly proportional to the length of XW. When ϕ becomes 90°, $XW = XY$. Therefore, the ratio of XW/XY in Figure 16-4 represents the ratio between the rate of cutting of lines of force at angle ϕ and the maximum rate of cutting.

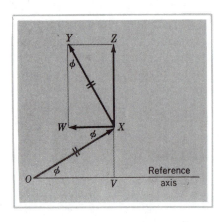

FIGURE 16-4
Geometric Representation of the Rate of Cutting of Magnetic Lines of Force.

From geometry:

Since $WX \parallel OV,$ $\therefore$ $\angle WXO = \angle XOV = \phi$

Since $\angle YXO = 90°,$ $\therefore$ $\angle WXY = 90° - \phi = \angle OXV$

Since $\angle XWY = 90°,$ $\therefore$ $\angle XYW = \phi$

Since $XY = OX,$ $\therefore$ $\triangle XOV \equiv \triangle XYW$

$$\therefore \quad \frac{XW}{XY} = \frac{XV}{OX}$$

As the loop rotates, the rate of cutting of magnetic lines of force at any instant is directly proportional to XV/OX. If OX remains constant, this ratio depends only on the magnitude of the angle ϕ. Since XV is the side of a right-angled triangle opposite the angle ϕ and OX is the hypotenuse, in trigonometry the ratio XV/OX is called the **sine** of the angle ϕ. We can determine accurately the ratio of XV/OX for any magnitude of the angle ϕ by using the **sin** key of an electronic calculator. Therefore,

*The voltage induced into the rotating conductor at any point is directly proportional to the **sine** of the angle through which the loop has rotated from the reference axis.*

We call this angle the **phase angle**.

In Figure 16-4, we determined by geometrical construction the relative magnitude of an induced voltage for an angle ϕ which is less than 90°. In its final form, the procedure amounted to drawing a perpendicular from the tip of the phasor (point X) to the horizontal reference axis (point V). As long as the length of the phasor OX remains constant, the magnitude of the induced voltage is directly proportional to the length of XV. Exactly the same geometrical construction applies to angular distances of more than 90° from the reference axis. In Figure 16-5(a), the phasor OX has rotated through an angle of 120°. When we draw a perpendicular from point X to the reference axis, we form a right-angled triangle in which the side XV is opposite an angle that is equal to $180° - \phi$ or 60°.

$$\therefore \sin 120° = \sin(180° - 120°) = \sin 60°$$

Similarly, in Figure 16-5(b), when $\phi = 210°$, the right-angled triangle has an angle opposite XV of $210° - 180° = 30°$. But, in this case, XV is *below* the reference axis.

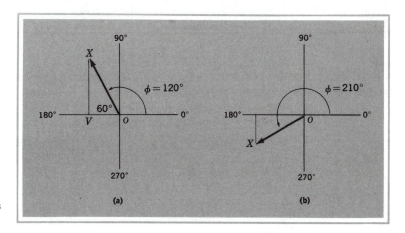

FIGURE 16-5
Determining the Sine of Angles Greater than 90°.

When we relate this diagram to the rotating loop of Figure 16-2, *XV* now represents a *negative* value of induced voltage, since the loop is moving in the opposite direction across the magnetic lines of force. Therefore, whenever *XV* is below the reference axis, it represents a negative quantity. So we must say, then, that sin 210° = −sin 30°.

The sine function is stored in the built-in memory of scientific calculators so that we can obtain the sine of *any* angle simply by pressing the **sin** key. BASIC language for personal computers also contains a SIN(X) function in its permanent memory. Most scientific calculators permit us to enter an angle in either degrees or radians. Since BASIC requires angles to be in radians, we must build a conversion step into any program providing for the input of angular data in degrees.

Since the voltage induced into a conductor rotating at a constant angular velocity in a uniform magnetic field is constantly varying, it is difficult for us to visualize its nature by a simple statement in words. Hence it is customary to plot a linear graph of the variation in the induced voltage over a complete cycle by expressing the angular distance ϕ as a linear distance along the horizontal axis of the graph and the induced voltage for each value of ϕ along the vertical axis of the graph, as in Figure 16-6. Since the induced voltage at any point during the rotation of the loop is directly proportional to the *sine* of the angle through which it has rotated from the reference axis, we can plot an accurate graph of the instantaneous variations in an alternating voltage by preparing a table of sine ϕ (at approximately 5° intervals) using a calculator **sin** key. We then plot these values on a sheet of graph paper and join the plots to form a smooth curve. A computer can calculate such a sine table automatically and then display the data in graph form on its output device. The resulting graph is called a **sine curve**. The alternating voltage that varies in accordance with this sine curve is called a **sine wave**.

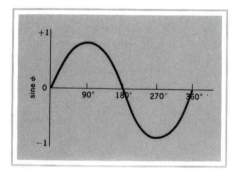

FIGURE 16-6
The Sine Curve.

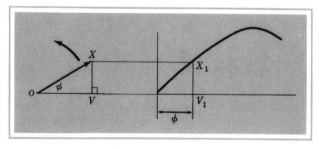

FIGURE 16-7
Preparing a Sine Curve by Geometric Projection.

Figure 16-7 shows another method of producing a sine curve. This method is a projection of the geometric construction we used in Figures 16-4 and 16-5. Since the length of phasor *OX* is constant, the altitude *VX* of the tip of the phasor is proportional to the sine of the angle ϕ, and, therefore, *VX* is directly proportional to the induced voltage. When we project *VX* to a position V_1X_1 on the linear graph at a distance along the horizontal axis representing the angular distance, the point X_1 will be a proper plot, as required to draw a sine curve. If we repeat this geometric projection at 5° intervals, we shall have sufficient plots to draw a complete sine curve.

16-4 THE PEAK VALUE OF A SINE WAVE

As we discovered in determining the nature of an alternating voltage, the greatest value that the induced voltage can attain occurs when $\phi = 90°$. At this point, the rate of cutting of magnetic lines of force is maximum. When $\phi = 270°$, the same maximum value is attained but the polarity of the induced voltage is reversed. The numerical value that the alternating voltage attains at these two angles is termed the **peak** value of the ac waveform.

The letter symbol for the peak value of an alternating source voltage is E_m ($_m$ for maximum).

This peak value is a numerical quantity in volts and does not have any polarity, since the same peak value in volts occurs at both 90° and 270°. The peak value is also independent of time, since, no matter where the rotating conductor may be at a certain instant, the peak value will appear as it passes the 90° and 270° points in its rotation. Therefore, the peak value is a constant for a given generator and does not have the sine-wave shape of the voltage induced from instant to instant.[†]

In the simple ac generator, the peak voltage will depend on just what the maximum rate of cutting of magnetic lines of force by the rotating loop happens to be. This rate, in turn, depends on (1) the total number of lines of force in the magnetic field, (2) the angular velocity of the loop, and (3) the number of turns of wire in the loop. From Faraday's law, when a conductor cuts across magnetic lines of force at the *rate* of one weber per second, an emf of one volt is induced in that conductor.

16-5 THE INSTANTANEOUS VALUE OF A SINE WAVE

The graphs of Figures 16-2, 16-6, and 16-7 show the manner in which the induced voltage changes from instant to instant as the loop rotates. Any value that is continually changing and thus is dependent on the exact instant in time for its numerical value is called an **instantaneous value.** Therefore, in representing the instantaneous value of the voltage induced in the rotating loop of the simple ac generator of Figure 16-1, we must use the lower case letter symbol *e*.

In the simple ac generator of Figure 16-1, during the first 180° of rotation of the loop, the voltage rose to a peak value in one direction and dropped back to zero; then, during the second 180° of rotation, it rose to a similar peak value but in the opposite direction and again dropped back to zero. Therefore, it took one complete revolution for the voltage to vary through the whole possible range of values. As the loop rotates past 360°, the same series of events recurs. We define this complete excursion of the instantaneous value of the induced voltage as one **cycle** of a **sine wave.** The time it takes for the instantaneous value to complete one cycle is called the **period** of the sine wave. The number of cycles completed in one second is called the **frequency** of the sine wave.

[†] In mathematical terminology, we refer to E_m as a **scalar** quantity, that is, a quantity which has a numerical magnitude but which does *not* have any direction associated with it.

The letter symbol for frequency is f (lower case).

Until 1965, frequency was expressed in cycles per second. But people became lazy and stated frequency in "cycles." Consequently, the SI unit is now preferred in technical writing.

The hertz is the SI unit of frequency. [†]
The unit symbol for hertz is **Hz**

Hertz and **cycles per second** are numerically synonymous.

Although the instantaneous voltage is never constant, its exact magnitude at any instant is directly proportional to sin ϕ. Since the instantaneous voltage reaches its peak value at 90° and since sin 90° = 1, we can determine the instantaneous voltage for any particular instant in time as the loop rotates from

$$e = E_{\mathrm{m}} \sin \phi \qquad (16\text{-}1)$$

In the simple **two-pole** generator of Figure 16-1, one complete mechanical revolution of the loop was required to generate one complete cycle of an electrical sine wave of voltage. But, in practical generators like the four-pole generator of Figure 16-8, one complete cycle of voltage will be generated as the loop moves through the magnetic flux under an N pole and then through the flux under an S pole. To use Equation (16-1), we must subdivide this cycle into 360 *electrical* degrees even though the loop has only rotated through 180 *mechanical* degrees. The relationship between *electrical* degrees and *mechanical* degrees will depend on the number of *pairs* of magnetic poles in the piece of rotating machinery. Sine waves of voltage can also be generated in **oscillator** circuits where there is no mechanical rotation. Again, to solve sin ϕ, a complete cycle must be subdivided into 360 *electrical* degrees.

[†] Named in honor of the German physicist Heinrich Hertz, who discovered "hertzian" or radio waves.

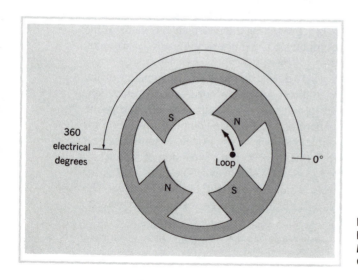

FIGURE 16-8
Relationship between Electrical and Mechanical Degrees in a Four-Pole Generator.

EXAMPLE 16-1
At what speed must the shaft of a six-pole 60-Hz alternator be turned?

SOLUTION
With three pairs of poles, one mechanical revolution generates three electrical cycles. Therefore, the shaft speed is

$$\frac{60}{3} = 20 \text{ r/s} = \mathbf{1200 \text{ r/min}}$$

Our next problem is how to determine the angle ϕ in electrical degrees. We are usually given the frequency of the sine wave. Since frequency can be expressed in cycles *per second*, if we multiply frequency by the time elapsed since the instantaneous value passed through zero, we have the angular position expressed in cycles and fractions of complete cycles. Multiplying this product by 360 then gives us an angle representing the instantaneous position of the induced voltage in its cycle in electrical degrees. Therefore, we can express Equation (16-1) in a more useful form as

$$e = E_{\mathrm{m}} \sin(360 ft)° \tag{16-2}$$

where e is the instantaneous source voltage at a given instant in time, E_{m} is the peak value of the sine wave in question, sin is the sine of the angle expressed in electrical degrees, f is the frequency of the sine wave in hertz, and t is the elapsed time in seconds.

EXAMPLE 16-2
What is the instantaneous value of a 60-Hz sine wave whose peak voltage is 150 V 10 ms after the instantaneous voltage passes through 0 V?

SOLUTION
$$\begin{aligned}
e &= E_{\mathrm{m}} \sin(360 ft)° \\
&= 150 \sin(360 \times 60 \times 0.01)° \\
&= 150 \sin 216°
\end{aligned}$$

To solve this equation with a calculator, the chain becomes

$$150 \boxed{\times} 216 \boxed{\mathbf{sin}} \boxed{=} \mathbf{-88.17 \ V}$$

SELF-CHECK

16-5:1. Referring to the instantaneous source voltage used in Example 16-2:

$$e = 150 \cdot \sin \phi \ \text{V}$$

at what angles (expressed in electrical degrees) does the instantaneous voltage equal:
(a) 0 V, (b) +26 V, and (c) −75 V?

16-5:2. At what times, in milliseconds, does the instantaneous source voltage used in Example 16-2 reach:

(a) 0 V, (b) +26 V, and (c) −75 V?

16-6 THE RADIAN

In dealing with the rotating conductor of Figure 16-1, we have been talking about angular *distance* and angular *velocity*. Although the Babylonian 360° system of angular measurement is very convenient for purposes of geometry and trigonometry, it is not suited to showing the relationship between the linear velocity of the conductor in Figure 16-1 as it moves around the circumference of a circle and its angular velocity. The linear velocity of the conductor governs the actual rate of cutting across magnetic lines of force. Therefore, for electrical purposes, it would be very useful to have a unit of angular distance which is related to the distance traveled by the free end of a rotating phasor. The **radian** is just such a unit.[†] By definition,

One radian is the angular distance through which the phasor travels when its free end (X in Figure 16-9) travels through a linear distance equal to the length (OX) of the phasor.

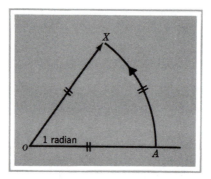

FIGURE 16-9
The Radian.

It is standard practice in electrical engineering to express the angular distance traveled by a rotating phasor in radians, and its angular velocity in radians per second. Since the rotating phasor in Figure 16-9 is the radius of a circle, and the circumference of a circle = $2\pi r$, there are 2π radians in one complete cycle. Hence, in marking the x-axis of a time graph (such as Figure 16-6) in radians,

$$90° = \pi/2 \text{ rad} \qquad 180° = \pi \text{ rad}$$
$$270° = 3\pi/2 \text{ rad} \qquad 360° = 2\pi \text{ rad}$$

We must also convert Equation (16-2) from angular distance in degrees to angular distance in radians. Most engineering calculators will allow us to solve trigonometric functions in radians without having to convert back to degrees. If we do have to convert

[†] The **radian** is the SI unit for a plane angle.

radians into degrees for some reason, the conversion factor becomes

$$2\pi \text{ rad} = 360°$$

$$\therefore \quad 1 \text{ rad} \doteq 57.3° \qquad (16\text{-}3)$$

As in Equation (16-2), we usually express angular distance as the product of angular velocity and elapsed time. In radian measure, angular velocity is expressed in **radians per second**.

The letter symbol for angular velocity in radians per second is the Greek letter ω (omega).

Expressed in radian measure, Equation (16-2) becomes

$$e = E_m \sin \omega t \qquad (16\text{-}4)$$

Since there are 2π radians in a cycle,

$$\omega = 2\pi f \qquad (16\text{-}5)$$

Therefore, we write Equation (16-4) as

$$e = E_m \sin 2\pi f t \qquad (16\text{-}6)$$

*In both Equation (16-4) and Equation (16-6), the angle is numerically expressed in radians. As we noted in Section 16-3, we call this angle the **phase angle** of the sine wave.*

EXAMPLE 16-3
Write the general equation for the instantaneous voltage of a 60-Hz generator whose peak voltage is 170 V.

SOLUTION

$$e = E_m \sin 2\pi f t$$

But
$$2\pi f = 2 \times 3.14 \times 60 = 377 \text{ rad/s}$$

$$\therefore e = \mathbf{170 \sin 377} t$$

Since 60 Hz is the standard power line frequency, we should memorize:

$$\mathbf{60 \text{ Hz} \doteq 377 \text{ rad/s}} \qquad (16\text{-}7)$$

EXAMPLE 16-2A
What is the instantaneous value of a 60-Hz sine wave whose peak voltage is 150 V 10 ms after the instantaneous voltage passes through 0 V?

$$e = E_m \sin 2\pi f t$$
$$= 150 \sin (2\pi \times 60 \times 0.01) \text{ rad}$$
$$= 150 \sin 1.2\pi \text{ rad}$$

We switch the calculator to the radian mode (usually by pressing a **d-r** key). The chain then becomes

$$150 \; \boxed{\times} \; \boxed{(} \; 1.2 \; \boxed{\times} \; \boxed{\pi} \; \boxed{)} \; \boxed{\textbf{sin}} \; \boxed{=} \; \textbf{--88.17 } V$$

SELF-CHECK

16-6:1. Using the equation of instantaneous voltage from Example 16-3:

$$e = 170 \cdot \sin 377t \text{ V}$$

find the angles (expressed in radians) when the instantaneous voltage is equal to $+144$ V.

16-6:2. Suppose the instantaneous voltage for a generator is given by:

$$e = 100 \cdot \sin 6283t \text{ V}$$

determine the following:
(a) the angular velocity of the instantaneous voltage.
(b) the frequency of the instantaneous voltage.
(c) the times when the voltage equals 100 V.

16-7 INSTANTANEOUS CURRENT IN A RESISTOR

Now that we are acquainted with the sinusoidal nature of the instantaneous voltage generated by a basic ac voltage source, we can determine the nature of the current that will flow when we connect resistance to the generator terminals in the simple ac circuit of Figure 16-10. At a particular instant in time, the instantaneous voltage developed by the generator has a certain magnitude and a certain polarity. If we consider only this one particular instant, there is no difference between the simple ac circuit of Figure 16-10 and

FIGURE 16-10
Simple Alternating-Current Circuit.

the equivalent dc circuit. The real difference between them is that, at the *next* instant in time, the magnitude of the instantaneous alternating voltage changes, but the applied voltage in the dc circuit continues on at a fixed magnitude. If we consider any particular *instant* in time, all the equations summarized at the end of Chapter 5 apply to the ac circuit of Figure 16-10.

Therefore, we can state that the instantaneous current through the resistance in Figure 16-10 will be

$$i = \frac{e}{R} \tag{16-8}$$

where i is the instantaneous current through the resistance, e is the instantaneous voltage applied to the resistance, and R is the resistance of the circuit.

Since resistance is determined by such physical factors as type of material, length, and cross section, it is a constant for a given circuit at a given temperature [indicated by an upper case R in Equation (16-8)]. Because R is a constant in Equation (16-8), whatever the variation in instantaneous voltage, the instantaneous current must stay right in step in order to make Ohm's law hold true. If the instantaneous voltage is a sine wave, the instantaneous current must also be a sine wave, as shown in Figure 16-11. The instantaneous current must reach its peak value at the same instant that the instantaneous voltage becomes maximum. It will become zero at the same instant that the instantaneous voltage becomes zero, and it will reverse its direction at the same instant that the instantaneous voltage across the resistance reverses its polarity. We can, therefore, substitute the general equation (16-4) describing a sine wave alternating applied voltage in Equation (16-8) giving

$$i = \frac{E_m}{R} \sin \omega t \tag{16-9}$$

Since Equations (16-4) and (16-9) both have the same phase angle (ωt), we say that the current in a resistance in an alternating-current circuit is **in phase** with the voltage across the resistance. And since the peak value of the instantaneous current depends on the peak

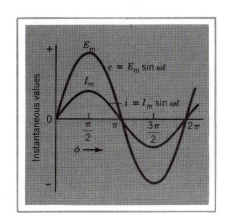

FIGURE 16-11
Instantaneous Current through a Resistor.

.value of the applied voltage,

$$I_{\mathrm{m}} = \frac{E_{\mathrm{m}}}{R} \tag{16-10}$$

$$\therefore i = I_{\mathrm{m}} \sin \omega t \tag{16-11}$$

where i is the instantaneous current through a resistor when a sine wave of voltage is applied, and I_{m} is the peak current as determined by the relationship $I_{\mathrm{m}} = E_{\mathrm{m}}/R$.

SELF-CHECK

16-7:1. For the circuit shown in Figure 16-10, suppose that the generator has the following instantaneous voltage waveform:

$$e = 170 \cdot \sin 377t \ \mathrm{V}$$

and that the resistor has a resistance of 850 Ω.
(a) Determine the peak current in the circuit.
(b) Write the equation for the instantaneous current in the circuit.
16-7:2. For the circuit described in Self-Check Problem 16-7:1, plot the curves of:
(a) instantaneous voltage versus phase angle (e versus ϕ) and
(b) instantaneous current versus phase angle (i versus ϕ) on one graph for the range $0 \leq \phi \leq 4\pi$.

16-8 INSTANTANEOUS POWER IN A RESISTOR

If we still think in terms of instantaneous values, we may apply Equations (5-2), (5-3), and (5-4) from Chapter 5 to an ac circuit. Therefore,

$$p = vi = i^2 R = \frac{v^2}{R} \tag{16-12}$$

where p is the instantaneous power in a resistor in watts, v is the instantaneous voltage drop across the resistance in volts, i is the instantaneous current through the resistance in amperes, and R is the resistance of the circuit in ohms.

As R is a constant for a given circuit, and since the instantaneous voltage and current in the basic circuit are *both* sine waves, the instantaneous power must be a sine-*squared* wave. To determine the nature of a sine-squared wave, we can plot a graph of instantaneous power in the same manner that we plotted the sine curve of Figure 16-6. If we calculate the *squares* of the sine ϕ plots on a linear graph, the instantaneous power in a resistor takes on the shape shown in Figure 16-12. Note that the instantaneous power in a resistor is always positive since squaring a negative quantity results in a positive quantity. We can interpret *positive* power as electric energy being converted into some other form of energy (heat in the case of a resistor). On this basis, *negative* power would represent some other form of energy being converted into electric energy. A resistor is not capable of doing this; therefore, the waveform of instantaneous power in a resistor has no negative component.

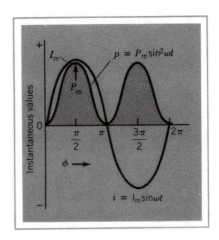

FIGURE 16-12
Instantaneous Power in a Resistor.

As Figure 16-12 indicates, the instantaneous power in the basic ac system is pulsating in nature, swinging from zero to maximum and back *twice* each cycle. If we consider an electric lamp as the resistance in Figure 16-10, as far as heating the filament is concerned, it does not matter which way the current flows through it. There will be one pulse of energy converted into heat and light on the one *half-cycle* of current and another similar pulse on the second half-cycle. If the lamp is operated from a 60-hertz source, it will increase in brilliance 120 times a second. Therefore, the instantaneous power pulsations are at *twice* the frequency of the voltage and current. This pulsating characteristic of instantaneous power is responsible for the flicker in 25-hertz lighting as used in some localities and for the vibration of small ac motors.

We can also show the pulsating nature of the power in an ac system by deriving an equation for instantaneous power.

By substituting $i = I_m \sin \omega t$ in Equation (16-12) we obtain

$$p = (I_m \sin \omega t)^2 \times R = I_m^2 R \times \sin^2 \omega t$$

And since peak values are maximum instantaneous values, from Equation (16-12)

$$P_m = I_m^2 R \tag{16-13}$$

$$\therefore p = P_m \sin^2 \omega t \tag{16-14}$$

where p is the instantaneous power in a resistance when a sine wave of current flows through it, and P_m is the peak power as determined by the relationship $P_m = I_m^2 R$.

From trigonometry, we can show that

$$\sin^2 \omega t = 0.5 (1 - \cos 2\omega t)^\dagger$$

$$\therefore p = 0.5 P_m - 0.5 P_m \cos 2\omega t \tag{16-15}$$

---

†

$$\cos (A - B) = \cos A \cos B + \sin A \sin B$$

$$\cos (A + B) = \cos A \cos B - \sin A \sin B$$

If $A = B$, then $\cos 0° = \cos^2 A + \sin^2 A$ and $\cos 2A = \cos^2 A - \sin^2 A$. Since $\cos 0° = 1$, subtracting gives $1 - \cos 2A = 2 \sin^2 A$, from which $\sin^2 A = 0.5(1 - \cos 2A)$.

This relationship tells us three things about the instantaneous power waveform which we can check by referring to Figure 16-12.

1. In any right-angled triangle, the cosine of one acute angle is also the sine of the other acute angle. Therefore, the general shape of the cosine curve and the sine curve are the same. Consequently, the fluctuations in instantaneous power are sinusoidal in nature.
2. Since 2ω is twice the angular velocity of ω, the frequency of the sinusoidal variation is twice as great as the frequency of the instantaneous voltage and current.
3. Since the limits of the value of a cosine are $+1$ when $\phi = 0°$ and -1 when $\phi = 180°$, the value of the expression $0.5(1 - \cos 2\omega t)$ can vary between zero and $+1$. Therefore, the instantaneous power waveform for a resistance is always positive.

SELF-CHECK

16-8:1. For the circuit shown in Figure 16-10, suppose that the generator has the following instantaneous voltage waveform:

$$e = 170 \cdot \sin 377t \ \text{V}$$

and that the resistor has a resistance of 850 Ω.
(a) Determine the peak resistor power.
(b) Write the equation for the instantaneous resistor power.
16-8:2. For the circuit described in Self-Check Problem 16-8:1, plot the curves of:
(a) instantaneous power versus phase angle (p versus ϕ) and
(b) instantaneous current versus phase angle (i versus ϕ) on one graph for the range $0 \leq \phi \leq 4\pi$.

16-9 PERIODIC WAVES

So far, we have considered only one form of alternating current—the sine wave. Although the sine wave is by far the most important ac waveform and the only one we shall consider in detail in this book, we should note that the sine wave is one of a larger group of **periodic waves**. In electric circuits, a periodic wave refers to any time-varying function such as voltage, current, or power which continually repeats exactly the same sequence of values after a given time interval known as the **period** of the waveform. Figure 16-13 shows the graphs of the manner in which the instantaneous values of six typical periodic waves vary as a function of time. The graphs of the sine wave and sine-squared wave are repetitions of the two waves shown in Figure 16-12. In this case, however, we have marked the horizontal axis of the graph in elapsed time (milliseconds), rather than in terms of the angular position of the rotating phasor. As shown in Figure 16-13(a), the **period** of the sine wave is the length of time it takes the instantaneous voltage or current to complete one **cycle** of events. As we noted in Section 16-8, the instantaneous power waveform in a resistor has a sinusoidal component which is twice

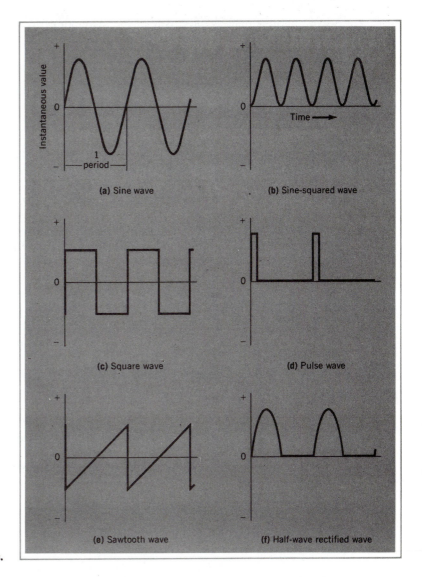

FIGURE 16-13
Typical Periodic Waves.

the frequency of the current and voltage waveforms. Hence, Figure 16-13(b) shows *four* complete cycles of a periodic wave. All other graphs in Figure 16-13 show *two* complete cycles.

The instantaneous values of a periodic wave change much too rapidly to be registered by the conventional electric meter movements of Chapter 10. However, the electron beam of the cathode-ray tube in an oscilloscope can quite readily respond to high frequency instantaneous voltages. If we apply the instantaneous voltage of a certain periodic wave [such as the sine wave of Figure 16-13(a)] to the vertical deflection system of an oscilloscope, the vertical motion of the electron beam will be exactly in step with these instantaneous values. This motion is still much too rapid for the eye to follow and all we would see is a vertical line proportional to the **peak to peak** value of the periodic wave. An important characteristic of the sawtooth periodic wave of Figure 16-13(e) is that the instantaneous voltage is a *linear* function of elapsed time. Hence, if we simultaneously apply to the horizontal deflection system of the oscilloscope a sawtooth wave having a period exactly twice that of the waveform fed to the vertical deflection system, the electron beam will trace out a duplicate of the graph of Figure 16-13(a).

16-10 THE AVERAGE VALUE OF A PERIODIC WAVE

By definition, an alternating current (or voltage) is one in which the *average* of the instantaneous values over a period of time is zero. Just by examining the waveforms of Figure 16-13, it is fairly obvious that the sine wave, square wave, and sawtooth wave are alternating currents (or voltages), since the instantaneous values are symmetrical about the horizontal axis of the graph. The sine-squared wave, pulse wave, and half-wave rectified wave do not qualify as alternating currents, since their instantaneous values are always of the same polarity.

We are quite familiar with the technique for *averaging* examination marks or determining batting averages in baseball. The same technique applies to determining the average value of any periodic wave. For complex periodic waves we can determine the average of the instantaneous values graphically by examining the *area* under a graph, as we did in Figures 12-19 and 15-12. But for the sine wave, we can obtain a more accurate average value by writing down the sine of 5°, 10°, 15°, and so on, for a complete cycle. To find the average value of any column of numbers, we add the column of figures and divide by the number of figures in the column. After going to all this trouble, the average value of a sine wave for a complete period is *zero*, since, for every positive value of instantaneous voltage or current during the one half-cycle, there is a similar negative value during the next half-cycle.

However, if we average *one-half cycle only*, we obtain a significant numerical constant. From the column of figures prepared as above from a sine table or calculator, we find that the average of sin ωt over one half-cycle is $22.9 \div 36 = 0.636$. If we average the half-cycle of a sine wave by integration (area under the curve method),[†] we arrive at a slightly more accurate result, expressing average values as $2/\pi$ or 0.637.

$$\text{Since } i = I_\text{m} \sin \omega t, \qquad \therefore I_\text{av} = I_\text{m} \times 0.637$$

Therefore, the *half*-cycle average value of a sine wave is

$$I_\text{av} = 0.637 I_\text{m} = \frac{2}{\pi} I_\text{m} \tag{16-16}$$

or

$$E_\text{av} = 0.637 E_\text{m} = \frac{2}{\pi} E_\text{m} \tag{16-17}$$

As we discovered in Section 16-8, the sine-squared wave of Figure 16-13(b) will have a *full*-cycle average value of one-half the peak value of the sine-squared wave. The

[†] See Appendix 2-6 for calculus solution.

pulse waveform will have an average value that depends on the ratio of the pulse duration to the time interval between pulses. The half-wave rectified sine wave of Figure 16-13(f) will have a full-cycle average value which, in turn, is the average of $0.637E_m$ for one half-cycle and zero for the next half-cycle, or $0.3183E_m$.

SELF-CHECK

16-10:1. For the circuit shown in Figure 16-10, suppose that the generator has the following instantaneous voltage waveform:

$$e = 170 \cdot \sin 377t \text{ V}$$

and that the resistor has a resistance of 850 Ω. Find:
(a) the *half*-cycle average voltage across the resistor
(b) the *half*-cycle average current in the resistor
(c) the *full*-cycle average voltage across the resistor

16-11 THE RMS VALUE OF A SINE WAVE

Instantaneous values of alternating current and voltage are not satisfactory for most measurement purposes since the pointers of meters cannot follow the rapid changes in polarity. Nor are they suitable for calculation purposes because by the time we complete the computation, the instant to which the calculation applied has long since passed. Full-cycle average values of sine waves are zero, and half-cycle average values are only useful for such applications as rectification where we are concerned primarily with a half-cycle of a sine wave.

These problems do not arise in dc circuits since the voltage, current, and power in a resistive circuit remain constant for a considerable period of time. As far as the end result of converting electric energy into light is concerned, it makes no difference whether the lamp is lighted by a direct current or an alternating current flowing through it. Perhaps, then, we can establish *equivalent steady-state* values for alternating current and voltage which will allow us to use the same interrelationships among voltage, current, resistance, power, work, etc. that we use in dc circuits. We can determine this **equivalent dc** or **effective** value of an alternating current experimentally by finding out what direct current must flow through a given resistance to produce in a certain time interval the same amount of heat energy as when the resistor is connected to a source of alternating voltage.

To determine this equivalent dc value of an alternating current, we can conduct the same experiment algebraically with pencil and paper. The basis for comparison in the experiment is *the same amount of heat energy in a certain time interval*. We can translate this into *average work per unit time or* simply **average power.** To determine average power in a resistive ac circuit, we start with Equation (16-12),

$$p = i^2R \tag{16-12}$$

Using the averaging technique we discussed in Section 16-10, we compute the *square* of the instantaneous current at small time intervals over a full cycle and then calculate the average or *mean* value of all these i^2 values. Hence,

$$P_{av} = \text{(full-cycle average of } i^2) \times R$$

In a dc circuit, $P = I^2R$. Since the effective values of an alternating current are to represent dc equivalent values, we can use the same letter symbols for both, that is, uppercase italic letters without subscripts. Therefore, in an ac circuit, $P = I^2R$ where P is average power and I is the dc equivalent value of the alternating current. Solving for I by combining the equations above,

$$I = \sqrt{\frac{P}{R}} = \sqrt{\text{full-cycle average of } i^2}$$

Note that R cancels out. We can define the dc equivalent value of an alternating current in terms of the *process* we go through to determine its value. We find the square *root* of the *mean* value of the *squares* of the instantaneous current over a full cycle. **Root-mean-square** (usually abbreviated to **rms**), **effective** and **equivalent dc** are synonymous terms. At present, **rms** is the preferred term, which we shall use for the remainder of the book.

For a sine wave of alternating current, we note that the instantaneous power in the graph of Figure 16-12 swings alternately and uniformly between zero and peak power. It appears quite reasonable to state, therefore, that *the average power in a resistor through which a sine-wave alternating current is flowing is one-half the peak power.*

We can also determine this relationship from the general Equation (16-14) for instantaneous power: $p = P_m \sin^2 \omega t$. As we have already shown, $\sin^2 \omega t = 0.5(1 - \cos 2\omega t)$. If we average a sine or cosine function over a *complete* cycle, the average must be zero, since for every positive value during the first half-cycle, there is an equivalent negative value during the second half-cycle. Hence, when averaged over a complete cycle, $\cos 2\omega t$ becomes zero and Equation (16-14) becomes

$$P = 0.5 P_m \tag{16-18}$$

For a sine-wave alternating current in an ac circuit,

$$P = I^2R \quad \text{and} \quad P_m = I_m^2 R$$

But
$$P = 0.5 P_m = 0.5 I_m^2 R$$

$$\therefore I^2R = 0.5 I_m^2 R, \text{ from which } I^2 = 0.5 I_m^2$$

and
$$I = \frac{I_m}{\sqrt{2}} = 0.707 I_m \tag{16-19}$$

The rms (effective) value of a sine wave of current is $1/\sqrt{2}$ or 0.707 of the peak value.

The rms value of a sine-wave voltage should be such that the *average* power is the product of the *rms* voltage across and *rms* current through the resistance of the circuit, just as in the equivalent dc circuit. Therefore,

$$P = VI$$

But when a sine-wave current flows through a resistor,

$$P = 0.5P_m \qquad (16\text{-}18)$$

Since peak instantaneous power occurs at the instant when both the instantaneous voltage across and the instantaneous current through a resistance circuit reach their peak values,

$$P_m = V_m I_m$$

As we have already determined,

$$I = \frac{I_m}{\sqrt{2}} \qquad (16\text{-}19)$$

Substituting in Equation (16-18)

$$V \times \frac{I_m}{\sqrt{2}} = 0.5V_m I_m$$

from which

$$V = \frac{V_m}{\sqrt{2}} = 0.707V_m \qquad (16\text{-}20)$$

Note also that
$$V_m = \sqrt{2}\,V = 1.414V$$

Since rms values are used for the majority of ac circuit measurements and computations, we shall assume that **all currents and voltages are stated as rms values unless it is specifically stated otherwise.**

EXAMPLE 16-4

What is the peak voltage of the customary 120-V 60-Hz electric service?

SOLUTION

$$E_m = 1.414E = 1.414 \times 120 \text{ V} = \mathbf{170\ V}$$

A term we may encounter occasionally in working with rms and average values of an ac waveform is its **form factor.** Form factor is the ratio of the rms value to the half-cycle average value of an ac wave. For a sine wave,

$$\text{form factor} = \frac{I}{I_{av}} = \frac{0.707I_m}{0.637I_m} = 1.11 \qquad (16\text{-}21)$$

SELF-CHECK

16-11:1. Suppose you wire a 1.0-kΩ resistor in series with an ac generator and that the ac in the resistor is given by:

$$i = 34 \cdot \sin 377t \text{ mA}.$$

Determine the following:
(a) the rms current in the resistor,
(b) the average power in the resistor,
(c) the peak instantaneous power in the resistor, and
(d) the effective voltage across the resistor.

PROBLEMS

16-5-1. If the peak voltage of a 60-Hz ac generator is 150 V, what is the instantaneous voltage after the loop has rotated through 135° from the reference axis?

16-5-2. If the peak voltage of a 1-kHz audio oscillator is 20 V, what is the instantaneous voltage when $\phi = 210°$?

16-5-3. A radio transmitter generates a sine-wave carrier with a frequency of 600 kHz. If the peak voltage fed to the antenna is 1 kV, what is the instantaneous voltage 2.0 μs after the instantaneous value passes through zero volts?

16-5-4. The instantaneous voltage from a 400-Hz aircraft generator is 95 V when $t = 1$ ms. What is the instantaneous voltage when $t = 2$ ms?

16-5-5. What is the instantaneous voltage after the loop in Problem 16-5-1 has rotated for 15 ms?

16-5-6. What is the instantaneous voltage from the audio oscillator of Problem 16-5-2 when $t = 600$ μs?

16-6-7. What is the instantaneous voltage after the loop of the generator in Problem 16-5-1 has rotated through 1 rad?

16-6-8. What is the instantaneous voltage from the audio oscillator in Problem 16-5-2 when $\phi = 5$ rad?

16-6-9. Write the general voltage equation for a 400-Hz source whose peak voltage is 200 V.

16-6-10. A certain alternating voltage is described by the equation

$$e = 170 \sin 1570t$$

(a) What is the frequency of the sine wave?
(b) What is the instantaneous voltage when $t = 5$ ms?

16-7-11. The instantaneous value of an 800-Hz sine-wave current is 5 A when $t = 0.5$ ms. What is the peak current?

16-7-12. If the instantaneous current drawn from a 50-Hz source is −263 mA when $t = 0.013$ s, what is the instantaneous current when $t = 0.017$ s?

16-7-13. Write the general equation for the instantaneous current in a 250-Ω resistor connected across the voltage source of Problem 16-5-1.

16-7-14. What is the instantaneous current in a 50-Ω dummy antenna resistance connected to the transmitter in Problem 16-5-3 at $t = 3.0$ μs?

16-8-15. What is the peak power input to the 250-Ω resistor in Problem 16-7-13?

16-8-16. What is the instantaneous power input to the dummy antenna in Problem 16-7-14?

16-10-17. What is the average value of the sine wave output of the voltage source in Problem 16-5-1?

16-10-18. If the peak value of the square wave in Figure 16-13(c) is 24 V, what is its average value?

16-10-19. If the average value of the sawtooth current waveform in Figure 16-13(e) is 120 mA, what is its peak value?

16-10-20. The conductor in the simple alternator shown in Figure 16-1 cuts across a total magnetic flux of 40 mWb during the first 180° of rotation. If the frequency of the ac output is 60 Hz, what is the half-cycle average induced voltage?

16-11-21. What is the rms value of the terminal voltage of the generator in Problem 16-5-1?

16-11-22. What is the rms current in the dummy antenna of Problem 16-7-14?

16-11-23. What is the peak voltage in the European 230-V, 50-Hz system?

16-11-24. What is the peak value of a 60-μV radio-frequency signal appearing at the input terminals of a radio receiver?

16-11-25. The equation of an alternating current is $i = 1.5 \sin 377t$. What rms voltage drop will appear across a 15-Ω resistor when this current flows through it?

16-11-26. Write an equation for the instantaneous voltage generated by a 117-V, 400-Hz alternator.

16-11-27. If the symmetrical square wave of current shown in Figure 16-13(c) has a peak value of 40 mA, what is its rms value?

16-11-28. If the symmetrical sawtooth voltage waveform in Figure 16-13(e) has a peak value of 30 V, what is its rms value?

16-11-29. The average power of a 250-Ω soldering iron is 55 W.
(a) What is the rms current through the iron?
(b) What is the peak voltage across the iron?

16-11-30. What current will an ac ammeter show when connected in series with a 16-Ω load whose peak power is 288 W?

16-11-31. How long will it take a 10-Ω heater connected to a 120-V, 60-Hz source to convert 1 kWh of electric energy into heat?

16-11-32. What is the "wattage" rating of a lamp whose hot resistance is 150 Ω when connected to a 117-V, 25-Hz source?

16-11-33. What is the rms value of the alternating voltage induced into the loop in Problem 16-10-20?

REVIEW QUESTIONS

16-1. What conditions govern the generation of a voltage by electromagnetic induction?

16-2. Why does the angular velocity of the rotating loop in Figure 16-1 tend to decrease when a load resistor is connected to the brushes?

16-3. Why is it not possible to maintain a constant polarity of induced voltage in the loop in Figure 16-1?

16-4. Why do we describe the instantaneous voltage waveform from the simple generator of Figure 16-1 as a **sine wave**?

16-5. .With a diagram similar to Figure 16-5, show that $\sin 3\pi/4$ rad $= \sin \pi/4$ rad.

16-6. What effect has the angular velocity of the loop in Figure 16-1 on (a) the frequency of the voltage waveform? (b) the peak value of the induced voltage?

16-7. Why does doubling the number of turns in the loop double the peak voltage?

16-8. Why is the peak value of the induced voltage never a negative quantity?

16-9. Why is the answer to Example 16-2 a negative quantity?

16-10. What is the significance of a negative value of instantaneous current?

16-11. What is the significance of a negative value of instantaneous power?

16-12. What is meant by angular *distance*?

16-13. Why is the radian measure preferred in writing the general equation for an alternating current?

16-14. What is the significance of the term **sine wave** in describing the behavior of an electric circuit?

16-15. What effect does a variation in the peak value of an alternating current have on the waveform?

16-16. Why can we substitute instantaneous ac values in the various equations we derived for use in dc circuits containing a voltage source and a load resistor?

16-17. Why must the instantaneous current through a resistor have the same waveform as the instantaneous voltage across it?

16-18. What is meant by the statement that the current through a resistor is in phase with the voltage across it?

16-19. What factors govern the instantaneous value of the current through a resistor in an ac circuit?

16-20. Why is the instantaneous power input to a resistor never negative?

16-21. Why does the instantaneous power input to a resistor pulsate at *twice* the frequency of the applied voltage?

16-22. Why is the average power input to a resistor operating from a sine-wave source one-half the peak power input?

16-23. Why is it necessary to consider average *power* when deriving an rms value for alternating current and voltage sine waves?

16-24. The equation for instantaneous voltage in an ac system is sometimes written in the form $e = \sqrt{2}\, E \sin \omega t$. Suggest a reason for using this form.

16-25. We obtain the average value of the alternating power input to a resistive load by averaging the instantaneous power input over a complete cycle. Why can we not determine the average value of an alternating current in the same manner?

*16-26. Using a personal computer, determine the rms value of a sine wave by the process of finding the square root of the mean values of sine squared for one complete cycle.

*16-27. Design a computer program to display a graph of the instantaneous power when a sine-wave alternating current flows in a resistor.

17

REACTANCE

As we discovered in Chapter 16, when we apply a sine wave of voltage to a resistor, the instantaneous current in the circuit must be an ac sine wave exactly in phase with the applied voltage. We established this relationship on the basis that current is the dependent variable in the circuit and that, in an alternating current circuit too, it must automatically take on a magnitude such that Ohm's law will hold true. As long as we keep in mind the continuously varying nature of a sine-wave alternating applied voltage, all the laws we have studied in dealing with dc circuits must also hold true for ac circuits. Any statement that would apply only for a special case and not for the general situation would not pass as a basic law of electric circuit theory.

Therefore, we were able to state Ohm's law in the form

$$i = \frac{v}{R} \tag{16-8}$$

where both v and i are instantaneous values. Although resistance is an electrical property of conductors, it is dependent on such physical factors as length, cross section, and type of material. Therefore, resistance cannot change in the short period of one cycle of a sine wave, and thus, as far as Equation (16-8) is concerned, resistance is a constant. Consequently, the instantaneous current must always be directly proportional to the instantaneous voltage.

Since the instantaneous current in a resistor reaches its peak value at the same instant as the instantaneous voltage across the resistance, we are also able to state Ohm's law in the form $I_m = V_m/R$, and, since the rms value of a sine wave is the dc equivalent value, which we will use for most of our ac circuit calculations, the most useful form of

Ohm's law is simply

$$I = \frac{V}{R} \qquad (17\text{-}1)$$

where I is the rms value of the sine-wave-current through a resistor, and V is the rms value of the sine-wave voltage across the resistor.

17-1 THE NATURE OF THE INSTANTANEOUS CURRENT IN AN IDEAL INDUCTOR

In setting out to determine the nature of the instantaneous current that must flow in the circuit of Figure 17-1, we assume that the resistance is so small that we can neglect it. In Chapter 15, we decided that we would think of the inductor as a load connected across a voltage source. Therefore, in order to satisfy Kirchhoff's voltage law, at every instant in time, the inductive voltage across the coil in Figure 17-1 must exactly equal the applied voltage. Hence,

$$v_L = e = E_m \sin \omega t$$

FIGURE 17-1
Inductance in an Alternating-Current Circuit.

If there is no resistance in the circuit, any voltage drop that appears across the terminals of an inductor must be due to the voltage induced in the coil by a *changing* current through it. From Equation (15-1), we can express the instantaneous voltage that must appear across the inductor in Figure 17-1 as

$$v_L = L \frac{di}{dt}$$

Therefore,

$$L \frac{di}{dt} = E_m \sin \omega t$$

from which

$$\frac{di}{dt} = \frac{E_m}{L} \sin \omega t \qquad (17\text{-}2)$$

By integrating Equation (17-2), we can establish a mathematical expression for the instantaneous current through an ideal inductor in an ac circuit in one fell swoop.[†] However, to help us *understand* the behavior of inductance in ac circuits, we can determine the nature of the instantaneous current through the inductor in Figure 17-1 by applying principles we have already learned. Since Kirchhoff's voltage law requires the voltage across the inductor in Figure 17-1 to be exactly equal to the applied voltage at every instant in time, we can represent the instantaneous voltage v_L across the inductor by the voltage sine curve of Figure 17-2. According to Faraday's law of electromagnetic induction, the magnitude of this inductive voltage at any instant is directly proportional to the *rate of change* of current through the coil at that instant. At the instant when we require maximum *positive* voltage across the inductor (at the $\pi/2$-rad point in Figure 17-2), the current must, therefore, be *changing* at the greatest *rate* in a *positive* direction. As we noted in discussing the rise of current through an inductor in Chapter 15, *rate of change of current* in a graph of current plotted against time is indicated by the *slope* of the current graph. Hence, maximum rate of change is indicated in the instantaneous current graph of Figure 17-2 by the steepest slope.

FIGURE 17-2
Instantaneous Current in an Inductor.

One-quarter cycle later in time (at the π-rad point in Figure 17-2), when the instantaneous voltage across the inductor must momentarily drop to zero, the current must stop *changing*. It does not matter how great the current may be, as long as it is neither rising nor falling, there will be no *change* in the number of magnetic lines of force linking the turns of the coil in order to induce a voltage in them. Therefore, at the π-rad point on the graph of Figure 17-2, the *slope* of the instantaneous current curve must become horizontal. Between $\pi/2$-rad and π-rad, the voltage across the inductor has to decrease along a sine curve, and thus the *rate of change* of current has to decrease in the same manner.

[†] See Appendix 2-7.

One-quarter cycle still later in time (at the $3\pi/2$-rad point in Figure 17-2), the instantaneous inductive voltage must become maximum negative, and, therefore, the instantaneous current must again change at a maximum rate but, this time, in the opposite direction. If we draw an instantaneous current graph that fulfills the conditions of having a maximum positive rate of change when the instantaneous voltage across the inductor is maximum positive, neither rising nor falling when the inductive voltage is zero, having a maximum negative rate of change when the voltage across the inductor is maximum negative, and so on, we find that the instantaneous current through the inductor in Figure 17-1 must have the *sine-wave* shape shown by the current curve in Figure 17-2. But the instantaneous current sine wave reaches its positive peak $\pi/2$ radians later in time than the instantaneous voltage drop across the inductor, and it passes through zero, going in a negative direction, $\pi/2$ radians later in time than the inductive voltage drop. Therefore,

For a sine-wave voltage drop to appear across an ideal inductor, the current through it must be a sine wave which lags the inductive voltage drop by $\pi/2$ radians.

When $t = 0$, $\sin \omega t = 0$. Therefore, the sine curve representing v_L in Figure 17-2 passes through zero at $t = 0$, increasing in a positive direction as time lapses and ωt increases from 0 to $\pi/2$ radians. However, the sine curve representing i_L in Figure 17-2 is at its maximum *negative* magnitude at $t = 0$. For this to occur, the phase angle ϕ of i_L must be $-\pi/2$ radians ($-90°$) so that sine $\phi = -1$ when $t = 0$. Hence, we must write the equation for the instantaneous current in the circuit of Figure 17-1 as

$$i_L = I_m \sin\left(\omega t - \frac{\pi}{2}\right) \text{ rad} \qquad (17\text{-}3)$$

We say that the instantaneous current in an ideal inductor in an alternating-current circuit is $\pi/2$ radians **out of phase** with the instantaneous voltage across it, with the current *lagging* the voltage by $\pi/2$ radians, or with the voltage *leading* the current by $\pi/2$ radians.

SELF-CHECK

17-1:1. Referring to Figure 17-1, suppose that the peak circuit current is 200 mA and that the instantaneous voltage of the generator is given by:

$$e = 170 \cdot \sin 377t \text{ V}$$

Write the equation that gives the instantaneous current in the circuit.

17-1:2. On one graph, plot the curves for $0 \leq \phi \leq 4\pi$ that represent:
  (a) instantaneous inductor voltage versus phase angle (v_L versus ϕ) and
  (b) the instantaneous inductor current versus phase angle (i_L versus ϕ)
    for the circuit described in Self-Check Problem 17-1:1.

17-2 INDUCTIVE REACTANCE

When we connected a dc voltage source to a circuit containing inductance but no resistance in Chapter 15, the current had to keep rising at a constant rate in order to maintain a constant induced voltage in the inductor. But when we connect a sine-wave voltage source to the circuit of Figure 17-1, the instantaneous current must take on a sine-wave form, lagging the applied voltage sine wave by $\pi/2$ radians in order to produce the required sine-wave voltage drop across the inductance. If the current in the inductor is a sine wave, it must have some definite rms value; thus we would discover a constant reading on the ac ammeter in Figure 17-1. Therefore, for a given alternating voltage source and a given inductance, the ratio V/I is a constant.

In discussing Ohm's law of constant proportionality, we decided that a constant V/I ratio represented the opposition of the circuit to electric current, and we called this opposition the **resistance** of the circuit. We can, therefore, say that the constant V_L/I_L ratio in Figure 17-1 represents the opposition of the inductance to *alternating* current. However, we cannot call this opposition **resistance,** since the alternating current in a resistor is in phase with the voltage drop across it, whereas the current in an ideal inductor lags the voltage across it by $\pi/2$ radians. Furthermore, an inductor does not convert electric energy into heat as does a resistor. Therefore, we must use a different term which suggests *opposition* to alternating current. Thus,

Inductive reactance is the opposition of inductance to alternating current.
The letter symbol for inductive reactance is X_L.

$$X_L = \frac{V_L}{I_L} \qquad (17\text{-}4)$$

Since inductive reactance is a V/I ratio just as is resistance, we can use the **ohm** as the unit of inductive reactance. Hence,

An ac circuit has an inductive reactance of one ohm when an alternating current having an rms value of one ampere flowing through the inductance creates an inductive voltage drop with an rms value of one volt across the inductance of the circuit.

SELF-CHECK

17-2:1. For the inductive circuit shown in Figure 17-1, suppose that the peak current is 200 mA and that the instantaneous voltage of the generator is given by:

$$e = 170 \cdot \sin 377t \quad \text{V}$$

Find the reactance of the inductor.

17-3 FACTORS GOVERNING INDUCTIVE REACTANCE

If we assume that the shape of the applied voltage in the circuit of Figure 17-1 is to remain sinusoidal, there are only three factors which we can vary independently: the peak voltage or **amplitude** of the sine wave of applied voltage (hence, also its rms value), the frequency of the applied voltage, and the inductance. Let us consider in turn, the effect of each independent variable on the current in the circuit.

If we double the peak value of the applied voltage leaving the frequency constant, the maximum rate of change of current must be doubled in order to develop the required inductive voltage drop. Since the frequency is unchanged, the current will reach its maximum value in the same length of time; yet, at each instant in time, its slope is twice as steep as its original values. Therefore, the peak value of the current must also be twice as great as originally. Consequently, doubling the amplitude of the applied voltage also doubles the current. As we would expect, since $X_L = V_L/I_L$, there is no change in inductive reactance. Hence,

The magnitude of the applied alternating voltage has no effect on the inductive reactance of an ac circuit.

If we double the frequency leaving the peak value of the applied voltage the same, the instantaneous current in the inductance must have the *same* maximum rate of change. But, because the frequency is doubled, the instantaneous current must cease rising in half the original time. Therefore, since the current rises with the same maximum slope for half as long, the amplitude of the current sine wave is now only half as great as originally, even though the amplitude of the voltage waveform is unchanged. Consequently, since $X_L = V_L/I_L$, the inductive reactance has been doubled. And

Inductive reactance is directly proportional to frequency.

If we double the inductance, leaving the source unchanged, according to Equation (17-2), the maximum rate of change of current will have to be cut in half to develop the same voltage. Therefore, the amplitude of the current is again only half as great for the same amplitude of voltage. And

Inductive reactance is directly proportional to the inductance.

We can avoid the calculus solution for Equation (17-2) by applying a technique we used in Section 15-6 to determine the energy stored by an inductance.[†] We can substitute the *average* rate of change of current for (di/dt) and the *average* value for sin ωt. Since the current must rise from zero to I_m in one quarter-cycle, and since the **period** of one cycle is $1/f$ seconds,

$$\text{average } \frac{di}{dt} = \frac{I_m}{\frac{1}{4} \times (1/f)} = 4fI_m$$

[†]See Appendix 2-7 for calculus solution.

Therefore, Equation (17-2) becomes

$$4fI_{m} = \frac{E_{m} \times 2/\pi}{L}$$

from which

$$\frac{E_{m}}{I_{m}} = 2\pi fL$$

and

$$X_{L} = 2\pi fL \qquad \text{(17-5)}$$

where X_L is inductive reactance in ohms, f is frequency in hertz, and L is inductance in henrys.

Since

$$\omega = 2\pi f \qquad \text{(16-5)}$$

$$X_{L} = \omega L \qquad \text{(17-6)}$$

where ω is angular velocity in radians per second.

SELF-CHECK

17-3:1. Given that an inductor has an inductance of 10 mH, find the reactance of the inductor at (a) 400 Hz and (b) 400 kHz. (c) Generally, if the frequency doubles, what happens to the reactance of an inductor?

17-3:2. Determine the inductance of the inductor in the circuit referred to in Self-Check Problem 17-2:1.

17-4 THE NATURE OF THE INSTANTANEOUS CURRENT IN A CAPACITOR

We can follow a similar procedure in order to determine the behavior of capacitance in ac circuits.* If we connect a capacitor across a sine-wave voltage source, as in Figure 17-3, Kirchhoff's voltage law requires the voltage across the capacitor at every instant in time to be exactly the same as the applied voltage at that instant. But the only way that a voltage across a capacitor can change is for the capacitor to charge or discharge. Consequently, the capacitor in Figure 17-3 must charge and discharge in such a manner that the voltage across it is a sine wave equal to the applied voltage at every instant in time.

Since

$$q = Cv \qquad \text{(11-7)}$$

$$\frac{dq}{dt} = C\frac{dv}{dt}$$

* See Appendix 2-8 for calculus solution.

But by definition, $i = dq/dt$. Therefore,

$$i = C \frac{dv}{dt} \qquad (17\text{-}7)$$

Since capacitance depends on such physical factors as the area of the plates and the dielectric constant, it cannot change in the short period of one cycle. Hence, as far as Equation (17-7) is concerned, capacitance is a constant. Accordingly, in determining the nature of the instantaneous current in Figure 17-3, we can state that it is always directly proportional to the *rate* at which the potential difference or voltage across the capacitor is changing.

FIGURE 17-3
Capacitance in an Alternating-Current Circuit.

Upon examining the sine curve of Figure 17-4 representing the instantaneous voltage across the capacitor, we note that the maximum voltage across the capacitor and the maximum *rate of change* of voltage occur $\pi/2$ radians apart in time. When the voltage across the capacitor is maximum, momentarily it is neither rising nor falling.

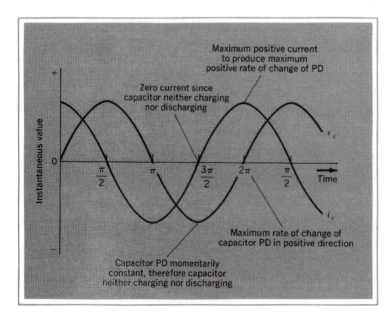

FIGURE 17-4
Instantaneous Current in a Capacitor.

Therefore, at this particular instant, the instantaneous current must be zero. Maximum rate of change of voltage occurs when the voltage sine curve of Figure 17-4 is steepest. The steepest slope of the instantaneous voltage sine wave occurs at the instant that the capacitor finishes discharging its stored charge and starts to recharge with the opposite polarity.[†] Therefore, the instantaneous current must be maximum in the positive direction when the voltage across the capacitor is changing from a negative polarity to a positive polarity. It will be maximum negative when the voltage changes from a positive polarity to a negative polarity, as shown in Figure 17-4.

When we plot an instantaneous current graph to fulfill the conditions of being zero when the voltage across the capacitance of the circuit is maximum and having a maximum positive value when the voltage is changing from a negative to a positive polarity, and so on, we find that the instantaneous current in the circuit of Figure 17-3 must follow a sine curve, as shown by the current graph of Figure 17-4. But the instantaneous current must be maximum positive at the instant that the voltage across the capacitance is just starting to build up in the positive direction. Then, $\pi/2$ radians later in time when the voltage across the capacitance has reached its positive peak, the instantaneous current has passed its positive peak and has dropped back to zero. Therefore,

For a sine-wave potential difference to be developed across a capacitor, the current through it must be a sine wave which leads the instantaneous voltage by $\pi/2$ radians.

For i_C to be maximum in a positive direction at $t = 0$, as indicated by the current sine curve in Figure 17-4, sin ϕ must be $+1$, and the phase angle ϕ must be $+\pi/2$ radians. Hence the equation for the instantaneous current "through" a capacitor in an alternating-current circuit becomes

$$i_c = I_\mathrm{m} \sin\left(\omega t + \frac{\pi}{2}\right) \text{ rad} \qquad (17\text{-}8)$$

Again, i_C is $\pi/2$ radians **out of phase** with v_C. But, in this case, i_C *leads* v_C by $\pi/2$ radians, or v_C *lags* i_C by $\pi/2$ radians.

SELF-CHECK

17-4:1. For the capacitive circuit shown in Figure 17-3, suppose that the peak circuit current is 25 mA and that the instantaneous voltage of the generator is given by:

$$e = 170 \cdot \sin 377t \text{ V}$$

Write the equation that gives the instantaneous current in the circuit.

[†]If we consider the manner in which a capacitor is discharged by electrons flowing into the plate which has a deficiency of electrons and flowing out of the plate which has a surplus of electrons, we realize that if this flow continues in the same direction after the original charges have been neutralized, the plate which was originally positively charged due to a deficiency of electrons will start to accumulate a surplus of electrons, thus becoming negatively charged. Similarly, the plate which was originally negatively charged will lose more electrons than are required to neutralize the original charge, thus starting to build up a positive charge. Therefore, there will be no change in the direction of the current as the polarity of the voltage across the capacitor reverses.

17-4:2. On one graph, plot the curves for $0 \leq \phi \leq 4\pi$ that represent:
  (a) the instantaneous capacitor voltage versus phase angle (v_c versus ϕ) and
  (b) the instantaneous capacitor current versus phase angle (i_c versus ϕ)
    for the circuit described in Self-Check Problem 17-4:1.

17-5 CAPACITIVE REACTANCE

When we connect a capacitor across a sine-wave voltage source, the instantaneous current must take on a sine wave form leading the voltage sine wave by $\pi/2$ radians in order to produce a sine wave of potential difference across the capacitor. If the current in the capacitor is a sine wave, it must have some definite rms value, and we would discover a steady reading on the ac ammeter of Figure 17-3. A steady reading on the ammeter does not mean that current is flowing *through* the insulating dielectric of the capacitor. In an ac circuit, instantaneous current flows in one direction for one half-cycle and then in the opposite direction for the next half-cycle. (The average of one complete cycle of a sine wave of current is zero.) Therefore, the capacitor can alternately charge, discharge, recharge with the opposite polarity, etc., with the required **displacement current** for the charging and discharging process forming a sine-wave alternating current in the circuit.[†]

For a given alternating voltage source and a given capacitance, both the sine-wave voltage across and the sine-wave current "through" the capacitor have definite rms values. Therefore, the ratio V/I is a constant. Just as in the case of resistance and inductive reactance, this constant V/I ratio represents the opposition of the capacitor to alternating current. Therefore,

Capacitive reactance is the opposition of capacitance to alternating current. The letter symbol for capacitive reactance is X_C.

$$X_C = \frac{V_C}{I_C} \tag{17-9}$$

An ac circuit has a capacitive reactance of one ohm when alternating current with an rms value of one ampere creates an alternating voltage with an rms value of one volt across the capacitance of the circuit.

SELF-CHECK

17-5:1. For the capacitive circuit shown in Figure 17-3, suppose that the peak circuit current is 25 mA and that the instantaneous voltage of the generator is given by:

$$e = 170 \cdot \sin 377t \ \text{V}$$

Find the capacitive reactance of the capacitor.

[†] Hence, the characteristic of capacitors is that they can "*pass*" alternating current but *block* direct current.

17-6 FACTORS GOVERNING CAPACITIVE REACTANCE

Now that we have established that the displacement current charging and discharging a capacitor in an ac circuit is a sine wave, we can discover the factors which govern capacitive reactance by again following the procedure of independently varying the amplitude and frequency of the applied voltage and the value of the capacitance in the circuit of Figure 17-3.

If we double the amplitude of the applied voltage, leaving the frequency constant, the capacitor will have to store twice as much charge in the same time interval. Therefore, the peak charging current must be twice as great. Since $X_C = V_C/I_C$, if both V_C and I_C are doubled, the capacitive reactance remains unchanged. And

The magnitude of the applied alternating voltage has no effect on the capacitive reactance of an ac circuit.

If we double the frequency, leaving the peak value of the applied voltage the same, the capacitor must store the same charge but in half the time. Therefore, the peak charging current must be twice as great. Consequently, although V_C is unchanged, I_C is doubled, and, since $X_C = V_C/I_C$, the magnitude of the capacitive reactance has been cut in half. Thus,

Capacitive reactance is **inversely** *proportional to frequency.*

Finally, if we double the capacitance, leaving the source unchanged, according to Equation (11-7), at the peak of the voltage sine wave, the capacitor must have stored twice as much charge. Since the frequency is unchanged, the capacitor must acquire this greater charge in the same length of time. Therefore, the peak charging current must be twice as great. Again, V_C is unchanged and I_C is doubled. Hence,

Capacitive reactance is **inversely** *proportional to the capacitance.*

Again we can solve for reactance without resorting to calculus by considering the *average* rate of voltage rise.[†] Rewriting Equation (17-7) in the form

$$\frac{dv}{dt} = \frac{I_m \sin \omega t}{C} \tag{17-10}$$

we can substitute the *average* rate of voltage rise for dv/dt and the *average* value for sin ωt. Since the voltage across the capacitor must rise from zero to E_M in one-quarter of a cycle,

$$\text{average } \frac{dv}{dt} = 4fE_m$$

[†]See Appendix 2-8 for calculus solution.

Therefore, Equation (17-10) becomes

$$4fE_m = \frac{I_m \times 2/\pi}{C}$$

from which

$$\frac{E_m}{I_m} = \frac{1}{2\pi fC}$$

and

$$X_c = \frac{1}{2\pi fC} \tag{17-11}$$

where X_C is capacitive reactance in ohms, f is frequency in hertz, and C is capacitance in farads.

Since

$$\omega = 2\pi f \tag{16-5}$$

when we substitute in Equation (17-11),

$$X_c = \frac{1}{\omega C} \tag{17-12}$$

SELF-CHECK

17-6:1. Given a capacitor that has a capacitance of 0.22 μF, find the reactance of the capacitor at (a) 400 Hz, (b) 400 kHz. (c) Generally, if the frequency doubles, what happens to the reactance of a capacitor?

17-6:2. Determine the capacitance of the capacitor in the circuit referred to in Self-Check Problem 17-5:1.

17-7 RESISTANCE, INDUCTIVE REACTANCE, AND CAPACITIVE REACTANCE

Resistance, inductive reactance, and capacitive reactance represent the opposition of the three basic properties of electric circuits to alternating current. Each is defined in terms of a V/I ratio and each is expressed in ohms. Although the instantaneous currents through all three components vary sinusoidally when a sine wave of voltage is applied across them, for a given circuit and a given frequency, the magnitude of this opposition (in ohms) is a *nonvarying* quantity which is independent of any particular instant in time. Resistance, inductive reactance, and capacitive reactance are *scalar* quantities represented by upper-case letter symbols.

However, there are significant differences in the behavior of resistance, inductance, and capacitance in ac circuits, which we must constantly keep in mind: (1) whereas resistance converts electric energy into heat, inductance and capacitance alternately *store* energy during the charging quarter-cycle, only to return this energy to the circuit during the next (discharging) quarter-cycle; (2) resistance is independent of frequency, but inductive reactance is *directly* proportional to frequency and capacitive reactance is *inversely* proportional to frequency; and (3) whereas the current through a resistor is in phase with the voltage across it, the current through an ideal inductor *lags* the voltage across it by 90° or $\pi/2$ radians and the alternating current "through" a capacitor *leads* the voltage across it by 90° or $\pi/2$ radians. We can summarize these characteristics in Table 17-1.

TABLE 17-1

R, L, AND C IN AC CIRCUITS		
Resistance	$R = \dfrac{V_R}{I_R} = R$ ohms	I_R is in phase with V_R
Inductive Reactance	$X_L = \dfrac{V_L}{I_L} = \omega L$ ohms	I_L lags V_L by 90°
Capacitive Reactance	$X_C = \dfrac{V_C}{I_C} = \dfrac{1}{\omega C}$ ohms	I_C leads V_C by 90°

It is important, therefore, that we identify clearly the *type* of opposition in making statements about ac circuits. For example, the statement that *the reactance of circuit A is* 200 *ohms* is a correct statement. But it provides only part of the necessary information. As shown in Table 17-1, there is a considerable difference between an ac circuit containing *inductive* reactance and one containing *capacitive* reactance. Careful attention to letter symbols and subscripts (R, X_L, and X_C) will provide the necessary identification. To maintain the identification along with a numerical value expressed in ohms, we sometimes include the appropriate *type* of ohms designation, as shown in the following examples.

EXAMPLE 17-1

What is the resistance of a 660-W toaster operated from a 110-V 60-Hz source?

SOLUTION

$$R = \frac{E^2}{P} = \frac{(110 \text{ V})^2}{660 \text{ W}} = \textbf{18.3 } \Omega \text{ (resistive)}$$

EXAMPLE 17-2

What inductive reactance develops a 40-V induced voltage when the current through it has an rms value of 80 mA?

SOLUTION

$$X_L = \frac{V_L}{I_L} = \frac{40 \text{ V}}{80 \text{ mA}} = \textbf{500 } \Omega \text{ (inductive)}$$

EXAMPLE 17-3

What is the reactance of a 68-pF capacitor when it is used in a 4.5-MHz circuit?

SOLUTION

$$X_C = \frac{1}{2\pi f C} = \frac{1}{2 \times \pi \times 4.5 \text{ MHz} \times 68 \text{ pF}} = \textbf{520 } \Omega \text{ (capacitive)}$$

Rather than considering dc and ac circuit-analysis techniques separately, some textbooks like to consider steady-state direct current as a special zero-frequency case of alternating current. At zero frequency, $\omega = 0$. Hence, $X_L = \omega L = 0$ and $X_C = 1/\omega C = \infty$; which only goes to show what we already know. The steady-state dc response of inductance is a short circuit, and the steady-state dc response of capacitance is an open circuit.

SELF-CHECK

17-7:1. Suppose you have a simple series circuit in which an ac generator is supplying a load and where the instantaneous voltage of the generator is given by

$$e = 170 \cdot \sin 377t \text{ V}$$

Determine the current in the load (rms value and *phase* with respect to the load voltage drop) if the load is:
(a) a 1.2-kΩ resistor
(b) a 4.7-H inductor
(c) a 3.3-μF capacitor

PROBLEMS

17-2-1. What is the inductive reactance of an ideal inductor which allows a 4-A current to flow when it is connected across a 117-V ac source?

17-2-2. What value of inductive reactance is required to limit current through it to 24 mA when the voltage across it is 35 V rms?

17-2-3. What rms ac voltage will appear across an inductive reactance of 48 Ω when the current through it is 300 mA?

17-2-4. What current must flow in an inductor having negligible resistance and a reactance of 2.2 kΩ to produce an inductive voltage drop of 120 V rms?

17-3-5. What is the 60-Hz reactance of a 4-H choke coil?

17-3-6. What is the reactance of a 250-mH radio-frequency choke at a frequency of 4.5 MHz?

17-3-7. At what frequency will a 0.5-H inductor have a reactance of 2 kΩ?

17-3-8. At what frequency will a 48-μH inductor have a reactance of 250 Ω?

17-3-9. What value of inductance will have a reactance of 1400 Ω at 475 kHz?

17-3-10. What value of inductance will draw a 160-mA rms current when connected to a 110-V 25-Hz source?

17-3-11. The solenoid of a magnetic contactor for starting an industrial electric motor draws 1.8 A from the 208-V 60-Hz source. Assuming negligible resistance, what is the inductance of the solenoid?

17-3-12. What value of inductance will draw a 1.5-μA current from a 4.5-MHz signal generator having a terminal voltage of 60 μV?

17-5-13. What value of capacitive reactance will allow a 3-mA alternating current to flow when it is connected to a 50-V 400-Hz source?

17-5-14. What is the reactance of a capacitor which allows a 7-μA current to flow when the voltage across it is 22 mV rms?

17-5-15. A capacitor has a reactance of 1200 Ω at a frequency of 465 kHz. What current is required at this frequency to maintain a voltage of 24 V rms across the capacitor?

17-5-16. What voltage will appear across a capacitor whose reactance is 250 Ω when it is "passing" a 180-mA alternating current?

17-6-17. What is the reactance of a 0.05-μF coupling capacitor at 400 Hz?

17-6-18. What is the reactance of a 27-pF capacitor at a frequency of 88 MHz?

17-6-19. At what frequency will an 8-μF capacitor have a reactance of 160 Ω?

17-6-20. At what frequency will a 0.002-μF capacitor have a reactance of 80 Ω?

17-6-21. What value of capacitance is required if its 400-Hz reactance is to be 50 Ω?

17-6-22. The input resistance to a transistor audio amplifier is 47 kΩ. What value of coupling capacitor should be used if its reactance is to equal the input resistance at 20 Hz?

17-6-23. What value of capacitance must be connected in parallel with a 0.25-μF capacitor to provide a total 1-kHz reactance of 400 Ω?

17-6-24. A capacitor is connected across a 6.6-kV 60-Hz transmission line for power factor correction. What capacitance is required for a 40-A reactive current?

17-7-25. What value of inductance is required in order that its reactance at 580 kHz be equal to the reactance of a 350-pF capacitor at that frequency?

17-7-26. At what frequency will the reactance of a 640-μH coil equal that of a 400-pF capacitor?

REVIEW QUESTIONS

17-1. Why must the instantaneous current sine wave be *exactly* $\pi/2$ radians out of phase with the applied voltage waveform across an ideal inductor?

17-2. Referring to the definition of inductance in Chapter 14, why must the current *lag* the voltage by $\pi/2$ radians, rather than lead it?

17-3. What is the advantage of expressing the phase difference between the voltage across and the current through an ideal inductor as $\pi/2$ radians, rather than 90°?

17-4. Why is it possible for us to label the *time* axis of the graph of Figure 17-2 in terms of *radians*?

17-5. What condition is imposed in expressing the elapse of time in radians?

17-6. Why does the meter in Figure 17-1 show a constant reading?

17-7. The term **resistance** is used in both dc and ac circuits. Why does the term **inductive reactance** not apply to dc circuits?

17-8. Why is it possible to express inductive reactance in *ohms*?

17-9. What is the significance of the symbol ω in the equation which states the factors governing inductive reactance?

17-10. With reference to the definition of capacitance in Chapter 11, why does the current *lead* the voltage across the capacitor by $\pi/2$ radians, rather than lag it?

17-11. When we connect the terminals of a capacitor to an ohmmeter, we obtain an infinitely high resistance reading. Nevertheless, the ammeter in Figure 17-3 can indicate a current of many amperes in the circuit. Explain.

17-12. The reactance of an inductor increases with an increase in frequency but the reactance of a capacitor decreases with an increase in frequency. Explain.

17-13. A single electrical component is sealed into a "black box" with a pair of terminals marked "1000 Ω." Suggest a laboratory procedure for determining whether the sealed box contains a resistor, a capacitor, or an inductor.

17-14. Explain how the polarity of the voltage across a capacitor can change without any change in the direction of the current "through" the capacitor.

*17-15. Design a PC program which will solve Problem 17-7-26 for any combination of inductance and capacitance. (In Chapter 23, we discover that the required frequency is called the **resonant frequency**.)

18

PHASORS

Now that we know how resistance, inductance, and capacitance behave individually in an alternating current circuit, we can proceed to ac circuits containing both resistance and reactance. We require a type of algebra that can handle numbers in *two* dimensions, rather than the simple *linear* algebra we have been using for dc circuits. Such algebra is called **complex algebra,** not because it is difficult, but rather because it deals with **real** and **imaginary** numbers that exist in the two-dimensional *complex plane*. There are several forms of complex numbers. In more advanced ac theory courses, we use the exponential form of complex numbers based on Euler's theorem. However, for this introductory course, we can limit our complex algebra to the polar and rectangular coordinates of complex numbers. Students who are quite familiar with conversion from polar to rectangular coordinates, and vice versa, can proceed to Chapter 19.

18-1 ADDITION OF SINE WAVES

Figure 18-1 shows a simple ac circuit containing resistance and inductance in series. The sine-wave voltage source will cause a sine wave of current to flow in the circuit. Since this is a simple series circuit, the current must be the same in all parts of the circuit. The current in the inductance must have the same magnitude as the current in the resistance, must reach its peak magnitude at the same instant, and become zero at the same instant. In fact, the sine-wave currents in the inductance and the resistance are one and the same thing, and the letter symbol for current in a series circuit requires no subscript.

The sine wave of current through the resistance will cause a sine-wave voltage drop across it. This sine wave of voltage is exactly *in phase* with the current sine wave. The

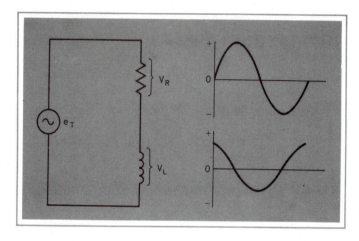

FIGURE 18-1
Simple AC Circuit.

instantaneous voltage across the resistance becomes maximum when the instantaneous current is maximum, etc.

In Chapter 17, we found that the instantaneous current through inductance has to *lag* the instantaneous voltage applied to it by $\pi/2$ radians. Conversely, then, the sine-wave voltage across the inductance portion of the circuit of Figure 18-1 must *lead* the current waveform through it by $\pi/2$ radians. Consequently, the sine-wave voltage across the inductance and the sine-wave voltage across the resistance are $\pi/2$ radians *out of phase*, with the instantaneous voltage across the inductance reaching its positive peak $\pi/2$ radians earlier in time than the instantaneous voltage across the resistance.

Kirchhoff's voltage law states that the sum of the voltage drops in a series circuit must equal the applied voltage. In dc circuits where the terminal voltage of the source is independent of *time*, we were able to find the sum of the voltage drops by simple arithmetic (linear algebra). Even when dealing with the charging of a capacitor in Chapter 12 and the rise of current in an inductor in Chapter 15, we were able to find the sum of the *instantaneous* voltage drops by simple arithmetic addition.

Our chore now is to find the *sum* of two out-of-phase sine-wave voltages appearing across the resistance and inductance of the circuit shown in Figure 18-1. As we discovered in Chapter 16, we can represent a sine-wave voltage either by its instantaneous value $v = V_\text{m} \sin \omega t$, or by its rms value $V = 0.707 \, V_\text{m}$.

The instantaneous value varies continuously and depends on an exact instant in *time*. Hence, we say that instantaneous values belong to the **time domain.** In the time domain, we use simple linear algebra and express angular distance (traveled by a rotating phasor in a given *time* interval) in radians.

We have already discovered that the rms value of a sine wave does not vary with the exact instant in time. However, we shall find that V_R and V_L also are 90° out of phase. To describe an alternating voltage by its rms value, we must include both magnitude and phase angle. $V \angle \phi$ is a complex number. We say that it belongs to the **frequency domain.** In the frequency domain, we use complex algebra and return to stating phase angle in *degrees*.

So that we may appreciate the advantages of being able to add rms values of sine waves, we shall first consider addition of their instantaneous values in the time domain.

18-2 ADDITION OF INSTANTANEOUS VALUES ON A LINEAR GRAPH

Since Kirchhoff's voltage law states that, at *every instant in time*, the algebraic sum of the various voltage drops must equal the algebraic sum of the applied voltages, we can then state that

$$e_T = v_R + v_L \qquad \text{(18-1)}$$

at each instant in time. Therefore, we can add the *instantaneous* values of sine waves by simple algebraic addition. Although such an addition for only one instant in time does not give us the overall picture of the addition of complete sine waves, we can repeat the calculation at small intervals in angular distance over a complete cycle to determine the nature of the resultant sum. To assist us in this time-consuming chore, we can draw the sine curves for the two sets of instantaneous values we wish to add on a common *time* axis, as in Figure 18-2. We indicate the passage of time by distance to the right along the horizontal axis of the graph, starting at the vertical axis with $t = 0$. And since angular velocity is constant for a given frequency, we can mark the horizontal axis of the graph in radians, rather than in milliseconds.

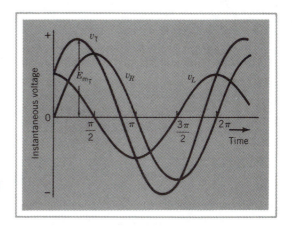

FIGURE 18-2
Addition of Instantaneous Values of Two Out-of-Phase Sine Waves on a Linear Graph.

Since the current sine wave is common to both the resistor and the ideal inductor in the circuit of Figure 18-1, we use its phase angle as a reference angle and let $i = I_m \sin \omega t$.* And since the voltage across the resistor is *in phase* with the common current, $v_R = I_m R \sin \omega t$. Therefore the sine curve for v_R in Figure 18-2 starts with zero magnitude at $t = 0$ and increases to maximum positive when $t = \pi/2$ radians. Since the sine-wave voltage across the ideal inductor must *lead* the common current by $\pi/2$ radians, it must already be at its positive peak at $t = 0$, dropping to zero at the $\pi/2$ mark on the scale. We indicate this phase relationship by writing the equation for the inductor

---

* Figure 18-2 omits the reference sine wave of current. It would have the same phase relationship as the v_R curve.

voltage as $v_L = I_m X_L \sin(\omega t + \pi/2)$. Thus, when $t = 0$, $\sin(\omega t + \pi/2) = \sin \pi/2 \text{ rad} = +1$.

Once we have drawn the two voltage sine waves with proper phase relationship and magnitudes (amplitudes) in Figure 18-2, we start to add instantaneous values graphically at small intervals along the horizontal axis. At $t = 0$, the instantaneous voltage across the resistance is zero; therefore, the instantaneous value of the total voltage is the same as the voltage across the inductance at that instant. As time elapses, the instantaneous voltage across the inductance decreases while the instantaneous voltage across the resistance increases. At approximately one radian their sum is a maximum, and the resultant waveform reaches its peak value at this point. At $\pi/2$ radians, the instantaneous voltage across the inductance is zero, and the instantaneous value of the total voltage becomes the same as the instantaneous voltage across the resistance. At approximately $(1 + \pi/2)$ radians, the instantaneous values of the voltages across the resistance and the inductance are equal but opposite in polarity. Therefore, at this point, the instantaneous value of the sum must become zero.

As we draw the v_T curve of Figure 18-2 through the plots we obtained by finding the simple algebraic sum of the instantaneous values at regular small intervals, we note that

The resultant sum of sine waves of the same frequency is also a sine wave of the same frequency.

We also note that, since the two sine waves we are adding do not reach their peak values at the *same* instant in time, the peak value of the resultant occurs at a time when the instantaneous values of the two component voltages are *less* than their peak values. Therefore,

$$E_{mT} \neq V_{mR} + V_{mL}$$

Since the rms value of any sine wave is 0.707 of the peak value,

$$|E_T| \neq |V_R| + |V_L|$$

The vertical bars enclosing the letter symbols indicate that the letter symbol represents the *magnitude* or *absolute value* of the quantity represented by the symbol.[†] The magnitude of a quantity is a simple positive numeral with no directional or angular data. Hence, the addition process represented by the above equations is simple algebraic addition. Therefore, although simple algebraic addition *does* apply to the instantaneous values of a sine wave, it does *not* apply to the peak and rms values if the sine waves being added are out of phase, as in Figure 18-2.

In the special case of sine waves that are exactly in phase, as in Figure 18-3, the resultant and all component sine waves do reach their peak values at the same instant in time. For *in-phase* sine waves,

$$E_{mT} = V_{m1} + V_{m2} \quad \text{and} \quad |E_T| = |V_1| + |V_2|$$

[†] The mathematical term for the absolute value of a quantity is the **modulus** of the quantity.

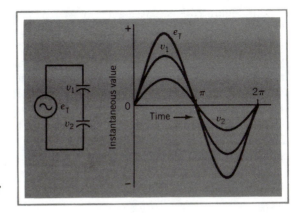

FIGURE 18-3
Addition of Instantaneous Values of Two In-Phase Sine Waves on a Linear Graph.

Therefore, we may add by simple arithmetic the rms values of sine waves that are exactly in phase.

With respect to the circuit of Figure 18-1, we can expand Equation (18-1) to

$$e_T = V_{mR} \sin \omega t + V_{mL} \sin(\omega t + \pi/2) \qquad (18\text{-}2)$$

SELF-CHECK

18-2:1. For the circuit shown in Figure 18-1, suppose that the components have the following values:

$$R = 4.7 \text{ k}\Omega, \quad L = 0.50 \text{ H}$$

and that the instantaneous current in the resistor is:

$$i_R = 5 \cdot \sin 6283t \quad \text{mA.}$$

Write the equations for the following instantaneous voltages: (a) v_R, (b) v_L and (c) e_T.

18-3 REPRESENTING A SINE WAVE BY A PHASOR DIAGRAM

Although we can determine the rms value of the resultant sine wave and its phase angle with respect to the component sine waves by means of the graphical addition of instantaneous values, as in Figures 18-2 and 18-3, the main feature of these graphs, which is apparent at a glance, is the sine-wave shape of the resultant. Once we accept that the resultant of the addition of sine waves of the same frequency is also a sine wave of the same frequency, the tedious procedure required to produce Figures 18-2 and 18-3 is not warranted. The original information from which we constructed the graphs of Figures 18-2 and 18-3 was simply their rms values and the phase angle between them. Rms value and phase angle will also serve to identify the resultant.

In Figure 16-3, we used a phasor as a means of drawing a simple sketch to indicate the position of a loop of wire as it cut across magnetic lines of force. The length of this phasor represented the maximum rate of cutting or E_m (which was a constant), and its position indicated the angular distance the loop had traveled in its cycle at that particular instant in time. If we draw phasors for the two sine-wave voltages in the circuit of Figure 18-1, the rotating phasor for the instantaneous voltage across the inductance will always stay exactly $\pi/2$ radians ahead of the rotating phasor for the instantaneous voltage across the resistance. Since the conventional direction for indicating this motion on a phasor diagram is counterclockwise, the rotating phasor for v_L will then be $\pi/2$ radians further counterclockwise than the phasor for v_R at every instant of time.

Because the phasors representing the voltages across the inductance and the resistance are always 90° apart, there is no motion of the one with respect to the other. Therefore,

We can represent sine waves of the same frequency by "stationary" phasors with the angle between the phasors representing the phase angle between the sine waves.

We can adopt a concept of stationary phasors by defining a **phasor quantity** as any quantity that is expressed in complex form, i.e., by an absolute magnitude and an angle. And since the rms value of a sine wave is always 0.707 of its peak value, we might as well make the length of the phasors (magnitude) proportional to the rms value of the sine wave they represent. These ac phasors belong to the frequency domain where we usually express angles in degrees.

In the simple series circuit of Figure 18-1, the current is common to all components. Therefore, in constructing the **phasor diagram** for a series circuit (Figure 18-4), we draw a phasor representing the rms value of the current along the reference axis of the diagram (in the three o'clock direction). Since the voltage across the resistance is in phase with the current through it, we mark V_R in the same direction as the current phasor. Since the voltage across the inductance *leads* the current through it by 90°, we draw the phasor for V_L 90° counterclockwise from the reference axis.

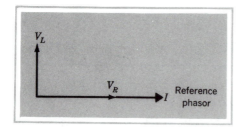

FIGURE 18-4
Phasor Diagram for the Circuit of Figure 18-1.

We prepared the phasor diagram of Figure 18-4 much more rapidly than the instantaneous voltage graphs of Figure 18-2. Nevertheless, we can see the pertinent data regarding rms values and phase angles at a glance in Figure 18-4, whereas we have to derive these data from the information shown in Figure 18-2. If we can now locate the phasor for the resultant voltage on the phasor diagram of Figure 18-4 without excessive calculations, we shall have a very powerful aid to the solution of ac circuits.

18-4 LETTER SYMBOLS FOR PHASOR QUANTITIES

In studying the behavior of electric and magnetic circuits, we encounter three distinct types of numerical quantities.

Scalars are quantities which possess magnitude but have no direction or angular information. Ohms, henrys, and farads are scalars. In dc circuits, where voltage and current are not time-varying quantities, they can be treated as scalars. In ac circuits, *instantaneous* values of voltage and current are scalars. Scalars are represented in equations by lightface italic letter symbols $(R, X_L, C, v, i,$ etc.). We add scalar quantities by simple algebraic addition.

Vectors are quantities which possess both magnitude and direction. Gravitational force, wind velocity, electric field intensity, magnetic flux density are all vector quantities. Although vector quantities may be represented by lightface italic letter symbols, if we wish to call attention to their *vector* characteristic, we use **boldface** italic letter symbols ($\boldsymbol{F}, \boldsymbol{v}, \boldsymbol{E}, \boldsymbol{B}$, etc.). The boldface type helps to remind us that vector quantities cannot be added by simple algebraic addition. Vector quantities are *complex* numbers representing magnitude and direction. Hence, vectors must be added by the rules for complex algebra.

Phasors are complex numbers also. The term **phasor quantity** refers specifically to steady-state values of quantities in alternating-current circuits which are complex numbers. The term **phasor** is used to avoid confusion with space *vectors* which are not time-varying functions. In technical journals, we are expected to know that rms ac quantities are phasor quantities—hence, the approved letter symbol for a phasor quantity is the familiar lightface italic letter symbol.[†] We should, therefore, represent an alternating voltage as a phasor quantity by the letter symbol V *where* $V = |V| \underline{/\phi}$.

To assist in distinguishing calculations requiring phasor algebra, many textbooks represent phasor quantities by boldface roman letter symbols. With this distinction, we can omit the vertical-bar solidus symbols of Section 18-2 and represent the magnitude of a phasor quantity by the standard lightface italic letter symbol. Hence, in the following chapters, we shall represent a phasor voltage by $\mathbf{V}$ *where* $\mathbf{V} = V \underline{/\phi}$.

Since we must use phasor algebra to add the two rms voltages shown in Figure 18-4, the equation that we could not write in Section 18-2 becomes

$$\mathbf{E}_T = \mathbf{V}_R + \mathbf{V}_L \tag{18-3}$$

18-5 PHASOR ADDITION BY GEOMETRICAL CONSTRUCTION

The resultant of two phasor quantities in an electrical phasor diagram (as in the vector diagram of forces in mechanics) is the diagonal from the origin to the opposite corner of a parallelogram with the two phasors as sides. If we draw the two phasors as accurately

---

[†]ANSI/IEEE Standard 280-1985, *IEEE Standard Letter Symbols for Quantities Used in Electrical Science and Electrical Engineering.*

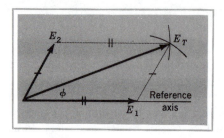

FIGURE 18-5
Geometrical Construction of the Resultant Phasor.

as possible on *polar* graph paper, we can draw the resultant with ruler and compass, as shown in Figure 18-5. Adjusting the compass to a radius equal to the length of E_1, we trace an arc using the tip of the E_2 phasor as the center. Then, adjusting the compass to a radius equal to the length of E_2, we trace an arc using the tip of E_1 as center. We then draw the resultant by joining the intersection of the two arcs to the origin with a straight line.

Applying this technique to the series circuit of Figure 18-1, we can quickly show the magnitude and phase relationships among **E**, **I**, V_R, and V_L. To illustrate the advantages of phasor diagrams, both the sine-wave graph of Figure 18-2 and the phasor diagram of Figure 18-4 are shown in Figures 18-6(a) and 18-6(b) respectively. Although we can determine (with some difficulty) the rms values and phase angles for **I**, V_R, V_L, and V_T from the graphical addition of sine waves in Figure 18-6(a), this information is clearly displayed by the phasor addition on the phasor diagram of Figure 18-6(b).

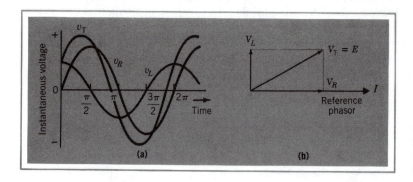

FIGURE 18-6
Sine-Wave Graph and Phasor Diagram for the *Same* Series Circuit of Figure 18-1.

If we perform the geometrical construction carefully, we can measure the phase angle between the applied voltage **E** and the current **I**, and the rms values of the voltages to two-figure accuracy directly from the phasor diagram. Although geometrical solutions are not considered accurate enough for ac circuit problems in later chapters, the phasor diagram is a valuable guide in performing the more accurate complex algebra solutions. If we draw the phasors in their proper proportions and mark off reasonably accurate phase angles, we can construct very quickly a free hand phasor diagram that will give us an approximate answer to the problem. The phasor diagram also assists us in checking sine, cosine, and tangent functions of angles larger than 90°. It also shows us whether we have the correct sign for the numerical quantities used in the algebraic solution. Therefore, we *always* construct a phasor diagram in solving ac circuits to show us what we have to do and to provide a quick check for the more accurate complex algebra procedures.

EXAMPLE 18-1

Three sources of sine-wave voltages of the same frequency have the rms values and phase angles listed. $\mathbf{E}_1 = 80$ V $\underline{/0°}$, $\mathbf{E}_2 = 100$ V $\underline{/150°}$, and $\mathbf{E}_3 = 40$ V $\underline{/45°}$. By geometrical construction on a phasor diagram solve for the total applied voltage when these sources are connected in series.

SOLUTION

Step I. Construct a phasor for each given voltage.

Step II. Construct a resultant phasor for any pair of phasors ($\mathbf{E}_1$, and $\mathbf{E}_3$ in Figure 18-7).

Step III. Construct a resultant phasor for the resultant of Step II and the remaining phasor.

Step IV. Measure the length and phase angle of the final resultant (approximately 79 V $\underline{/80°}$, in Figure 18-7).

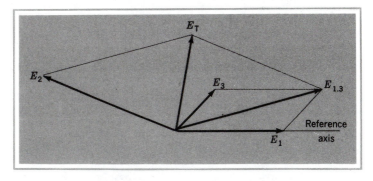

FIGURE 18-7
Solving Example 18-1 by Geometrical Construction on a Phasor Diagram.

SELF-CHECK

18-5:1. Referring to the circuit shown in Figure 18-1 and given that:

$$\mathbf{V}_R = 16.6 \text{ V} \underline{/0°} \text{ and } \mathbf{V}_L = 11.1 \text{ V} \underline{/90°}$$

use phasor addition by geometrical construction to find $\mathbf{E}_T = \mathbf{V}_R + \mathbf{V}_L$.
(Note that this is the same circuit that was analyzed in Self-Check Problem 18-2:1.)

18-6 ADDITION OF PHASORS WHICH ARE AT RIGHT ANGLES

We have already noted that, for the specific case of phasor quantities having a zero-degree angle between them, we can accomplish phasor addition by simple arithmetical addition (Figure 18-3). Similarly, when the angle between a pair of phasor quantities is exactly 180°, we can determine the phasor sum by simple arithmetical subtraction. In the specific case of two phasor quantities with an angle of exactly 90° between them, we can determine the phasor sum very accurately by making use of the relationships among the various sides and angles of a right-angled triangle as defined in trigonometry. Geometry also supplies a very useful relationship which applies to this specific case of two quantities with a 90° angle between them. Pythagoras' theorem states that the square of the hypotenuse of a right-angled triangle is equal to the sum of the squares of the other two sides.

In the simple series circuit of Figure 18-1, since the voltage drop across the resistance must be exactly in phase with the current and the voltage across the inductance must lead the current by exactly 90°, the two voltage phasors we wish to add are exactly 90° out of phase. If we solve for the resultant sum by geometrical construction, as in Figure 18-8(a) the parallelogram becomes a rectangle with side *AC* exactly equal to the phasor *OB* which represents the voltage across the inductance. Since the triangle of Figure 18-8(b) contains exactly the same data as the conventional phasor diagram of Figure 18-8(a), which has all phasors starting from the same point of origin, we are allowed to use triangular diagrams in solving problems in phasor addition if we find that we can show data more clearly in this form.

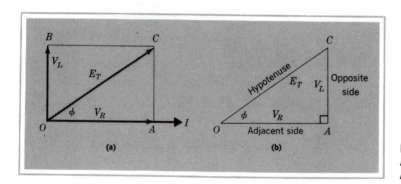

FIGURE 18-8
Addition of Phasors at Right Angles.

Since $\mathbf{E_T}$ is a phasor quantity, it must possess angle information as well as magnitude information. Solving first for the angle between the total voltage and the current, from trigonometry,

$$\tan \phi = \frac{\text{opposite side}}{\text{adjacent side}} = \frac{AC}{OA} = \frac{V_L}{V_R}$$

Therefore, if we know the magnitude of the voltage across the inductance and the voltage drop across the resistance, we can determine the phase angle ϕ from trigonometric tangent tables or calculators with tan^{-1} keys.

$$\therefore \phi = \arctan \frac{V_L}{V_R} \tag{18-4}$$

where **arctan** means *"the angle whose tangent is."* [†]

We now have a choice of two methods of solving for the magnitude of the total voltage phasor OC. From trigonometry,

$$\sin \phi = \frac{\text{opposite side}}{\text{hypotenuse}} = \frac{AC}{OC}$$

from which
$$E_T = \frac{V_L}{\sin \phi} \tag{18-5}$$

or
$$\cos \phi = \frac{\text{adjacent side}}{\text{hypotenuse}} = \frac{OA}{OC}$$

from which
$$E_T = \frac{V_R}{\cos \phi} \tag{18-6}$$

The second method for determining the magnitude of the resultant is based on Pythagoras' theorem. In Figure 18-8(b),

$$OC^2 = OA^2 + AC^2$$
$$\therefore E_T = \sqrt{V_R^2 + V_L^2} \tag{18-7}$$

Note that it is conventional to state the resistance component before the reactive component in Equation (18-7). Since we require both magnitude and angle information to identify the exact nature of the total voltage, we may combine Equations (18-7) and (18-4) in the form

$$\mathbf{E_T} = \sqrt{V_R^2 + V_L^2} \Big/ \arctan \frac{V_L}{V_R} \tag{18-8}$$

> **EXAMPLE 18-2**
> If the voltage drop across the resistance in the circuit of Figure 18-1 is 15 V rms and the voltage across the inductance is 10 V rms, what are the magnitude of the total voltage and its phase angle with respect to the current?
>
> **TRIGONOMETRY SOLUTION**
>
> $$\phi = \arctan \frac{V_L}{V_R} = \arctan \frac{10}{15} = 33.7°$$

[†] **Arctan** usually appears on calculator keys as tan^{-1}.

$$E_T = \frac{V_R}{\cos \phi} = \frac{15}{0.832} = 18 \text{ V}$$

or

$$E_T = \frac{V_L}{\sin \phi} = \frac{10}{0.555} = 18 \text{ V}$$

or

$$E_T = \sqrt{V_R^2 + V_L^2} = \sqrt{225 + 100} = 18 \text{ V}$$

$$\therefore \mathbf{E_T} = \mathbf{18 \text{ V} \underline{/+33.7°}}$$

The positive angle shows that the voltage *leads* the current.

CALCULATOR SOLUTION

We can solve for $\mathbf{E_T}$ with a calculator using trigonometry function keys according to the above steps. It is worth the time involved just to become familiar with the use of these keys.

However, calculators with a $\rightarrow$P key can give us the solution in a single step when the two phasor quantities are at right angles. In Figure 18-8, V_R and V_L are the rectangular coordinates of $\mathbf{E_T}$. Entering 15 as the value for x, we then press the $x \leftrightarrow y$ key and enter 10 as the value for y. When we press the $\rightarrow$P key, the display shows the magnitude of the resultant voltage 18.027 756. Pressing the $x \leftrightarrow y$ key once more then displays the phase angle 33.690 067°. We will understand this coordinate conversion procedure better after we have studied the next section.

SELF-CHECK

18-6:1. Referring to the circuit shown in Figure 18-1 and given that:

$$\mathbf{V_R} = 16.6 \text{ V} \underline{/0°} \text{ and } \mathbf{V_L} = 11.1 \text{ V} \underline{/90°},$$

calculate $\mathbf{E_T} = \mathbf{V_R} + \mathbf{V_L}$.
(Note that this is the same circuit that was analyzed in Self-Check Problems 18-2:1 and 18-5:1)

18-6:2. For the series circuit shown in Figure 18-1, suppose that $\mathbf{E_T} = 120 \text{ V} \underline{/60°}$. Find (a) $\mathbf{V_R}$ and (b) $\mathbf{V_L}$.

18-6:3. For the series circuit shown in Figure 18-1 suppose that the following voltage *magnitudes* are known:

$$V_R = 40 \text{ V} \text{ and } E_T = 50 \text{ V}.$$

Find the following voltage phasors:
(a) $\mathbf{V_R}$, (b) $\mathbf{V_L}$, and (c) $\mathbf{E_T}$.

18-7 EXPRESSING A PHASOR QUANTITY IN RECTANGULAR COORDINATES

When we describe a phasor by stating its magnitude and phase angle with respect to the reference axis, as we did in Example 18-2, we are stating its **polar coordinates;** that is, the coordinates that identify the phasor are its length along a radius from the origin and its angular distance from the reference axis. Polar coordinates are desirable for expressing the final answer to electric circuit calculations since the numerical quantities involved in this method of representation are those which we obtain in the laboratory with measuring instruments such as voltmeters, ammeters, and wattmeters.

By applying the basic trigonometric relationships to the phasor diagram of Figure 18-8, we were able to determine the polar coordinates of the total voltage from two components, one of which lies along the reference axis and one of which is perpendicular to the reference axis. Therefore, if we are given a phasor in polar form ($E \underline{/\phi}$ in Figure 18-9), we can use the same basic trigonometry functions to express it in terms of one component along the reference axis and one component at right angles to the reference axis. We find the reference axis component (called the **real** component in mathematics) by multiplying the magnitude of the phasor by the *cosine* of its phase angle. We find the quadrature (perpendicular) component (called the **imaginary** component in mathematics) by multiplying the magnitude of the phasor by the *sine* of its phase angle.

FIGURE 18-9
Determining the Rectangular Coordinates of a Phasor.

We can, therefore, identify a phasor by two coordinates, one of which is the reference axis (or real) component and one of which is the quadrature (or imaginary) component. These two components are called the **rectangular coordinates** of the phasor. A rectangular coordinate must fall along a boundary between the four quadrants of the complex plane. In polar coordinate representation, we identify the boundaries between the quadrants in terms of Babylonian degrees, as shown in Figure 18-10(a). In rectangular coordinate representation, we identify the four possible directions for the coordinates by the complex plane *operators* shown in Figure 18-10(b).[†] Hence, the rectangular coordinates of a phasor consist of a real number identified simply by + or − signs, and an imaginary number identified by either +j or −j operators. In writing the rectangular coordinates of a phasor, it is conventional to write the reference axis (real) component before the j (imaginary) component.

With reference to Figure 18-9, we can state the procedure for converting a phasor from polar coordinates to rectangular coordinates in equation form as

$$E \underline{/\phi} = +E \cos \phi + jE \sin \phi \qquad (18\text{-}9)$$

[†] In the complex plane, multiplying a vector quantity along the reference axis by the j operator *rotates* it 90° in a counterclockwise direction. To rotate it another 90°, we simply multiply once more by the j operator. Hence, multiplying a vector quantity along the reference axis by j^2 changes its direction by 180°. In polar coordinates, reversing the direction of a phasor is the same as multiplying by −1. Hence $j^2 = -1$. It follows that $j = \sqrt{-1}$, which has no solution—thus the name *imaginary* for quantities preceded by the +j or −j operator. Multiplying the original vector quantity once more by j rotates it a further 90° to the 270° location in polar coordinates. Since $j^3 = j^2 \times j$ and $j^2 = -1$, then the complex plane equivalent of 270° is −j.

FIGURE 18-10
Boundaries Between the Quadrants in (a) Polar Coordinates, and (b) Rectangular Coordinates.

EXAMPLE 18-3
Express 18 V $\underline{/33.7°}$ in rectangular coordinates.

SOLUTION
Step I. Always draw a freehand phasor diagram as a quick means of identifying the correct operators to use with the rectangular coordinates (Figure 18-11).

FIGURE 18-11
Phasor Diagram for Example 18-3.

Step IIA. *Using* sin *and* cos *keys of a calculator:*

$$\mathbf{E} = +E \cos \phi + jE \sin \phi = +18 \cos 33.7° + j18 \sin 33.7°$$

We calculate each coordinate separately.

$$18 \times 33.7 \; \boxed{\cos} = 15 \quad \text{and} \quad 18 \times 33.7 \; \boxed{\sin} = 10$$
$$\therefore \mathbf{E} = 15 + j10 \; \mathbf{V}$$

Step IIB. *Using the* →R *coordinate conversion key:* For coordinate conversion, the calculator can handle two unknowns. In polar coordinates, x represents the magnitude and y the phase angle. In rectangular coordinates, x represents the real component and y the imaginary or j component. The calculator automatically enters or displays the x component first. Hence, we enter

$$18 \quad x \leftrightarrow y \quad 33.7$$

The $x \leftrightarrow y$ key moves the x component into storage so that the y component can be entered. We have now entered the polar coordinates of **E**. Pressing the →**R** key automatically converts x and y to rectangular coordinates and displays the x or real component. Pressing the $x \leftrightarrow y$ key then displays the j component.

EXAMPLE 18-4
State 4 A $\underline{/240°}$ in rectangular coordinates.

SOLUTION
Step I. Again a reasonably accurate freehand phasor diagram (Figure 18-12) shows the proper operators for the solution as a check on the signs displayed by the calculator.

FIGURE 18-12
Phasor Diagram for Example 18-4.

Step II. Whether we use the sin and cos keys or the →**R** key, the calculator can automatically handle *any* angle and provide the correct + or − signs for the rectangular coordinates. Hence,

$$4 \quad x \leftrightarrow y \quad 240 \quad →R \quad \text{displays} \quad -2.0$$

and pressing $x \leftrightarrow y$ again displays -3.464

$$\therefore \mathbf{I} = -2 - \mathbf{j}3.46 \text{ A}$$

EXAMPLE 18-5
State 48 $\underline{/-45°}$ in rectangular coordinates.

SOLUTION
A *negative* angle is simply one measured *clockwise* from the reference axis. From Figure 18-13,

$$48 \underline{/-45°} = +48 \cos 45° - \mathbf{j}48 \sin 45° = \mathbf{+33.9 - j33.9}$$

FIGURE 18-13
Phasor Diagram for Example 18-5.

Again the calculator will handle the negative angle automatically if we enter

$$48 \quad \boxed{x \leftrightarrow y} \quad 45 \quad \boxed{+/-} \quad \boxed{\to R}$$

The procedure for converting from rectangular coordinates into polar coordinates is similar to that indicated by Equation (18-8). The magnitude of the polar quantity is simply the root of the sum of the squares of the reference axis rectangular coordinate and the j or quadrature rectangular coordinate. We find the proper phase angle for the resultant by drawing a phasor diagram of the rectangular coordinates. The angle contained within the triangle constructed on the horizontal axis is numerically equal to

$$\arctan \frac{\text{quadrature coordinate}}{\text{reference axis coordinate}}$$

From the position of the resultant in the phasor diagram, we can then determine the actual phase angle with respect to the reference axis.

EXAMPLE 18-6
State $-60 + j30$ volts in polar form.

TRIGONOMETRY SOLUTION
If we use the trigonometry function keys of the calculator, we must pay special attention to the phasor diagram of Figure 18-14. Since the $\tan^{-1}$ key of the calculator

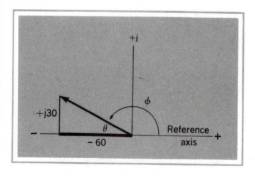

FIGURE 18-14
Phasor Diagram for Example 18-6.

cannot distinguish between $30/{-}60$ and $-30/60$, it will display an angle of $-26.565°$. From Figure 18-14, we realize that the angle we want for the polar coordinates is

$$\phi = 180° - 26.6° = +153.4°$$

$$V \text{ is } 60 \boxed{x^2} + 30 \boxed{x^2} = \boxed{\sqrt{x}}$$

which displays 67.1 V or

$$V \text{ is } 60 \boxed{\div} 153.4 \boxed{\cos} = -67.1 \text{ V}$$

In polar coordinates, the magnitude of the phasor is always a positive quantity. Hence, we ignore the $-$ sign that cos 153.4° gives in the second solution.

$$\therefore \mathbf{V} = \mathbf{67 \text{ V}} \underline{/\mathbf{+153.4°}}$$

CALCULATOR SOLUTION
Since the calculator knows that x is the real coordinate and y is the imaginary coordinate, when we use the $\boxed{\rightarrow P}$ key, the calculator will automatically locate the correct angle for polar coordinates.

$$60 \boxed{+/-} \boxed{x \leftrightarrow y} 30 \boxed{\rightarrow P}$$

displays 67.082 V. Pressing the $\boxed{x \leftrightarrow y}$ key then displays 153.43°.

SELF-CHECK

18-7:1. Express each of the following phasors in rectangular coordinate form:
  (a) $\mathbf{E_T} = 75 \text{ V} \underline{/60°}$
  (b) $\mathbf{V} = 16 \text{ V} \underline{/{-}15°}$
  (c) $\mathbf{I} = 50 \text{ mA} \underline{/120°}$
18-7:2. Express each of the following phasors in polar coordinate form:
  (a) $\mathbf{E_T} = 10 + \text{j}20 \text{ V}$
  (b) $\mathbf{V} = -16 - \text{j}38 \text{ V}$
  (c) $\mathbf{I} = 5.5 - \text{j}2.5 \text{ A}$

18-8 PHASOR ADDITION BY RECTANGULAR COORDINATES

Although there are trigonometric procedures for adding phasors having angles between them of other than 0° or 90°, now that we can express phasor quantities in rectangular coordinates, we seldom need such procedures. When we are required to add several phasors with various phase angles, we first convert each one to its rectangular coordinates. We can now add all components in a horizontal ($+$ and $-$) direction by simple algebraic addition. Similarly, we can add all components falling in the vertical ($+$j and

−j) direction by simple algebraic addition. This leaves us with one + or − coordinate and one +j or −j coordinate. These then are the rectangular coordinates of the sum of the individual phasors. All that remains is to convert these final rectangular coordinates into polar form. This procedure, as set forth in Example 18-7, is the conventional method of phasor addition for alternating-current circuit problems.

EXAMPLE 18-7

What is the sum of $\mathbf{E}_1 = 80$ V $\underline{/+60°}$ and $\mathbf{E}_2 = 80$ V $\underline{/-135°}$?

TRIGONOMETRY SOLUTION

Step I. Although we do not rely on geometrical construction for an accurate solution, we *always* construct a freehand phasor diagram (Figure 18-15), so we can check on any error in entering data into the calculator.

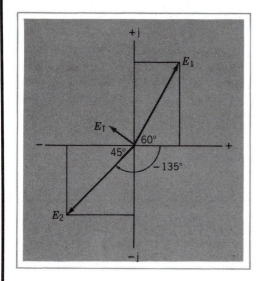

FIGURE 18-15
Phasor Diagram for Example 18-7.

Step II. Converting each voltage into its rectangular coordinates,

$$\mathbf{E}_1 = 80 \cos 60° \quad + j80 \sin 60° \quad = +40 \quad + j69.28 \text{ V}$$
$$\mathbf{E}_2 = 80 \cos -135° + j80 \sin -135° = -56.57 - j56.57 \text{ V}$$

Step III. Adding gives $\qquad \mathbf{E}_T = -16.57 + j12.71 \text{ V}$

Step IV. Converting the resultant into polar coordinates,

$$\theta = \arctan \frac{12.71}{-16.57} = -37.49°$$

From Figure 18-15,

$$\phi = 180° - 37.49° = +142.51°$$

$$E_T = \sqrt{16.57^2 + 12.71^2} = 20.88 \text{ V}$$

$$\therefore \mathbf{E_T} = \mathbf{20.9} \text{ V } \underline{/+142.5°}$$

CALCULATOR SOLUTION

Most pocket calculators do not have enough storage to keep the solution to each step in the calculator. Hence, we jot down the coordinates after each conversion.

$$80 \text{ V}\underline{/+60°} = +40 \qquad + \text{j}69.2820 \text{ V}$$

$$80 \text{ V}\underline{/-135°} = -56.5685 - \text{j}56.5685 \text{ V}$$

Adding gives
$$\mathbf{E_T} = -16.5685 + \text{j}12.7135 \text{ V}$$

Converting to polar coordinates,

$$\mathbf{E_T} = \mathbf{20.884} \text{ V } \underline{/+142.5°}$$

COMPUTER SOLUTION

Unless we have ready access to a previously prepared program for converting polar to rectangular coordinates and vice versa, a personal computer cannot compete with an engineering calculator for this type of example. BASIC computer language is quite capable of performing the mathematics required to find $\mathbf{E_T}$. However, BASIC does not provide for automatic coordinate conversion like the $\rightarrow$P and $\rightarrow$R keys of a calculator. We can use the computer in the calculator mode to perform the steps of the Trigonometry Solution, or we can write a simple program to add two phasor quantities. In the following sample program note that Line 160 recognizes that BASIC solves trigonometric functions in radians and does not store a value for π. Line 230 takes care of the dilemma in the Trigonometry Solution as to which quadrant arctan ATN(X) belongs.

```
100 REM SUM OF TWO PHASORS
110 PRINT "ENTER MAGNITUDE AND ANGLE IN DEGREES OF PHASOR 1"
120 INPUT M1, A1
130 PRINT "ENTER MAGNITUDE AND ANGLE IN DEGREES OF PHASOR 2"
140 INPUT M2, A2
150 REM CONVERT ANGLES TO RADIANS
160 PI = 3.1416: R1=A1*PI/180: R2=A2*PI/180
170 REM ADD REAL X AND IMAGINARY Y COORDINATES
180 LET  XR=M1*COS(R1) + M2*COS(R2)
190 LET  YR=M1*SIN(R1) + M2*SIN(R2)
200 REM CONVERT RESULTANT TO POLAR COORDINATES
210 LET  MR=SQR(XR^2 + YR^2)
220 LET  AR=180/PI*ATN(YR/XR)
230 IF  XR < 0 THEN AR=AR + 180
240 PRINT "MAGNITUDE OF RESULTANT IS ";MR
250 PRINT "ANGLE OF RESULTANT IS ";AR;" DEGREES"
260 END
```

SELF-CHECK

18-8:1. For the following phasors:

$$\mathbf{V}_1 = 24 + j15 \text{ V}$$
$$\mathbf{V}_2 = 30 - j42 \text{ V}$$
$$\mathbf{V}_3 = 35 \text{ V} \underline{/70°}$$
$$\mathbf{V}_4 = 24 \text{ V} \underline{/-135°}$$

Use phasor addition to compute the following sums:
(a) $\mathbf{V}_1 + \mathbf{V}_2$
(b) $\mathbf{V}_1 + \mathbf{V}_3$
(c) $\mathbf{V}_2 + \mathbf{V}_3$
(d) $\mathbf{V}_3 + \mathbf{V}_4$

18-9 SUBTRACTION OF PHASOR QUANTITIES

When expressed in rectangular coordinates, subtraction of two phasor quantities requires only simple algebraic subtraction of the reference axis and the quadrature components. We then convert resultant rectangular coordinates into polar form.

EXAMPLE 18-8
Subtract $4 - j5$ from $3 + j6$.

SOLUTION

$$3 + j6$$
$$\underline{4 - j5}$$

Subtracting: $\quad\quad -1 + j11$

Draw a simple phasor diagram of $-1 + j11$ to help us determine the phase angle of the polar form (Figure 18-16).

$$-1 + j11 = \sqrt{1^2 + 11^2} \underline{/180 - \arctan 11/1}$$
$$= 11.05 \underline{/+95.2°}$$

We can also subtract by reversing the $+$ and $-$ signs in the subtrahend and then performing an algebraic *addition*.

$$3 + j6$$
$$\underline{-4 + j5}$$

Adding gives $\quad\quad -1 + j11$

FIGURE 18-16
Phasor Diagram for Example 18-8.

Again, the calculator chain 1 $+/-$ $x \leftrightarrow y$ 11 $\rightarrow$ P $x \leftrightarrow y$ automatically provides the correct angle.

The second procedure in Example 18-8 gives us a clue as to how to go about subtracting phasors on a phasor diagram. If we locate $+4 - j5$ and $-4 + j5$ on a phasor diagram, as in Figure 18-17, we note that they represent phasors which are exactly equal in length but 180° apart in angular location. Therefore, we can use geometrical construction to subtract two phasor quantities by simply reversing the direction of the subtrahend phasor and then carrying out the customary geometrical construction for the addition of phasors on a phasor diagram.

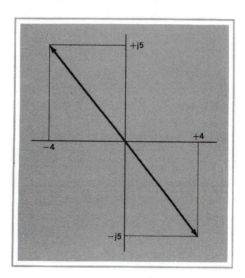

FIGURE 18-17
Reversing the Signs of Rectangular Coordinates.

EXAMPLE 18-9
Subtract $\mathbf{E}_1 = 40 \underline{/30°}$ from $\mathbf{E}_2 = 50 \underline{/135°}$ by geometrical construction on a phasor diagram.

SOLUTION
See Figure 18-18.

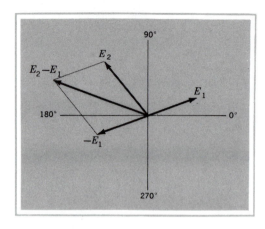

FIGURE 18-18
Subtraction of Phasors by Geometrical Construction.

SELF-CHECK

18-9:1. For the following phasors:

$$\mathbf{V}_1 = 24 + j15 \text{ V}$$
$$\mathbf{V}_2 = 30 - j42 \text{ V}$$
$$\mathbf{V}_3 = 35 \text{ V} \underline{/70°}$$
$$\mathbf{V}_4 = 24 \text{ V} \underline{/-135°}$$

Use phasor subtraction to compute the following differences:
(a) $\mathbf{V}_1 - \mathbf{V}_2$
(b) $\mathbf{V}_1 - \mathbf{V}_3$
(c) $\mathbf{V}_2 - \mathbf{V}_4$

18-10 MULTIPLICATION AND DIVISION OF PHASOR QUANTITIES

Multiplication and division of phasor quantities in *polar* form is the simplest of all the complex algebra processes.

To multiply phasor quantities expressed in polar form, multiply their magnitudes and algebraically add their phase angles.

$$\therefore |E_1|\underline{/\phi_1} \times |E_2|\underline{/\phi_2} = |E_1||E_2|\underline{/\phi_1 + \phi_2} \qquad (18\text{-}10)$$

To divide phasor quantities expressed in polar form, divide their magnitudes and algebraically subtract their phase angles.

$$\therefore |E_1|\underline{/\phi_1} \div |E_2|\underline{/\phi_2} = \frac{|E_1|}{|E_2|}\underline{/\phi_1 - \phi_2} \qquad (18\text{-}11)$$

EXAMPLE 18-10

Solve $80\,\underline{/40°} \times 11\,\underline{/-15°}$ and $80\,\underline{/40°} \div 11\,\underline{/-15°}$.

SOLUTION

$$80\,\underline{/40°} \times 11\,\underline{/-15°} = 80 \times 11\,\underline{/40 + (-15)} = \mathbf{880\,\underline{/25°}}$$
$$80\,\underline{/40°} \div 11\,\underline{/-15°} = \tfrac{80}{11}\,\underline{/40 - (-15)} = \mathbf{7.27\,\underline{/55°}}$$

Although the procedure for multiplying and dividing phasor quantities expressed in *rectangular* coordinates is quite straightforward, it does require skill in manipulating algebraic terms. Therefore, it is usually safer to carry out phasor multiplication and division by the simple polar method even though this often requires conversion from rectangular to polar coordinates. If our calculator has $\rightarrow$ P and $\rightarrow$ R keys, such conversion is the quickest way to carry out phasor multiplication and division. However, to illustrate the procedure for multiplication and division in rectangular form, we can refer to Example 18-11.

EXAMPLE 18-11

Solve $(8 + j6) \times (2 - j4)$ and $(8 + j6) \div (2 - j4)$.

SOLUTION

$$8 + j6$$
$$\underline{2 - j4}$$

Multiplying by 2, $16 + j12$

Multiplying by $-j4$, $\underline{\quad -j32 - j^2 24}$

Adding, $16 - j20 - j^2 24$

The next step is to take care of the term j^2. Since the operator j represents a rotation of 90° from the reference axis, j^2 represents a rotation of 180° from the reference axis. Therefore, the operator j^2 is the same as the operator $-$. If we wish, we can say $j^2 = -1$.

$$\therefore 16 - j20 - j^2 24 \quad \text{becomes} \quad 16 - j20 + 24$$

and the product becomes $\mathbf{40 - j20.}$

The procedure for division is more involved. First, we set up the phasor quantities involved in fraction form. If we multiply both numerator and denominator of a fraction by the same quantity, we do not change the value of the fraction. In choosing this extra factor, we wish to get rid of any j term in the denominator. To eliminate the j term, the extra factor must have the same reference axis coordinate as the denominator and the same magnitude of quadrature coordinate as the denominator *but with the opposite sign.* [†]

[†] Such a term is called the **conjugate** of the denominator.

$$\therefore (8 + j6) \div (2 - j4) = \frac{8 + j6}{2 - j4} = \frac{(8 + j6) \times (2 + j4)}{(2 - j4) \times (2 + j4)}$$

Completing the multiplication gives

$$\therefore (8 + j6) \div (2 - j4) = \frac{-8 + j44}{20}$$

Now that there is no j term in the denominator, we can carry out the division quite readily.

$$\therefore (8 + j6) \div (2 - j4) = \mathbf{-0.4 + j2.2}$$

SELF-CHECK

18-10:1. For the following phasors:

$$V_1 = 24 + j15 \text{ V}$$
$$V_2 = 30 - j42 \text{ V}$$
$$V_3 = 35 \text{ V } \underline{/70°}$$
$$V_4 = 24 \text{ V } \underline{/-135°}$$

Use phasor algebra to compute the results of the following operations:
(a) $V_1 \times V_3$
(b) $V_1 \div V_3$
(c) $V_1 \times (V_3 + V_4)$
(d) $\dfrac{V_3 \cdot V_4}{(V_1 + V_2)}$

PROBLEMS

Solve Problems 18-2-1 to 18-2-4 by drawing reasonably accurate freehand sine curves on a sheet of graph paper and determining the resultant by addition of instantaneous values.

18-2-1. A 200-V sine wave lags a 100-V sine wave of the same frequency by 45°. Find the rms value of the resultant.

18-2-2. A 2-A sine-wave current flows through a capacitor and an inductor in series. The reactance of the capacitor is 200 Ω and the reactance of the inductor is 150 Ω. Find the rms value of the total voltage.

18-2-3. The equations for the two voltage sine waves of Figure 18-2 are $v_R = 180 \sin 1508t$ and $v_L = 120 \sin(1508t + \pi/2)$. Find the equation for v_T. What is the frequency of these sine waves?

18-2-4. The currents in the two branches of a parallel circuit are given as $i_1 = 80 \sin 377t$ and $i_2 = 50 \sin(377t - \pi/3)$. Solve graphically for the total current.

18-5-5. Draw phasor diagrams for Problems 18-2-1 and 18-2-2 and solve by geometrical construction.

18-5-6. Solve Problem 18-2-4 by geometrical construction on a phasor diagram.

18-5-7. Three sources of alternating voltage each generate 120 V rms, but E_2 leads E_1 by 120°, and E_3 lags E_1 by 120°. Draw a phasor diagram and determine the resultant voltage when all three sources are connected in series.

18-5-8. Draw a phasor diagram and determine the resultant voltage if the leads to source E_1 in Problem 18-5-7 are reversed.

18-5-9. Find the resultant of the following phasors by geometrical construction: $80\underline{/+45°}$, $60\underline{/-135°}$, $40\underline{/-\pi/3}$ rad, $20\underline{/2}$ rad.

18-5-10. An induction motor draws a 10-A current from a 120-V 60-Hz source. This current lags the applied voltage by 30°. A capacitor in parallel with the motor draws a 5-A current from the source. Draw a schematic diagram, a phasor diagram, and solve for the total current drawn from the source by geometrical construction.

18-7-11. Convert the following phasor quantities into rectangular coordinates.
(a) $25\underline{/36.9°}$
(b) $36\underline{/140°}$
(c) $178\underline{/300°}$
(d) $60\underline{/-120°}$
(e) $14.2\underline{/2}$ rad
(f) $6.8\underline{/\pi/5}$ rad

18-7-12. Convert the following to polar coordinates.
(a) $3 + j4$
(b) $-14 - j10$
(c) $12.7 + j0$
(d) $-0.8 + j1.2$
(e) $0 - j18$
(f) $0.67 - j0.43$

Express all the following answers in polar coordinates.

18-8-13. Add $13 - j7$ and $11 + j18$.

18-8-14. Add $1.4 + j0.8$, $-0.9 + j2.1$, and $0 - j3.9$.

18-8-15. Add $170\underline{/200°}$ and $88\underline{/-75°}$.

18-8-16. Add $1.8\underline{/125°}$, $2.7\underline{/-157°}$, and $1.3\underline{/-66°}$.

18-8-17. Add $40 + j72$ and $60\underline{/100°}$.

18-8-18. Add $120 + j0$, $120\underline{/-120°}$, and $-60 + j104$.

18-9-19. Subtract $2.8\underline{/60°}$ from $3.1\underline{/45°}$.
(a) By geometrical construction.
(b) By complex algebra.

18-9-20. Add $120 + j0$, $-120\underline{/-120°}$, and $-60 + j104$.

18-10-21. Solve $4.1\underline{/-64°} \times 13\underline{/13°}$ and $4.1\underline{/-64°} \div 13\underline{/13°}$.

18-10-22. Solve $(3 + j4) \times (5 - j6)$ and $(3 + j4) \div (5 - j6)$.

18-10-23. Solve $\mathbf{Z}_{eq} = \dfrac{\mathbf{Z}_1\mathbf{Z}_2}{\mathbf{Z}_1 + \mathbf{Z}_2}$ if $\mathbf{Z}_1 = 40\ \Omega\underline{/30°}$ and $\mathbf{Z}_2 = 12\ \Omega\underline{/-45°}$.

18-10-24. Solve $\dfrac{(47 + j13)(-21 - j32)}{51\underline{/-111°} + 19\underline{/+70°}}$

REVIEW QUESTIONS

18-1. Rms values in an ac circuit are defined as dc equivalent values. Why then can we not add rms values by simple algebra as we can in a dc circuit?

18-2. Why is it possible to add instantaneous values in an ac circuit by simple algebra?

18-3. When writing an equation for the instantaneous value of the sum of two sine waves, how is the phase difference between them indicated?

18-4. What advantage does the phasor diagram have over the linear graph of instantaneous values plotted against time?

18-5. What information does a linear graph show that is not shown by a phasor diagram?

18-6. Why is a knowledge of complex algebra required in solving ac circuit problems?

18-7. What condition is required to be able to add rms values in an ac circuit either by phasor construction or by complex algebra?

18-8. Is it possible to add the instantaneous values of two sine waves if their frequencies are different? Explain.

18-9. What is the significance of an equation such as $|I_T| = |I_1| + |I_2|$?

18-10. Show with a phasor diagram that $\sin 210° = -\sin 30°$.

18-11. Show with a phasor diagram that $\cos 135° = -\cos 45°$.

18-12. What is meant by **algebraic addition?**

18-13. What is meant by **Pythagorean addition?**

18-14. What is meant by **polar coordinates?**

18-15. What is meant by **rectangular coordinates?**

18-16. Why is it possible to convert phasor quantities from one set of coordinates to the other?

18-17. What is the significance of the operator j in dealing with phasor quantities?

18-18. What is the advantage of expressing phasor quantities in rectangular coordinates?

18-19. Explain the procedure for subtracting phasor quantities by geometrical construction on a phasor diagram.

18-20. Is it possible to determine the product of two phasor quantities by geometrical construction on a phasor diagram? Explain.

*18-21. Given two phasors $I_1 = 40 \text{ mA} \underline{/+120°}$ and $I_2 = 60 \text{ mA} \underline{/-30°}$, either run the computer program for Example 18-7 or go through the steps indicated in the program to determine the resultant sum.

*18-22. Write a computer program to express in rectangular coordinates the quotient of two phasor quantities which are stated in rectangular coordinates.

*18-23. Modify the computer program of Example 18-7 so that the input data can be in the form of the general equations for two sine waves of the same frequency $[v_1 = E_{m_1} \sin(\omega t + \phi_1)$ and $v_2 = E_{m_2} \sin(\omega t + \phi_2)]$ and the solution printed out in the form of the general equation for the sum of v_1 and v_2.

19

IMPEDANCE

19-1 RESISTANCE AND INDUCTANCE IN SERIES

In Chapter 18, we noted that all rules and laws which apply to dc circuits must also apply to ac circuits, provided that we keep in mind the sinusoidally varying nature of the voltage source in an ac system. The most important of the rules of series circuits is the one by which we define a series circuit.

The same current must flow in all components in a series circuit.

Therefore, the starting point in determining the behavior of all series ac circuits is to represent this common current by the reference phasor of a phasor diagram, as in Figure 19-1.

A second characteristic of series circuits is that the sum of the various voltage drops must equal the applied voltage. If we use instantaneous values, this becomes a simple algebraic sum. However, using instantaneous values is a tedious means of solving an ac circuit. To use the more practical *rms* values of voltage and current, we can restate this second characteristic of series circuits thus:

In a series ac circuit, the **phasor** *sum of the various voltage drops must equal the applied voltage.*

Since the voltage drop across the resistance must be in phase with the current, its phasor falls along the + rectangular coordinate axis in Figure 19-1. Similarly, since the voltage across the inductance must lead the current by 90°, its phasor falls along the +j rectangular coordinate axis. Therefore, we may express the relationship among the

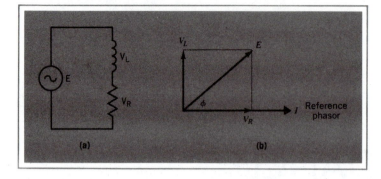

FIGURE 19-1
Phasor Diagram of Voltage and Current Relationships in a Series Circuit Containing Resistance and Inductance.

voltages in a simple ac series circuit quite readily in rectangular coordinates as

$$\mathbf{E} = V_R + jV_L \tag{19-1}$$

Drawing in the resultant phasor on the phasor diagram of Figure 19-1, we can also express this relationship in polar form.

$$\mathbf{E} = \sqrt{V_R^2 + V_L^2} \Big/ \arctan \frac{V_L}{V_R} \tag{19-2}$$

Because current is the only parameter common to all components in a series circuit, in constructing the phasor diagram of Figure 19-1, we started with current as the reference phasor and solved by geometrical construction for the location of the phasor representing the total voltage or applied voltage. In the actual circuit, when the inductance and resistance are connected in series to a given source, the current (the dependent variable) must automatically take on a sine-wave shape and a magnitude and phase such that it will develop a voltage drop across the resistance and voltage across the inductance with a phasor sum equal to the applied voltage. Nevertheless, we always think of current as the reference phasor for series circuits, and the angle in Equation (19-2) is the **phase angle** of the applied voltage *with respect to* the common current.

The letter symbol for phase angle is the Greek letter ϕ (phi) or the Greek letter θ (theta). [†]

SELF-CHECK

19-1:1. Referring to the circuit shown in Figure 19-1, suppose that $V_R = 100$ V and $V_L = 66$ V. Determine the value of the source voltage phasor, **E**,
(a) in rectangular coordinates and
(b) in polar form.

[†] The international standard symbol for phase angle is the Greek letter ϕ. North American textbooks have, in the past, shown a preference for the Greek letter θ. At one point in time, one of several North American standards used θ to represent the phase angle of the voltage *with respect to* the current, and ϕ to represent the phase angle of the current *with respect to* the voltage. Current North American standards permit the use of either ϕ or θ to represent the phase angle *between* current and voltage. In line with the present North American trend to adopt international standards as rapidly as is feasible, we shall use the Greek letter ϕ to represent phase angle in this text.

19-2 IMPEDANCE

Since the current in the series circuit of Figure 19-1 must automatically take on a sine-wave form of certain magnitude and phase with respect to the applied voltage, for a given ac circuit the ratio between the applied voltage and the current is a constant. Just as V_R/I represents the opposition of resistance to alternating current and V_L/I represents the opposition of inductance to alternating current, the ratio V_T/I or E/I represents the *total* opposition of the circuit to alternating current. In resistance the voltage and current are in phase, and in reactance the voltage and current are exactly 90° out of phase. But as we can see from the phasor diagram of Figure 19-1, the angle ϕ between the applied voltage and the current is between 0° and +90°. Therefore, we cannot call this total opposition either resistance or reactance.

The total opposition to current in an ac circuit is called impedance.
The letter symbol for impedance is Z.

$$\mathbf{Z} = \frac{\mathbf{E}}{\mathbf{I}} \tag{19-3}$$

where $\mathbf{Z}$ *is the impedance of an ac circuit in ohms,* $\mathbf{E}$ *is the applied voltage in volts, and* $\mathbf{I}$ *is the current drawn from the source in amperes.*

Since current is the dependent variable in the circuit, we must have some means of determining the impedance before we can solve for the current. Once we know the impedance, we can use Equation (19-3) to determine the current in the circuit. Because the current is common to all components in a series circuit, we can divide every term in Equation (19-1) by I. Hence,

$$\frac{\mathbf{E}}{I} = \frac{V_R}{I} + \mathbf{j}\frac{V_L}{I}$$

from which

$$\mathbf{Z} = R + \mathbf{j}X_L \tag{19-4}$$

Also, when we divide each term in Equation (19-2) by I,

$$\mathbf{Z} = \sqrt{R^2 + X_L^2} \Big/ \arctan \frac{X_L}{R} \tag{19-5}$$

Therefore,

The impedance of a series circuit is the **phasor** *sum of the resistance and reactance.*

Like resistance and reactance, impedance is a V/I ratio expressed in ohms. Like resistance and reactance, impedance is not a sinusoidally varying quantity. But unlike R, X_L, and X_C which do indicate the exact phase angle between V and I, the letter symbol Z gives no such indication. The statement $Z = 5$ kilohms tells us only the magnitude of the total opposition to electric current. There is no indication as to whether the impedance is largely inductive (with the voltage leading the current) or capacitive (with the voltage lagging the current). To provide this information we must recognize that impedance is a complex number which fits the definition of a phasor quantity. Hence, Equation (19-3) can be expanded into polar coordinates as

$$\mathbf{Z} = \frac{|E|\underline{/\phi}}{|I|\underline{/0°}} = |Z|\underline{/\phi} \tag{19-6}$$

We can represent impedance in the customary common-origin type of phasor diagram of Figure 19-2(a). However, to preserve a distinction between sinusoidal phasor quantities such as voltage and current on the one hand and nonsinusoidal phasor quantities measured in ohms on the other, we reserve the common-origin type of phasor diagram for voltage and current. In Section 18-6, we found that we could perform the same graphical addition of phasors with a triangular type of diagram. Hence, we use the format of Figure 19-2(b) and call it an **impedance diagram.** As indicated by Equation (19-4), resistance is always drawn in the reference direction and inductive reactance in the +j direction. It follows that capacitive reactance is drawn in the −j direction.

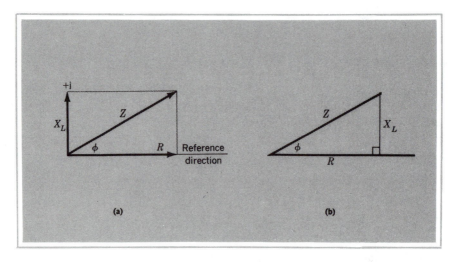

FIGURE 19-2
Impedance Diagram for a Series Circuit Containing Resistance and Inductance.

EXAMPLE 19-1
A solenoid having an inductance of 0.5 H and a resistance of 100 Ω is connected to a 120-V 60-Hz source. What is the magnitude and phase of the current with respect to the applied voltage?

SOLUTION
   Step 1.   In an ac circuit, we are more concerned with the *reactance* of the solenoid than its inductance. Therefore, the first step is to express its reactance at 60 Hz.

$$X_L = \omega L = 377 \times 0.5 = 188.5 \ \Omega \ (\text{inductive})$$

Step II. Always draw a circuit diagram, labeling it with all pertinent data, and draw either a voltage phasor diagram or an impedance diagram to show the relationships that exist in the given circuit (Figure 19-3).

FIGURE 19-3
Impedance Diagram for Example 19-1.

Step III. From the impedance diagram, the rectangular coordinates of the impedance are

$$\mathbf{Z} = R + jX_L = 100 + j188.5 \ \Omega$$

Step IVA. Solve for the impedance in polar form.

$$\phi = \arctan \frac{X_L}{R} = \arctan \frac{188.5}{100} = 62°$$

$$Z = \sqrt{R^2 + X_L^2} = \sqrt{100^2 + 188.5^2} = 213 \ \Omega$$

$$\therefore \mathbf{Z} = 213 \ \Omega \underline{/+62°}$$

Step IVB. With a calculator, we can bypass the trigonometry of rectangular to polar coordinate conversion and let the calculator do all the work.

$$100 \ \boxed{x \leftrightarrow y} \ 188.5 \ \rightarrow \text{P} \ 213.383 \ \boxed{x \leftrightarrow y} \ 62.054$$

Step VA. Solve for the magnitude of the current.

$$I = \frac{E}{Z} = \frac{120 \text{ V}}{213 \ \Omega} = 0.563 \text{ A}$$

And from Figure 19-3, this current *lags* the applied voltage by 62°.

Step VB. Treating the terms of Equation (19-6) as phasor quantities with the angle of the applied voltage as 0°, we can solve for both magnitude and angle of the current with respect to the applied voltage by phasor division. Therefore,

$$\mathbf{I} = \frac{\mathbf{E}}{\mathbf{Z}} = \frac{120 \text{ V}\underline{/0°}}{213 \text{ }\Omega\underline{/62°}} = \mathbf{0.563 \text{ A}}\underline{\mathbf{/-62°}}$$

SELF-CHECK

19-2:1. A motor has an impedance of $635 + j300 \text{ }\Omega$. What current (phasor form) will the motor draw when connected to a 120-V 60-Hz source?

19-2:2. For the series circuit shown in Figure 19-1, suppose that the source is a 230-V 50-Hz generator and that the resistance is 1.5 kΩ and the inductance is 1.8 H. Determine
(a) the impedance of the circuit and
(b) the current in the circuit.

19-2:3. A series circuit similar to that shown in Figure 19-1, contains a 160-Ω resistor and a 50-mH inductor. Using an ammeter, the magnitude of the circuit current is found to be 2.0 A. Determine the source voltage if the source frequency is:
(a) 100 Hz
(b) 1.0 kHz.

19-3 PRACTICAL INDUCTORS

In practical circuits, we are not able to obtain inductance alone since the wire used in the winding possesses some resistance. Therefore, the practical inductor is not a reactance with a +90° angle. With only two leads from the choke in Figure 19-4(a), we cannot connect a voltmeter to read only the inductive voltage drop across its inductance or the *IR* drop across its resistance. We can read only the total voltage drop between the two

FIGURE 19-4
Practical Inductor and Resistor in Series.

leads. This voltage leads the current through the choke by something *less* than 90°. If we are given the *impedance* of the choke in polar form with its proper angle, we can determine what the resistance and inductance components of the practical inductor are by conversion to rectangular coordinates.

EXAMPLE 19-2

The choke in the circuit of Figure 19-4(a), has an impedance of 80 Ω /80°. What is the total impedance when it is connected in series with a 75-Ω resistor?

SOLUTION

Since we do not directly tackle the addition of two phasors with an 80° angle between them as in the phasor diagram of Figure 19-4(b), we must first express the impedance of the choke in rectangular coordinates.

$$\therefore \mathbf{Z}_{coil} = 80 \cos 80° + j80 \sin 80° = 14 + j79 \; \Omega$$

The total impedance of the circuit in Figure 19-4(a) now becomes

$$\mathbf{Z}_T = 75 + 14 + j79 \; \Omega$$

Since the two resistance components are both along the + axis of an impedance diagram, we may add them by simple arithmetic.

$$\therefore \mathbf{Z}_T = 89 + j79 \; \Omega$$

Changing back to polar form,

$$\phi = \arctan \frac{79}{89} = 41.6°$$

$$Z_T = \sqrt{89^2 + 79^2} = 119 \; \Omega$$

$$\therefore \mathbf{Z}_T = \mathbf{119 \; \Omega \; /+41.6°}$$

Since the $\rightarrow$ P and $\rightarrow$ R keys of a calculator conceal the actual process of converting from rectangular to polar coordinates, and vice versa, the solutions to examples will continue to show the "longhand" steps. Students have the option of using calculator short cuts and checking these results with the solutions shown. Now that we are familiar with the Pythagorean relationship among the magnitudes of resistance, reactance, and impedance, we might consider the shorter trigonometry procedure for determining Z with calculators not possessing $\rightarrow$ P and $\rightarrow$ R keys.

$$Z = \frac{R}{\cos \phi} \quad \text{or} \quad \frac{X}{\sin \phi} \tag{19-7}$$

While the angle ϕ is still displayed by the calculator from having solved arctan X/R, press cos $1/x$ $\times$ and enter R. Pressing = then displays Z.

The following example doesn't lend itself to shortcuts. An alternative solution can be based on the cosine rule of trigonometry rather than on a Pythagorean equation.

EXAMPLE 19-3

A coil and a resistor each have an 80-V drop across them when connected in series to a 120-V 60-Hz source. If they draw a 0.5-A current from the source, what is the resistance of the coil?

SOLUTION
  Step I.

$$Z_T = \frac{E_T}{I} = \frac{120 \text{ V}}{0.5 \text{ A}} = 240 \text{ } \Omega$$

$$Z_{coil} = \frac{V_{coil}}{I} = \frac{80 \text{ V}}{0.5 \text{ A}} = 160 \text{ } \Omega$$

$$R = \frac{V_R}{I} = \frac{80 \text{ V}}{0.5 \text{ A}} = 160 \text{ } \Omega$$

Step II.   Since the coil possesses both resistance and inductance, the angle of its impedance ϕ must be something less than $+90°$, as shown in Figure 19-5(b).

$$\therefore \mathbf{Z}_{coil} = 160 \text{ } \Omega \underline{/\phi}$$

We can express this in rectangular coordinates as

$$\mathbf{Z}_{coil} = 160 \cos \phi + j160 \sin \phi$$

(a)   (b)

FIGURE 19-5
Impedance Diagram for Example 19-3.

Step III.

$$\therefore \mathbf{Z}_T = 160 + 160 \cos \phi + j160 \sin \phi = 240 \; \Omega$$

In polar form, this becomes

$$240 = \sqrt{(160 + 160 \cos \phi)^2 + (160 \sin \phi)^2}$$

Squaring both sides of the equation,

$$57\,600 = 25\,600 + 51\,200 \cos \phi + 25\,600 \cos^2 \phi + 25\,600 \sin^2 \phi$$

$$\therefore \cos \phi = \frac{32\,000 - 25\,600(\cos^2 \phi + \sin^2 \phi)}{51\,200} = \frac{5 - 4(\cos^2 \phi + \sin^2 \phi)}{8}$$

Step IV. But from triangle ABC in Figure 19-5(b),

$$\cos^2 \phi = \frac{AC^2}{AB^2} \quad \text{and} \quad \sin^2 \phi = \frac{BC^2}{AB^2}$$

$$\therefore \cos^2 \phi + \sin^2 \phi = \frac{AC^2}{AB^2} + \frac{BC^2}{AB^2} = \frac{AC^2 + BC^2}{AB^2}$$

But from Pythagoras' theorem, $AC^2 + BC^2 = AB^2$

$$\therefore \cos^2 \phi + \sin^2 \phi = 1$$

and

$$\cos \phi = \frac{5 - 4}{8} = 0.125$$

Step V. The resistance of the coil then becomes

$$R_{\text{coil}} = Z_{\text{coil}} \cos \phi = 160 \times 0.125 = 20 \; \Omega$$

SELF-CHECK

19-3:1. Find the impedance at 60 Hz of the series combination of a 220-Ω resistor and a practical coil which has 50 Ω of resistance and 600 mH of inductance.

19-3:2. Suppose the series combination of resistor and practical coil from Self-Check Problem 19-3:1 is connected to a 60-Hz source which has a voltage of 120 V $\underline{/0°}$.
   (a) Find the current in the circuit.
   (b) Find the reading that would appear on a voltmeter connected across the resistor.
   (c) Find the reading that would appear on a voltmeter connected across the practical coil.

19-4 RESISTANCE AND CAPACITANCE IN SERIES

In the circuit of Figure 19-6(a), the current again is common to all components. Since the potential difference across the capacitor *lags* the current through it by 90°, the phasor diagram shows the total-voltage phasor located in quadrant IV. From this phasor diagram, we can state that

$$\mathbf{E} = V_R - jV_C \tag{19-8}$$

or
$$\mathbf{E} = \sqrt{V_R^2 + V_C^2} \Big/ -\arctan \frac{V_C}{V_R} \tag{19-9}$$

and
$$\mathbf{Z} = R - jX_C \tag{19-10}$$

or
$$\mathbf{Z} = \sqrt{R^2 + X_C^2} \Big/ -\arctan \frac{X_C}{R} \tag{19-11}$$

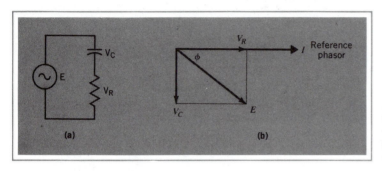

FIGURE 19-6
Phasor Diagram of Voltage and Current Relationships in a Series Circuit Containing Resistance and Capacitance.

Since *inductive* reactance is always drawn in a +j direction in impedance diagrams and *capacitive* reactance in a −j direction, in polar form, an inductive impedance always has a + angle between 0 and +90° and a capacitive impedance always has a − angle between 0 and −90°.

EXAMPLE 19-4

What capacitance when connected in series with a 500-Ω resistor will limit the current drawn from a 48-mV 465-kHz source to 20 μA?

SOLUTION

$$Z = \frac{E}{I} = \frac{48 \text{ mV}}{20 \text{ μA}} = 2.4 \text{ k}\Omega$$

From Pythagoras' theorem with respect to Figure 19-7,

$$X_C = \sqrt{Z^2 - R^2} = \sqrt{2400^2 - 500^2} = 2350 \text{ } \Omega$$

FIGURE 19-7
Impedance Diagram for Example 19-4.

since $\quad X_C = \dfrac{1}{2\pi f C}$

$$C = \frac{1}{2\pi f X_C} = \frac{1}{2 \times \pi \times 465 \text{ kHz} \times 2350 \ \Omega}F = \textbf{146 pF}$$

SELF-CHECK

19-4:1. For the circuit shown in Figure 19-6, suppose that the source is 50 V 400 Hz, the resistance is 2.2 kΩ and the capacitance is 0.047 μF.
(a) Find the circuit impedance.
(b) Find the circuit current.

19-4:2. For the circuit shown in Figure 19-6, suppose that the source is 50 V 400 Hz and the capacitance is 0.047 μF.
(a) What circuit resistance would limit the circuit current to 2.0 mA?
(b) What would be the phase angle of this 2.0 mA current?

19-5 RESISTANCE, INDUCTANCE, AND CAPACITANCE IN SERIES

We can express the sum of the various voltages in the series circuit of Figure 19-8 in rectangular form as

$$\mathbf{E} = V_R + jV_L - jV_C \tag{19-12}$$

The voltage across the inductance *leads* the common current by 90°, and the voltage across the capacitance *lags* the current by 90°. Therefore, these two voltages are exactly

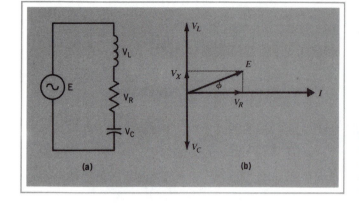

FIGURE 19-8
Phasor Diagram of Voltage and Current Relationships in a Series Circuit Containing Resistance, Inductance, and Capacitance.

180° out of phase. We can then add these two phasors by simple arithmetical subtraction, and Equation (19-12) becomes

$$\mathbf{E} = V_R + j(V_L - V_C) \tag{19-13}$$

Unless the two reactive voltages are exactly equal, this leaves the circuit with an **equivalent reactive voltage** (V_X in Figure 19-8), which will be either inductive or capacitive depending on which of the original reactive voltages is the larger. Note that as far as a fixed-frequency generator is concerned in the circuit of Figure 19-8, the circuit appears to contain only inductance since V_L is greater than V_C. The overall circuit cannot be both inductive and capacitive at the same time. The current must *either* lead *or* lag the applied voltage; it cannot do both.

Therefore, converting Equation (19-13) into polar coordinates gives

$$\mathbf{E} = \sqrt{V_R^2 + V_X^2} \bigg/ \arctan \frac{V_X}{V_R} \tag{19-14}$$

where $V_X = V_L - V_C$ (the net reactive voltage).

If we divide every term in Equations (19-13) and (19-14) by the common current, the impedance of series circuits containing R, L, and C becomes

$$\mathbf{Z} = R + jX_L - jX_C = R + j(X_L - X_C) \tag{19-15}$$

or

$$\mathbf{Z} = \sqrt{R^2 + X_{eq}^2} \bigg/ \arctan \frac{X_{eq}}{R} \tag{19-16}$$

where $X_{eq} = X_L - X_C$ (the net equivalent reactance).

EXAMPLE 19-5
What is the total impedance at 20 kHz of a series circuit consisting of a 1.5-mH inductance, a 100-Ω resistance, and an 0.08-μF capacitance?

SOLUTION
    Step I.   Again, in ac circuits we are interested primarily in the *reactance* of the

inductor and the capacitor. Hence,

$$X_L = 2\pi f L = 2 \times \pi \times 20 \text{ kHz} \times 1.5 \text{ mH}$$

$$= 188.5 \ \Omega \text{ (inductive)}$$

and

$$X_C = \frac{1}{2\pi f C} = \frac{1}{2 \times \pi \times 20 \text{ kHz} \times 0.08 \ \mu\text{F}}$$

$$= 99.47 \ \Omega \text{ (capacitive)}$$

Step II.   Draw a schematic diagram and either a phasor diagram or an impedance diagram, as in Figure 19-9(b).

FIGURE   19-9
Impedance Diagram for Example 19-5.

Step III.

$$\mathbf{Z} = R + jX_L - jX_C = 100 + j188.5 - j99.5 = 100 + j89 \ \Omega$$

Step IV.

$$\phi = \arctan \frac{X_{eq}}{R} = \arctan \frac{89}{100} = +41.7°$$

$$Z = \frac{R}{\cos \phi} = \frac{100 \ \Omega}{\cos 41.7°} = 134 \ \Omega$$

$$\therefore \mathbf{Z} = 134 \ \Omega \ \underline{/+41.7°}$$

SELF-CHECK

19-5:1. A series RLC circuit, similar to the one shown in Figure 19-8, is to operate at 1.0 kHz. The circuit components are as follows:

$$R = 100 \ \Omega, \qquad L = 30 \ \text{mH} \qquad \text{and} \ . \qquad C = 1.6 \ \mu F.$$

Find the impedance of this series circuit.

19-5:2. In a series RLC circuit similar to the one shown in Figure 19-8, the capacitive reactance is known to be 100 Ω. When a 100-V 1-kHz source is used, the circuit current equals 0.8 A and lags the source voltage by 28°.
   (a) Determine the resistance of the resistor.
   (b) Determine the inductance of the inductor.

19-6 RESISTANCE, INDUCTANCE, AND CAPACITANCE IN PARALLEL

On examining Figure 19-10(a), we recall that the characteristic by which we define a parallel circuit is that

The same voltage appears across all parallel branches.

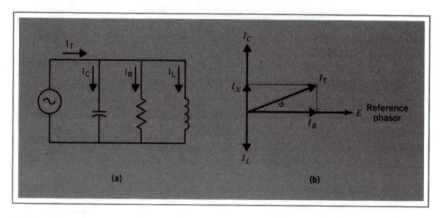

FIGURE 19-10
Phasor Diagram of Voltage and Current Relationships in a Parallel Circuit Containing Resistance, Inductance, and Capacitance.

Therefore, in preparing a phasor diagram for any parallel circuit, we always use this common voltage as the reference phasor. If we are given the numerical value of this voltage, we can solve for the individual branch currents by Ohm's law. Since

$$I_R = \frac{E \ \underline{/0°}}{R \ \underline{/0°}}$$

then I_R is in phase with the reference phasor.

Since

$$I_L = \frac{E \underline{/0°}}{X_L \underline{/+90°}}$$

then I_L lags the reference voltage by 90°.

And since

$$I_C = \frac{E \underline{/0°}}{X_C \underline{/-90°}}$$

I_C leads the reference voltage by 90°, as shown in Figure 19-10. The total current in the parallel circuit then becomes the *phasor* sum of the branch currents. Therefore, from Figure 19-10,

$$\mathbf{I_T} = I_R - jI_L + jI_C = I_R + j(I_C - I_L) \tag{19-17}$$

Converting Equation (19-17) to polar coordinates gives

$$\mathbf{I_T} = \sqrt{I_R^2 + I_X^2} \underline{\Big/ \arctan \frac{I_X}{I_R}} \tag{19-18}$$

where $I_X = I_C - I_L$ (*the net reactive current*). We can now determine the equivalent impedance of a parallel circuit by using Equation (19-6).

$$\mathbf{Z_{eq}} = \frac{E}{\mathbf{I_T}} \tag{19-6}$$

This method of solving for the equivalent impedance of a parallel circuit is called the *total-current method.* Even if the exact value of the applied voltage is not known, we may assume any convenient value in order to solve for the equivalent impedance.

⌐ EXAMPLE 19-6
What is the equivalent impedance at 20 kHz of a circuit consisting of a 1.5-mH inductance, a 100-Ω resistance, and an 0.08-μF capacitance all connected in parallel?

SOLUTION
    Step I.

$$X_L = 2\pi f L = 2 \times \pi \times 20 \text{ kHz} \times 1.5 \text{ mH}$$
$$= 188.5 \ \Omega \text{ (inductive)}$$

$$X_C = \frac{1}{2\pi f C} = \frac{1}{2 \times \pi \times 20 \text{ kHz} \times 0.08 \ \mu\text{F}}$$
$$= 99.5 \ \Omega \text{ (capacitive)}$$

Step II. Assume a convenient value of applied voltage. For example, let $E = 200 \text{ V} \underline{/0°}$.

Then $\qquad I_R = \dfrac{E\underline{/0°}}{R} = \dfrac{200\underline{/0°}}{100\underline{/0°}} = 2 \text{ A}\underline{/0°}$

$$I_L = \dfrac{E\underline{/0°}}{+jX_L} = \dfrac{200\underline{/0°}}{188.5\underline{/+90°}} = 1.06 \text{ A}\underline{/-90°}$$

and $\qquad I_c = \dfrac{E\underline{/0°}}{-jX_C} = \dfrac{200\underline{/0°}}{99.5\underline{/-90°}} = 2.01 \text{ A}\underline{/+90°}$

To facilitate the phasor division, we expressed the direction associated with R, X_L, and X_C in polar form rather than rectangular coordinate operators. Note also that dividing a quantity having the same direction as the reference phasor by a $+j$ quantity creates a $-j$ quantity, and vice versa.

Step III. Prepare a circuit diagram and phasor diagram, as in Figure 19-11.

Step IV.

$$\mathbf{I_T} = I_R - jI_L + jI_C = 2 - j1.06 + j2.01 = 2 + j0.95 \text{ A}$$

Step V.

$$\phi = \arctan \frac{I_X}{I_R} = \arctan \frac{0.95}{2} = +25.4°$$

$$I_T = \frac{I_R}{\cos \phi} = \frac{2.0 \text{ A}}{\cos 25.4°} = 2.214 \text{ A}$$

$$\therefore \mathbf{I_T} = 2.21 \text{ A}\underline{/+25.4°}$$

FIGURE 19-11
Phasor Diagram for Example 19-6.

Step VI.

$$Z_{eq} = \frac{E}{I_T} = \frac{200\,\underline{/0°}}{2.21\,\underline{/+25.4°}} = 90.5\ \Omega\,\underline{/-25.4°}$$

Note that the equivalent impedance is capacitive, since the capacitive branch draws a greater *current* from the source than does the inductive branch. When the *same* three components were connected in series in Example 19-5, the total impedance was inductive, since there was a greater *voltage* across the inductance than across the capacitance.

SELF-CHECK

19-6:1. Consider an *RLC* parallel circuit similar to that shown in Figure 19-10, where the source voltage is 120 V, the resistance is 1.0 kΩ, the inductance is 0.10 H and the capacitance is 1.0 μF. If the source frequency is 800 Hz, find:
(a) the total current
(b) the total impedance for this parallel circuit.
19-6:2. Referring to the *RLC* parallel circuit shown in Figure 19-10, suppose that the source has a voltage of 120 V $\underline{/0°}$ and has a frequency of 60 Hz. Measurements give the total circuit current as $I_T = 4 - j3$ A and the capacitance as 42 μF. Determine:
(a) the resistance and
(b) the inductance of this circuit.

19-7 CONDUCTANCE, SUSCEPTANCE, AND ADMITTANCE

Although we listed the total-current method in solving for the equivalent resistance of parallel dc circuits, we preferred to use a method that did not require the in-between step of solving for total current. To do this, we found it necessary to change our point of view in dealing with parallel circuits. Rather than thinking in terms of the ability of a circuit to *oppose* current, we found that in dealing with parallel circuits we could obtain a better picture of the circuit behavior if we thought in terms of the ability of the various branches to *pass* current. We call this reciprocal property of resistance **conductance**.

> *Conductance is a measure of the ability of resistance to* **pass** *electric current.*
> *The letter symbol for conductance is G.*

The unit of conductance is the **siemens** as defined by the equation $G = 1/R$.
If we wish, we can write Equation (19-17) as

$$\frac{E}{Z} = \frac{E}{R} - j\frac{E}{X_L} + j\frac{E}{X_C}$$

Since the *same* voltage appears across all components in a simple parallel circuit, this becomes

$$\frac{1}{\mathbf{Z}} = \frac{1}{R} - j\frac{1}{X_L} + j\frac{1}{X_C} \qquad (19\text{-}19)$$

We already have a term and a letter symbol for the reciprocal of resistance, namely, **conductance**. We could certainly clean up the appearance of Equation (19-19) if we also had such reciprocals for impedance and reactance. We defined impedance as the total *opposition* to current in an ac circuit. Hence, its reciprocal,

> *Admittance is the overall ability of an electric circuit to **pass** alternating current. The letter symbol for admittance is **Y**.*

where $\mathbf{Y} = 1/\mathbf{Z}$ *siemens*. Similarly, the reciprocal of reactance,

> *Susceptance is the ability of inductance or capacitance to **pass** alternating current. The letter symbol for susceptance is **B**.*

where $B = 1/X$ *siemens*.

Therefore, Equation (19-19) becomes

$$\mathbf{Y} = G - jB_L + jB_C = G + j(B_C - B_L) \qquad (19\text{-}20)$$

Note that when we divide a phasor quantity with a $+90°$ angle into $1\underline{/0°}$, the quotient has a $-90°$ angle, and vice versa. Therefore, the reciprocal of $+jX_L$ is $-jB_L$, and the reciprocal of $-jX_C$ is $+jB_C$. Using the now familiar relationship, we can express the polar coordinates of admittance as

$$\mathbf{Y} = \sqrt{G^2 + B_{eq}^2} \underline{\bigg/ \arctan \frac{B_{eq}}{G}} \qquad (19\text{-}21)$$

where $B_{eq} = B_C - B_L$ *(the net equivalent susceptance)*. Since $B_L = 1/X_L$,

$$B_L = \frac{1}{2\pi f L} = \frac{1}{\omega L} \qquad (19\text{-}22)$$

and

$$B_C = 2\pi f C = \omega C \qquad (19\text{-}23)$$

EXAMPLE 19-6A

What is the equivalent impedance at 20 kHz of a circuit consisting of a 1.5-mH inductance, a 100-Ω resistance, and an 0.08-μF capacitance all connected in parallel?

SOLUTION

Step I.

$$G = \frac{1}{R} = \frac{1}{100 \ \Omega} = 0.01 \text{ siemens}$$

$$B_C = 2\pi f C = 2 \times \pi \times 20 \text{ kHz} \times 0.08 \ \mu\text{F}$$

$$= 0.01 \text{ S (capacitive)}$$

$$B_L = \frac{1}{2\pi f L} = \frac{1}{2 \times \pi \times 20 \text{ kHz} \times 1.5 \text{ mH}}$$

$$= 0.0053 \text{ S (inductive)}$$

Step II.   Prepare an admittance diagram for the circuit of Figure 19-11(a). (Figure 19-12.) Note that it is not possible to draw an *impedance* diagram using the actual values given in the circuit of Figure 19-11(a), since the equivalent impedance of a parallel circuit must be *smaller* than the resistance, and it is impossible for the hypotenuse of a right-angled triangle to be smaller than either of the two sides.

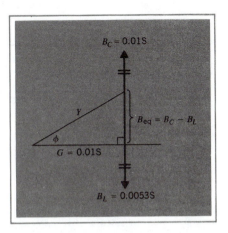

FIGURE   19-12
Admittance Diagram for Example 19-6(a).

Step III.

$$\mathbf{Y} = G + j(B_C - B_L) = 0.01 + j(0.01 - 0.0053)$$

$$= 0.01 + j0.0047 \text{ S}$$

Step IV.

$$\phi = \arctan \frac{B_{eq}}{G} = \arctan \frac{0.0047}{0.01} = +25.2°$$

$$Y = \frac{G}{\cos \phi} = \frac{0.01 \text{ s}}{\cos 25.2°} = 0.011\,05 \text{ S}$$

$$\therefore \mathbf{Y} = 11.05 \text{ mS} \underline{/+25.2°}$$

Step V.

$$\mathbf{Z}_{eq} = \frac{1}{\mathbf{Y}} = \frac{1 \underline{/0°}}{11.05 \text{ mS} \underline{/+25.2°}} = \mathbf{90.5 \ \Omega \underline{/-25.2°}}$$

SELF-CHECK

19-7:1. A 2.2-μF capacitor and a 2.0-H inductor are to be used in a 60-Hz circuit. Find:
    (a) the susceptance of the capacitor
    (b) the susceptance of the inductor
    (c) the admittance of the parallel combination of the inductor and the capacitor

19-7:2. A 120-V 60-Hz source is connected to a parallel combination of a 3.3-kΩ resistor, a 2.2-μF capacitor and a 2.0-H inductor.
    (a) Determine the admittance of the parallel combination.
    (b) Find the current drawn from the source.

19-7:3. Consider a parallel circuit similar to the one shown in Figure 19-10. When connected to a 90 V $\underline{/0°}$ source the circuit draws a total current of 20 mA $\underline{/-20°}$. When the inductor branch is removed, the same source drives a total current of 25 mA magnitude. Find R, B_C and B_L for the original circuit.

19-8 IMPEDANCE AND ADMITTANCE

In the simple parallel circuit of Figure 19-10, we made sure that each branch contained only one type of opposition to alternating current. Hence, we were able to substitute G for $1/R$ and B for $1/X$. Suppose, however, that one of the branches had been a practical inductor possessing both resistance and inductive reactance, as in Figure 19-13. The current through this branch would now be inversely proportional to its **impedance**. Hence, if we want to use a reciprocal unit, it would have to be **Y** for $1/\mathbf{Z}$.

FIGURE 19-13
Finding the Conductance and Susceptance of a Practical Inductor.

Using the resistance and reactance values given for the practical inductor in Figure 19-13, we *cannot* say that the conductance of this branch is

$$G = \frac{1}{R} = \frac{1}{30} = 0.333 \text{ S}$$

and the susceptance of the branch is *not*

$$B = \frac{1}{X} = \frac{1}{40} = 0.025 \text{ S}$$

Doing so would leave us with a conductance and susceptance in *series*, just as in the original circuit of Figure 19-13. But in Equation (19-20), conductance and susceptance are the rectangular coordinates of admittance, and we developed Equation (19-20) from *parallel* branches. We can illustrate the procedure for finding the conductance and susceptance of a practical inductor from the following example.

EXAMPLE 19-7

What is the conductance and susceptance of a practical inductor whose resistance is 30 Ω and whose inductive reactance is 40 Ω?

SOLUTION

    Step I. Find the impedance of the inductor.

$$\mathbf{Z} = \sqrt{R^2 + X^2}\ \Big/\ \arctan \frac{X}{R}$$

$$= \sqrt{30^2 + 40^2}\ \arctan \frac{40}{30} = 50\ \Omega\underline{/+53.1°}$$

    Step II. From the definition of admittance,

$$\mathbf{Y} = \frac{1}{\mathbf{Z}} = \frac{1}{50\ \Omega\underline{/+53.1°}} = 0.02\ \mathrm{S}\underline{/-53.1°}$$

    Step III. Since $\mathbf{Y} = G - jB$, we split $\mathbf{Y}$ into its rectangular coordinates.

$$G = Y \cos \phi = 0.02 \cos 53.1° = 0.02 \times 0.6$$

$$\therefore G = \textbf{0.012 siemens}$$

And

$$B_L = Y \sin \phi = 0.02 \sin 53.1° = 0.02 \times 0.8$$

$$\therefore B_L = \textbf{0.016 siemens}$$

Compare these answers with the *wrong* answers on page 516. With a calculator, we can make the double conversion with a single chain as follows:

    30 $\boxed{x \leftrightarrow y}$ 40 $\rightarrow$ P displays 50 = Z

    Press $\boxed{1/x}$ to display 0.02 = Y

    Press $\boxed{x \leftrightarrow y}$ to display 53.13° = ϕ

    Press $\boxed{+/-}$ to change the sign of ϕ

    Press $\boxed{\rightarrow R}$ and then $\boxed{x \leftrightarrow y}$ to display the rectangular coordinates of $\mathbf{Y}$

$$0.012 - j0.016\ \mathrm{S}$$

Note that the rectangular coordinates of an **impedance** are the resistance and reactance components which always represent the *series* equivalent circuit, while the rectangular coordinates of an **admittance** are the conductance and susceptance com-

.ponents which always represent the *parallel* equivalent circuit. We first encountered these "similar but opposite" characteristics of series and parallel circuits when we compared series and parallel resistances in dc circuits in Section 6-8. Extending this comparison to ac circuits, we can summarize the reciprocal characteristics of series and parallel circuits in Table 19-1.

SELF-CHECK

19-8:1. Determine the conductance and susceptance at 60 Hz of a practical coil which has the following series specifications:

$$R = 100 \ \Omega \quad \text{and} \quad L = 1.0 \text{ H}.$$

19-8:2. Determine the series-equivalent resistance and reactance of a capacitor which has an admittance of

$$\mathbf{Y} = 10 + \text{j}300 \ \mu\text{S}.$$

PROBLEMS

Draw a circuit diagram and either a phasor diagram or an impedance diagram (series circuits) or admittance diagram (parallel circuits) with the solution for each problem.

19-1-1. Calculate the magnitudes of the missing rms voltages in the ac circuits shown in Figure 19-14.

(a)   (b)

FIGURE 19-14

19-1-2. What are the phase angles of the source voltages with respect to the current in the ac circuits shown in Figure 19-14?

19-1-3. The voltages across an ideal inductor and a resistor connected in series are 24 V and 30 V rms, respectively. What is the magnitude and phase angle of the total voltage with respect to the common current?

19-1-4. An inductor connected to a 120-V 60-Hz voltage source has an *IR* drop of 72 V across the resistance of its winding. What is the magnitude and phase angle of the voltage across its inductance with respect to the applied voltage?

19-2-5. Calculate the total impedance (polar coordinates) at 60 Hz of a 1.8-H inductance in series with a 500-Ω resistance.

TABLE 19-1

SERIES AND PARALLEL AC CIRCUIT CHARACTERISTICS

Characteristic	Series components	Parallel components
Phasor diagram		
Sum of phasors	$\mathbf{V_T} = \mathbf{V_1} + \mathbf{V_2} + \mathbf{V_3} +$ etc.	$\mathbf{I_T} = \mathbf{I_1} + \mathbf{I_2} + \mathbf{I_3} +$ etc.
Defining equation	$\mathbf{Z} = \dfrac{\mathbf{V_T}}{\mathbf{I}}$	$\mathbf{Y} = \dfrac{\mathbf{I_T}}{\mathbf{V}}$
Impedance diagram		Not possible
Admittance diagram	Not possible	
Rectangular form	$\mathbf{Z} = R + jX_{eq}$	$\mathbf{Y} = G + jB_{eq}$
Resistive component	R ohms	G siemens
Reactive component	$+jX_{eq} = +j(X_L - X_C)$ where $X_L = \omega L$ $X_C = \dfrac{1}{\omega C}$	$+jB_{eq} = +j(B_C - B_L)$ where $B_C = \omega C$ $B_L = \dfrac{1}{\omega L}$
Polar form	$\mathbf{Z} = \lvert Z \rvert \; \underline{/\phi}$	$\mathbf{Y} = \lvert Y \rvert \; \underline{/\phi}$
Magnitude	$\lvert Z \rvert = \sqrt{R^2 + X_{eq}^2}$	$\lvert Y \rvert = \sqrt{G^2 + B_{eq}^2}$
Phase angle	$\phi = \arctan \dfrac{X_{eq}}{R}$	$\phi = \arctan \dfrac{B_{eq}}{G}$

19-2-6. .At what frequency will an inductance of 500 mH in series with a resistance of 250 Ω have an impedance of 300 Ω?

19-2-7. When the operating coil of a contactor is connected to a 120-V 60-Hz source, it draws a 0.9-A current which lags the applied voltage by 72°. What is its impedance
(a) In polar coordinates?
(b) In rectangular coordinates?

19-2-8. When a 400-Hz sine-wave voltage of 4 V $\underline{/0°}$ is applied to the voice coil of a loudspeaker, the resulting current is 0.5 A $\underline{/-20°}$. Express the impedance of the loudspeaker in both polar and rectangular coordinates.

19-2-9. What is the resistance of a 4-H choke whose impedance at 25 Hz has a magnitude of 750 Ω? What is the angle of the impedance?

19-2-10. What two components in series will have a 2-kHz impedance of 5 kΩ $\underline{/+60°}$?

19-2-11. Determine the series equivalent circuit of a device that has an impedance of 2.4 kΩ $\underline{/+30°}$ at a frequency of 200 Hz.

19-2-12. Determine the series equivalent circuit of a device which draws a current of 200 μA $\underline{/-30°}$ from a 1-kHz voltage source of 50 mV $\underline{/0°}$.

19-2-13. When connected to a 12-V battery, a certain choke draws a current of 0.25 A. When connected to a 120-V 60-Hz source, it draws a current of 1.0 A. Determine the resistance and inductance of the choke.

19-2-14. When connected to a 117-V 25-Hz source, a certain inductor draws a 5.0-A current. When connected to a 120-V 60-Hz source, the same inductor draws a 3.0-A current. Determine the resistance and inductance of the inductor.

19-3-15. A 15-W fluorescent lamp operates with a ballast inductance in series with it. The voltage across the lamp is measured as 56 V rms, and the total voltage across the ballast inductor is measured as 100 V rms when the total applied voltage is 120 V 60 Hz. What are the resistance and inductance of the ballast? (The lamp itself represents resistance only.)

19-3-16. A projector using a special 120-V 300-W lamp is operated on the European 230-V 50-Hz system by connecting a choke in series with the lamp. If the resistance of the choke is 20 Ω, what must its inductance be in order to provide the correct voltage drop across the lamp?

19-4-17. Calculate the magnitudes of the missing rms voltages in the ac circuits shown in Figure 19-15.

(a)   (b)

FIGURE 19-15

19-4-18. The voltages across a capacitor and a resistor connected in series in an ac circuit are measured as 60 V and 80 V rms, respectively. What is the magnitude and phase angle of the total voltage with respect to the common current?

19-4-19. What is the 1.5-MHz impedance (polar coordinates) of a 200-pF capacitor in series with a 1.2-kΩ resistor?

19-4-20. Calculate the total impedance (polar coordinates) at 1 kHz of a 0.5-μF capacitor and a 500-Ω resistor in series.

19-4-21. What are the magnitude and phase angle of the current that an $80 - j95$ Ω impedance will draw from a 117-V ac source?

19-4-22. The input circuit of a radio receiver draws a current of 3 μA$\underline{/+15°}$ from a 1-MHz signal generator when the terminal voltage is 225 μV$\underline{/0°}$. Express the input impedance of the receiver in rectangular coordinates.

19-4-23. At what frequency will a 0.02-μF capacitor and a 120-kΩ resistor in series have an impedance of 200 kΩ?

19-4-24. Determine the series equivalent circuit of a 4.5-MHz impedance of 150 $\Omega\underline{/-75°}$.

19-4-25. What value of capacitor must be connected in series with a 560-Ω resistor to limit its dissipation to 5 W when connected to a 120-V 60-Hz source?

19-5-26. Calculate the magnitude and angle of the open-circuit output voltages in the coupling circuits shown in Figure 19-16 if the applied voltage in both cases is 150 μV$\underline{/0°}$. [There are two possible answers for the circuit in Figure 19-16(b).]

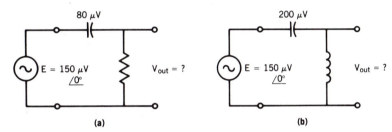

(a)   (b)

FIGURE 19-16

19-5-27. What is the total 60-Hz impedance (polar form) of a 1-H choke whose resistance is 100 Ω in series with a 10-μF capacitor?

19-5-28. Calculate the impedance (polar form) of an ac circuit consisting of a 680-Ω resistor, a 200-mH inductor, and a 0.1-μF capacitor in series at frequencies of (a) 400 Hz, (b) 1 kHz, and (c) 3 kHz.

19-5-29. At a frequency of 400 Hz, what value of capacitor must be connected in series with an impedance of 75 $\Omega\underline{/+60°}$ to form a total impedance of 75 $\Omega\underline{/-60°}$?

19-5-30. What is the total impedance when $\mathbf{Z}_1 = 240$ $\Omega\underline{/+15°}$, $\mathbf{Z}_2 = 480$ $\Omega\underline{/-45°}$, and $\mathbf{Z}_3 = 1$ k$\Omega\underline{/+60°}$ are all connected in series?

19-5-31. A relay coil requires a 100-mA current through it in order to close the contacts of the relay. If it is operated from a dc source, the required voltage is 24 V. If it is operated from a 60-Hz source, the required voltage is 160 V. What value of capacitance in series with the relay coil will allow its operation from a 120-V 60-Hz supply? (Two answers.)

19-5-32. At a frequency of 400 Hz, what value of capacitance must be connected in series with an impedance of 75 $\Omega\underline{/+60°}$ to form
(a) An inductive impedance of 50 Ω?
(b) A capacitive impedance of 50 Ω?

19-5-33. At what frequency will the total impedance of the series circuit of Problem 19-5-28 have an angle of $\underline{/0°}$?

19-5-34. A coil and capacitor are connected in series across a 400-Hz source. The voltage across the coil is measured as 48 V, across the capacitor as 52 V, and across the whole circuit as 6 V. The current in the circuit is 100 mA. Determine

(a) The inductance of the coil.
(b) The resistance of the coil.
(c) The capacitance of the capacitor.
(d) The total impedance in polar form.

19-6-35. Calculate the magnitude and the phase angle with respect to the applied voltage of the unknown current in the ac circuits shown in Figure 19-17.

(a)

(b)

FIGURE 19-17

19-6-36. If the magnitudes of the applied voltages in the circuits shown in Figure 19-17 are 75 V rms, calculate the equivalent impedance of each circuit.

19-6-37. A 60-W incandescent lamp and a 44.2-μF capacitor are connected in parallel to a 120-V 60-Hz ac source.
(a) What is the total current drawn from the source?
(b) What is the equivalent impedance of the load?

19-6-38. The bias circuit of an audio amplifier consists of a 2200-Ω resistor in parallel with a 0.2μF capacitor. What is the equivalent impedance of this bias circuit at a frequency of 1 kHz?

19-7-39. Calculate the total admittance and equivalent impedance (polar coordinates) at 60 Hz of a 20-mS conductance in parallel with a 40-mS capacitive susceptance.

19-7-40. Calculate the total admittance and equivalent impedance (polar coordinates) at 45 MHz of a 300-μS conductance in parallel with a 30-pF capacitor.

19-7-41. Calculate the 60-Hz admittance of a 0.5-H inductance in parallel with a 12-mS conductance.

19-7-42. What is the equivalent impedance of a device whose admittance is $0.05 - j0.015$ S? Is this device inductive or capacitive?

19-7-43. What values of conductance and susceptance must be connected in parallel to form an admittance of 500 mS$\underline{/-30°}$?

19-7-44. Calculate the admittance (polar form) of an ac circuit consisting of a 680-Ω resistor, a 200-mH inductor, and a 0.1-μF capacitor in parallel at frequencies of (a) 400 Hz, (b) 1 kHz, and (c) 3 kHz.

19-7-45. Calculate the impedance (polar form) of the circuit in Problem 19-7-44. (Compare with the answers to Problem 19-5-28.)

19-7-46. What is the total admittance (in polar coordinates) when an ac circuit consists of two parallel branches whose admittances are $36 + j24$ mS and $42 - j20$ mS?

19-7-47. What is the equivalent impedance of an ac circuit consisting of two parallel branches whose admittances are 64 μS$\underline{/+15°}$ and 48 μS$\underline{/-45°}$?

19-7-48. A given impedance draws a 5.0-A current from a 120-V 60-Hz source. The current through the impedance lags the voltage across it by 60°. What is the total current drawn from the source when a 100-μF capacitor is connected in parallel with the given impedance?

19-7-49. What is the equivalent impedance (polar form) when $\mathbf{Z}_1 = 240 \ \Omega\underline{/+15°}$, $\mathbf{Z}_2 = 480 \ \Omega\underline{/-45°}$, and $\mathbf{Z}_3 = 1 \ k\Omega\underline{/+60°}$ are all connected in parallel? (Compare with the answer to Problem 19-5-30.)

19-8-50. What values of resistance and capacitance in parallel have a 465-kHz admittance of $40 \ \mu S\underline{/+60°}$?

19-8-51. Determine the parallel equivalent circuit of a device that has an impedance of $2.4 \ k\Omega\underline{/+30°}$ at a frequency of 200 Hz. (Compare with the answer to Problem 19-2-11.)

19-8-52. Determine the parallel equivalent circuit at 4.5 MHz of an impedance of $150 \ \Omega \ \underline{/-75°}$. (Compare with the answer to Problem 19-4-24.)

19-8-53. Determine the series components which will have an equivalent admittance at 60 Hz of $20 - j80$ mS.

19-8-54. Determine the series components which will have an equivalent admittance at 25 MHz of $120 + j90$ S.

19-8-55. What value of capacitance must be connected in parallel with a 75-Ω resistor to form a total admittance at 500 Hz of 20 mS?

19-8-56. What value of resistance must be connected in parallel with an inductance of 1.5 H to produce a 60-Hz impedance of 400 Ω?

REVIEW QUESTIONS

19-1. Why is the additional term **impedance** required in dealing with the opposition of a circuit to alternating current?

19-2. Why must we add the resistance and reactance of a series circuit by the Pythagorean method in order to find the total impedance?

19-3. Why is it necessary to state an angle as well as a magnitude when an impedance is given in polar form?

19-4. Why does a + angle with an impedance always represent a circuit possessing some inductive reactance?

19-5. How is the angle of an impedance related to the voltage across and the current through the impedance?

19-6. Why is it not necessary to state an angle when an impedance is stated in rectangular coordinates?

19-7. What is the significance of a total impedance having a 0° angle?

19-8. Why must the angle of the impedance of a practical inductor be something less than +90°?

19-9. Expressing the impedance of a practical inductor in rectangular coordinates gives us the actual resistance and reactance components which form the practical inductor. Explain.

19-10. How do we go about determining the total impedance of a practical inductor and a resistor in series when the angle between them is less than 90°?

19-11. Under what circumstances is it possible for the total impedance to decrease when a capacitor is added in series with a given impedance?

19-12. Under what circumstances is it possible for the rms current drawn from a certain source to have the same magnitude when an inductance is connected in series with a given impedance?

19-13. Explain the statement that an ac circuit containing both inductance and capacitance appears to the source only as either an inductive circuit or a capacitive circuit.

19-14. What is meant by **equivalent reactance**?

19-15. Why is quadrant I the capacitive quadrant of a parallel circuit phasor diagram, whereas quadrant IV is the capacitive quadrant of a series circuit phasor diagram?

19-16. Why can we not prepare an impedance diagram for a parallel circuit?

19-17. What is the advantage of thinking in terms of admittance, conductance, and susceptance in dealing with parallel ac circuits?

19-18. Why does B_L have a $-j$ operator for phasor addition, whereas X_L has a $+j$ operator?

19-19. What is meant by the **equivalent impedance** of a parallel ac circuit?

19-20. Explain why $\mathbf{Z} = 20 + j40\ \Omega$ is *not* the equivalent impedance of $\mathbf{Y} = 50 - j25$ mS.

*19-21. Write a computer program which will permit entering the rectangular coordinates of an impedance and then print out the rectangular coordinates of the equivalent admittance, or vice versa. Verify the program by using it to solve Example 19-7.

20

POWER IN ALTERNATING-CURRENT CIRCUITS

Most of the examples in this chapter deal with power in ac circuits at the standard North American power-line frequency of 60 hertz. This is not surprising when we recall that we defined **power** as the *rate of doing work* or the *rate of converting energy from one form to another*. Much of the terminology we encounter in this chapter has come from power-systems applications and has resisted a trend toward increased use of the mathematical terminology of complex algebra.

20-1 POWER IN A RESISTOR

In Chapter 16, we investigated the nature of the instantaneous power in a resistor in order in establish rms values for voltage and current in ac circuits. We can set a pattern for dealing with power in ac circuits in general by reviewing the procedure we used to determine the power (or rate of conversion of energy) in a resistor.

If we consider only one particular instant in time, all the power equations we used in direct-current circuits (Chapter 5) must apply to alternating-current circuits. Therefore,

$$p = vi \qquad (20\text{-}1)$$

And when we plot the sine-wave voltage drop across and current through a resistor, as shown by the shaded curves of Figure 20-1, we can place a series of plots on a linear graph at small intervals, representing the product of v and i at those particular instants.

FIGURE 20-1
Instantaneous Power in a Resistor.

Joining these plots gives us the black curve of Figure 20-1 representing power input to a resistor.

Since the voltage drop across a resistor must be exactly in phase with the current through it, the instantaneous voltage and current reach their peak values simultaneously. Whenever i is a positive quantity, v is also a positive quantity; and whenever i is a negative quantity, v is also a negative quantity. Since the product of two negative quantities is a positive quantity, the instantaneous-power graph for a resistor is always positive. This indicates that a resistor must always convert electric energy into heat, never vice versa.

We also noted that the instantaneous power in a resistor is pulsating in nature, fluctuating sinusoidally between zero and a peak value of $P_m = V_m I_m$ twice during each complete cycle of voltage and current. Because of this smooth sinusoidal variation between zero and P_m, we concluded that, as far as doing work is concerned, we can consider the power in a resistive ac circuit as having an average, or dc equivalent value equal to $0.5\,P_m$.

We verified this conclusion by substituting the sine-wave form for the instantaneous voltage and current in Equation (20-1), giving

$$p = V_m I_m \sin^2 \omega t = P_m \sin^2 \omega t \tag{16-14}$$

From trigonometric half-angle relationships, this became

$$p = 0.5\,P_m(1 - \cos 2\omega t) \tag{16-15}$$

Since the *average* of a cosine wave over a complete cycle is zero, from Equation (16-15) the average power $P = 0.5\,P_m$.

In Chapter 16, we used this average value of power in a resistor to establish the rms value of a sine wave of voltage and current as 0.707 of its peak value. Therefore, it follows that the average power input to the resistance of an ac circuit is simply the product of the rms value of the voltage drop across and the rms value of the current through the resistance, and

$$P = V_R I_R \tag{20-2}$$

Because a resistor can convert energy only from an electric form into heat or light but never vice versa, the average power in watts determined by Equation (20-2) used to be called the **true power** input to the resistance of an ac circuit. More recent terminology refers to the average power input to the resistance of an ac circuit as **real power** or **active power.** We shall appreciate the choice of such designations when we consider power in an ac circuit containing both resistance and reactance in Section 20-4. Note that the letter symbol for average power in an ac circuit in Equation (20-2) is the same as that for power in a dc circuit.

*The product of the rms voltage across and the current in the **resistance** of an ac circuit is called the active power of the circuit.*
The letter symbol for active power is P.

Since $R = V_R/I_R$, just as in dc circuits,

$$P = V_R I_R = I_R^2 R = \frac{V_R^2}{R} \tag{20-3}$$

where V_R is the rms voltage across only the resistance portion of the circuit and I_R is the rms current through only the resistance portion of the circuit.

EXAMPLE 20-1

A wattmeter shows an average power of 144 W in an ac circuit in which the rms current is 2 A. What is the resistance of this circuit?

SOLUTION

$$R = \frac{P}{I^2} = \frac{144\ \text{W}}{(2\ \text{A})^2} = \textbf{36}\ \boldsymbol{\Omega}$$

SELF-CHECK

20-1:1. A 360-Ω resistor is connected across a 120-V 60-Hz source. Find the average power dissipated in the resistor.

20-1:2. For the resistor in Problem 20-1:1, determine the following:
   (a) the maximum instantaneous power and
   (b) the minimum instantaneous power.

20-2 POWER IN AN IDEAL INDUCTOR

We can determine the nature of the instantaneous power in an ideal inductor by the same technique of plotting the product of the instantaneous voltage and current on a linear graph. But in preparing the shaded sine curves of Figure 20-2, we must note that the instantaneous current in an ideal inductor *lags* the voltage across it by $\pi/2$ radians. (Refer to Figure 17-2.) Hence, although the graph of the instantaneous voltage in Figure 20-2

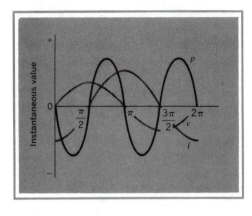

FIGURE 20-2
Instantaneous Power in an Ideal Inductor.

reaches its positive peak at $\pi/2$ radians, the instantaneous current graph does not reach its positive peak until $\pi/2$ radians later—at π radians.

At the zero-radian point on the graph of Figure 20-2, the instantaneous voltage across the inductance of the circuit is zero. Therefore, between 0 and $\pi/2$ radians, the instantaneous voltage is increasing in a positive direction. To generate this voltage, the *rate of change* of current is also increasing in a positive direction as the instantaneous current decreases toward zero. During this interval the instantaneous current is actually flowing *against* the applied voltage as the magnetic field around the inductor collapses, thus returning the energy stored in its magnetic field to the source. As we would expect, the product of a positive v and a negative i gives a *negative* instantaneous power as energy is transferred from the inductance to the source. At $\pi/2$ radians, the current momentarily becomes zero; thus the instantaneous power is also zero.

Between $\pi/2$ and π radians, the instantaneous voltage across the inductor is still positive but decreasing in magnitude. The instantaneous current goes positive as it maintains its positive rate of change in order to induce a positive voltage in the inductance. The product $v \times i$ is now a positive quantity as the instantaneous current increases and energy is stored in the magnetic field of the inductor. At π radians when the instantaneous voltage becomes zero, the instantaneous power also momentarily becomes zero.

The instantaneous power again becomes negative between π and $3\pi/2$ radians when the instantaneous current is decreasing and the collapsing magnetic field again returns energy to the system. Between $3\pi/2$ and 2π radians both v and i have negative polarities creating a positive $v \times i$ product as the source supplies energy to rebuild the magnetic field of the inductor.

The resulting black curve of Figure 20-2 for instantaneous power in the inductance of an ac circuit is sinusoidal in shape at twice the frequency of the instantaneous voltage and current, just as in the resistance of an ac circuit. But in the inductance, the instantaneous power fluctuates equally between a positive maximum and a negative maximum as the ideal inductor alternately stores and then returns equal amounts of energy in the circuit. Therefore, **the *average* power in the inductance of an ac circuit is zero.**

Again we can substitute for v and i in Equation (20-1) using the expression for instantaneous current in an ideal inductor given in Equation (17-3). Hence,

$$p = v \times i = V_m \sin \omega t \times I_m \sin\left(\omega t - \frac{\pi}{2}\right)$$

From trigonometry, $\sin \theta - 90°$ is the same as $-\cos \theta$. Hence,

$$p = -V_m I_m \sin \omega t \cos \omega t$$

From the half-angle relationship, this becomes[†]

$$p = -0.5 \, V_m I_m \sin 2\omega t \tag{20-4}$$

Hence, the instantaneous power in an ideal inductor is a sine wave at twice the frequency of the instantaneous voltage and current and starting off in the *negative* direction from 0°. But when we average $\sin 2\omega t$ over a complete cycle, there is no nonvarying term in Equation (20-4) as there was in Equation (16-15). Hence, for an ideal inductor, the *average* power is zero.

For the resistance of an ac circuit, we can determine average power simply by multiplying the rms values of current and voltage [Equation (20-2)]. In the circuit of Figure 20-3, the meters show us the rms values of the voltage across and the current through an ideal inductor. But since the average power in the inductance is zero, $V_L I_L$ does *not* represent average power. However, this product is directly proportional to the amount of energy stored and returned by the inductor every time the current changes direction. Therefore,

The product of the rms voltage and current in an ideal inductor is called the reactive power of the inductor.

The letter symbol for reactive power is Q.[‡]

$$\therefore Q = V_L I_L \tag{20-5}$$

[†]
$$\sin (A + B) = \sin A \cos B + \cos A \sin B$$
$$\sin (A - B) = \sin A \cos B - \cos A \sin B$$

If $A = B$, since $\sin 0° = 0$, $\sin 2A = 2 \sin A \cos A$.

[‡] Prior to 1968, reactive power was represented by the more descriptive letter symbol P_q.

FIGURE 20-3
Reactive Power in an Ideal Inductor.

We can now see why the preferred term for the average power in the resistance of an ac circuit is **active power.** Since the average power in the inductance of an ac circuit is zero, to avoid the possibility of mistaking reactive power for active power, we reserve the **watt** exclusively for active power in an ac circuit. Reactive power is simply the product of the voltage across and the current through the reactance. Therefore,

The voltampere (reactive), which is usually abbreviated to **var,** *is the unit of reactive power.*

Since

$$X_L = \frac{V_L}{I_L} \tag{17-4}$$

$$\therefore Q = V_L I_L = I_L^2 X_L = \frac{V_L^2}{X_L} \qquad \text{vars} \quad (20\text{-}6)$$

EXAMPLE 20-2

What is the reactive power of an ideal inductor with an inductance of 0.5 H drawing 0.5 A from a 60-Hz source?

SOLUTION

$$X_L = \omega L = 377 \times 0.5 \text{ H} = 188.5 \ \Omega \text{ (inductive)}$$

$$Q = I_L^2 X_L = (0.5 \text{ A})^2 \times 188.5 \ \Omega = \textbf{47.1 vars}$$

SELF-CHECK

20-2:1. For the circuit shown in Figure 20-3, suppose that the source is a 120-V 60-Hz generator and that the inductance is 1 H.
    (a) Write the equation for the instantaneous power in the inductor.
    (b) Determine the average (real) power in the inductor.
20-2:2. For the circuit described in Self-Check Problem 20-2:1, find the reactive power of the inductor.

20-3 POWER IN A CAPACITOR

Since the instantaneous current in a capacitor leads the instantaneous voltage across it by $\pi/2$ radians, we again find that the instantaneous power graph of Figure 20-4, although a sine wave at twice the frequency of the instantaneous voltage and current, goes alternately positive and negative for equal time intervals. As we would expect, on examining Figure 20-4 we find that the instantaneous power is positive while the potential difference across the capacitor is rising and the capacitor is taking energy from the source to store up a charge on its plates. While the voltage across the capacitor is decreasing,

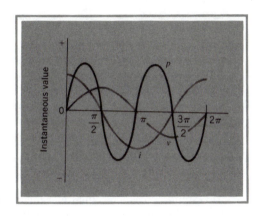

FIGURE 20-4
Instantaneous Power in a Capacitor.

the capacitor is discharging the stored energy back into the system. During this $\pi/2$-radian interval, the instantaneous power is, therefore, a negative quantity.

Because an ideal capacitor must discharge as much energy between $\pi/2$ and π radians as it stored between 0 and $\pi/2$ radians, **the *average* power in the capacitance of an ac circuit is zero.** In this case, substitution in Equation (20-1) gives

$$p = 0.5\, V_m I_m \sin 2\omega t \qquad (20\text{-}7)$$

Again we find that the product of the rms voltage across and current in the capacitance of an ac circuit does *not* represent average power.

> *The product of the rms voltage and current in a capacitor is called the reactive power of the capacitor.*

$$\therefore Q = V_C I_C = I_C^2 X_C = \frac{V_C^2}{X_C} \qquad \text{vars} \quad (20\text{-}8)$$

EXAMPLE 20-3
What value of capacitance will have a reactive power of 50 vars when connected across a 120-V 60-Hz source?

SOLUTION
Since $Q = V_C^2/X_C$,

$$X_C = \frac{V_C^2}{Q} = \frac{(120 \text{ V})^2}{50 \text{ vars}} = 288 \ \Omega \text{ (capacitive)}$$

Since $X_C = 1/\omega C$,

$$C = \frac{1}{\omega X_C} = \frac{1}{377 \times 288 \ \Omega} = 9.2 \times 10^{-6} \text{ F} = \mathbf{9.2 \ \mu F}$$

SELF-CHECK

20-3:1. Suppose that a capacitor draws 100 mA from a 120-V 60-Hz source:
    (a) Write the equation for the instantaneous power in the capacitor.
    (b) Determine the average (real) power in the capacitor.
20-3:2. Find the reactive power of the capacitor described in Self-Check Problem 20-3:1.

20-4 POWER IN A CIRCUIT CONTAINING RESISTANCE AND REACTANCE

If an electric circuit consists of equal resistance and inductive reactance in series, as in Figure 20-5, the instantaneous current lags the applied instantaneous voltage by $\pi/4$ radians (45°). Therefore, when we plot the instantaneous power on the graph of Figure 20-6, it is a negative quantity between 0 and $\pi/4$ radians as the inductance is returning more energy to the system than the resistance is converting to heat. For the longer time interval between $\pi/4$ and π radians, the instantaneous power is positive. At the peak of this instantaneous power wave, both the resistance and the inductance are taking energy from the source.

(a)

(b)

FIGURE 20-5
Apparent Power in an AC Circuit Containing Resistance and Inductive Reactance in Series.

The instantaneous-power graph for a circuit containing both resistance and reactance is again a sine wave at twice the frequency of the voltage and current. But, for the circuit of Figure 20-5, the instantaneous-power graph is neither all positive, as in the case of resistance, nor equally positive and negative, as in the case of reactance. Since the instantaneous-power graph is more positive than negative, there is an average power component which represents the active power input to the resistance of the circuit. Part of the positive instantaneous power is offset by the negative instantaneous power to account for the reactive power of the reactance of the circuit. In the general case,

$$p = V_m I_m \sin \omega t \sin (\omega t + \phi) \tag{20-9}$$

where ϕ is the phase angle of the current with respect to the applied voltage. The phase angle of the rms current with respect to the rms voltage in the phasor diagram of Figure

FIGURE 20-6
Instantaneous Power in an AC Circuit Containing Resistance and Inductive Reactance.

20-5(b) is $-45°$. In the time graph of instantaneous values in Figure 20-6, the same phase angle is shown as $-\pi/4$ radians.

If we use the $\sin(A + B)$ relationship to substitute for $\sin(\omega t + \phi)$ in Equation (20-9),

$$p = V_m I_m \sin \omega t (\sin \omega t \cos \phi + \cos \omega t \sin \phi)$$
$$= \cos \phi (V_m I_m \sin^2 \omega t) + \sin \phi (V_m I_m \sin \omega t \cos \omega t)$$

We should recognize the expressions in parentheses as the instantaneous power in the resistance and reactance of the circuit, respectively. Hence,

$$p = \cos \phi \, p_A + \sin \phi \, p_R \tag{20-10}$$

where p_A *is the instantaneous power in the resistance of the circuit and* p_R *is the instantaneous power in the reactance of the circuit.*

In the example above, the product of total rms voltage and total rms current represents neither active power in **watts** nor reactive power in **vars.** If we have only the voltmeter and ammeter readings to go by in the circuit of Figure 20-5, we have no means of determining the active and reactive power components of the circuit. Therefore,

The product of the total rms voltage and the total rms current in an ac circuit is called **the** *apparent power of the circuit.*
The letter symbol for apparent power is S.[†]

$$\therefore S = V_T I_T \tag{20-11}$$

At this point, we can only talk about the magnitude of the apparent power. In the next section, we find that, like impedance and admittance, we can treat S as a complex number with an angle as well as a magnitude. Because apparent power is neither active power in **watts** nor reactive power in **vars,** we must always express apparent power simply as

[†]Prior to 1968, apparent power was represented by the more descriptive letter symbol P_s.

the product of volts and amperes. Therefore,

The voltampere (VA) is the unit of apparent power. [†]

Since

$$\frac{V_T}{I_T} = Z \qquad (19\text{-}3)$$

$$\therefore S = V_T I_T = I_T^2 Z = \frac{V_T^2}{Z} \qquad (20\text{-}12)$$

EXAMPLE 20-4

The input impedance of a given transmitting antenna system is 500 $\Omega\underline{/+20°}$. A radio-frequency ammeter shows an input current of 8 A. What apparent power is the transmitter supplying to the antenna system?

SOLUTION

$$S = I^2 Z = (8 \text{ A})^2 \times 500 \ \Omega = 32\,000 \text{ VA} = \textbf{32 kVA}$$

SELF-CHECK

20-4:1. Determine the magnitude of the apparent power supplied by a 15-V $\underline{/0°}$ source that supplies a current of 2 A $\underline{/-45°}$.

20-4:2. In the circuit shown in Figure 20-5, suppose that the load's inductive reactance is 300 Ω, the load's resistance is 400 Ω, and the source voltage is 120 V.
 (a) Determine the magnitude of the apparent power supplied to the load.
 (b) Determine the real power supplied to the load.
 (c) Determine the reactive power supplied to the load.

20-5 THE POWER TRIANGLE

Although total voltage times total current does not give us any indication of the active power in the circuit of Figure 20-5, we can find the active power from the product of V_R and I and the reactive power from the product of V_L and I. Since V_L leads V_R by 90° (Figure 20-5), we can state that

$$V_T = \sqrt{V_R^2 + V_L^2} \qquad (19\text{-}2)$$

Multiplying each side of equation (19-2) by the common current,

$$V_T I = \sqrt{(V_R I)^2 + (V_L I)^2}$$

[†]**Load** in an electric circuit is defined as either a device that receives power or the power or apparent power to such a device. Hence, we sometimes find voltamperes referred to as the **load** on a source. **Voltampere** and **var** do not follow the customary pattern for naming SI units after pioneers in science. However, lacking other names for the SI units of apparent and reactive power, the International Electrotechnical Commission accepts the names **voltampere** and **var**.

from which
$$S = \sqrt{P^2 + Q^2} \qquad \text{(20-13)}$$

Therefore, we can use the sides of a right-angled triangle to represent the relationship among the three types of power in an ac circuit.

Similarly, in the parallel circuit of Figure 20-7, since the current in the capacitor branch must lead the current in the resistor branch by exactly 90°,

$$I_\mathrm{T} = \sqrt{I_R^2 + I_C^2} \qquad \text{(19-17)}$$

FIGURE 20-7
Apparent Power in an AC Circuit Containing Resistance and Capacitive Reactance in Parallel.

Multiplying each side of Equation (19-17) by the common voltage,

$$V\,I_\mathrm{T} = \sqrt{(VI_R)^2 + (VI_C)^2}$$

and again
$$S = \sqrt{P^2 + Q^2} \qquad \text{(20-13)}$$

Note that Equation (20-13) applies to both *series* and *parallel* circuits. Since power is the product of current (the common phasor of series circuits) and voltage (the common phasor of parallel circuits), we find that the *same* equations for power apply to series, parallel, and also combination series-parallel ac circuits. This knowledge will provide us with a simple method of solving series-parallel circuits without having to reduce them to series-equivalent or parallel-equivalent circuits, as we shall discover in Section 20-7.

Although instantaneous power does vary sinusoidally, it does so at *twice* the frequency of the voltage and current. Consequently, we cannot draw power phasors on the same phasor diagram with voltage and current. Therefore, we prefer not to represent the relationship among active power, reactive power, and apparent power by the conventional phasor diagram of the type shown in Figures 20-5(b) and 20-7(b). It is customary to represent this Pythagorean relationship of Equation (20-13) by means of right-angled triangles, as shown in Figure 20-8.

As we would expect, we draw that side of the triangle which represents active power horizontally. We draw the side representing reactive power in a vertical direction, with the right angle at the *right*-hand end of the side representing active power. Since Equation (20-13) applies to either a series or a parallel circuit, we can develop our **power triangle** from either an impedance diagram (Figure 19-2) or an admittance diagram (Figure 19-12). If we start with an impedance diagram, in which inductive reactance is drawn in the +j direction, the power triangle appears as shown in Figure 20-8(a). But if we start from an admittance diagram for parallel circuits, in which inductive sus-

FIGURE 20-8
Power Triangle: (a) Based on Common Current as Reference Axis;
(b) Based on Voltage as Reference Axis.

ceptance is drawn in the $-j$ direction, the power triangle has the form shown in Figure 20-8(b).

To avoid confusion between inductive and capacitive reactive power in our power triangles, we must select *one* of these formats for *all* ac circuits, whether series, parallel, or series-parallel. The International Electrotechnical Commission, which establishes international standards for electrical units and symbols, prefers inductive reactive power to be shown in the $-j$ direction, as in Figure 20-8(b). Hence, it is conventional to think of the *voltage* of the source as the reference phasor in dealing with power in alternating-current circuits.

We can now add to the magnitude of the apparent power the angle ϕ shown in Figure 20-8(b). With the convention adopted for drawing power diagrams, it turns out that **S** has the same angle as the admittance of the circuit and the angle of the conjugate impedance (opposite sign) of the circuit. When we treat **S** as a complex number, we call it the **phasor power** of the circuit. Thus,

$$\mathbf{S} = P + jQ \qquad (20\text{-}14)$$

We can now see where the terms **real power** for active power and **quadrature** (imaginary) **power** for reactive power originated. To satisfy the convention for the sign of ϕ, capacitive-reactive power has a $+j$ operator and inductive-reactive power has a $-j$ operator.

Since the voltage across the capacitance in a series circuit is 180° out of phase with the voltage across the inductance, the net reactive voltage is the *difference* between the two reactive voltages. Similarly, in a simple parallel circuit, the net reactive current is the *difference* between the capacitive and inductive branch currents. Therefore, since $Q = V_X I_X$, the net reactive power in any alternating-current circuit (either series or parallel) is the *difference* between the capacitive reactive power and the inductive reactive power.

If we examine Figure 20-2 closely, we note that the instantaneous power in an inductance is positive when the current is rising and building up a magnetic field around the inductor. But Figure 20-4 shows that, when the current is rising, the capacitance is discharging its stored energy back into the system. Therefore, in an ac circuit containing both inductance and capacitance, the capacitance always returns energy to the circuit when the inductance takes energy from the circuit, and vice versa. Consequently, a certain amount of energy can be traded back and forth between the inductance and the capacitance of the circuit and, as far as the source is concerned, the net reactive power required by the circuit is the *difference* between the capacitive reactive power and the inductive reactive power.

EXAMPLE 20-5

A circuit consists of two branches connected in parallel to a 120-V 60-Hz source. Branch I consists of a 75-Ω resistance and a 100-Ω inductive reactance in series. Branch II consists of a 200-Ω capacitive reactance. Determine the apparent power of the circuit.

SOLUTION 1

Step I. In the circuit diagram of Figure 20-9,

$$\mathbf{Z}_1 = R + jX_L = 75 + j100 = 125 \ \Omega \underline{/+53.1°}$$

$$\therefore I_1 = \frac{E}{Z_1} = \frac{120 \text{ V}}{125 \ \Omega} = 0.96 \text{ A}$$

FIGURE 20-9
Diagrams for Example 20-5.

Step II. Active power of the circuit is

$$P = I_1^2 R = 0.96^2 \times 75 = 69.1 \text{ W}$$

Reactive power of the inductance is

$$Q = I_1^2 X_L = 0.96^2 \times 100 = 92.2 \text{ vars}$$

Reactive power of the capacitance is

$$Q = \frac{E^2}{X_C} = \frac{120^2}{200} = 72 \text{ vars}$$

Net reactive power of the circuit is

$$Q = 92.2 - 72 = 20.2 \text{ vars (inductive)}$$

Step III.

$$S = \sqrt{P^2 + Q^2} = \sqrt{69.1^2 + 20.2^2} = \mathbf{72 \text{ VA}}$$

SOLUTION 2
A somewhat similar procedure involves solving Steps II and III in terms of current rather than power.

Step I.

$$\mathbf{Z_1} = R + jX_L = 75 + j100 = 125 \text{ } \Omega\underline{/+53.1°}$$

$$\therefore \mathbf{I_1} = \frac{E}{\mathbf{Z_1}} = \frac{120 \underline{/0°}}{125 \underline{/+53.1°}} = 0.96 \text{ A}\underline{/-53.1°}$$

$$\therefore \mathbf{I_1} = 0.58 - j0.77 \text{ A}$$

Step II.

$$I_C = \frac{E}{-jX_C} = \frac{120 \underline{/0°}}{200 \underline{/-90°}} = 0.6 \text{ A}\underline{/+90°}$$

$$= 0 + j0.6 \text{ A}$$

$$\therefore \mathbf{I_T} = I_1 + I_C = (0.58 - j0.77) + (0 + j0.6)$$

$$= 0.58 - j0.17 \text{ A}$$

and

$$I_T = \sqrt{0.58^2 + 0.17^2} = 0.6 \text{ A}$$

Step III.

$$\therefore S = E_T \text{ } I_T = 120 \times 0.6 = \mathbf{72 \text{ VA}}$$

SELF-CHECK

20-5:1. For the parallel *RC* circuit shown in Figure 20-7, suppose that the resistance is 1200 Ω and the capacitive reactance is 500 Ω. Determine the phasor power, **S**, supplied by the source which has $E = 60$ V.

20-5:2. An *RLC* series circuit has resistance of 500 Ω, inductive reactance of 600 Ω, and capacitive reactance of 400 Ω. Determine the phasor power supplied by a 120-V source.

20-6 POWER FACTOR

Since the hypotenuse of a right-angled triangle must be greater than either of the other two sides, the apparent power that a generator must supply to a reactive load is always greater than the active power that the load can convert into some other form of energy. Since most industrial loads possess appreciable inductive reactance, this relationship between the active power that the load can use and the apparent power that the source must supply is of considerable importance.

> *The ratio between the active power and the apparent power of a load in an ac circuit is called the power factor of the load.*

Examining the power triangle of Figure 20-8(b), we note that the ratio of active power to apparent power is the cosine of the **power factor angle** ϕ between active power and apparent power. Tracing the construction of the power triangle back through admittance diagram to the phasor diagram for current and voltage, we find that the **power factor angle** is the same as the **phase angle** between voltage across and current through the load.

From the definition of power factor,

$$\text{Power factor} = \frac{P}{S} = \cos \phi \qquad (20\text{-}15)$$

In keeping with international practice, North American standards have dropped a letter symbol for power factor.[†] In equations,

> *The symbol for power factor is cos ϕ.*

In keeping with our decision to show inductive reactive power in a $-j$ direction in power triangles, we distinguish between inductive and capacitive loads by stating that inductive loads always have a **lagging power factor** and that capacitive loads always have a **leading power factor.** Since the active power cannot be greater than the apparent power, the power factor cannot have a numerical value greater than unity. Since it is a simple ratio, power factor is always a positive dimensionless number. We may express power factor either as a decimal fraction or as a percentage.

EXAMPLE 20-6
What are the active power and power factor of a load whose impedance is 60 $\Omega \underline{/+60°}$ when connected to a 120-V 60-Hz source?

SOLUTION 1
Step I. Expressing the impedance in rectangular coordinates produces the equivalent of a series circuit consisting of resistance and reactance in series. Therefore, as shown in Figure 20-10,

$$\mathbf{Z} = 60 \cos 60° + j60 \sin 60°$$

[†] Prior to 1968, power factor was represented by the letter symbol, F_p.

FIGURE 20-10
Schematic Diagram for Example 20-6.

from which $R = 30\ \Omega$.

Step II.

$$I = \frac{E}{Z} = \frac{120\ V}{60\ \Omega} = 2.0\ A$$

$$\therefore P = I^2R = 4 \times 30 = \mathbf{120\ W}$$

Step III.

$$S = I^2Z = 4 \times 60 = 240\ VA$$

$$\therefore \cos \phi = \frac{P}{S} = \frac{120}{240} = \mathbf{0.5\ lagging}$$

Although Solution 1 follows the definition of power factor, it is a roundabout way of tackling the problem.

SOLUTION 2
Step I. Power Factor $= \cos \phi = \cos 60° = \mathbf{0.5\ lagging}$

Step II.

$$S = \frac{E^2}{Z} = \frac{120^2}{60} = 240\ VA$$

Step III. Since $\cos \phi = P/S$ (Figure 20-11),

$$P = S \cos \phi = 240 \times 0.5 = \mathbf{120\ W}$$

From the second solution, we note that we can express active power in terms of power factor and apparent power. This leads to the following very useful expression for active power in ac circuits.

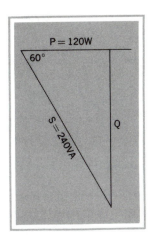

FIGURE 20-11
Power Triangle for Example 20-6.

$$P = E I \cos \phi \tag{20-16}$$

Once we know the active power and apparent power of an ac load, we can determine reactive power by

$$Q = \sqrt{S^2 - P^2} \tag{20-13}$$

Since $P = S \cos \phi$, this equation becomes

$$Q = S\sqrt{1 - \cos^2 \phi} \tag{20-17}$$

Although our main concern in dealing with ac loads is the ratio between active power and apparent power, there are some instances when we wish to solve directly for the reactive power of the load. In these cases, it is useful to be able to express the ratio between reactive power and apparent power.

> *Reactive factor is the ratio between the reactive power and the apparent power of an ac load.*

Once again, as shown by the power triangle of Figure 20-8(b), from the definition of reactive factor,

$$\text{Reactive factor} = \frac{Q}{S} = \sin \phi \tag{20-18}$$

The symbol for reactive factor is sin ϕ.[†]

[†] Prior to 1968, reactive factor was represented by the letter symbol F_q.

SELF-CHECK

20-6-1. Suppose that the circuit shown in Figure 20-7 has a resistance of 1200 Ω and a capacitive reactance of 500 Ω and is supplied by a 120-V source. Determine the power factor of this circuit.

20-6:2. Determine the power factor for an *RLC* series circuit where the resistance is 500 Ω, the inductive reactance is 600 Ω, and the capacitive reactance is 400 Ω.

20-7 POWER FACTOR CORRECTION

We can appreciate the consequences of a low power factor in an industrial load by considering the state of affairs illustrated by Figure 20-12. Since the motor in Figure 20-12(a) has a 70% lagging power factor, the apparent power, as far as the source is concerned, is

$$S = \frac{P}{\cos \phi} = \frac{840}{0.7} = 1200 \text{ VA}$$

and, therefore, the current drain on the source is

$$I = \frac{S}{V} = \frac{1200 \text{ VA}}{120 \text{ V}} = 10 \text{ A}$$

(a)   (b)

FIGURE 20-12
Effect of Load Power Factor.

If we now replace the original motor by one capable of developing an active power of 840 watts at 100% power factor, as in Figure 20-12(b), the power apparently drawn from the source will now be 840 VA. The current drawn by the second motor is only 7 amperes. Therefore, the 70% power-factor motor requires a current of 10 amperes to do the same work that the unity-power-factor motor can do by drawing only 7 amperes from the source.

The wire size used in the windings of the generator and the conductors connecting the load to the generator depends on the *current* they have to pass, and the copper losses in the system depend on the square of the *current*. Therefore, it is more economical for an electrical utility company to feed an 840-W 100% power-factor load than to feed an

840-W 70% lagging-power-factor load. Since a great many industrial loads possess considerable inductive reactance, these electrical utility companies are very much concerned with maintaining a high overall power factor in their systems.

As we have noted, an 840-W 70% lagging-power-factor motor draws three amperes more from the source than an equivalent 840-W 100% power-factor motor. As a power triangle shows, this additional current is involved in supplying the reactive power required by the inductive reactance of the motor. But, as shown in Figure 20-2, this portion of the current alternately transfers energy from the system into the magnetic field of the inductance and then returns the stored energy back to the system. Although the average power used by the inductance during this process is zero, this charge and discharge current has to flow through the generator windings and connecting conductors, thus forcing us to use a larger wire size for a given load power.

We can illustrate this charge and discharge or *reactive* current component by expressing the current of each of the motors in Figure 20-12 in reactangular coordinates. In the case of the 70% lagging-power-factor motor, the current lags the applied voltage by arccos 0.7 = 45.6°.

$$\therefore \mathbf{I} = 10 \cos 45.6° - j10 \sin 45.6° = 7 - j7.14 \text{ A}$$

In the case of the 100% power-factor motor, the current is in phase with the applied voltage.

$$\therefore \mathbf{I} = 7 - j0 \text{ A}$$

When we express the load current in this manner, we note that both have the same *reference axis* component. It is the component that contributes to the active power of the load. The 70% lagging-power-factor load requires the additional *quadrature* component to take care of the inductance of the load.

We recall that in an ac circuit containing both inductance and capacitance, the capacitance returns energy to the system while the inductance takes energy from the system; and vice versa. If we then place a capacitor in the circuit containing the 70% lagging-power-factor motor, it should be possible for the capacitor and the inductance of the load to trade their reactive power back and forth without the reactive component of the current having to travel all the way from the source to the load and back. In practice, this would allow us to use lower power factor loads yet obtain the advantages of a high system power factor.

For the moment, we connect a 308-μF capacitor in *series* with the 70% lagging-power-factor motor, as in Figure 20-13. The impedance of the motor is $Z_M = V/I = 120/10 = 12 \ \Omega$.

$$\phi = \arccos 0.7 = 45.6°$$
$$\therefore \mathbf{Z_M} = 12 \ \Omega \underline{/+45.6°}$$
$$= 12 \cos 45.6° + j12 \sin 45.6° = 8.4 + j8.6 \ \Omega$$
$$X_C = \frac{1}{\omega C} = \frac{1}{377 \times 308 \ \mu F} = 8.6 \ \Omega \text{(capacitive)}$$

FIGURE 20-13
Effect of a Series Capacitor on a Lagging Power-Factor Load.

Because the motor and capacitor are in series,

$$\mathbf{Z}_T = 8.4 + j8.6 - j8.6 = 8.4 + j0 \ \Omega$$

$$\therefore \mathbf{Z}_T = 8.4 \ \Omega \underline{/0°}$$

Since the phase angle of the total impedance is now 0°, the current drawn from the generator is in phase with the source voltage. But

$$I = \frac{V}{Z_T} = \frac{120 \ V}{8.4 \ \Omega} = 14.3 \ A$$

Therefore, connecting a capacitor in *series* with an inductive load to reduce the net reactive power does not *reduce* the current drain on the source. It *increases* it. Moreover, since the impedance of the motor itself is still 12 ohms, the voltage across the motor in the circuit of Figure 20-13 becomes

$$V_M = IZ_M = 14.3 \ A \times 12 \ \Omega = 172 \ V$$

Although connecting a capacitor in *series* with the motor has raised the system power factor to 100%, it has served only to aggravate the original problem by increasing the current drawn from the source by applying excessive voltage to the motor. Therefore, it is not practical to correct for a low power factor by connecting a capacitor in *series* with an inductive load.

> *Power factor correction or improvement consists of adding a capacitive reactive power to an ac circuit in such a manner that the apparent power drawn from the source is reduced without altering the current through or the voltage across the load itself.*

One of the characteristics of a parallel circuit is that, although a change in one branch will affect the *total* current, it will *not* affect the current in the other branches. Therefore, in the circuit of Figure 20-14, we have connected a capacitor in parallel with the 840-W 70% lagging-power-factor motor. Since the motor is still connected directly across the 120-V source, its current is still 10 amperes, lagging the applied voltage by 45.6°. Because of its lagging power factor, the motor has a reactive power of $Q = VI \sin \phi = 120 \times 10 \times 0.714 = 860$ vars. If the capacitor in parallel with the

FIGURE 20-14
Power-Factor Correction.

motor also has a reactive power of 860 vars, they will be able to trade this reactive power back and forth between them, and, as far as the generator is concerned, the net reactive power in the circuit is zero. Since the voltage across and the current through the motor are unchanged, the active power is still 840 watts. Therefore, the overall apparent power that the generator must supply is 840 voltamperes, and the current drawn from the generator becomes seven amperes. Note that adding the capacitor to the circuit has decreased the current to seven amperes, while the motor current still remains at ten amperes. As far as the generator is concerned, the 70% lagging-power-factor motor and capacitor in parallel are the equivalent of a unity-power-factor motor.

Although we can solve for the required capacitance by noting that the capacitor current must equal the reactive component of the load current or that the susceptance of the capacitor must equal the susceptance component of the load admittance, the simplest procedure is to work in terms of reactive power, since power equations are independent of series or parallel connection of the circuit components.

To obtain an overall unity power factor with a lagging power factor in a load, a capacitor having a reactive power equal to the reactive power of the load is connected in parallel with the load.

EXAMPLE 20-7
What value of capacitance must be connected in parallel with a motor drawing 10 A at 70% lagging power factor from a 120-V 60-Hz source in order for the generator current to be minimum?

SOLUTION
Step I. The reactive power of the load is

$$Q = VI\sqrt{1 - \cos^2 \phi} = 120 \times 10\sqrt{1 - 0.7^2} = 857 \text{ vars}$$

Therefore, we must select a capacitor with a reactive power of 857 vars when connected across a 120-V 60-Hz source.

Step II. Since $Q = V^2/X_C$,

$$\therefore X_C = \frac{V^2}{Q} = \frac{(120 \text{ V})^2}{857 \text{ vars}} = 16.8 \ \Omega \text{ (capacitive)}$$

$$C = \frac{1}{\omega X_C} = \frac{1}{377 \times 16.8 \ \Omega} = 1.58 \times 10^{-4} \text{ F} = \textbf{158 } \boldsymbol{\mu}\textbf{F}$$

Note: As a check that *adding* an extra branch does *reduce* the total current drawn from the source without affecting the load current, draw an accurate phasor diagram for the circuit of Figure 20-14.

EXAMPLE 20-8

A fluorescent lamp and its ballast inductance draw a 1.0-A current at a 50% lagging power factor from a 120-V 60-Hz source. What is the overall power factor when a 26.5-μF capacitor is connected across the fixture?

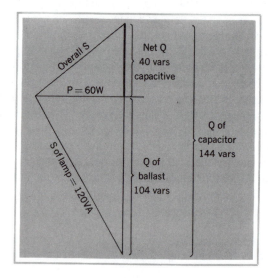

FIGURE 20-15
Power Triangle for Example 20-8.

SOLUTION

Step I. Prepare a power triangle for the lamp and its ballast inductance.

$$S = VI = 120 \text{ V} \times 1.0 \text{ A} = 120 \text{ VA}$$

$$P = S \cos \phi = 120 \text{ VA} \times 0.5 = 60 \text{ W}$$

$$Q = \sqrt{S^2 - P^2} = \sqrt{120^2 - 60^2} = 104 \text{ vars (inductive)}$$

Step II.

$$X_C = \frac{1}{\omega C} = \frac{1}{377 \times 26.5 \ \mu\text{F}} = 100 \ \Omega \text{ (capacitive)}$$

and
$$Q_C = \frac{V^2}{X_C} = \frac{120^2}{100} = 144 \text{ vars (capacitive)}$$

Step III. From the power triangle of Figure 20-15,

$$\text{Final } Q = 144 - 104 = 40 \text{ vars (capacitive)}$$
$$\text{Final } S = \sqrt{P^2 + Q^2} = \sqrt{60^2 + 40^2} = 72 \text{ VA}$$

Step IV.

$$\text{Overall } \cos \phi = \frac{P}{S} = \frac{60}{72} = \textbf{83.2\% leading}$$

EXAMPLE 20-9

What value of capacitance must be connected in parallel with a load drawing 1 kW at 70.7% lagging power factor from a 208-V 60-Hz source in order to raise the overall power factor to 91% lagging?

SOLUTION

Step I.

$$\text{Original } S = \frac{P}{\cos \phi} = \frac{1000 \text{ W}}{0.707} = 1.41 \text{ kVA}$$

$$\text{Load } Q = \sqrt{S^2 - P^2} = \sqrt{1.41^2 - 1^2} = 1 \text{ kvar (inductive)}$$

Step II.

$$\text{Final } S = \frac{P}{\cos \phi} = \frac{1000 \text{ W}}{0.91} = 1.1 \text{ kVA}$$

$$\text{Overall } Q = \sqrt{S^2 - P^2} = \sqrt{1.1^2 - 1^2} = 456 \text{ vars (inductive)}$$

Step III. Therefore, from the power triangle of Figure 20-16, the reactive power

FIGURE 20-16
Power Triangle for Example 20-9.

of the capacitor must be

$$Q_C = 1000 - 456 = 544 \text{ vars (capacitive)}$$

$$\therefore X_C = \frac{V^2}{Q_C} = \frac{208^2}{544} = 80 \ \Omega \text{ (capacitive)}$$

and

$$C = \frac{1}{\omega X_C} = \frac{1}{377 \times 80 \ \Omega} = 3.3 \times 10^{-5} \text{ F} = \mathbf{33 \ \mu F}$$

SELF-CHECK

20-7:1. What additional element must be added to the circuit of Example 20-6 in order to make the final power factor equal to one?

20-7:2. Starting with the circuit shown in Figure 20-7, where the resistance is 1200 Ω, the capacitive reactance is 500 Ω, and the source is a 120-V 60-Hz generator, determine what additional component must be added to modify the circuit so that the overall power factor becomes 95% lagging.

20-8 MEASURING POWER IN AC CIRCUITS

In a dc system, we seldom require wattmeters to indicate power input to a load, since we can obtain this information simply by multiplying the voltmeter reading by the ammeter reading. However, in an ac system, the product of the voltmeter and ammeter readings gives us only the *apparent*-power input to a load. Unless we have further data as to the power factor of the load or the phase angle between load voltage and load current, we cannot determine the *active*-power input to the load in this manner.

In Chapter 10, we discovered that an electrodynamometer movement produces a torque which is proportional to the product of the currents in the moving coil and the stationary coils. If we connect the electrodynamometer as shown in Figure 20-17, the *load current* flows through the stationary coils. The current through the moving coil and its series multiplier resistor is proportional to the *voltage across the load*. Hence the pointer displacement of the resulting wattmeter is proportional to the *product* of load current and load voltage, i.e., load *power*.

FIGURE 20-17
Measuring AC Power with a Wattmeter.

PART IV   ALTERNATING CURRENT

In direct-current circuits, this wattmeter reading is a real-power reading, since reactive power does not exist. To determine what the wattmeter will read in an alternating-current circuit, we must examine the nature of the instantaneous torque produced by the sine-wave currents in the stationary and moving coils. We can do this by referring first to the voltage and current relationships for a resistance load, as shown in Figure 20-1.

The current through the current coil is represented by the shaded current curve, and the current through the voltage coil is represented by the shaded voltage curve. The instantaneous power curve in Figure 20-1 represents the product $v \times i$ and also represents the instantaneous torque developed by the electrodynamometer movement. Since the current in both coils reverses at the same instant, the instantaneous torque (like the instantaneous power) is always positive. However, the inertia of the pointer and voltage coil assembly will not allow the pointer to follow the variations in instantaneous torque due to the alternating current in the coils. Therefore, the pointer will take up a position which represents the average between zero and maximum instantaneous torque. The position of the pointer then represents the *average* power into the resistor.

If we now connect the electrodynamometer to an inductor, the currents in the two coils will be 90° out of phase, as indicated by the shaded curves in Figure 20-2. Therefore, the product of the two instantaneous fluxes will produce a clockwise torque for one quarter-cycle and a counterclockwise torque for the next quarter-cycle. As a result, the net meter deflection is zero, just as the average of the instantaneous power (the active power) in inductance is zero. If we connect the electrodynamometer to the circuit of Figure 20-5(a), the instantaneous torque can be represented by the instantaneous power curve in Figure 20-6. Now the clockwise torque is greater than the counterclockwise torque and the meter will give a reading. We can consider that the instantaneous torques due to the inductance average to zero, leaving a net positive torque representing the active power into the resistance of the circuit.

Therefore, when we connect an electrodynamometer movement with its current coils in series with a load and its voltage terminals across the load in an ac circuit, the inertia of the movement causes it to indicate the *average* instantaneous power or the *active* power input to the load. Consequently, we call this instrument a **wattmeter.** We should also note that the wattmeter reading can be represented by

$$P = EI \cos \phi \qquad (20\text{-}16)$$

where E is the voltage applied to the voltage terminals, I is the current flowing through the current terminals, and ϕ is the phase angle between the voltage and current applied to the wattmeter.

In the wattmeter movement, the required current through the moving voltage coil is determined by a *multiplier resistor* in series with the coil. If we replace this resistor with a *capacitor* having the required capacitive reactance at the power line frequency, the current through the voltage coil will have the same magnitude but now will be 90° out of phase with the voltage applied to the voltage terminals of the meter (since the capacitor current leads the applied voltage by 90°). If we connect this meter to a resistive circuit, the net torque is zero, since the voltage-coil current is now 90° out of phase with the current through the current coil. Similarly, the meter will now show a reading when connected to an inductive or capacitive load. This modification to the electrodynamometer has produced a meter which indicates *reactive power*. Hence, we call this meter a **varmeter.**

20-9 EFFECTIVE RESISTANCE

We can determine the resistance of an electric circuit by measuring the voltage across and the current through it. However, for the results to be accurate, we must be sure that the voltage across the resistor is due only to *IR* drop. We recall that, as the current through an electric conductor increases, closed loop magnetic lines of force expand outward from the center of the conductor. As a result, in addition to the *IR* drop, a small counter emf is induced in the copper that is being linked by the expanding magnetic field. In a dc circuit, where the current is not changing, these induced voltages do not appear. Therefore, when we make *direct* current measurements of the voltage across and current through a practical resistor, we are reasonably sure that the value of resistance determined from $R = V/I$ is the actual **ohmic resistance** of the resistor.

If we wish, we can also determine the resistance of a dc circuit by using a wattmeter and an ammeter. This is more in keeping with the techniques we use in establishing the rms or dc equivalent value of an alternating current. Since $R = V/I$, multiplying numerator and denominator by I gives

$$R = \frac{V}{I} = \frac{VI}{I^2} = \frac{P}{I^2}$$

Because the wattmeter reads the *product* of the voltage across and current through a dc circuit, we are essentially measuring the same parameters by both techniques.

However, in an ac circuit, the ratio between the voltage across and the current through a circuit component is, by definition, its **impedance**, rather than just its resistance. Since it is not possible for us to connect an ac voltmeter in such a manner that it indicates only the *IR* drop across the resistance of the component, we cannot use the voltmeter-ammeter method to determine the **effective resistance** of an ac circuit. But, as we noted in Chapter 19, the *same* current flows through both the resistance and the reactance of practical impedances. We can then accurately measure I_R in an ac circuit, even though we cannot accurately measure V_R. As we noted earlier in this chapter, the active power in an ac circuit represents the power input to the resistance of the circuit. Since a wattmeter measures only the active power in an ac circuit, we can determine the effective resistance of an ac circuit by the wattmeter-ammeter method, even though the voltmeter-ammeter technique is not valid. Hence, in both ac and dc circuits, we are correct in expressing resistance by the relationship

$$R = \frac{P}{I^2} \tag{20-19}$$

After going to this trouble to obtain a true resistance value in a practical ac circuit (rather than the total impedance of the circuit), we would discover that the **effective** or **ac resistance** of the circuit is appreciably higher than the plain **ohmic resistance** of the same circuit with only direct current through it. In addition to thinking of resistance as the property of an electric circuit that opposes electric current, we must now think of **resistance as the property of an electric circuit that dissipates electric energy.** There are several mechanisms for dissipating energy in an ac circuit that are not present in a dc circuit.

One such factor is apparent if we again consider the expanding and collapsing magnetic lines of force around an electric conductor in an ac circuit. There will be more changes in flux linkage at the center of the conductor than at its surface. Therefore, there is more opposition to alternating current in the center of the conductor than there is at the surface. As the frequency of the alternating current is increased, the rate of change of flux increases, which, in turn, increases the induced voltage at the center of the conductor. This effect is so pronounced that at radio frequencies almost all the current flows along the surface of the conductor. This is called **skin effect.** Since there is practically no current in the center of the conductor at radio frequencies, its effective cross-sectional area is greatly reduced, and, therefore, the effective resistance of the conductor is much greater than its dc ohmic resistance.

A second factor governing the effective resistance of an ac circuit is the dissipation of energy through radiation of electromagnetic waves from the magnetic and electric fields associated with the ac circuit. As we have discovered, these fields build up and collapse twice each cycle. Under ideal conditions, as much reactive power is returned to the electric circuit during the collapsing of the field as is taken from the circuit during the building up of the field. But as the frequency is increased, some of the energy escapes as a radiated electromagnetic wave; hence, less energy is returned to the circuit during the collapse of the field than was needed to build it up. These **radiation losses** increase the value of P in Equation (20-19) and, therefore, increase the effective resistance of the ac circuit.

Suppose that we now connect an ammeter and a wattmeter so that we can determine the effective resistance of a coil of wire in an ac circuit. Having made the necessary calculations, we now insert an iron core into the coil. Due to the increase in inductive reactance, both meter readings decrease. But when we calculate the effective resistance, we find that it is greater than before. We can blame this increase on the hysteresis of the iron core and also on eddy current in the core. Both hysteresis and eddy current cause energy supplied by the electric source to be converted into heat. This requires an increase in the active power input to the coil for a certain coil current. Since this increase in active power is registered by the wattmeter, the effect of hysteresis and eddy current shows up in our calculation of the effective resistance of the iron-core coil. Again, the higher the frequency, the greater the discrepancy between ohmic resistance and effective resistance.

Similarly, a capacitor which, when measured in a dc circuit has an infinitely high resistance, can cause an appreciable wattmeter reading in an ac circuit. This produces the effect of a resistance in parallel with the capacitor. The wattmeter, in this case, is indicating the transfer of energy into heat as the dielectric of the capacitor is stressed, first in one direction and then in the other, many times a second. Since this loss of energy by conversion to heat is similar to the hysteresis loss in a magnetic circuit, it is called **dielectric hysteresis.** It is interesting to note that, because magnetic and dielectric hysteresis losses and eddy current losses all tend to raise the temperature of the ac circuit, there will also be an additional increase in the ohmic resistance if the circuit has an appreciable positive temperature coefficient of resistance.

SELF-CHECK

20-9:1. What is the effective resistance of a motor that draws 10 A at 70% lagging power factor from a 120-V 60-Hz source?

PROBLEMS

20-1-1. A toaster draws a 6-A current from a 110-V 60-Hz source.
(a) What is the active power of the toaster?
(b) What is the peak value of the instantaneous power input to the toaster?

20-1-2. The voltage drop across the heater of a certain cathode-ray tube is 6.3 V rms when the current through it is 0.3 A. What is the average rate of conversion of electric energy into heat?

20-1-3. What is the average power dissipation of a 75-Ω resistor in which the sine-wave current reaches a peak value of 2.0 A?

20-1-4. What power rating must a 300-Ω dummy antenna resistor possess if the rms voltage drop across it is 96 V?

20-2-5. An inductance of 3.0 mH passes a 1.0-kHz sine-wave current of 20 mA.
(a) What is the average power input to the inductor?
(b) What is the reactive power?
(c) What is the peak rate at which it can store energy?

20-2-6. The voltage across and the current through an inductor having negligible resistance are measured as 60 V and 80 mA rms. If the frequency of the voltage source is 60 Hz,
(a) What is the reactive power of the inductor?
(b) What is its inductance?

20-3-7. A 26.5-μF capacitor is connected across a 120-V 60-Hz source.
(a) What is the reactive power input to the capacitor?
(b) What is the peak rate at which it can discharge its stored energy back into the circuit?

20-3-8. An industrial capacitor used in power factor correction is rated at 7.5 kvars when connected to a 208-V 60-Hz circuit. What is its capacitance?

Draw fully labeled power triangles for remaining problems.

20-4-9. A solenoid having an inductance of 0.5 H and a resistance of 24 Ω is connected to a 120-V 60-Hz source.
(a) What is the apparent power input to the solenoid?
(b) What is the active power input to the solenoid?
(c) What is the reactive power input to the solenoid?

20-4-10. The operating coil of a relay draws 100 mA from a 24-V 400-Hz source. If the active power input is 360 mW,
(a) What is the resistance of the coil?
(b) What is the inductance of the coil?

20-4-11. What is the average power input to the transmitting antenna in Example 20-4?

20-4-12. The reactive power input to an impedance of 480 $\Omega \underline{/-60°}$ is 300 mvars. At what rate is energy being dissipated by the impedance?

20-4-13. A 100-pF capacitor is connected in series with a coil having an inductance of 100 μH and a resistance of 20 Ω. At what frequency will this network have zero net reactive power?

20-4-14. The ratio between the reactive power and the active power of the coil in Problem 20-4-13 is called its Q. What is the Q of the coil at a frequency of 1 MHz?

20-6-15. What is the power factor of the solenoid in Problem 20-4-9?

20-6-16. What is the power factor of the relay coil in Problem 20-4-10?

20-6-17. What is the reactive factor of the solenoid in Problem 20-4-9?

20-6-18. What is the reactive factor of the relay coil in Problem 20-4-10?

20-6-19. What value of resistance and inductance in series will draw the same current from a 120-V 60-Hz source as a 500-W 60% lagging power factor load?

20-6-20. What value of resistance and inductance in parallel will draw the same current from a 120-V 60-Hz source as a 500-W 60% lagging power factor load?

20-6-21. (a) What current must a 110-V alternator supply to operate a 2-kW 75% lagging-power-factor load?
(b) What active power could the alternator supply for the same magnitude of current in its windings to a unity-power-factor load?

20-6-22. A resistor and capacitor in parallel draw a total current of 50 mA from a 208-V 60-Hz source at an 80% leading power factor. Calculate the resistance and capacitance.

20-6-23. An apparent power input of 25 kVA must be provided to the "work coil" of a radio-frequency induction heater in order to generate heat in the steel work piece at the rate of 5 kW. What is the reactive factor of the load?

20-6-24. Part of the "crossover" network used to feed the "woofer" and "tweeter" in a loudspeaker system consists of an 8-Ω resistor and a capacitor in parallel. The power input to these two components at 1 kHz is 500 mW at a 20% leading power factor. What is the capacitance of the capacitor?

20-6-25. An impedance coil having an 0.2 lagging power factor is connected in series with a 300-W lamp in order to supply the lamp with 120 V from a 208-V 60-Hz source. What is the voltage across the terminals of the impedance coil?

20-6-26. An induction motor which draws 2.0 A from a 120-V 60-Hz source at 0.8 lagging power factor and a 100-W lamp are connected in parallel. What is their overall power factor?

20-7-27. An induction motor draws 6.0 A at 0.8 lagging power factor from a 208-V 60-Hz source.
(a) What value of capacitance must be placed in parallel with the motor to raise the overall power factor to unity?
(b) What are the magnitudes of the final motor current, capacitor current, and line current?

20-7-28. What value of capacitance is required to produce an overall power factor of 0.96 lagging with the motor of Problem 20-7-27?

20-7-29. What is the overall power factor when a 50-μF capacitor is connected in parallel with the motor of Problem 20-7-27?

20-7-30. A synchronous motor capable of operating with a leading power factor draws 15 kW from a distribution transformer while driving an air compressor. The remainder of the load on the transformer is 80 kW at 0.85 lagging power factor.
(a) How many kilovars of capacitive reactive power must the synchronous motor produce to raise the overall power factor to 0.96 lagging?
(b) What is the reactive factor of the synchronous motor operating in this manner?

20-7-31. The power factor of a load on a 120-V 60-Hz source is raised from 0.707 lagging to 0.866 lagging by connecting a 53-μF capacitor across the load. What is the active power of the load?

20-7-32. The power factor of a load on a 120-V 60-Hz source is raised from 0.866 lagging to 0.966 leading by connecting a $110\frac{1}{2}$-μF capacitor in parallel with the load. What is the rms load current?

20-9-33. What is the effective resistance of the transmitting antenna system in Example 20-4?

20-9-34. An 0.0005-μF radio transmitter capacitor has a 10% leading power factor at 500 kHz. What is the effective resistance of the capacitor (in parallel with its capacitance) at this frequency? At what rate is heat generated in this capacitor when the rms radio-frequency current "through" it is 3.0 A?

20-9-35. The ac resistance of an air-core solenoid is measured by the ammeter-wattmeter method. When the solenoid is connected to a 120-V 60-Hz source, the ammeter indicates 2.0 A and the wattmeter reads 48 W. When an iron core is inserted into the coil, the readings are 180 mA and 500 mW, respectively.

(a) What is the effective resistance and inductance of the solenoid without the iron core?

(b) What is the effective resistance and inductance of the coil with the iron core inserted?

20-9-36. We can also calculate effective resistance by employing a precision noninductive resistor instead of a wattmeter. An iron-core inductor draws a current of 1.0 A from a 120-V 60-Hz source. When a noninductive 100-Ω precision resistor is connected in series with the coil, the current becomes 0.6 A. What is the effective resistance and inductance of the iron-core inductor?

REVIEW QUESTIONS

20-1. Why can we state that $p = vi$ in any ac circuit?

20-2. Why is the instantaneous power in resistance always a positive quantity?

20-3. Why can we state that the average power in the resistance in a sine-wave ac system is one-half the peak power?

20-4. Why is the product $V_R \times I_R$ sometimes called the **real power?**

20-5. Given an ac voltmeter and an ac ammeter, how would you determine the peak value of the instantaneous power as an ideal inductor builds up its magnetic field in an ac circuit?

20-6. Why is the average power of an ideal inductor zero?

20-7. What is the significance of the term **reactive power?**

20-8. What is the significance of the term **apparent power?**

20-9. What is the apparent power of an ideal resistor?

20-10. What is the apparent power of an ideal capacitor?

20-11. Why should we not express apparent power in watts?

20-12. Why can we say that the net reactive power of an ac circuit is the *difference* between the inductive reactive power and the capacitive reactive power?

20-13. Why is the total active power of a network always the sum of the individual active powers regardless of series or parallel connection of the components?

20-14. Why is the apparent power of an ac circuit the root of the sum of the squares of the active and reactive powers rather than the simple sum?

20-15. Why is the angle between the active power and apparent power in a power triangle the same as the angle between the total current and total voltage of a load?

20-16. What is the significance of the term **power factor** in dealing with industrial loads?

20-17. What is meant by a **leading power factor?**

20-18. Why are ac generators and transformers rated in terms of kilovoltamperes rather than kilowatts?

20-19. Of what practical use is the term **reactive factor?**

20-20. Derive an expression for reactive factor in terms of power factor.

20-21. Although placing capacitance in series with an inductive load raises the overall power factor, why can we not consider this to be a satisfactory means of power-factor correction?

20-22. What is the purpose of power-factor correction?

20-23. Why does adding capacitance in parallel with an inductive load achieve this purpose?

20-24. In determining the rms value of a sine-wave by laboratory experiment, we determine the equivalent direct current that will produce the same lamp brilliance. Why do we work with current rather than voltage?

20-25. Why does the effective resistance of an electric conductor increase as the frequency of the alternating current through it increases?

20-26. Some radio coils are wound from "Litz" wire, which consists of many strands of fine, separately insulated wire. Suggest why Litz wire is more suitable for this purpose than a solid conductor of the same gauge.

20-27. How do you account for the change in effective resistance of the solenoid in Problem 20-9-35? What resistance would you expect an ohmmeter to indicate for the iron-core solenoid?

*20-28. Design a personal computer program which asks for the following input data for a two-terminal ac load: voltmeter reading, ammeter reading, and wattmeter reading. On receiving these input data, the computer should print out a table showing the following data on the load: total impedance, ac resistance, reactance, apparent power, true power, reactive power, power factor, and reactive factor. This program cannot indicate whether the power factor is leading or lagging. Why?

*20-29. Design a personal computer program which asks for the following input data for a two-terminal inductive ac load: rms input voltage, rms input current, power factor, and frequency. The program should then direct the computer to print out the parallel capacitance required to raise the overall power factor to unity.

PART V

IMPEDANCE NETWORKS

ALTERNATING CURRENTS IN IMPEDANCE NETWORKS must obey all the basic laws of electric circuit behavior we encountered in the direct-current circuits in Part 2. To qualify as fundamental principles of electric circuit behavior, all dc circuit-analysis formats and theorems must also hold true for alternating-current circuits. The only difference is that steady-state ac quantities possess angles as well as magnitudes. Hence, the solution of numerical examples usually requires using the rules of complex algebra. This added dimension simply requires a little extra caution in performing the various algebraic routines on a calculator or personal computer.

In Chapters 21 and 22, we shall review the basic laws and circuit-analysis techniques we first encountered in Part 2 and adapt them to alternating-current examples. The remaining chapters deal with electric circuit characteristics which are unique to ac systems.

21

SERIES AND PARALLEL IMPEDANCES

21-1 RESISTANCE AND IMPEDANCE

Resistance and **impedance** both represent opposition to electric current. Both are expressed in **ohms.** However, whereas resistance is a specific property of both direct- and alternating-current circuits, we define impedance as the opposition of a portion of a circuit to *alternating* current. We commenced our consideration of circuit-analysis techniques with resistance networks because all numerical quantities in direct-current circuits are *scalar* quantities, which are limited to a numerical magnitude and, in some instances, a + or − polarity. Consequently, all numerical computation in Chapters 8 and 9 is based on *linear* algebra operations. As we discovered in Chapter 19, impedance is a phasor quantity in alternating-current circuits. Therefore, numerical computation for impedance networks is based on *phasor* algebra operations.

As long as we perform all numerical computation by **phasor** *algebra, we may apply to impedance networks all the relationships we developed in Part 2 for resistance networks.*

In this chapter, we shall consider the basic series and parallel impedance relationships as a means of becoming acquainted with the characteristics and limitations of phasor algebra before we tackle the more elaborate and multisource networks which require application of the network theorems we encountered in Chapters 8 and 9.

21-2 IMPEDANCES IN SERIES— KIRCHHOFF'S VOLTAGE LAW

In section 18-2, we noted that, if we consider only a particular *instant* in time (if time could stand still), that the circuit relationships in Figure 21-1(a) for that instant are exactly the same as in a direct-current circuit with the numerical values that prevail at that instant. Consequently, from Kirchhoff's voltage law, we are able to state that

$$e = v_T = v_R + v_L \tag{18-1}$$

where we perform the numerical operations by *linear* algebra.

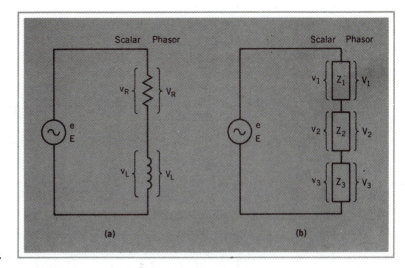

FIGURE 21-1
Kirchhoff's Voltage Law
Applied to an
Alternating-Current Circuit.

To extend Kirchhoff's voltage law to the much more useful rms values for alternating-current circuit parameters, we must note that rms values are *phasor* quantities. We can write the Kirchhoff's voltage-law relationship for the circuit of Figure 21-1(a) in the form

$$\mathbf{E}_T = \mathbf{V}_R + \mathbf{V}_L \tag{18-3}$$

provided that we note that each term is a phasor and that we must perform the numerical computation by phasor algebra. (In Figure 21-1, we have indicated that both the time domain relationships of Equation (18-1) and the frequency domain relationships of Equation (18-3) apply to the same ac circuit by placing the lowercase symbol v to the left of the components and the boldface $\mathbf{V}$ to the right of the components.)

We can now proceed one step further and apply Kirchhoff's voltage law to the series circuit of Figure 21-1(b). Although we can write the Kirchhoff's voltage-law equation in terms of instantaneous values, we are concerned mainly with rms values; hence, we write the equation in the form

$$\mathbf{E} = \mathbf{V}_T = \mathbf{V}_1 + \mathbf{V}_2 + \mathbf{V}_3 \tag{21-1}$$

where (unlike the equivalent equation for a dc circuit) all terms are phasor quantities.

We can divide every term in Equation (21-1) by the common current in the series circuit of Figure 21-1(b), (using phasor division, of course), giving

$$\frac{\mathbf{V_T}}{\mathbf{I}} = \frac{\mathbf{V_1}}{\mathbf{I}} + \frac{\mathbf{V_2}}{\mathbf{I}} + \frac{\mathbf{V_3}}{\mathbf{I}}$$

From the definition of impedance, this becomes

$$\mathbf{Z_T} = \mathbf{Z_1} + \mathbf{Z_2} + \mathbf{Z_3} \qquad (21\text{-}2)$$

and we may state that

*The total impedance of several impedances in series is their **phasor sum**.*

EXAMPLE 21-1

What is the total impedance when $\mathbf{Z_1} = 60 \ \Omega \underline{/+60°}$ and $\mathbf{Z_2} = 80 \ \Omega \underline{/-45°}$ are connected in series?

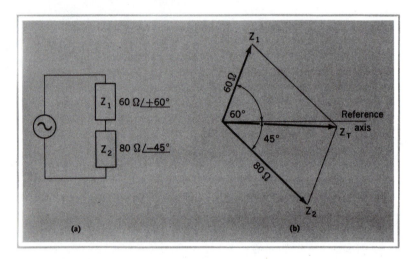

FIGURE 21-2
Impedances in Series.

SOLUTION

Step I. Draw a circuit diagram on which we can show all pertinent data for ready reference, and a fairly accurate freehand impedance diagram to indicate the approximate solution.

Step II. Express the impedances in rectangular form so that we can add their coordinates.

$$\mathbf{Z_1} = 60 \cos 60° + j60 \sin 60° = 30 \quad + j52$$
$$\mathbf{Z_2} = 80 \cos 45° - j80 \sin 45° = \underline{56.6 - j56.6}$$

Adding gives
$$\mathbf{Z_T} = \overline{86.6 - \quad j4.6 \ \Omega}$$

Step III. Depending on the nature of any further computations, convert the total impedance into polar form. [For example, if the next step is to determine the current in the circuit of Figure 21-2(a) for a given E.]

$$\phi = \arctan \frac{-4.6}{86.6} = -3°$$

$$Z = \frac{86.6}{\cos -3°} = 86.7 \ \Omega$$

$$\therefore \mathbf{Z} = \mathbf{86.7 \ \Omega \underline{/-3°}}$$

EXAMPLE 21-2
Given $\mathbf{Z}_1 = 10\,000 + j15\,000 \ \Omega$, $\mathbf{Z}_2 = 4700 + j0 \ \Omega$, and $\mathbf{Z}_3 = 25\,000 - j10\,000 \ \Omega$. Express the total impedance of these three impedances in series in rectangular coordinates.

SOLUTION
All we need to do in this case is to perform the addition of rectangular coordinates.

$$\mathbf{Z}_1 = 10\,000 + j15\,000$$
$$\mathbf{Z}_2 = \ \ \ 4700 + j0$$
$$\underline{\mathbf{Z}_3 = 25\,000 - j10\,000}$$

and
$$\mathbf{Z}_T = \mathbf{39\,700 + j5000 \ \Omega}$$

We might be inclined to suggest from the simple solution of Example 21-2 that we should always express impedance in rectangular coordinates. However, we must not allow ourselves to be misled by first appearances. As we noted in Section 19-8 (see Table 19-1), the rectangular coordinates of an impedance represent a *series* resistance and reactance. Hence, the addition process of Example 21-2 applies only to *series* impedances.

As an application of Kirchhoff's voltage law to series ac circuits, we can now extend the **voltage-divider principle** of Section 7-4 to series impedances. Therefore, Equation (7-1) becomes

$$\mathbf{V}_x = \mathbf{E}\frac{\mathbf{Z}_x}{\mathbf{Z}_T} \tag{21-3}$$

where every term is treated as a phasor quantity.

EXAMPLE 21-3
Given an applied voltage in the circuit of Figure 21-2(a) of 120 V $\underline{/0°}$, what is the voltage drop across $\mathbf{Z}_1$?

SOLUTION 1

By Ohm's law:

Step I.   Having determined $\mathbf{Z}_T$ in Example 21-1,

$$\mathbf{I} = \frac{\mathbf{E}}{\mathbf{Z}_T} = \frac{120 \text{ V} \underline{/0°}}{86.7 \text{ }\Omega \underline{/-3°}} = 1.38 \text{ A} \underline{/+3°}$$

Step II.

$$\mathbf{V}_1 = \mathbf{IZ}_1 = 1.38 \text{ A} \underline{/+3°} \times 60 \text{ }\Omega \underline{/+60°}$$
$$= \mathbf{83 \text{ V} \underline{/+63°}}$$

SOLUTION 2

By the voltage-divider principle: We can solve for $\mathbf{V}_1$ in a single step without first determining $\mathbf{I}$.

$$\mathbf{V}_1 = \mathbf{E}\frac{\mathbf{Z}_1}{\mathbf{Z}_T} = 120 \text{ V} \underline{/0°} \times \frac{60 \text{ }\Omega \underline{/+60°}}{86.7 \text{ }\Omega \underline{/-3°}}$$
$$= \mathbf{83 \text{ V} \underline{/+63°}}$$

SELF-CHECK

21-2:1. For the series circuit shown in Figure 21-1(b), compute the total impedance using the following individual impedances:

$\mathbf{Z}_1 = 250 + \text{j}500 \text{ }\Omega$
$\mathbf{Z}_2 = 400 \text{ }\Omega \underline{/-40°}$
$\mathbf{Z}_3 = -\text{j}100 \text{ }\Omega$

21-2:2. For the circuit used in Self-Check Problem 21-2:1 with $\mathbf{E} = 60 \text{ V} \underline{/0°}$, use the voltage-divider principle to find the voltage across each of the individual impedances.

21-3 IMPEDANCES IN PARALLEL— KIRCHHOFF'S CURRENT LAW

In working with resistors in parallel in Chapter 6, we noted that the *more* resistors we connect in parallel, the *less* is the equivalent resistance. We, therefore, found it to our advantage to switch our point of view in dealing with parallel circuits and to think in terms of the ability of the various branches to *conduct* electric current. Consequently, the solution of parallel circuits is a bit more involved than the solution of series circuits, because we must first make this conversion from *opposition* into *ability to conduct*.

In the solution of parallel ac circuits, it is even more important that we do our computation in numerical quantities that represent *ability to conduct*. (See Table 19-1.) We can use the same procedures in parallel ac circuits that we used in parallel dc circuits. There are two sets of parameters which are proportional to the ability of a circuit to conduct: (1) the total of all branch *currents,* and (2) the total of all branch *admittances.* In ac circuits, these must of course be *phasor* totals.

For parallel branches in an ac circuit, Kirchhoff's current law gives

$$\mathbf{I_T} = \mathbf{I_1} + \mathbf{I_2} + \mathbf{I_3} + \text{etc.} \tag{21-4}$$

where all terms are phasor quantities. Dividing every term in Equation (21-4) by the *common* voltage across *parallel* branches, gives

$$\frac{\mathbf{I_T}}{\mathbf{V}} = \frac{\mathbf{I_1}}{\mathbf{V}} + \frac{\mathbf{I_2}}{\mathbf{V}} + \frac{\mathbf{I_3}}{\mathbf{V}}$$

From the definition of admittance, this equation becomes

$$\mathbf{Y_T} = \mathbf{Y_1} + \mathbf{Y_2} + \mathbf{Y_3} + \text{etc.} \tag{21-5}$$

and we may state that

> *The total admittance of several parallel branches is the* **phasor** *sum of the branch admittances.*

We can determine the equivalent *impedance* of a parallel ac circuit either from Equation (21-4) by the total-current method,

$$\mathbf{Z_{eq}} = \frac{\mathbf{E}}{\mathbf{I_T}} \tag{21-6}$$

or from Equation (21-5) by the total-admittance method,

$$\mathbf{Z_{eq}} = \frac{1}{\mathbf{Y_T}} \tag{21-7}$$

EXAMPLE 21-4
What is the equivalent impedance when $\mathbf{Z_1} = 60~\Omega \underline{/+60°}$ and $\mathbf{Z_2} = 80~\Omega \underline{/-45°}$ are connected in parallel?

SOLUTION 1
Total-current method:

Step I. Draw a schematic diagram and a current phasor diagram, as in Figure 21-3.

Step II. Assume a convenient value of applied voltage, e.g., $\mathbf{E} = 240~\text{V} \underline{/0°}$.

Step III. Determine the branch currents.

$$\mathbf{I_1} = \frac{\mathbf{E}}{\mathbf{Z_1}} = \frac{240 \underline{/0°}}{60 \underline{/+60°}} = 4~\text{A} \underline{/-60°}$$

$$\mathbf{I_2} = \frac{\mathbf{E}}{\mathbf{Z_2}} = \frac{240 \underline{/0°}}{80 \underline{/-45°}} = 3~\text{A} \underline{/+45°}$$

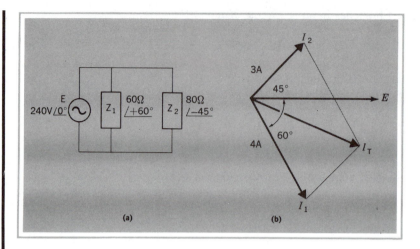

FIGURE 21-3
Impedances in Parallel.

Step IV. Express the branch currents in rectangular coordinates so they can be added.

$$\mathbf{I}_1 = 4 \cos 60° - j4 \sin 60° = 2 \qquad - j3.464$$

$$\mathbf{I}_2 = 3 \cos 45° + j3 \sin 45° = \underline{2.121 + j2.121}$$

Adding gives
$$\mathbf{I}_T = 4.121 - j1.343 \text{ A}$$

Step V. Convert the total current to polar form.

$$\phi = \arctan \frac{-1.343}{4.121} = -18°$$

$$I_T = \frac{4.121}{\cos -18°} = 4.334 \text{ A}$$

$$\therefore \mathbf{I}_T = 4.334 \text{ A} \underline{/-18°}$$

Step VI.

$$\mathbf{Z}_{eq} = \frac{\mathbf{E}}{\mathbf{I}_T} = \frac{240 \text{ V} \underline{/0°}}{4.334 \text{ A} \underline{/-18°}} = \mathbf{55.4 \ \Omega \underline{/+18°}}$$

SOLUTION 2
Total-admittance method:

Step I. as shown in Figure 21-4,

$$\mathbf{Y}_1 = \frac{1}{\mathbf{Z}_1} = \frac{1}{60 \ \Omega \underline{/+60°}} = 0.0167 \text{ S} \underline{/-60°}$$

$$\mathbf{Y}_2 = \frac{1}{\mathbf{Z}_2} = \frac{1}{80 \ \Omega \underline{/-45°}} = 0.0125 \text{ S} \underline{/+45°}$$

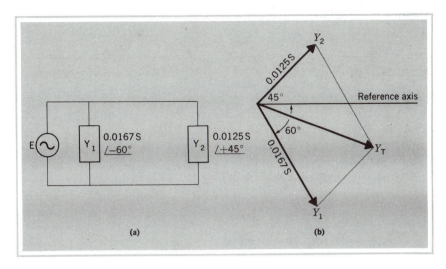

FIGURE 21-4
Admittances in Parallel.

Step II.

$$\mathbf{Y}_1 = 0.0167 \cos 60° - j0.0167 \sin 60° = 0.008\,33 - j0.014\,43 \text{ S}$$

$$\mathbf{Y}_2 = 0.0125 \cos 45° + j0.0125 \sin 45° = \underline{0.008\,84 + j0.008\,84 \text{ S}}$$

Adding gives
$$\mathbf{Y}_T = 0.017\,17 - j0.005\,59 \text{ S}$$

Step III.

$$\phi = \arctan \frac{-0.005\,59}{0.017\,17} = -18°$$

$$Y_T = \frac{0.017\,17}{\cos -18°} = 0.018\,05 \text{ S}$$

$$\therefore \ \mathbf{Y}_T = 0.018 \text{ S } \underline{/{-18°}}$$

Step IV.

$$\mathbf{Z}_{eq} = \frac{1}{\mathbf{Y}_T} = \frac{1}{0.018 \text{ S } \underline{/{-18°}}} = \mathbf{55.4 \ \Omega \underline{/{+18°}}}$$

Although Solution 1 may seem easier, since we can avoid switching our thinking to quantities expressed in siemens, Solution 2 is much preferred as we encounter more elaborate ac networks. We shall now drop the total-current procedure as a means to an end and work as much as possible in terms of admittance in parallel circuits. For the special case of *two* branches in parallel,

$$\mathbf{Y}_T = \mathbf{Y}_1 + \mathbf{Y}_2 = \frac{1}{\mathbf{Z}_1} + \frac{1}{\mathbf{Z}_2} = \frac{\mathbf{Z}_2 + \mathbf{Z}_1}{\mathbf{Z}_1 \mathbf{Z}_2}$$

Inverting this equation,

$$\mathbf{Z}_{eq} = \frac{\mathbf{Z}_1 \mathbf{Z}_2}{\mathbf{Z}_1 + \mathbf{Z}_2} \qquad (21\text{-}8)$$

where all terms are treated as phasor quantities. Applying Equation (21-8) to Example 21-4, we obtain a shorter numerical computation *without adjusting our thinking to the admittance concept of parallel branches.*

SOLUTION 3

Step I. Solving the numerator,

$$\mathbf{Z}_1 \mathbf{Z}_2 = 60\underline{/+60°} \times 80\underline{/-45°} = 4800\underline{/+15°}$$

Step II. Solving the denominator,

$$\mathbf{Z}_1 = 60 \ \Omega\underline{/+60°} = 30 \quad + j51.96$$
$$\mathbf{Z}_2 = 80 \ \Omega\underline{/-45°} = 56.57 - j56.57$$

Adding gives $\qquad \mathbf{Z}_1 + \mathbf{Z}_2 = 86.57 - j4.61 \ \Omega$

from which $\qquad \mathbf{Z}_1 + \mathbf{Z}_2 = 86.69 \ \Omega\underline{/-3°}$

Step III.

$$\frac{\mathbf{Z}_1 \mathbf{Z}_2}{\mathbf{Z}_1 + \mathbf{Z}_2} = \frac{4800\underline{/+15°}}{86.69\underline{/-3°}}$$

$$\therefore \ \mathbf{Z}_{eq} = 55.4 \ \Omega\underline{/+18°}$$

EXAMPLE 21-5

Given $\quad \mathbf{Z}_1 = 10\,000 + j15\,000 \ \Omega$, $\mathbf{Z}_2 = 4700 + j0 \ \Omega$, $\quad$ and $\quad \mathbf{Z}_3 = 25\,000 - j10\,000 \ \Omega$. What equivalent impedance (rectangular form) does the source "see" when it feeds $\mathbf{Z}_1$, $\mathbf{Z}_2$, and $\mathbf{Z}_3$ all connected in parallel?

SOLUTION

Step I. Since we cannot work with the rectangular coordinates of impedance in dealing with parallel branches, we first convert to polar form.

$$\mathbf{Z}_1 = 10\,000 + j15\,000 = 18.03 \ k\Omega\underline{/+56.3°}$$
$$\mathbf{Z}_2 = \quad 4700 + j0 \quad = 4.7 \ k\Omega\underline{/0°}$$
$$\mathbf{Z}_3 = 25\,000 - j10\,000 = 26.93 \ k\Omega\underline{/-21.8°}$$

Step II. Taking the reciprocal of each impedance,

$$\mathbf{Y}_1 = 55.47 \ \mu S\underline{/-56.3°}$$

$$\mathbf{Y}_2 = 212.8 \ \mu S \underline{/0°}$$

$$\mathbf{Y}_3 = 37.14 \ \mu S \underline{/+21.8°}$$

Step III. Converting to rectangular coordinates,[†]

$$\mathbf{Y}_1 = 30.77 - j46.15 \ \mu S$$
$$\mathbf{Y}_2 = 212.8 + j0 \mu S$$
$$\mathbf{Y}_3 = 34.48 + j13.79 \ \mu S$$
$$\overline{\mathbf{Y}_T = 278.1 - j32.36 \ \mu S}$$

Step IV. Converting $\mathbf{Y}_T$ to polar form,

$$\mathbf{Y}_T = 280 \ \mu S \underline{/-6.64°}$$

Step V. Taking the reciprocal of $\mathbf{Y}_T$,

$$\mathbf{Z}_{eq} = \frac{1}{\mathbf{Y}_T} = \frac{1}{280 \ \mu S \underline{/-6.64°}} = 3572 \ \Omega \underline{/+6.64°}$$

Step VI. Converting to rectangular coordinates,

$$\mathbf{Z}_{eq} = \mathbf{3548 + j413} \ \Omega$$

Since the *same* voltage appears across parallel branches, we can establish the **current-divider principle** for ac circuits from the relationship

$$\mathbf{V} = \mathbf{I}_1\mathbf{Z}_1 = \mathbf{I}_2\mathbf{Z}_2$$

from which

$$\frac{\mathbf{I}_1}{\mathbf{I}_2} = \frac{\mathbf{Z}_2}{\mathbf{Z}_1} \tag{21-9}$$

Hence, as we would expect, the split in current between two parallel branches in an ac circuit is *inversely* proportional to the impedance of the two branches. However, we find it more convenient in dealing with parallel branches in ac circuits to express the current-divider principle in terms of branch admittances. Starting with the common voltage for all branches in parallel,

$$\mathbf{V} = \frac{\mathbf{I}_x}{\mathbf{Y}_x} = \frac{\mathbf{I}_T}{\mathbf{Y}_T}$$

[†] As we noted in Section 19-8, the rectangular coordinates of the admittance are *not* the reciprocals of the rectangular coordinates of the impedance.

from which

$$I_x = I_T \frac{Y_x}{Y_T} \qquad (21\text{-}10)$$

where every term is treated as a phasor quantity.

Note that when we switch our thinking to admittance, the current-divider Equation (21-10) becomes the exact dual of the voltage-divider Equation (21-3). For the special case of only *two* parallel branches, we can substitute the reciprocal of Equation (21-8) for Y_T and express the current-divider principle in the form

$$I_1 = I_T \frac{Z_2}{Z_1 + Z_2} \qquad (21\text{-}11)$$

Equation (21-11) again allows us to minimize numerical computation in dealing with two parallel branches without adjusting our thinking to the admittance concept of parallel branches.

EXAMPLE 21-6

What current will a 10-k$\Omega\underline{/-45°}$ load draw from the Norton-equivalent source shown in Figure 21-5?

FIGURE 21-5
Diagrams for Example 21-6.

SOLUTION 1
By Ohm's law:

Step I.

$$Z_{eq} = \frac{Z_1 \times Z_2}{Z_1 + Z_2} = \frac{5\ k\Omega\underline{/+30°} \times 10\ k\Omega\underline{/-45°}}{(4330 + j2500) + (7071 - j7071)}$$

$$= \frac{50 \times 10^6 \underline{/-15°}}{(11\ 401 - j4571)} = \frac{50 \times 10^6 \underline{/-15°}}{12\ 283 \underline{/-21.85°}}$$

$$= 4.07\ k\Omega \underline{/+6.85°}$$

Step II.

$$\mathbf{V} = \mathbf{I_T Z_{eq}} = 50\ mA\underline{/-30°} \times 4.07\ k\Omega\underline{/+6.85°}$$

$$= 203.5\ V\underline{/-23.15°}$$

Step III.

$$\mathbf{I_L} = \frac{\mathbf{V}}{\mathbf{Z_L}} = \frac{203.5\ V\underline{/-23.15°}}{10\ k\Omega\underline{/-45°}} = \mathbf{20.35\ mA\underline{/+21.85°}}$$

SOLUTION 2

By the current-divider principle:

Step I.

$$\mathbf{Y_L} = \frac{1}{\mathbf{Z_L}} = \frac{1}{10\ k\Omega\underline{/-45°}} = 100\ \mu S\underline{/+45°} = 70.71 + j70.71\ \mu S$$

$$\mathbf{Y_N} = \frac{1}{\mathbf{Z_N}} = \frac{1}{5\ k\Omega\underline{/+30°}} = 200\ \mu S\underline{/-30°} = \underline{173.2 - j100\ \mu S}$$

$$\mathbf{Y_T} = 243.91 - j29.29\ \mu S$$

From which $\mathbf{Y_T} = 245.7\ \mu S\underline{/-6.85°}$

Step II. From Equation (21-10)

$$\mathbf{I_L} = \mathbf{I_T}\frac{\mathbf{Y_L}}{\mathbf{Y_T}} = 50\ mA\underline{/-30°} \times \frac{100\ \mu S\underline{/+45°}}{245.7\ \mu S\underline{/-6.85°}}$$

$$= \mathbf{20.35\ mA\underline{/+21.85°}}$$

SELF-CHECK

21-3:1. A generator with source voltage $\mathbf{E} = 60\ V\underline{/0°}$ supplies a circuit that contains three parallel branches. The branch impedances are as follows:

$$\mathbf{Z_1} = 400\ \Omega$$
$$\mathbf{Z_2} = -j250\ \Omega$$
$$\mathbf{Z_3} = 400 + j500\ \Omega$$

Determine the total admittance of the circuit.

21-3:2. For the circuit described in Self-Check Problem 21-3:1, use the current-divider principle to determine the currents in each of the three branches.

21-4 SERIES-PARALLEL IMPEDANCES

As long as we remember that we are working with phasor quantities, the procedure for solving series-parallel impedance combinations is essentially the same as the procedure we developed for dc circuits in Section 7-1.

EXAMPLE 21-7
What total load is seen by the source in the circuit diagram of Figure 21-6(a)?

FIGURE 21-6
Circuit Diagrams for Example 21-7.

SOLUTION
   Step I.   Determine the single Z_{eq} which can replace the parallel combination of Z_2 and Z_3.

$$Z_{eq} = \frac{Z_2 \times Z_3}{Z_2 + Z_3} = \frac{5 \text{ k}\Omega\underline{/+30°} \times 10 \text{ k}\Omega\underline{/-45°}}{(4330 + j2500) + (7071 - j7071)}$$

$$= \frac{50 \times 10^6\underline{/-15°}}{11\,401 - j4571} = \frac{50 \times 10^6\underline{/-15°}}{12\,283\underline{/-21.85°}}$$

$$= 4.07 \text{ k}\Omega\underline{/+6.85°}$$

   Step II.   Convert Z_1 and Z_{eq} to rectangular coordinates. For the simple series circuit of Figure 21-6(b),

$$Z_1 = 1000 - j1732 \ \Omega$$

$$Z_{eq} = \underline{4041 + \ \ j485} \ \Omega$$

$$Z_T = 5041 - j1247 \ \Omega = \mathbf{5193 \ \Omega\underline{/-13.9°}}$$

EXAMPLE 21-8

If the output terminals of the π-network in Figure 21-7 are left open-circuit, what is the input impedance at 10 kHz?

FIGURE 21-7
Pi-Network for Example 21-8.

SOLUTION

Step I. First we must determine the single impedance that can replace the resistor, inductor, and right-hand capacitor, all of which form a simple series circuit when no load is connected to the output terminals.

$$X_L = 2 \times \pi \times 10 \text{ kHz} \times 200 \text{ mH} = 12\ 566\ \Omega$$

$$X_C = \frac{1}{2 \times \pi \times 10 \text{ kHz} \times 0.001\ \mu\text{F}} = 15\ 915\ \Omega$$

$$\mathbf{Z}_s = R + j(X_L - X_C) = 4700 + j(12\ 566 - 15\ 915)\ \Omega$$

$$= 4700 - j3349\ \Omega = 5771\ \Omega\underline{/-35.47°}$$

Step II. We can now use any of the methods for determining the equivalent impedance of two impedances in parallel.

$$B_C = \omega C = 2 \times \pi \times 10 \text{ kHz} \times 0.001\ \mu\text{F} = 62.83\ \mu\text{S (capacitive)}$$

$$\mathbf{Y}_s = \frac{1}{\mathbf{Z}_s} = \frac{1}{5771\ \Omega\underline{/-35.47°}} = 173.3\ \mu\text{S}\underline{/+35.47°}$$

Step III.

$$\mathbf{Y}_T = B_C + \mathbf{Y}_s = +j62.83\ \mu\text{S} + (141.2 + j100.4)\ \mu\text{S}$$

$$= 141.2 + j163.23\ \mu\text{S} = 215.8\ \mu\text{S}\underline{/+49.1°}$$

Step IV.

$$\therefore\ \mathbf{Z}_{in} = \frac{1}{\mathbf{Y}_T} = \frac{1}{215.8\ \mu\text{S}\underline{/+49.1°}} = \mathbf{4634}\ \Omega\underline{/-49.1°}$$

SELF-CHECK

21-4:1. In the series-parallel circuit shown in Figure 21-8(a), the individual devices have the following values:

$$\mathbf{E} = 100 \text{ V} \underline{/0°}$$
$$\mathbf{Z_1} = 400 - j500 \text{ }\Omega$$
$$\mathbf{Z_2} = -j500 \text{ }\Omega$$
$$\mathbf{Z_3} = 400 + j500 \text{ }\Omega$$

(a) Find the total impedance for this circuit.
(b) Find the voltage drop across $\mathbf{Z_1}$.

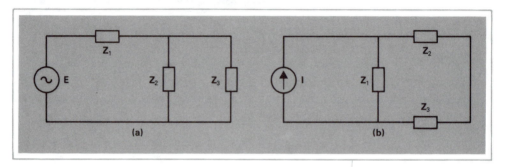

FIGURE 21-8

21-4:2. In the series-parallel circuit shown in Figure 21-8(b), the individual devices have the following values:

$$\mathbf{I} = 100 \text{ mA} \underline{/0°}$$
$$\mathbf{Z_1} = 400 - j500 \text{ }\Omega$$
$$\mathbf{Z_2} = -j500 \text{ }\Omega$$
$$\mathbf{Z_3} = 400 + j500 \text{ }\Omega$$

(a) Find the total impedance for this circuit.
(b) Find the current in $\mathbf{Z_1}$.

21-5 EQUIVALENT CIRCUITS

We shall encounter many impedance networks which can be solved more readily if we can replace an elaborate network by a simple equivalent network. This essentially was the basis for the circuit-simplification pattern of solving series-parallel impedance networks in Section 21-4. The equivalent-circuit technique becomes particularly important in dealing with equivalent circuits for electronic amplifiers.

The complex nature of an impedance provides a procedure (in addition to the techniques we considered in Chapter 8) for substituting an equivalent parallel circuit for a series circuit, and vice versa. This procedure had no particular significance in dc circuits. Given an impedance sealed in a "black box" as in Figure 21-9(a), the source cannot distinguish between the possible circuits of Figures 21-9(b) and 21-9(c). Hence, the circuit of Figure 21-9(b) is the equivalent of the circuit of Figure 21-9(c), and vice versa. As we noted in Section 19-8 (Table 19-1), Figure 21-9(b) represents the rectangular coordinates of the impedance of the black box in ohms; but to find the parallel equivalent circuit of Figure 21-9(c), we must first find the rectangular coordinates of the equivalent admittance in siemens.

FIGURE 21-9
Series- and Parallel-Equivalent Circuits for an Inductive Load in an AC Circuit.

When we solve impedance network problems by plugging numbers into formulas, one of the most common sources of error is selecting the wrong equivalent circuit for the rectangular coordinates we intend to use. Again, we should note and understand thoroughly that

Whenever we express an impedance (ohms) in rectangular coordinates, we are automatically specifying an equivalent circuit consisting of a resistance (ohms) and reactance (ohms) in series, as shown in Figure 21-10.

FIGURE 21-10
Series-Equivalent Circuit of an Impedance.

Applying basic trigonometry functions to the impedance diagram of Figure 21-10,

$$R_s = Z \cos \phi \tag{21-12}$$

and
$$X_s = Z \sin \phi \tag{21-13}$$

Likewise,

Whenever we express an admittance (siemens) in rectangular coordinates, we are automatically specifying an equivalent circuit consisting of a conductance (siemens) and susceptance (siemens) in parallel, as shown in Figure 21-11.

FIGURE 21-11
Parallel-Equivalent Circuit of an Admittance.

In schematic diagrams for practical circuits, we find that it is customary to state the values of resistors and the reactances of chokes in **ohms,** even though they may actually be connected in *parallel*. Therefore, in Figure 21-11, we would like to express the R_p and the X_p of the two parallel branches in ohms. From the admittance diagram of Figure 21-11,

$$G_p = Y \cos \phi \tag{21-14}$$

But
$$G = \frac{1}{R} \quad \text{and} \quad Y = \frac{1}{Z}$$

$$\therefore R_p = \frac{Z}{\cos \phi} \tag{21-15}$$

Similarly,
$$B_p = Y \sin \phi \tag{21-16}$$

and
$$X_p = \frac{Z}{\sin \phi} \tag{21-17}$$

The next step is to be able to replace the given series circuit of Figure 21-12(a) with an exact parallel-equivalent circuit with R_p and X_p in ohms, without having to perform all the in-between calculations every time. For a parallel circuit to be the exact equivalent of a given series circuit as far as the source is concerned, their impedances must be exactly the same both in magnitude and in angle.

From Equation (21-12),
$$\frac{R_s}{Z} = \cos \phi$$

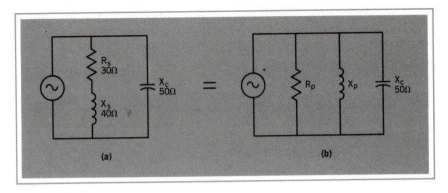

FIGURE 21-12
Parallel Equivalent of a Series Branch.

and from Equation (21-15), $\qquad \dfrac{Z}{R_p} = \cos \phi$

$$\therefore \quad \dfrac{R_s}{Z} = \dfrac{Z}{R_p} \tag{21-18}$$

from which $\qquad R_p = \dfrac{Z^2}{R_s} = \dfrac{(R_s^2 + X_s^2)}{R_s} \tag{21-19}$

Similarly $\qquad X_p = \dfrac{Z^2}{X_s} = \dfrac{(R_s^2 + X_s^2)}{X_s} \tag{21-20}$

EXAMPLE 21-9
An ac circuit consists of two parallel branches, one consisting of a 30-Ω resistance and 40-Ω inductive reactance in series. The other branch consists of a 50-Ω capacitive reactance. Find the equivalent parallel circuit.

SOLUTION

$$R_p = \dfrac{R_s^2 + X_s^2}{R_s} = \dfrac{900 + 1600}{30} = 83.3 \ \Omega$$

$$X_p = \dfrac{R_s^2 + X_s^2}{X_s} = \dfrac{2500}{40} = 62.5 \ \Omega \ \text{(inductive)}$$

The equivalent susceptance of the parallel inductance and capacitance branches of the equivalent circuit of Figure 21-12(b) is

$$B_{eq} = B_C - B_L = \dfrac{1}{X_C} - \dfrac{1}{X_L} = \dfrac{1}{50} - \dfrac{1}{62.5} = 0.02 - 0.016$$

$$= 0.004 \ \text{S (capacitive)}$$

$$\therefore \ X_{eq} = \dfrac{1}{B_{eq}} = \dfrac{1}{0.004 \ \text{S}} = 250 \ \Omega \ \text{(capacitive)}$$

Therefore, the equivalent parallel circuit consists of an 83.3-Ω resistance in parallel with a 250-Ω capacitive reactance.

We shall find this type of conversion very useful in dealing with practical parallel-resonant circuits in Chapter 23.

Returning to Equation (21-18), we can derive the following equations for the series-equivalent components of a given parallel circuit.

$$R_s = \frac{R_p X_p^2}{R_p^2 + X_p^2} \tag{21-21}$$

and

$$X_s = \frac{R_p^2 X_p}{R_p^2 + X_p^2} \tag{21-22}$$

In Section 20-5, we discovered that the *same* power relationship [Equation (20-13)] applies to series, parallel, and also series-parallel networks. We used this knowledge to advantage in establishing a procedure for solving power-factor correction problems.

We can also add this useful relationship to our collection of equivalent-circuit techniques. Given the active power and reactive power for any ac network, as shown in Figure 21-13(a), P/I^2 and Q/I^2 automatically give us the *series*-equivalent circuit in ohms where

$$R_s = \frac{P}{I^2} \tag{21-23}$$

and

$$X_s = \frac{Q}{I^2} \tag{21-24}$$

Likewise, as shown in Figure 21-13(b), V^2/P and V^2/Q automatically give us the *parallel*-equivalent circuit in ohms where

$$R_p = \frac{V^2}{P} \tag{21-25}$$

and

$$X_p = \frac{V^2}{Q} \tag{21-26}$$

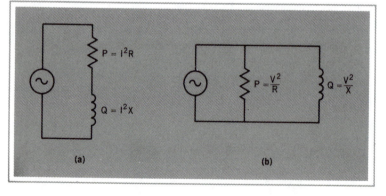

(a)

(b)

FIGURE 21-13
Series- and Parallel-Equivalent Circuits from Power Relationships.

EXAMPLE 21-9A

An ac circuit consists of two parallel branches. One branch consists of a 30-Ω resistance and a 40-Ω inductive reactance in series. The other branch consists of a 50-Ω capacitive reactance [Figure 21-12(a)]. Find the equivalent parallel circuit.

SOLUTION

Assume a convenient source voltage. Let $E = 100$ V. The impedance of the series branch is

$$Z_s = \sqrt{30^2 + 40^2} = 50 \ \Omega$$

Hence, the current in each branch has a magnitude of

$$I = \frac{100 \text{ V}}{50 \ \Omega} = 2 \text{ A}$$

The active power is

$$P = I_s^2 R = 2^2 \times 30 = 120 \text{ W}$$

The inductive reactive power is

$$Q_L = I_s^2 X_L = 2^2 \times 40 = 160 \text{ vars}$$

The capacitive reactive power is

$$Q_C = \frac{V^2}{X_C} = \frac{100^2}{50} = 200 \text{ vars}$$

As shown in the power triangle of Figure 21-14, the net reactive power is $200 - 160 = 40$ vars capacitive from which,

$$R_p = \frac{V^2}{P} = \frac{100^2}{120} = \textbf{83.3} \ \boldsymbol{\Omega}$$

and

$$X_p = \frac{V^2}{Q} = \frac{100^2}{40} = \textbf{250} \ \boldsymbol{\Omega} \text{ (capacitive)}$$

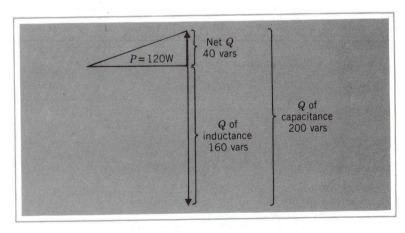

FIGURE 21-14
Power Triangle for Example 21-9A.

Note that, in both of the equivalent-circuit techniques we have developed in this section, we have escaped *phasor* algebra because the relationships are based on *magnitudes* only.

SELF-CHECK

21-5:1. Determine the 400-Hz parallel equivalent resistance and reactance for a practical coil whose series resistance is 50 Ω and series inductance is 120 mH.
21-5:2. Find the series equivalent for the parallel circuit in Figure 21-4.

21-6 SOURCE CONVERSION

Another useful equivalent-circuit technique is the conversion of a constant-voltage source to an equivalent constant-current source, and vice versa, which we developed in Section 8-4. Although *instantaneous* voltages and currents vary continuously in ac circuits, rms values are represented by steady-state phasors. Hence, we can still speak of constant-voltage and constant-current sources in alternating-current networks. For any "black-box" ac source, Equation (8-1) becomes

$$\mathbf{Z}_{int} = \frac{\text{open-circuit rms voltage}}{\text{short-circuit rms current}} \tag{21-27}$$

If we are given a constant-voltage ac source, as in Figure 21-15(a), the equivalent constant-current source of Figure 21-15(b) has the *same* internal impedance and a short-circuit current of

$$\mathbf{I}_{sc} = \frac{\mathbf{E}}{\mathbf{Z}_{int}} \tag{21-28}$$

Similarly, if we are given the constant-current ac source of Figure 21-15(b), the equivalent constant-voltage source of Figure 21-15(a) has the *same* internal impedance and an open-circuit voltage of

$$\mathbf{E}_{oc} = \mathbf{I}\,\mathbf{Z}_{int} \tag{21-29}$$

(a)   (b)

FIGURE 21-15
Alternating-Current Source Conversion.

In both cases, the phasor algebra provides us with an angle, as well as a magnitude, for the equivalent source.

Source conversion provides a third solution for Example 21-6.

EXAMPLE 21-6A

What current will a 10-kΩ $\underline{/-45°}$ load draw from the Norton-equivalent source shown in Figure 21-5?

SOLUTION

We can replace the original constant-current source shown again in Figure 21-16(a) with the equivalent constant-voltage source of Figure 21-16(b).

$$\mathbf{E}_x = \mathbf{IZ}_x = 50 \text{ mA}\underline{/-30°} \times 5 \text{ k}\Omega\underline{/+30°} = 250 \text{ V}\underline{/0°}$$

$$\mathbf{Z}_T = \mathbf{Z}_x + \mathbf{Z}_L = 5 \text{ k}\Omega\underline{/+30°} + 10 \text{ k}\Omega\underline{/-45°}$$

$$= (4330 + j2500) + (7071 - j7071)$$

$$= (11\,401 - j4571) = 12\,283 \text{ }\Omega\underline{/-21.85°}$$

$$\mathbf{I}_L = \frac{\mathbf{E}_x}{\mathbf{Z}_T} = \frac{250 \text{ V}\underline{/0°}}{12\,283 \text{ }\Omega\underline{/-21.85°}} = \mathbf{20.35 \text{ mA}\underline{/+21.85°}}$$

FIGURE 21-16
Source Conversion for Example 21-6A.

SELF-CHECK

21-6:1. A voltage source has an open-circuit voltage of 150 V and an internal impedance of 300 + j400 Ω. Convert this source to an equivalent constant-current source.

21-7 MAXIMUM POWER TRANSFER

In Section 6-7, we considered the selection of a load resistance which would permit maximum power input to the load from a source having a fixed open-circuit voltage and a fixed internal resistance. We discovered that maximum power input occurs when

$R_L = R_{int}$. We can now extend our discussion of maximum power transfer to alternating-current circuits in which the internal **impedance** of the source is partially reactive.

From our discussion of active power in Chapter 20, the power input to the load in the circuit in Figure 21-17 is

$$P_L = I^2 R_L$$

But
$$I = \frac{E}{Z_T} = \frac{E}{\mathbf{Z}_{int} + \mathbf{Z}_L} = \frac{E}{\sqrt{(R_{int} + R_L)^2 + (X_{int} + X_L)^2}}$$

Hence,
$$P_L = \frac{E^2 R_L}{(R_{int} + R_L)^2 + (X_{int} + X_L)^2} \qquad (21\text{-}30)$$

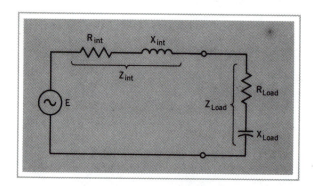

FIGURE 21-17
Maximum Power Transfer in an AC Circuit.

R_{int} and X_{int} remain fixed, depending on the characteristics of the source. For the moment, we leave R_L fixed and vary X_{Load}. From Equation (21-30) P_L will be maximum when the magnitude of the denominator of the expression is minimum. If we select a value of load reactance that is equal in magnitude but *opposite* in sign to the internal reactance, $(X_{int} + X_L)^2$ in Equation (21-30) becomes zero. Consequently, P_L will be maximum for the given value of R_L. Thus, as shown in Figure 21-17, if the source has an *inductive* internal impedance, we select a *capacitive* load reactance such that the net reactance of the circuit is zero.

Having cancelled out the reactance of the source, we now vary R_L just as we did in Section 6-7. Maximum power input to the load again requires that we select $R_L = R_{int}$. Hence, in an ac circuit,

Maximum power transfer occurs when the load impedance is the **conjugate** *of the internal impedance of the source.*

$\mathbf{Z}_L$ is the *conjugate* of $\mathbf{Z}_{int}$ when $R_L = R_{int}$ and the reactance of the load is equal in magnitude but *opposite* in sign to the reactance of the source.

EXAMPLE 21-10
A 50-MHz RF signal generator has an internal impedance of $60 + j45\ \Omega$. What value of load resistance and capacitance in parallel with R_L are required to obtain maximum power output from the signal generator?

SOLUTION

For maximum power transfer,

$$\mathbf{Z}_L = 60 - j45 \ \Omega$$

60 Ω and 45 Ω are the resistance and reactance, respectively, of a *series* circuit. From Equation (21-19),

$$R_p = \frac{R_s^2 + X_s^2}{R_s} = \frac{60^2 + 45^2}{60} = \mathbf{93.75 \ \Omega}$$

and from Equation (21-20),

$$X_p = \frac{R_s^2 + X_s^2}{X_s} = \frac{5625}{45} = 125 \ \Omega$$

$$\therefore C = \frac{1}{\omega X_C} = \frac{1}{2 \times \pi \times 50 \ \text{MHz} \times 125 \ \Omega} = \mathbf{25.5 \ pF}$$

SELF-CHECK

21-7:1. What series combination of ideal components will draw maximum power from a 400-Hz voltage source with the following characteristics:

$$E = 100 \ \text{V}$$
$$\mathbf{Z}_{\text{int}} = 30 + j40 \ \Omega?$$

21-7:2. What is the maximum power that can be supplied to a circuit by a 60-Hz voltage source that has an open-circuit voltage of 120-V and an internal impedance of 100 $\Omega\underline{/-25°}$?

21-8 CIRCLE DIAGRAMS

When we wish to examine the effect of varying one of the parameters of an impedance network, for example, the reactance of the load in the circuit of Figure 21-17, we can avoid having to draw a whole series of phasor diagrams by resorting to a bit of basic geometry.

Since the current in a series circuit is common to all components, we have been using current as the reference phasor for series-circuit phasor diagrams, as in Figure 21-18(b). We then draw the total voltage as the resultant of the voltages across the resistance and the reactance of the circuit.

But, in practice, the applied voltage is a constant, and as either the resistance or reactance of the circuit is changed, the current (the dependent variable of most electric circuits) must change in such a manner that

$$V_R + jV_L = \mathbf{E}$$

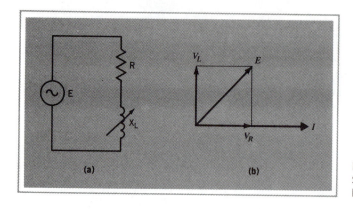

(a)

(b)

FIGURE 21-18
Simple Series Circuit with a Variable Reactance.

We can redraw the phasor diagram with the applied voltage as the reference phasor, as shown in Figure 21-19. Since the applied voltage is constant, and since the voltage across the inductance must always lead the voltage across the resistance by exactly 90°, this phasor diagram will always be in the form of a right-angled triangle with a constant hypotenuse. We can show that, under these circumstances, the junction of the V_R and V_L phasors must be on the circumference of a circle with the **E** phasor as a diameter.[†]

In the circuit of Figure 21-18(a), the resistance remains constant but the inductive reactance can change. Since $I = V_R/R$, and since R is constant, the current phasor in Figure 21-19 will have the same angle and same proportionate magnitude as the V_R phasor. Therefore, as shown in Figure 21-21, in a series circuit containing a fixed

[†]$\triangle ABC$ in Figure 21-20,

$$\angle CAB + \angle ABC + \angle BCA = 180°$$
$$\therefore \angle CAB + \angle ABD + \angle DBC + \angle BCA = 180°$$

But since $AD = DB = DC$,

$$\angle CAB = \angle ABD$$

Similarly,

$$\angle BCA = \angle DBC$$
$$\therefore 2(\angle ABD + \angle DBC) = 180°$$

from which

$$\angle ABC = 90°$$

Conversely, whenever

$$\angle ABC = 90°$$
$$AD = DB = DC = \text{radii of a circle}$$

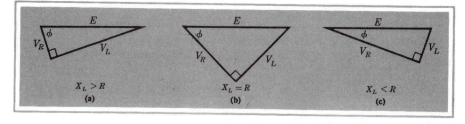

FIGURE 21-19
Effect of Varying X_L in a Series Circuit Connected to a Fixed Source Voltage.

FIGURE 21-20

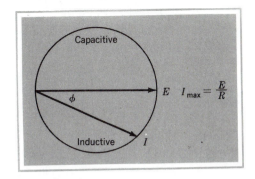

FIGURE 21-21
Circle Diagram for Current in a Series Circuit with a Fixed Resistance and a Variable Reactance.

resistance and a variable reactance, the tip of the current phasor must also lie on the circumference of a circle having the applied voltage as the reference phasor and a diameter of $I_{max} = E/R$, because the current will be greatest when the reactance drops to zero. Such is the situation when the source has a constant magnitude but a variable frequency. We use the circular locus of the current phasor as the frequency changes in the analysis of resonant circuits and transmission lines.

Figure 21-22 shows a simple series circuit in which the inductive reactance remains constant and the resistance varies. This condition will be encountered in analyzing the operating characteristics of induction motors as the mechanical load on the motor is changed. The current will be maximum when the resistance is zero. Hence, $I_{max} = E/X_L$, lagging the applied voltage by 90° as shown. As the resistance increases, the current decreases and, at the same time, the phase angle decreases. Again, since the applied voltage is constant, and since V_R and V_L must always be at right angles, the tip of the current phasor must be on the circumference of a circle, with E/X_L as the diameter.

From Figure 21-22(b),

$$\frac{I}{I_{max}} = \cos(90° - \phi) = \sin\phi \qquad (21\text{-}31)$$

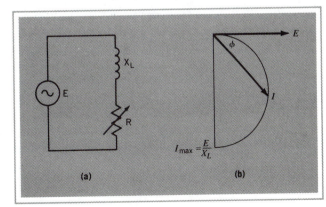

FIGURE 21-22
Circle Diagram for Current in a Series Circuit with a Fixed Inductive Reactance and a Variable Resistance.

EXAMPLE 21-11

A series circuit with a fixed reactance has an 0.5 lagging power factor when it draws a 10-A current from a 120-V 60-Hz source. What is the maximum active power for this circuit?

SOLUTION

$$\phi = \text{arccos } 0.5 = 60°$$

$$I_{max} = \frac{I}{\sin 60°} = \frac{10}{0.866} = 11.55 \text{ A}$$

$$P = EI \cos \phi$$

where E is the applied voltage and I cos ϕ is the in-phase rectangular coordinate of the current. As the circle diagram of Figure 21-22(b) shows, *I cos ϕ will be maximum when $\phi = 45°$.* Therefore, when $\phi = 45°$,

$$I = I_{max} \sin \phi = 11.55 \times 0.7071 = 8.165 \text{ A}$$

$$\therefore P_{max} = EI \cos \phi = 120 \times 8.165 \times 0.7071 = \textbf{693 W}$$

PROBLEMS

Draw schematic and phasor diagrams wherever applicable.

21-2-1. What is the total impedance when 400 $\Omega\,/+15°$ and 800 $\Omega\,/+60°$ are connected in series?

21-2-2. What is the total impedance when $\mathbf{Z}_1 = 5 \text{ k}\Omega\,/+30°$, $\mathbf{Z}_2 = 4.7 \text{ k}\Omega\,/-45°$, and $\mathbf{Z}_3 = 2.2 \text{ k}\Omega\,/+75°$ are connected in series?

21-2-3. A coil having a resistance of 100 Ω and an inductance of 0.5 H, a 60-μF capacitor, and a 30-Ω resistor are all connected in series to a 60-Hz source whose open-circuit voltage is 208 V $/0°$.
(a) What are the magnitude and phase of the current through the coil?
(b) What are the magnitude and phase of the voltage across the coil?

21-2-4. What value of impedance must be connected in series with a 60-$\Omega\,/+45°$ impedance for the current drawn from a 120-V source to be 1 A lagging the source voltage by 30°?

21-3-5. What is the total admittance when the two components of Problem 21-2-1 are connected in parallel?

21-3-6. What is the equivalent impedance when the three impedances of Problem 21-2-2 are connected in parallel?

21-3-7. What is the overall power factor when the three components of Problem 21-2-3 are connected in parallel?

21-3-8. What value of impedance must be connected in parallel with a 10-k$\Omega\,/-60°$ impedance for the current from a 1.5-V 100-kHz source to be 150 μA $/0°$?

21-4-9. Determine the magnitude and phase of the two capacitor currents in the circuit of Figure 21-23(a).

21-4-10. Find the magnitude and phase of the voltage across the capacitor in the circuit of Figure 21-23(b).

21-4-11. Find the magnitude and phase of the current in the output resistor in the circuit of Figure 21-24(a).

(a)   **(b)**

FIGURE   21-23

(a)   **(b)**

FIGURE   21-24

21-4-12. Find the magnitude and phase of the current drawn from the voltage source in the circuit of Figure 21-24(b).
    (a) With the output terminals open-circuit.
    (b) With the output terminals short-circuit.

21-4-13. Find the magnitude and phase of the open-circuit output voltage in the circuit of Figure 21-24(b).

21-4-14. Find the magnitude and phase of the voltage source required in the circuit of Figure 21-25(a) for the total drain on the source to be 50 mA$\underline{/+45°}$.

21-4-15. Find the magnitude and phase of the open-circuit voltage between terminals A and B in the Hay bridge circuit shown in Figure 21-25(b), if the source voltage is 50 V$\underline{/+30°}$.

21-4-16. If the source voltage in Figure 21-25(b) is 15 V$\underline{/0°}$, what is the total source current with terminals A and B (a) open-circuit, and (b) short-circuit?

(a)   **(b)**

FIGURE   21-25

21-5-17. What values of capacitance and resistance in series will form a 1-kHz impedance of 600 $\Omega\underline{/-30°}$?

21-5-18. What values of capacitance and resistance in parallel will form a 1-kHz impedance of 600 $\Omega\underline{/-30°}$?

21-5-19. What values of inductance and resistance in parallel will be equivalent at 60 Hz to a 0.5-H inductance and a 120-Ω resistance in series?

21-5-20. What value of capacitance and resistance in series will be equivalent at 400 Hz to a 0.02-μF capacitor and a 6800-Ω resistor in parallel?

21-5-21. Reduce the circuit of Figure 21-26 to the simplest possible series circuit containing a resistance and a reactance for (a) $f = 60$ Hz, and (b) $f = 100$ Hz.

21-5-22. Reduce the circuit of Figure 21-26 to the simplest possible parallel circuit for $f = 50$ Hz.

21-5-23. Reduce the circuit of Figure 21-27 to the simplest possible series circuit.

FIGURE 21-26

FIGURE 21-27

FIGURE 21-28

21-5-24. Reduce the circuit of Figure 21-27 to the simplest possible parallel circuit.

21-5-25. Determine the input impedance (rectangular coordinates) for the network of Figure 21-28.

21-5-26. Determine the input admittance (rectangular coordinates) for the network of Figure 21-28.

21-6-27. Determine the equivalent constant-current sources for the voltage sources shown in Figure 21-29.

21-6-28. Determine the equivalent constant-voltage sources for the current sources shown in Figure 21-30.

21-6-29. Given a constant-voltage source with an open-circuit voltage of 110 V$\underline{/-45°}$ and an internal impedance of 30 + j40 Ω, determine the equivalent constant-current source and the rectangular coordinates of its internal admittance.

21-6-30. Given a constant-current source with a short-circuit current of 400 mA$\underline{/+60°}$ and an internal admittance of 20 + j80 mS, determine the equivalent constant-voltage source and its internal impedance in rectangular form.

FIGURE 21-29

FIGURE 21-30

21-7-31. If the open-circuit voltage of the RF signal generator in Example 21-10 is 240 mV, what is its maximum power output?

21-7-32. What is the power output in Problem 21-7-31 if we remove the 25-pF capacitor?

21-7-33. A television signal generator has an internal impedance of 50 $\Omega\underline{/-30°}$ at 83.25 MHz. Determine the series components of a dummy antenna load designed for maximum power transfer.

21-7-34. A transmitter operating on an assigned frequency of 1850 kHz delivers maximum power to an antenna when the antenna is tuned to represent a load consisting of a 200-pF capacitance in parallel with a 160-Ω resistance. Express the internal impedance "seen" at the output terminals of the transmitter in rectangular coordinates..

21-8-35. A circuit consisting of a 0.75-H inductance, a 40-Ω resistance, and a 26.5-μF capacitance in series is connected to a source which remains constant at 120 V, but whose frequency can vary from 25 Hz to 60 Hz. Draw an accurate diagram showing the manner in which the magnitude and phase of the current vary as the frequency is varied through its range.

21-8-36. Draw an accurate circle diagram showing how the open-circuit output voltage E_0 varies as the rheostat is varied from zero to 500 kΩ in the phase shifter shown in Figure 21-31.

FIGURE 21-31
Phase Shifter.

21-8-37. In an ac circuit, $\mathbf{Z}_1$ consists of a 50-Ω capacitive reactance and $\mathbf{Z}_2 = 50\ \Omega\underline{/+30°}$. Determine the equivalent impedance of $\mathbf{Z}_1$ and $\mathbf{Z}_2$ in series and of $\mathbf{Z}_1$ and $\mathbf{Z}_2$ in parallel. Determine the relationship that must exist between any two impedances so that their parallel equivalent impedance is the same as their series equivalent impedance.

REVIEW QUESTIONS

21-1. Write a statement for Kirchhoff's voltage law as it applies to alternating-current circuits.

21-2. Given the magnitudes of two impedances in ohms, why can we not calculate the total impedance of these two impedances in series?

21-3. Under what conditions could the total impedance of two impedances in series be less than that of either one by itself?

21-4. Under what circumstances is it (a) more convenient and (b) less convenient, to state impedances in rectangular coordinates?

21-5. Write a statement for the voltage-divider principle as it applies to alternating-current circuits.

21-6. What is the advantage in thinking in terms of total **admittance** when dealing with parallel ac circuits?

21-7. Show how Equation (21-8) is derived.

21-8. Derive Equation (21-11) from Equation (21-10).

21-9. What conditions must be fulfilled for a series circuit to be the exact equivalent of a given parallel circuit?

21-10. What is the significance of equivalent circuits in analyzing the performance of electric and electronic circuits?

21-11. Given a resistance and reactance in series with $R_s = 4X_s$, in the equivalent parallel circuit, $X_p = 4\,R_p$. Explain this inversion.

21-12. A coil and capacitor connected in series to a given source represent an inductive impedance. When the same two components are connected in parallel to the same source, the total impedance becomes capacitive. Explain.

21-13. If we vary the frequency of the source in the circuit of Figure 21-9(a), we can in fact determine whether the "black box" contains circuit (b) or circuit (c). Explain how this is achieved.

21-14. Why is it possible to derive either a series- or a parallel-equivalent circuit from the *same* power triangle?

21-15. Show that Equation (21-23) is the same as $R_s = Z \cos \phi$.

21-16. Show that Equation (21-26) is the same as

$$X_p = \frac{Z}{\sin \phi}$$

21-17. What is meant by the **conjugate** of an impedance?

21-18. What is the significance of conjugate impedance in ac networks?

21-19. Why does the tip of the current phasor move around the circumference of a circle when the resistance alone is varied in an ac circuit containing resistance and reactance in series.

21-20. Draw a circle diagram for the output voltage of the phase shifter of Figure 21-32 as the variable resistance is increased from zero to 500 kΩ.

FIGURE 21-32
Phase Shifter.

*21-21. Design a computer program that will ask for three pieces of input data: the value of a pure resistance, the value of a pure reactance, and whether they are connected in series or in parallel. When the third piece of information has been entered, the computer should then print out the values of resistance and reactance for the opposite equivalent circuit. Why does this program not need to know whether the reactance is inductive or capacitive?

* 21-22. Circle diagrams are alternatives to tedious repetitive calculations in ac networks containing a variable impedance. Computers can do such repetitive calculations rapidly and accurately. Write a computer program to print out a table of magnitudes and phase angles which will verify the circle diagram you produced for the phase shifter in Problem 21-8-36.

22

IMPEDANCE NETWORKS

In Chapter 8, we encountered resistance networks which we could not solve by simple series-parallel circuit relationships. We developed two general lines of attack on such networks: (1) we can write a set of simultaneous **loop** or **nodal** equations for the networks based on Kirchhoff's voltage and current laws, respectively; or (2) we can use network theorems to simplify the original networks to a level where we can tackle them by simple series-parallel relationships.

As long as we perform all numerical computations by *phasor* algebra, we follow exactly the same procedures in dealing with impedance networks in ac circuits. Unfortunately, this switch from linear algebra to phasor algebra greatly increases the amount of numerical computation. To minimize the possibility of errors in performing these calculations on an electronic calculator, we need to develop an understanding of the various network analysis techniques so that we can select the one that is most straightforward and involves the minimum of computation for a particular network.

The advantage of a personal computer is that errors are limited to incorrectly entering requested data—once a suitable program has been verified. The disadvantage is that too much time is required to write (and debug) a special program for each example involving complex algebra. From this chapter onward, for a personal computer to be a useful aid in solving alternating-current networks, we require access to a suitable circuit-analysis software package. Due to the variations in such programs, we must leave specific instruction in their application to the manuals provided with the software.

22-1 LOOP EQUATIONS

We can restate Kirchhoff's voltage law from Section 8-5, worded to suit the phasor quantities we encounter in ac circuits.

In any complete ac circuit (closed loop), the phasor sum of the voltage drops must equal the **phasor** *sum of the source voltages.*

This leads to equations in the form

$$\mathbf{V_1} + \mathbf{V_2} + \mathbf{V_3} = \mathbf{E} \qquad (22\text{-}1)$$

where every term is a phasor quantity. As in direct-current circuits, we have a choice of two methods of writing Kirchhoff's voltage-law equations: **loop** equations and **mesh** equations. The loop procedure more closely follows our statement of Kirchhoff's voltage law and is probably easier to *understand*. Mesh equations provide a format for setting up essentially the same Kirchhoff's voltage-law equations by following a few simple rules.

In writing *loop* equations, we substitute an **IZ** voltage drop across each impedance for each **V** term in Equation (22-1). Since the various values of **Z** are usually known, we end up with a set of simultaneous equations in which loop currents are the unknown quantities.

If we could omit $\mathbf{Z_4}$ in the network shown in Figure 22-1, we would have a simple T-network which we could solve by the series-parallel circuit techniques of Chapter 21. The addition of $\mathbf{Z_4}$ turns the network into a bridge circuit in which it is quite likely that

FIGURE 22-1
Bridged-T Network.

there will be a different current through each of the five impedances. We can avoid having to write *five* simultaneous equations by constructing a set of tracing loops from the + terminal of the source to its − terminal. Hence, instead of writing I_L for the current through Z_L in Figure 22-1, we immediately write $Z_L(I_2 + I_3)$ as our substitution for V_L. We write the *sum* of I_2 and I_3 since both tracing loops pass in the same direction through Z_L.

Because the current in a dc circuit always flows in the same direction, establishing the direction of the tracing loops presents no problem. We simply assumed a direction from the + terminal of the source, through the network, and back to the − terminal of the source. However, in an ac circuit, the instantaneous current reverses its polarity twice each cycle. Nevertheless, we must adopt some method of marking the direction of an ac source, since reversing the leads of an ac source will change the angle of the source voltage by 180°, just as reversing the leads of a dc source will invert the polarity of the source voltage. In Figure 22-1, the lower lead is common to both the source and the load. In nodal analysis, we call this the **reference node.** It is customary to express as a positive number the voltage appearing at the upper terminal of the source *with respect to* the reference node (which is sometimes grounded).

To make this choice quite clear, particularly in multisource networks, there are two methods of marking the direction for ac sources. We can use + and − signs, as shown in Figure 22-1. In an ac circuit, these signs tell us that the angle given for **E** is measured from the + terminal *with respect to* the − terminal. Hence, we should mark the positive direction for tracing loops through the external network from the + terminal to the − terminal of the source. Obviously, these + and − signs do not indicate instantaneous polarity.

The second method is to draw an arrow alongside the source, also shown in Figure 22-1. This arrow tells us that the angle given for **E** is measured from the head of the arrow *with respect to* the other terminal. We can also take the direction for tracing loops through the external network by projecting the arrow around the loop. To avoid possible confusion from using + and − signs in an ac circuit diagram, we shall use the arrow method in the following chapters.

There are several possible combinations of tracing loops which we can draw for the network of Figure 22-1. Using those shown in Figure 22-1, we can write the three loop equations from Kirchhoff's voltage law.

For Loop I_1,

$$Z_1 (I_1 + I_2) + Z_3 I_1 = E$$

For Loop I_2,

$$Z_1(I_1 + I_2) + Z_2 I_2 + Z_L (I_2 + I_3) = E$$

For Loop I_3,

$$Z_4 I_3 + Z_L (I_2 + I_3) = E$$

Collecting terms,

$$(\mathbf{Z}_1 + \mathbf{Z}_3)\mathbf{I}_1 + \qquad\qquad \mathbf{Z}_1\mathbf{I}_2 \qquad\qquad = \mathbf{E}$$

$$\mathbf{Z}_1\mathbf{I}_1 + (\mathbf{Z}_1 + \mathbf{Z}_2 + \mathbf{Z}_L)\mathbf{I}_2 + \qquad \mathbf{Z}_L\mathbf{I}_3 = \mathbf{E}$$

$$\mathbf{Z}_L\mathbf{I}_2 + (\mathbf{Z}_4 + \mathbf{Z}_L)\mathbf{I}_3 = \mathbf{E}$$

EXAMPLE 22-1

Given $\mathbf{E} = 20\text{ V}\underline{/0°}$, $\mathbf{Z}_1 = \mathbf{Z}_2 = -\text{j}5\text{ k}\Omega$, $\mathbf{Z}_3 = \mathbf{Z}_4 = (20 + \text{j}0)\text{ k}\Omega$, and $\mathbf{Z}_L = (5 + \text{j}5)\text{ k}\Omega$ for the bridged-T network of Figure 22-1, find the current through $\mathbf{Z}_L$.

SOLUTION

Although we could proceed to substitute in the equations we wrote for the tracing loops of Figure 22-1, we would have to solve for both $\mathbf{I}_2$ and $\mathbf{I}_3$ in order to determine $\mathbf{I}_L$. If we select a different set of tracing loops, as shown in Figure 22-2, we need solve only for $\mathbf{I}_3$.

FIGURE 22-2
Circuit Diagram for Example 22-1.

Step I. Writing the three loop equations,

$$\mathbf{Z}_1\mathbf{I}_1 + \mathbf{Z}_3(\mathbf{I}_1 + \mathbf{I}_2) = \mathbf{E}$$

$$\mathbf{Z}_4(\mathbf{I}_2 + \mathbf{I}_3) + \mathbf{Z}_2\mathbf{I}_2 + \mathbf{Z}_3(\mathbf{I}_2 + \mathbf{I}_1) = \mathbf{E}$$

$$\mathbf{Z}_4(\mathbf{I}_3 + \mathbf{I}_2) + \mathbf{Z}_L\mathbf{I}_3 = \mathbf{E}$$

Collecting terms,

$$(\mathbf{Z}_1 + \mathbf{Z}_3)\mathbf{I}_1 + \qquad\qquad \mathbf{Z}_3\mathbf{I}_2 \qquad\qquad = \mathbf{E}$$

$$\mathbf{Z}_3\mathbf{I}_1 + (\mathbf{Z}_2 + \mathbf{Z}_3 + \mathbf{Z}_4)\mathbf{I}_2 + \qquad \mathbf{Z}_4\mathbf{I}_3 = \mathbf{E}$$

$$\mathbf{Z}_4\mathbf{I}_2 + (\mathbf{Z}_4 + \mathbf{Z}_L)\mathbf{I}_3 = \mathbf{E}$$

Step II. If we substitute for $\mathbf{Z}$ in *kilohms* and for $\mathbf{E}$ in *volts*, the current will be in *milliamperes*. And since the loop equations show that we shall be adding impedances, we might as well make our substitutions in rectangular coordinates. Therefore,

$$(20 - j5)I_1 + (20 + j0)I_2 \qquad\qquad = 20 + j0$$

$$(20 + j0)I_1 + (40 - j5)I_2 + (20 + j0)I_3 = 20 + j0$$

$$(20 + j0)I_2 + (25 + j5)I_3 = 20 + j0$$

Step III. Since we now have to *multiply* the diagonals of the determinant matrices, we use the $\boxed{\rightarrow \text{P}}$ key of our calculator to write the impedances in the matrices in polar form.

$$I_3 = \dfrac{\begin{vmatrix} 20.61\underline{/-14.04°} & 20\underline{/0°} & 20\underline{/0°} \\ 20\underline{/0°} & 40.31\underline{/-7.13°} & 20\underline{/0°} \\ 0 & 20\underline{/0°} & 20\underline{/0°} \end{vmatrix}}{\begin{vmatrix} 20.61\underline{/-14.04°} & 20\underline{/0°} & 0 \\ 20\underline{/0°} & 40.31\underline{/-7.13°} & 20\underline{/0°} \\ 0 & 20\underline{/0°} & 25.5\underline{/+11.31°} \end{vmatrix}}$$

$$I_3 = \frac{16\,616\underline{/-21.17°} + 0 + 8000\underline{/0°} - 0 - 8244\underline{/-14.04°} - 8000\underline{/0°}}{21\,185\underline{/-9.86°} + 0 + 0 - 0 - 8244\underline{/-14.04°} - 10\,196\underline{/+11.31°}}$$

$$= \frac{(15\,495 - j6000) - (7998 - j2000)}{(20\,872 - j3628) - (7998 - j2000) - (9998 + j2000)}$$

$$= \frac{7497 - j4000}{2876 - j3628} = \frac{8497\underline{/-28.1°}}{4630\underline{/-51.6°}} = 1.835 \text{ mA}\underline{/+23.5°}$$

$$\therefore \mathbf{I}_L = \mathbf{1.84 \text{ mA}\underline{/+23.5°}}$$

In the single-source network of Example 22-1, we started all tracing loops from the "positive" terminal of the source. This keeps the number of negative terms in the loop equations to a minimum. In Example 22-1, we needed three loops to include all circuit elements in such a way that we could combine loop currents to determine six different currents in the six different branches of the network. If there is more than one voltage source in a network, at least one tracing loop starts and ends with each source. Again, the source direction determines the loop current direction and there must be sufficient loop equations to determine actual currents in all branches. In Example 22-2, there are two sources and only three different branches. Hence, we can determine all actual currents from two loop current equations.

EXAMPLE 22-2

Two 60-Hz alternators are connected in parallel to feed a load whose impedance is $20 \, \Omega \underline{/-15°}$. Alternator 1 develops an open-circuit voltage of 120 V and has an internal impedance of $10 \, \Omega \underline{/+30°}$. Alternator 2 has an open-circuit voltage of 117 V which lags the voltage of alternator 1 by a constant 15°. Alternator 2 has an internal impedance of $8 \, \Omega \underline{/+45°}$. Find the load current.

SOLUTION

Step I. Writing the two loop equations for the circuit of Figure 22-3,

FIGURE 22-3
Circuit Diagram for Example 22-2.

For the $\mathbf{I}_1$ loop,

$$\mathbf{Z}_1 \mathbf{I}_1 + \mathbf{Z}_L (\mathbf{I}_1 + \mathbf{I}_2) = \mathbf{E}_1$$

And for the $\mathbf{I}_2$ loop,

$$\mathbf{Z}_2 \mathbf{I}_2 + \mathbf{Z}_L (\mathbf{I}_2 + \mathbf{I}_1) = \mathbf{E}_2$$

Collecting terms,

$$(\mathbf{Z}_1 + \mathbf{Z}_L)\mathbf{I}_1 + \mathbf{Z}_L \mathbf{I}_2 = \mathbf{E}_1$$
$$\mathbf{Z}_L \mathbf{I}_1 + (\mathbf{Z}_2 + \mathbf{Z}_L)\mathbf{I}_2 = \mathbf{E}_2$$

$$\mathbf{I}_1 = \frac{\begin{vmatrix} \mathbf{E}_1 & \mathbf{Z}_L \\ \mathbf{E}_2 & (\mathbf{Z}_2 + \mathbf{Z}_L) \end{vmatrix}}{\begin{vmatrix} (\mathbf{Z}_1 + \mathbf{Z}_L) & \mathbf{Z}_L \\ \mathbf{Z}_L & (\mathbf{Z}_2 + \mathbf{Z}_L) \end{vmatrix}} \quad \text{and} \quad \mathbf{I}_2 = \frac{\begin{vmatrix} (\mathbf{Z}_1 + \mathbf{Z}_L) & \mathbf{E}_1 \\ \mathbf{Z}_L & \mathbf{E}_2 \end{vmatrix}}{\begin{vmatrix} (\mathbf{Z}_1 + \mathbf{Z}_L) & \mathbf{Z}_L \\ \mathbf{Z}_L & (\mathbf{Z}_2 + \mathbf{Z}_L) \end{vmatrix}}$$

Step II.

$$\mathbf{Z_1} + \mathbf{Z_L} = 10\underline{/+30°} + 20\underline{/+15°} = (8.66 + j5) + (19.32 + j5.176)$$

$$= 27.98 + j10.176 = 29.77\ \Omega\underline{/+19.99°}$$

$$\mathbf{Z_2} + \mathbf{Z_L} = 8\underline{/+45°} + 20\underline{/+15°} = (5.657 + j5.657) + (19.32 + j5.176)$$

$$= 24.98 + j10.833 = 27.23\ \Omega\underline{/+23.45°}$$

$$\mathbf{I_1} = \frac{\begin{vmatrix} 120\underline{/0°} & 20\underline{/+15°} \\ 117\underline{/-15°} & 27.23\underline{/+23.45°} \end{vmatrix}}{\begin{vmatrix} 29.77\underline{/+19.99°} & 20\underline{/+15°} \\ 20\underline{/+15°} & 27.23\underline{/+23.45°} \end{vmatrix}} = \frac{3268\underline{/+23.45°} - 2340\underline{/0°}}{810.6\underline{/+43.44°} - 400\underline{/+30°}}$$

$$= \frac{(2998 + j1301) - (2340 + j0)}{(588.6 + j557.4) - (346.4 + j200)} = \frac{658 + j1301}{242.2 + j357.4}$$

$$= \frac{1458\underline{/+63.17°}}{431.7\underline{/+55.88°}} = 3.377\ A\underline{/+7.29°}$$

$$\mathbf{I_2} = \frac{\begin{vmatrix} 29.77\underline{/+19.99°} & 120\underline{/0°} \\ 20\underline{/+15°} & 117\underline{/-15°} \end{vmatrix}}{D} = \frac{3483\underline{/+4.99°} - 2400\underline{/+15°}}{D}$$

$$= \frac{(3470 + j303) - (2318 + j621)}{D} = \frac{1152 - j318}{D}$$

$$= \frac{1195\underline{/-15.43°}}{431.7\underline{/+55.88°}} = 2.768\ A\underline{/-71.31°}$$

Step III. From Figure 22-3,

$$\mathbf{I_L} = \mathbf{I_1} + \mathbf{I_2} = 3.377\ A\underline{/+7.29°} + 2.768\ A\underline{/-71.31°}$$

$$= (3.35 + j0.429) + (0.887 - j2.622)$$

$$= 4.237 - j2.193 = \mathbf{4.77\ A\underline{/-27.4°}}$$

SELF-CHECK

22-1:1. For the circuit shown in Figure 22-4:
    (a) Write the loop equations traced by $\mathbf{I_1}$ and $\mathbf{I_2}$.
    (b) Solve the loop equations for $\mathbf{I_1}$ and $\mathbf{I_2}$.
    (c) Find the actual currents in the inductor and both voltage sources.

FIGURE 22-4

22-2 MESH EQUATIONS

The rules for writing the optional mesh equations are the same for both dc and ac circuits. However, in ac networks, both the dependent and independent variables are phasor quantities.

1. We must convert any current sources in the network to equivalent constant-voltage sources before laying out the circuit diagram.

2. We cannot use mesh equations unless we can lay out the circuit diagram so that there are no electrical components *within* any closed loop "mesh" (a *planar* circuit diagram).

3. We draw and identify a current loop for every mesh so that mesh currents are either all clockwise or all counterclockwise. (We shall choose clockwise.) As a result, when we write the mesh equations, the current for the identified mesh is a *positive* quantity and all currents in adjacent meshes with mutual components are *negative* quantities.

4. If the direction of the mesh current is the *same* as the direction indicated for a source in the mesh, the source voltage is a *positive* quantity. If the directions are opposite, the source voltage is a *negative* quantity. (There may be no source in some meshes.)

5. Since the object of these rules is to set up the simultaneous equations directly in matrix format (as in the collecting-of-terms step in the *loop* procedure), the coefficient for the identified positive mesh current is the *sum* of the impedances around the mesh. The coefficients for the negative adjacent mesh currents are the impedances of the mutual components. Thus the mesh equation for the $\mathbf{I}_A$ mesh in Figure 22-5 has the form

$$(\mathbf{Z}_1 + \mathbf{Z}_3)\mathbf{I}_A - \mathbf{Z}_3\mathbf{I}_B - \mathbf{Z}_1\mathbf{I}_C = +\mathbf{E} \qquad (22\text{-}2)$$

> *EXAMPLE 22-1A*
> Given $\mathbf{E} = 20 \text{ V}\underline{/0°}$, $\mathbf{Z}_1 = \mathbf{Z}_2 = -j5 \text{ k}\Omega$, $\mathbf{Z}_3 = \mathbf{Z}_4 = (20 + j0) \text{ k}\Omega$, and $\mathbf{Z}_L = (5 + j5) \text{ k}\Omega$ for the bridged-T network of Figures 22-1 and 22-5, find the current through $\mathbf{Z}_L$.

FIGURE 22-5
Circuit Diagram for Example 22-1A.

SOLUTION

Step I. Substituting for **Z** in kilohms and solving for **I** in milliamperes, and following the mesh format rules, the three mesh equations for $\mathbf{I_A}$, $\mathbf{I_B}$, and $\mathbf{I_C}$, respectively, are

$$(20 - j5)\mathbf{I_A} - (20 + j0)\mathbf{I_B} - (0 - j5)\mathbf{I_C} = +20\underline{/0°}$$

$$-(20 + j0)\mathbf{I_A} + (25 + j0)\mathbf{I_B} - (0 - j5)\mathbf{I_C} = 0$$

$$-(0 - j5)\mathbf{I_A} - (0 - j5)\mathbf{I_B} + (20 - j10)\mathbf{I_C} = 0$$

Step II. Having written the required mesh equations in a single step, solution of the three simultaneous equations now follows the same procedure as the loop method. To find $\mathbf{I_L}$, we need solve only for $\mathbf{I_B}$. Using the $\boxed{\rightarrow \text{P}}$ key of the calculator, the matrices become

$$\mathbf{I_B} = \cfrac{\begin{vmatrix} +20.62\underline{/-14.04°} & +20\underline{/0°} & -5\underline{/-90°} \\ -20\underline{/0°} & 0 & -5\underline{/-90°} \\ -5\underline{/-90°} & 0 & +22.36\underline{/-26.57°} \end{vmatrix}}{\begin{vmatrix} +20.62\underline{/-14.04°} & -20\underline{/0°} & -5\underline{/-90°} \\ -20\underline{/0°} & +25\underline{/0°} & -5\underline{/-90°} \\ -5\underline{/-90°} & -5\underline{/-90°} & 22.36\underline{/-26.57°} \end{vmatrix}}$$

$$= \cfrac{0 + 500\underline{/-180°} + 0 - 0 - 0 + 8944\underline{/-26.57°}}{11\,524\underline{/-40.61°} - 500\underline{/-180°} - 500\underline{/-180°} - 625\underline{/-180°} - 515.4\underline{/-194.04°} - 8944\underline{/-26.57°}}$$

$$= \cfrac{(-500 + j0) + (8000 - j4000)}{(8750 - j7500) - (-1625 + j0) - (-500 + j125) - (8000 - j4000)}$$

$$= \cfrac{7500 - j4000}{2875 - j3625} = \cfrac{8500\underline{/-28.1°}}{4627\underline{/-51.6°}}$$

$$\mathbf{I_L} = \mathbf{I_B} = \mathbf{1.84 \ mA}\underline{/+23.5°}$$

EXAMPLE 22-2A

Two 60-Hz alternators are connected in parallel to feed a load whose impedance is $20\ \Omega\underline{/-15°}$. Alternator 1 develops an open-circuit voltage of 120 V and has an internal impedance of $10\ \Omega\underline{/+30°}$. Alternator 2 has an open-circuit voltage of 117 V which lags the voltage of alternator 1 by a constant 15°. Alternator 2 has an internal impedance of $8\ \Omega\underline{/+45°}$. Find the load current.

SOLUTION

We can use the circuit diagram of Figure 22-3 for mesh equations simply by reversing the direction of the I_2 loop.

Step I. The mesh equations for I_1 and I_2, respectively, are

$$(\mathbf{Z}_1 + \mathbf{Z}_L)\mathbf{I}_1 - \mathbf{Z}_L\,\mathbf{I}_2 = +\mathbf{E}_1$$

$$-\mathbf{Z}_L\,\mathbf{I}_1 + (\mathbf{Z}_2 + \mathbf{Z}_L)\mathbf{I}_2 = -\mathbf{E}_2$$

$$\mathbf{I}_1 = \frac{\begin{vmatrix} \mathbf{E}_1 & -\mathbf{Z}_L \\ -\mathbf{E}_2 & \mathbf{Z}_2 + \mathbf{Z}_L \end{vmatrix}}{\begin{vmatrix} \mathbf{Z}_1 + \mathbf{Z}_L & -\mathbf{Z}_L \\ -\mathbf{Z}_L & \mathbf{Z}_2 + \mathbf{Z}_L \end{vmatrix}} \qquad \mathbf{I}_2 = \frac{\begin{vmatrix} \mathbf{Z}_1 + \mathbf{Z}_L & +\mathbf{E}_1 \\ -\mathbf{Z}_L & -\mathbf{E}_2 \end{vmatrix}}{\begin{vmatrix} \mathbf{Z}_1 + \mathbf{Z}_L & -\mathbf{Z}_L \\ -\mathbf{Z}_L & \mathbf{Z}_2 + \mathbf{Z}_L \end{vmatrix}}$$

Step II. Comparing these mesh equations with the loop equations of Example 22-2 after collecting terms, we find the same coefficients but with some − signs due to reversing the direction of I_2. Solving for I_1 and I_2,

$$\mathbf{I}_1 = 3.377\ A\underline{/+7.29°} \qquad \mathbf{I}_2 = 2.768\ A\underline{/+108.69°}$$

Step III. From Figure 22-3, with both mesh currents clockwise,

$$\mathbf{I}_L = \mathbf{I}_1 - \mathbf{I}_2 = 3.377\ A\underline{/+7.29°} - 2.768\ A\underline{/+108.69°}$$

$$= (3.35 + j0.429) - (-0.887 + j2.622)$$

$$= 4.237 - j2.193 = \mathbf{4.77\ A}\underline{\mathbf{/-27.4°}}$$

SELF-CHECK

22-2:1. For the circuit shown in Figure 22-6(a):
(a) Write the mesh equations traced by I_1 and I_2.
(b) Solve the mesh equations for I_1 and I_2.
(c) Find the actual currents in the inductor and both voltage sources.
(Your answers should agree with those obtained in Self-Check Problem 22-1:1.)

22-2:2. For the circuit shown in Figure 22-6(b):
(a) Write the mesh equations traced by I_1, I_2, and I_3.
(b) Solve the mesh equations for I_1, I_2, and I_3.
(c) Find the actual voltages at points A and B.

FIGURE 22-6

22-3 SUPERPOSITION THEOREM

In networks containing more than one source, we can avoid the simultaneous equations of the loop procedure by applying the **superposition theorem.** We can restate the theorem to embrace the phasor quantities encountered in ac circuits.

The current that flows in any branch of a network of impedances resulting from the simultaneous application of a number of voltage sources distributed in any manner throughout the network is the phasor sum of the component currents in that branch that would be caused by each source acting independently in turn while the others are replaced in the network by their respective internal impedances.

Since Example 22-1 has only one source, the superposition theorem cannot help us with the bridged-T network problem. However, we can use the superposition theorem to avoid the simultaneous equations of Example 22-2.

EXAMPLE 22-2B

Two 60-Hz alternators are connected in parallel to feed a load whose impedance is $20\ \Omega/{+}15°$. Alternator 1 develops an open-circuit voltage of 120 V and has an internal impedance of $10\ \Omega/{+}30°$. Alternator 2 has an open-circuit voltage of 117 V which lags the voltage of alternator 1 by a constant 15°. Alternator 2 has an internal impedance of $8\ \Omega/{+}45°$. Find the load current.

SOLUTION

Step I. We consider $\mathbf{E}_1$ to be acting alone in the network. All that remains of $\mathbf{E}_2$ in the circuit of Figure 22-7(a) is its internal impedance, $\mathbf{Z}_2$.

$$\mathbf{Z}_T = \mathbf{Z}_1 + \frac{\mathbf{Z}_2\mathbf{Z}_L}{\mathbf{Z}_2 + \mathbf{Z}_L}$$

$$= (8.66 + j5) + \frac{(5.657 + j5.657) \times (19.319 + j5.176)^\dagger}{(5.657 + j5.657) + (19.319 + j5.176)}$$

$$= (8.66 + j5) + \frac{(80.01 + j138.57)}{(24.98 + j10.83)}$$

$$= (8.66 + j5) + (4.72 + j3.5) = (13.38 + j8.5)\ \Omega$$

$$= 15.85\ \Omega\underline{/+32.43°}$$

$$\therefore \mathbf{I}_1 = \frac{\mathbf{E}_1}{\mathbf{Z}_T} = \frac{120\underline{/0°}}{15.85\underline{/+32.43°}} = 7.571\ \text{A}\underline{/-32.43°}$$

FIGURE 22-7
Circuit Diagram for Example 22-2B.

Using the current-divider principle to find the first component of the load current, from Figure 22-7(a),

$$\mathbf{I}_L = \mathbf{I}_1\frac{\mathbf{Z}_2}{\mathbf{Z}_2 + \mathbf{Z}_L} = 7.571\underline{/-32.43°} \times \frac{8\underline{/+45°}}{27.23\underline{/+23.44°}}$$

$$= 2.224\ \text{A}\underline{/-10.87°}$$

---

† As an alternative to repeated use of the $\boxed{\rightarrow P}$ and $\boxed{\rightarrow R}$ keys, we can, if we wish, carry out the phasor multiplication and division in rectangular coordinates, using the procedure shown in Example 18-11.

Step II. We now consider $\mathbf{E}_2$ to be acting alone in the network giving the circuit of Figure 22-7(b),

$$\mathbf{Z}_T = \mathbf{Z}_2 + \frac{\mathbf{Z}_1 \mathbf{Z}_L}{\mathbf{Z}_1 + \mathbf{Z}_L}$$

$$= (5.657 + j5.657) + \frac{(8.66 + j5) \times (19.319 + j5.176)}{(8.66 + j5) + (19.319 + j5.176)}$$

$$= (5.657 + j5.657) + \frac{(141.42 + j141.42)}{(27.98 + j10.18)}$$

$$= (5.657 + j5.657) + (6.088 + j2.839) = (11.745 + j8.496)\ \Omega$$

$$= 14.50\ \Omega\underline{/+35.88^\circ}$$

$$\therefore \mathbf{I}_2 = \frac{\mathbf{E}_2}{\mathbf{Z}_T} = \frac{117\underline{/-15^\circ}}{14.5\underline{/+35.88^\circ}} = 8.069\ \text{A}\underline{/-50.88^\circ}$$

The second component of the load current is

$$\mathbf{I}_L = \mathbf{I}_2\,\frac{\mathbf{Z}_1}{\mathbf{Z}_1 + \mathbf{Z}_L} = 8.069\underline{/-50.88^\circ} \times \frac{10\underline{/+30^\circ}}{29.77\underline{/+19.99^\circ}}$$

$$= 2.710\ \text{A}\underline{/-40.87^\circ}$$

Step III. The actual load current is the phasor sum of the two component currents. Hence,

$$\mathbf{I}_L = 2.224\underline{/-10.87^\circ} + 2.710\underline{/-40.87^\circ}$$

$$= (2.184 - j0.419) + (2.049 - j1.773)$$

$$= (4.233 - j2.192)\ \text{A} = \mathbf{4.77\ A}\underline{\mathbf{/-27.4^\circ}}$$

The superposition theorem was developed as a circuit-analysis tool long before the concept of equivalent constant-current sources became popular. Nevertheless, we can still use the superposition principle in networks containing current sources. To illustrate the procedure without a lot of complex algebra, we can consider the simple network of Figure 22-8(a).

EXAMPLE 22-3
Assuming that the ac milliammeter in the network of Figure 22-8(a) has zero impedance, calculate its reading.

SOLUTION
Step I. For $\mathbf{E}_A$ to act alone in the network, $\mathbf{I}_B$ becomes zero by leaving an *open* circuit in place of its symbol, as shown in Figure 22-8(b). The first component of the

FIGURE 22-8
Circuit Diagram for Example 22-3.

meter current is

$$\mathbf{I}_1 = \frac{\mathbf{E}_A}{\mathbf{Z}_1 + \mathbf{Z}_2} = \frac{20 \ V\underline{/0°}}{(2 + j4) \ k\Omega} = 4.472 \ mA\underline{/-63.43°}$$

Step II. For $\mathbf{I}_B$ to act alone, $\mathbf{E}_A$ becomes zero by placing a short circuit in place of its symbol, as shown in Figure 22-8(c). Using the current-divider principle, the second component of the meter current is

$$\mathbf{I}_2 = \frac{\mathbf{Z}_1}{\mathbf{Z}_1 + \mathbf{Z}_2} \times \mathbf{I}_B = \frac{2 \ k\Omega\underline{/0°}}{4.472 \ k\Omega\underline{/+63.43°}} \times 40 \ mA\underline{/+30°}$$

$$= 17.89 \ mA\underline{/-33.43°}$$

Step III. From the directions of the component currents in Figures 22-8(b) and (c),

$$\mathbf{I}_M = \mathbf{I}_1 + \mathbf{I}_2 = (2 - j4) + (14.93 - j9.86) = 16.93 - j13.86 \ mA$$

$$\therefore I_M = \mathbf{21.88 \ mA}$$

SELF-CHECK

22-3:1. For the circuit shown in Figure 22-6(a):
   (a) Draw the two single-source circuits that will be used to find the component currents.
   (b) Solve the single-source circuits for the component currents.

(c) Combine the component currents to determine the actual currents in the inductor and both voltage sources.

(Your answers should agree with those obtained in Self-Check Problem 22-2:1.)

22-4 THÉVENIN'S THEOREM

Because of the considerable amount of numerical computation involved when we solve networks in their original form, techniques which will permit us to replace the original circuit with a simplified equivalent circuit are particularly significant in the analysis of alternating-current circuit behavior. Since Thévenin's theorem allows us to replace a fairly complex impedance network containing one or more voltage sources with a single source and its internal impedance, this theorem has become one of the most useful of our circuit-analysis techniques. When we apply it to ac circuits, we may state Thévenin's theorem thus,

> *Any two-terminal network of fixed impedances and voltage sources may be replaced by a single voltage source having an equivalent voltage equal to the open-circuit voltage at the terminals of the original network and having an internal impedance equal to the imped-ance looking back into the network from the two terminals with all voltage sources replaced by their internal impedances.*

Thévenin's theorem is particularly useful in analyzing the performance of elec-tronic amplifiers. We can represent the *active* components of the circuit (transistors) by Thévenin-equivalent sources. Then, as shown in Figure 22-9, we can incorporate the coupling network into the signal source by a Thévenin transformation.

FIGURE 22-9
AC Circuit Simplification by Thévenin's Theorem.

EXAMPLE 22-4

Given the signal source and four-terminal network of Figure 22-9(a), determine the Thévenin-equivalent source.

SOLUTION

Step I. With no load connected to the output terminals, the impedance across the

original source is

$$\mathbf{Z} = 50 \text{ k}\Omega + 50 \text{ k}\Omega - j50 \text{ k}\Omega = 111.8 \text{ k}\Omega\underline{/-26.56°}$$

$$\mathbf{I} = \frac{\mathbf{E}}{\mathbf{Z}} = \frac{10 \text{ V}\underline{/0°}}{111.8 \text{ k}\Omega\underline{/-26.56°}} = 89.44 \text{ }\mu\text{A}\underline{/+26.56°}$$

Therefore, the open-circuit voltage is

$$\mathbf{E}_{th} = \mathbf{V}_C = \mathbf{I}\mathbf{X}_C = 89.44 \text{ }\mu\text{A}\underline{/+26.56°} \times 50 \text{ k}\Omega\underline{/-90°}$$

$$= \mathbf{4.47 \text{ V}\underline{/-63.44°}}$$

Step II. Shorting out the alternator symbol leaving only its 50-kΩ internal re-sistance in the circuit, we now look back into the open-circuit output terminals to find $\mathbf{Z}_{th}$. Since all impedances are in kilohms, we can simplify the following calculation by working in kilohms.

$$\mathbf{Z}_{th} = (50 + j0) + \frac{(100 + j0) \times (0 - j50)}{(100 + j0) + (0 - j50)}$$

$$= (50 + j0) + \frac{-j5000}{(100 - j50)} = (50 + j0) + (20 - j40)$$

$$= (70 - j40) \text{ k}\Omega = \mathbf{80.62 \text{ k}\Omega\underline{/-29.74°}}$$

By being able to remove the load from the source network, Thévenin's theorem provides an alternate solution for Example 22-1.

EXAMPLE 22-1B
Given the source and bridged-T network of Figure 22-10(a), what current will flow through a $(5 + j5)$-kΩ load connected to this source?

FIGURE 22-10
Circuit Diagram for Example 22-1B.

SOLUTION

Step I. With the load removed from the original network, we can solve for the total impedance now seen by the source using series-parallel impedance techniques. To assist us in writing the necessary equations, we can rearrange the circuit of Figure 22-10(a) into the form shown in Figure 22-11(a) and again work directly in kilohms.

$$Z_T = Z_3 + \frac{Z_1(Z_4 + Z_2)}{Z_1 + (Z_4 + Z_2)} = (20 + j0) + \frac{-j5(20 - j5)}{-j5 + (20 - j5)}$$

$$= (20 + j0) + \frac{-25 - j100}{20 - j10} = (20 + j0) + (1 - j4.5)$$

$$Z_T = (21 - j4.5) \text{ k}\Omega = 21.48 \text{ k}\Omega\underline{/-12.09°}$$

$$I_T = \frac{E}{Z_T} = \frac{20 \text{ V}\underline{/0°}}{21.48 \text{ k}\Omega\underline{/-12.09°}} = 931.2 \text{ } \mu A\underline{/+12.09°}$$

FIGURE 22-11
Determining Thévenin-Equivalent Voltage and Impedance.

From the current-divider principle, the current through Z_4 and Z_2 in Figure 22-11(a) is

$$I_2 = I_T \frac{Z_1}{Z_1 + (Z_4 + Z_2)} = I_T \frac{-j5 \text{ k}\Omega}{(20 - j10)\text{k}\Omega}$$

$$= 931.2 \text{ } \mu A\underline{/+12.09°} \times \frac{5 \text{ k}\Omega\underline{/-90°}}{22.36 \text{ k}\Omega\underline{/-26.57°}}$$

$$= 208.2 \text{ } \mu A\underline{/-51.34°}$$

Step II. Calculating the voltage drops across Z_2 and Z_3,

$$V_2 = I_2 Z_2 = 208.2 \text{ } \mu A\underline{/-51.34°} \times 5 \text{ k}\Omega\underline{/-90°}$$

$$= 1.041 \text{ V} \underline{/-141.34°}$$

$$\mathbf{V}_3 = \mathbf{I}_T \mathbf{Z}_3 = 931.2 \ \mu\text{A} \underline{/+12.09°} \times 20 \text{ k}\Omega \underline{/0°}$$

$$= 18.624 \text{ V} \underline{/+12.09°}$$

$$\therefore \mathbf{E}_{th} = \mathbf{V}_2 + \mathbf{V}_3 = 1.041 \text{ V} \underline{/-141.34°} + 18.624 \text{ V} \underline{/+12.09°}$$

$$= (-0.813 - j0.650) + (18.211 + j3.901)$$

$$= (17.398 + j3.251) \text{ V} = 17.7 \text{ V} \underline{/+10.58°}$$

Step III. From Figure 22-11(b), we can first find the equivalent of $\mathbf{Z}_3$ and $\mathbf{Z}_1$ in parallel. Working in kilohms,

$$\mathbf{Z}_{eq} = \frac{\mathbf{Z}_3 \times \mathbf{Z}_1}{\mathbf{Z}_3 + \mathbf{Z}_1} = \frac{20 \times (-j5)}{20 - j5} = \frac{-j100}{20 - j5}$$

$$= (1.176 - j4.706) \text{ k}\Omega$$

$$\mathbf{Z}_{th} = \frac{\mathbf{Z}_4 \times (\mathbf{Z}_2 + \mathbf{Z}_{eq})}{\mathbf{Z}_4 + (\mathbf{Z}_2 + \mathbf{Z}_{eq})} = \frac{20 \times (-j5 + 1.176 - j4.706)}{20 + (-j5 + 1.176 - j4.706)}$$

$$= \frac{23.53 - j194.12}{21.176 - j9.706} = (4.39 - j7.15) \text{ k}\Omega$$

Step IV. Returning to the Thévenin-equivalent circuit of Figure 22-10(b),

$$\mathbf{I} = \frac{\mathbf{E}_{th}}{\mathbf{Z}_{th} + \mathbf{Z}_L} = \frac{17.7 \text{ V} \underline{/+10.58°}}{(4.39 - j7.15) \text{ k}\Omega + (5 + j5) \text{ k}\Omega}$$

$$= \frac{17.7 \text{ V} \underline{/+10.58°}}{(9.39 - j2.15) \text{ k}\Omega} = \frac{17.7 \text{ V} \underline{/+10.58°}}{9.64 \text{ k}\Omega \underline{/-12.92°}}$$

$$= \mathbf{1.84 \text{ mA}} \underline{\mathbf{/+23.5°}}$$

Thévenin's theorem also allows us to escape the simultaneous equations in the multisource network of Example 22-2. When we remove $\mathbf{Z}_L$ in the circuit of Figures 22-3 and 22-12(a), we are left with a simple series circuit from which we can determine the Thévenin-equivalent source feeding $\mathbf{Z}_L$ in Figure 22-12(b).

FIGURE 22-12
Circuit Diagrams for Example 22-2C.

EXAMPLE 22-2C

Two 60-Hz alternators are connected in parallel to feed a load whose impedance is $20\ \Omega/\underline{-15°}$. Alternator 1 develops an open-circuit voltage of 120 V and has an internal impedance of $10\ \Omega/\underline{+30°}$. Alternator 2 has an open-circuit voltage of 117 V which lags the voltage of alternator 1 by a constant 15°. Alternator 2 has an internal impedance of $8\ \Omega/\underline{+45°}$. Find the load current.

SOLUTION

Step I. Writing a single Kirchhoff's voltage law equation for the remaining loop in Figure 22-12(a),

$$(\mathbf{Z}_1 + \mathbf{Z}_2)\mathbf{I} = \mathbf{E}_1 - \mathbf{E}_2$$

$$\mathbf{I} = \frac{120\ \text{V}/\underline{0°} - 117\ \text{V}/\underline{-15°}}{10\ \Omega/\underline{+30°} + 8\ \Omega/\underline{+45°}}$$

$$= \frac{(120 + j0) - (113 - j30.28)}{(8.66 + j5) + (5.657 + j5.657)}$$

$$= \frac{7 + j30.28}{14.317 + j10.657} = \frac{31.078/\underline{+76.98°}}{17.85/\underline{+36.66°}} = 1.741\ \text{A}/\underline{+40.32°}$$

The voltage drop across $\mathbf{Z}_1$ is

$$\mathbf{V}_1 = \mathbf{IZ}_1 = 1.741\ \text{A}/\underline{+40.32°} \times 10\ \Omega/\underline{+30°}$$

$$= 17.41\ \text{V}/\underline{+70.32°}$$

$$\mathbf{V}_{AB} = \mathbf{E}_1 - \mathbf{V}_1 = 120\ \text{V}/\underline{0°} - 17.41\ \text{V}/\underline{+70.32°}$$

$$= (120 + j0) - (5.863 + j16.393)$$

$$\therefore\ \mathbf{V}_{th} = \mathbf{V}_{AB} = 114.137 - j16.393 = 115.3\ \text{V}/\underline{-8.173°}$$

Step II. The Thévenin-equivalent resistance as "seen" looking back into terminals A and B in Figure 22-12(a) consists of $\mathbf{Z}_1$ and $\mathbf{Z}_2$ in parallel. Hence,

$$\mathbf{Z}_{th} = \frac{\mathbf{Z}_1 \times \mathbf{Z}_2}{\mathbf{Z}_1 + \mathbf{Z}_2} = \frac{80\ \Omega^2/\underline{+75°}}{17.85\ \Omega/\underline{+36.66°}}$$

$$= 4.482\ \Omega/\underline{+38.34°} = 3.515 + j2.780\ \Omega$$

Step III. In the Thévenin-equivalent circuit of Figure 22-12(b),

$$\mathbf{I}_L = \frac{\mathbf{E}_{th}}{\mathbf{Z}_{th} + \mathbf{Z}_L} = \frac{115.3\ \text{V}/\underline{-8.173°}}{(3.515 + j2.780) + (19.319 + j5.176)\ \Omega}$$

$$= \frac{115.3\ \text{V}/\underline{-8.173°}}{22.834 + j7.956\ \Omega} = \frac{115.3\ \text{V}/\underline{-8.173°}}{24.18\ \Omega/\underline{+19.21°}}$$

$$= \mathbf{4.77\ A}/\underline{\mathbf{-27.4°}}$$

FIGURE 22-13

SELF-CHECK

22-4:1. Referring to the circuit shown in Figure 22-13:
   (a) Find the Thévenin equivalent load "seen" by source E_1 by applying Thévenin's theorem across terminals a and b.
   (b) Use the Thévenin equivalent circuit to solve for the current in source E_1.
   (Your answer should agree with that obtained in Self-Check Problem 22-2:1.)

22-5 NORTON'S THEOREM

In Section 8-2, we used a "black-box" technique to find the open-circuit terminal voltage and internal resistance of a two-terminal source network. Starting with the same source network, Norton showed that the black box could contain an equivalent constant-current source. The same black-box "measurements" of E_{oc} and I_{sc} yield either a constant-voltage source of E_{oc} and $R_x = E_{oc}/I_{sc}$, or a constant-current source of I_{sc} and $R_x = E_{oc}/I_{sc}$.

After we find the Thévenin-equivalent constant-voltage source for a two-terminal network of voltage sources and fixed impedances, we can carry circuit simplification one more step by converting to the Norton-equivalent constant-current source. We combine these two steps into a single procedure called Norton's theorem. For alternating-current circuits, Norton's theorem becomes

*Any two-terminal network of fixed impedances and voltage sources may be replaced by a single **constant-current** source whose current is equal to the current drawn by a short circuit across the terminals of the original network and having in **parallel** with the constant-current source an impedance equal to the impedance looking back into the network from the two terminals.*

Converting a constant-voltage source to an equivalent constant-current source is an application of Norton's theorem. Norton's theorem is particularly useful when a selected branch in a network is fed from several voltage sources in parallel. Hence, Norton's theorem provides an even shorter solution for Example 22-2C than Thévenin's theorem.

EXAMPLE 22-2D
Two 60-Hz alternators are connected in parallel to feed a load whose impedance is $20\ \Omega/\underline{+15°}$. Alternator 1 develops an open-circuit voltage of 120 V and has an

internal impedance of 10 $\Omega/+30°$. Alternator 2 has an open-circuit voltage of 117 V which lags the voltage of alternator 1 by a constant 15°. Alternator 2 has an internal impedance of 8 $\Omega/+45°$. Find the load current.

SOLUTION
    Step I. Determine the short-circuit current of each alternator.

$$\mathbf{I_1} = \frac{\mathbf{E_1}}{\mathbf{Z_1}} = \frac{120 \text{ V}/0°}{10 \ \Omega/+30°} = 12 \text{ A}/-30°$$

$$\mathbf{I_2} = \frac{\mathbf{E_2}}{\mathbf{Z_2}} = \frac{117 \text{ V}/-15°}{8 \ \Omega/+45°} = 14.63 \text{ A}/-60°$$

In drawing Figure 22-14, we made sure that the tracing directions for the constant-current sources matched the tracing directions for the alternators in Figure 22-3. Hence, the *sum* of the two Norton currents divides among the three impedance branches in the equivalent circuit of Figure 22-14.

FIGURE 22-14
Circuit Diagram for Examples 22-2D and 22-2E.

$$\therefore \mathbf{I_T} = \mathbf{I_1} + \mathbf{I_2} = 12 \text{ A}/-30° + 14.6 \text{ A}/-60°$$
$$= (10.39 + j6) + (7.31 - j12.67) = (17.7 - j18.67) \text{ A}$$
$$= 25.73 \text{ A}/-46.53°$$

Step II.

$$\mathbf{Y_T} = \mathbf{Y_1} + \mathbf{Y_2} + \mathbf{Y_3} = 0.1 \text{ S}/-30° + 0.05 \text{ S}/-15° + 0.125 \text{ S}/-45°$$
$$= (0.0866 - j0.05) + (0.0483 - j0.0129) + (0.0884 - j0.0884)$$
$$= (0.2233 - j0.1513) \text{ S} = 0.2697 \text{ S}/-34.12°$$

From the current-divider principle [Equation (21-10)],

$$\mathbf{I_L} = \mathbf{I_T} \frac{\mathbf{Y_L}}{\mathbf{Y_T}} = 25.73 \text{ A}/-46.53° \times \frac{0.05 \text{ S}/-15°}{0.2697 \text{ S}/-34.12°}$$
$$= \mathbf{4.77 \text{ A}/-27.41°}$$

We used Example 22-3 to show that the superposition theorem applies to both voltage and current sources. However, when we encounter mixed sources in a network, as in Figures 22-8(a) and 22-15(a), it is usually simpler to convert to either one or the other. In the network of Figure 22-15(a), if we convert the constant-current source into a constant-voltage source, we have to include Z_2 and the milliammeter in the transformation. Hence, the meter current in the new circuit would not be the same as in the original. But we can replace the constant-voltage source on the left of Figure 22-15(a) with its Norton-equivalent constant-current source, as shown in Figure 22-15(b).

FIGURE 22-15
Circuit Diagrams for Example 22-3A.

EXAMPLE 22-3A
Assuming that the ac milliammeter in the network of Figure 22-15(a) has zero imped-ance, calculate its reading.

SOLUTION
   Step I.   The Norton-equivalent source on the left of Figure 22-15(b) has the *same* internal impedance Z_1 and

$$\mathbf{I_A} = \frac{\mathbf{E_A}}{\mathbf{Z}_1} = \frac{20 \text{ V}\underline{/0°}}{2 \text{ k}\Omega\underline{/0°}} = 10 \text{ mA}\underline{/0°}$$

   Step II.   Writing a single Kirchhoff's current-law equation for the network of Figure 22-15(b),

$$(\mathbf{Y}_1 + \mathbf{Y}_2)\mathbf{V} = \mathbf{I_A} + \mathbf{I_B}$$

$$\mathbf{V} = \frac{10 \text{ mA}\underline{/0°} + 40 \text{ mA}\underline{/+30°}}{500 \text{ } \mu\text{S}\underline{/0°} + 250 \text{ } \mu\text{S}\underline{/-90°}}$$

$$= \frac{(10 + j0) + (34.64 + j20)}{(500 + j0) + (0 - j250)} = \frac{44.64 + j20}{500 - j250}$$

$$\therefore V = \frac{48.92 \text{ mA}}{559 \text{ } \mu\text{S}} = 87.5 \text{ V}$$

Step III.

$$I_{\mathrm{M}} = \frac{V}{Z_2} = \frac{87.5 \text{ V}}{4 \text{ k}\Omega} = \mathbf{21.88 \text{ mA}}$$

SELF-CHECK

22-5:1. Referring to the circuit shown in Figure 22-16:
(a) Convert both voltage sources to current sources and draw the Norton-equivalent circuit.
(b) Use the Norton-equivalent circuit to solve for the current in the inductor.
(Your answer should agree with that obtained in Self-Check Problem 22-2:1.)

FIGURE 22-16

22-6 NODAL ANALYSIS

Adapting Kirchhoff's current law to alternating-current circuits, we can state that

The phasor sum of the currents **leaving** *any junction point or node in an ac circuit must equal the phasor sum of the currents* **entering** *that node.*

The nodal equivalent of the loop procedure requires us to mark the directions of all branch currents leaving and entering independent nodes, as shown in Figure 22-14. From inspection, we can then write a Kirchhoff's current-law equation.

$$\mathbf{I}_{Z_1} + \mathbf{I}_{Z_L} + \mathbf{I}_{Z_2} = \mathbf{I}_1 + \mathbf{I}_2$$

We then replace impedance currents with $\mathbf{I} = \mathbf{V} \times \mathbf{Y}$, the product of the voltage across each impedance and its admittance. Writing the Kirchhoff's current-law equation is quite straightforward in the circuit of Figure 22-14, since all currents leaving the upper node are impedance currents and all currents entering the node are source currents. This is not the case in the circuit of Figure 22-17, and it is a bit more difficult to write suitable nodal equations by inspection. Consequently, we usually use a nodal analysis *format* which is the *dual* of mesh equations.

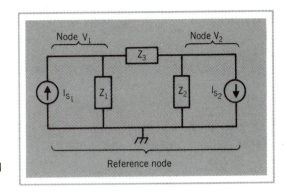

FIGURE 22-17
Circuit Diagram for Writing Nodal Equations.

For the nodal analysis format, we can restate Kirchhoff's current law in the form,

The phasor sum of the impedance *currents leaving any node in an ac circuit must equal the phasor sum of the* source *currents entering that node.*

Regardless of the actual directions of source currents, Kirchhoff's current-law nodal equations always have the format,

$$\mathbf{I}_{Z_1} + \mathbf{I}_{Z_2} + \mathbf{I}_{Z_3} + \cdots = \mathbf{I}_{S_1} + \mathbf{I}_{S_2} + \cdots \tag{22-3}$$

If an *actual* source current *leaves* the node, we enter it as a *negative* quantity in Equation (22-3). The signs for impedance currents are taken care of automatically by the rules for substituting voltages in the **VY** products which replace each impedance current.

1. For nodal equations to form the *dual* of mesh equations, we start by converting all voltage sources to equivalent constant-current sources to provide the source-current terms for the right-hand side of Equation (22-3).
2. We select a point in the network which is common to as many branches as possible and label it "reference node" in the circuit diagram. We label all other independent nodes with a **V** symbol and a suitable subscript. (The dependent variables in nodal analysis are the voltages between independent nodes and the reference node.) We can redraw our diagram to show junction point nodes, as in Figure 8-25, or mark the nodes with brackets, as in Figure 22-17.
3. In writing a Kirchhoff's current-law equation for node $\mathbf{V}_1$ in Figure 22-17, the voltage across an impedance connected between that node and the reference node ($\mathbf{Z}_1$ in Figure 22-17), is simply the node voltage $\mathbf{V}_1$ (with respect to the reference node). For an impedance connected between the node in question and an adjacent independent node ($\mathbf{Z}_3$ in Figure 22-17), the voltage across such an impedance always has the form $(\mathbf{V}_x - \mathbf{V}_A)$ where $\mathbf{V}_x$ is the node voltage in question and $\mathbf{V}_A$ is the adjacent node voltage (with respect to reference node). Following this format automatically takes care of the sign for the impedance current *leaving* the node in question. We can now write Kirchhoff's current-law equations for the two independent nodes in the circuit of Figure 22-17.

For node V_1,

$$V_1 Y_1 + (V_1 - V_2)Y_3 = I_{S_1} \qquad (1)$$

For node V_2,

$$V_2 Y_2 + (V_2 - V_1)Y_3 = -I_{S_2} \qquad (2)$$

Collecting the terms into matrix form,

$$(Y_1 + Y_3)V_1 - \qquad Y_3 V_2 = I_{S_1} \qquad (22\text{-}4)$$
$$-Y_3 V_1 + (Y_2 + Y_3)V_2 = -I_{S_2} \qquad (22\text{-}5)$$

4. By noting the structure of Equations (22-4) and (22-5), we can skip Step 3 and write nodal equations directly in matrix form. The left-hand side of nodal equations consists of one positive term and any number of negative terms. The positive term is the node voltage in question and its coefficient is the phasor sum of *all* admittances connected to that node. Each negative term is an adjacent node voltage with the admittance between that adjacent node and the node in question as its coefficient. The right-hand side of the nodal equation is the phasor sum of all source currents *entering* the node in question.

EXAMPLE 22-2E
Two 60-Hz alternators are connected in parallel to feed a load whose impedance is $20 \ \Omega/\underline{+15°}$. Alternator 1 develops an open-circuit voltage of 120 V and has an internal impedance of $10 \ \Omega/\underline{+30°}$. Alternator 2 has an open-circuit voltage of 117 V which lags the voltage of alternator 1 by a constant 15°. Alternator 2 has an internal impedance of $8 \ \Omega/\underline{+45°}$. Find the load current.

SOLUTION
Step I. Replacing each source by its equivalent constant-current source and maintaining the same direction for the currents produces the equivalent circuit of Figure 22-14 where

$$I_1 = \frac{E_1}{Z_1} = \frac{120 \ V/\underline{0°}}{10 \ \Omega/\underline{+30°}} = 12 \ A/\underline{-30°}$$

and

$$I_2 = \frac{E_2}{Z_2} = \frac{117 \ V/\underline{-15°}}{8 \ \Omega/\underline{+45°}} = 14.63 \ A/\underline{-60°}$$

Step II. If we take the lower conductor in the circuit diagram of Figure 22-14 as the reference node, the upper conductor represents a single independent node. Hence we need only one nodal equation.

$$(Y_1 + Y_L + Y_2)V = I_1 + I_2$$

from which

$$V = \frac{12\ A\underline{/-30°} + 14.63\ A\underline{/-60°}}{0.1\ S\underline{/-30°} + 0.05\ S\underline{/-15°} + 0.125\ S\underline{/-45°}}$$

$$= \frac{(10.39 - j6) + (7.31 - j12.67)}{(0.0866 - j0.05) + (0.0483 - j0.0129) + (0.0884 - j0.0884)}$$

$$= \frac{17.7 - j18.67}{0.2233 - j0.1513} = \frac{25.73\ A\underline{/-46.53°}}{0.2697\ S\underline{/-34.12°}}$$

$$= 95.4\ V\underline{/-12.41°}$$

Step III.

$$I_L = \frac{V}{Z_L} = \frac{95.4\ V\underline{/-12.41°}}{20\ \Omega\underline{/+15°}} = \mathbf{4.77\ A\underline{/-27.41°}}$$

There are much simpler methods than nodal analysis for solving the network of Figure 22-17. If we convert both current sources to equivalent constant-voltage sources, we have a simple single-loop circuit which we can easily solve by Ohm's law. However, as a further example, we shall complete the solution by nodal analysis.

EXAMPLE 22-5
Given the following parameters for the network of Figure 22-17, determine the magnitude of the current in Z_3 by nodal analysis.

$$I_{S_1} = 50\ mA\underline{/0°} \qquad I_{S_2} = 100\ mA\underline{/+90°}$$

$$Z_1 = +j50\ \Omega \qquad Z_2 = -j25\ \Omega \qquad Z_3 = 50\text{-}\Omega\ \text{resistor}$$

SOLUTION
Step I. Converting impedances to admittances,

$$Y_1 = -j20\ mS \qquad Y_2 = +j40\ mS \qquad Y_3 = 20\text{-mS conductance}$$

(These are the values we mark on the circuit diagram of Figure 22-17.)

$$Y_1 + Y_3 = (20 - j20)mS = 28.3\ mS\underline{/-45°}$$

$$Y_2 + Y_3 = (20 + j40)mS = 44.7\ mS\underline{/+63.4°}$$

Step II. The nodal equation for node V_1 becomes

$$28.3\underline{/-45°}\ V_1 - 20\underline{/0°}\ V_2 = 50\underline{/0°}†$$

The nodal equation for node V_2 becomes

$$44.7\underline{/+63.4°}\ V_2 - 20\underline{/0°}\ V_1 = -100\underline{/+90°}†$$

---

† Since milliamperes/millisiemens = volts, the nodal equations balance without repeated use of the prefix *milli* when entering admittance and current coefficients into the calculator.

Step III.

$$V_1 = \frac{\begin{vmatrix} 50\underline{/0°} & -20\underline{/0°} \\ -100\underline{/+90°} & 44.7\underline{/+63.4°} \end{vmatrix}}{\begin{vmatrix} 28.3\underline{/-45°} & -20\underline{/0°} \\ -20\underline{/0°} & 44.7\underline{/+63.4°} \end{vmatrix}} = \frac{2236\underline{/+63.4°} - 2000\underline{/+90°}}{1265\underline{/+18.4°} - 400\underline{/0°}}$$

$$= \frac{1000\underline{/0°}}{894\underline{/+26.6°}} = 1.12 \text{ V}\underline{/-26.6°}$$

$$V_2 = \frac{\begin{vmatrix} 28.3\underline{/-45°} & 50\underline{/0°} \\ -20\underline{/0°} & -100\underline{/+90°} \end{vmatrix}}{D} = \frac{-2828\underline{/+45°} + 1000\underline{/0°}}{D}$$

$$= \frac{2236\underline{/-116.6°}}{894\underline{/+26.6°}} = 2.5 \text{ V}\underline{/-143.1°}$$

Step IV.

$$\mathbf{V}_1 - \mathbf{V}_2 = (-2 - j1.5) - (1 - j0.5) = (-3 - j1) \text{ V}$$
$$= 3.16 \text{ V } \underline{/-161.6°}$$

Step V.

$$I_{Z_3} = \frac{\mathbf{V}_1 - \mathbf{V}_2}{Z_3} = \frac{3.16 \text{ V}}{50 \text{ } \Omega} = \mathbf{63 \text{ mA}}$$

If we attempt to solve the bridged-T network of Example 22-1 by nodal analysis, we would discover that there are *three* independent nodes. Therefore, we would require three nodal equations requiring third-order determinants. Second, we would find that we cannot convert the voltage source to an equivalent constant-current source, since the internal impedance of the given source is zero.

SELF-CHECK

22-6:1. For the circuit shown in Figure 22-16 and using the source conversions from Self-Check Problem 22-5:1:
  (a) Write the nodal equation that will give the voltage across the inductor.
  (b) Solve the nodal equation for the inductor voltage and current.
  (Your answers should agree with those obtained in Self-Check Problem 22-2:1.)

22-6:2. For the circuit shown in Figure 22-6(b):
  (a) Convert the voltage sources to current sources.
  (b) Write the nodal equations for V_A and V_B.
  (c) Solve for V_A, V_B, and the capacitor current.
  (Your answers should agree with those obtained in Self-Check Problem 22-2:2.)

22-7 DELTA-WYE TRANSFORMATION

By substituting the impedance letter symbol for the resistance symbol, we can apply to ac circuits the equations we developed in Section 9-6 for converting a given delta network into an equivalent wye network, and vice versa. Since the derivation of these equations is the same as for resistance networks, we shall simply state the equations at this point. In using these transformation equations, we must remember that we must treat each $\mathbf{Z}$ symbol as a *phasor* quantity.

$$\mathbf{Z}_A = \frac{\mathbf{Z}_Y \mathbf{Z}_Z}{\mathbf{Z}_X + \mathbf{Z}_Y + \mathbf{Z}_Z} \tag{22-6}$$

$$\mathbf{Z}_B = \frac{\mathbf{Z}_X \mathbf{Z}_Z}{\mathbf{Z}_X + \mathbf{Z}_Y + \mathbf{Z}_Z} \tag{22-7}$$

$$\mathbf{Z}_C = \frac{\mathbf{Z}_X \mathbf{Z}_Y}{\mathbf{Z}_X + \mathbf{Z}_Y + \mathbf{Z}_Z} \tag{22-8}$$

$$\mathbf{Z}_X = \frac{\mathbf{Z}_A \mathbf{Z}_B + \mathbf{Z}_B \mathbf{Z}_C + \mathbf{Z}_C \mathbf{Z}_A}{\mathbf{Z}_A} \tag{22-9}$$

$$\mathbf{Z}_Y = \frac{\mathbf{Z}_A \mathbf{Z}_B + \mathbf{Z}_B \mathbf{Z}_C + \mathbf{Z}_C \mathbf{Z}_A}{\mathbf{Z}_B} \tag{22-10}$$

$$\mathbf{Z}_Z = \frac{\mathbf{Z}_A \mathbf{Z}_B + \mathbf{Z}_B \mathbf{Z}_C + \mathbf{Z}_C \mathbf{Z}_A}{\mathbf{Z}_C} \tag{22-11}$$

In Section 9-6, we developed an alternate form for the wye to delta transformation equations based on the *dual* format of the delta to wye equations. This alternate form is particularly useful in ac circuits, where we must use *phasor* algebra, since they involve fewer computations.

$$\mathbf{Z}_X = \mathbf{Z}_B \mathbf{Z}_C (\mathbf{Y}_A + \mathbf{Y}_B + \mathbf{Y}_C) \tag{22-12}$$

$$\mathbf{Z}_Y = \mathbf{Z}_A \mathbf{Z}_C (\mathbf{Y}_A + \mathbf{Y}_B + \mathbf{Y}_C) \tag{22-13}$$

$$\mathbf{Z}_Z = \mathbf{Z}_A \mathbf{Z}_B (\mathbf{Y}_A + \mathbf{Y}_B + \mathbf{Y}_C) \tag{22-14}$$

These transformation equations were developed originally for use in three-phase ac power-distribution networks having the wye and delta forms of Figure 22-18(a) and (b). However, many of the four-terminal coupling networks encountered in electronic circuitry have a *common* input and output terminal (which is often grounded). In such coupling circuits, we can often identify the wye network in the form of the T-network of Figure 22-18(c). Similarly, the π-network of Figure 22-18(d) is essentially the same as the delta network of Figure 22-18(b). Therefore, we may also use the Equations (22-6) to (22-14) for T to π transformations, and vice versa. This transformation provides an additional means of solving the bridged-T network of Example 22-1. If we replace the T-network consisting of $\mathbf{Z}_1$, $\mathbf{Z}_2$, and $\mathbf{Z}_3$ in Figure 22-19(a) with the equivalent π-network, we can solve the resulting circuit of Figure 22-19(b) as a series-parallel impedance network.

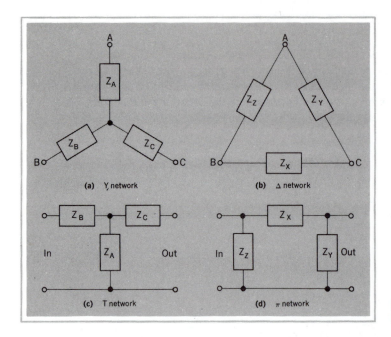

(a) Y network   **(b)** Δ network

(c) T network   **(d)** π network

FIGURE 22-18
Delta-Wye Transformation.

(a)   **(b)**

FIGURE 22-19
Circuit Diagram for Example 22-1C.

EXAMPLE 22-1C
Given $\mathbf{E} = 20$ V $\underline{/0°}$, $\mathbf{Z}_1 = \mathbf{Z}_2 = -j5$ kΩ, $\mathbf{Z}_3 = \mathbf{Z}_4 = (20 + j0)$ kΩ, and $\mathbf{Z}_L = (5 + j5)$ kΩ for the bridged-T network of Figure 22-19(a), find the current through $\mathbf{Z}_L$.

SOLUTION
    Step l. Labeling $\mathbf{Z}_1$, $\mathbf{Z}_2$, and $\mathbf{Z}_3$ as $\mathbf{Z}_A$, $\mathbf{Z}_B$, and $\mathbf{Z}_C$, respectively, in Figure 22-19(a), we can then label the π-network of Figure 22-19(b) with the appropriate

subscripts. These must correspond according to the exact relationships shown in Figure 22-18; that is, Z_X does *not* connect to the same terminal as Z_A, and so forth around the network.

Note that we need not solve for Z_Y, since it is directly across the source and does not affect the current through the load.

Step II. Working out the common numerator for Equations (22-9) and (22-11) and again working directly in kilohms,

$$Z_A Z_B + Z_B Z_C + Z_C Z_A = -j5 \times (-j5) + (-j5) \times 20 + 20 \times (-j5)$$

$$= -25 + (-j100) + (-j100) = -25 - j200$$

$$= 201.56 \underline{/-97.12°}$$

$$\therefore Z_X = \frac{Z_A Z_B + Z_B Z_C + Z_C Z_A}{Z_A} = \frac{201.56 \underline{/-97.12°}}{5 \underline{/-90°}} = 40.31 \text{ k}\Omega \underline{/-7.12°}$$

and $$Z_Z = \frac{Z_A Z_B + Z_B Z_C + Z_C Z_A}{Z_C} = \frac{201.56 \underline{/-97.12°}}{20 \underline{/0°}} = 10.08 \text{ k}\Omega \underline{/-97.12°}$$

Note that Z_Z appears to have a *negative* resistance component. Therefore, in this case, we cannot construct an *actual* equivalent π-network, using only passive circuit elements. Nevertheless, we can still use the pencil-and-paper transformation as a means of solving a network problem.

Step III. Excluding Z_Y, the load on the source in Figure 22-19(b) is

$$Z_T = \frac{Z_4 \times Z_Z}{Z_4 + Z_Z} + \frac{Z_X \times Z_L}{Z_X + Z_L}$$

$$= \frac{20 \underline{/0°} \times 10.08 \underline{/-97.12°}}{(20 + j0) + (-1.25 - j10.0)} + \frac{40.31 \underline{/-7.12°} \times 7.07 \underline{/+45°}}{(40 - j5) + (5 + j5)}$$

$$= \frac{201.56 \underline{/-97.12°}}{18.75 - j10.0} + \frac{284.99 \underline{/+37.88°}}{45 + j0}$$

$$= \frac{201.56 \underline{/-97.12°}}{21.25 \underline{/-28.07°}} + \frac{284.99 \underline{/+37.88°}}{45 \underline{/0°}}$$

$$= 9.49 \underline{/-69.05°} + 6.33 \underline{/+37.88°}$$

$$= (3.39 - j8.86) + (5.0 + j3.89) = 8.39 - j4.97$$

$$= 9.75 \text{ k}\Omega \underline{/-30.64°}$$

Step IV. The current drawn from the source by these four impedances is

$$\mathbf{I_T} = \frac{\mathbf{E}}{\mathbf{Z_T}} = \frac{20 \text{ V } \underline{/0^\circ}}{9.75 \text{ k}\Omega \ \underline{/-30.64^\circ}} = 2.05 \text{ mA } \underline{/+30.64^\circ}$$

Step V. Using the current-divider principle,

$$\mathbf{I_L} = \mathbf{I_T} \frac{\mathbf{Z}_X}{\mathbf{Z}_X + \mathbf{Z}_L} = 2.05 \text{ mA } \underline{/+30.64^\circ} \times \frac{40.31 \text{ k}\Omega \ \underline{/-7.12^\circ}}{45 \text{ k}\Omega \ \underline{/0^\circ}}$$

$$= \mathbf{1.84 \text{ mA } \underline{/+23.5^\circ}}$$

SELF-CHECK

22-7:1. The following impedances are Y-connected:

$$\mathbf{Z_A} = 30 + j40 \ \Omega$$
$$\mathbf{Z_B} = 35 - j35 \ \Omega$$
$$\mathbf{Z_C} = 50 \ \Omega$$

Solve for the Δ-equivalents.

22-8 ALTERNATING-CURRENT BRIDGES

If we replace the resistance symbols of the basic Wheatstone bridge with impedance symbols, we obtain the general ac bridge circuit of Figure 22-20. A brief consideration of ac bridges at this point will serve the dual purpose of acquainting us with the method by which we can obtain precision measurement of ac circuit parameters, and of providing us with a source of practical ac network problems.

FIGURE 22-20
General Form of an AC Bridge.

For the null indicator in Figure 22-20 to read zero, the voltage across its terminals must be zero. For this to be so, the voltage across $\mathbf{Z}_1$ and $\mathbf{Z}_3$ must be exactly the same both in magnitude and phase. Therefore,

$$\mathbf{V_1} = \mathbf{V_3} \quad \text{and} \quad \mathbf{I_1Z_1} = \mathbf{I_3Z_3}$$

Because no current flows through the null indicator when the bridge is balanced,

$$\mathbf{I}_1 = \frac{\mathbf{E}}{\mathbf{Z}_1 + \mathbf{Z}_2} \quad \text{and} \quad \mathbf{I}_3 = \frac{\mathbf{E}}{\mathbf{Z}_3 + \mathbf{Z}_4}$$

Hence,

$$\frac{\mathbf{E}\mathbf{Z}_1}{\mathbf{Z}_1 + \mathbf{Z}_2} = \frac{\mathbf{E}\mathbf{Z}_3}{\mathbf{Z}_3 + \mathbf{Z}_4}$$

from which

$$\mathbf{Z}_1\mathbf{Z}_4 = \mathbf{Z}_2\mathbf{Z}_3 \qquad (22\text{-}15)$$

This is known as the **general bridge equation** and must hold true for any special form of an ac bridge when it is properly balanced to give zero voltage across the null indicator.

The bridge arrangement of Figure 22-21 is called a **capacitance comparison bridge** since the unknown capacitance C_x and its loss resistance R_x are to be balanced against the standard capacitor C_s. In this particular case, the terms of the general bridge equation become

$$\mathbf{Z}_1 = R_1 \qquad \mathbf{Z}_2 = R_2 - j\frac{1}{\omega C_s}$$

$$\mathbf{Z}_3 = R_3 \qquad \mathbf{Z}_4 = R_x - j\frac{1}{\omega C_x}$$

FIGURE 22-21
Capacitance-Comparison Bridge.

Substituting in Equation (22-15) gives

$$R_1 R_x - j\frac{R_1}{\omega C_x} = R_2 R_3 - j\frac{R_3}{\omega C_s}$$

As an impedance diagram will show, altering the magnitude of a reactance can have no effect on the magnitude of the total resistance of the circuit because these two quantities are at right angles. Therefore, the j term on the left of the above equation must equal the j term on the right of the equation. Similarly, the two horizontal or *real* components must be equal. Hence, we can separate equations expressed in rectangular coordinates into two equations.

$$R_1 R_x = R_2 R_3$$

from which
$$R_x = \frac{R_2 R_3}{R_1} \qquad \text{(22-16)}$$

and
$$\frac{R_1}{\omega C_x} = \frac{R_3}{\omega C_s}$$

from which
$$C_x = C_s \frac{R_1}{R_3} \qquad \text{(22-17)}$$

Note that in deriving Equations (22-16) and (22-17), the angular velocity term disappeared. Therefore, the accuracy of the capacitance comparison bridge is not dependent on the accuracy of the frequency of the source. None of the ac bridges is dependent on the magnitude of the applied voltage for an accurate reading when balanced, since this term disappeared in deriving Equation (22-15).

Although we can construct a similar inductance comparison bridge to determine accurately an unknown inductance, standard inductors are expensive and bulky. The **Maxwell bridge** circuit of Figure 22-22 allows us to balance the bridge by using a standard capacitance for comparison purposes. In this case, Branch 1 contains parallel components. Thus, it is more convenient to identify the four arms as

$$\mathbf{Y}_1 = \frac{1}{R_1} + j\omega C \qquad \mathbf{Z}_2 = R_2$$

$$\mathbf{Z}_3 = R_3 \qquad \mathbf{Z}_4 = R_x + j\omega L_x$$

Substituting $1/\mathbf{Y}_1$ for $\mathbf{Z}_1$ in Equation (22-15) gives

$$\mathbf{Z}_4 = \mathbf{Y}_1 \mathbf{Z}_2 \mathbf{Z}_3$$

Substituting the given values for the Maxwell bridge gives

$$R_x + j\omega L_x = \frac{R_2 R_3}{R_1} + j\omega C_s R_2 R_3$$

FIGURE 22-22
Maxwell Bridge.

Again separating the real and j terms into separate equations,

$$R_x = \frac{R_2 R_3}{R_1}$$ (22-18)

$$L_x = C_s R_2 R_3$$ (22-19)

The Maxwell bridge is best suited to coils in which R_x is an appreciable fraction of ωL_x. As Equation (22-18) indicates, if R_x is very small, R_1 must be very large in order to obtain balance. For such coils, the Hay bridge of Figure 22-23 is preferred, even though the equations turn out to be frequency dependent.

In this case,

$$\mathbf{Z}_1 = R_1 - j\frac{1}{\omega C_s} \qquad \mathbf{Z}_2 = R_2$$

$$\mathbf{Z}_3 = R_3 \qquad\qquad \mathbf{Z}_4 = R_x + j\omega L_x$$

FIGURE 22-23
Hay Bridge.

Substituting in Equation (22-15),

$$\left(R_1 - j\frac{1}{\omega C_s}\right)(R_x + j\omega L_x) = R_2 R_3$$

$$R_1 R_x + j\omega L_x R_1 - \frac{jR_x}{\omega C_s} + \frac{L_x}{C_s} = R_2 R_3$$

Separating the real and j terms,

$$R_1 R_x + \frac{L_x}{C_s} = R_2 R_3$$ (1)

$$\omega L_x R_1 = \frac{R_x}{\omega C_s}$$ (2)

From Equation (2), $$R_x = \omega^2 L_x C_s R_1 \qquad (3)$$

Substituting in Equation (1) gives

$$\omega^2 L_x C_s R_1^2 + \frac{L_x}{C_s} = R_2 R_3$$

from which

$$L_x = \frac{C_s R_2 R_3}{1 + \omega^2 C_s^2 R_1^2} \qquad (22\text{-}20)$$

Substituting Equation (22-20) in Equation (3) gives

$$R_x = \frac{\omega^2 C_s^2 R_1 R_2 R_3}{1 + \omega^2 C_s^2 R_1^2} \qquad (22\text{-}21)$$

EXAMPLE 22-6

The Maxwell bridge of Figure 22-22 balances when $C_s = 0.01 \ \mu F$, $R_1 = 470 \ k\Omega$, $R_2 = 5.1 \ k\Omega$, and $R_3 = 100 \ k\Omega$. Determine the inductance and resistance of the inductor being measured.

SOLUTION

$$L_x = C_s R_2 R_3 = 0.01 \ \mu F \times 5.1 \ k\Omega \times 100 \ k\Omega = \mathbf{5.1 \ H}$$

$$R_x = \frac{R_2 R_3}{R_1} = \frac{5.1 \ k\Omega \times 100 \ k\Omega}{470 \ k\Omega} = \mathbf{1085 \ \Omega}$$

SELF-CHECK

22-8:1. The capacitance-comparison bridge shown in Figure 22-21 balances when:

$$R_1 = 33 \ k\Omega \qquad R_2 = 6.6 \ \Omega$$
$$R_3 = 1.0 \ k\Omega \qquad C_s = 0.10 \ \mu F.$$

Determine series capacitance and resistance for the unknown capacitor.

PROBLEMS

Problems 22-1 to 22-20 are drill problems in network solution. Solve odd-numbered problems by one method and the even-numbered problem for the same network by a different method. Accordingly, most problems in this assignment do not include section numbers.

22-1. Determine the source current drawn from each source in Figure 22-24(a).

22-2. Determine the voltage across the 50-Ω resistor in Figure 22-24(a).

22-3. Determine the capacitor current in Figure 22-24(b).

FIGURE 22-24

22-4. Determine the voltage drop across each 100-Ω resistor in Figure 22-24(b).

22-5. Determine the voltage drop across the $(10 - j10)$-Ω branch in Figure 22-25(a).

22-6. Determine both source currents in Figure 22-25(a).

22-7. Determine all three branch currents in Figure 22-25(b).

22-8. Reverse the polarity of the 20 V$\underline{/+90°}$ source and determine all three branch currents in Figure 22-25(b).

FIGURE 22-25

22-9. Determine the currents in $\mathbf{Z}_1$ and $\mathbf{Z}_4$ in Figure 22-26(a).

22-10. Determine the currents in $\mathbf{Z}_2$ and $\mathbf{Z}_3$ in Figure 22-26(a).

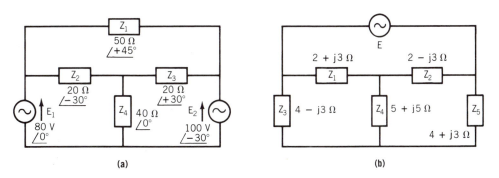

FIGURE 22-26

22-11. If the source voltage in Figure 22-26(b) is 10 V $\underline{/0°}$, determine the magnitude of the source current.

22-12. If the source voltage in Figure 22-26(b) is 25 V $\underline{/-45°}$, determine the magnitude of the voltage across $\mathbf{Z}_4$.

22-13. Determine the magnitude of the current through the inductor in Figure 22-27(a).

22-14. Determine the magnitude of the voltage across the capacitor in Figure 22-27(a).

22-15. Determine the magnitude of the voltage across each constant-current source in Figure 22-27(b).

22-16. Determine the magnitude of the current through the capacitor in Figure 22-27(b).

22-17. Determine the load current in Figure 22-28(a).

(a)　　　　　　　　　　(b)

FIGURE 22-27

(a)　　　　　　　　　　(b)

FIGURE 22-28

22-18. Determine the load voltage in Figure 22-28(a) if the direction of the 100 V $\underline{/+30°}$ source is reversed.

22-19. Determine the voltage drop across the 5-kΩ resistor in Figure 22-28(b).

22-20. Determine the magnitude of the two capacitor currents in Figure 22-28(b) if the direction of the 200-V source is reversed.

Use the method you think is most appropriate in solving the remaining applications networks.

22-21. Find the current in the common neutral lead of the three-wire, single-phase system shown in Figure 22-29.

22-22. The three-wire, single-phase system of Problem 22-21 can be converted into a two-phase system if we make $\mathbf{E}_2 = 110$ V $\underline{/90°}$. Find the current in $\mathbf{Z}_1$ under these circumstances.

FIGURE 22-29

22-23. Figure 22-30 shows one of the equivalent circuits for a transistor. It includes a current-controlled voltage source whose magnitude equals the mutual resistance r_m of the transistor × the current through r_e. If the signal source has an open-circuit voltage $E_g = 2$ V rms, and an internal resistance of $R_g = 200$ Ω, find the rms output voltage across a load resistor $R_L = 2$ kΩ.

22-24. Figure 22-31 shows an equivalent output circuit of a transistor amplifier. To determine its output impedance, we short circuit the input terminals and feed an ac signal E_o back into the output terminals and calculate the current I_o drawn from this source. From Ohm's law, the output impedance is simply $Z_o = E_o/I_o$. Set up an equation for the output impedance in terms of the circuit parameters given in Figure 22-31.

22-4-25. Find the Thévenin-equivalent source for the voltage source shown in Figure 22-32 when **E** has a frequency of 400 Hz.

22-5-26. Find the Norton-equivalent source at 400 Hz for the source in Figure 22-32.

22-4-27. Using Thévenin's theorem, determine an equivalent source for the network of Figure 22-32 that

$r_e = 20$ ohms $r_c = 2$ megohms
$r_b = 200$ ohms $r_m = 2$ megohms

FIGURE 22-30
Transistor Equivalent Circuit.

FIGURE 22-31
Output Impedance of a Common-
Collector Transistor Amplifier.

FIGURE 22-32

will be applicable at *any* frequency. (*Hint:* Do not include the 33-kΩ resistor in the Thévenin transformation.)

22-5-28. Determine the parameters for a Norton-equivalent source that will be applicable for any frequency for the network of Figure 22-32. Compare with the solution for Problem 22-5-26.

22-4-29. A 26.5-μF capacitor is connected in series with a 100-Ω resistor to a 120-V 60-Hz source. What value of resistance must be connected in parallel with the capacitor to make the current in the capacitor have a magnitude of 0.5 A?

22-30. At what frequency will the output voltage of the network of Figure 22-33 be in phase with the input voltage?

22-4-31. A voltage amplifier and its *CR* output coupling network can be represented by the source shown in Figure 22-34. $E = 10$ V $\underline{/180°}$; $r_a = 20$ kΩ; $R_b = 80$ kΩ; $C_c = 0.02$ μF; $R_g = 470$ kΩ. Using 1 kHz as a reference frequency, determine by how many degrees the phase of the output voltage will shift as the frequency decreases to 100 Hz.

22-4-32. An amplifier and its coupling network can be represented by the signal source shown in Figure 22-35. At what frequency will the output voltage drop to 70.7% of its magnitude at 1 kHz as a result of the increase in the reactance of the coupling capacitor as the frequency decreases?

FIGURE 22-33

FIGURE 22-34
Amplifier Coupling Circuit.

FIGURE 22-35
Equivalent Circuit of a Transistor Amplifier Coupling Circuit.

22-33. If the two capacitors in the circuit of Figure 22-36 are equal, what must their capacitance be for the open-circuit output voltage of this network to be 90° out of phase with the input voltage at 60 Hz?

22-4-34. Solve Example 22-2 by combining Thévenin's theorem with the superposition theorem—that is, for each of the circuit diagrams in Figure 22-7, find the *component* current through $\mathbf{Z}_L$ by Thévenin's theorem rather than by series-parallel impedance network techniques.

22-7-35. Find the equivalent T-network for the circuit of Figure 22-37 when the input frequency is 400 Hz.

22-7-36. Determine the equivalent π-network for the transistor equivalent circuit of Figure 22-30.

22-7-37. If the ammeter in Figure 22-38 has negligible internal resistance, solve for the ammeter reading, using the wye-delta transformation equations.

FIGURE 22-36

FIGURE 22-37

FIGURE 22-38

22-7-38. Find the source current in Problem 22-11 using wye-delta transformation.

22-8-39. The bridge circuit of Figure 22-20 balances when $\mathbf{Z}_1 = 470 \ \Omega \underline{/0°}$, $\mathbf{Z}_2$ is a $0.01 \mu F$ capacitor in series with an 860-kΩ resistor, $\mathbf{Z}_3 = 5 \ k\Omega \underline{/0°}$, and $f = 1$ kHz. Determine the components that constitute $\mathbf{Z}_4$.

22-8-40. The bridge circuit of Figure 22-20 balances when $\mathbf{Z}_2 = 10 \ k\Omega \underline{/0°}$, $\mathbf{Z}_3 = 5.6 \ k\Omega \underline{/0°}$, $\mathbf{Z}_4$ consists of a 0.01-μF capacitor in parallel with a 1.23-MΩ resistor, and $f = 1$ kHz. Determine the components that constitute $\mathbf{Z}_1$.

22-8-41. The bridge circuit of Figure 22-20 balances when $\mathbf{Z}_1 = 10 \ k\Omega \underline{/0°}$, $\mathbf{Z}_3$ is a 0.01-μF capacitor in series with a 27-Ω resistor, $\mathbf{Z}_4 = 2700 \ \Omega \underline{/0°}$, and $f = 1$ kHz. Determine the value of $\mathbf{Z}_2$ in rectangular coordinates.

22-8-42. The bridge circuit of Figure 22-20 balances when $\mathbf{Z}_1 = 475 \ \Omega \underline{/0°}$, $\mathbf{Z}_2 = 1780 \ \Omega \underline{/+75°}$, $\mathbf{Z}_3 = 1000 \ \Omega \underline{/0°}$, and $f = 1$ kHz. Determine the components that constitute $\mathbf{Z}_4$.

22-8-43. The bridge circuit of Figure 22-20 balances when $\mathbf{Z}_1 = 200 \ \Omega \underline{/0°}$, $\mathbf{Z}_2 = 100 \ \Omega \underline{/0°}$, $\mathbf{Z}_3 = 20 \ \Omega \underline{/0°}$, and $\mathbf{Z}_4$ consists of a 200-pF capacitor in series with a coil whose inductance and resistance are unknown. $f = 1$ MHz. Determine L_x and R_x.

22-8-44. In the bridge circuit of Figure 22-20, $\mathbf{Z}_1 = 200 \ \Omega \underline{/-45°}$, $\mathbf{Z}_2 = 1000 \ \Omega \underline{/0°}$, $\mathbf{Z}_3 = 10 \ k\Omega \underline{/0°}$, and $\mathbf{Z}_4 = 500 \ \Omega \underline{/+60°}$. The null indicator has an impedance of $5 \ k\Omega \underline{/0°}$. What is the total current drawn from a 50-V 1-kHz source?

REVIEW QUESTIONS

22-1. In writing loop equations for ac networks with more than one source, why is it necessary to identify which terminal of the source is which?

22-2. Compare the relative merits of the six solutions for Example 22-2 and select the method which you think is most appropriate. State your reasons for your choice.

22-3. What are the advantages and disadvantages of the superposition theorem as a general network theorem?

22-4. Thévenin's theorem is widely used in electronic circuit-analysis techniques. What is the main feature of Thévenin's theorem that would account for this popularity?

22-5. To be quite precise in stating Thévenin's theorem, we should have specified a network of fixed *linear, bilateral* impedances. In an ac system, what is the significance of (a) a *linear* impedance, and (b) a *bilateral* impedance?

22-6. We can solve Example 22-2 in one fell swoop by using the algebraic statement of Millman's theorem,

$$V_L = \frac{E_1 Y_1 + E_2 Y_2}{Y_1 + Y_L + Y_2} \tag{22-22}$$

Show that Equation (22-22) is valid for the network of Figure 22-2. (*Hint:* The solution for Example 22-2D is essentially a procedure based on Millman's theorem.)

22-7. One factor governing the choice between loop or nodal equations in solving an impedance network is the number of simultaneous equations required by each method. In some networks (such as Example 22-1), the same number of equations is required for each method. The choice then depends on whether we are asked to find one or more branch *voltages* or one or more branch *currents*. What would be your choice on this second basis?

22-8. Under what circumstances can we apply the equations that we developed for wye-delta transformations in *three*-terminal networks to *four*-terminal coupling networks?

22-9. Why do we not need to take the impedance of the null indicator into account in a balanced ac bridge circuit?

22-10. Why does a variation in the magnitude of the source voltage not affect the accuracy of a balanced bridge measurement?

22-11. How will the sensitivity of the null indicator affect the accuracy of an ac bridge measurement?

22-12. Why can we split the general bridge Equation (22-15) into two separate equations in deriving equations for the unknown components in the various ac bridge circuits?

22-13. After assembling a bridge in the form shown in Figure 22-39, we find that it cannot be balanced. Show why by deriving equations for L_x and R_x from the general bridge equation.

22-14. Why is the Hay bridge more suitable than the Maxwell bridge for measuring coils with a very low resistance?

FIGURE 22-39

FIGURE 22-40
Schering Bridge.

22-15. Derive equations for C_x and R_x in the Schering bridge of Figure 22-40.

With electronic calculators, we no longer need to learn how to determine sines and tangents from trigonometry tables. Similarly, if personal computers and circuit-analysis software become as commonplace as calculators, we can solve complex electrical networks without learning the circuit-analysis techniques of this chapter. We don't need to know how a computer goes about circuit-analysis calculations. All we need to know is how to enter the requested network parameters.

Since the primary objective of this course is to understand why electric circuits must perform in a predictable manner, we have avoided circuit-analysis software in working with examples in this chapter. Now that we are familiar with circuit-analysis techniques, we are able to appreciate the computer's advantage of speed and accuracy.

*22-16.** If you have access to a computer and circuit-analysis software, solve some of the problems in this chapter using the available circuit-analysis program.

*22-17.** If you have access to a personal computer but not to circuit-analysis programs, produce your own programs for storage on diskette (or cassette) to perform the time-consuming and error-prone tasks of solving second- and third-order simultaneous equations in which both coefficients and unknowns are phasor quantities.

*22-18.** Produce for storage on diskette (or cassette) programs to carry out delta-wye and wye-delta transformations involving impedance phasors.

23

RESONANCE

23-1 EFFECT OF VARYING FREQUENCY IN A SERIES *RLC* CIRCUIT

In most of the ac network examples we have considered, the frequency of the source has remained constant. This is the case in electric power systems. The standard power-line frequency in North America is 60 Hz.* The standard frequency in Europe is 50 Hz. To reduce the weight of iron required for the magnetic circuits of motors and transformers, aircraft electrical systems use frequencies of 400 Hz or 1 kHz. In radio circuitry, however, frequency is an important variable. Therefore, we must now devote our attention to the behavior of electric circuits in which we can consider that the rms voltage remains constant but the frequency varies.

If the circuit connected to a variable-frequency source contains only resistance, the frequency of the source will have negligible effect on the magnitude or phase of the current drawn from the source, since $I = E/R$. But if we connect an inductor to a variable-frequency source, because $X_L = 2\pi f L$, its inductive reactance is directly proportional to the frequency. Similarly, the reactance of a capacitor *decreases* as the frequency increases.

In the circuit of Figure 23-1(a), resistance, inductance, and capacitance are all connected in series to a variable-frequency source. When the frequency is quite low, the inductive reactance is quite small and the capacitive reactance is quite large. Consequently, the net reactance is a large capacitive reactance, much greater in magnitude than the resistance, and thus the total impedance is practically equal to $-jX_C$. As we increase the frequency, not only does the capacitive reactance become smaller, but the

---

* A few older generating stations still produce electric energy as a 25-Hz sine wave.

FIGURE 23-1
Series *RLC* Circuit Connected to a Variable-Frequency Source.

inductive reactance increases. We can show the effect of increasing frequency on the three components of the circuit of Figure 23-1(a) by plotting graphs of their resistance or reactance as a function of frequency, as shown in Figure 23-1(b). As we have noted, resistance is not frequency-sensitive. Hence, its graph is a horizontal line. And since X_L is *directly* proportional to frequency, its graph is a straight line through the origin with a slope proportional to the inductance of the coil. To show that we *subtract* X_C from X_L in determining the net reactance, we plot X_C along the negative Y-axis of the graph. Since X_C is *inversely* proportional to the frequency, its graph has the reciprocal-type curve as shown.

As we increase the frequency, the net reactance $(X_L - X_C)$ decreases until we reach a frequency, marked by the brackets in Figure 23-1(b), at which $X_L = X_C$. The net reactance at this frequency is zero, and, therefore, the total impedance of the circuit is simply the resistance of R. As we increase the frequency past this point, the inductive reactance exceeds the capacitive reactance, and the total impedance becomes inductive, approaching the condition of being practically the same as the inductive reactance of the circuit when the frequency is quite high.

To examine the effect of varying frequency on the impedance of the series *RLC* circuit of Figure 23-1(a) in a more quantitative manner, we can program a computer to make repetitive calculations by increasing frequency in steps over a range of approximately 160 Hz to 16 kHz.[†] Table 23-1 contains selected calculations from such a computer printout. When the frequency of the source is 159 Hz, the capacitive reactance of the series circuit is 100 times as great as the inductive reactance, and the series *RLC* circuit appears to consist of a 100-ohm resistor in a series with a *capacitance* having a reactance of 9900 ohms. As we noted in Figure 23-1(b), there is a frequency at which the inductive and capacitive reactances are numerically equal and cancel. This frequency is known as the **resonant frequency** of the series *RLC* circuit and is identified by the letter symbol f_r. For

[†] See the computer program we used in Section 6-7.

TABLE 23-1

EFFECT OF VARYING FREQUENCY IN A SERIES *RLC* CIRCUIT

f (Hz)	R (Ω)	X_L (Ω)	X_C (Ω)	$X_L - X_C$ (Ω)	$Z = \sqrt{R^2 + X^2}$	ϕ Arctan X/R	$I = E/Z$ (A)
159	100	100	10 K	$-j9900$	9900	$-89.4°$	0.0101
660	100	415	2415	$-j2000$	2002	$-87.2°$	.0499
985	100	618	1618	$-j1000$	1005	$-84.3°$	.0995
1245	100	781	1281	$-j500$	509	$-78.7°$	.1965
1440	100	905	1105	$-j200$	223	$-63.4°$	.448
1515	100	952	1052	$-j100$	141	$-45°.4$	.707
1590	100	1000	1000	$-j0$	100	$-40°$	1.00
1675	100	1052	952	$+j100$	141	$+45°.4$	.707
1760	100	1105	905	$+j200$	223	$+63.4°$	.448
2040	100	1281	781	$+j500$	509	$+78.7°$	.1965
2575	100	1618	618	$+j1000$	1005	$+84.3°$	.0995
3840	100	2415	415	$+j2000$	2002	$+87.3°$	.0499
15 900	100	10 000	100	$+j9900$	9900	$+89.4°$	0.0101

the circuit of Figure 23-1(a), the resonant frequency is 1590 Hz. For a frequency of 15 900 Hz, Table 23-1 shows that X_L now greatly exceeds X_C and the series *RLC* circuit behaves as if it consisted of a 100-ohm resistor in series with an *inductance* having a reactance of 9900 ohms.

We can display the data of Table 23-1 graphically for each frequency by preparing a set of impedance diagrams similar to Figure 19-9. Figure 23-2 shows the impedance diagrams for the series *RLC* circuit of Figure 23-1(a) for the resonant frequency of 1590 Hz and for two specific frequencies: (a) when the frequency of the source is *lower* than the resonant frequency and (c) when the frequency is *higher* than the resonant frequency.

We would require thirteen separate impedance diagrams to show all the situations given in Table 23-1. Although these impedance diagrams show the impedance re-

FIGURE 23-2
Impedance of a Series *RLC* Circuit when (a) $f < f_r$, (b) $f = f_r$, and (c) $f > f_r$.

lationships for each *specific* frequency, they do not really show clearly the manner in which the impedance of a series *RLC* circuit varies as we vary the frequency of the source. Since we have chosen to make frequency the independent variable in this chapter, we adopt the type of graph we used in Figure 23-1(b) where we plot frequency along the X-axis of the graph. We can program the computer to display or print out such a graph directly from the calculations it made to produce Table 23-1. Before we mark a frequency scale along the X-axis of the graph, we should note the nonlinear relationship between frequency (left-hand column) and current (right-hand column) in Table 23-1.* For the current to rise from 0.0101 A to 1.00 A, the frequency must increase by $1590 - 159 = 1431$ Hz. But for the current to drop from 1.00 A back down to 0.0101 A, the frequency of the source must increase by $15\,900 - 1590 = 14\,310$ Hz—ten times as great a range!† If we use the usual *linear* scale to represent frequency, the resulting **resonance curve** will be cramped and skewed toward the left-hand side of the graph.

To restore the symmetry that is apparent in Table 23-1, we note that there is a *ten-fold* increase in frequency from 159 Hz to 1590 Hz, and another *ten-fold* increase in frequency from 1590 Hz to 15 900 Hz. Hence, we should lay out a frequency scale such that each unit of horizontal distance along the X-axis represents a *ten-fold* increase in frequency. Since $\log 1 = 0$, $\log 10 = 1$, $\log 100 = 2$, $\log 1000 = 3$, etc., we call such a scale a logarithmic scale. Graph paper is available in a **semi-log** format where only the X-axis is laid out in logarithmic divisions, or **log-log** format with logarithmic divisions along both axes. In plotting the graphs of Figure 23-3 from the last three columns of Table 23-1, we select *semi-log* scales for phase angle and *log-log* scales for both current and impedance. To focus attention on how these three dependent variables respond to changes in frequency, we have left out the scale markings of the original graph paper.

If the magnitude of the source voltage remains at 100 volts rms in spite of the frequency variation, the magnitude of the current in Table 23-1 and Figure 23-3 is inversely proportional to the impedance. Therefore, whereas the impedance becomes minimum at 1590 hertz, the current will be maximum at this frequency. As we have noted, at very low frequencies, the impedance of the series *RLC* circuit of Figure 23-1(a) is almost completely capacitive reactance, and thus the angle of the impedance is almost $-90°$. Consequently, at these low frequencies, the current *leads* the applied voltage by almost 90°. At 1590 hertz, the net reactance becomes zero and the total impedance is simply the resistance of the circuit with a 0° angle. Then, at very high frequencies, the impedance becomes almost completely inductive reactance, under which circumstances the current *lags* the applied voltage by almost 90°. If we wish to show the effect of frequency on both the magnitude and phase of the current, we can prepare a circle diagram similar to Figure 21-20.

SELF-CHECK

23-1:1. Find the frequency for a series combination of a 20-mH coil and a 0.033-μF capacitor at which the two reactances will be equal in magnitude.

---

* Although X_L and X_C individually are linear functions of frequency, as shown by Column 5 of Table 23-1, the *net* reactance $(X_L - X_C)$ is not.

† We have already encountered a similar situation in transcribing the data from Table 6-1 to the graph showing power output as a function of a variable load resistance in Figure 6-9.

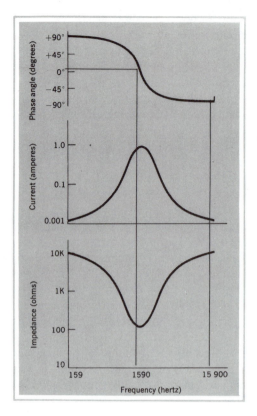

FIGURE 23-3
Effect of Varying Frequency on Impedance, Current, and Phase Angle of the Current in a Series *RLC* Circuit.

23-2 SERIES RESONANCE

In the series *RLC* circuit of Figure 23-1(a), when the frequency of the source is 1590 Hz:

1. The inductive reactance and the capacitive reactance of the circuit are equal.
2. Therefore, the impedance is minimum and is equal to the resistance of the circuit.
3. Hence, the current in the circuit is maximum and is equal to E/R.
4. The current through the circuit is in phase with the applied voltage.

This condition is known as **resonance.** The series *RLC* circuit of Figure 23-1(a) is called a **resonant circuit,** and the frequency at which resonance occurs is called the **resonant frequency** of the circuit.

The letter symbol for resonant frequency is f_r.

Since at resonance $X_L = X_C$, then $2\pi f_r L = 1/2\pi f_r C$, from which

$$f_r = \frac{1}{2\pi\sqrt{LC}} \tag{23-1}$$

where f_r is in hertz, L is in henrys, and C is in farads.

Although we have considered only the case of fixed values of L and C with the frequency varying to obtain resonance, rearranging Equation (23-1) will show that we can bring a given RLC circuit into resonance at a certain frequency by varying either the inductance or capacitance of the circuit. Inductance is usually varied by changing the reluctance of a coil's magnetic circuit. This is accomplished by moving a powdered-iron core on a threaded shaft in or out of the coil. Variable inductance is usually used for preset tuning adjustments. Continuously variable **tuning** is usually accomplished by rotating one set of plates of an air-dielectric capacitor between a set of fixed plates. More recently, electronic tuning varies the capacitance of a resonant circuit by varying the dc reverse-bias voltage applied to a specially designed pn-junction called a **varactor.**

We note from Equation (23-1) that the resistance of the series-resonant circuit has no bearing on the resonant frequency. The resistance governs the minimum impedance of the circuit which occurs at resonance. As a result, the resistance governs the steepness of the "skirts" of the resonance curve. In the series-resonant circuit of Figure 23-4(a), the inductance and capacitance are the same as in the circuit of Figure 23-1(a). Therefore, its resonant frequency is still 1590 hertz. However, the resistance has been reduced from 100 ohms to 50 ohms. If we prepare a table similar to Table 23-1 for the circuit of Figure 23-4(a), we can then plot the resonance curve shown in Figure 23-4(b). Off resonance, the total impedance has not been affected very much by the change in resistance. But at the resonant frequency, the impedance is only half that of the circuit of Figure 23-1(a), and thus the circuit current at resonance is twice as great.

(a)

(b)

FIGURE 23-4
Effect of Resistance on the Shape of the Resonance Curve.

We also note from Equation (23-1) that as long as the product of L and C is constant, the resonant frequency of a series-resonant circuit will remain constant. In the circuit of Figure 23-5(a) we have doubled the inductance and cut the capacitance in half. Therefore, the LC product and the resonant frequency are unchanged. However, we have increased the L/C ratio by a factor of $4:1$. This means that at resonance, $X_L = X_C = 2000$ ohms instead of 1000 ohms, as in the circuit of Figure 23-1(a). Hence, at a frequency of 1515 Hz, $X_L - X_C = 200$ ohms instead of 100 ohms, as in the circuit of Figure 23-1(a). This increase in the L/C ratio will, therefore, steepen the skirts of the resonance curve, as shown in Figure 23-5(b).

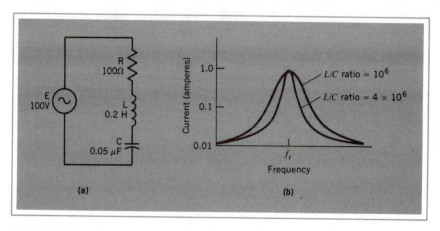

FIGURE 23-5
Effect of *L/C* Ratio on the Shape of the Resonance Curve.

EXAMPLE 23-1

A coil in a tuned circuit in a radio receiver has an inductance of 300 μH and a resistance of 15 Ω. What value of capacitance must be connected in series with the coil for the circuit to be series resonant at 840 kHz?

SOLUTION

$$X_L = 2\pi f L = 2 \times \pi \times 840 \text{ kHz} \times 300 \ \mu\text{H} = 1583 \ \Omega$$

At resonance,
$$X_C = X_L = 1583 \ \Omega$$

$$\therefore C = \frac{1}{2 \times \pi \times 840 \text{ kHz} \times 1583 \ \Omega} = 120 \text{ pF}$$

SELF-CHECK

23-2:1. How will the resonant frequency of the circuit shown in Figure 23-5 change if:
(a) *R* is made ten times larger?
(b) *L* is made ten times larger?
23-2:2. What inductance, when connected in series with a 100-pF capacitor, will produce a series resonant frequency of 20 kHz?

23-3 Q OF RESONANT CIRCUITS

Decreasing the resistance in a series resonant circuit and increasing the *L/C* ratio both have the effect of steepening the skirts of the resonance curve. Perhaps we can combine both these variables into a single factor relating to resonant circuits. As we noted when we considered power factor correction in Section 20-7, in an ac circuit containing both inductance and capacitance, the capacitance *stores* energy while the inductance *returns* energy stored in its magnetic field, and vice versa. Therefore, ac circuits containing both inductance and capacitance *exchange* reactive power between the inductance and capacitance. Hence, the source is required to supply only the difference between the reactive power of the inductance and the reactive power of the capacitance. This accounts for the net reactance of a series circuit being the *difference* between the inductive and capacitive reactance, and the reactive voltage of a series circuit being the *difference* between V_L and V_C.

At resonance, the reactive power of the inductance and capacitance are equal, and the source has to supply only the active power required by the resistance of the circuit.

The ratio between the reactive power of either the inductance or the capacitance at resonance and active power of a resonant circuit is called the Q factor of that resonant circuit.

At this point, standard letter symbols become rather confusing since the letter Q represents both reactive power and quality factor (and also electric charge). Therefore, in stating the definition of **quality factor** (more commonly called Q factor) in equation form, we shall revert to the secondary symbol for reactive power P_q.

$$\therefore Q = \frac{P_q}{P} \tag{23-2}$$

Since the resistance in a resonant circuit is usually the resistance associated with the coil, we often refer to the Q *of the coil.*

Because $\qquad P_q = I^2 X_L \quad \text{and} \quad P = I^2 R, \quad \text{then} \quad Q = \frac{I^2 X_L}{I^2 R}.$

But the current is common to all components in a series circuit.

$$\therefore Q = \frac{X_L}{R} = \frac{\omega L}{R} \tag{23-3}$$

Equation (23-3) provides us with a simpler form for the calculation of Q than Equation (23-2). It is sometimes given as a definition of Q.
Since $\omega L = 2\pi f L$ and since $f_r = 1/2\pi\sqrt{LC}$,

$$\therefore Q = \frac{1}{R}\sqrt{\frac{L}{C}} \tag{23-4}$$

The Q factor of a resonant circuit is, therefore, the single factor which takes into account the effect of both the resistance and the L/C ratio on the shape of a resonance curve. The higher the Q, the steeper the skirts of the resonance curve.

EXAMPLE 23-2
A series resonant circuit consists of a 50-μH inductance whose resistance is 5 Ω, and a 200-pF capacitor. What is its Q?

SOLUTION 1

$$Q = \frac{1}{R}\sqrt{\frac{L}{C}} = \frac{1}{5}\sqrt{\frac{50\ \mu H}{200\ pF}} = \frac{1}{5} \times 500 = \mathbf{100}$$

SOLUTION 2
Since it is quite likely that we shall need to know the resonant frequency of a circuit, this solution may be quicker in the long run.

$$f_r = \frac{1}{2\pi\sqrt{LC}} = \frac{1}{2 \times \pi \times \sqrt{50\ \mu H \times 200\ pF}} = 1.592\ MHz$$

$$X_L = 2\pi f L = 2 \times \pi \times 1.592\ MHz \times 50\ \mu H = 500\ \Omega$$

$$Q = \frac{X_L}{R} = \frac{500}{5} = \mathbf{100}$$

SELF-CHECK

23-3:1. Determine the Q of the resonant circuit in Figure 23-4 for both:
(a) $R = 50\ \Omega$ and
(b) $R = 100\ \Omega$
23-3:2. Determine the Q of the resonant circuit in Example 23-1.

23-4 RESONANT RISE OF VOLTAGE

We have not as yet considered the effect on the voltage distribution in a series-resonant circuit of varying the frequency of the source. With respect to the circuit of Figure 23-1(a), we can prepare Table 23-2.

TABLE 23-2

RESONANT RISE OF VOLTAGE			
$f(Hz)$	$V_R = IR$	$V_L = IX_L$	$V_C = IX_C$
159	1.01 V	1.01 V	101 V
660	4.99	20.7	120
985	9.95	61.5	161
1245	19.65	153.5	252
1440	44.8	406	495
1515	70.7	675	747
1590	100	1000	1000
1675	70.7	747	675
1760	44.8	495	406
2040	19.65	252	153.5
2575	9.95	161	61.5
3840	4.99	120	20.7
15 900	1.01	101	1.01

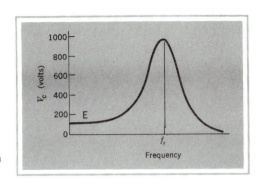

FIGURE 23-6
Effect of Varying Frequency on the Voltage across the Capacitance of a Series-Resonant Circuit.

Plotting the data from the right-hand column of Table 23-2, we obtain the graph of Figure 23-6, which shows the manner in which the voltage across the capacitor varies as we vary the frequency of the source. At frequencies well below the resonant frequency of the circuit, the voltage across the capacitor is practically the same as the applied voltage, since the total impedance of the circuit at these frequencies is mainly the reactance of the capacitor. As the frequency approaches the resonant frequency, even though the reactance of the capacitor is decreasing, the increasing current due to the decreasing total impedance causes the voltage across the capacitance to exceed the applied voltage. Because V_L and V_C are equal but $180°$ out of phase at resonance, V_R is the same as the applied voltage, and V_C is much greater than the applied voltage. As we increase the frequency to many times the resonant frequency, the impedance of the circuit is practically the same as the inductive reactance of the circuit and the voltage across the capacitance becomes very small.

Since capacitive reactance decreases as frequency increases, the product IX_C reaches a maximum at a frequency just slightly *lower* than the resonant frequency of the circuit, as shown in Figure 23-6. This frequency difference is usually so small that we can neglect it. In Table 23-2, the frequency readings are not close enough to show the exact frequency at which V_C is maximum. We can plot a similar graph showing the rise in the voltage across the inductance of the series-resonant circuit as we vary the frequency of the source through resonance. In this case, the exact frequency at which V_L is a maximum is slightly higher than the resonant frequency of the circuit. This increase in the voltage across the capacitance (and the inductance) of a series-resonant circuit at resonance, as shown in Figure 23-6, is called **resonant rise of voltage.**

$$V_C = IX_C$$

But $I = E/Z$ and at resonance, $Z = R$.

$$\therefore V_C = \frac{EX_C}{R}$$

But $X_C/R = Q$. Therefore, the resonant rise of voltage in a series-resonant circuit is

$$V_C = QE \tag{23-5}$$

where E is the source voltage at the resonant frequency. Because Q is usually much greater than unity, a voltage much greater than the source voltage appears across the capacitance and inductance of a series-resonant circuit at resonance. However, since the voltage drops across the inductance and capacitance are equal and opposite at resonance and the voltage drop across the resistance is equal to the applied voltage, the phasor sum of the voltage drops does equal the applied voltage, as required by Kirchhoff's voltage law. Equation (23-5) is sometimes given as a definition of Q.

Resonant rise of voltage can be put to practical use in radio receivers as a means of *increasing* the input voltage of a desired signal to which the circuit is tuned. Figure 23-7(a) shows a typical antenna-input circuit of a radio receiver. Antenna current in the primary coil induces a voltage in the tuned secondary coil. At first glance, it appears that the coil and tuning capacitor are in *parallel*. Actually, the induced voltage acts in *series* with the coil and capacitor, as shown by the equivalent circuit of Figure 23-7(b). At resonance, the voltage across the tuning capacitor (which becomes the input to the amplifier) is Q times the voltage induced into the coil by the antenna signal. As indicated by Equation (23-5), the higher the Q of the tuned circuit, the greater its **sensitivity.**

SELF-CHECK

23-4:1. For the resonant circuit in Figure 23-4, determine the voltage across the capacitor at resonance for both:
(a) $R = 50 \ \Omega$ and
(b) $R = 100 \ \Omega$

23-4:2. Determine the voltage (magnitude and phase), at resonance, across each of the devices in the circuit shown in Figure 23-5.

23-5 SELECTIVITY

Since the capacitance of the circuit of Figure 23-7 is variable, we can *select* the frequency at which the resonant rise of voltage provides us with a voltage gain of Q times the input voltage. The shape of the resonance curve of Figure 23-6 is such that frequencies close

FIGURE 23-7
Resonant Rise of Voltage in a Radio Receiver Input Circuit: (a) Actual Circuit; (b) Equivalent Circuit.

to the resonant frequency receive just about as great an increase in voltage as the resonant frequency itself. Therefore, we can say that a resonant circuit selects a certain *band* of frequencies. This is desirable in radio communications since, according to Hartley's bandwidth law, any signal carrying intelligence requires a band of frequencies whose bandwidth is directly proportional to the rate at which intelligence is transmitted.

However, the skirts of the resonance curve are not steep enough for us to be able to draw a definite boundary between those frequencies which will be selected and those frequencies which will be rejected. We must arbitrarily select a point on the resonance curve on either side of the resonant frequency, which we shall consider to be the practical boundary between selected and rejected frequencies.

Since the current is common to all components in a series circuit, we return to the graph of current versus frequency in a series-resonant circuit to determine the bandwidth of a tuned circuit. As the frequency of the source is decreased from the resonant frequency of a given tuned circuit, the impedance of the circuit will increase and the current will decrease, as shown in Figure 23-8(a). We consider the frequency f_1 at which the current is diminished to $1/\sqrt{2}$, or 0.707, of its value at resonance as the limit of the band of frequencies below the resonant frequency which the tuned circuit will accept. Similarly, the frequency f_2 in Figure 23-8(a), at which the current is again diminished to 0.707 of the maximum current at resonance, is considered to be the limit of the band of frequencies above the resonant frequency which the tuned circuit will accept. Since $P = I^2 R$, the active power input to the tuned circuit at frequencies f_1 and f_2 is

$$P = (0.707\ I_{\text{max}})^2 R = 0.5\ P_{\text{max}}$$

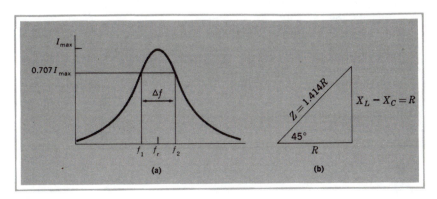

FIGURE 23-8
Bandwidth of a Resonant Circuit.

Consequently, we call frequencies f_1 and f_2 on the resonance curve the **half-power points.** And we refer to the frequency range Δf between f_1 and f_2 as the **bandwidth** of the resonant circuit.

The bandwidth of a resonant circuit is the range of frequencies between upper and lower half-power frequencies.
The standard letter symbol for bandwidth is B.

By definition,
$$B = f_2 - f_1 \qquad \text{(23-6)}*$$

For the current to drop to 0.707 of the current at resonance, the impedance of the tuned circuit must be $1/0.707$ or 1.414 times the impedance at resonance. At resonance $Z = R$; hence, the circuit impedance at the half-power frequencies must be $1.414R$. As indicated by Figure 23-8(b), the net reactance at the half-power points must be equal to the resistance of the tuned circuit for this to be the case. Since inductive reactance is directly proportional to frequency and capacitive reactance is inversely proportional to frequency, at frequency f_2, the value of X_L must have increased by an amount approximately equal to $0.5R$ and X_C must have decreased by an amount approximately equal to $0.5R$. From this, it follows that

$$2\pi f_2 L - 2\pi f_1 L = R, \quad \text{or} \quad 2\pi L(f_2 - f_1) = R$$

$$\therefore B = \frac{R}{2\pi L} \qquad \text{(23-7)}$$

Dividing both sides of Equation (23-7) by f_r,

$$\frac{B}{f_r} = \frac{R}{2\pi f_r L} = \frac{R}{X_L} = \frac{1}{Q}$$

$$\therefore B = \frac{f_r}{Q} \qquad \text{(23-8)}$$

Therefore, an increase in the Q of a resonant circuit not only increases the **sensitivity** by increasing the resonant rise of voltage, but it also increases the **selectivity** by decreasing the bandwidth of the tuned circuit.

EXAMPLE 23-3
(a) Determine the bandwidth of the resonant circuit of Example 23-2.
(b) If this circuit is connected to a 40-μV signal source, what is the voltage across the tuning capacitor at resonance?

SOLUTION

(a)
$$B = \frac{f_r}{Q} = \frac{1.592 \text{ MHz}}{100} = \textbf{15.92 kHz}$$

(b)
$$V_C = QE = 100 \times 40 \ \mu\text{V} = \textbf{4 mV}$$

SELF-CHECK

23-5:1. Determine the bandwidth of the resonant circuit shown in Figure 23-4 for both:
    (a) $R = 50 \ \Omega$
    (b) $R = 100 \ \Omega$
23-5:2. Find the bandwidth of the resonant circuit in Example 23-1.

    * We sometimes use Δf as a more descriptive symbol for bandwidth.

23-6 THEORETICAL PARALLEL-RESONANT CIRCUIT

The theoretical parallel-resonant circuit of Figure 23-9 consists of an *ideal* inductor, a capacitor, and a *high* resistance, all connected in parallel to a voltage source. Since the current in the inductive branch lags the applied voltage by 90° and the current in the capacitive branch leads the applied voltage by 90°,

$$I_T = \sqrt{I_{R_p}^2 + (I_C - I_L)^2}$$

Because $I_L = E/X_L$ and $I_C = E/X_C$, the current in the inductive branch will decrease, and the current in the capacitive branch will increase as the frequency of the source increases. Therefore, there will be a particular frequency at which $I_L = I_C$. At this frequency, the current drawn from the source will be a minimum and will be equal to the current in the resistance branch of the theoretical parallel-resonant circuit.

For the parallel *RLC* circuit of Figure 23-9, we can plot a series of admittance diagrams similar to Figure 19-12 or a series of current phasor diagrams similar to Figure 19-11. Using current phasor diagrams, at the resonant frequency as shown in Figure 23-10(b), the currents in the inductance and capacitance branches cancel out and the total

FIGURE 23-9
Theoretical Parallel-Resonant Circuit.

FIGURE 23-10
Total Current in a Parallel *RLC* Circuit when (a) $f < f_r$, (b) $f = f_r$, and (c) $f > f_r$.

current is minimum and is equal to the current in the R_p branch. At a frequency below the resonant frequency, X_L is less than X_C and the total current is greater than I_R and is *inductive,* as shown in Figure 23-10(a). Similarly, as shown in Figure 23-10(c), at a frequency above the resonant frequency, the total current in the theoretical parallel-resonant circuit is *capacitive.* Note that the parallel-resonant circuit is the exact opposite of the series-resonant circuit in this respect.

If we carefully plot a graph of total current versus frequency for the circuit of Figure 23-9, we note (lower graph of Figure 23-11) that we produce a resonance curve which has the same shape but is inverted in comparison with the current versus frequency graph for a series-resonant circuit. But, by now, we should be accustomed to the "similar but opposite" characteristics of parallel and series circuits. For a constant source voltage, total current is directly proportional to the total admittance of the circuit. Hence, the total-admittance graph will have the same shape as the total-current resonance curve of Figure 23-11. And since the equivalent impedance of the theoretical parallel-resonant circuit is the reciprocal of the total admittance, it has the shape of the upper curve in Figure 23-11. Again, we should note that the symmetrical appearance of these resonance curves is due to using a *logarithmic* frequency scale, as in Figure 23-3. Because the resonance graphs for a theoretical parallel-resonant circuit are the reciprocal curves of the equivalent series-resonant circuit, parallel resonance is sometimes called **antiresonance.**

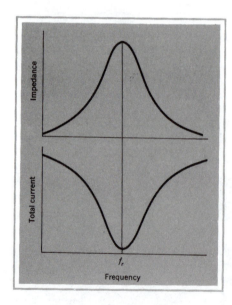

FIGURE 23-11
Parallel-Resonance Curves.

Since the current in the inductance branch and the current in the capacitance branch are equal at resonance, it follows that the resonant frequency of the theoretical parallel-resonant circuit is, as in the case of series resonance,

$$f_r = \frac{1}{2\pi\sqrt{LC}}$$

(23-1)

Although the *total* current in a parallel-resonant circuit is minimum at resonance, if the reactances of the inductive and capacitive branches are considerably less than the resistance of the high-resistance branch, the current in the inductance and capacitance is many times as great as the line current. Remembering that the capacitance is charging when the magnetic field around the inductance is collapsing, and vice versa, we can consider this current as being the medium which carries the energy back and forth between the capacitance and the inductance. Therefore, at resonance, we can say that a considerable **tank current** flows in the closed loop consisting of the inductance and capacitance branches, while the resistance branch draws current from the source.

Again, we are concerned with the relationship between the resistance and reactance of the resonant circuit. We define the Q factor of the parallel-resonant circuit as the ratio between the reactive power of either the inductance or the capacitance at resonance and the active power of the resistance. Hence,

$$Q = \frac{P_q}{P} = \frac{E^2/X_L}{E^2/R_p}$$

$$Q = \frac{R_p}{X_L} \tag{23-9}$$

Note that this results in a ratio between resistance and reactance which is inverted as compared with the Q of a series-resonant circuit. But note also that the *parallel* resistance is quite high, 100 000 ohms or so, whereas the *series* resistance is only a few ohms.

At resonance, the tank current in the theoretical parallel-resonant circuit will be the same as I_L or I_C, and the total current will be the same as I_{R_p}. Since $E = I_R R_p = I_L X_L$,

$$\text{tank current } = Q \times \text{ line current} \tag{23-10}$$

We can think of Equation (23-10) as representing a **resonant rise of tank current** similar to the resonant rise of voltage in a series-resonant circuit. However, when we examine the current graph of Figure 23-11, we find that there is a *decrease* in the current drawn from the source rather than a resonant rise in tank current. The problem with the circuit of Figure 23-9 is that we have neglected the internal resistance of the source. When we use a parallel-resonant circuit as the output load for a transistor amplifier, we find that the comparatively high internal resistance of the transistor makes it possible for us to translate the resonant rise of impedance of the parallel-resonant circuit into a resonant rise in output voltage.

In the equivalent circuit of Figure 23-12(a), from the voltage-divider principle,

$$\mathbf{V}_{\text{out}} = \mathbf{E} \times \frac{\mathbf{Z}_p}{R_{\text{int}} + \mathbf{Z}_p} \tag{7-1}$$

When the frequency of the signal source is considerably off resonance, the internal resistance of the source is much greater than the impedance of the parallel-resonant circuit. Consequently, most of the signal voltage generated by the constant-voltage

FIGURE 23-12
Resonant Rise in Output Voltage due to Resonant Rise of Impedance of Parallel-Resonant Circuit.

source is lost across the internal resistance of the source. But, at the resonant frequency, the impedance of the parallel-resonant circuit is usually greater than the internal impedance of the signal source. Hence, the output voltage in Figure 23-12(a) is much greater at the resonant frequency than for frequencies off resonance.

In preparing equivalent circuits for transistor amplifiers, we usually find it more convenient to convert the constant-voltage source to its equivalent constant-current source of Figure 23-12(b). From Ohm's law,

$$\mathbf{V}_{\text{out}} = \mathbf{IZ}_{\text{T}}$$

And since $\mathbf{Z}_{\text{T}}$ consists of the internal resistance of the source and the impedance of the tuned circuit in parallel, $\mathbf{V}_{\text{out}}$ will be greatest at resonance when Z_p is maximum (Figure 23-11) and will approach the open-circuit voltage of the constant-voltage source of Figure 23-12. Off resonance, the output voltage will be considerably less than the open-circuit voltage, as shown in Figure 23-13.

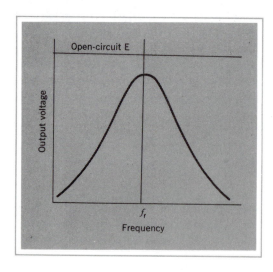

FIGURE 23-13
Output Voltage Across a Parallel-Resonant Circuit when the Source Has Appreciable Internal Resistance.

SELF-CHECK

23-6:1. Using the theoretical parallel-resonant circuit shown in Figure 23-9 and the values:

$$R_p = 2.0 \text{ k}\Omega$$
$$L = 50 \ \mu\text{H}$$
$$C = 0.33 \ \mu\text{F}$$

Determine the following:
(a) the resonant frequency
(b) the resonant Q

23-7 PRACTICAL PARALLEL-RESONANT CIRCUITS

The losses in the dielectric of a practical capacitor are usually so small that we can neglect them. If we are required to take these losses into consideration, we can represent them as a high resistance in *parallel* with the capacitance. But, since capacitor losses are so small in comparison to the resistance of practical inductors, we can omit such parallel resistance from the equivalent circuit of Figure 23-14. However, we must represent the resistance of the coil as acting in *series* with its inductance, as shown in Figure 23-14.

FIGURE 23-14
Practical Parallel-Resonant Circuit.

Since the practical parallel-resonant circuit is not a simple parallel circuit, we must be careful how we define its resonant frequency. We can define the resonant frequency of a practical parallel-resonant circuit by any of the following statements, all of which specify essentially the same conditions. **The resonant frequency of a practical parallel-resonant circuit is the frequency at which**

1. **The reactive power of the inductance and capacitance are equal.**
2. **The circuit has unity power factor.**
3. **The total impedance is completely resistive.**
4. **The line current is in phase with the source voltage.**

If we assume that the frequency of the voltage applied to the circuit of Figure 23-14 is such that $X_L = X_C$, the impedance of the inductive branch will be slightly greater than the reactance of the capacitive branch due to the resistance in series with the inductive reactance. Therefore, we must *decrease* the frequency of the source slightly to allow sufficient current in the inductive branch to make the reactive power of the inductive branch equal to the reactive power of the capacitance.

To assist us in solving for the actual resonant frequency of a practical parallel-resonant circuit, we can apply a technique we developed in Chapter 21. We can replace the series LR branch by an equivalent parallel circuit, as shown in Figure 23-15, where

$$R_p = \frac{Z^2}{R_s} = \frac{R_s^2 + X_s^2}{R_s} \tag{21-19}$$

$$X_p = \frac{Z^2}{X_s} = \frac{R_s^2 + X_s^2}{X_s} \tag{21-20}$$

FIGURE 23-15
Parallel Equivalent of a Practical Parallel-Resonant Circuit.

At resonance, the reactances of the inductive and capacitive branches of the *equivalent* parallel circuit must be equal.

$$\therefore \frac{1}{2\pi f_r C} = \frac{R_s^2 + (2\pi f_r L)^2}{2\pi f_r L} \tag{23-11}$$

$$\frac{L}{C} = R_s^2 + (2\pi f_r L)^2$$

from which

$$f_r^2 = \frac{L - CR_s^2}{4\pi^2 L^2 C}$$

$$\therefore f_r = \frac{1}{2\pi\sqrt{LC}} \times \sqrt{1 - \frac{CR_s^2}{L}} \tag{23-12}$$

Examining the factor under the square root sign in Equation (23-12), we can make several important observations about the resonant frequency of practical parallel-resonant circuits.

1. The magnitude of the factor under the square root sign is less than unity. Hence, the resonant frequency of a practical parallel-resonant circuit is slightly *lower* than the frequency given by Equation (23-1).

2. Unlike in series-resonant circuits or theoretical parallel-resonant circuits, the resonant frequency of a practical parallel-resonant circuit *is* dependent on the resistance of the circuit.

3. If the series resistance is large enough so that $CR_s^2/L > 1$, there is no parallel-resonant frequency for that particular circuit.

To evaluate the extent of the reduction in the resonant frequency of the practical parallel-resonant circuit of Figure 23-14, we first of all investigate the Q of a practical parallel-resonant circuit. We have already determined that the Q of a theoretical parallel-resonant circuit is

$$Q = \frac{R_p}{X_L} \qquad (23\text{-}9)$$

Applying Equation (23-9) to the parallel equivalent of a practical parallel-resonant circuit shown in Figure 23-15.

$$\therefore Q = \frac{Z^2}{R_s} \times \frac{X_s}{Z^2} = \frac{X_s}{R_s}$$

Hence,

The Q of a practical parallel-resonant circuit is the same as the Q of the coil itself.

By substituting $2\pi f_r L/Q = R_s$ in Equation (23-11),

$$\frac{1}{2\pi f_r C} = \frac{(2\pi f_r L/Q)^2 + (2\pi f_r L)^2}{2\pi f_r L} = 2\pi f_r L\left(\frac{1 + Q^2}{Q^2}\right)$$

from which
$$f_r = \frac{1}{2\pi\sqrt{LC}} \times \sqrt{\frac{Q^2}{1 + Q^2}} \qquad (23\text{-}13)$$

From Equation (23-13), if the Q of the coil is reasonably high ($Q > 10$), the difference between the resonant frequency produced by Equation (23-13) and Equation (23-1) is small enough for us to consider the resonant frequency of most practical parallel-resonant circuits to be simply

$$f_r = \frac{1}{2\pi\sqrt{LC}} \qquad (23\text{-}1)$$

We have noted that the impedance of a parallel-resonant circuit is maximum at resonance and is

$$Z_P = R_p = \frac{R_s^2 + X_s^2}{R_s} = R_s + \frac{X_s^2}{R_s}$$

There are several useful forms of this equation. Factoring out R_s and substituting $Q = X_s/R_s$,

$$Z_p = R_s(1 + Q^2) \qquad (23\text{-}14)$$

Or simply substituting for X_s/R_s,

$$Z_p = R_s + QX_s \qquad (23\text{-}15)$$

Since R_s is usually very small compared to QX_s,

$$\therefore Z_p \approx QX_L \qquad (23\text{-}16)$$

Substituting for X_L from Equation (23-3),

$$Z_p \approx Q^2 R_s \qquad (23\text{-}17)$$

Or substituting for Q from Equation (23-3),

$$Z_p \approx \frac{X_L^2}{R_s} \qquad (23\text{-}18)$$

and since

$$X_L^2 \approx X_L \times X_C = \frac{\omega L}{\omega C}$$

then

$$Z_p \approx \frac{L}{CR} \qquad (23\text{-}19)$$

Again, if $Q > 10$, for practical purposes we can replace the "approximately-equals" signs in the above equations by "equals" signs.

EXAMPLE 23-4

A coil in a tuned circuit in a radio receiver has an inductance of 50 μH and a resistance of 25 Ω. It forms a parallel-resonant circuit with a 200-pF capacitor.
(a) At what frequency is the total impedance completely resistive?
(b) What is the magnitude of the parallel impedance at this frequency?

SOLUTION

(a)
$$f_r = \frac{1}{2\pi\sqrt{LC}} \times \sqrt{1 - \frac{CR_s^2}{L}}$$

$$= \frac{1}{2 \times \pi \times \sqrt{50 \ \mu H \times 200 \ pF}} \times \sqrt{1 - \frac{CR_s^2}{L}}$$

$$= 1.5915 \ \text{MHz} \times \sqrt{1 - \frac{200 \ pF \times (25 \ \Omega)^2}{50 \ \mu H}}$$

$$\therefore f_r = 1.5915 \ \text{MHz} \times 0.99875 = \mathbf{1.5895 \ MHz}$$

(b)

$$X_L = 2\pi fL = 2 \times \pi \times 1.5895 \text{ MHz} \times 50 \ \mu\text{H} \approx 500 \ \Omega$$

$$Q = \frac{X_L}{R_s} \approx \frac{500 \ \Omega}{25 \ \Omega} \approx 20$$

$$\therefore Z_p = QX_L \approx 20 \times 500 \ \Omega \approx \mathbf{10 \ k\Omega}$$

SELF-CHECK

23-7:1. Using the practical parallel-resonant circuit shown in Figure 23-14 and given the following values:

$$E = 100 \text{ mV}$$
$$R_{\text{int}} = 1.0 \text{ k}\Omega, \ R_s = 3.0 \ \Omega$$
$$L = 50 \ \mu\text{H}, \quad C = 0.33 \ \mu\text{F}$$

Determine the following:
(a) the resonant frequency
(b) the resonant Q

23-7:2. Change the R_s to 1.0 Ω and recalculate the results of Self-Check Problem 23-7:1.

23-8 SELECTIVITY OF PARALLEL-RESONANT CIRCUITS

As we noted in Section 23-5, the **selectivity** of a resonant circuit is a measure of its ability to differentiate between a desired frequency and undesired interference from other frequencies. We usually express selectivity in terms of the **bandwidth** between the **half-power points** either side of the resonant frequency. At these points, the voltage across the equivalent parallel-resonant circuit in Figure 23-15 will be 0.707 of its magnitude at resonance. With a constant-current source, this means that the equivalent impedance will have dropped to 0.707 of its value at resonance. Following the reasoning we used in Section 23-5, and since we are working with a *parallel* circuit, the *susceptance* of L and C must each change by an amount equal to $1/R_p$ between these two frequencies. Taking $B_C = \omega C$,

$$2\pi C (f_2 - f_1) = \frac{1}{R_p}$$

from which

$$B = \frac{1}{2\pi C R_p} = \frac{f_r}{2\pi f_r C R_p} = f_r \times \frac{X}{R_p}$$

And if $Q > 10$, this becomes

$$B = \frac{f_r}{Q} \qquad\qquad (24\text{-}8)^\dagger$$

† Note that this is the same equation as for a *series*-resonant circuit. This shouldn't surprise us when we remember that Q is defined as the ratio of reactive *power* to active *power*.

As we discovered in Figure 23-12, the internal resistance of the signal source has a significant effect on the performance of a parallel-resonant circuit. The effective Q of a practical parallel-resonant circuit is seldom determined by R_p alone. In the equivalent circuit of Figure 23-15, the internal resistance of the source is directly in parallel with R_p. This reduces the parallel Q of the circuit and thus increases Δf, resulting in a reduction in selectivity.

EXAMPLE 23-5

The parallel-resonant circuit of Example 23-4 is connected across a signal source having an internal resistance of 40 kΩ. What is the bandwidth of the tuned circuit?

SOLUTION

The additional *parallel* resistance does not enter into the equation for the resonant frequency. Hence, the reactance of either branch at resonance is still 500 Ω. The parallel resistance due to the coil is

$$R_p = \frac{X^2}{R_s} = \frac{(500 \ \Omega)^2}{25 \ \Omega} = 10 \ \text{k}\Omega$$

The equivalent parallel resistance becomes

$$R_{eq} = \frac{R_{int} \times R_p}{R_{int} + R_p} = \frac{40 \ \text{k}\Omega \times 10 \ \text{k}\Omega}{40 \ \text{k}\Omega + 10 \ \text{k}\Omega} = 8 \ \text{k}\Omega$$

$$Q = \frac{R_{eq}}{X} = \frac{8 \ \text{k}\Omega}{500 \ \Omega} = 16$$

$$\therefore B = \frac{f_r}{Q} = \frac{1.5895 \ \text{MHz}}{16} \approx \textbf{99 kHz}$$

Returning to Equation (23-12), we noted that if R_s is large enough, there is no parallel-resonant frequency for a given coil and capacitor combination. Purposely increasing the series resistance (or reducing the Q) of a tuned circuit is called **damping.** When R_s is large enough for parallel resonance to be impossible, the result is **critical damping.** Just as we damp meter movements to keep the pointers from oscillating, we add resistance to peaking coils in television receivers to keep them from **ringing** at a frequency determined by their inductance and distributed capacitance.

SELF-CHECK

23-8:1. Determine the bandwidth of the parallel-resonant circuit in Self-Check Problem 23-7:1.
23-8:2. Determine the bandwidth of the parallel-resonant circuit in Self-Check Problem 23-7:2.

PROBLEMS

23-2-1. What is the resonant frequency of a coil with an inductance of 400 mH and a resistance of 100 Ω in series with a 0.016 μF capacitor?

23-2-2. At what frequency is a 27-pF capacitor and a 48-μH inductor with a resistance of 20 Ω series resonant?

23-2-3. What value of capacitance is required to tune a 500-μH coil to series resonance at 465 kHz?

23-2-4. What value of inductance in series with a 12-pF capacitor is resonant at 45 MHz?

23-2-5. A variable tuning capacitor in a radio receiver has a maximum capacitance of 365 pF and a minimum capacitance of 30 pF.
(a) What inductance is required for the lowest frequency to which the circuit can tune to be 540 kHz?
(b) What is the highest frequency to which the circuit can be tuned?

23-2-6. At what frequency is a circuit consisting of a 40-pF capacitor and a 90-μH coil with a Q of 80 series resonant?

23-3-7. What is the Q of the resonant circuit of Problem 23-2-1?

23-3-8. What is the Q of the resonant circuit of Problem 23-2-2?

23-3-9. What is the resistance of the coil in Problem 23-2-6?

23-3-10. (a) What is the L/C ratio of the tuned circuit of Problem 23-2-3?
(b) If the resistance of the coil is 30 Ω, what is its Q?

23-4-11. If the circuit of Problem 23-2-6 is connected to a 500-μV source, what voltage will appear across the capacitor at resonance?

23-4-12. What current will flow in the circuit of Problem 23-2-6 at resonance?

23-5-13. What is the bandwidth of the resonant circuit of Problem 23-2-2?

23-5-14. What is the bandwidth of the circuit of Problem 23-2-6?

23-5-15. A tuned circuit consisting of a 40-μH inductance and a 100-pF capacitance in series has a bandwidth of 25 kHz. What is its Q?

23-5-16. What is the resistance of the coil in Problem 23-5-15?

23-5-17. The coil and capacitor of a tuned circuit have an L/C ratio of 10^5. The Q of the tuned circuit is 80 and its bandwidth is 5.8 kHz.
(a) Calculate the upper and lower frequency limits of the bandpass.
(b) Calculate the inductance and resistance of the coil.

23-5-18. A series resonant circuit has a resistance of 20 Ω, and a bandwidth of 20 kHz at a resonant frequency of 1 MHz. Determine the required values of L and C.

23-7-19. What is the resonant frequency of a tuned circuit consisting of the coil and capacitor of Problem 23-2-6 connected in parallel?

23-7-20. What is the impedance of the parallel-tuned circuit of Problem 23-7-19 at resonance?

23-7-21. If the resonant circuit of Problem 23-7-19 is connected to a 500-μV source, what is the tank current at resonance?

23-7-22. What will the parallel-resonant frequency be if loading the inductive branch reduces the Q of the tuned circuit of Problem 23-7-19 to 20?

23-7-23. A coil with a Q of 90 and a capacitor form a parallel-resonant circuit tuned to 4.5 MHz. The total impedance of the circuit at resonance is 60 kΩ. Find
   (a) The inductance of the coil.
   (b) The capacitance of the capacitor.

23-7-24. A tuned circuit in a radio transmitter is composed of a tank coil with an inductance of 80 μH tuned by an 800-pF capacitor. The RF voltage across this tank circuit is 1200 V rms. The overall Q of the circuit is 30.
   (a) Calculate the tank current.
   (b) At what rate must RF energy be supplied to this circuit to maintain this amplitude?

REVIEW QUESTIONS

23-1. Using the data from Table 23-1, plot accurate graphs of current as a function of frequency (a) on linear graph paper, and (b) on log-log graph paper.

23-2. Using the data from the two right-hand columns of Table 23-1, plot a circle diagram showing how both current and phase angle change with a change in source frequency over the range from 159 Hz to 15.9 kHz.

23-3. What is meant by the general term **resonance?**

23-4. What property of an electric circuit makes resonance possible?

23-5. Derive an equation to determine the capacitance required to tune a given inductance to series resonance at a given frequency.

23-6. Why does the resistance in a series-resonant circuit have no bearing on the resonant frequency?

23-7. What is the significance of the resistance in a series-resonant circuit?

23-8. What is the significance of the term L/C **ratio** in resonant circuits?

23-9. What is the significance of the Q factor of resonant circuits?

23-10. What is the significance of the term **resonant rise of voltage** in discussing resonant circuits?

23-11. What is the significance of the term **half-power points** in discussing resonant circuits?

23-12. How does a parallel-resonant circuit differ in circuit behavior from a series-resonant circuit?

23-13. Why are parallel-resonant circuits capacitive at frequencies above the resonant frequency, whereas series-resonant circuits are inductive at frequencies above the resonant frequency?

23-14. Why does *decreasing* the resistance of the coil of a parallel-resonant circuit *increase* the total impedance at resonance?

23-15. Why does a change in the resistance of a coil in a parallel-resonant circuit affect the resonant frequency?

23-16. If a resistor were placed in parallel with the capacitor of a series-resonant circuit, would the resonant frequency be affected? Explain.

23-17. Why does the resonant frequency of a parallel-resonant circuit tend to drift as the **loading** on the circuit is increased?

23-18. What is the significance of the term **tank current** in discussing resonant circuits?

*23-19. Using either computer or calculator, prepare a table similar to Table 23-1 for the circuit of Figure 23-4.

*23-20. Use the same program to prepare a table similar to Table 23-1 for the circuit of Figure 23-5.

24

TRANSFORMERS

24-1 MUTUAL INDUCTION

We encountered mutual induction briefly in Chapter 14. Figure 14-2 showed a circuit in which *changing* the current in one winding (**primary**) induced a voltage in another winding (**secondary**), which was *linked* by the magnetic lines of force created by the current in the primary winding. The generation of a voltage in one winding by a *changing* current in another winding is called **mutual induction.** In Section 14-1, we considered mutual induction only to the extent that it illustrated Faraday's and Lenz's laws. Since mutual induction is based on *changing* current, we did not pursue its application to dc circuits. However, since the current in an ac circuit is continually changing, mutual induction plays a very important role in ac circuits.

24-2 TRANSFORMER ACTION

A practical device which depends on mutual induction for its operation is the **transformer.** In schematic diagrams, it is customary to connect the source to the *left*-hand coil (primary) and the load to the *right*-hand coil (secondary), as in Figure 24-1. For the moment, we shall assume that the secondary terminals are left open-circuit as shown.

According to Kirchhoff's voltage law, a voltage must appear across the primary coil such that it is exactly equal to the applied voltage at every instant in time. As we discovered in Section 17-1, if the resistance of the primary coil is negligible, this voltage must be induced into the coil by a changing flux in the core. Since the applied voltage is a sine wave, and since, according to Faraday's law, the induced voltage is proportional

FIGURE 24-1
Transformer Action.

to the *rate of change* of flux, the flux in the transformer core must be a sine wave lagging the applied voltage by 90°.

Since we define the weber as the magnetic flux which must cut (or link) an electric conductor in one second in order to induce an average voltage of one volt, and because *each* turn of a coil is linked by the same magnetic field, it follows that

$$E_{av} = \frac{N\Phi_m}{t} \qquad (24\text{-}1)$$

where E_{av} is the average voltage induced into a coil, N is the number of turns in the coil, and t is the time (seconds) it takes the flux to rise from zero to Φ_m webers.

Since the flux in the transformer must be a sine wave, it must rise from zero to Φ_m in one-quarter of a cycle.

$$\therefore E_{av} = \frac{N\Phi_m}{1/4f} = 4fN\Phi_m$$

But the *average* value of a sine wave is 0.637 of the peak value, and the *rms* value of a sine wave is 0.707 of the peak value; hence,

$$E = 1.11E_{av} \qquad (24\text{-}2)$$

$$\therefore E = 4.44fN\Phi_m \qquad (24\text{-}3)$$

where f is the frequency of the applied voltage in hertz, N is the number of turns in the coil, and Φ_m is the peak value of the flux in webers. We call Equation (24-3) the **general transformer equation.**[†]

Applying the general transformer equation to the primary winding of the simple transformer of Figure 24-1, we obtain

$$E_p = 4.44fN_p\Phi_m \qquad (24\text{-}4)$$

To create this sine wave of flux in the core, an alternating current must flow in the primary winding. And since

---

[†] See Appendix 2-9 for calculus derivation of the general transformer equation.

$$\Phi_p = \frac{F_m}{R_m} = \frac{N_p I_p}{R_m} \qquad \text{(13-2) and (13-1)}$$

it follows that the peak value of the primary current is

$$I_{pm} = \Phi_m \frac{R_m}{N_p}$$

from which
$$I_p = \frac{\Phi_m}{\sqrt{2}} \times \frac{R_m}{N_p} \qquad \text{(24-5)}$$

To apply Equations (24-4) and (24-5) to a practical transformer, we must know something about the *magnetic* circuit of the transformer core. First, for both the magnetic flux in the core and the primary current which induces it to be pure sine waves, the flux density in the core must remain in the linear region of the magnetization curve for the type of iron used. From the *BH* curve of Figure 13-21, B_m for sheet steel should not be greater than 1.1 teslas to stay below the upper knee (saturation) area of sheet steel. Maximum flux Φ_m will then depend on the cross-sectional area of the core. Once we decide on the maximum flux density for the core we plan to use, we can calculate the number of turns we need to use in our primary winding.

If we use Equation (24-5) to solve for the primary current, we must determine the reluctance of the magnetic circuit of the transformer core. If we use the magnetization curve of Figure 13-21, we can start with the maximum magnetizing force needed to create B_m. From Equation (13-6),

$$F_m = H_m l$$

And from Equation (13-1),

$$F_m = N I_m$$

Hence,
$$I_m = \frac{H_m l}{N}$$

and
$$I = \frac{H_m l}{\sqrt{2}\,N} \qquad \text{(24-6)}$$

If, for the moment, we can neglect core losses, even though the secondary of the transformer is an open circuit, there must be a sine wave of primary current in phase with the flux and lagging the applied voltage by 90°. We call this primary current the **exciting** or **magnetizing** current of the transformer. In good transformers, we expect this magnetizing current to be less than 5% of the total primary current when we connect full load to the secondary winding.

EXAMPLE 24-1
The sheet-steel core of a transformer has a cross-sectional area of 5.0 cm² and an average path length of 25 cm. If the peak flux density in the core is to be 1.0 T (see the *BH* curves of Figure 13-21),

(a) How many turns are required on a primary winding which is to be connected to a 120-V 60-Hz source?

(b) What is the rms value of the exciting current in the primary winding?

SOLUTION

(a) $\Phi_m = 5.0 \text{ cm}^2 \times 1.0 \text{ T} = 0.0005 \text{ Wb}$

$$N_p = \frac{E_p}{4.44\,f\Phi_m} = \frac{120 \text{ V}}{4.44 \times 60 \text{ Hz} \times 0.0005 \text{ Wb}} = \textbf{901 turns}$$

(b) From the BH curve for sheet steel in Figure 13-21, when $B_m = 1.0$ T, $H_m = 425$ At/m.

$$\therefore I_p = \frac{H_m l}{\sqrt{2}\,N_p} = \frac{425 \text{ At/m} \times 25 \text{ cm}}{\sqrt{2} \times 901 \text{ turns}} = \textbf{83.4 mA}$$

SELF-CHECK

24-2:1. A transformer has a primary voltage of 200-V 400-Hz. The cast-steel core of the transformer has a cross-sectional area of 2.0 cm^2 and an average flux length of 16 cm. A 500-turn primary coil is wound on the core.
(a) Determine the peak value of the flux within the core.
(b) Determine the magnetizing current in the primary (see Figure 13-21).

24-3 TRANSFORMATION RATIO

In the ideal transformer, all the flux created by primary current will link the secondary winding. Therefore, a voltage will be induced in the secondary winding,

$$E_s = 4.44\,fN_s\Phi_m \tag{24-7}$$

Dividing Equation (24-7) into Equation (24-4) gives

$$\frac{E_p}{E_s} = \frac{N_p}{N_s} = a \tag{24-8}$$

Consequently, when all the primary flux links the secondary winding, the ratio of the primary induced voltage to the secondary induced voltage is the same as the ratio of the number of primary turns to the number of secondary turns. These ratios are called the **transformation ratio,** which we represent by the letter symbol a.

Equation (24-8) shows us why transformers are so widely used in power systems. They provide a means of transforming a given source of alternating voltage into any desired voltage with minimum loss. If we arrange to produce a secondary voltage higher than the source voltage, we have a **step-up** transformer. Similarly, a **step-down** transformer is one in which the secondary voltage is less than the primary voltage.

EXAMPLE 24-2

The secondary of the transformer of Example 24-1 is to be wound to supply 6.3 V for the heater of a TV picture tube. How many turns must there be on the secondary winding?

SOLUTION

$$a = \frac{E_p}{E_s} = \frac{120}{6.3} = 19$$

$$N_s = \frac{N_p}{a} = \frac{901}{19} \approx \textbf{48 turns}$$

If we now connect the load impedance to the secondary winding in the simple transformer circuit of Figure 24-1, the voltage induced into the secondary winding by the primary flux will cause current to flow in the secondary circuit. This current flowing in the secondary winding will create a magnetomotive force of $N_s I_s$ ampere-turns, which, according to Lenz's law, tends to *reduce* the amplitude of the sine wave of flux in the core. But, in a lossless transformer, the amplitude of the flux sine wave must not change, since it is responsible for always inducing a voltage equal to the applied voltage into the primary winding. Therefore, when a load is connected to the secondary, *additional* primary current must flow in order to develop a primary mmf component which exactly cancels the secondary mmf. Hence,

$$N_p I_p' = N_s I_s \qquad (24\text{-}9)$$

where I_p' is the additional or load component of the primary current. We can rearrange Equation (24-9) to give

$$\frac{I_s}{I_p'} = \frac{N_p}{N_s} = a \qquad (24\text{-}10)$$

If the load connected to the secondary terminals of the transformer is a pure resistance, the secondary current will be in phase with the secondary voltage. Since E_p and E_s are both induced by the same changing flux, it follows that the additional I_p' is in phase with the applied voltage. Because the magnetizing current, I_{mag} in Figure 24-2, lags the applied voltage by almost 90°, the total primary current is the *phasor* sum of these two components.

FIGURE 24-2
Total Primary Current of a Transformer with a Resistance Secondary Load.

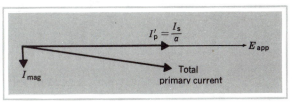

SELF-CHECK

24-3:1. An ideal transformer has a transformation ratio of 0.030. The primary of the transformer has 60 turns and is designed for a voltage of 120 V. Determine:
(a) The secondary voltage and
(b) The number of turns on the secondary coil.

24-3:2. An ideal transformer has a 1200-turn 60-V primary. When operating at full load, the primary draws 0.50 A. Determine the following for the 300-turn secondary coil:
(a) The secondary voltage and
(b) The secondary current.

24-4 IMPEDANCE TRANSFORMATION

When we connect a load to the secondary winding of the simple transformer of Figure 24-1, the primary current increases and shifts in phase to the position shown in the phasor diagram of Figure 24-2. But, as far as the source is concerned, we can treat the total primary current as being the phasor sum of two separate *parallel* branch currents. Hence, the right-hand circuit of Figure 24-3 is the equivalent circuit for the actual left-hand circuit. The imaginary parallel branch carrying I_p' appears to have an impedance of

$$Z_p = \frac{E_p}{I_p'}$$

FIGURE 24-3
Reflected Impedance.

From Equation (24-8), $$E_p = \frac{N_p}{N_s} E_s = aE_s$$

and from Equation (24-10), $$I_p' = \frac{N_s}{N_p} I_s = \frac{I_s}{a}$$

$$\therefore Z_p = \frac{E_s}{I_s}\left(\frac{N_p}{N_s}\right)^2$$

But E_s/I_s is dependent on the total impedance of the secondary circuit. In a good transformer, the impedance of the secondary winding will be small compared with the load impedance;

$$\therefore Z_p \approx \left(\frac{N_p}{N_s}\right)^2 Z_L \approx a^2 Z_L \qquad (24\text{-}11)$$

where Z_p is called the **reflected impedance.** It is approximately equal to the *square* of the turns ratio times the load impedance.

Therefore, we can use transformers as a means of transforming a given load impedance into a different value with minimum loss. There are many circuit applications in which a transformer is used as an **impedance matching** device. One example is in coupling a given load to an electronic amplifier which requires a specific value of load impedance for maximum power output with minimum distortion, as illustrated in the following example.

EXAMPLE 24-3
An audio transformer has 1200 turns in the primary winding. How many turns must be wound in the secondary winding to make a 4-Ω loudspeaker appear as a 1-kΩ load to the amplifier connected to the primary?

SOLUTION

$$a^2 = \frac{Z_p}{Z_L} = \frac{1000}{4} = 250$$

$$\therefore a = 15.81$$

$$N_s = \frac{N_p}{a} = \frac{1200}{15.81} = \textbf{76 turns}$$

EXAMPLE 24-4
The audio output transformer of Figure 24-4 presents the correct load impedance to a transistor amplifier when *either* an 8-Ω load is connected to terminals *A* and *B or* a 16-Ω load is connected to terminals *A* and *C*. What load may be connected to terminals *B* and *C* alone to present the same reflected load?

SOLUTION
We can solve this problem either by assuming convenient values for the number of

FIGURE 24-4
Tapped Audio Output Transformer.

primary turns and reflected impedance or with algebraic symbols. Let N_1 represent the number of secondary turns between terminals A and B, and N_2 the total number of turns between terminals A and C.

Then

$$Z_p = 8\left(\frac{N_p}{N_1}\right)^2 = 16\left(\frac{N_p}{N_2}\right)^2 = Z_x\left(\frac{N_p}{N_2 - N_1}\right)^2$$

from which

$$\frac{\sqrt{8}}{N_1} = \frac{\sqrt{16}}{N_2} = \frac{\sqrt{Z_x}}{N_2 - N_1}$$

from which

$$\sqrt{Z_x} = \sqrt{16} - \sqrt{8}, \text{ and } Z_x = (\sqrt{16} - \sqrt{8})^2 = \mathbf{1.37 \ \Omega}$$

Note that the required impedance is *not* simply the difference between the two given loads.

SELF-CHECK

24-4:1. A transformer has a transformation ratio of 12 and a load impedance of 50 Ω. Determine the reflected impedance for this case.

24-4:2. Determine the proper load impedance for the transformer described in Self-Check Problem 24-3:2.

24-5 LEAKAGE REACTANCE

In establishing the transformation ratio in Section 24-3, we assumed that *all* the flux created by the magnetizing current in the primary linked the secondary winding. However, since there is no magnetic insulator, some of the magnetic lines of force produced by the magnetizing current do not link the secondary, even in the best transformers. These lines of force are called the primary **leakage flux,** as shown in Figure 24-5. The remainder of the total primary flux does link the secondary and is called the **mutual flux.** Similarly, when load current flows in the secondary, some of the secondary magnetic lines of force do not link the primary. These lines of force are called the secondary leakage flux.

FIGURE 24-5
Leakage Flux.

FIGURE 24-6
Effect of Leakage Reactance. R_{CL}, Core Losses; L_m, Primary Self-Inductance Governing Magnetizing Current; R_p, Primary Winding Resistance; X_p, Primary Winding Leakage Reactance; T, Ideal Transformer with Transformation Ratio a; X_s, Secondary Winding Leakage Reactance; R_s, Secondary Winding Resistance; Z_L, Secondary Load Impedance.

The effect of leakage flux is the same as if Φ_m/Φ_p of the total primary turns were developing lines of force that link the secondary winding and Φ_L/Φ_p of the primary turns act as a self-inductance in series with the primary. The reactance of these ineffective primary turns is called the primary **leakage reactance.** As far as the source is concerned, the transformer takes on the appearance shown in Figure 24-6.

In most iron-core transformers, the primary and secondary leakage reactances represent approximately the same percentage of their respective windings; thus, the transformation ratio of the perfect transformer portion of the equivalent circuit of Figure 24-6 is still essentially N_p/N_s. Since this represents a perfect transformer, upon looking into its primary, the source "sees" a reflected load of

$$\mathbf{Z}_{ref} = a^2(R_s + jX_s + \mathbf{Z}_L) \tag{24-12}$$

We can, therefore, reduce the transformer still further to the equivalent circuit of Figure 24-7(a). If the transformer is fully loaded, the magnetizing current in the branch of the

(a) Complete equivalent network for Figure 24-6. (b) Simplified equivalent network.

FIGURE 24-7
Equivalent Networks for an Iron-Core Transformer. $R_e = R_p + a^2R_s$ and $X_e = X_p + a^2X_s$.

equivalent circuit containing R_{CL} and L_m is so small compared with the load current that it creates negligible voltage drop across R_p and X_p. Consequently, when we wish to determine the effects of loading a transformer, we can omit the R_{CL} and L_m branch and simplify the equivalent circuit still further to the simple series circuit of Figure 24-7(b).

SELF-CHECK

24-5:1. A 120-V 60-Hz transformer has 1200 turns on the primary and 400 turns on the secondary. The impedance of the secondary coil is $3 + j2 \ \Omega$. Determine the reflected impedance when a 50-Ω resistive load is used.

24-6 OPEN-CIRCUIT AND SHORT-CIRCUIT TESTS

We can determine experimentally the transformer constants for the equivalent networks of Figure 24-7 from two simple tests. If we leave the secondary terminals open-circuit, $\mathbf{Z}_L$ in Figure 24-7(a) is infinitely large, and current will flow only in the R_{CL} and $j\omega L_m$ branch of the equivalent network. If the transformer under test is designed with a 120-volt primary winding, we probably use the 120-volt winding as the primary winding for the test. However, if the transformer is designed to step a 550-volt source down to 110 volts, to protect the instruments, we use the 110-volt winding as the primary for the test, taking adequate steps to ensure that no one can come in contact with the open-circuit high-voltage winding in the setup of Figure 24-8(a).

We feed the winding we are using as the primary for the test with the exact rated voltage for the winding through a group of instruments consisting of wattmeter, voltmeter, and ammeter, connected as shown in Figure 24-8(a). Since the magnetizing current is only a very small fraction of the full-load primary current, we can assume that the I^2R loss due to magnetizing current alone in the primary winding is negligible. Therefore, the wattmeter shows the **core losses.** These core losses are due to eddy currents in the core and to the hysteresis loop of the core iron. The eddy currents produce a small counterflux in the core just as if we had connected a resistor across the secondary winding. The hysteresis loop of the iron causes the exciting current to lag the applied voltage by slightly less than 90°. The in-phase component of the magnetizing current contributes to the active power reading of the wattmeter.

From the voltmeter and ammeter readings, we can determine the total impedance of the magnetizing-current and core-loss branch of the equivalent network of Figure 24-7(a). As this diagram indicates, an increase in load current will slightly increase the voltage drop across R_p and X_p. This, in turn, will slightly reduce the current in the R_{CL} and L_m branch. Consequently, there is a slight reduction in magnetizing current as the transformer is fully loaded. For practical purposes, we can assume that the core losses remain constant as the load is increased from zero to full rated load.

For the short-circuit test, we short-circuit the low-voltage winding of the transformer, as shown in Figure 24-8(b). In the case of the step-down transformer, the low-voltage winding will be the normal secondary winding. The short circuit is equivalent to making $\mathbf{Z}_L$ zero in Figure 24-7(b). Since $R_e + jX_e$ is fairly small, we must greatly reduce the voltage applied to the high-voltage winding to keep the currents in the transformer windings from exceeding their rated values. With this reduction in applied voltage, we obtain maximum protection for personnel and equipment by using the

FIGURE 24-8
Meter Connections: (a) Open-Circuit Test; (b) Short-Circuit Test.

high-voltage winding as the primary, as shown in Figure 24-8(b). Rheostat R allows us to adjust the applied voltage to a value which will permit rated current to flow in the windings of the transformer. An applied voltage of less than 5% of the rated voltage for the winding being used as the primary is required to produce rated current in the windings. Since reducing the applied voltage reduces the exciting current to less than 5% of its normal value, the core losses are now less than 0.25% of their normal value, and we can assume that the wattmeter is showing only the copper losses from which we can determine the value of R_e.

EXAMPLE 24-5

When the secondary winding of a small step-down transformer is short-circuited, the applied primary voltage is 6 V, the primary current is 4 A, and the wattmeter reading is 12 W. Determine R_e and X_e.

SOLUTION

$$R_e = \frac{P}{I^2} = \frac{12 \text{ W}}{(4 \text{ A})^2} = \textbf{0.75 } \boldsymbol{\Omega}$$

$$Z_e = \frac{E}{I} = \frac{6 \text{ V}}{4 \text{ A}} = 1.5 \text{ } \Omega$$

$$\therefore X_e = \sqrt{Z_e^2 - R_e^2} = \sqrt{1.5^2 - 0.75^2} = \textbf{1.3 ohms (inductive)}$$

SELF-CHECK

24-6:1. A transformer has a transformation ratio of 5. On a short-circuit test, the following meter readings were obtained:

$$\text{wattmeter} = 2.8 \text{ W}$$
$$\text{ammeter} = 2.0 \text{ A}$$
$$\text{voltmeter} = 21 \text{ V}$$

Determine R_e and X_e used in Self-Check problem 24-6:1.

24-6:2. When the same transformer undergoes a 60-Hz open-circuit test the following readings are obtained:

$$\text{wattmeter} = 2.4 \text{ W}$$
$$\text{ammeter} = 0.1 \text{ A}$$
$$\text{voltmeter} = 120 \text{ V}$$

Determine R_{CL} and L_m.

24-7 EFFICIENCY

The core-loss data obtained from the open-circuit test are required in calculating the efficiency of a transformer under load. From Equation (5-6),

$$\eta = \frac{P_{out}}{P_{in}} = \frac{P_L}{P_L + \text{copper loss} + \text{core loss}} \tag{24-13}$$

The core loss remains constant and is equal to the wattmeter reading from the open-circuit test. But the copper loss is proportional to the *square* of current in the equivalent network of Figure 24-7(b). For practical purposes, the current, in turn, is proportional to the *apparent* power into the load. Hence, we can rewrite Equation (24-13),

$$\eta = \frac{S_L \cos \phi}{S_L \cos \phi + I^2 R_e + P_{CL}} \tag{24-14}$$

EXAMPLE 24-6

The transformer tested in Example 24-5 has a transformation ratio of 10 and develops a full-load secondary voltage of 12 V when the load current is 40 A at 0.866 lagging power factor. The open-circuit test indicates core losses of 5 W. Calculate the efficiency of the transformer with this load.

SOLUTION

Step I.

$$S_L \cos \phi = 12 \text{ V} \times 40 \text{ A} \times 0.866 = 416 \text{ W}$$

Step II. In the equivalent network of Figure 24-7(b),

$$I = \frac{I_s}{a} = \frac{40 \text{ A}}{10} = 4 \text{ A}$$

$$\text{Copper loss} = I^2 R_e = (4 \text{ A})^2 \times 0.75 \ \Omega = 12 \text{ W}$$

Step III.

$$\therefore \eta = \frac{416 \text{ W}}{416 \text{ W} + 12 \text{ W} + 5 \text{ W}} = 0.96 = \mathbf{96\%}$$

Since transformers are usually operated at somewhat less than full load, it is customary to design them so that *maximum* efficiency also occurs at somewhat less than full load. As we would expect from earlier consideration of the maximum power transfer theorem,

For maximum efficiency, copper loss = core loss. [†]

EXAMPLE 24-7
What secondary current must be drawn from the transformer of Example 24-6 for maximum efficiency?

SOLUTION
    Step I.

$$I^2 R_e = P_{\text{CL}} = 5 \text{ W}$$

from which

$$I = \sqrt{\frac{P}{R_e}} = \sqrt{\frac{5 \text{ W}}{0.75 \ \Omega}} = 2.58 \text{ A}$$

Step II. The secondary current is

$$I_s = aI = 10 \times 2.58 \text{ A} = \mathbf{25.8 \text{ A}}$$

SELF-CHECK

24-7:1. A transformer has a transformation ratio of 5. On a short-circuit test and an open-circuit test, the following meter readings were obtained:

Short-circuit test	Open-circuit test
wattmeter = 2.8 W	wattmeter = 2.4 W
ammeter = 2.0 A	ammeter = 0.1 A
voltmeter = 21 V	voltmeter = 120 V

[†] See Appendix 2-10.

A 2.0-Ω resistive load is found to draw 11 A. Determine the efficiency of this transformer with this load.

24-7:2. For the transformer described in Self-Check Problem 24-7:1, determine what load current would produce maximum efficiency.

24-8 EFFECT OF LOADING A TRANSFORMER

One important effect of the leakage reactance and resistance of the windings in Figure 24-6 is that the load voltage V_L is not the same as the voltage induced into the secondary since, to satisfy Kirchhoff's voltage law,

$$E_s = I_s R_s + j I_s X_s + V_L$$

Similarly, the voltage induced into the primary by the mutual flux must be slightly less than the total applied voltage V_p. Therefore, as the current drawn by the load is changed, the voltage across the load must also change, even though the applied voltage may remain constant.

We can show the effect of load current on the load voltage by preparing phasor diagrams for the equivalent network of Figure 24-7(b). The current flowing in this equivalent circuit is the primary load current, which is $1/a$ times the actual load current. Likewise, the voltage across $a^2 Z_L$ in Figure 24-7 is a times the actual load voltage. Because the current is common to all components, we use it as a reference phasor in the phasor diagrams of Figure 24-9. The total applied primary voltage will be the phasor sum of three voltages: $a V_L$, which will have an angle with respect to the current depending on the power factor of the load (power-factor angle), $I_p R_e$ (or $I_L/a \times R_e$), which will be in phase with the current, and $I_p X_e$ (or $I_L/a \times X_e$), which will lead the current by 90°. Hence, when we reflect the secondary circuit elements back into the primary circuit to form the simple equivalent circuit of Figure 24-7(b),

$$V_p = a V_L + I_p R_e + I_p X_e \qquad (24\text{-}15)$$

Although we could add these phasor voltages by the customary geometrical construction we learned in Chapter 18 (where all phasors are drawn from a common origin),

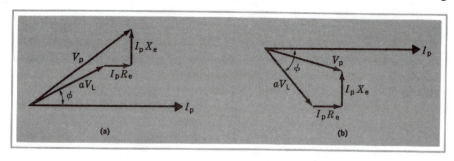

FIGURE 24-9
Phasor Diagrams for a Transformer Under Load: (a) Lagging Power Factor Load; (b) Leading Power Factor Load.

our transformer phasor diagrams will be less confusing if we add each phasor, with its proper direction and magnitude, to the tip of the preceding phasor, as shown in Figure 24-9. Such phasors are called **funicular** (chain-like) phasors.

Since we are not including magnetizing current in our equivalent circuit, when we disconnect the load, I_p becomes zero. Hence, I_pR_e and I_pX_e in Figure 24-9 shrink to zero and aV_L and V_p are identical. But, as shown in Figure 24-9(a), when the load has a lagging power factor so that V_L leads the current, the load voltage will drop appreciably under load. However, with a leading power-factor load, as in Figure 24-9(b), the full-load voltage and no-load voltage are almost the same. The percentage change in secondary voltage from no load to full load with respect to the full-load voltage is called the percentage **voltage regulation** of the transformer.

$$\text{Voltage regulation} = \frac{V_{NL} - V_{FL}}{V_{FL}} \times 100\% \qquad (24\text{-}16)$$

where V_{NL} is the no-load secondary voltage, and V_{FL} is the full-load secondary voltage.

EXAMPLE 24-8

Calculate the percentage voltage regulation for the transformer of Example 24-6.

SOLUTION

$$aV_L = 10 \times 12 \text{ V} = 120 \text{ V}$$

Since the power factor is 0.866 lagging, this voltage will lead the current by arccos 0.866 or 30°, as shown in Figure 24-10.

$$I_pR_e = \frac{40 \text{ A}}{10} \times 0.75 \text{ }\Omega = 3 \text{ V}$$

$$I_pX_e = \frac{40 \text{ A}}{10} \times 1.3 \text{ }\Omega = 5.2 \text{ V}$$

As shown in Figure 24-10, we must express the load voltage in rectangular coordinates in order to carry out the phasor addition.

$$aV_L = 120 \cos 30° + j120 \sin 30° = 104 + j60 \text{ V}$$

$$V_p = (104 + 3) + j(60 + 5.2) \text{ V}$$

$$\therefore V_p = \sqrt{107^2 + 65.2^2} = 125.3 \text{ V}$$

$$\text{No-load } V_L = \frac{V_p}{a} = \frac{125.3 \text{ V}}{10} = 12.53 \text{ V}$$

$$\text{Voltage regulation} = \frac{V_{NL} - V_{FL}}{V_{FL}} \times 100\%$$

FIGURE 24-10
Phasor Diagram for Example 24-8.

$$\therefore \text{voltage regulation} = \frac{12.53 - 12.0}{12.0} \times 100\%$$

$$= \mathbf{4.4\%}$$

SELF-CHECK

24-8:1. A transformer has a transformation ratio of 5 and operates on a 120-V 60-Hz primary voltage. At full load the voltage across a 2-Ω load is 23 V. Determine the voltage regulation of the transformer.

24-9 AUTOTRANSFORMERS

In many applications where we do not need the electrical isolation of separately insulated primary and secondary transformer windings, we can obtain the desired high-efficiency, voltage-changing ability of the transformer at reduced cost by using the **autotransformer** construction shown in Figure 24-11.

Instead of separate primary and secondary windings, the autotransformer of Figure 24-11 has one continuous tapped winding. Primary and secondary windings in a trans-

FIGURE 24-11
Autotransformer.

former must have separate load currents to produce equal and opposite mmfs. Since the lower 100 turns in Figure 24-11 are connected to the load, this becomes the secondary winding. Note that the primary consists of only the top 200 turns. Hence, the primary voltage is *not* the same as the source voltage. Since the source voltage is connected across all 300 turns, from Equation (24-8), the "primary" voltage is 80 V and the secondary voltage is 40 V.

The load in Figure 24-11 draws 15 A at 40 V for an apparent power of 600 VA. If we assume negligible core and copper losses, the apparent power drawn from the source is also 600 VA, from which $I'_p = 600/120 = 5$ A. From Kirchhoff's current law, the secondary current must be 10 A with a direction as shown in Figure 24-11. This agrees with Equation (24-10). The direction of I_s is *opposite* to the direction of I'_p, which agrees with Lenz's law requirement that the two load components of the currents in the transformer windings must produce equal and opposite mmfs.

The apparent power delivered to the load by the secondary winding through transformer action is

$$S_s = V_s \times I_s = 40 \text{ V} \times 10 \text{ A} = 400 \text{ VA}$$

The remaining 200 VA reaches the load by direct conduction due to the primary current forming one third of the load current. Hence, an autotransformer used to reduce the voltage applied to a 600-VA load can be considerably smaller than a 600-VA separate-winding transformer.

We can connect a separate-winding transformer as an autotransformer as long as we arrange the two windings to act as a single continuous winding, as shown in Figure 24-12. We can also use an autotransformer in a step-up mode.

FIGURE 24-12
Separate-Winding Transformer Connected as a Step-up Autotransformer.

EXAMPLE 24-9
We want to raise the available 110-V 60-Hz power-line voltage to provide approximately 120 V to a 10-A load. Our junk box provides a 120-V to 12-V two-winding filament transformer rated at 120 VA. Show that this transformer can accomplish the task of feeding the 1200-VA load when it is connected as in Figure 24-12.

SOLUTION
From the transformer specifications,

$$a = \frac{V_p}{V_s} = \frac{120 \text{ V}}{12 \text{ V}} = 10$$

As connected in Figure 24-12, $V_p = E_{\text{source}}$.

$$\therefore V_s = \frac{110 \text{ V}}{10} = 11 \text{ V}$$

and
$$V_L = 110 \text{ V} + 11 \text{ V} = 121 \text{ V}$$

Since I_L flows through the secondary winding,

$$I'_p = \frac{I_s}{a} = \frac{10 \text{ A}}{10} = 1 \text{ A}$$

Therefore, the current drawn from the source is

$$I_{\text{source}} = 1 \text{ A} + 10 \text{ A} = 11 \text{ A}$$
$$S_L = 121 \text{ V} \times 10 \text{ A} = 1210 \text{ VA}$$

But the apparent power delivered to the load by transformer action is only

$$S_p = 110 \text{ V} \times 1 \text{ A} = \textbf{110 VA} \quad \text{or} \quad S_s = 11 \text{ V} \times 10 \text{ A} = \textbf{110 VA}$$

SELF-CHECK

24-9:1. Referring to the autotransformer shown in Figure 24-11, suppose that the load is changed so that it is connected across the 200-turn coil and draws 9.0 A. If the 120-V source is not changed, determine:
(a) the voltage across the load
(b) the apparent power delivered to the load
(c) the apparent power delivered by the secondary coil.

24-10 AUDIO TRANSFORMERS

The most extensive application of the iron-core transformer is the transforming of an ac source voltage to higher or lower rms magnitudes in power transmission systems and power supplies for electronic devices. For such applications, transformers are designed to operate at a single, fixed power-line frequency. In Examples 24-3 and 24-4, we noted that transformers also serve a useful function in *matching* low-impedance loudspeakers to the output impedances of audio amplifiers. In addition, we can use transformers with center-tapped secondary windings to produce *two* output voltages equal in magnitude but

180° out of phase. By grounding the center tap, such transformers provide a *push-pull* output voltage from a *single-ended* input voltage.

Unlike power transformers, audio transformers must have as uniform as possible a response over a frequency range of at least 20 Hz to 20 kHz. To determine the effect of this range of input frequencies on the performance of an audio transformer, we must consider two additional circuit parameters in the equivalent circuit of Figure 24-13—the internal resistance of the source R_{int}, and, at higher frequencies, the equivalent shunt capacitance C_e of the distributed capacitance between the turns.

FIGURE 24-13
Equivalent Circuit of an Audio Transformer.

As the frequency of the source decreases from a reference frequency of 1 kilohertz, the primary exciting current must increase in order to maintain the required mutual flux. Another way of saying the same thing is that, as the frequency of the source decreases, the reactance of the primary inductance decreases. This means that as the frequency decreases there will be more of the source voltage lost across R_{int} and less applied to the equivalent transformer circuit. We can consider the low-frequency limit of the transformer to be at the frequency at which the inductive reactance of the primary drops to the same value as the internal resistance of the source. Good low-frequency response, therefore, requires sources with low internal resistance and transformers with high primary inductance.

As the frequency increases, the voltage across the equivalent leakage reactance increases, which tends to reduce the output voltage, as shown by the shaded curve of Figure 24-14. But if the transformer is designed so that the equivalent capacitance and

FIGURE 24-14
Frequency Response of a Typical Audio Transformer.

the leakage inductance are resonant at approximately 20 kilohertz, the resonant rise of voltage across this capacitance can be used to advantage to offset the natural roll-off at the high-frequency end of the audio spectrum. Excessive resonant rise would be undesirable, but this rise can be controlled in the transformer design by choosing the proper ratio between the reactance and resistance. At frequencies above the resonant frequency of X_e and C_e, the output voltage is very small. Since *pitch* response of the human ear relates to *doubling* the frequency of a sound wave (one octave), we use a logarithmic frequency scale in the graph of Figure 24-14.

In earlier network analysis chapters, we learned how to analyze a network as a whole using such techniques as mesh and nodal equations. Equivalent-circuit theorems (Thévenin's and Norton's) showed us how to split networks into voltage and current sources with two output terminals, and loads with two input terminals. Transformers belong to a third class of networks connected between two-terminal sources and two-terminal loads. Such coupling networks (which are the topic of the next chapter) are four-terminal networks with two input and two output terminals.*

PROBLEMS

24-2-1. A small filament transformer has a laminated sheet steel core with an average path length of 20 cm and a cross-sectional area of 5 cm². How many turns must there be on a primary winding to be connected to a 120-V 60-Hz source if the flux density must not exceed 1.1 T?

24-2-2. A 2300/230-V 15-kVA 60-Hz transformer is designed to have an induced voltage of 2.2 V per turn. Determine the number of primary turns and the full-load primary current.

24-2-3. Assuming the magnetizing current to be a pure sine wave, calculate the rms value of the primary current in the transformer of Problem 24-2-1 when the secondary winding is open-circuit. (Use the *BH* curves of Figure 13-21.)

24-2-4. The primary of the transformer of Problem 24-2-1 is connected to a 110-V 25-Hz source. What is the rms value of the exciting current?

24-2-5. What cross-sectional area must the core of the transformer of Problem 24-2-2 have if the flux density is not to exceed 1.1 T?

24-2-6. With the secondary open-circuit, the primary current in the transformer of Problem 24-2-2 is measured as 400 mA. What is the average path length of the magnetic circuit of the transformer (approximately)?

24-3-7. What is the transformation ratio of a transformer having 40 turns on the secondary winding and 680 turns on the primary?

24-3-8. What is the transformation ratio of a transformer which has an open-circuit secondary voltage of 300 V rms when the primary is connected to a 120-V 60-Hz source?

24-3-9. What rms voltage must be applied to the primary of the transformer in Problem 24-3-7 to develop an open-circuit secondary voltage of 6.3 V?

24-3-10. What is the load component of the primary current when a 2-kΩ resistor is connected across the secondary winding of the transformer in Problem 24-3-8?

24-3-11. The secondary of the transformer in Problem 24-3-8 has 1400 turns. How many primary turns are required?

*A popular "buzz word" for a two-terminal network is a **one-port** network. Transformers, then, become **two-port** networks.

24-3-12. The primary of a certain transformer has 2.7 turns per volt of applied voltage. How many turns must there be on a 5-V secondary winding?

24-3-13. How many turns are there on a 7.5-V secondary winding on the transformer of Problem 24-2-1?

24-3-14. When an 0.8-Ω load is connected to the 24-V secondary of a step-down transformer, the primary current is measured as 8.0 A. What is the primary voltage of the transformer?

24-4-15. An audio output transformer has 1500 primary turns and 40 secondary turns. What is the reflected value of a 4-Ω load connected to its secondary?

24-4-16. An audio transformer is listed as 7-kΩ primary reflected impedance with a 4-Ω load at 1000 Hz. This transformer is to be used with an amplifier requiring a 4500-Ω load. What value of load must be connected to the transformer secondary?

24-4-17. An audio output transformer has 1200 primary turns and two secondary windings of 50 turns each. If the circuit feeding the transformer must see a reflected load of 5 kΩ, what load must be connected to the secondary windings when they are connected
(a) in series?
(b) in parallel?

24-4-18. An audio transformer has a 6-kΩ primary winding and an 8-Ω secondary with a tap at 4 Ω with respect to a common secondary terminal. Through error, a 4-Ω load is connected from the 4-Ω tap to the 8-Ω terminal. What is the reflected impedance?

24-6-19. A short-circuit test on a 2300/208-V 500-kVA 60-Hz distribution transformer yields the following data: $V_p = 95$ V, $I_p = 218$ A, and $P_{in} = 8.2$ kW. Calculate R_e and X_e for the equivalent circuit of Figure 24-7(b).

24-6-20. The open-circuit test on the transformer of Problem 24-6-19, with the high-voltage winding open and the low-voltage winding used as the primary, yields the following data: $V = 208$ V, $I = 84$ A, and $P = 1.8$ kW. Calculate the magnitude and phase angle of the exciting current when the transformer is used with the high-voltage winding as the primary.

Draw detailed and accurately scaled phasor diagrams with the remaining problems.

24-7-21. Calculate the efficiency and power output for maximum efficiency conditions with a 90% lagging power-factor load connected to the transformer of Problems 24-6-19 and 24-6-20.

24-7-22. Given the following data on a 550/220-V 50-kVA 60-Hz transformer: $R_p = 0.03$ Ω, $X_P = 0.05$ Ω, $R_s = 0.005$ Ω, and $X_s = 0.01$ Ω. If the power input to this transformer is 200 W when the secondary is open-circuit, calculate the full-load efficiency with a unity power-factor load.

24-8-23. Calculate the percentage voltage regulation for the transformer in Problem 24-6-19 with a 70.7% lagging power-factor load.

24-8-24. Calculate the percentage voltage regulation for the transformer in Problem 24-6-19 with an 86.6% leading power-factor load.

24-8-25. Calculate the percentage voltage regulation of the transformer in Problem 24-7-22 with a 50% lagging power-factor load.

REVIEW QUESTIONS

24-1. Why is mutual induction more significant in ac circuits than in dc circuits?

24-2. What is a **transformer?**

24-3. What is the purpose of an iron core in power and audio transformers?

24-4. What problems would be encountered in using iron cores in radio-frequency transformers?

24-5. What is meant by the **general transformer equation?** What is its application?

24-6. What is the significance of the numerical value 1.11 in Equation (24-2)? [Compare Equation (24-2) with Equation (16-21).]

24-7. What is meant by **exciting current?**

24-8. What factors determine the magnitude of the exciting current?

24-9. The exciting current in a certain transformer lags the applied voltage by 87°. Account for this angle.

24-10. Account for the results obtained in Problem 24-2-4 and relate these results to a practical transformer.

24-11. What would be the effect of operating a 110-V 25-Hz transformer on a 120-V 60-Hz supply?

24-12. What is the relationship between the turns ratio of a transformer and its transformation ratio?

24-13. What is the relationship between the load current ratio and the turns ratio of a transformer?

24-14. By combining Equations (24-9) and (24-11), determine the power input to power output ratio of a transformer.

24-15. What is meant by **reflected impedance?**

24-16. Why is the impedance ratio of an audio transformer equal to the *square* of the turns ratio?

24-17. What is the significance of impedance ratio in practical transformers?

24-18. Complete the individual steps in reducing the initial equation in Example 24-4 to its final form.

24-19. What is meant by **mutual flux** and **leakage flux?**

24-20. Define **leakage reactance.**

24-21. Why is the total core loss of a transformer essentially independent of load current?

24-22. Why can we assume that the wattmeter reading on an open-circuit transformer test is essentially only the core loss?

24-23. Why is the core loss on a short-circuit transformer test less than 1/400 of its normal value?

24-24. Explain why we can state that $R_e = R_p + a^2 R_s$ and $X_e = X_p + a^2 X_s$ in Figure 24-7(b).

24-25. Draw a phasor diagram showing the relationship between primary terminal voltage and secondary terminal voltage for a unity power-factor load.

24-26. What factor is primarily responsible for the difference between the no-load and full-load secondary voltages of a power transformer?

24-27. Why does a leading power-factor load give better voltage regulation than an equivalent lagging power-factor load?

24-28. Draw a circuit diagram for an autotransformer feeding a 208-V load from a 120-V source. Explain its operation with a numerical example.

24-29. By using the numerical value of load you selected in Question 24-28, show why an autotransformer can be appreciably smaller than the equivalent two-winding transformer.

24-30. What factors limit the lowest frequency that an audio transformer can handle satisfactorily?

24-31. What factors limit the highest frequency that an audio transformer can handle satisfactorily?

***24-32.** Assuming a constant power factor of 0.866 lagging for the load connected to the transformer of Examples 24-5 and 24-6, and constant primary and secondary voltages, program a computer to prepare a table of percent efficiency (four significant digits) for load currents from 4 to 40 A in 2-A steps. In addition, display these data in graph form.

25

COUPLED CIRCUITS

25-1 DETERMINING COUPLING NETWORK PARAMETERS

There is an old saying about there being more than one way of skinning a cat. We have certainly discovered that this is true in solving electrical networks and in reducing them to simple equivalent circuits. We have already found several ways of solving the impedance network of Figure 25-1(a). Since this network is not too elaborate, we can view the network as a whole and write a set of Kirchhoff's voltage-law (loop) equations or Kirchhoff's current-law (nodal) equations from which we can solve for I_1 and I_2.

Thévenin's theorem provides us with a more powerful tool for working on this network. Thévenin's theorem views all networks as consisting of only two components, a two-terminal load and a two-terminal constant-voltage source. This technique is very useful for two reasons: (1) particularly in dealing with electric power systems, our basic notion of a simple electric circuit is that of connecting a load across a voltage source; and (2) Thévenin's theorem provides us with a laboratory method for determining the parameters of the Thévenin-equivalent source by laboratory measurements, even though we have no idea what the sealed two-terminal "black box" actually contains.

With the increasing significance of the signal-transmitting circuits of electronic communications and control systems, we need still another way of viewing the circuit of Figure 25-1(a). Suppose that we replace E_2 by a load impedance, as in Figure 25-1(b). The circuit now becomes a T-network, coupling the signal source E_1 to the load Z_L. Our experience with such networks tells us that I_2 will turn out to be a negative quantity if

FIGURE 25-1
Coupling Networks: (a) Between
Voltage Sources; (b) Between a
Source and a Load; (c) Of
Unknown Internal Composition.

we do not change its direction. However, to standardize our conventions, we think of the input current I_1 and the output current I_2 of the generalized coupling network of Figure 25-1(c) as having a direction *into* the network at both upper or "live" terminals of the network.

If we knew that the network of Figure 25-1(c) actually had the form of Figure 25-1(b), we could solve for I_2 (and I_1 if needed) by simple series-parallel impedance techniques. But if the coupling network is sealed up in a four-terminal "black box," as in Figure 25-1(c), we need some way of stating the network parameters in a form that we can obtain by the "black-box" measurement techniques we first used in Chapter 8. The simplest generalized coupling network will have one component or parameter which is common to both the input current I_1 and the output current I_2, one component through which only I_1 flows, and one through which only I_2 flows. We can, therefore, represent the coupling network, whatever its internal composition may be, by a simple T-network.

To determine the parameters of an unknown network, we connect the network in the test circuit shown in Figure 25-2, together with voltmeters whose internal resistances are high enough that we can neglect any voltmeter current. To find out what we "see" when looking back into the output terminals, we require E_2 rather than the actual load impedance.

FIGURE 25-2
Determining the Open-Circuit Impedance Parameters of a Coupling Network.

25-2 OPEN-CIRCUIT IMPEDANCE PARAMETERS

To define the composition of a four-terminal, two-port network such as shown in Figure 25-1(c), we require four parameters. The test circuit of Figure 25-2 will give us one fairly common set of parameters called the **open-circuit impedance parameters**[†] of the network.

We start by opening the right-hand switch in Figure 25-2 so that $I_2 = 0$. Killing I_2 removes the possibility of the input to the network containing any component dependent on the output circuit. When we close the left-hand switch, the readings from voltmeter V_1 and ammeter I_1 give us a V/I ratio which represents only the open-circuit input *impedance* of the network. (Volts ÷ amperes = ohms.)

We now open the left-hand switch and close the right-hand switch. Since I_1 becomes zero, we won't be able to "see" the input impedance because there will be no voltage drop across it. However, it shouldn't surprise us to find a reading on voltmeter V_1 as a part of I_2 is coupled back to the input side of the network. We can create a parameter to express the amount of output-to-input coupling by recording the ratio of voltmeter V_1 reading to ammeter I_2 reading. Again we express a V/I ratio in ohms. Since it is the ratio between the voltage created at the *input* terminals (1,1) and the current flowing into the *output* terminals (2,2) to create it, we call this second parameter the **reverse-transfer impedance** of the coupling network. The same "measurement" technique provides similar output parameters. The complete set of open-circuit impedance parameters[‡] consists of

$$\mathbf{z}_{11} = \frac{\mathbf{E}_1}{\mathbf{I}_1} \qquad (\mathbf{I}_2 = 0) \quad \text{is the **open-circuit input impedance**} \qquad (25\text{-}1)$$

[†] "Open-circuit" because we prevent the output circuit from affecting the measurement of input impedance by *opening* the right-hand switch in Figure 25-2.

[‡] We use lowercase letter symbols for impedance parameters because they are equivalent-circuit parameters, rather than actual circuit elements. Note carefully the subscript convention we observe.

$$\mathbf{z}_{12} = \frac{\mathbf{V}_1}{\mathbf{I}_2} \qquad (\mathbf{I}_1 = 0) \quad \text{is the \textbf{reverse-transfer impedance}} \qquad (25\text{-}2)$$

$$\mathbf{z}_{21} = \frac{\mathbf{V}_2}{\mathbf{I}_1} \qquad (\mathbf{I}_2 = 0) \quad \text{is the \textbf{forward-transfer impedance}} \qquad (25\text{-}3)$$

$$\mathbf{z}_{22} = \frac{\mathbf{E}_2}{\mathbf{I}_2} \qquad (\mathbf{I}_1 = 0) \quad \text{is the \textbf{open-circuit output impedance}} \qquad (25\text{-}4)$$

Having determined the four **z**-parameters, we now close *both* switches in Figure 25-2 so that the network actually does couple the output and input circuits. Consequently, *both* $\mathbf{z}_{11}$ and $\mathbf{z}_{12}$ are now present in the input circuit and $\mathbf{V}_1$ consists of *two* voltages in series: the *IZ* drop due to $\mathbf{I}_1$ flowing in $\mathbf{z}_{11}$, plus a coupled voltage representing $\mathbf{I}_2$ flowing through the reverse-transfer impedance $\mathbf{z}_{12}$. We can write a Kirchhoff's voltage-law equation for the input terminals (1,1) of the network as

$$\mathbf{z}_{11}\mathbf{I}_1 + \mathbf{z}_{12}\mathbf{I}_2 = \mathbf{E}_1 \qquad (25\text{-}5)$$

Similarly, for the output terminals (2,2),

$$\mathbf{z}_{21}\mathbf{I}_1 + \mathbf{z}_{22}\mathbf{I}_2 = \mathbf{E}_2 \qquad (25\text{-}6)$$

The next step is to determine the equivalent circuit represented by the two Kirchhoff's voltage-law Equations (25-5) and (25-6). Using our "black-box" reasoning, we can investigate the two voltages we "see" looking into the input terminals of the coupling network. Obviously, the sum of two voltages represents a *series* circuit. With the left-hand switch open and the right-hand switch closed in Figure 25-2, we discovered a reading on voltmeter V_1 due to output current $\mathbf{I}_2$. Hence, we can argue that we are looking into a Thévenin-equivalent voltage source which is *dependent* on the current in some other part of the circuit. The open-circuit voltage of this current-controlled voltage source is $\mathbf{z}_{12}\mathbf{I}_2$. From the way we "measured" it, $\mathbf{z}_{11}$ appears to be the internal impedance of the dependent source. Looking into the output terminals, we see a similar Thévenin-equivalent dependent voltage source. Hence, Figure 25-3 becomes the equivalent circuit for the open-circuit impedance parameters of a four-terminal coupling network.

We can now perform the same "measurements" on paper with the dotted T-network in position in Figure 25-2. With the left-hand switch closed and the right-hand switch

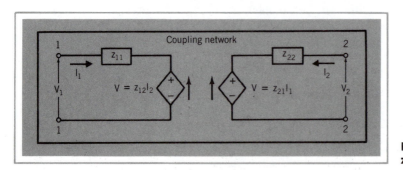

FIGURE 25-3
z-Parameter Equivalent Circuit.

open,

$$I_1 = \frac{E_1}{Z_p + Z_m}$$

from which
$$Z_p + Z_m = \frac{E_1}{I_1} = z_{11}$$
(25-7)

While the left-hand switch is still closed and the right-hand switch open, voltmeter V_2 reads

$$V_2 = Z_m I_1$$

Hence,
$$Z_m = \frac{V_2}{I_1} = z_{21}$$
(25-8)

Similarly, with the left-hand switch in Figure 25-2 open and the right-hand switch closed,

$$Z_s + Z_m = \frac{E_2}{I_2} = z_{22}$$
(25-9)

and
$$Z_m = \frac{V_1}{I_2} = z_{12}$$
(25-10)

Hence, the z-parameters for a T-network are

$$z_{11} = Z_p + Z_m \quad (25\text{-}7) \qquad z_{22} = Z_s + Z_m \quad (25\text{-}9)$$

$$z_{21} = Z_m \quad (25\text{-}8) \qquad z_{12} = Z_m \quad (25\text{-}10)$$

We can also rearrange Equations (25-7) to (25-10) to express the three T-network components in terms of z-parameters.

$$Z_m = z_{12} = z_{21}$$
(25-11)

$$Z_p = z_{11} - z_{12}$$
(25-12)

and
$$Z_s = z_{22} - z_{21}$$
(25-13)

If we know the z-parameters for a passive coupling network, we can readily establish the equivalent T-network, or vice versa. To use a T-network, note that both forward- and reverse-transfer impedances must be the same. This is characteristic of any *passive* network containing only resistance and reactance elements.[†] Consequently, we call Z_m the **mutual impedance** of the coupling network. In *active* networks containing transistors, z_{12} and z_{21} must be different for amplification to take place. The z-parameter

---

[†] The reciprocity theorem of Section 9-8 depends on this equal forward- and reverse-transfer impedance characteristic of coupling networks.

equivalent circuit of Figure 25-3 can represent both passive and active coupling networks.

EXAMPLE 25-1
Check Equations (25-5) and (25-6) by writing Kirchhoff's voltage-law equations for the T-network in Figure 25-2.

SOLUTION

$$\mathbf{I_1}\mathbf{Z_p} + (\mathbf{I_1} + \mathbf{I_2})\mathbf{Z_m} = (\mathbf{Z_p} + \mathbf{Z_m})\mathbf{I_1} + \mathbf{Z_m}\mathbf{I_2} = \mathbf{E_1}$$

$$\therefore (\mathbf{z_{11}} - \mathbf{z_{12}} + \mathbf{z_{12}})\mathbf{I_1} + \mathbf{z_{12}}\mathbf{I_2} = \mathbf{E_1} \qquad (25\text{-}5)$$

$$\mathbf{I_2}\mathbf{Z_s} + (\mathbf{I_1} + \mathbf{I_2})\mathbf{Z_m} = (\mathbf{Z_s} + \mathbf{Z_m})\mathbf{I_2} + \mathbf{Z_m}\mathbf{I_{1'}} = \mathbf{E_2}$$

$$\therefore (\mathbf{z_{22}} - \mathbf{z_{21}} + \mathbf{z_{21}})\mathbf{I_2} + \mathbf{z_{21}}\mathbf{I_1} = \mathbf{E_2} \qquad (25\text{-}6)$$

EXAMPLE 25-2
Write Equations (25-5) and (25-6) as they apply to the network in Figure 25-4.

SOLUTION
In this example, $\mathbf{Z_L}$ becomes the internal impedance of a source producing zero voltage. As Figure 25-4 shows, this internal impedance acts in series with $\mathbf{Z_s}$. There is no change in the equation for $\mathbf{E_1}$:

$$\mathbf{z_{11}}\mathbf{I_1} + \mathbf{z_{12}}\mathbf{I_2} = \mathbf{E_1} \qquad (25\text{-}5)$$

FIGURE 25-4
T-Coupling Network.

But the second equation is

$$(\mathbf{Z_L} + \mathbf{Z_s} + \mathbf{Z_m})\mathbf{I_2} + \mathbf{Z_m}\mathbf{I_1} = 0$$

from which

$$\mathbf{z_{21}}\mathbf{I_1} + (\mathbf{z_{22}} + \mathbf{Z_L})\mathbf{I_2} = 0 \qquad (25\text{-}14)$$

Equations (25-5) and (25-14) provide us with a means of developing a simpler expression for the total input impedance of the T-coupling network of Figure 25-4 (and any generalized coupling network) than an equation for $\mathbf{Z_{in}}$ in terms of series and parallel impedances. From Equation (25-14),

$$I_2 = \frac{-z_{21} I_1}{z_{22} + Z_L}$$ (25-15)

Substituting in Equation (25-5),

$$z_{11} I_1 - \frac{z_{12} z_{21} I_1}{z_{22} + Z_L} = E_1$$

Hence,

$$Z_{in} = \frac{E_1}{I_1} = z_{11} - \frac{z_{12} z_{21}}{z_{22} + Z_L}$$

But from Equation (25-11),

$$z_{12} z_{21} = Z_m^2$$

Therefore,

$$Z_{in} = z_{11} - \frac{Z_m^2}{z_{22} + Z_L}$$ (25-16)

The second term in Equation (25-16) is a significant parameter in coupled-circuit techniques; hence, we give it a special name:

$$\textbf{Coupled impedance} = \frac{Z_m^2}{z_{22} + Z_L}$$ (25-17)

Equation (25-16) represents an equivalent circuit consisting of two impedances connected in *series*. The negative sign with the coupled impedance means that the coupled impedance tends to *reduce* the input impedance of the network so that energy may be coupled to the load. Thus, our coupled-circuit technique gives us a *series* **coupled impedance** in contrast to the *parallel* **reflected impedance** which we developed with the transformation-ratio technique for dealing with iron-core transformers in Chapter 24. However, as far as the source is concerned, both techniques produce the same results. In Chapter 24, disconnecting the load made the reflected impedance in *parallel* with the primary impedance infinitely large so that Z_{in} was simply the open-circuit input impedance of the transformer. With **z**-parameters, disconnecting the load so that Z_L in Equation (25-17) goes to infinity results in the coupled impedance in *series* with the open-circuit input impedance z_{11} becoming zero. Again, Z_{in} is simply the open-circuit input impedance.

⌐ EXAMPLE 25-3
What is the input impedance of the H-attenuator in Figure 25-5 when a 600-Ω resistive load is connected to its output terminals? (Check the results by series-parallel resistance calculations.)

FIGURE 25-5
H-Attenuator.

SOLUTION

We may regard the H-attenuator as a balanced form of a T-network. The equivalent T-network has $\mathbf{Z}_p = \mathbf{Z}_s = 200\ \Omega$. Hence,

$$z_{11} = z_{22} = 200 + 800 = 1000\ \Omega$$

and
$$\mathbf{Z}_m = 800\ \Omega \qquad (\text{all } \underline{/0°})$$

$$\mathbf{Z}_{in} = z_{11} - \frac{\mathbf{Z}_m^2}{z_{22} + \mathbf{Z}_L} = 1000 - \frac{640\,000}{1000 + 600}$$

$$\therefore Z_{in} = \mathbf{600\ \Omega}$$

EXAMPLE 25-4

Calculate the power into a 75-Ω load connected to the high-pass filter circuit in Figure 25-6.

SOLUTION

$$z_{11} = z_{22} = -j\,\frac{1}{\omega C} + j\omega L$$

$$= -j\,\frac{1}{2 \times \pi \times 1\ \text{kHz} \times 2\ \mu F} + j(2 \times \pi \times 1\ \text{kHz} \times 6\ \text{mH})$$

$$= -j79.58 + j37.70 = -j41.88\ \Omega$$

$$\mathbf{Z}_m^2 = (+j37.7)^2 = -1421^{\dagger}$$

Hence, $$\mathbf{Z}_{in} = z_{11} - \frac{\mathbf{Z}_m^2}{z_{22} + \mathbf{Z}_L} \qquad\qquad (25\text{-}16)$$

$$= -j41.88 + \frac{1421}{75 - j41.88} = 14.44 - j33.82$$

and $$Z_{in} = 36.77\ \Omega$$

† Remember that $(+j)^2 = (-j)^2 = -1$.

FIGURE 25-6
High-Pass Filter Network.

(We do not need to record the angle of $\mathbf{Z}_{in}$ to determine the magnitude of the load current.)

$$I_{in} = \frac{E}{Z_{in}} = \frac{10 \text{ V}}{36.77 \text{ } \Omega} = 272 \text{ mA}$$

$$\mathbf{I}_2 = \frac{-\mathbf{z}_{21}\mathbf{I}_1}{\mathbf{z}_{22} + \mathbf{Z}_L} \quad (25\text{-}15)$$

$$\therefore I_2 = \frac{37.7}{85.9} \times 272 \text{ mA} = 119.4 \text{ mA}$$

Since the load contains only resistance,

$$P_L = I_2^2 R_L = (119.4 \text{ mA})^2 \times 75 \text{ } \Omega = \mathbf{1.07 \text{ W}}$$

SELF-CHECK

25-2:1. Given the T-coupling circuit in Figure 25-4 and the following values:

$$\mathbf{Z}_p = -j15 \text{ } \Omega$$
$$\mathbf{Z}_m = 100 + j20 \text{ } \Omega$$
$$\mathbf{Z}_s = -j12 \text{ } \Omega$$
$$\mathbf{Z}_L = 50 \text{ } \Omega$$

(a) Determine the coupled impedance.
(b) Determine the input impedance.

25-3 SHORT-CIRCUIT ADMITTANCE PARAMETERS

In Figure 25-7, we want to represent the generalized coupling network by the π-network shown in dotted lines. If we follow the rules associated with Equations (25-1) to (25-4) carefully, we can still derive equations relating the three π-network impedances to **z**-parameters for the network.[†] But by now we should be sufficiently familiar with the

[†] See Review Question 25-15.

FIGURE 25-7
Determining the Short-Circuit Admittance Parameters of a Coupling Network.

dual form of network equations to suspect that it might be simpler to work with *admittances* when we encounter a coupling network in the form of a π-network (which is a dual for a T-network). Although the resulting **short-circuit admittance parameters** are not often used in modern equivalent-circuit techniques, we shall derive the **y**-parameters of a generalized four-terminal coupling network as an introduction to the hybrid parameters of Section 25-4.

In the test circuit of Figure 25-7, to ensure that the source $\mathbf{E}_2$ does not interfere with the input meter readings when we are "measuring" the input admittance, we place the right-hand switch in the position which short-circuits the output terminals (2,2) through ammeter I_2. Similarly, when we apply $\mathbf{E}_2$ to the output terminals, the left-hand switch must short-circuit the input terminals (1,1) through ammeter I_1.

Starting with $\mathbf{E}_1$ applied to the input terminals, we write the input meter readings as an I/V ratio. Hence, we are stating the input *admittance* of the network with the output terminals short-circuit. Again we should not be surprised to find a reading on ammeter I_2 short-circuiting the output terminals as a portion of the input current is coupled through to the output. Looking back into the short-circuited output terminals, we see a Norton-equivalent source containing a *dependent* current source controlled by the voltage applied to the input terminals. To find the constant that relates $\mathbf{I}_2$ to $\mathbf{E}_1$, we simply divide ammeter I_2 reading by voltmeter V_1 reading. And since amperes ÷ volts = siemens, we call this constant the **forward-transfer admittance** of the network. Reversing the position of the switches allows us to "measure" a **short-circuit output admittance** and a **reverse-transfer admittance.** Hence, the four admittance parameters of any four-terminal, two-port network are

$$\mathbf{y}_{11} = \frac{\mathbf{I}_1}{\mathbf{E}_1} \quad (\mathbf{E}_2 = 0) \quad \text{is the } \textbf{short-circuit input admittance} \quad (25\text{-}18)$$

$$\mathbf{y}_{12} = \frac{\mathbf{I}_1}{\mathbf{E}_2} \quad (\mathbf{E}_1 = 0) \quad \text{is the } \textbf{reverse-transfer admittance} \quad (25\text{-}19)$$

$$y_{21} = \frac{I_2}{E_1} \quad (E_2 = 0) \quad \text{is the \textbf{forward-transfer admittance}} \qquad (25\text{-}20)$$

$$y_{22} = \frac{I_2}{E_2} \quad (E_1 = 0) \quad \text{is the \textbf{short-circuit output admittance}} \qquad (25\text{-}21)$$

Since a Norton-equivalent current source has to be in *parallel* with the input admittance, when we look into the input terminals of the network, we see *two* paths for I_1 to follow when it enters the network. Writing the Kirchhoff's current-law equation for I_1 entering the input terminals (1,1),

$$y_{11}\,E_1 + y_{12}\,E_2 = I_1 \qquad (25\text{-}22)$$

Similarly, for the output terminals (2,2),

$$y_{21}\,E_1 + y_{22}\,E_2 = I_2 \qquad (25\text{-}23)$$

Drawing the equivalent circuit which satisfies Equations (25-22) and (25-23) gives us the equivalent circuit for the short-circuit admittance parameters of a four-terminal coupling network, as shown in Figure 25-8.

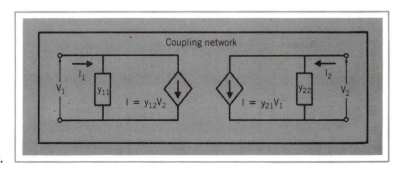

FIGURE 25-8
y-Parameter Equivalent Circuit.

If we now assume that the coupling network consists of the π-network shown in dotted lines in Figure 25-7, we can repeat the "measurements" of Equations (25-18) to (25-21) using values from the π-network. Starting with ammeter I_2 placing a short circuit across the output terminals,

$$I_1 = E_1(Y_p + Y_m)$$

$$\therefore Y_p + Y_m = \frac{I_1}{E_1} = y_{11} \qquad (25\text{-}24)$$

Similarly, $\qquad\qquad Y_s + Y_m = \dfrac{I_2}{E_2} = y_{22} \qquad (25\text{-}25)$

With the left-hand switch in the short-circuit position, there can be no voltage drop across $\mathbf{Y}_p$—hence, no current through $\mathbf{Y}_p$. All the current through $\mathbf{Y}_m$ must flow through ammeter I_1. As a result, $\mathbf{I}_1$ will have the same *magnitude* as the current through $\mathbf{Y}_m$. But the current through $\mathbf{Y}_m$ is caused by $\mathbf{E}_2$ and, as we can see from Figure 25-7, has the *opposite* direction to $\mathbf{I}_1$. We would find that ammeter I_1 reads *backwards* in the short-circuit position of the switch. Hence,

$$\mathbf{I}_1 = -\mathbf{I}_{Y_m} = -\mathbf{Y}_m \mathbf{E}_2$$

Therefore,

$$\mathbf{Y}_m = \frac{-\mathbf{I}_1}{\mathbf{E}_2} = -\mathbf{y}_{12} \tag{25-26}$$

Similarly

$$\mathbf{Y}_m = \frac{-\mathbf{I}_2}{\mathbf{E}_1} = -\mathbf{y}_{21} \tag{25-27}$$

Hence, the y-parameters for a π-network are

$$\mathbf{y}_{11} = \mathbf{Y}_p + \mathbf{Y}_m \quad (25\text{-}24) \qquad \mathbf{y}_{22} = \mathbf{Y}_s + \mathbf{Y}_m \quad (25\text{-}25)$$

$$\mathbf{y}_{12} = -\mathbf{Y}_m \quad (25\text{-}26) \qquad \mathbf{y}_{21} = -\mathbf{Y}_m \quad (25\text{-}27)$$

Rearranging Equations (25-24) to (25-27), we can now state the equivalent π-network in terms of the y-parameters of the coupling network as

$$\mathbf{Y}_m = -\mathbf{y}_{12} = -\mathbf{y}_{21} \tag{25-28}$$

$$\mathbf{Y}_p = \mathbf{y}_{11} + \mathbf{y}_{12} \tag{25-29}$$

and

$$\mathbf{Y}_s = \mathbf{y}_{22} + \mathbf{y}_{21} \tag{25-30}$$

EXAMPLE 25-5

Determine the short-circuit admittance parameters for the low-pass filter circuit in Figure 25-9.

SOLUTION

$$\mathbf{Y}_p = \mathbf{Y}_s = +j\omega C = +j(2 \times \pi \times 1 \text{ kHz} \times 38 \ \mu\text{F}) = +j238.8 \text{ mS}$$

$$\mathbf{Y}_m = -j\frac{1}{\omega L} = -j\frac{1}{2 \times \pi \times 1 \text{ kHz} \times 1.4 \text{ mH}} = -j113.7 \text{ mS}$$

FIGURE 25-9
Low-Pass Filter Network.

The y-parameters are

$$\mathbf{y}_{11} = \mathbf{y}_{22} = \mathbf{Y}_p + \mathbf{Y}_m = +j238.8 \text{ mS} - j113.7 \text{ mS} = \mathbf{+j125.1 \text{ mS}}$$

$$\mathbf{y}_{12} = \mathbf{y}_{21} = -\mathbf{Y}_m = -(-j113.7 \text{ mS}) = \mathbf{+j113.7 \text{ mS}}$$

SELF-CHECK

25-3:1. The manufacturer of a sealed low-pass filter module specifies its 20-kHz admittance parameters as:

$$\mathbf{y}_{11} = +j100 \text{ mS}$$
$$\mathbf{y}_{12} = +j150 \text{ mS}$$
$$\mathbf{y}_{22} = +j90 \ \text{ mS}$$

(a) By modeling the filter as the coupling network in Figure 25-7, determine: $\mathbf{Y}_p$, $\mathbf{Y}_m$, and $\mathbf{Y}_s$.

(b) Assuming the module has internal circuitry similar to that shown in Figure 25-9, determine: C_p, C_s and L_m.

25.4 HYBRID PARAMETERS

Instead of being made up of all impedances or all admittances, as the name indicates, hybrid parameters use a mixture of units. In analyzing circuits containing active devices such as transistors, we find it more convenient to think of the input terminals of a four-terminal coupling network as a Thévenin-equivalent voltage source and the output terminals as a Norton-equivalent current source. We can still use the same lab measurement technique to determine the four hybrid parameters for any four-terminal coupling network.

To find the open-circuit voltage of the Thévenin-equivalent source at input terminals (1,1) in Figure 25-10(a), we feed V_2 into the output terminals (2,2). In this case, we consider the Thévenin-equivalent source to be a *voltage*-controlled voltage source. Hence the parameter which represents the fraction of the output voltage appearing at the input terminals is $\mathbf{h}_{12} = \mathbf{V}_1/\mathbf{V}_2$. Since volts ÷ volts = a ratio without units, $\mathbf{h}_{12}$ is the **reverse-voltage *ratio*** parameter.

Since we are treating the dependent source as a *voltage*-controlled voltage source, to kill V_2 while we measure the input voltage and current, we short-circuit the output terminals, as in Figure 25-10(b). By recording the ratio $\mathbf{V}_1/\mathbf{I}_1$, the parameter $\mathbf{h}_{11}$ is expressed in ohms and represents the **short-circuit input impedance** of the network. Since $\mathbf{h}_{12}\mathbf{V}_2$ is a *voltage* source, the equivalent input circuit for the coupling network shows the dependent voltage source and input impedance in series, as in Figure 25-10(c).

To determine the short-circuit current of the Norton-equivalent source at the output terminals (2,2) in Figure 25-11(a), we feed I_1 into the input terminals and short-circuit the output terminals through ammeter I_2. Because the network is linear, I_2 will be a constant fraction of the input current I_1. In this case, amperes ÷ amperes = a ratio without units and $\mathbf{h}_{21}$ becomes the **forward-current ratio** parameter.

FIGURE 25-10
Finding the Thévenin-Equivalent Input Circuit of a Four-Terminal Network: (a) Open-Circuit Reverse Voltage; (b) Internal Input Resistance; (c) Network Input Parameters.

Since the output impedance of a Norton-equivalent source is in *parallel* with the current source, it is more convenient to think in terms of output admittance. And since we are treating this dependent source as a *current*-controlled current source, to kill I_1 while we measure I_2 and V_2 in Figure 25-11(b), we leave the input terminals of the network open-circuit. By recording the ratio I_2/V_2, the parameter h_{22} is expressed in siemens and represents the **open-circuit output admittance.** Summarizing the four sets of measurements to determine the hybrid parameters of a four-terminal coupling network,

$$h_{11} = \frac{V_1}{I_1} \quad (V_2 = 0) \quad \text{is the **short-circuit input impedance**} \quad (25\text{-}31)$$

$$h_{12} = \frac{V_1}{V_2} \quad (I_1 = 0) \quad \text{is the **reverse-voltage ratio**} \quad (25\text{-}32)$$

$$h_{21} = \frac{I_2}{I_1} \quad (V_2 = 0) \quad \text{is the **forward-current ratio**} \quad (25\text{-}33)$$

$$h_{22} = \frac{I_2}{V_2} \quad (I_1 = 0) \quad \text{is the **open-circuit output admittance**} \quad (25\text{-}34)$$

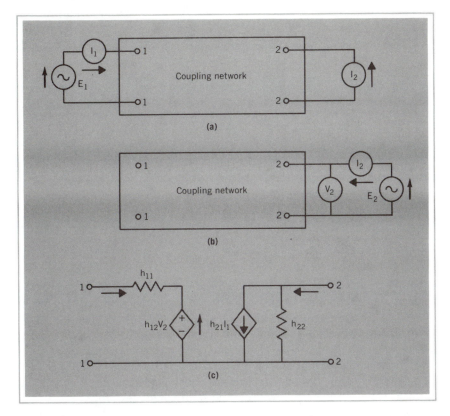

FIGURE 25-11
Finding the Norton-Equivalent Output Circuit of a Four-Terminal Network: (a) Short-Circuit Forward Current; (b) Output Conductance; (c) Complete Hybrid Parameters.

For the Thévenin-equivalent source we see looking into the input terminals (1,1) of the network, we can write a Kirchhoff's voltage-law equation, as we did for **z**-parameters. For the Norton-equivalent source looking back into the output terminals (2,2), we write a Kirchhoff's current-law equation, as we did for **y**-parameters. The two unknowns in this case are I_1 and V_2.

$$\mathbf{h}_{11}\mathbf{I}_1 + \mathbf{h}_{12}\mathbf{V}_2 = \mathbf{E}_1 \qquad (25\text{-}35)$$

$$\mathbf{h}_{21}\mathbf{I}_1 + \mathbf{h}_{22}\mathbf{V}_2 = \mathbf{I}_2 \qquad (25\text{-}36)$$

The transistor amplifier equivalent circuit of Figure 25-12 is a typical example of hybrid parameters.

EXAMPLE 25-6
Given the following parameters for the transistor amplifier equivalent circuit of Figure 25-12, calculate the voltage amplification when a 2.2-kΩ resistance load is connected across the output terminals.

FIGURE 25-12
Hybrid Parameters of a Simple Transistor Amplifier.

$$\mathbf{h}_{11} = 50 \ \Omega \qquad \mathbf{h}_{12} = 3 \times 10^{-4}$$

$$\mathbf{h}_{21} = -0.95 \qquad \mathbf{h}_{22} = 2 \times 10^{-5} \ \text{S}$$

SOLUTION

We define voltage amplification A as V_2/E_1. One solution is to derive a formula for amplification by rearranging Equations (25-35) and (25-36) and then substituting for the parameters. We shall substitute directly and solve for V_2. Since there are no reactance elements in this equivalent circuit, we can solve Equations (25-35) and (25-36) by linear algebra. There are four variables in Equations (25-35) and (25-36). With only two equations, we must make two of them independent. We can pick some convenient value for E_1, say 1 V rms. And from Ohm's law,

$$I_2 = \frac{-V_2}{R_{\text{L}}}$$

The negative sign, as we have noted, is because the $\mathbf{I}_2$ direction in Equations (25-35) and (25-36) was established by a generator in place of R_{L}. For the same reason, $\mathbf{h}_{21}$ ($\mathbf{I}_2/\mathbf{I}_1$) in the above data is also a negative quantity. Substituting these values in Equations (25-35) and (25-36),

$$50 \ I_1 + 3 \times 10^{-4} \ V_2 = 1 \tag{1}$$

$$-0.95 \ I_1 + 2 \times 10^{-5} \ V_2 = \frac{-V_2}{2.2 \ \text{k}\Omega} \tag{2}$$

From (1),
$$I_1 = \frac{1 - 3 \times 10^{-4} V_2}{50}$$

Substituting in (2),

$$-0.95 \left(\frac{1 - 3 \times 10^{-4} V_2}{50} \right) + 2 \times 10^{-5} V_2 = \frac{-V_2}{2.2 \ \text{k}\Omega}$$

from which
$$V_2 = 39.9 \ \text{V}$$

Hence, the voltage amplification is

$$A = \frac{V_2}{E_1} = 40$$

In Section 9-5, we discovered that we cannot use the method stated in Thévenin's theorem to obtain the Thévenin-equivalent internal resistance of a dependent source when the controlling element is included in the transformation. Since this is the case when we want to determine input and output impedances of coupling networks, we must "measure" the V/I ratio. We determined $\mathbf{h}_{11}$ with a short-circuit across the output terminals of the network. There will be a slightly different value for $\mathbf{Z}_{in}$ in the circuit of Figure 25-12 where reverse coupling occurs.

EXAMPLE 25-7
The source $\mathbf{E}_1$ in the equivalent circuit of Figure 25-12 has an internal resistance of 50 Ω. Determine the input and output resistances for the transistor amplifier circuit of Example 25-6.

SOLUTION
For this example, we shall derive formulas and then substitute the given parameters.
(a) In Example 25-6, we noted that

$$I_2 = \frac{-V_2}{R_L}$$

Substituting this expression for $\mathbf{I}_2$ in Equation (25-36) and collecting terms gives

$$V_2 = \frac{-\mathbf{h}_{21}\,\mathbf{Z}_L\,\mathbf{I}_1}{1 + \mathbf{h}_{22}\,\mathbf{Z}_L}$$

Substituting for $\mathbf{V}_2$ in Equation (25-35) and collecting terms,[†]

$$V_1 = \mathbf{h}_{11}\,\mathbf{I}_1 - \frac{\mathbf{h}_{12}\mathbf{h}_{21}\,\mathbf{Z}_L\,\mathbf{I}_1}{1 + \mathbf{h}_{22}\,\mathbf{Z}_L}$$

$$\therefore \mathbf{Z}_{in} = \frac{\mathbf{V}_1}{\mathbf{I}_1} = \mathbf{h}_{11} - \frac{\mathbf{h}_{12}\mathbf{h}_{21}\,\mathbf{Z}_L}{1 + \mathbf{h}_{22}\,\mathbf{Z}_L} \qquad (25\text{-}37)$$

Substituting for the parameters in Equation (25-37),

$$R_{in} = \mathbf{Z}_{in} = 50 - \frac{-0.95 \times 3 \times 10^{-4} \times 2.2 \times 10^3}{1 + 2 \times 10^{-5} \times 2.2 \times 10^3}$$

$$= 50 + 0.601 = 50.6\ \Omega$$

[†] We use $\mathbf{V}_1$ rather than $\mathbf{E}_1$ because $\mathbf{V}_1$ will not be the same as $\mathbf{E}_1$ if the source has any internal resistance. The source resistance is *not* included in the input resistance of the network.

(b) To obtain $\mathbf{h}_{22}$, we killed $\mathbf{I}_1$ by leaving the input terminals $(1,1)$ in Figure 25-11(b) open-circuit. To "measure" the output impedance, we kill source $\mathbf{E}_1$ but leave its internal resistance $\mathbf{Z}_s$ connected across terminals $(1,1)$. The Thévenin-equivalent source in Figure 25-11(c) creates a current through $\mathbf{h}_{11}$ and $\mathbf{Z}_s$ in the *opposite* direction to $\mathbf{I}_1$. Hence,

$$\mathbf{I}_1 = \frac{-\mathbf{h}_{12}\mathbf{V}_2}{\mathbf{h}_{11} + \mathbf{Z}_s}$$

Substituting for $\mathbf{I}_1$ in Equation (25-36),

$$\mathbf{I}_2 = \frac{-\mathbf{h}_{12}\mathbf{h}_{21}\mathbf{V}_2}{\mathbf{h}_{11} + \mathbf{Z}_s} + \mathbf{h}_{22}\mathbf{V}_2$$

$$= \mathbf{V}_2 \frac{\mathbf{h}_{22}(\mathbf{h}_{11} + \mathbf{Z}_s) - \mathbf{h}_{12}\mathbf{h}_{21}}{\mathbf{h}_{11} + \mathbf{Z}_s}$$

from which

$$\frac{\mathbf{V}_2}{\mathbf{I}_2} = \frac{\mathbf{h}_{11} + \mathbf{Z}_s}{\mathbf{h}_{22}(\mathbf{h}_{11} + \mathbf{Z}_s) - \mathbf{h}_{12}\mathbf{h}_{21}}$$

Dividing both numerator and denominator by $(\mathbf{h}_{11} + \mathbf{Z}_s)$,

$$\mathbf{Z}_{\text{out}} = \frac{1}{\mathbf{h}_{22} - \dfrac{\mathbf{h}_{12}\mathbf{h}_{21}}{\mathbf{h}_{11} + \mathbf{Z}_s}} \tag{25-38}$$

Substituting for the parameters in Equation (25-38),

$$\mathbf{Z}_{\text{out}} = \frac{1}{2 \times 10^{-5} - \dfrac{-0.95 \times 3 \times 10^{-4}}{50 + 50}} = \textbf{43.76 k}\boldsymbol{\Omega}$$

SELF-CHECK

25-4:1. For the transistor amplifier equivalent circuit in Example 25-6, determine the current amplification which is defined:

$$A_i = \frac{\text{load current}}{\text{source current}}$$

25-4:2. For the transistor amplifier equivalent circuit in Example 25-6, determine the power amplification which can be defined:

$$A_p = \frac{\text{power to the load}}{\text{power from the source}}$$

25-5 AIR-CORE TRANSFORMERS

We encounter a good example of the coupled-circuit technique of circuit analysis in dealing with the magnetically coupled circuits used in radio circuitry. The equivalent circuit we developed in Chapter 24 for iron-core power and audio transformers based on the **transformation ratio** is valid only because the leakage flux is very small compared to the mutual flux. We cannot use the transformation-ratio technique with loosely coupled air-core transformers, since the mutual flux in these transformers is only a fraction of the total primary flux. This fraction is called the **coefficient of coupling.** Therefore, by definition,

$$k = \frac{\Phi_m}{\Phi_p} \tag{25-39}$$

where k is the coefficient of coupling.

The value of k can range from unity when all the primary flux links the secondary, as in the ideal transformer of Figure 24-1, down to zero when the secondary winding is completely outside the magnetic field of the primary or is purposely positioned at right angles to the primary flux [Figure 25-13(c)]. The physical arrangement of Figure 25-13(c) is used in radio circuits where magnetic coupling between tuned circuits is to be kept to a minimum and the physical dimensions are rather restricted. In order to obtain *tight* coupling with k close to unity in radio-frequency coils where iron cores are not practical, the two strands of wire are interwound on the coil form as a **bifilar** winding [Figure 25-13(a)]. Magnetic coupling of *tuned* coils requires a very *loose* coupling with a coefficient of coupling of approximately 0.01, as shown by the physical placement of the two coils on the coil form of Figure 25-13(b).

FIGURE 25-13
Variations in Magnetic Coupling: (a) Tight Coupling; (b) Loose Coupling;
(c) Minimum Coupling.

25-6 MUTUAL INDUCTANCE

To develop an equivalent coupling network for an air-core transformer, we require one parameter which is common to both primary and secondary circuits. This parameter is the **mutual inductance** of the pair of coils. In Chapter 14, we found that a coil has a *self*-inductance of one henry when current in that coil *changing* at a rate of one ampere per second induces a voltage of one volt in the coil. Applying the same line of thought to the process of mutual induction, we can say that

A pair of magnetically coupled coils has the mutual inductance of one henry when current changing at the rate of one ampere per second in one coil induces an average voltage of one volt in the other coil.
The letter symbol for mutual inductance is M.

Therefore, by definition,

$$E_{s_{av}} = \frac{MI_{p_m}}{t} \tag{25-40}$$

Since the instantaneous current must rise from zero to maximum in one-quarter of a cycle of a sine wave,

$$E_{s_{av}} = \frac{MI_{p_m}}{1/4f} = 4fMI_{p_m}$$

But

$$E_{av} = \frac{2}{\pi} E_m \tag{16-17}$$

Thus,

$$\frac{2}{\pi} E_{s_m} = 4fMI_{p_m}$$

from which

$$E_s = 2\pi fMI_p \tag{25-41}$$

Rearranging Equation (25-41) gives

$$\frac{E_s}{I_p} = 2\pi fM = X_M \tag{25-42}$$

where X_M is called the **mutual reactance** of the magnetically coupled windings.

┌ EXAMPLE 25-8
120 V, 60 Hz is applied to the primary of a transformer whose primary inductance is 5 H. The open-circuit secondary voltage is 40 V. Neglecting losses, determine the mutual inductance between the two windings.

SOLUTION

$$E_s = \omega MI_p$$

But

$$I_p = \frac{E_p}{\omega L_p}$$

$$\therefore E_s = \frac{M}{L_p} E_p$$

L and $$M = L_p \frac{E_s}{E_p} = 5 \times \frac{40}{120} = \mathbf{1.67\ H}$$

Combining Equations (24-1) and (25-40)

$$\frac{N\Phi_m}{t} = \frac{MI_{Pm}}{t}$$

But from Equation (25-39) $$\Phi_m = k\Phi_p$$

$$\therefore MI_{Pm} = N_s k \Phi_p \qquad (1)$$

If we now reverse the windings and use the original secondary winding as the primary, and vice versa, it follows that

$$MI_{s_m} = N_p k \Phi_s \qquad (2)$$

Multiplying Equations (1) and (2),

$$M^2 = k^2 \left(\frac{N_p \Phi_p}{I_{Pm}}\right)\left(\frac{N_s \Phi_s}{I_{s_m}}\right) \qquad (3)$$

But from the definition of self-inductance,

$$E_{av} = \frac{LI_m}{t}$$

and from the definition of the weber,

$$E_{av} = \frac{N\Phi}{t}$$

$$\therefore L = \frac{N\Phi}{I_m}$$

Substituting in Equation (3) gives

$$M = k\sqrt{L_p L_s} \qquad (25\text{-}43)$$

EXAMPLE 25-9

If the secondary winding of the transformer of Example 25-8 has a self-inductance of 0.8 H, what is the coefficient of coupling between the windings?

SOLUTION

$$k = \frac{M}{\sqrt{L_p L_s}} = \frac{1.67}{\sqrt{5 \times 0.8}} = \mathbf{0.835}$$

FIGURE 25-14
Determining Mutual Inductance.

One method for experimentally determining mutual inductance is shown in Figure 25-14. We can measure the total inductance of the series-connected coils on an inductance bridge. If we connect the windings as shown and check by our hand rule, the flux produced by the current in one winding is in the same direction around the core as the flux produced by the current in the other winding. This increases the total flux, which increases the voltage induced by a given alternating current and, in turn, increases the total inductance. All the induced voltages will be in phase. There are four induced voltages: the self-induced voltage in the primary, the voltage mutually induced in the primary by current in the secondary, the voltage mutually induced in the secondary by current in the primary, and the self-induced voltage in the secondary:

$$\therefore E = I(\omega L_p + \omega M + \omega M + \omega L_s)$$

from which $L_T = L_p + L_s + 2M$.

If, however, we reverse the leads to the secondary, the mutually induced voltages in each coil are 180° out of phase with the self-induced voltages, resulting in

$$L_T = L_p + L_s - 2M$$

Therefore, we can extend our original Equation (14-6) for two inductances in series to include magnetic coupling between them;

$$\therefore L_T = L_p + L_s \pm 2M \tag{25-44}$$

The mutual inductance of the magnetically coupled coils in Figure 25-14 will be one-quarter of the difference between the total inductance readings with the coils connected series-aiding and then series-opposing.

SELF-CHECK

25-6:1. An air-core transformer has the following specifications:

$$L_p = 100 \ \mu H$$

$$L_s = 50 \ \mu H$$
$$k = 0.20$$

Find the mutual inductance.

25-6:2. If the transformer specified in Self-Check Problem 25-6:1 is connected to a 250-mV 10-kHz source, determine the open-circuit secondary voltage.

25-7 COUPLED IMPEDANCE

Before we set up the equivalent coupling network for a loosely coupled transformer, we should note a useful practical advantage that transformer coupling has over other forms of coupling networks. For a given direction of E_1 in Figure 25-15, we can reverse the phase of the output voltage by reversing the direction of one of the windings, or simply by reversing the leads to one of the windings. Where it is necessary to keep track of this phase relationship in a circuit diagram, we mark one end of each winding with a dot, as shown in Figure 25-15(a) and (b). If, at a certain moment, changing mutual flux induces an instantaneous voltage into the primary winding with a polarity such that the dotted end of the winding is positive with respect to the undotted end, then the same mutual flux change must induce an instantaneous voltage into the secondary winding with a polarity such that the dotted end of the secondary winding is positive with respect to the undotted end.

Although reversing the secondary polarity will reverse the current through the external circuit, note that I_2 in Figure 25-15(b) still must flow into the dot end of the

(a)

(b)

(c)

FIGURE 25-15
Equivalent Circuit of an Air-Core Transformer.

secondary winding, just as it does in Figure 25-15(a). Hence, the primary circuit is not affected by any phase reversal we obtain by reversing the secondary leads. In selecting the direction for these current arrows, we must remember that the *same* mutual flux (which is produced by these currents) induces voltages into both primary and secondary windings. So if we draw I_1 pointing into the dot end of the primary winding, we must also draw I_2 pointing into the dot end of the secondary winding.

When we write the Kirchhoff's voltage-law (loop) equation for the primary loop, there are three voltages which we add as phasors to equal E_1: an *IR* drop across the resistance of the primary circuit, a *self*-induced voltage due to I_1 flowing through the primary winding, and a *mutually*-induced voltage due to I_2 flowing in the secondary circuit. Figure 25-15(c) shows one form of equivalent circuit for this transformer, in which the self-induced voltage is represented by an inductance symbol and the mutually induced voltage by a CCVS symbol. To satisfy Lenz's law, the polarity of the CCVS must be such that it opposes changes in I_1 due to the applied voltage E_1. Hence, changing the polarity of the CCVS in the secondary circuit will *not* change the polarity of the voltage I_s $(+j\omega M)$ in the primary circuit. Consequently, Kirchhoff's voltage-law equation for the primary loop becomes

$$\mathbf{I}_p(R_p + j\omega L_p) + \mathbf{I}_s(+j\omega M) = \mathbf{E}_p$$

or

$$\mathbf{I}_p\mathbf{Z}_p + \mathbf{I}_s(+j\omega M) = \mathbf{E}_p \qquad (25\text{-}45)$$

where $\mathbf{Z}_p$ is the open-circuit impedance of the primary circuit by itself, and ωM is the mutual reactance of the two windings. Note that Lenz's law gives us the same direction for I_2 that we chose for our generalized coupling network. [Compare Equation (25-45) with Equation (25-5).] Hence, we write the loop equation for the secondary circuit of Figure 25-15(c) as

$$\mathbf{I}_s(\mathbf{Z}_L + R_s + j\omega L_s) + \mathbf{I}_p(+j\omega M) = 0$$

or

$$\mathbf{I}_s(\mathbf{Z}_L + \mathbf{Z}_s) + \mathbf{I}_p(+j\omega M) = 0 \qquad (25\text{-}46)$$

where $\mathbf{Z}_s$ is the open-circuit impedance of the secondary winding by itself. [Compare Equations (25-46) and (25-14).] From Equation (25-46),

$$\mathbf{I}_s = \frac{-\mathbf{I}_p(+j\omega M)}{\mathbf{Z}_s + \mathbf{Z}_L} \qquad (25\text{-}47)$$

Substituting in Equation (25-45),

$$\mathbf{I}_p\mathbf{Z}_p - \mathbf{I}_p\frac{(+j\omega M)^2}{\mathbf{Z}_s + \mathbf{Z}_L} = \mathbf{E}_p \qquad (25\text{-}48)$$

Equation (25-48) is the equivalent of Equation (25-16). In this case, we can go one step further by substituting -1 for j^2. Thus,

$$\mathbf{E}_p = \mathbf{I}_p\mathbf{Z}_p + \mathbf{I}_p\frac{(\omega M)^2}{\mathbf{Z}_s + \mathbf{Z}_L} \qquad (25\text{-}49)$$

and, dividing through by I_p,

$$Z_{in} = Z_p + \frac{(\omega M)^2}{Z_s + Z_L} \qquad (25\text{-}50)$$

where $(\omega M)^2/(Z_s + Z_L)$ *is the* **coupled impedance** *for transformer coupling*. Note that when we substitute -1 for j^2, $(\omega M)^2$ has now become an impedance with a $0°$ angle.

We can check Equation (25-50) by considering the effect of loading the secondary winding of a pair of magnetically coupled coils. If the secondary is left open-circuit, Z_L is infinitely large and, therefore, the coupled impedance in Equation (25-50) becomes zero. As a result, the total impedance is simply the primary impedance alone as we would expect. If we connect a resistance across the secondary winding, Z_L will have a $0°$ angle. Since Z_s is largely the inductive reactance of the secondary winding, the total secondary-circuit impedance is inductive. Consequently, when we carry out the phasor division, dividing an impedance with a $+$ angle into $(\omega M)^2$ with its $0°$ angle results in a *capacitive* coupled impedance. Because Z_p is largely *inductive*, the *capacitive* impedance coupled into the primary circuit via the mutual inductance adds the coupled resistance component in series with the primary resistance component but *reduces* the total primary reactance. As a result, the total impedance becomes smaller as the secondary is loaded with a resistance, thus allowing more primary current to flow in order to transfer energy to the secondary circuit. This then checks with our first approach to transformer action in which we discovered that primary current must increase when a transformer is loaded in order to keep the amplitude of the mutual-flux sine wave constant.

EXAMPLE 25-10

If the resistance of the windings can be neglected, determine the total input impedance to the transformer of Examples 25-8 and 25-9, (a) with the secondary open-circuit, and (b) with a 50-Ω resistance connected to the secondary.

SOLUTION

(a) $\qquad Z_{in} = Z_p = \omega L_p = 377 \times 5 = +j1885 \ \Omega$

(b) $\qquad$ Coupled $Z = \dfrac{(\omega M)^2}{Z_s + Z_L} = \dfrac{(377 \times 1.667)^2}{+j(377 \times 0.8) + 50}$

$$= \frac{395\,000 \underline{/0°}}{305.7 \underline{/80.59°}} = 1292 \ \Omega \underline{/-80.59°} = 211.2 - j1275 \ \Omega$$

$$\therefore Z_{in} = (+j1885) + (211.2 - j1275) = 211.2 + j610 \ \Omega$$

$$= \mathbf{645.5 \ \Omega \underline{/+70.9°}}$$

SELF-CHECK

25-7:1. Suppose the windings of the transformer in Example 25-10 each have 25-Ω of resistance. Find the input impedance when the 50-Ω resistive load is connected to the secondary.

25-8 TUNED TRANSFORMERS

While we have been getting used to the concept of representing any four-terminal coupling network by standardized input and output parameters, we have chosen to ignore the question of the **frequency response** of such networks. With many coupling networks, such as the H-attenuator of Figure 25-5 and the transistor equivalent circuit of Figure 25-12, the *frequency* of source voltage E_1 has little effect on output voltage V_2. But with other coupling networks, such as the high-pass filter of Figure 25-6 and the low-pass filter of Figure 25-9, the magnitude (and phase) of the output voltage is quite dependent on the *frequency* of E_1. We call such frequency-dependent coupling circuits **filter networks** (or simply **filters**).

We can convert the loose-coupled air-core transformer of Figure 25-15 into a filter by *tuning* its self-inductance (L_p and/or L_s) to resonance at a desired frequency. If we consider the source to be a constant-voltage source, as in Figure 25-16(a), the primary winding is designed to form a series-resonant circuit with the capacitive reactance of C_p equal to the inductive reactance of the primary winding at the desired resonant frequency. This will result in maximum current in the primary winding at resonance (neglecting for the moment any coupled impedance due to secondary current). Since the transistor amplifiers we use to feed the primary of a tuned transformer have a relatively high internal resistance, they behave as constant-current sources. Accordingly, C_p must be connected in *parallel* with the primary winding so that the resonant rise of impedance will increase the voltage across (and the tank current in) the primary at resonance, as in Figure 25-16(b).

FIGURE 25-16
Tuned Transformer: (a) Fed from a Low-Impedance Constant-Voltage Source; (b) Fed from a High-Impedance Constant-Current Source.

When we consider the effect of tuning the secondary circuit, the total secondary-circuit impedance [$\mathbf{Z}_s + \mathbf{Z}_L$ in Equation (25-50)] will be minimum and equal to R_s at resonance. Therefore, the secondary circuit behaves like a series-resonant circuit. Since $(\omega M)^2$ is essentially constant over the small range of frequencies near resonance, when $\mathbf{Z}_s + \mathbf{Z}_L$ becomes a *minimum,* the coupled impedance becomes a *maximum,* as shown in Figure 25-17. When the source frequency is slightly lower than the resonant frequency,

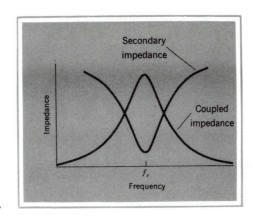

FIGURE 25-17
Effect of a Tuned Secondary Circuit.

X_{C_s} is greater than X_{L_s} and the secondary impedance becomes capacitive. But dividing a *capacitive* impedance into $(\omega M)^2$ with its $0°$ angle results in an *inductive* coupled impedance. Thus, as far as the signal source is concerned, the secondary behaves as if it were a *parallel*-resonant circuit in *series* with the primary winding.

The primary impedance by itself in the circuit of Figure 25-16(a) is attempting to become a minimum at resonance in order to allow maximum primary current to flow. But the coupled impedance is attempting to raise the primary impedance at resonance, thus limiting the maximum primary current. The extent to which the coupled secondary-tuned circuit affects the primary resonance curve depends on the degree of coupling between the coils. When the coupling is very loose, the mutual inductance is very small and, even at resonance, the coupled impedance is *smaller* than the resistance of the primary circuit. Under these circumstances, the only effect that the coupled secondary has on the shape of the primary-current resonance curve is to limit its peak value slightly. The secondary current tends to have the usual series resonance curve. But, with loose coupling, since the primary current rises to a maximum at resonance and since the secondary induced voltage depends on the primary current, the secondary resonance curve is much sharper than that of a single-tuned circuit having the same Q.[†]

We recall from earlier studies that maximum transfer of energy occurs when the load resistance is equal to the resistance of the source. If we apply this principle to magnetically coupled tuned circuits, maximum energy transfer will take place when the coupled resistance is equal to the primary-circuit resistance. Since the coupled resistance depends on the mutual inductance, which in turn depends on the coefficient of coupling, there is a **critical coupling** for a given pair of tuned circuits at which maximum energy transfer from primary to secondary takes place.

At resonance, $$\mathbf{Z}_p = R_p$$

and the coupled $$\mathbf{Z} = \frac{(\omega M)^2}{R_s}$$

[†] The bandwidth is approximately $B \approx kf_r$.

Therefore, for critical coupling,

$$M^2 = R_p R_s / \omega^2$$

But

$$M^2 = k^2 L_p L_s \qquad (25\text{-}43)$$

Hence,

$$k^2 = \frac{R_p}{\omega L_p} \times \frac{R_s}{\omega L_s}$$

from which

$$k_c = \frac{1}{\sqrt{Q_p Q_s}} \qquad (25\text{-}51)$$

With critical coupling, the secondary current attains its greatest value. But, with the coupled impedance rising at resonance, the primary current is no longer maximum at resonance. As we have already noted, at a frequency slightly below resonance, the coupled impedance is *inductive*, whereas the impedance of the series-tuned primary is *capacitive*. Therefore, there are two frequencies, one slightly above the resonant frequency and one slightly below where the coupled reactance tunes the primary reactance to give minimum total primary impedance and thus maximum primary current. This double hump in the primary current tends to flatten the peak of the secondary current response curve because the voltage induced into the secondary at the resonant frequency is not quite as great as at the two hump frequencies either side of resonance.

If the tuned circuits are overcoupled, the increase in coupled impedance moves the primary-current humps further apart. This, in turn, causes such a decrease in primary current at resonance that the secondary current also starts to show a double-humped curve, as shown in Figure 25-18. In practice, a coefficient of coupling of 1.5 times the critical coefficient of coupling produces such a slight dip in the secondary current at resonance that the resulting secondary-current resonance curve has a very desirable flat top with steep skirts.

25-9 RESONANT FILTER NETWORKS

The transformer coupling of Section 25-8 is not essential to the design of **band-pass** filters. We can also design four-terminal band-pass filters using the simple resonant circuits of Chapter 23. We can redraw the series-resonant circuit of Figure 23-7 and the parallel-resonant circuit of Figure 23-12 as the four-terminal band-pass filters of Figure 25-19(a) and (b), respectively.

We can also use resonant circuits to *reject* a specific frequency without affecting the response of the circuit to all other frequencies. Such an application is called a **wavetrap.** Figure 25-20 illustrates two wavetrap circuits. In both cases, the four-terminal network forms a voltage divider across the input. If we assume a very high-impedance load connected to the output terminals, then the output voltage of the circuit of Figure 25-20(a) will be

$$V_{out} = E_{in} \times \frac{Z_s}{R + Z_s} \qquad (25\text{-}52)$$

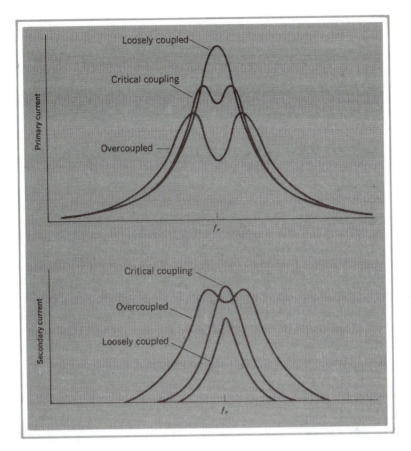

FIGURE 25-18
Effect of Coefficient of Coupling on the Resonance Curves of a Tuned Transformer.

FIGURE 25-19
(a) Series-Resonant, and (b) Parallel-Resonant Bandpass Filters.

FIGURE 25-20
Wavetraps.

At resonance, Z_s will be equal to the resistance of the coil, which is quite small compared with the resistance R. With the values given in Figure 25-20(a), the output voltage will be only 1% of the input voltage at the resonant frequency. With a high-Q tuned circuit, at frequencies other than the resonant frequency, Z_s will be much greater than R and the fraction $Z_s/(R + Z_s)$ will approach 100%.

In the wavetrap circuit of Figure 25-20(b), the load voltage will be

$$V_{out} = E_{in} \times \frac{R_L}{R_L + Z_p} \tag{25-53}$$

At resonance the impedance of the parallel-resonant circuit will be many times as great as the resistance of the load. This results in a very small voltage across the load at this frequency. At frequencies off resonance, Z_p becomes quite small, and, therefore, $R_L/(R_L + Z_p)$ approaches 100%.

In many communications circuits, it is desirable to include the selection of one frequency and the rejection of another frequency into a combined **double-resonant** circuit such as those shown in Figure 25-21.

FIGURE 25-21
Double-Resonant Circuits.

EXAMPLE 25-11

If the capacitance in the circuit of Figure 25-21(a) is 200 pF, determine the inductances of L_p and L_s in order to reject a signal at 456 kHz and to accept a signal at 1200 kHz.

SOLUTION

Step I. L_p and C form a parallel-resonant circuit which will represent a high impedance in series with R_L at resonance. Hence, L_p and C form a wavetrap and must be tuned to the frequency which is to be rejected, 456 kHz.

$$X_C = \frac{1}{2\pi fC} = \frac{1}{2 \times \pi \times 456 \text{ kHz} \times 200 \text{ pF}} = 1745 \ \Omega$$

$$L_p = \frac{X_L}{2\pi f} = \frac{1745 \ \Omega}{2 \times \pi \times 456 \text{ kHz}} = \textbf{609} \ \boldsymbol{\mu}\textbf{H}$$

Step II. At 1200 kHz,

$$B_C = 2\pi fC = 2 \times \pi \times 1200 \text{ kHz} \times 200 \text{ pF} = 1.508 \text{ mS}$$

$$B_{L_p} = \frac{1}{2\pi fL} = \frac{1}{2 \times \pi \times 1200 \text{ kHz} \times 609 \ \mu\text{H}} = 217.8 \ \mu\text{S}$$

Therefore, the net susceptance of the parallel circuit at 1200 kHz is

$$B = 1.508 \text{ mS} - 217.8 \ \mu\text{S} = 1.2902 \text{ mS capacitive}$$

For the whole circuit to be *series* resonant at 1200 kHz,

$$B_{L_s} = 1.2902 \text{ mS}$$

$$\therefore L_s = \frac{1}{2\pi fB_{L_s}} = \frac{1}{2 \times \pi \times 1200 \text{ kHz} \times 1.2902 \text{ mS}}$$

$$= \textbf{102.8} \ \boldsymbol{\mu}\textbf{H}$$

Figure 25-21(b) illustrates an alternate circuit arrangement to achieve the results required in Example 25-11. Here L_s and C are selected first to form a series-resonant circuit at the desired frequency of 1200 kHz, thus comprising a low impedance in series with R_L at 1200 kHz. At 456 kHz, this series branch has a net capacitive reactance which can be tuned with L_p as a parallel-resonant circuit at 456 kHz. If the selected frequency is lower than the rejected frequency, L_s in the circuit in Figure 25-21(a) and L_p in the circuit of Figure 25-21(b) must be replaced with capacitors.

SELF-CHECK

25-9:1. Design a double-resonant filter that uses a 0.033-μF capacitor and the circuit shown in Figure 25-21(b) and passes a 35-kHz signal while rejecting a 19-kHz signal.

25-10 HIGH-PASS AND LOW-PASS FILTERS

We can also use inductors and capacitors in four-terminal filter networks which are designed to pass all frequencies above a certain **cutoff frequency,** and to reject all frequencies below the cutoff frequency. We call these **high-pass** filters. Similarly, a **low-pass** filter rejects all frequencies above the cutoff frequency and passes all frequencies below it. In Example 25-4, we worked with a symmetrical LC high-pass filter, but only at a single frequency of 1 kHz. In the circuit of Figure 25-22(a), we have connected the same filter to a variable-frequency source.

As we might expect, the cutoff frequency for the filter of Figure 25-22(a) is the frequency at which $X_L = X_C$ of the input capacitor ($\mathbf{z}_{11} = 0$). Hence,

$$f_c = \frac{1}{2\pi\sqrt{LC}} = 1453 \text{ Hz} \qquad\qquad (23\text{-}1)$$

FIGURE 25-22
LC High-Pass Filter.

As we increase the source frequency above the cutoff frequency, the reactance of the two capacitors in series with R_L becomes quite small and the reactance of the inductance in parallel with R_L becomes quite large. Consequently, V_L is almost equal to E_1. But for frequencies below f_c, X_L is small while X_C is large. From the voltage divider principle, the voltage drop across R_L, therefore, becomes very small.

As we discovered in Chapter 23, when we plot a frequency-response graph for a network, we vary the source frequency in steps of one *octave*. Hence, the frequency column of Table 25-1 consists of frequencies which are *twice* the preceding frequency. We can now put a computer to work repeating all the calculations of Example 25-4 for each frequency in order to print out the V_L column of Table 25-1. Plotting these data on the logarithmic X-axis scale of Figure 25-22(b) gives us a graphical representation of the

TABLE 25-1

OUTPUT VOLTAGE *vs* FREQUENCY FOR A HIGH-PASS FILTER	
Frequency (Hz)	Output Voltage (V)
91	0.003
182	0.027
363	0.23
727	2.39
*1000	8.96
f_c 1453	13.69
2096	10.15
5812	10.0
11 624	10.0
23 248	10.0

*Fixed frequency of Example 25-4.

frequency response of the high-pass filter of Figure 25-22(a). Note the slight resonant rise of voltage at the cutoff frequency.

In the low-pass filter of Figure 25-23(a), we have used the same capacitors and inductor as in Figure 25-22(a), but we have reversed their positions. We have also reduced the load resistance to equal X_L or X_C at f_c. If we go through a similar routine of multiple calculations to determine V_L for various source frequencies, we obtain the frequency-response graph of Figure 25-23(b). Note the smaller resonant rise of output voltage at a frequency slightly lower than f_c.

Figure 25-24 shows the simplest of four-terminal, high- and low-pass filter networks, consisting of a resistor and a capacitor. We define their cutoff frequency as the

FIGURE 25-23
LC Low-Pass Filter.

FIGURE 25-24
(a) A Simple *RC* High-Pass Filter; (b) a Simple *RC* Low-Pass Filter.

frequency at which

$$R = X_C = \frac{1}{2\pi f C}$$

from which

$$f_c = \frac{1}{2\pi RC} \tag{25-54}$$

For the moment, we assume that the variable-frequency source in Figure 25-24 has a constant amplitude of 10 V rms and negligible internal resistance. By using the same resistor and capacitor for both filters, we can show all the necessary data in one table. The cutoff frequency for both filters is

$$f_c = \frac{1}{2\pi RC} = \frac{1}{2\pi \times 10^4 \times 10^{-8}} = 1592 \text{ Hz}$$

At this frequency,

$$X_C = R = 10 \text{ k}\Omega$$

and

$$V_R = \mathbf{E}_1 \times \frac{R}{R - jX_C} = 10 \text{ V} \underline{/0°} \times \frac{10^4 \underline{/0°}}{10^4 - j10^4}$$

$$= 10 \text{ V} \underline{/0°} \times \frac{10\,000 \underline{/0°}}{14\,140 \underline{/-45°}} = 7.07 \text{ V} \underline{/+45°}$$

Similarly,

$$V_C = 10 \text{ V} \underline{/0°} \times \frac{10\,000 \underline{/-90°}}{14\,140 \underline{/-45°}} = 7.07 \text{ V} \underline{/-45°}$$

Note that we are including the phase angle of the output voltage with respect to the applied voltage. Both magnitude and phase angle will change when we set out to do our multiple calculations for Table 25-2.

Comparing the graphs of Figure 25-25 with those of Figures 25-22 and 25-23, we note that *RC* filters do not have as sharp a boundary between the pass band and reject band as do *LC* filters. We have to arbitrarily set the boundary by defining the cutoff frequency as the frequency at which $X_C = R$. From the amplitude response graphs of Figure 25-25, we note that the amplitude of the output voltage has dropped off to $1/\sqrt{2}$ or 0.707 of

TABLE 25-2

Frequency (Hz)	V_R (V)	V_C (V)
50	$0.31\underline{/+88.2°}$	$9.99\underline{/-1.8°}$
100	$0.62\underline{/+86.4°}$	$9.98\underline{/-3.6°}$
199	$1.24\underline{/+82.9°}$	$9.92\underline{/-7.1°}$
398	$2.43\underline{/+76°}$	$9.70\underline{/-14°}$
796	$4.47\underline{/+63.4°}$	$8.94\underline{/-26.6°}$
f_c 1592	$7.07\underline{/+45°}$	$7.07\underline{/-45°}$
3184	$8.94\underline{/+26.6°}$	$4.47\underline{/-63.4°}$
6368	$9.70\underline{/+14°}$	$2.43\underline{/-76°}$
12736	$9.92\underline{/+7.1°}$	$1.24\underline{/-82.9°}$
25472	$9.98\underline{/+3.6°}$	$0.62\underline{/-86.4°}$
50944	$9.99\underline{/+1.8°}$	$0.31\underline{/-88.2°}$

VOLTAGE DISTRIBUTION IN AN RC FILTER

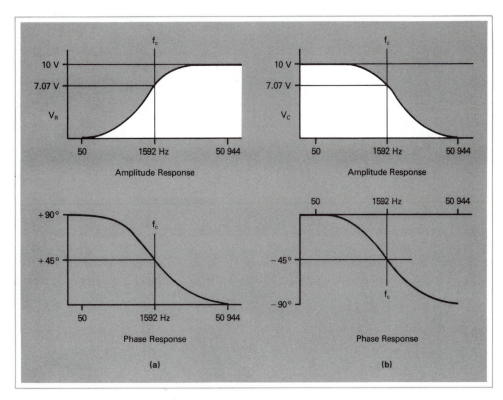

FIGURE 25-25
(a) *RC* High-Pass Filter; (b) *RC* Low-Pass Filter.

the input voltage at the cutoff frequency. We sometimes define cutoff frequency as the frequency at which the output voltage drops off to 0.707 of the pass band amplitude. Note also that the phase of the output voltage shifts quite rapidly as the source frequency moves from pass band to reject band.[†]

We sometimes design *RC* filters into audio systems to act as "tone" controls. More often than not, we discover that normal *RC* coupling circuits between amplifier stages produce unintended high- or low-pass filter characteristics. Figure 25-26 is an equivalent circuit for the capacitive coupling between an amplifier output and the open-circuit input to the next stage. Since amplifiers have appreciable internal resistance, we must include it in the equivalent circuit. We have also shown capacitor C_s representing all the stray shunt capacitance between the stages. The equivalent circuit indicates that the interstage coupling network contains both a low-pass and a high-pass filter with different cutoff frequencies. Rather than repeat the laborious calculations of Tables 25-1 and 25-2, if we can determine the two cutoff frequencies, we can pattern a frequency response graph after the basic models of Figure 25-25.

FIGURE 25-26
Equivalent Circuit for Interstage Capacitive Coupling.

EXAMPLE 25-12
Determine the upper and lower cutoff frequencies of the pass band of the coupling network of Figure 25-26.

SOLUTION
   Step I. At the upper cutoff frequency, the reactance of the 0.2-μF coupling capacitor will be very small compared to the other resistances and reactance. As far as circuit impedances are concerned, we can replace it with a short circuit in the equivalent circuit of Figure 25-27(a). Considering the shunt capacitance to be a "load," we produce the Thévenin-equivalent source of Figure 25-27(b). The upper cutoff frequency then becomes

$$f_c = \frac{1}{2\pi \times 5 \text{ k}\Omega \times 1 \text{ nF}} \approx \mathbf{31.8 \text{ kHz}}$$

[†] The phase response of coupling networks is important in designing feedback amplifiers.

FIGURE 25-27
Equivalent Circuit for Upper Cutoff Frequency.

FIGURE 25-28
(a) Equivalent Circuit for Lower Cutoff Frequency; (b) Pass Band of Coupling Network.

Step II. At the lower cutoff frequency, the reactance of the shunt capacitance will be very large compared to the other resistances and reactance. Hence we can neglect it in the equivalent circuit of Figure 25-28(a). In calculating the lower cutoff frequency, we must include the internal resistance of the source when we determine the frequency at which $X_C = R$ and the output drops to 0.707 of the mid-band amplitude. Hence, the lower cutoff frequency becomes

$$f_c = \frac{1}{2\pi \times 20 \text{ k}\Omega \times 0.2 \text{ }\mu\text{F}} \approx \textbf{40 Hz}$$

SELF-CHECK

25-10:1. Design an *LC* low-pass filter that uses a 0.33-μF capacitor and has a cutoff frequency of 8.0 kHz.

25-10:2. Design an *RC* low-pass filter that uses a 0.33-μF capacitor and has a cutoff frequency of 8.0 kHz.

25-11 THE DECIBEL

To draw the frequency response graphs of Sections 25-9 and 25-10, we had to maintain a constant magnitude of source voltage for all frequencies. Such a situation is not very realistic in practical networks. An alternative method of drawing a frequency-response graph for a coupling network is to calibrate the Y-axis in terms of the *ratio* between output voltage and input voltage.

With most *passive* coupling networks, the output voltage is less than the input voltage. A transformer, however, is a passive coupling network that can produce an output voltage greater than the source voltage. But, in both cases, the output *power* of the passive coupling device must be less than the input *power*. Hence, particularly in communications systems, we prefer to talk about the **power loss** in passive coupling networks, and the **power gain** of active coupling devices, such as amplifiers. The terms power gain and power loss infer a *ratio* between power output and power input. And since determination of power gain and power loss was particularly important in the early days of the telephone, it was desirable to express these ratios as *logarithms* in order to duplicate the nonlinear response of the human ear to sound waves. Consequently, a unit called the **bel**[†] was defined as the common logarithm (base of ten) of the ratio of output power to input power. The bel proved to be too large a unit for practical purposes. Hence, we usually express power ratio in tenths of a bel, or **decibels.**

> *Power gain (or loss) is the ratio of power output to power input in a four-terminal coupling network, usually expressed in decibels.*

By definition,

$$\text{Power gain (in dB)} = 10 \log_{10} \frac{P_{\text{out}}}{P_{\text{in}}} \qquad (25\text{-}55)^{\ddagger}$$

Since the logarithm of a quantity less than 1 is a negative number, a negative number of decibels automatically indicates a power *loss*.

EXAMPLE 25-13

What is the power gain of an amplifier in which an input power of 50 μW results in an output power of 20 W?

SOLUTION

$$\text{Power gain} = 10 \log_{10} \frac{20 \text{ W}}{50 \text{ μW}} = +56 \text{ dB}$$

[†] Named in honor of Alexander Graham Bell.

[‡] The standard letter symbol for gain is *G*. However, to avoid confusion with the letter symbol for conductance, and also to emphasize that **gain** is a *power* ratio, we shall use the full term **power gain** in this section.

On a calculator,

$$10 \ \boxed{\times} \ \boxed{(} \ 20 \ \boxed{\div} \ 50 \ \boxed{EE} \ \boxed{+/-} \ 6 \ \boxed{)} \ \boxed{\log} \ \boxed{=}$$

EXAMPLE 25-14

What is the power loss of the attenuator shown in Figure 25-29 when it is connected between a signal source and a 600-Ω load?

FIGURE 25-29
Attenuator Circuit.

SOLUTION

Step I. Assume an rms load current of 1 mA. Then

$$P_{\text{out}} = I^2 R_L = (1 \text{ mA})^2 \times 600 \ \Omega = 600 \ \mu\text{W}$$

Step II. Input resistance (see Example 25-3),

$$R_{\text{in}} = 600 \ \Omega$$

From the current-divider principle if $I_L = 1$ mA, then

$$I_{\text{in}} = 2 \text{ mA}$$

Hence, $$P_{\text{in}} = I^2 R_{\text{in}} = (2 \text{ mA})^2 \times 600 \ \Omega = 2.4 \text{ mW}$$

Step III.

$$\text{Power loss} = 10 \log_{10} \frac{600 \ \mu\text{W}}{2.4 \text{ mW}} = \mathbf{-6 \ dB}$$

Since decibels represent *power* ratios, if (and only if) the input and output resistances of a four-terminal network are equal, we can substitute V^2/R for P in Equation (25-55) with R canceling out. Hence,

$$\text{Power gain (dB)} = 10 \log_{10} \frac{P_{\text{out}}}{P_{\text{in}}} = 10 \log_{10} \frac{V_{\text{out}}^2}{V_{\text{in}}^2}$$

In logarithms, the log of a square is twice the log of the root quantity. Therefore,

$$\text{Power gain (in dB)} = 20 \log_{10} \frac{V_{\text{out}}}{V_{\text{in}}} \quad (R_{\text{out}} = R_{\text{in}}) \tag{25-56}$$

Sometimes we wish to compare output voltage (or current) with input voltage (or current) simply in ratio form ($V_{\text{out}}/V_{\text{in}}$). Strictly speaking, the term **gain** refers only to *power* gain. The term for the voltage (or current) ratio is **amplification** which has the letter symbol A. Hence,

$$A_V = \frac{V_{\text{out}}}{V_{\text{in}}} \tag{25-57}$$

The logarithmic nature of the decibel makes it possible for us to *add* decibels whereas we would have to *multiply* a chain of power gains and losses using only the basic power ratios. Every time we *double* power gain (or power loss), the ratio expressed in decibels changes by 3. For example, a $2:1$ power ratio $= 3$ dB; a $4:1$ power ratio $= 2^2 = 6$ dB; and an $8:1$ power ratio $= 2^3 = 9$ dB. And since decibels are logarithmic units, we can now use *semi-log* graph paper with linear Y-axis divisions for the frequency response graphs of the next section.

SELF-CHECK

25-11:1. What is the gain of a circuit which has an input of 12 W and a useful output of 6 W?
25-11:2. What is the power output of an amplifier that has an input of 20 mW and a gain of 23 dB?

25-12 BODE DIAGRAMS

When Hendrik W. Bode was designing feedback amplifiers at the Bell Telephone Laboratories circa 1940, he developed a simple method of determining the amplitude and phase response of four-terminal coupling networks without having to resort to the tedious calculations we needed to produce Tables 25-1 and 25-2. (They didn't have personal computers in those days.) Bode diagrams use semi-log graph paper to plot the loss (or gain) of a coupling network in decibels against frequency on the logarithmic X-axis scale.

In Figures 12-5 and 15-4, we discovered that we could draw a reasonably accurate graph for the exponential charge of a capacitor or inductor by using straight lines called **asymptotes.** We can also use asymptotes to position the roll-off of the high- and low-pass *RC* coupling networks of Figure 25-24. To demonstrate the accuracy of the technique, we first convert the voltage output calculations of Table 25-2 to voltage ratios in decibels. We also note that 1593 Hz is the cutoff frequency and identify the other frequencies as multiples and submultiples of f_c, thus obtaining Table 25-3. A sample calculation becomes

At f_c,
$$A = 20 \log \frac{7.07 \text{ V}}{10.0 \text{ V}} = -3 \text{ dB}$$

TABLE 25-3

VOLTAGE RATIOS IN AN RC FILTER		
Frequency	High-Pass	Low-Pass
$f_c/32$	−30 dB	0 dB
$f_c/16$	−24 dB	−0.02 dB
$f_c/8$	−18 dB	−0.07 dB
$f_c/4$	−12.3 dB	−0.26 dB
$f_c/2$	−7 dB	−1 dB
f_c = 1592 Hz	−3 dB	−3 dB
$2 f_c$	−1 dB	−7 dB
$4 f_c$	−0.26 dB	−12.3 dB
$8 f_c$	−0.07 dB	−18 dB
$16 f_c$	−0.02 dB	−24 dB
$32 f_c$	0 dB	−30 dB

The lower curve of Figure 25-30 is the plot of actual response figures from Table 25-3. Note that it passes through −3 dB at f_c. However, Table 25-3 also indicates that, for frequencies below $f_c/8$, the amplitude response drops off at a fixed rate of 6 dB per octave. (We could also say 20 dB per decade.) Hence we can draw a straight line with a slope of 6 dB per octave passing through 0 dB at f_c to form one asymptote for our "idealized" amplitude response. Table 25-3 also indicates that, for frequencies above 8 f_c, the actual amplitude response is very close to 0 dB. Hence, the other asymptote is a horizontal straight line representing 0 dB. The upper graph of Figure 25-30 becomes

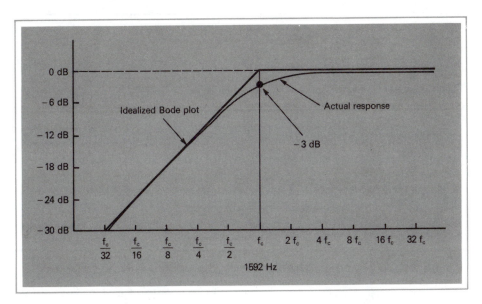

FIGURE 25-30
Bode Diagram for the High-Pass Coupling Network of Figure 25-24(a).

the idealized Bode plot for the amplitude vs frequency response of the high-pass RC coupling network of Figure 25-24(a).

EXAMPLE 25-15

Construct a Bode diagram for the low-pass RC coupling network of Figure 25-24(b).

SOLUTION

Step I. Calculate $f_c = \dfrac{1}{2\pi RC} = \dfrac{1}{2\pi \times 10^4 \times 10^{-8}} = 1592$ Hz

Step II. Using 3-cycle semi-log graph paper, locate 1592 Hz in the center cycle of the X-axis and place a plot at 0 dB. Draw a horizontal line along the 0-dB line from 100 Hz to 1592 Hz.

Step III. Draw a sloping line from 0 dB at 1592 Hz through -20 dB at 15 920 Hz. These two lines form the Bode diagram of Figure 25-31.

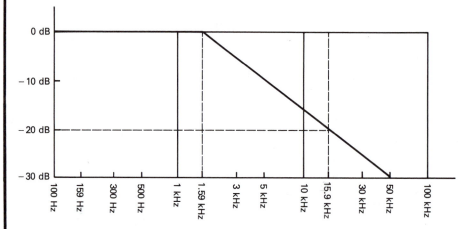

FIGURE 25-31
Bode Diagram for the Low-Pass Coupling Network of Figure 25-24(b).

We can also construct a Bode diagram for the phase response graphs of Figure 25-25 by noting that the actual phase response graph is quite linear between $f_c/4$ and $4\,f_c$. In the Bode phase response diagram of Figure 25-32, we draw a straight line from $+90°$ at $f_c/10$ through 45° at f_c to 0° at 10 f_c. The heavier "curve" of Figure 25-32 then becomes the Bode plot for the phase response of the high-pass coupling network of Figure 25-24(a). The lighter curve shows the actual phase response.

SELF-CHECK

25-12:1. Draw a Bode diagram for the RC low-pass filter designed in Self-Check Problem 25-10:2.

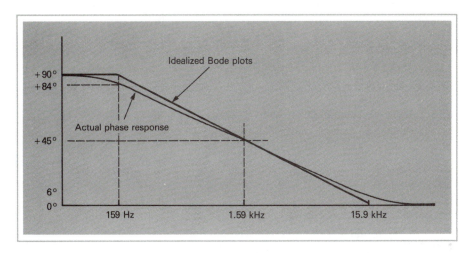

FIGURE 25-32
Bode Diagram for Phase Response of the High-Pass Coupling Network of Figure 25-24(a).

PROBLEMS

*In Problems 25-2-1 to 25-2-4, calculate the open-circuit **z**-parameters for each T-network. Use the subscript notation of Figure 25-4.*

25-2-1. $\mathbf{Z}_p = R_p = 357\ \Omega$, $\mathbf{Z}_S = R_S = 36\ \Omega$, $\mathbf{Z}_m = R_m = 286\ \Omega$.

25-2-2. $\mathbf{Z}_p = R_p = 2500\ \Omega$, $\mathbf{Z}_s = -j500\ \Omega$, $\mathbf{Z}_m = R_m = 10\ k\Omega$.

25-2-3. $\mathbf{Z}_p = \mathbf{Z}_s = 20 + j800\ \Omega$, $\mathbf{Z}_m = -j800\ \Omega$.

25-2-4. $\mathbf{Z}_p = \mathbf{Z}_s =$ a 50-μH inductance in series with a 200-pF capacitance, $\mathbf{Z}_m =$ a 200-μH inductance, $f = 1$ MHz.

*In Problems 25-2-5 to 25-2-8, find the equivalent T-networks for the "black boxes" having the given **z**-parameters.*

25-2-5. $\mathbf{z}_{11} = 600\ \Omega\,\underline{/0°}$, $\mathbf{z}_{12} = \mathbf{z}_{21} = 400\ \Omega\,\underline{/0°}$, $\mathbf{z}_{22} = 1200\ \Omega\,\underline{/0°}$.

25-2-6. $\mathbf{z}_{11} = 500\ k\Omega - j5\ k\Omega$, $\mathbf{z}_{12} = \mathbf{z}_{21} = \mathbf{z}_{22} = 500\ k\Omega + j0\ \Omega$.

25-2-7. $\mathbf{z}_{11} = \mathbf{z}_{22} = 50 + j0\ \Omega$, $\mathbf{z}_{12} = \mathbf{z}_{21} = -j1200\ \Omega$.

25-2-8. $\mathbf{z}_{11} = 100 + j1500\ \Omega$, $\mathbf{z}_{12} = \mathbf{z}_{21} = 40 + j500\ \Omega$, $\mathbf{z}_{22} = 80 + j1200\ \Omega$.

25-2-9. What is the coupled impedance when a 1500-Ω resistive load is connected to the output terminals of the network in Problem 25-2-5?

25-2-10. What is the coupled impedance when a capacitive reactance of 1240 Ω is connected to the output terminals of the network in Problem 25-2-8?

25-2-11. What is $\mathbf{Z}_{in}$ when a 250-Ω resistive load is connected to the output terminals of the T-network in Problem 25-2-1?

25-2-12. What is $\mathbf{Z}_{in}$ when a 1-kΩ resistive load is connected to the output terminals of the T-network in Problem 25-2-4?

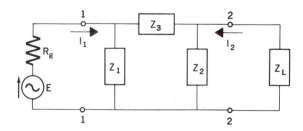

FIGURE 25-33
Pi-Coupling Network.

In Problems 25-2-13 to 25-2-16, find the open circuit z-parameters for each π-network. Use the subscript notation of Figure 25-33.

25-2-13. $Z_1 = R_1 = 3470\ \Omega$, $Z_2 = R_2 = 350\ \Omega$, $Z_3 = R_3 = 440\ \Omega$.

25-2-14. $Z_1 = 80\ k\Omega\ \underline{/0°}$, $Z_2 = 470\ k\Omega\ \underline{/0°}$, Z_3 is a 0.02-μF capacitor, and $f = 100$ Hz.

25-2-15. $Z_1 = Z_2 = -j3600\ \Omega$, $Z_3 = +j3600\ \Omega$.

25-2-16. $Z_1 = Z_2 = 150\ \mu$H, $Z_3 = 600\ \Omega$, $f = 500$ kHz.

25-2-17. Find Z_{in} when a 300-Ω resistive load is connected to the output terminals of the π-network in Problem 25-2-13.

25-2-18. Find Z_{in} when a load of $300 - j400\ \Omega$ is connected to the output terminals of the π-network in Problem 25-2-16.

25-2-19. With the techniques used in deriving Equations (25-7) to (25-10), and noting that the internal resistance of the internal source symbol is zero, find the z-parameters for the transistor in Figure 22-26.

25-2-20. By rearranging Equation (25-15), calculate the ratio of output current to input current when a 250-Ω resistive load is connected to the output terminals of the T-network of Problem 25-2-1. Attempt to derive a single equation for the ratio of output voltage V_2 to input voltage E_1 in terms of z-parameters and Z_L and use it to find V_2/E_1 for this network and load.

In Problems 25-3-21 and 25-3-22, find the short-circuit y-parameters for each π-network. Use the subscript notation of Figure 25-7.

25-3-21. $Y_p = G_p = 2$ mS, $Y_s = G_s = 4$ mS, $Y_m = G_m = 5$ mS.

25-3-22. $Y_p = Y_s = +j400\ \mu$S, $Y_m = -j400\ \mu$S.

In Problems 25-3-23 and 25-3-24, find the equivalent π-networks for the "black boxes" having the given y-parameters.

25-3-23. $y_{11} = 400\ \mu S\ \underline{/0°}$, $y_{12} = y_{21} = 200\ \mu S\ \underline{/180°}$, $y_{22} = 600\ \mu S\ \underline{/0°}$.

25-3-24. $y_{11} = y_{22} = 10\ mS\ \underline{/-45°}$, $y_{12} = y_{21} = 10\ mS\ \underline{/-135°}$.

25-3-25. Find the y-parameters for the network of Problem 25-2-16.

25-3-26. Find the z-parameters for the network of Problem 25-3-23.

25-4-27. Hybrid parameters are not used for simple resistance coupling networks. However, they may readily be derived. What would the h-parameters for the network of Problem 25-2-1 look like?

25-4-28. Find the hybrid parameters for the transistor in Figure 22-26.

25-4-29. For a common-base configuration, a certain transistor has the following parameters:

$$h_{11} = 40 \ \Omega \qquad h_{12} = 4 \times 10^{-4}$$

$$h_{21} = -0.96 \qquad h_{22} = 10^{-5} \ S$$

The transistor feeds into a 4.7-kΩ resistive load and is fed from a 1-kHz signal source having an open-circuit voltage of 10 mV and an internal resistance of 250 Ω. Calculate the voltage gain (the ratio of output to input voltage).

25-4-30. For a common-emitter configuration, the same transistor as used in Problem 25-4-29 has the following parameters:

$$h_{11} = 1000 \ \Omega \qquad h_{12} = 9.6 \times 10^{-3}$$

$$h_{21} = 24 \qquad h_{22} = 2.5 \times 10^{-4} \ S$$

For the same signal source and load as in Problem 25-4-29, calculate the voltage gain for this circuit configuration.

25-6-31. An air-core transformer with adjustable coupling has a primary self-inductance of 200 mH and a secondary self-inductance of 240 mH. If a 500-μA current at $f = 1$ kHz is flowing in the primary, what coefficient of coupling is required to induce a secondary voltage of 50 mV?

25-6-32. Two coils wound on a common core have self-inductances of 1.6 and 3.2 H, respectively. When connected in series, their total inductance is 8 H.
(a) What is the coefficient of coupling?
(b) What is the total inductance when the leads of one of the coils are reversed?

25-7-33. An air-core transformer has a primary inductance of 200 mH, a secondary inductance of 50 mH, and a coefficient of coupling of 0.1. If the resistance of the windings can be neglected, what is the coupled impedance at 100 kHz when a 10-kΩ resistor is connected across the secondary?

25-7-34. If the voltage applied to the primary winding of the transformer in Problem 25-7-33 has an rms value of 20 mV, what voltage will appear across the 10-kΩ resistor in the secondary circuit?

25-8-35. The primary and secondary windings of an intermediate-frequency transformer in a radio receiver each have a self-inductance of 360 μH. The total equivalent primary-circuit series resistance is 30 Ω, and the total equivalent secondary-circuit series resistance is 24 Ω.
(a) What value of tuning capacitors are required to tune each winding to 456 kHz?
(b) What is the critical coupling coefficient?
(c) What is the approximate bandwidth of the IF transformer if its coils are critically coupled?

25-8-36. An IF transformer in a television receiver has an untuned primary winding and a tuned secondary. The primary self-inductance is 2.4 μH, the secondary inductance is 3.0 μH, and the coefficient of coupling is 0.15. The secondary circuit has an effective resistance of 20 Ω. If the primary current contains a 42-MHz component with an rms value of 50 μA, what voltage will appear across the tuning capacitance in the secondary when the secondary is tuned to resonance?

25-9-37. A 4-H filter choke with a resistance of 20 Ω is connected in series with a pulsating dc source and a resistive load. What value of capacitance must be connected across the choke to provide the most effective filtering of a 120-Hz ripple component in the output of the pulsating source?

25-9-38. An audio filter for reducing the 10-kHz heterodyne between broadcast stations contains a parallel-resonant circuit whose capacitance branch is 0.02 μF. The inductance branch has a resistance of 100 Ω. What inductance should this branch possess?

25-9-39. Solve Example 25-11 with reference to the circuit of Figure 25-21(b).

25-9-40. Design a filter which can be used to pass the output of a 200-kHz powerline carrier transmitter whose bandwidth is 6 kHz but reject any third harmonic content in the transmitter output. This filter will contain an 0.5-mH coil.

25-10-41. (a) What is the cutoff frequency of the low-pass filter network of Example 25-5?
(b) What is the voltage across R_L at the cutoff frequency if E remains at 10 V rms?
(c) Draw a freehand sketch of the amplitude vs frequency response for this filter.
(d) What value of resistance should be connected in parallel with R_L to minimize the resonant rise of voltage at f_c?

25-10-42. An LC high-pass filter consists of a series capacitor, and an inductor in parallel with the output terminals. The filter feeds a 50-Ω resistive load. Determine the values of L and C for a cutoff frequency of 500 kHz with critical damping.

25-10-43. Given $C = 0.5\ \mu$F and $R = 22$ kΩ for the high-pass filter of Figure 25-34(a), and assuming a signal source with a constant rms output of 5 V and negligible internal resistance,
(a) What is the cutoff frequency if the output terminals remain open-circuit?
(b) What is the magnitude and phase of the open-circuit output voltage at f_c?
(c) What is the magnitude and phase of the open-circuit output voltage at $f = 10$ Hz?

(a)　　　　　　　　　　(b)

FIGURE 25-34

25-10-44. If the source in Problem 25-10-43 has an internal resistance of 5 kΩ,
(a) What is the cutoff frequency of the system?
(b) What is the open-circuit output voltage at the original cutoff frequency of Problem 25-10-43?
(c) If a 10-kΩ resistor is now connected across the output terminals, what is the cutoff frequency?

25-10-45. Assuming that the source connected to the low-pass filter of Figure 25-34(b) has an rms output of 15 mV with negligible internal resistance,
(a) What value of capacitance is required for a cutoff frequency of 30 kHz if $R = 4.7$ kΩ?
(b) At what frequency will the open-circuit output voltage lag the source voltage by 60°?
(c) What is the rms magnitude of the open-circuit output voltage at this frequency?

25-10-46. The low-pass filter of Figure 25-34(b) consists of $R = 33$ kΩ and $C = 0.0005\ \mu$F. It is fed from a constant-voltage source with an internal resistance of 10 kΩ. A 10-kΩ resistive load is connected across the output terminals.
(a) What is the cutoff frequency of the system?
(b) If the open-circuit voltage of the source is 1 V rms, what is the output voltage at the cutoff frequency?
(c) Draw a freehand amplitude vs frequency response graph.

25-11-47. An input power of 12 mW is required to produce a power output of 30 W from an audio power amplifier. What is its power gain in decibels?

25-11-48. A broadcast studio feeds a 6-mW signal into a telephone line between the studio and the transmitter. If the signal power arriving at the transmitter is 950 μW, what is the power loss of the telephone line in dB?

25-11-49. What power must a TV signal generator feed into a 12-dB attenuator if the power output is to be 50 mW?

25-11-50. What signal voltage must be applied to the 10-kΩ input terminals of an amplifier having a power gain of 40 dB to produce a power output of 15 W?

25-11-51. By how many dB must the gain of an audio amplifier be increased if the output power is to be increased from 5 W to 25 W?

25-11-52. By how many times must the input voltage to a fixed-gain amplifier be increased in order to increase the output power by 15 dB?

25-12-53. Draw an accurate Bode diagram for the amplitude vs frequency response of the coupling network of Problem 25-10-43.

25-12-54. Draw an accurate Bode diagram for the amplitude vs frequency response of the coupling network of Problem 25-10-45.

25-12-55. Draw accurate Bode diagrams for both amplitude and phase response of the complete networks of Problem 25-10-44(c).

25-12-56. Draw accurate Bode diagrams for both amplitude and phase response of the complete network of Problem 25-10-46. On these diagrams, draw freehand graphs of the actual response.

REVIEW QUESTIONS

25-1. Why will I_2 in the circuit of Figure 25-1(b) turn out to be a negative quantity?

25-2. What is meant by the "black-box" technique of determining the equivalent circuit of a network?

25-3. Why do we require at least three components in the equivalent circuit of a generalized coupling network?

25-4. Give a practical example (with a circuit diagram) in which one of these components has an impedance of 0 Ω.

25-5. Why are z_{12} and z_{21} of the open-circuit impedance parameters called **transfer impedances?**

25-6. With reference to Equations (25-5) and (25-6), discuss the merit of our choice of direction for I_1 and I_2.

25-7. What is the **mutual impedance** of a coupling network?

25-8. Under what circumstances will a coupling network have no mutual impedance?

25-9. Figure 22-26 shows the (older) R-parameters of a transistor equivalent circuit. What is the reason for the internal voltage source symbol in this circuit?

25-10. Account for the $-$ sign in Equation (25-15).

25-11. Account for the $-$ sign in Equation (25-16).

25-12. At a given frequency, it would make no difference to Z_{in} in Figure 25-4 if Z_m were a pure capacitive reactance of X ohms or a pure inductive reactance of X ohms. Explain.

25-13. Why is the H-attenuator network of Figure 25-5 called a "balanced" network, whereas the equivalent T-network is not?

25-14. Under what circumstances must we use the short-circuit parameters of a coupling network, rather than the open-circuit parameters?

25-15. By carefully following the rules for Equations (25-1) to (25-4), determine the open-circuit **z**-parameters for the π-network of Figure 25-19.

25-16. From the **z**-parameters derived in Question 25-15, find the equivalent T-network.

25-17. Derive an equation for $\mathbf{Y}_{in}$ when a load admittance replaces $\mathbf{E}_2$ in Figure 25-7 which will be the dual of Equation (25-16).

25-18. When we connect a load to a four-terminal coupling network, $I_2 = -V_2 Y_L$. Rearrange Equation (25-23) to provide a formula for the output voltage to input voltage ratio of a network in terms of its **y**-parameters and load admittance. Compare with Equation (25-15) for **z**-parameters and with the equation for V_2/E_1 in terms of **z**-parameters you developed in Problem 25-2-20.

25-19. Why is the term **hybrid** used to describe the parameters of Figure 25-12?

25-20. State the units in which each of the four hybrid parameters is expressed.

25-21. Is it possible to apply the Thévenin-equivalent circuit technique used in developing the hybrid parameters to a *passive* coupling network? Explain with numerical examples.

25-22. Define **coefficient of coupling.**

25-23. What is meant by **loose** coupling and **tight** coupling?

25-24. Express coefficient of coupling in terms of mutual flux and leakage flux.

25-25. What effect has the coefficient of coupling on the mutual inductance of a pair of magnetically coupled coils?

25-26. What is meant by **mutual reactance?**

25-27. How does the coupled impedance of loose-coupled coils differ from the reflected impedance of an audio transformer?

25-28. How is the coupled impedance related to the coefficient of coupling?

25-29. Why does an inductive secondary circuit appear as a capacitive coupled impedance?

25-30. Define **critical coupling.**

25-31. Why is the secondary current greater at frequencies slightly off resonance than at the resonant frequency in an overcoupled tuned transformer?

25-32. Why does increasing the coupling between two tuned circuits increase their bandwidth?

25-33. Does resonant rise of voltage take place in the secondary circuit of a tuned transformer? Explain.

25-34. The input capacitor in the network of Figure 25-9 has no effect on the magnitude of the output voltage. What is its purpose?

25-35. What is the significance of the answer to Problem 25-9-38?

25-36. In the construction of the filter circuit of Figure 25-21(b), L_p and L_s are both made variable to allow the filter to be tuned. In which order must these inductances be adjusted? Why?

25-37. The filter network of Figure 25-21(a) performs equally well if the input and output terminals are reversed. Why does the same situation not apply to the filter network of Figure 25-19(a)?

25-38. Why does the low-pass filter of Figure 25-23 have a more distinct boundary between pass-band and reject-band than the low-pass filter of Figure 25-25(b)?

25-39. What factors govern the pass-band of a typical audio amplifier?

25-40. What are the purpose and advantage of expressing power (and voltage) ratios in a communications system in logarithmic terms?

25-41. When the power *output* of an amplifier is stated in decibels, it is also necessary to state a reference power (usually 1 mW). Explain.

25-42. What is the overall power gain of a system composed of a tape cassette head feeding a preamplifier having a gain of 20 dB, which in turn feeds a tone-control network with an insertion loss of 10 dB, which in turn feeds a power amplifier with a gain of 40 dB, which in turn feeds a loudspeaker?

25-43. What is the purpose of Bode diagrams?

25-44. What is the advantage of using decibels for the Y-axis of amplitude response Bode diagrams?

*25-45. Design a computer program to print out the ratio of V_{out} to E_{in} for any four-terminal coupling network when the input data consist of its **z**-parameters and load impedance. (Problem 25-2-20.)

*25-46. Write a computer program which will print out the voltage gain of a transistor amplifier when you enter the four hybrid parameters of the amplifier and R_L.

*25-47. Develop a computer program to display or print out a graph of the output/input voltage ratio (in decibels) when $R_L = 100 \ \Omega$ in Example 25-11.

*25-48. Write a computer program which we could use to produce Table 25-2.

26

THREE-PHASE SYSTEMS

26-1 ADVANTAGES OF POLYPHASE SYSTEMS

In most of the ac networks we have considered up to this point, we have encountered only one source of alternating voltage. Such circuits are referred to as **single-phase** circuits. In Chapter 16, we were introduced to the sinusoidal nature of alternating voltages by considering the instantaneous voltage induced in an electric conductor rotating at a constant angular velocity in a uniform magnetic field. We also noted that most of our electric energy is generated at the expense of mechanical energy by **alternators** based on the principle of a loop rotating in a magnetic field. However, when we consider practical systems for the generation and distribution of large amounts of electric energy, the single-phase system has several disadvantages.

If we add a second loop to the simple alternator of Chapter 16, as shown in Figure 26-1(a), two separate sine-wave voltages will be induced in the one machine. Since both loops are mounted on the same rotor assembly, both rotate at the same angular velocity. Therefore, both voltages have the same frequency. If both loops have the same number of turns, both voltages will have the same rms value. But what is most important is that loop A is mounted on the rotor 90° ahead (in the direction of rotation) of loop B. Consequently, loop A always cuts a certain magnetic line of force 90° ahead (in time) of loop B. Therefore, although we can use the two voltages separately if we wish, as in Figure 26-2(a), the voltage in loop A is always 90° out of phase with the voltage in loop B, as shown in Figure 26-1(b). This simple alternator is referred to as a **two-phase** alternator.

In Figure 26-2, we have connected identical loads to the two windings of the simple two-phase alternator. Note the customary procedure of using a coil to represent the alternator winding rather than the general ac generator symbol. This procedure enables

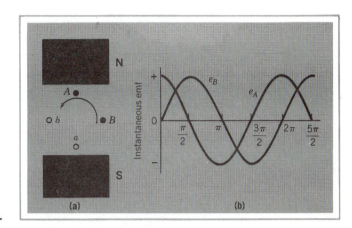

FIGURE 26-1
Simple Two-Phase Generator.

FIGURE 26-2
Simple Two-Phase System.

us to draw the coils at right angles in a circuit diagram to indicate that their voltages are 90° out of phase. Even though we can connect each winding independently to its load, as in Figure 26-2(a), it is customary to use a common or **neutral** lead for the two circuits, as in Figure 26-2(b). This neutral conductor reduces the number of conductors required from four to three. Although, as the circuit diagram of Figure 26-2(b) indicates, the neutral current is the sum of the two load currents, this is a *phasor* sum. Therefore, if the two loads are identical, the neutral current is not twice the current in each of the other two conductors, but, as shown in the phasor diagram of Figure 26-3, it is only $\sqrt{2}$, or 1.414, times the current in each of the other conductors. Therefore, the total copper cross section to feed the two 120-ohm loads at 120 volts in the two-phase system is only the copper cross section required to handle one ampere, 1.414 amperes, and one ampere, a total of 3.414 amperes. The equivalent single-phase system to supply the two 120-ohm loads in parallel at 120 volts would require two conductors, each carrying two amperes, or a total copper cross section to handle four amperes. This then is the first advantage of any polyphase system. **Less copper is required to supply a given load power at a given voltage with a polyphase system than with a single-phase system.**

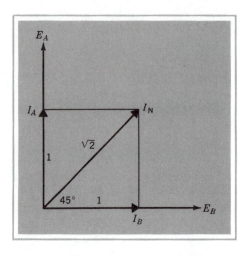

FIGURE 26-3
Phasor Diagram for the Neutral Current in a Simple Two-Phase System.

We can illustrate the second advantage of a polyphase system by plotting on a common graph, as in Figure 26-4, the graphs of the instantaneous power to the two identical loads in the circuits of Figure 26-2. The total instantaneous power supplied by the alternator at any instant is the sum of the instantaneous power to the two loads at that instant. Since the two voltages are 90° out of phase, the instantaneous power to one load becomes maximum as the other becomes zero. Even though the instantaneous power in each load is the customary pulsating sine-squared wave, if we check carefully, we find that at every instant in time the sum of the two instantaneous powers is constant. Therefore, **if the load on each phase of a polyphase source is identical, the instantaneous power output of the alternator is constant.** This is an important advantage in large machines since it allows a steady conversion of mechanical energy into electric energy.

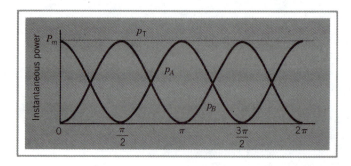

FIGURE 26-4
Instantaneous Power in a Balanced Two-Phase System.

The third advantage can be demonstrated by connecting the two-phase alternator to a set of coils placed at right angles, as shown in Figure 26-5(a). Each coil will pass a sine wave of current, but the current in coil B will be 90° out of phase with the current in coil A. The magnetic flux produced by the two currents at any instant depends on the magnitude at that instant of the two instantaneous currents. At the 0° position, the current in coil B is zero and the current in coil A is maximum in the positive direction. This

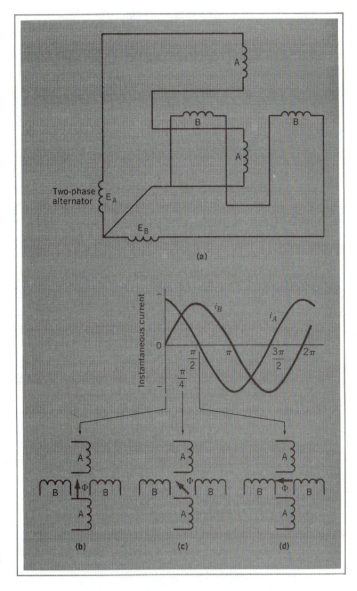

FIGURE 26-5
Producing a Rotating Magnetic Field in a Two-Phase System.

produces a magnetic field in which the direction of the magnetic lines of force is that indicated by the arrow in Figure 26-5(b). At 45°, both coils pass $0.707I_m$, resulting in a net flux which is the *phasor* sum of these two components. This results in a total flux with the *same* total flux density but with the direction of the magnetic lines of force being that shown in Figure 26-5(c). At the 90° point in the cycle, coil A current is zero and coil B current is maximum positive, resulting in a magnetic field which is now at right angles to its original direction at 0° in the cycle.

If we continue the diagrams commenced in Figure 26-5, we find that the coil arrangement of Figure 26-5(a) connected to a two-phase source develops **a magnetic field which has a constant flux density and rotates at the frequency of the applied**

sine wave. If we place a compass needle in the center of the coils in Figure 26-5(a), it will rotate with the magnetic field at a **synchronous speed,** which, in this case, is the same as the frequency of the two-phase sine wave. This rotating magnetic field, characteristic of all polyphase systems, greatly simplifies ac motor construction. A single-phase system produces only a magnetic field with increases and decreases in flux density and *reverses* its direction each 180° but does *not* rotate.

26-2 GENERATION OF THREE-PHASE VOLTAGES

The simple two-phase system that we used to show the advantages of polyphase electric power generation has been used to a limited extent in the distribution of electric energy. However, its main application in present-day equipment is in servomechanisms. The auto pilot of an aircraft is an example of a servomechanism. If the aircraft deviates from its course, a gyrocompass senses this error and develops voltages which operate a servomotor, which, in turn, operates the rudder to bring the aircraft back on course. Two-phase motors are ideal for servosystem applications because their rotation depends on the presence of an error voltage and their direction of rotation depends on whether this sine-wave error voltage leads or lags a reference voltage by 90°.

In search of an improvement on the two-phase system for the distribution of electric energy, let us suppose that we place two coils 180° apart on the rotor of a simple ac generator and connect them to two identical loads with a common neutral lead, as shown in Figure 26-6. The neutral current is now the phasor sum of two equal currents which are 180° out of phase. Therefore, with identical loads, the neutral current in this system is zero. Consequently, this system requires less copper than the two-phase system of Figure 26-2(b) to supply each load with 115 V at a given power. However, the total voltage from line *A* to line *B* is greater than in the two-phase system. And the instantaneous power in each coil of the generator in Figure 26-6 reaches its peak at the *same* instant, and thus the total instantaneous power *is* pulsating in nature. Also, with two currents 180° out of phase, we cannot produce a rotating magnetic field. Therefore, this system is *not* a two-phase system. It is called the Edison three-wire, single-phase system

FIGURE 26-6
Edison Three-Wire Single-Phase System.

and is widely used for distributing electric energy to residences because of the reduction in line current (and the resulting reduction in wire size required) for a given load power and voltage, as compared with the simple two-wire, single-phase system.

The system which is universally used for electric power distribution is a *three*-phase system, based on the simple generator of Figure 26-7(a), which has three coils mounted on the rotor at 120° intervals. This alternator generates three sine-wave voltages which are 120° out of phase with one another, as shown in Figure 26-7(b). Figure 26-8(a) shows one method of connecting these coils to a load. With a **balanced** load, the currents in each arm of the load will be equal in magnitude and 120° out of phase with one another, as shown in the phasor diagram of Figure 26-8(b).[†] The neutral current is the phasor sum of these three load currents. As the phasor diagram of Figure 26-8(b) indicates, this neutral current will be zero with a balanced load. As a result, with a balanced load, even less copper is required to convey energy at a given rate and given voltage in the three-phase system than in an equivalent two-phase system.

[†] A balanced load is one in which the impedance of each arm of the load is identical, both in magnitude and phase (power factor) angle. A balanced load need not be only resistance.

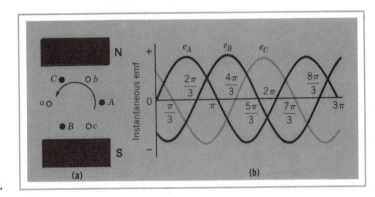

FIGURE 26-7
Simple Three-Phase Generator.

FIGURE 26-8
Simple Three-Phase System.

With identical load resistors in Figure 26-8(a),

$$p_T = \frac{e_A^2 + e_B^2 + e_C^2}{R}$$

Careful inspection of Figure 26-7(b) shows us that $e_A^2 + e_B^2 + e_C^2$ is always constant. Therefore, the instantaneous power in a balanced three-phase system is a constant. We also find that we can use the voltages generated by a three-phase alternator to produce a rotating magnetic field of constant flux density, just as we did with the two-phase system. Consequently, the three-phase system has all the advantages of the two-phase system plus others that we shall discover as we investigate the behavior of three-phase circuits.

26-3 DOUBLE-SUBSCRIPT NOTATION

The three-phase (3ϕ) circuit of Figure 26-8 represents an impedance network with three voltage sources acting simultaneously. Reversing the leads to one of the coils in an ac generator has the same effect on an ac circuit as reversing the leads to a battery in a dc circuit containing several dc voltage sources. Hence, we cannot take the six leads from the coils in Figure 26-8(a), connect them in a random manner, and then assume that we obtain the current relationships shown by the phasor diagram of Figure 26-8(b). When we consider that we can reverse the leads to each coil of the simple 3ϕ generator, thus changing the angle of the source voltage by 180°, we can obtain quite a few possible combinations.

In the circuit of Figure 26-8(a), we indicate the manner in which we connected the leads to each winding by drawing an arrow alongside the symbol representing the alternator winding. Using the angle given with the source voltage, we must consider the tracing direction around the loop to be in the direction indicated by the arrow. However, the arrow system by itself does not allow us to show the connection of the leads to each coil when we write a Kirchhoff's voltage-law equation for the loop. Neither does the letter symbol $\mathbf{E}_A$ give us any indication of which lead is which.

This problem can be solved by using a **double-subscript notation.** In one of the common systems of double-subscript notation, one end of each coil is indicated by a lowercase letter and the other by an uppercase letter, as in Figure 26-9(a). In any complete electric circuit, these coils will be included in tracing loops around the circuit. If, as we follow one of these tracing loops, it enters coil A by the terminal marked with the lowercase letter and leaves by the terminal marked with the uppercase letter, we would represent the induced voltage of that coil as $\mathbf{E}_{aA}$. **The first of the two subscripts always represents the end of a circuit component at which the tracing loop enters the component, and the second subscript represents the end of the component from which the tracing loop leaves the component.**

When we use double-subscript notation to identify voltages, we must also be careful to distinguish between a **voltage rise** *inside* the source and a **voltage drop** in the *external* circuit. In the simplified diagram of Figure 26-10, we have connected a load directly across an alternator winding whose terminals are marked a and A. The direction of the induced voltage for this winding, as shown by the arrow in Figure 26-10, is from

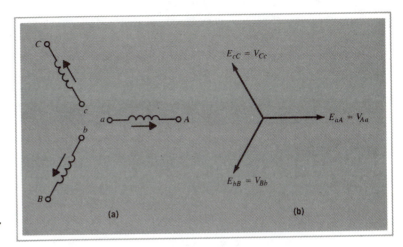

FIGURE 26-9
Double-Subscript Notation for Three-Phase Voltages.

FIGURE 26-10
Double-Subscript Notation for Voltage Rise and Voltage Drop.

a to *A* *inside* the alternator. Therefore, we identify the voltage rise inside the alternator as $\mathbf{E}_{aA}$. To be consistent with the tracing loop, the voltage drop across the load $\mathbf{V}_{Aa}$ must reverse the order of the subscripts. From Kirchhoff's voltage law,

$$\mathbf{E}_{aA} = \mathbf{V}_{Aa} \tag{26-1}$$

Hence, the same phasors in Figure 26-9(b) can be used to represent either the voltage rises *inside* the source or the voltage drops in the external loads.

When we write loop equations based on Kirchhoff's voltage law, we include both $\mathbf{E}$ and $\mathbf{V}$ in the same equations [as in Equation (26-1)]. Hence, proper designation of letter symbols and subscripts is quite important. But when we write Kirchhoff's current-law equations for nodal analysis, we use only the external voltage drop between the two nodes and encounter a further convention for subscripts. **In the circuit external to the source, the *second* subscript with the voltage drop symbol V is the *reference* node.** Hence, in Figure 26-10, terminal *a* is the reference node and the voltmeter is measuring the voltage from terminal *A* with respect to terminal *a*.

In the following sections, when we are concerned with the manner in which we connect alternator windings, we use the internal voltage rise symbol **E**. If we reverse the leads to coil *A* so that the tracing loop in a given circuit enters the coil at end *A* and leaves at end *a*, we would represent the source voltage in Kirchhoff's voltage-law equations as $\mathbf{E}_{Aa}$, where $\mathbf{E}_{Aa}$ has the same rms magnitude as $\mathbf{E}_{aA}$ but its angle is displaced by an additional 180°. From our study of subtraction of phasors in Chapter 18, since $\mathbf{E}_{Aa}$ is 180° out of phase with $\mathbf{E}_{aA}$,

$$\therefore \mathbf{E}_{Aa} = -\mathbf{E}_{aA} \tag{26-2}$$

When we make calculations in the *external* circuit, we usually use the appropriate voltage drop symbol $\mathbf{V}$ with conventional double subscripts.

26-4 FOUR-WIRE WYE-CONNECTED SYSTEM

We can assume that, in winding the coils on the simple 3ϕ generator of Figure 26-7, we were methodical enough that if we take the tracing direction from the lowercase letter to the uppercase letter in each coil, the three voltages will be 120° out of phase. We assume that these three generated voltages for the alternator of Figure 26-7 are

$$\mathbf{E}_{aA} = 120 \text{ V } \underline{/0°}$$

$$\mathbf{E}_{bB} = 120 \text{ V } \underline{/-120°}$$

$$\mathbf{E}_{cC} = 120 \text{ V } \underline{/+120°}$$

Since one of the reasons for using a three-phase system is to cut down on the conductors required to carry electric energy to a load, we can connect one end of each generator coil to a common neutral. **It is conventional to consider the tracing direction as going out from the generator to the load along the individual lines and returning along the common neutral lead;** consequently, we connect the lowercase letter ends of the three windings to the neutral, as shown in Figure 26-11. To show that points a, b, and c are all connected to the neutral, we can replace these lowercase letters with the letter o. Hence, the three source voltages become $\mathbf{E}_{oA} = \mathbf{E}_{aA}$, $\mathbf{E}_{oB} = \mathbf{E}_{bB}$, and $\mathbf{E}_{oC} = \mathbf{E}_{cC}$. As shown in Figure 26-11(b), we designate the equivalent voltage drops from line to neutral as $\mathbf{V}_{AN}$, $\mathbf{V}_{BN}$, and $\mathbf{V}_{CN}$. Connecting the alternator coils in this manner produces a

FIGURE 26-11
Wye-Connected Alternator Windings.

wye- (Y)-connected three-phase source, so called because of the appearance of the circuit diagram and the phasor diagram of Figure 26-11.[†] The voltage measured from line to neutral with a wye-connected source is called the **phase voltage** of the source.

We could also form a wye-connected 3ϕ source by connecting all the uppercase letter ends of the coils to the common neutral. However, we must avoid the situation shown in Figure 26-12. Coils A and C have their lowercase letter ends connected to the neutral, but coil B leads have been reversed. Since the tracing direction must be out along the line and back along the neutral, we must reverse the subscripts and represent the voltage induced in coil B as $\mathbf{E}_{Bb}$ in Figure 26-12(b), which is 180° displaced from $\mathbf{E}_{bB}$ in Figure 26-11(b). If we connect such a source to three equal resistors, as the phasor diagram of Figure 26-12(b) suggests, there will be considerable neutral current.

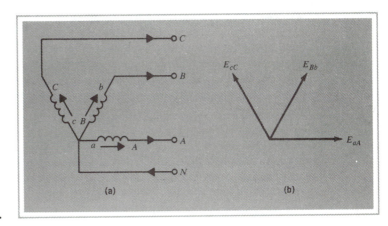

FIGURE 26-12
Incorrect Wye Connection.

Figure 26-13 shows three impedances connected as a four-wire wye load. If we trace the wiring carefully, we note that $\mathbf{Z}_A$ is connected directly across coil A of the generator. We also find that the current in $\mathbf{Z}_A$, the current in line A, and the current in coil A must all be the same since they are all part of the same series loop. Because it is

[†]A system in which one terminal of each phase is connected to a common point (star point) can also be called a **star**-connected system.

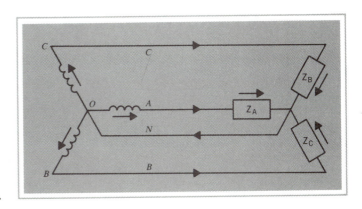

FIGURE 26-13
Four-Wire Wye-Connected Load.

conventional to consider the tracing direction as outward from the alternator along the individual lines and back toward the alternator along the neutral, we do not need double subscripts to represent the current in a four-wire wye system. Therefore, $\mathbf{I}_A$ represents the current in $\mathbf{Z}_A$, line A, and generator coil A. Hence,

$$\mathbf{I}_A = \frac{\mathbf{E}_{oA}}{\mathbf{Z}_A} = \frac{\mathbf{V}_{AN}}{\mathbf{Z}_A}, \ \mathbf{I}_B = \frac{\mathbf{E}_{oB}}{\mathbf{Z}_B} = \frac{\mathbf{V}_{BN}}{\mathbf{Z}_B}, \ \text{ and } \ \mathbf{I}_C = \frac{\mathbf{E}_{oC}}{\mathbf{Z}_C} = \frac{\mathbf{V}_{CN}}{\mathbf{Z}_C} \qquad (26\text{-}3)$$

and

$$\mathbf{I}_N = \mathbf{I}_A + \mathbf{I}_B + \mathbf{I}_C \qquad (26\text{-}4)$$

where all quantities are phasors.

The three load impedances of the four-wire wye load may be independent single-phase loads. For example, $\mathbf{Z}_A$ may be a bank of lamps, $\mathbf{Z}_B$ may be a heater, and $\mathbf{Z}_C$ a single-phase motor. It is customary in dividing the total single-phase load among the three phases to make the line currents as nearly equal in magnitude as possible. Also, $\mathbf{Z}_A$, $\mathbf{Z}_B$, and $\mathbf{Z}_C$ may represent the windings of a 3ϕ motor, in which case $\mathbf{Z}_A = \mathbf{Z}_B = \mathbf{Z}_C$ and the load is a balanced load. In such cases, the three line currents will be equal in magnitude, and the neutral current will be zero. For a four-wire balanced wye load,

$$I_L = \frac{V_P}{Z} \qquad (26\text{-}5)$$

where I_L is the magnitude of the current in each line, V_P is the magnitude of the phase voltage, and Z is the impedance from each line to neutral. The load may also be made up of a combination of single-phase and three-phase loads.

EXAMPLE 26-1

In the circuit of Figure 26-13, $\mathbf{Z}_A = \mathbf{Z}_B = \mathbf{Z}_C = 30 \ \Omega \underline{/+30°}$. The phase voltages are as shown in Figure 26-11(b).
(a) Find the current in each line.
(b) Find the neutral current.
(c) Draw a phasor diagram showing phase voltages and line currents.

SOLUTION

(a)
$$\mathbf{I}_A = \frac{\mathbf{V}_{AN}}{\mathbf{Z}_A} = \frac{120 \ \text{V} \underline{/0°}}{30 \ \Omega \underline{/+30°}} = 4 \ \text{A} \underline{/-30°}$$

$$\mathbf{I}_B = \frac{\mathbf{V}_{BN}}{\mathbf{Z}_B} = \frac{120 \ \text{V} \underline{/-120°}}{30 \ \Omega \underline{/+30°}} = 4 \ \text{A} \underline{/-150°}$$

$$\mathbf{I}_C = \frac{\mathbf{V}_{CN}}{\mathbf{Z}_C} = \frac{120 \ \text{V} \underline{/+120°}}{30 \ \Omega \underline{/+30°}} = 4 \ \text{A} \underline{/+90°}$$

(b)
$$\mathbf{I}_N = \mathbf{I}_A + \mathbf{I}_B + \mathbf{I}_C = 4 \ \text{A} \underline{/-30°} + 4 \ \text{A} \underline{/-150°} + 4 \ \text{A} \underline{/+90°}$$
$$= (3.464 - j2) + (-3.464 - j2) + (0 + j4) = \mathbf{0 \ A}$$

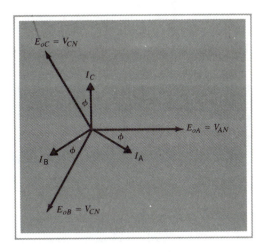

FIGURE 26-14
Phasor Diagram for Example 26-1.

(c) Figure 26-14 shows the phasor diagram for the phase voltage and line currents. Once we realize that the neutral current with a balanced wye load is zero, it is not necessary to go through the calculations of step (b). Note that each load current lags the voltage applied to that load by 30°. This is called the **power-factor angle, ϕ.**

EXAMPLE 26-2
In the circuit of Figure 26-13, $\mathbf{Z}_A$ is a 360-W lamp, $\mathbf{Z}_B$ passes a current of 4 A at a 0.966 lagging power factor, and $\mathbf{Z}_C = 60 \ \Omega \underline{/-30°}$. Find the neutral current.

SOLUTION
Because the lamp possesses only resistance, $I_A = P_A/V_{AN} = 360 \text{ W}/120 \text{ V} = 3$ A in phase with $\mathbf{V}_{AN}$ (Figure 26-15).

$$\therefore \mathbf{I}_A = 3 \text{ A} \underline{/0°}$$

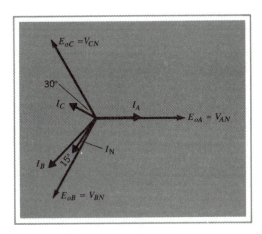

FIGURE 26-15
Phasor Diagram for Example 26-2.

Since $\mathbf{Z}_B$ has a 0.966 lagging power factor, the power-factor angle is arccos 0.966 = 15°. Therefore, $\mathbf{I}_B$ lags $\mathbf{V}_{BN}$ by 15° and

$$\therefore \mathbf{I}_B = 4 \text{ A}\underline{/-135°}$$

$$\mathbf{I}_C = \frac{\mathbf{V}_{CN}}{\mathbf{Z}_C} = \frac{120 \text{ V}\underline{/+120°}}{60 \text{ }\Omega\underline{/-30°}} = 2 \text{ A}\underline{/+150°}$$

$$\mathbf{I}_N = \mathbf{I}_A + \mathbf{I}_B + \mathbf{I}_C$$

$$= 3 \text{ A}\underline{/0°} + 4 \text{ A}\underline{/-135°} + 2 \text{ A}\underline{/+150°}$$

$$= (3 + j0) + (-2.828 - j2.828) + (-1.732 + j1)$$

$$= -1.56 - j1.828 \text{ A} = \mathbf{2.4 \text{ A}\underline{/-130.5°}}$$

SELF-CHECK

26-4:1. A 4-wire wye-connected load, similar to that shown in Figure 26-13, has the following impedances:

$$\mathbf{Z}_A = 30 + j40 \text{ }\Omega$$
$$\mathbf{Z}_B = 35 - j35 \text{ }\Omega$$
$$\mathbf{Z}_C = 50 \text{ }\Omega$$

Using the phase voltages from Figure 26-11, find:
(a) The line currents
(b) The neutral current

26-4:2. Two phases of a 4-wire wye-connected load are fixed at:

load A: 1200-VA motor with an 80% lagging power factor,
load B: 1450-W electric heater

What additional load, when used on the third phase, will reduce the neutral current to zero?

26-5 DELTA-CONNECTED SYSTEM

The phasor diagram of Figure 26-9(b) suggests that, when properly connected, the **phasor sum of the voltages generated in the three coils of a three-phase source is zero.**[†] To add voltages by circuit connection, we connect the coils in series (like adding batteries in series). We start by connecting end b of coil B to end A of coil A and end c of coil C to end B of coil B, as shown in Figure 26-16(a). Finally, we connect a voltmeter between the remaining two terminals to read the phasor total voltage.

[†] The three-phase source need not be a rotating machine. The secondary windings of three transformers fed from a three-phase system can be considered as representing the coils of a three-phase source.

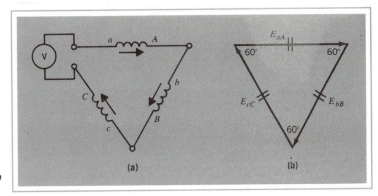

FIGURE 26-16
Delta Connection of a Three-Phase Source.

When we are required to add several phasors diagrammatically, we can cut down on the geometrical construction and also obtain a less cluttered phasor diagram by switching to funicular phasors (Section 24-8). Since the head end of the phasor for E_{aA} represents the uppercase letter end of the coil [Figure 26-16(a)], we can duplicate the circuit connection of Figure 26-16(a) by drawing the phasor $\mathbf{E}_{bB}$ with its proper magnitude and direction from the tip of the phasor $\mathbf{E}_{aA}$. Similarly, we construct the phasor for $\mathbf{E}_{cC}$ from the tip of the phasor $\mathbf{E}_{bB}$. This returns the tip of the phasor $\mathbf{E}_{cC}$ back to the origin, forming an equilateral triangle. If we have actually connected the coils as we have specified, the voltmeter in Figure 26-16(a) will read zero. Therefore, it is safe to connect terminal C to terminal a without any short-circuit current flowing around the loop. In practice, we would always check with a voltmeter before making this final connection, because reversing the leads to any one of the coils (coil C in Figure 26-17) results in a voltmeter reading of twice the phase voltage. This would cause a very heavy current, limited only by the impedance of the three coils, to flow around the loop.

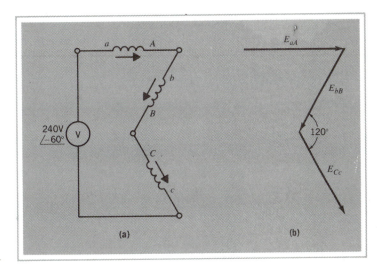

FIGURE 26-17
Incorrect Delta Connection.

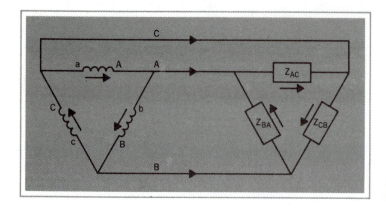

FIGURE 26-18
Delta-Connected Load.

Since we can connect the windings of a three-phase source into a delta without any short-circuit current flowing, we require only three wires to connect the three sets of voltages to a load which can also be connected in delta fashion, as in Figure 26-18.[†] In the **delta** (Δ)-connected system, each arm of the load is connected directly across one of the generator coils. We can label the three lines according to the *uppercase* or second subscript letters of the various phase voltages connected to each line. Thus, $\mathbf{E}_{aA}$ becomes $\mathbf{E}_{CA}$, $\mathbf{E}_{bB}$ becomes $\mathbf{E}_{AB}$, and $\mathbf{E}_{cC}$ becomes $\mathbf{E}_{BC}$. The equivalent voltage drops become $\mathbf{V}_{AC}$, $\mathbf{V}_{BA}$, and $\mathbf{V}_{CB}$.

In the wye-connected load where each arm of the load is connected from line to neutral, we found that a single subscript for each impedance was quite satisfactory. In the delta load, each arm is connected from line to line. Consequently, to identify each impedance properly, we use two subscripts, these being the two lines between which the arm in question is connected. The order in which we write the subscripts is taken from the tracing direction around each loop. Let us examine carefully the loop between lines A and B. The tracing direction through generator coil B is from b to B. Therefore, the voltage generated in that coil is written $\mathbf{E}_{AB}$. This loop then passes through the load impedance from line B to line A. Hence, the impedance is identified as $\mathbf{Z}_{BA}$. The subscripts are in the same order as the subscripts for the voltage drop $\mathbf{V}_{BA}$. Hence, when working with delta-connected loads, it will be less confusing if we work with $\mathbf{V}$ rather than $\mathbf{E}$, since this allows us to write the voltage, current, and impedance subscripts at the load all with the same order.

The phase current of the source is defined as the current flowing in a particular coil of the source, and the phase current of a load is the current flowing in a particular arm of the load. In the wye-connected system, the source phase current, load phase current, and line current are all one and the same. But careful inspection of the delta system of Figure 26-18 shows us that each line has to carry current for *two* arms of the load. Closer inspection shows us that the tracing direction for one of these currents is *away* from the source and the other is *toward* the source. Therefore, the line current to a delta load must be the phasor *difference* between the two load phase currents flowing in that line. (However, if two phasor quantities have an angle between them of greater than 90°, their phasor difference is *greater* than either one by itself.)

[†] So called because of the similarity to the Greek letter Δ.

Hence, the load phase currents in Figure 26-18 are

$$\mathbf{I}_{AC} = \frac{\mathbf{V}_{AC}}{\mathbf{Z}_{AC}} = \frac{\mathbf{E}_{CA}}{\mathbf{Z}_{AC}} \qquad \text{(26-6A)}$$

$$\mathbf{I}_{CB} = \frac{\mathbf{V}_{CB}}{\mathbf{Z}_{CB}} = \frac{\mathbf{E}_{BC}}{\mathbf{Z}_{CB}} \qquad \text{(26-6B)}$$

$$\mathbf{I}_{BA} = \frac{\mathbf{V}_{BA}}{\mathbf{Z}_{BA}} = \frac{\mathbf{E}_{AB}}{\mathbf{Z}_{BA}} \qquad \text{(26-6C)}$$

Even in a three-wire system, it is conventional to consider the direction of the line current as being *outward* along all *three* lines *away* from the source, as shown in Figure 26-18. Therefore, with reference to Figure 26-18,

$$\mathbf{I}_A = \mathbf{I}_{AC} - \mathbf{I}_{BA} \qquad \text{(26-7A)}$$

$$\mathbf{I}_B = \mathbf{I}_{BA} - \mathbf{I}_{CB} \qquad \text{(26-7B)}$$

$$\mathbf{I}_C = \mathbf{I}_{CB} - \mathbf{I}_{AC} \qquad \text{(26-7C)}$$

In a 3ϕ, three-wire system, there is no neutral for a return tracing direction from the load back to the generator. Consequently, even with an unbalanced load, in any 3ϕ, three-wire system, the phasor sum of the three line currents must be zero. We can show this by adding the three Equations (26-7).

$$\mathbf{I}_A + \mathbf{I}_B + \mathbf{I}_C = 0 \qquad \text{(26-8)}$$

This relationship is useful in checking the solution of 3ϕ, three-wire ac circuit problems.

EXAMPLE 26-3
The impedances of the branches of the delta load in Figure 26-18 are $\mathbf{Z}_{AC} = 60~\Omega\underline{/0°}$, $\mathbf{Z}_{CB} = 30~\Omega\underline{/-30°}$, and $\mathbf{Z}_{BA} = 30~\Omega\underline{/+45°}$. Determine the three line currents. Use the alternator voltages given in Figure 26-11(b).

SOLUTION
Step I. Start to prepare a phasor diagram, adding phasors as we determine their data. The complete phasor diagram is shown in Figure 26-19.

Step II. Determine the phase currents.

$$\mathbf{I}_{AC} = \frac{\mathbf{V}_{AC}}{\mathbf{Z}_{AC}} = \frac{120~V\underline{/0°}}{60~\Omega\underline{/0°}} = 2~A\underline{/0°}$$

$$\mathbf{I}_{CB} = \frac{\mathbf{V}_{CB}}{\mathbf{Z}_{CB}} = \frac{120~V\underline{/+120°}}{30~\Omega\underline{/-30°}} = 4~A\underline{/+150°}$$

$$\mathbf{I}_{BA} = \frac{\mathbf{V}_{BA}}{\mathbf{Z}_{BA}} = \frac{120~V\underline{/-120°}}{30~\Omega\underline{/+45°}} = 4~A\underline{/-165°}$$

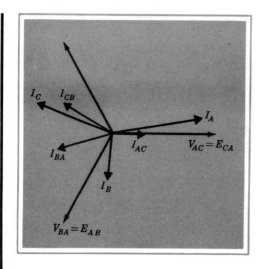

FIGURE 26-19
Phasor Diagram for Example 26-3.

Step III. Determine the line currents.

$$\mathbf{I}_A = \mathbf{I}_{AC} - \mathbf{I}_{BA} = 2 \text{ A} \underline{/0°} - 4 \text{ A} \underline{/-165°}$$
$$= (2 + j0) - (-3.864 - j1.035)$$
$$= 5.864 + j1.035 \text{ A} = \mathbf{5.96 \text{ A} \underline{/+10°}}$$

$$\mathbf{I}_B = \mathbf{I}_{BA} - \mathbf{I}_{CB} = 4 \text{ A} \underline{/-165°} - 4 \text{ A} \underline{/+150°}$$
$$= (-3.864 - j1.035) - (-3.464 + j2)$$
$$= -0.4 - j3.035 \text{ A} = \mathbf{3.06 \text{ A} \underline{/-97.5°}}$$

$$\mathbf{I}_C = \mathbf{I}_{CB} - \mathbf{I}_{AC} = 4 \text{ A} \underline{/+150°} - 2 \text{ A} \underline{/0°}$$
$$= (-3.464 + j2) - (2 + j0)$$
$$= -5.464 + j2 \text{ A} = \mathbf{5.82 \text{ A} \underline{/+159.9°}}$$

Check.

$$\mathbf{I}_A + \mathbf{I}_B + \mathbf{I}_C = (5.864 + j1.035) + (-0.4 - j3.035) + (-5.464 + j2)$$
$$= 0 + j0 \text{ A}$$

Perhaps we can understand the relationship between line current and phase current in a delta system better if we carry out the phasor subtraction by geometrical construction. From the solution for Example 26-3,

$$\mathbf{I}_B = \mathbf{I}_{BA} - \mathbf{I}_{CB} = 4 \underline{/-165°} - 4 \underline{/+150°}$$

As we noted in Chapter 18, we can draw a phasor for $-\mathbf{I}_{CB}$ in the phasor diagram of Figure 26-20 by drawing a phasor equal in length to the $\mathbf{I}_{CB}$ phasor but in the opposite

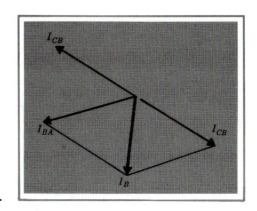

FIGURE 26-20
Phasor Subtraction of Phase Currents.

direction from the origin. Having constructed the $-\mathbf{I}_{CB}$ phasor, we can now add the $\mathbf{I}_{BA}$ phasor and the $-\mathbf{I}_{CB}$ phasor by the usual geometrical construction to complete the parallelogram.

In a balanced delta load, the impedances of all three arms are equal in magnitude and angle. Since the phase voltages are equal and 120° out of phase, it follows that the phase currents will also be equal and 120° out of phase, as shown in Figure 26-21. Therefore, $\mathbf{I}_{CB}$ and $-\mathbf{I}_{AC}$ are equal in magnitude and have a 60° angle between them.

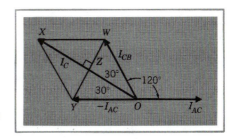

FIGURE 26-21
Showing the Relationship between Line Current and Phase Current with a Balanced Delta Load.

We can label the parallelogram by which we construct the line-current phasor as shown in Figure 26-21. Since $\mathbf{I}_{CB}$ and $-\mathbf{I}_{AC}$ are equal in magnitude and have a 60° angle between them, OWY is an equilateral triangle which is bisected by the diagonal OX which bisects WY at right angles at point Z

$$\therefore OW = 2WZ$$

From Pythagoras' theorem,

$$OZ = \sqrt{OW^2 - WZ^2} = WZ\sqrt{2^2 - 1^2} = \sqrt{3}\,WZ = \frac{\sqrt{3}}{2}OW$$

Because the diagonals of a parallelogram bisect each other,

$$OX = 2OZ = \sqrt{3}\,OW$$

Hence,

The line current to a balanced delta load has a magnitude of $\sqrt{3}$ times the phase current in each arm of the load and is displaced 30° from the phase current.

SELF-CHECK

26-5:1. A 3ϕ delta-connected system similar to the one shown in Figure 26-18 has the following loads:

$$\mathbf{Z}_{AC} = 30 + j40 \ \Omega$$
$$\mathbf{Z}_{BA} = 35 - j35 \ \Omega$$
$$\mathbf{Z}_{CB} = 50 \ \Omega$$

(a) Using the voltages given in Example 26-3, find the line currents.
(b) Use Equation (26-8) to check your line currents.

26-5:2. Determine the line currents in a 120-V delta-connected balanced 3ϕ system (similar to Figure 26-18) where the loads are:

$$\mathbf{Z} = 12 \ \Omega \underline{/25°}$$

Use the voltages given in Example 26-3.

26-6 WYE-DELTA SYSTEM

In discussing the four-wire wye-connected three-phase system, we considered only the source voltages as measured from line to neutral. Since the coils of the wye-connected source are connected from line to neutral, this voltage is the phase voltage of the wye-connected source. However, in working with three-phase systems, it is customary to work in terms of the voltage measured between lines and called the **line voltage.** In fact, unless it is specified otherwise, the voltages given for a three-phase system are **line** voltages. When the source is delta-connected, this distinction poses no problem. Because the coils of the source are connected from line to line, the phase voltage and line voltage of a delta-connected source are one and the same thing.

In the wye-connected source of Figure 26-22, each voltmeter reads a line voltage which depends on the voltages generated in *two* coils. In writing the subscripts for the three line voltages, we must decide whether to read clockwise or counterclockwise around the phasor diagram for the source. As long as we follow the same direction for all three line voltages, they will have the same relative angular positions. In Figure 26-22, we have drawn the tracing arrows for the source in the direction which will give us the same subscripts for the line voltages in this wye-connected source as we had for the delta-connected source of Figure 26-18.

Following the tracing arrow from line A to line B through the wye-connected source in Figure 26-22 shows us that

$$\mathbf{E}_{AB} = \mathbf{E}_{Aa} + \mathbf{E}_{bB} = -\mathbf{E}_{aA} + \mathbf{E}_{bB} \tag{26-9A}$$

FIGURE 26-22
Line Voltages of a Wye-Connected
Three-Phase Source.

Therefore, the line voltage in a wye-connected source is the phasor *difference* between two phase voltages. Since the angle between the phase voltages is greater than 90°, this phasor difference is greater than either phase voltage by itself. Using the values given with Figure 26-11, we get

$$\mathbf{E}_{AB} = -120 \text{ V}\underline{/0°} + 120 \text{ V}\underline{/-120°}$$
$$= -(120 + \text{j}0) + (-60 - \text{j}104)$$
$$= -180 - \text{j}104 \text{ V} = 208 \text{ V}\underline{/-150°}$$

We can also solve for $\mathbf{E}_{AB}$ by geometrical construction, as shown in Figure 26-23. First, we construct phasors for the given phase voltages $\mathbf{E}_{oA}$ and $\mathbf{E}_{oB}$. Then we construct the phasor for $-\mathbf{E}_{oA}$ by reversing the direction of the $\mathbf{E}_{oA}$ phasor. $\mathbf{E}_{AB}$ then becomes the phasor sum of $-\mathbf{E}_{oA}$ and $\mathbf{E}_{oB}$.

FIGURE 26-23
Determining the Line Voltages of a
Wye-Connected Source by
Geometrical Construction.

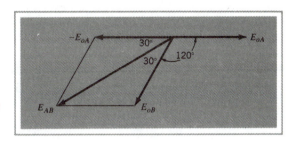

If the coils of the wye-connected source are properly connected, the phase voltages are always equal in magnitude and 120° out of phase. Consequently, just as in the case of the line current to a balanced delta load, the line voltage from a wye must be

$$V_\text{L} = \sqrt{3} \, V_\text{P} = 1.732 \, V_\text{P} \qquad (26\text{-}10)$$

For the phase voltages given with the wye-connected source of Figure 26-11, the three line voltages are

$$\mathbf{V}_{BA} = \mathbf{E}_{AB} = \mathbf{E}_{Aa} + \mathbf{E}_{bB} = \mathbf{E}_{bB} - \mathbf{E}_{aA} = 208 \text{ V}\underline{/-150°} \qquad \text{(26-9A)}$$

$$\mathbf{V}_{CB} = \mathbf{E}_{BC} = \mathbf{E}_{Bb} + \mathbf{E}_{cC} = \mathbf{E}_{cC} - \mathbf{E}_{bB} = 208 \text{ V}\underline{/+90°} \qquad \text{(26-9B)}$$

$$\mathbf{V}_{AC} = \mathbf{E}_{CA} = \mathbf{E}_{Cc} + \mathbf{E}_{aA} = \mathbf{E}_{aA} - \mathbf{E}_{cC} = 208 \text{ V}\underline{/-30°} \qquad \text{(26-9C)}$$

The complete phasor diagram for both line and phase voltages of a wye-connected three-phase source is shown in Figure 26-24(a). The geometrical construction lines have been omitted in order to keep the diagram as simple as possible. Note that the line voltages are always displaced 30° from the phase voltages. Since these are generated voltages, changes in the load connected to a wye-connected source cannot alter the phasors of Figure 26-24.

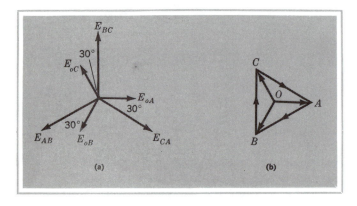

FIGURE 26-24
Phasor Diagrams for a
Wye-Connected Three-Phase Source.

Because the opposite sides of a parallelogram are equal in length and parallel, a line joining the tips of the $\mathbf{E}_{oA}$ and $\mathbf{E}_{oB}$ phasors in Figure 26-24 would have the same magnitude and direction as $\mathbf{E}_{AB}$. Therefore, we can check rapidly for the magnitude and direction of the line-voltage phasors with respect to the phase-voltage phasors for a wye-connected three-phase source without special geometrical construction by joining the tips of the phase-voltage phasors, as in Figure 26-24(b).

An important feature of the wye-connected three-phase source is that two different values of three-phase source voltage are available in the one system. This is why a 3ϕ source consisting of wye-connected transformer secondary windings rated at 120/208 volts is probably the most widely used distribution system in North America. If we wish to apply 120 volts, 3ϕ to the load, we connect $\mathbf{Z}_A$, $\mathbf{Z}_B$, and $\mathbf{Z}_C$ as a four-wire wye load, as in Figure 26-13. But if we wish to operate the same load from a 208-V 3ϕ source, we can connect the three arms of the load in delta fashion, as shown in Figure 26-25. Since the various load impedances are connected from line to line, we use the same double-subscript identification that we used with the simple delta-connected system.

EXAMPLE 26-4
A balanced load, each arm of which has an impedance of 52 $\Omega\underline{/+45°}$, is delta-connected to a 120/208-V 3ϕ wye-connected source, as in Figure 26-25. What is the magnitude of the current in each coil of the source?

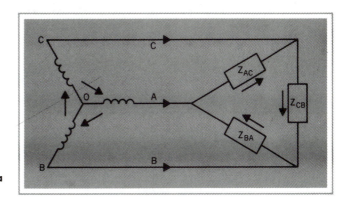

FIGURE 26-25
Wye-Connected Source Feeding a
Delta-Connected Load.

SOLUTION
Since each arm of the load is connected from line to line, the load voltage is the line voltage.

$$\therefore I_{BA} = I_{CB} = I_{AC} = \frac{208 \text{ V}}{52 \text{ }\Omega} = 4 \text{ A}$$

The load is a balanced load; hence,

$$I_L = 1.732 I_P = 1.732 \times 4 \text{ A} = 6.93 \text{ A}$$

As Figure 26-25 indicates, the current in each winding of a wye-connected source is the same as the current in the line connected to that particular coil. Therefore, the current in each winding of the source is **6.93 A.**

When a delta load is connected to a wye-connected source, as in Figure 26-25, the phase voltages of the *source* are not used. Thus, once we have determined the line voltages, we can disregard the phase voltages completely. Hence, the procedure for solving an unbalanced wye-delta circuit is exactly the same as that given in Example 26-3.

SELF-CHECK

26-6:1. Suppose that the line currents must not exceed 15 A in a 120/208-V Y-Δ system similar to that shown in Figure 26-25. What Δ-connected balanced load will produce line currents equal to 15 A?

26-6:2. In a 120/208-V Y-Δ system similar to that shown in Figure 26-25, the loads are as follows:

$$\mathbf{Z}_{AC} = 30 + j40 \text{ }\Omega$$
$$\mathbf{Z}_{BA} = 35 - j35 \text{ }\Omega$$
$$\mathbf{Z}_{CB} = 50 \text{ }\Omega$$

Determine the current in each coil of the source.

26-7 POWER IN A BALANCED THREE-PHASE SYSTEM

In a balanced three-phase load, the power in each of the three arms will be identical. Consequently the total power is three times the power in each phase. Since, as far as the load is concerned, *phase* voltage is the voltage across each arm of the load and *phase* current is the current through each arm of the load,

$$P_T = 3V_P I_P \cos \phi \qquad (26\text{-}11A)$$

Because the junction points on three-phase loads such as three-phase motors are not always accessible, it is more convenient to express the total power in terms of *line* voltage and *line* current, which we can readily measure. In the wye-connected system of Figure 26-13,

$$V_L = \sqrt{3}\, V_P \qquad \text{and} \qquad I_L = I_P$$

In the balanced delta-connected load of Figure 26-18,

$$V_L = V_P \qquad \text{and} \qquad I_L = \sqrt{3}\, I_P$$

In both cases, substitution in Equation (26-11A) gives

$$P_T = \sqrt{3}\, V_L I_L \cos \phi \qquad (26\text{-}11B)$$

> EXAMPLE 26-5
> What is the total power input to the load in Example 26-4?
>
> SOLUTION
> Since each impedance has a power-factor angle of $+45°$,
>
> $$\cos \phi = \cos 45° = 0.707 \text{ lagging}$$
>
> $$\therefore P_T = 1.732 \times V_L I_L \cos \phi = 1.732 \times 208 \text{ V} \times 6.93 \text{ A} \times 0.707 = \mathbf{1765 \ W}$$

> EXAMPLE 26-6
> A 550-V 3ϕ motor delivers 15 hp to a mechanical load with an 80% efficiency. If the power factor of the motor is 0.9 lagging, what is the line current?
>
> SOLUTION
>
> $$P_T = \frac{\text{hp} \times 746 \text{ W}}{\eta} = \frac{15 \times 746 \text{ W}}{0.8} = 14 \text{ kW}$$
>
> From Equation (26-11),
>
> $$I_L = \frac{P_T}{1.732 V_L \cos \phi} = \frac{14 \text{ kW}}{1.732 \times 550 \text{ V} \times 0.9} = \mathbf{16.32 \ A}$$

SELF-CHECK

26-7:1. What is the power delivered to the load in Self-Check Problem 26-5:2?

26-7:2. A Δ-connected 3ϕ motor draws a line current of 5 A at an 80% lagging power factor when supplied by a 120/208-V 3ϕ source. What power is delivered to the motor?

26-8 MEASURING POWER IN A THREE-PHASE SYSTEM

One method of measuring the total power in a three-phase load is to connect three wattmeters so that each meter measures the voltage across and the current through one of the arms of the load, as in Figure 26-26. The total power is, therefore, the sum of the three individual phase powers. This system applies to either balanced or unbalanced loads. However, if we know that the load is balanced (all wattmeters read the same), we can leave out two of the meters and take the total power as three times the reading of the remaining wattmeter.

FIGURE 26-26
Measuring Individual Phase Powers.

As we noted in Section 20-8, reversing the leads to one of the coils of a wattmeter results in a negative torque, which drives the pointer off scale to the left. To assist us in connecting the wattmeter correctly in the first instance, one terminal of each wattmeter coil is marked with a ± identification. As shown in Figure 26-26, we connect the identified terminal of the *current* coil to the source and the identified terminal of the *voltage* coil to the line containing the current coil. Where we use a separate wattmeter for each arm of the load, as in Figure 26-26, a negative wattmeter reading indicates an incorrect connection which we can readily correct. However, with the two-wattmeter method of measuring three-phase power, which we shall consider in Section 26-9, it is important that we start out with the wattmeters properly connected.

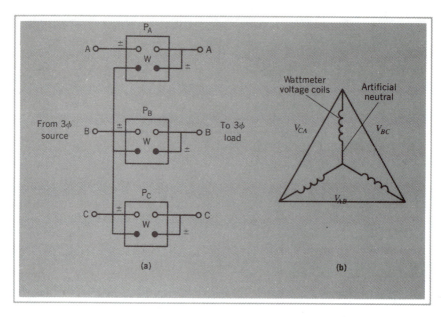

FIGURE 26-27
Three-Wattmeter Measurement of Three-Phase Power.

With many three-phase loads, the junction points in both wye- and delta-connected loads are not accessible for measuring phase currents and phase voltages independently, as in Figure 26-36. We can, however, switch to the wattmeter connections shown in Figure 26-27 to measure the total power to either a wye- or a delta-connected load. By connecting the identical voltage coils of the three wattmeters as a wye load, each wattmeter will have a voltage equivalent to the phase voltage of a wye-connected system $(V_L/\sqrt{3})$, as shown in Figure 26-27(b). On this basis, regardless of load connections, the line current will appear to the wattmeters as if it were the same as the phase current of a wye-connected load. Therefore, the total power is the sum of the three wattmeter readings.

Again, if the load is balanced, all wattmeters will read alike. But, in this case, we cannot remove two wattmeters without disturbing the voltage to the voltage coil of the remaining wattmeter. Consequently, when we remove two wattmeters, we must provide some means of restoring the proper voltage to the remaining wattmeter voltage coil. One method is to complete the wye connection with two impedance coils whose impedances are exactly equal to the impedance of the voltage coil of the remaining meter. A second method is to create an artificial neutral to which to connect the voltage coil, by connecting three equal resistors as a wye load on the three-phase system.

A third method of using a single wattmeter to indicate total power in a balanced three-phase load is shown in Figure 26-28. The unidentified terminal of the voltage coil of the wattmeter is returned to a center-tapped resistor across the two lines not containing the current coil. With this connection, the voltage coil receives a voltage of

$$\mathbf{V}_M = \mathbf{V}_{AB} + 0.5\mathbf{V}_{BC}$$

FIGURE 26-28
Single-Wattmeter Measurement of Three-Phase Power in a Balanced Load.

Constructing the resultant phasor in Figure 26-28(b), we find that the voltage coil receives a voltage having the same phase position as if it were connected to a neutral, that is, the same angle as a phase voltage in an equivalent wye system. But, unlike the phase voltage of the equivalent wye, which has a magnitude of $V_L/\sqrt{3}$, the wattmeter voltage, as Figure 26-28(b) shows, is $V_L \cos 30°$, which is $(\sqrt{3}/2)V_L$. Therefore, since $P_T = \sqrt{3} V_L I_L \cos \phi$, the wattmeter reads *one-half* of the total power to a balanced three-phase load.

26-9 TWO-WATTMETER MEASUREMENT OF THREE-PHASE POWER

The most widely used method of measuring three-phase power is based on the two-wattmeter connections of Figures 26-29 and 26-30. Its advantages are: (1) the same connections apply to both wye- and delta-connected loads; (2) it shows the total power in both balanced and unbalanced loads with one less meter than the three-wattmeter systems; (3) with balanced loads, we can determine load power factor from the two wattmeter readings. To determine what each wattmeter reads, we shall consider first the lagging power-factor balanced wye load of Figure 26-29.

Examining the connections to wattmeter P_1, we find that the current coil carries the current $\mathbf{I}_A$, which lags the voltage $\mathbf{V}_{AN}$ across $\mathbf{Z}_A$ by the power-factor angle ϕ. However, the voltage coil is connected across the *line* voltage $\mathbf{V}_{AB}$. As we have already noted, the line voltage in a wye-connected three-phase system is displaced 30° from the phase voltage. Locating $\mathbf{V}_{AB}$ on the phasor diagram (from $\mathbf{V}_{AB} = \mathbf{V}_{AN} + \mathbf{V}_{NB} = \mathbf{V}_{AN} - \mathbf{V}_{BN}$), we find that the total angle between $\mathbf{I}_A$ and $\mathbf{V}_{AB}$ is $(30° + \phi)$. Thus, wattmeter P_1 reads

$$P_1 = V_{AB}I_A \cos(30° + \phi)$$

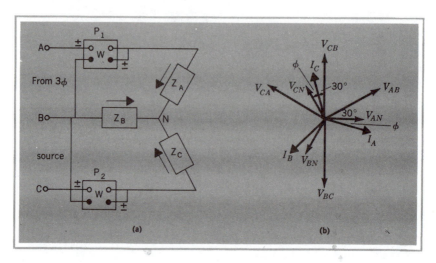

FIGURE 26-29
Two-Wattmeter Power Measurement with a Balanced Wye Load.

FIGURE 26-30
Two-Wattmeter Power Measurement with a Balanced Delta Load.

The current coil of wattmeter P_2 carries the current $\mathbf{I}_C$. But note that the voltage coil is connected to read $\mathbf{V}_{CB}$ rather than $\mathbf{V}_{BC}$ (the systematically arranged line voltage). Therefore, locating $\mathbf{V}_{CB}$ on the phasor diagram of Figure 26-29(b) (colored line), we find that the angle between $\mathbf{I}_C$ and $\mathbf{V}_{BC}$ is $(30° - \phi)$. Hence, in general terms,

$$P_1 = V_L I_L \cos(30° + \phi) \qquad (26\text{-}12)$$

$$P_2 = V_L I_L \cos(30° - \phi) \qquad (26\text{-}13)$$

In the balanced delta-connected load of Figure 26-30, each phase current lags the line voltage by the power-factor angle ϕ. As we have already noted, with a *balanced* load, the line current is displaced 30° from the phase current in each arm of the load. Therefore, locating $\mathbf{I}_A$ on the phasor diagram (from $\mathbf{I}_A = \mathbf{I}_{AB} - \mathbf{I}_{CA}$), we find that the angle between $\mathbf{I}_A$ and $\mathbf{V}_{AB}$ is $(30° + \phi)$. To find the angle between $\mathbf{I}_C$ and the colored phasor $\mathbf{V}_{CB}$, we can draw in the $-\mathbf{I}_{BC}$ phasor. With a balanced load, there is a 30° angle between $-\mathbf{I}_{BC}$ and $\mathbf{I}_C$. The angle between $\mathbf{V}_{CB}$ and $-\mathbf{I}_{BC}$ is the same as the angle between $\mathbf{V}_{BC}$ and $\mathbf{I}_{BC}$, the power-factor angle ϕ. Consequently, the angle between $\mathbf{I}_C$ and $\mathbf{V}_{CB}$ is $(30° - \phi)$, and again

$$P_1 = V_L I_L \cos (30° + \phi) \tag{26-12}$$

$$P_2 = V_L I_L \cos (30° - \phi) \tag{26-13}$$

Adding Equations (26-12) and (26-13) gives

$$P_1 + P_2 = V_L I_L [\cos (30 + \phi) + \cos (30 - \phi)]$$

To expand the terms in brackets, we substitute the trigonometric expansion for $\cos (A + B)$ and $\cos (A - B)$. Therefore,

$$P_1 + P_2 = V_L I_L (\cos 30 \cos \phi - \sin 30 \sin \phi + \cos 30 \cos \phi + \sin 30 \sin \phi)$$

$$= V_L I_L \, 2 \cos 30 \cos \phi$$

But $2 \cos 30° = \sqrt{3}$.

$$\therefore \; P_1 + P_2 = \sqrt{3} \, V_L I_L \cos \phi$$

which is the total power in a balanced 3ϕ system.

We can also show that the sum of the two wattmeter readings is the total power for an unbalanced load by returning to the circuit of Figure 26-30(a) and considering *instantaneous* power values. In the delta-connected load,

$$p_t = (v_{AB} i_{AB}) + (v_{BC} i_{BC}) + (v_{CA} i_{CA}) \tag{1}$$

But, as we noted when we delta-connected a 3ϕ source,

$$v_{AB} + v_{BC} + v_{CA} = 0 \tag{2}$$

from which

$$v_{CA} = -v_{AB} - v_{BC}$$

Substituting in Equation (1) gives

$$p_t = (v_{AB} i_{AB}) + (v_{BC} i_{BC}) - (v_{AB} i_{CA}) - (v_{BC} i_{CA})$$

Collecting the voltage terms and substituting $-v_{CB}$ for v_{BC},

$$p_t = v_{AB} (i_{AB} - i_{CA}) + v_{CB} (i_{CA} - i_{BC}) = (v_{AB} i_A) + (v_{CB} i_C)$$

Inspection of the circuit of Figure 26-30(a) shows us that this expression indicates precisely what the two wattmeters are connected to read. By a similar algebraic procedure, we can show that the total instantaneous power in *any* wye-connected load can also be reduced to the sum of the two wattmeter readings. Therefore, with the two wattmeters connected as in Figures 26-29 and 26-30, **total power input to any three-phase load, balanced or unbalanced, is the algebraic sum of the two wattmeter readings.**

The third advantage we claimed for the two-wattmeter system of three-phase power measurement was that with balanced loads we can determine the load power factor from the two wattmeter readings. First, let us consider the effect of various load power factors on the meter readings. With unity power factor, Equations (26-12) and (26-13) both become

$$P_1 = P_2 = V_L I_L \cos 30° = \frac{\sqrt{3}}{2} V_L I_L$$

Hence, **each wattmeter reads half of the total power.**

If the power-factor angle is $+30°$,

$$P_1 = V_L I_L \cos 60° = 0.5\, V_L I_L$$
$$P_2 = V_L I_L \cos 0° = V_L I_L$$

One wattmeter reads one-third of the total power and the other reads two-thirds. A power-factor angle of $-30°$ simply reverses the wattmeter readings.

If the power-factor angle is $60°$,

$$P_1 = V_L I_L \cos 90° = 0$$
$$P_2 = V_L I_L \cos - 30° = \frac{\sqrt{3}}{2} V_L I_L$$

One wattmeter reads zero, and the other then registers the total power. Since this represents a power factor of 0.5, note that the total power is half that at unity power factor for the same line voltages and currents.

If the power factor drops below 50%, the angle $(30° + \phi)$ exceeds 90°, and the cosine becomes a negative quantity. Therefore, in the wattmeter movement, the counterclockwise torque exceeds the clockwise torque, and the meter tries to read off scale to the left. To read the total active power to this low power-factor load, we must reverse the wattmeter voltage *or* current coil leads to bring the reading onto the scale and then *subtract* this reading from that of the other meter.

[†] If we encounter a negative wattmeter reading when conducting a laboratory assignment on power input to a three-phase load, it can mean either a load power factor of less than 50%, or incorrect connection of the wattmeter coils in the first instance. Conversely, if one of the wattmeters shows a small positive power reading, we do not know for sure that we have not already reversed the wattmeter and are really reading a power which should be subtracted due to a low power-factor load. There are several ways of checking if we are not sure that the original meter connections were correct. Perhaps the simplest test is to "single-phase" the system by momentarily opening the line containing the current coil of the *other* wattmeter. If the wattmeter in question reads negative, it was incorrectly connected in the first place. However, if it shows a positive power reading under single-phase conditions, any negative reading when three-phase source voltages are restored to the load is due to low power factor. We must then reverse its voltage coil to obtain a reading which we then *subtract* from the other wattmeter reading.

When we know the power factor of a balanced load, we can express the ratio between the two wattmeter readings by dividing Equation (26-13) into Equation (26-12). This gives

$$\frac{P_1}{P_2} = \frac{\cos (30° + \phi)}{\cos (30° - \phi)} \qquad (26\text{-}14)$$

To use the two wattmeter readings to determine the load power factor, we start by subtracting Equation (26-12) from Equation (26-13).

$$\therefore P_2 - P_1 = V_L I_L \cos (30° - \phi) - V_L I_L \cos (30° + \phi)$$

When we substitute the trigonometric expansions for $\cos (30° - \phi)$ and $\cos (30° + \phi)$ and simplify,

$$P_2 - P_1 = V_L I_L \, 2 \sin 30° \sin \phi$$

But $\sin 30° = 0.5$,

$$\therefore P_2 - P_1 = V_L I_L \sin \phi$$

Dividing this expression by the sum of Equations (26-12) and (26-13),

$$\frac{P_2 - P_1}{P_2 + P_1} = \frac{V_L I_L \sin \phi}{\sqrt{3} \, V_L I_L \cos \phi} = \frac{\tan \phi}{\sqrt{3}}$$

$$\therefore \tan \phi = \sqrt{3} \frac{P_2 - P_1}{P_2 + P_1} \qquad (26\text{-}15)$$

Knowing the two wattmeter readings, we can solve for the power-factor angle ϕ and then determine the power factor of the load from $\cos \phi$.

EXAMPLE 26-7

The total power input to the load in Example 26-4 is measured by the two-wattmeter method. What are the two wattmeter readings?

SOLUTION

$$P_1 = V_L I_L \cos (30° + 45°) = 208 \text{ V} \times 6.93 \text{ A} \times \cos 75° = \textbf{373 W}$$

$$P_2 = V_L I_L \cos (30° - 45°) = 208 \text{ V} \times 6.93 \text{ A} \times \cos - 15° = \textbf{1392 W}$$

For laboratory tests on variable power-factor three-phase loads such as induction motors, we can avoid repeated calculation of power factor from Equation (26-15) by plotting an accurate graph based on Equation (26-14). This power-factor graph has the shape shown in Figure 26-31.

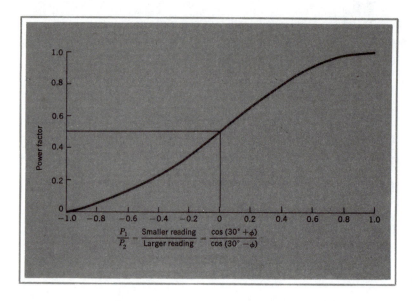

FIGURE 26-31
Power Factor from Ratio of
Two-Wattmeter Readings.

SELF-CHECK

26-9:1. The two-wattmeter method is used to measure the power delivered to an unbalanced Δ-load as shown in Figure 26-30. The readings on the wattmeters are as follows:

$$\text{wattmeter one} = 300 \text{ W}$$
$$\text{wattmeter two} = 500 \text{ W}$$

What total power does the load draw?

26-9:2. The two-wattmeter method is used to measure the power delivered to a balanced Y-load. The readings on the wattmeters are as follows:

$$\text{wattmeter one} = 300 \text{ W}$$
$$\text{wattmeter two} = 500 \text{ W}$$

(a) What total power does the load draw?
(b) What is the load's power factor?

26-10 PHASE SEQUENCE

In the three-phase circuit problems we have encountered so far, we have been given data from which we could determine the exact angle (with respect to the reference axis) of all the phase and line voltages. When a three-phase service is supplied by an electrical utility company, we know that all three line voltages have the same magnitude and are displaced from one another by a 120° angle. We can measure the *magnitude* of the voltages quite readily. However, this does not tell us whether V_{BN} *lags* V_{AN} by 120° or *leads* by 120°. This information is called the **phase sequence** (or phase rotation) of a three-phase system. It is important information, since it governs the direction of rotation of three-phase motors and also governs the division of the current among the three lines feeding an unbalanced load. It is also important that the electrical utility maintain the same phase sequence.

In the 3ϕ source of Figure 26-32(a), we are given information that shows that V_{BN} lags V_{AN} by 120° and V_{CN} lags V_{BN} by 120°. Therefore, the instantaneous voltage to line A passes through its positive peak value first; then the instantaneous voltage to line B passes through its positive peak value, followed by the instantaneous voltage to line C. As a result, we say that the system in Figure 26-32(a) has an ABC phase sequence. If we reverse the leads to two of the generator coils, as in Figure 26-32(b), V_{BN} now *leads* V_{AN} by 120° and V_{CN} *leads* V_{BN} by 120°. Thus the phase sequence has been reversed and is stated as a CBA phase sequence, because the instantaneous voltage to line C passes a certain point in its cycle 120° before the instantaneous voltage to line B, which in turn is 120° ahead of the instantaneous voltage to line A.

Once we have prepared a phasor diagram for the voltages in a 3ϕ system, we can read the phase sequence from the phasor diagram quite readily. Since the direction of rotation of a phasor is counterclockwise, the phase sequence is the order in which the

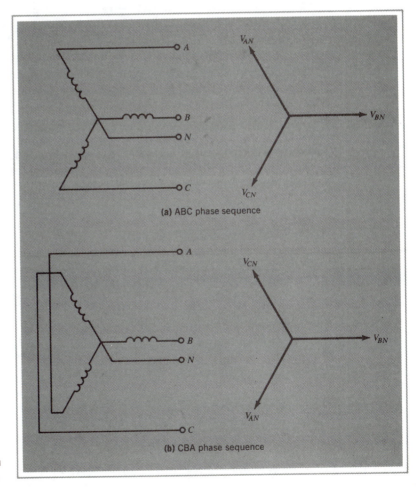

FIGURE 26-32
Phase Sequence of a
Three-Phase Source.

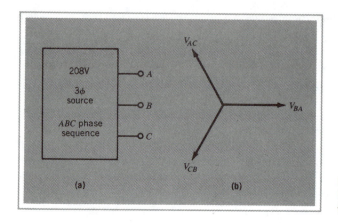

FIGURE 26-33
Determining Phase Sequence from a Phasor Diagram.

voltage phasors would pass the reference axis if they were rotated in a counterclockwise direction. If we prefer to leave the phasors stationary, we can read the phase sequence from a phasor diagram by reading *clockwise* around the diagram.

When we are given the line voltage and phase sequence of a three-phase source, as in Figure 26-33(a), we can prepare a phasor diagram by placing one phasor (usually $\mathbf{V}_{BA}$) along the reference axis. To identify the other two voltage phasors, as we read *clockwise* around the phasor diagram of Figure 26-33(b), *either* all first subscript letters *or* all second subscript letters must agree with the specified phase sequence.

We can show the effect of phase sequence on the line currents to an unbalanced load by considering the following example.

EXAMPLE 26-8
$\mathbf{Z}_{AB} = 52\ \Omega\underline{/-30°}$, $\mathbf{Z}_{BC} = 52\ \Omega\underline{/+45°}$, and $\mathbf{Z}_{CA} = 104\ \Omega\underline{/0°}$, are connected as a delta load to a 208-V 3ϕ source. Find the magnitude of the current in each line if the phase sequence of the source is (a) *ABC* and (b) *CBA*.

SOLUTION
(a) With *ABC* phase sequence.

    Step I.

$$\mathbf{I}_{AB} = \frac{\mathbf{V}_{AB}}{\mathbf{Z}_{AB}} = \frac{208\ \text{V}\underline{/0°}}{52\ \Omega\underline{/-30°}} = 4\ \text{A}\underline{/+30°}$$

$$\mathbf{I}_{BC} = \frac{\mathbf{V}_{BC}}{\mathbf{Z}_{BC}} = \frac{208\ \text{V}\underline{/-120°}}{52\ \Omega\underline{/+45°}} = 4\ \text{A}\underline{/-165°}$$

$$\mathbf{I}_{CA} = \frac{\mathbf{V}_{CA}}{\mathbf{Z}_{CA}} = \frac{208\ \text{V}\underline{/+120°}}{104\ \Omega\underline{/0°}} = 2\ \text{A}\underline{/+120°}$$

    Step II. From the circuit diagram of Figure 26-34(a),

$$\mathbf{I}_A = \mathbf{I}_{AB} - \mathbf{I}_{CA} = 4\ \text{A}\underline{/+30°} - 2\ \text{A}\underline{/+120°}$$

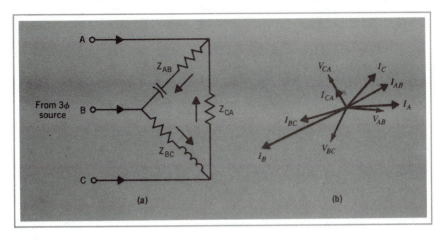

FIGURE 26-34
Phasor Diagram for Example 26-8(a).

$$\mathbf{I}_R = (3.464 + j2) - (-1 + j1.732) = 4.464 + j0.268 \text{ A}$$

$$\therefore I_A = \mathbf{4.47 \text{ A}}$$

$$\mathbf{I}_B = \mathbf{I}_{BC} - \mathbf{I}_{AB} = 4 \text{ A}\underline{/-165°} - 4 \text{ A}\underline{/+30°}$$

$$= (-3.863 - j1.035) - (3.464 + j2) = -7.327 - j3.035 \text{ A}$$

$$\therefore I_B = \mathbf{7.93 \text{ A}}$$

$$\mathbf{I}_C = \mathbf{I}_{CA} - \mathbf{I}_{BC} = 2 \text{ A}\underline{/+120°} - 4 \text{ A}\underline{/-165°}$$

$$= (-1 + j1.732) - (3.863 - j1.035) = 2.863 + j2.767 \text{ A}$$

$$\therefore I_C = \mathbf{3.98 \text{ A}}$$

(b) With *CBA* phase sequence (as shown in Figure 26-35).

Step I.

$$\mathbf{I}_{AB} = \frac{\mathbf{V}_{AB}}{\mathbf{Z}_{AB}} = \frac{208 \text{ V}\underline{/0°}}{52 \text{ }\Omega\underline{/-30°}} = 4 \text{ A}\underline{/+30°}$$

$$\mathbf{I}_{BC} = \frac{\mathbf{V}_{BC}}{\mathbf{Z}_{BC}} = \frac{208 \text{ V}\underline{/+120°}}{52 \text{ }\Omega\underline{/+45°}} = 4 \text{ A}\underline{/+75°}$$

$$\mathbf{I}_{CA} = \frac{\mathbf{V}_{CA}}{\mathbf{Z}_{CA}} = \frac{208 \text{ V}\underline{/-120°}}{104 \text{ }\Omega\underline{/0°}} = 2 \text{ A}\underline{/-120°}$$

Step II.

$$\mathbf{I}_A = \mathbf{I}_{AB} - \mathbf{I}_{CA} = 4 \text{ A}\underline{/+30°} - 2 \text{ A}\underline{/-120°}$$

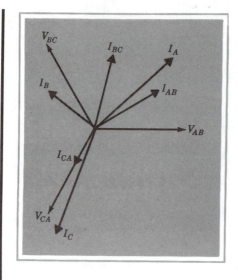

FIGURE 26-35
Phasor Diagram for Example 26-8(b).

$$\mathbf{I}_A = (3.464 + j2) - (-1 - j1.732) = 4.464 + j3.732 \text{ A}$$

$$\therefore I_A = \mathbf{5.82 \text{ A}}$$

$$\mathbf{I}_B = \mathbf{I}_{BC} - \mathbf{I}_{AB} = 4 \text{ A}\underline{/+75°} - 4 \text{ A}\underline{/+30°}$$

$$= (1.035 + j3.863) - (3.464 + j2) = -2.429 + j1.863 \text{ A}$$

$$\therefore I_B = \mathbf{3.06 \text{ A}}$$

$$\mathbf{I}_C = \mathbf{I}_{CA} - \mathbf{I}_{BC} = 2 \text{ A}\underline{/-120°} - 4 \text{ A}\underline{/+75°}$$

$$= (-1 - j1.732) - (1.035 + j3.863) = -2.035 - j5.595 \text{ A}$$

$$\therefore I_C = \mathbf{5.95 \text{ A}}$$

SELF-CHECK

26-10:1. The following loads are Δ-connected to a 120/208-V 3ϕ source:

$$\mathbf{Z}_{AB} = 30 + j40 \text{ } \Omega$$
$$\mathbf{Z}_{BC} = 35 - j35 \text{ } \Omega$$
$$\mathbf{Z}_{CA} = 50 \text{ } \Omega$$

Using $\mathbf{V}_{AB}$ as the reference voltage, find the line currents if the phase sequence is ABC.
26-10:2. Repeat Self-Check Problem 26-10:1 for a phase sequence of CBA.

26-11 UNBALANCED THREE-WIRE WYE LOADS

We have been able to solve unbalanced four-wire wye loads and delta loads with no particular difficulty since, in both cases, we know the phase voltage across each arm of the load. But with the unbalanced three-wire wye load of Figure 26-36, we do not know what the phase voltages will be, since the wye junction point is *floating*. There are two

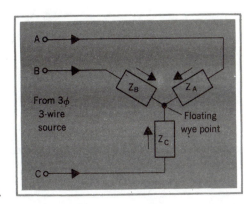

FIGURE 26-36
Three-Wire Wye Load.

fairly straightforward procedures based on the ac network solutions we studied in Chapter 22. The first is based on writing Kirchhoff's-law equations for the wye load of Figure 26-36.

Following the conventional tracing arrows through the wye load in Figure 26-36, we can write three loop equations,

$$I_A Z_A - I_B Z_B = E_{BA} = V_{AB} \tag{1}$$

$$I_B Z_B - I_C Z_C = E_{CB} = V_{BC} \tag{2}$$

$$I_C Z_C - I_A Z_A = E_{AC} = V_{CA} \tag{3}$$

Although we have three equations for three unknown line currents, if we try to solve the determinant, we discover that the three equations are not independent, due to the symmetrical relationship of the left-hand sides. Hence, we require an additional equation. Because there is no neutral lead to this wye load, from Kirchhoff's current law,

$$I_A + I_B + I_C = 0 \tag{4}$$

Omitting Equation (2) which does not contain I_A, we can rearrange Equations (1), (3), and (4) in the form

$$Z_A I_A - Z_B I_B \qquad\qquad = V_{AB} \tag{1}$$

$$-Z_A I_A + \qquad\qquad + Z_C I_C = V_{CA} \tag{3}$$

$$I_A + \qquad I_B + \qquad I_C = 0 \tag{4}$$

from which,

$$I_A = \frac{\begin{vmatrix} V_{AB} & -Z_B & \\ V_{CA} & & Z_C \\ 1 & 1 \end{vmatrix}}{\begin{vmatrix} Z_A & -Z_B & \\ -Z_A & & Z_C \\ 1 & 1 & 1 \end{vmatrix}} = \frac{0 + (-0) + 0 - 0 - V_{AB} Z_C - (-V_{CA} Z_B)}{0 + (-Z_B Z_C) + 0 - 0 - Z_A Z_C - Z_A Z_B}$$

$$\therefore \mathbf{I}_A = \frac{\mathbf{V}_{AB}\,\mathbf{Z}_C - \mathbf{V}_{CA}\,\mathbf{Z}_B}{\mathbf{Z}_A\,\mathbf{Z}_B + \mathbf{Z}_B\,\mathbf{Z}_C + \mathbf{Z}_C\,\mathbf{Z}_A} \tag{26-16A}$$

Similarly,
$$\mathbf{I}_B = \frac{\mathbf{V}_{BC}\,\mathbf{Z}_A - \mathbf{V}_{AB}\,\mathbf{Z}_C}{\mathbf{Z}_A\,\mathbf{Z}_B + \mathbf{Z}_B\,\mathbf{Z}_C + \mathbf{Z}_C\,\mathbf{Z}_A} \tag{26-16B}$$

and
$$\mathbf{I}_C = \frac{\mathbf{V}_{CA}\,\mathbf{Z}_B - \mathbf{V}_{BC}\,\mathbf{Z}_A}{\mathbf{Z}_A\,\mathbf{Z}_B + \mathbf{Z}_B\,\mathbf{Z}_C + \mathbf{Z}_C\,\mathbf{Z}_A} \tag{26-16C}$$

EXAMPLE 26-9

The phase sequence tester shown in Figure 26-37 is connected to a 208-V 3ϕ 60-Hz source whose phase sequence is *ABC*. The choke has an inductance of 0.5 H with negligible resistance, and the lamps each have a resistance of 100 Ω. (We shall assume that the resistance remains constant even though the current through the lamp changes). Using Equations (26-16), determine the magnitude of the current in each lamp.

SOLUTION

$$\mathbf{Z}_C = +j\omega L = +j377 \times 0.5\ \mathrm{H} = 188.5\ \Omega\,\underline{/+90°}$$

$$\mathbf{I}_A = \frac{\mathbf{V}_{AB}\,\mathbf{Z}_C - \mathbf{V}_{CA}\,\mathbf{Z}_B}{\mathbf{Z}_A\,\mathbf{Z}_B + \mathbf{Z}_B\,\mathbf{Z}_C + \mathbf{Z}_C\,\mathbf{Z}_A}$$

$$= \frac{(208\,\underline{/0°} \times 188.5\,\underline{/+90°}) - (208\,\underline{/+120°} \times 100\,\underline{/0°})}{(100\,\underline{/0°} \times 100\,\underline{/0°}) + (100\,\underline{/0°} \times 188.5\,\underline{/90°}) + (188.5\,\underline{/90°} \times 100\,\underline{/0°})}$$

$$= \frac{39\,208\,\underline{/+90°} - 20\,800\,\underline{/120°}}{10\,000\,\underline{/0°} + 18\,850\,\underline{/+90°} + 18\,850\,\underline{/+90°}}$$

$$= \frac{(0 + j39\,208) - (-10\,400 + j18\,013)}{10\,000 + j37\,700} = \frac{10\,400 + j21\,195}{10\,000 + j37\,700}$$

$$\therefore \mathbf{I}_A = \frac{23\,609}{39\,003} = \mathbf{0.605\ A}$$

Similarly,

$$\mathbf{I}_B = \frac{(208\,\underline{/-120°} \times 100\,\underline{/0°}) - (208\,\underline{/0°} \times 188.5\,\underline{/+90°})}{39\,003}$$

$$= \frac{20\,800\,\underline{/-120°} - 39\,208\,\underline{/+90°}}{39\,003}$$

$$= \frac{(-10\,400 - j18\,013) - (0 + j39\,208)}{39\,003} = \frac{-10\,400 - j57\,221}{39\,003}$$

FIGURE 26-37
Phase-Sequence Tester.

and

$$I_B = \frac{58\,158}{39\,003} = \mathbf{1.49\ A}$$

Therefore, with an *ABC* phase sequence, lamp *B* is brighter than lamp *A*.

The appearance of Equations (26-16A), (26-16B), and (26-16C) may suggest to us the second method for solving three-phase circuits with an unbalanced three-wire wye load. In Chapter 22, we set down equations which would allow us to substitute an equivalent delta circuit for the original wye-connected load. Once we know the impedance of the delta arms, we can solve for the line currents in the usual manner. Therefore, in the circuit of Figure 26-38,

$$\mathbf{Z}_{BC} = \frac{\mathbf{Z}_A \mathbf{Z}_B + \mathbf{Z}_B \mathbf{Z}_C + \mathbf{Z}_C \mathbf{Z}_A}{\mathbf{Z}_A} \tag{22-5}$$

$$\mathbf{Z}_{CA} = \frac{\mathbf{Z}_A \mathbf{Z}_B + \mathbf{Z}_B \mathbf{Z}_C + \mathbf{Z}_C \mathbf{Z}_A}{\mathbf{Z}_B} \tag{22-6}$$

$$\mathbf{Z}_{AB} = \frac{\mathbf{Z}_A \mathbf{Z}_B + \mathbf{Z}_B \mathbf{Z}_C + \mathbf{Z}_C \mathbf{Z}_A}{\mathbf{Z}_C} \tag{22-7}$$

EXAMPLE 26-10
If the phase sequence of the 208-V 3ϕ 60-Hz source in Example 26-9 is changed to *CBA*, determine the magnitude of the current in each lamp, using the wye-delta transformation equations.

SOLUTION
Since we have already solved the numerator of the wye-delta transformation equations in solving Example 26-9,

FIGURE 26-38
Using a Wye-Delta Transformation to Solve Unbalanced Three-Wire Wye Loads.

Step I.

$$\mathbf{Z}_{AB} = \frac{39\,003\,\underline{/+75.14°}}{188.5\,\underline{/+90°}} = 206.9\ \Omega\,\underline{/-14.86°}$$

$$\mathbf{Z}_{BC} = \frac{39\,003\,\underline{/+75.14°}}{100\,\underline{/0°}} = 390\ \Omega\,\underline{/+75.14°}$$

$$\mathbf{Z}_{CA} = \frac{39\,003\,\underline{/+75.14°}}{100\,\underline{/0°}} = 390\ \Omega\,\underline{/+75.14°}$$

Step II.

$$\mathbf{I}_{AB} = \frac{\mathbf{V}_{AB}}{\mathbf{Z}_{AB}} = \frac{208\ \text{V}\,\underline{/0°}}{206.9\ \Omega\,\underline{/-14.86°}} = 1.005\ \text{A}\,\underline{/+14.86°}$$

$$\mathbf{I}_{BC} = \frac{\mathbf{V}_{BC}}{\mathbf{Z}_{BC}} = \frac{208\ \text{V}\,\underline{/+120°}}{390\ \Omega\,\underline{/+75.14°}} = 0.533\ \text{A}\,\underline{/+44.86°}$$

$$\mathbf{I}_{CA} = \frac{\mathbf{V}_{CA}}{\mathbf{Z}_{CA}} = \frac{208\ \text{V}\,\underline{/-120°}}{390\ \Omega\,\underline{/+75.14°}} = 0.533\ \text{A}\,\underline{/+164.86°}$$

Step III.   From Figure 26-38,

$$\mathbf{I}_A = \mathbf{I}_{AB} - \mathbf{I}_{CA} = 1.005\ \text{A}\,\underline{/+14.86°} - 0.533\ \text{A}\,\underline{/+164.86°}$$

$$= (0.9714 + j0.2577) - (-0.5145 + j0.1392)$$

$$= 1.486 + j0.1185 \text{ A}$$

$$\therefore I_A = \mathbf{1.49 \text{ A}}$$

and
$$\mathbf{I}_B = \mathbf{I}_{BC} - \mathbf{I}_{AB} = 0.533 \text{ A}\underline{/+44.86°} - 1.005 \text{ A}\underline{/+14.86°}$$

$$= (0.3778 + j0.3760) - (0.9714 + j0.2577)$$

$$= -0.5936 + j0.1183 \text{ A}$$

$$\therefore I_B = \mathbf{0.605 \text{ A}}$$

Hence, with a *CBA* phase sequence, lamp *A* is brighter than lamp *B*.

Once we have solved for line currents, we can return to the original wye-connected circuit and solve for the various phase voltages on the basis that $\mathbf{V}_{AN} = \mathbf{I}_A \mathbf{Z}_A$, etc.

SELF-CHECK

26-11:1. The following loads are 3-wire Y-connected to a 120/208-V 3ϕ source as shown in Figure 26-36:

$$\mathbf{Z}_A = 30 + j40 \ \Omega$$
$$\mathbf{Z}_B = 35 - j35 \ \Omega$$
$$\mathbf{Z}_C = 50 \ \Omega$$

Use $\mathbf{V}_{AB}$ as the reference voltage, an *ABC* phase sequence, and Y-Δ transformation equations to find the line currents.

PROBLEMS

26-4-1. Given a 60-Hz 3ϕ four-wire wye-connected source consisting of $\mathbf{E}_{oA} = 120 \text{ V}\underline{/0°}$, $\mathbf{E}_{oB} = 120 \text{ V}\underline{/+120°}$, and $\mathbf{E}_{oC} = 120 \text{ V}\underline{/-120°}$. A balanced four-wire wye-connected load consists of three 10-Ω resistors connected to this source.
  (a) Draw a fully labeled schematic diagram showing tracing arrows.
  (b) Determine algebraically the magnitude and phase of the current in each line.
  (c) Draw a phasor diagram of the phase voltages and currents.
  (d) Determine algebraically the neutral current.

26-4-2. The load in Problem 26-4-1 is replaced with a balanced four-wire wye-connected load each arm of which is rated at 200 W with an 80% lagging power factor. Repeat all four parts of Problem 26-4-1. What will the neutral current be if line *A* becomes open-circuit?

26-4-3. The load in Problem 26-4-1 is replaced with an unbalanced four-wire wye-connected load consisting of $\mathbf{Z}_A = 30 \ \Omega\underline{/+30°}$, $\mathbf{Z}_B = 40 \ \Omega\underline{/-30°}$, and $\mathbf{Z}_C = 50 \ \Omega\underline{/-90°}$. Repeat all four parts of Problem 26-4-1.

26-4-4. The load in Problem 26-4-1 is replaced with an unbalanced four-wire wye-connected load in which $\mathbf{Z}_A$ is a 30-Ω heater, $\mathbf{Z}_B$ is a 150-W lamp, and $\mathbf{Z}_C$ is an 0.1-H inductor whose resistance is 5 Ω. Repeat all four parts of Problem 26-4-1.

26-5-5. Given a 60-Hz 3ϕ delta-connected source in which $\mathbf{E}_{CA} = 550$ V$\underline{/0°}$, $\mathbf{E}_{AB} = 550$ V$\underline{/+120°}$, and $\mathbf{E}_{BC} = 550$ V$\underline{/-120°}$. A balanced delta-connected load consisting of three 11-Ω resistors is connected to this source.
(a) Draw a fully labeled schematic diagram showing tracing arrows.
(b) Determine the phase currents algebraically.
(c) Draw a phasor diagram of the phase voltages and currents.
(d) Determine the line currents by geometrical construction on a phasor diagram.
(e) Determine the line currents by phasor algebra.

26-5-6. The load in Problem 26-5-5 is replaced by a balanced delta load which has a total power rating of 15 kW at 96.6% leading power factor. Repeat all five parts of Problem 26-5-5.

26-5-7. The load in Problem 26-5-5 is replaced by a delta load consisting of $\mathbf{Z}_{AC} = 5\ \Omega\underline{/+45°}$, $\mathbf{Z}_{CB} = 10\ \Omega\underline{/-36.9°}$, and $\mathbf{Z}_{BA} = 11\ \Omega\underline{/0°}$. Repeat all five parts of Problem 26-5-5.

26-5-8. The load in Problem 26-5-5 is replaced by a delta-connected load in which $\mathbf{Z}_{AC} = 10\ \text{k}\Omega\underline{/0°}$, $\mathbf{Z}_{CB}$ consists of a 10-H choke with a resistance of 50 Ω, and $\mathbf{Z}_{BA}$ is a 0.5-μF capacitor. Repeat all five parts of Problem 26-5-5.

26-6-9. The load of Problem 26-5-5 is connected to the source of Problem 26-4-1.
(a) Draw a fully labeled schematic diagram showing tracing arrows.
(b) Determine the line voltages by phasor construction.
(c) Determine the line voltages by phasor algebra.
(d) Determine the phase currents.
(e) Determine the line currents by phasor algebra.
(f) Draw a complete phasor diagram for the system.

26-6-10. The load of Problem 26-5-6 is connected to the source of Problem 26-4-1. Repeat all six parts of Problem 26-6-9.

26-6-11. The load of Problem 26-5-7 is connected to the source of Problem 26-4-1. Repeat all parts of Problem 26-6-9.

26-6-12. The load of Problem 26-5-8 is connected to the source of Problem 26-4-1. Repeat all parts of Problem 26-6-9.

26-7-13. What is the total power input to the load in Problem 26-4-1?

26-7-14. What is the total power input to the load in Problem 26-5-5?

26-7-15. What is the total power input to the load in Problem 26-4-3?

26-7-16. What is the total power input to the load in Problem 26-5-7?

26-8-17. A wattmeter is connected with its current coil in line B and its voltage coil from line A to line B in the circuit of Problem 26-4-1. What is its reading?

26-8-18. A wattmeter is connected with its current coil in line C and its voltage coil from line A to line B in the circuit of Problem 26-4-1. What is its reading?

26-9-19. The total power input to the load in Problem 26-5-5 is checked by the two-wattmeter method. What is the reading on each wattmeter?

26-9-20. The total power input to the load in Problem 26-4-2 is checked by the two-wattmeter method. What is the reading on each wattmeter?

26-9-21. A balanced delta load connected to a 208-V 3ϕ source is checked by the two-wattmeter method whose readings are 2 kW and 1 kW, respectively.
(a) What is the total three-phase power?
(b) What is the power factor of the load?
(c) What is the magnitude of the line current?
(d) What is the impedance (magnitude and angle) of each arm of the load?
(e) What is the angle between the line voltage and the line current?

26-9-22. Repeat the five parts of Problem 26-9-21 when one wattmeter reads 3 kW and the other reads zero.

26-10-23. What is the phase sequence of the source in Problem 26-4-1?

26-10-24. What is the phase sequence of the source in Problem 26-5-5?

26-10-25. If the phase sequence of Problem 26-8-17 is reversed, what is the wattmeter reading?

26-10-26. If the phase sequence of Problem 26-8-18 is reversed, what is the wattmeter reading?

26-11-27. The load of Problem 26-4-1 is connected to the source of Problem 26-5-5. Repeat the four parts of Problem 26-4-1.

26-11-28. The neutral lead in Problem 26-4-3 becomes open circuit. Determine the current in each line.

26-11-29. Determine the three line currents in Problem 26-11-28 if the phase sequence of the source is reversed.

26-11-30. The load of Problem 26-5-7 is connected to a 208-V 3ϕ source with a *CBA* phase sequence through three conductors, each of which has an impedance of $2 + j1$ Ω. Determine the magnitude of the current in each line.

REVIEW QUESTIONS

26-1. What are the three main advantages of a polyphase system compared to a single-phase power distribution system?

26-2. Why is it desirable to arrange that the total instantaneous power input to a polyphase system be free from the customary sine-squared fluctuations?

26-3. Draw a graph of the total instantaneous power if one of the resistors in Figure 26-2 is changed to 180 Ω.

26-4. Continue the vector representation of the magnetic flux in Figure 26-5 at 45° intervals for the remainder of the cycle.

26-5. Why does the flux density of the rotating magnetic field in Figure 26-5 remain constant?

26-6. What is the significance of the term **synchronous speed?**

26-7. How could we reverse the direction of rotation of the magnetic field in Figure 26-5?

26-8. Why can we not produce a rotating magnetic field with the Edison three-wire system of Figure 26-6?

26-9. What is the advantage of using the system shown in Figure 26-6 for electric power distribution?

26-10. What voltage would appear across a load connected between line *A* and line *B* in the two-phase system of Figure 26-2(b)?

26-11. What is meant by a **balanced load** in a polyphase system?

26-12. Why does a load not have to have unity power factor to be classed as a balanced load?

26-13. Draw a graph showing that the total instantaneous power input to a balanced three-phase load is constant.

26-14. Why can we state that, in a balanced three-phase load, the currents in the three arms are equal in magnitude and displaced 120° in phase from one another regardless of the power factor of the load?

26-15. Show (with diagrams) how a rotating magnetic field can be produced when a three-phase source is available.

26-16. What is meant by the terms **phase voltage** and **line voltage** in a three-phase system? How do these terms apply to: (a) a wye-connected source? (b) a delta-connected source?

26-17. What is meant by the terms **phase current** and **line current** in a three-phase system? How do these terms apply to: (a) a wye-connected load? (b) a delta-connected load?

26-18. Draw a phasor diagram for a wye-connected source in which the upper-case letter ends of the windings in Figure 26-11 are connected to the neutral.

26-19. What is the function of the voltmeter when delta-connecting the coils of a three-phase source?

26-20. How would you test for incorrect wye connection of a three-phase source with an ac voltmeter?

26-21. Three transformers whose primary windings are fed from a three-phase source have the following secondary voltages: $E_{aA} = 240$ V$\underline{/0°}$, $E_{bB} = 240$ V$\underline{/-120°}$, and $E_{cC} = 240$ V$\underline{/+120°}$. Each secondary is center-tapped, and these center taps are connected to a common neutral, as shown in Figure 26-39. Draw a fully labeled phasor diagram for the six phase source available from these **star**-connected secondary windings.

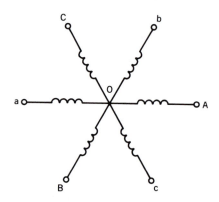

FIGURE 26-39
Star Connection of Three-Phase
Transformer Secondary Windings.

26-22. Show with a phasor diagram why the line current to a balanced delta load is greater than the phase current even though we *subtract* one phase current from the other to obtain the line current.

26-23. Why must the phasor sum of the three line currents in any three-phase, three-wire system be zero?

26-24. Show with a phasor diagram that the line voltage from a wye-connected source is $\sqrt{3}$ times the voltage generated in each winding and displaced 30° from the voltages in the windings.

26-25. $I_L = \sqrt{3}\,I_p$ in a delta-connected load depends on the load being a balanced load. However, $V_L = \sqrt{3}\,V_p$ in a wye-connected source is not dependent on a balanced load. Explain.

26-26. What is the advantage of expressing power input to a balanced three-phase load in terms of *line* voltage and current rather than *phase* voltage and current?

26-27. Why do the wattmeters in Figure 26-26 read the *active* power of each arm of the load?

26-28. What advantage does the three-wattmeter system of Figure 26-27 have over that of Figure 26-26?

26-29. With reference to Figure 26-27, what is meant by an **artificial neutral?**

26-30. What advantage does the two-wattmeter system of three-phase power measurement have over a single wattmeter?

26-31. With the two-wattmeter system of three-phase power measurement, why can the two wattmeters have different readings even though a balanced load is used?

26-32. Show that the total instantaneous power in *any* **wye**-connected load is equal to the sum of the two wattmeter readings in Figure 26-29.

26-33. The horizontal axis scale of the power factor graph of Figure 26-31 ranges from -1.0 through 0 to $+1.0$. Explain the reason for these particular limits.

26-34. Why is it necessary for an electrical utility company to maintain the same phase sequence?

26-35. What effect will reversing the tracing arrows when writing the voltage subscripts in the wye-connected source of Figure 26-22 have on the phase sequence of the source? Explain.

26-36. Why are we instructed to read *clockwise* around a phasor diagram when determining the phase sequence?

26-37. Explain the principle of operation of the phase sequence tester of Figure 26-37(a).

26-38. Account for the changes in load phase voltages when the neutral lead becomes open-circuit in Problem 26-11-28.

26-39. What is meant by a **floating wye point** in a three-phase system?

26-40. Derive Equation (26-16B) from the basic Kirchhoff's-law equations.

26-41. We can solve three-phase networks with the superposition theorem. The individual calculations are simpler, but there are many repetitive component-current calculations. However, we must use *phase* voltages for the source rather than *line* voltages. Using diagrams, explain this restriction.

*26-42. Write a computer program which will print out the overall power factor of a three-phase load when the input data are the readings for the two-wattmeter method of power measurement.

*26-43. Write a program to print out the magnitude of the three line currents I_A, I_B, and I_C when any three-wire wye load is connected to a 208-V 3ϕ source. The program should ask the operator to select phase sequence *ABC* or *CBA* for the source and enter the magnitude and angle for Z_A, Z_B, and Z_C.

APPENDICES

In solving simultaneous equations, we require as many equations as there are unknowns. For the network examples in this text, we can limit our discussion to second- and third-order simultaneous equations and write them in the general form shown below, arranging the order of the terms so that the *same* unknown appears in a vertical column.

Second-Order Equations	Third-Order Equations

$$a_1 x + b_1 y = k_1 \quad (1)$$

$$a_2 x + b_2 y = k_2 \quad (2)$$

$$a_1 x + b_1 y + c_1 z = k_1 \quad (A)$$

$$a_2 x + b_2 y + c_2 z = k_2 \quad (B)$$

$$a_3 x + b_3 y + c_3 z = k_3 \quad (C)$$

where x, y, and z are the unknowns and a, b, c, and k are numerical coefficients which can be either $+$ or $-$ quantities or even zero in some cases.

In our search for a systematic procedure, we can solve the generalized second-order simultaneous equations above by the more familiar elimination method.

$$b_2 \times (1) \text{ gives} \quad a_1 b_2 x + b_1 b_2 y = k_1 b_2$$

$$b_1 \times (2) \text{ gives} \quad a_2 b_1 x + b_1 b_2 y = k_2 b_1$$

Subtracting,

$$x(a_1 b_2 - a_2 b_1) = k_1 b_2 - k_2 b_1$$

from which

$$x = \frac{k_1 b_2 - k_2 b_1}{a_1 b_2 - a_2 b_1} \tag{3}$$

Similarly,

$$y = \frac{a_1 k_2 - a_2 k_1}{a_1 b_2 - a_2 b_1} \tag{4}$$

We can, therefore, solve *any* second-order simultaneous equations by *memorizing* Equations (3) and (4). The numerators and denominators of these generalized equations are called **determinants.** Note that the *same* determinant appears in the denominators of both Equation (3) and Equation (4). Hence, we can rewrite Equations (3) and (4) as

$$x = \frac{D_x}{D} \quad \text{and} \quad y = \frac{D_y}{D}$$

These equations are called **Cramer's rule.**

We now need a simple way of memorizing the formulas for D, D_x, and D_y. We can achieve this by writing the numerical coefficients which make up the determinants in matrix form. We first write the simultaneous equations in the systematic form of Equations (1) and (2). Directly below, we write the matrices for the determinants. Each determinant matrix contains as many vertical columns and horizontal rows as there are unknowns. The denominator matrix D simply reproduces the numerical coefficients of the unknowns in the *same* matrix position as in the original equations.

$$
\begin{array}{ccccc}
\textit{Column} & & \textit{Column} & & \textit{Substitution} \\
\textit{1} & & \textit{2} & & \textit{column} \\
a_1 x & + & b_1 y & = & k_1 & \quad (1) \\
a_2 x & + & b_2 y & = & k_2 & \quad (2)
\end{array}
$$

$$x = \frac{\begin{vmatrix} k_1 & b_1 \\ k_2 & b_2 \end{vmatrix}}{\begin{vmatrix} a_1 & b_1 \\ a_2 & b_2 \end{vmatrix}} \qquad (5)$$

$$y = \frac{\begin{vmatrix} a_1 & k_1 \\ a_2 & k_2 \end{vmatrix}}{\begin{vmatrix} a_1 & b_1 \\ a_2 & b_2 \end{vmatrix}} \qquad (6)$$

The numerator matrix for the determinant D_x replaces the coefficients of x (a_1 and a_2) with the numerical constants k_1 and k_2 from the right-hand side of Equations (1) and (2). Similarly, the numerator matrix for the determinant D_y replaces the coefficients of y (b_1 and b_2) with the same numerical constants k_1 and k_2.

To evaluate the three determinants so as to arrive at the same result as Equations (3) and (4), we use the following procedure. Using the denominator determinant D as an example, we require the product $a_1 b_2$ *minus* the product $a_2 b_1$. This can be accomplished in the D matrix by multiplying coefficients along *diagonals,* always moving diagonally to the right. To obtain the required solution ($a_1 b_2 - a_2 b_1$) we always *subtract* the *upward* diagonal from the *downward* diagonal.

Second operation—subtractive diagonal
Reverse sign of the product

$$\begin{vmatrix} a_1 & b_1 \\ a_2 & b_2 \end{vmatrix} = a_1 b_2 - a_2 b_1$$

First operation—additive diagonal

Checking against the numerator determinants in Equations (3) and (4), we find that the same rule applies to the evaluation of the matrices for D_x and D_y. The best way to memorize the procedure is to work numerical examples.

$$\begin{vmatrix} 4 & 2 \\ 3 & 5 \end{vmatrix} = (4 \times 5) - (3 \times 2) = 14$$

$$\begin{vmatrix} 3 & 5 \\ 4 & 2 \end{vmatrix} = 6 - 20 = -14$$

$$\begin{vmatrix} -2 & -4 \\ -3 & 5 \end{vmatrix} = (-2 \times 5) - (-3 \times -4) = -10 - 12 = -22$$

Using the simultaneous equations from Step V of Example 8-2,

$$60I_1 + 10I_2 = 80$$

$$10I_1 + 35I_2 = 80$$

$$I_1 = \dfrac{\begin{vmatrix} 80 & 10 \\ 80 & 35 \end{vmatrix}}{\begin{vmatrix} 60 & 10 \\ 10 & 35 \end{vmatrix}} = \dfrac{2800 - 800}{2100 - 100} = \dfrac{2000}{2000} = 1\ \text{A}$$

$$I_2 = \dfrac{\begin{vmatrix} 60 & 80 \\ 10 & 80 \end{vmatrix}}{\begin{vmatrix} 60 & 10 \\ 10 & 35 \end{vmatrix}} = \dfrac{4800 - 800}{2000} = \dfrac{4000}{2000} = 2\ \text{A}$$

We can check our solution by substituting the answers into the original equations:

$$60 \times 1 + 10 \times 2 = 80$$

$$10 \times 1 + 35 \times 2 = 80$$

Adding a third unknown greatly increases the amount of numerical computation required in the solution of the third-order simultaneous equations. However, the procedure for evaluating third-order determinants is exactly the same as the one we have developed for evaluating second-order determinants. If we take the time to solve the three generalized Equations (A), (B), and (C) for x, we obtain

$$x = \frac{k_1 b_2 c_3 + b_1 c_2 k_3 + c_1 k_2 b_3 - k_3 b_2 c_1 - b_3 c_2 k_1 - c_3 k_2 b_1}{a_1 b_2 c_3 + b_1 c_2 a_3 + c_1 a_2 b_3 - a_3 b_2 c_1 - b_3 c_2 a_1 - c_3 a_2 b_1}$$

The equations for y and z are similar, all with the same denominator determinant. Our third order determinant matrices must provide us with three downward additive diagonals and three upward subtractive diagonals, each diagonal being the product of three coefficients. First, we set up the required matrices from Equations (A), (B), and (C).

Column 1		Column 2		Column 3		Substitution column	
$a_1 x$	+	$b_1 y$	+	$c_1 z$	=	k_1	(A)
$a_2 x$	+	$b_2 y$	+	$c_2 z$	=	k_2	(B)
$a_3 x$	+	$b_3 y$	+	$c_3 z$	=	k_3	(C)

$$x = \frac{\begin{vmatrix} k_1 & b_1 & c_1 \\ k_2 & b_2 & c_2 \\ k_3 & b_3 & c_3 \end{vmatrix}}{\begin{vmatrix} a_1 & b_1 & c_1 \\ a_2 & b_2 & c_2 \\ a_3 & b_3 & c_3 \end{vmatrix}}$$

$$y = \frac{\begin{vmatrix} a_1 & k_1 & c_1 \\ a_2 & k_2 & c_2 \\ a_3 & k_3 & c_3 \end{vmatrix}}{D}$$

$$z = \frac{\begin{vmatrix} a_1 & b_1 & k_1 \\ a_2 & b_2 & k_2 \\ a_3 & b_3 & k_3 \end{vmatrix}}{D}$$

To obtain the required diagonals from the 3×3 matrices, we can think of them as being printed on a vertical cylinder, so that there are three rows to each diagonal, no matter where we start. To achieve the same effect on flat paper, we repeat the first two columns in sequence on the right-hand side of the original matrix.

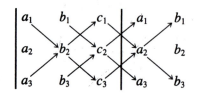

Second operation—three subtractive diagonals
Reverse sign of these three products

First operation—three additive diagonals

$$\therefore D = a_1 b_2 c_3 + b_1 c_2 a_3 + c_1 a_2 b_3 - a_3 b_2 c_1 - b_3 c_2 a_1 - c_3 a_2 b_1$$

EXAMPLE

$$D = \begin{vmatrix} 4 & 3 & -2 \\ 5 & -2 & 4 \\ 6 & 0 & 1 \end{vmatrix}$$

SOLUTION

$$D = \begin{vmatrix} 4 & 3 & -2 \\ 5 & -2 & 4 \\ 6 & 0 & 1 \end{vmatrix} \begin{matrix} 4 & 3 \\ 5 & -2 \\ 6 & 0 \end{matrix}$$

$$= -8 + 72 + 0 - 24 - 0 - 15 = +25$$

APPENDIX 2 CALCULUS SOLUTIONS

2-1 Maximum Power-Transfer Theorem

Given a voltage source with a fixed internal resistance R_{int}, we wish to determine the value of load resistance R_L in the circuit of Fig. A2-1 which will permit maximum power in the load resistance.

FIGURE A2-1

First, we set up an equation for power output in terms of the variable load resistance:

$$P_L = I^2 R_L$$

but $I = E/(R_{int} + R_L)$; hence,

$$P_L = \frac{E^2 R_L}{(R_{int} + R_L)^2} \tag{7}$$

Maximum power transfer will occur when the value of R_L is such that the derivative of P_L with respect to R_L is zero:

$$\frac{dP_L}{dR_L} = \frac{E^2[(R_{int} + R_L)^2 - R_L(2R_{int} + 2R_L)]}{(R_{int} + R_L)^4} = 0$$

Since $E \neq 0$, the factor in square brackets must equal zero, from which

$$R_{int} + R_L = 2R_L$$

Therefore, maximum power output occurs when $R_L = R_{int}$. $\tag{8}$

2-2 Instantaneous PD in a CR Circuit

When we throw the switch to position 1 in the circuit of Fig. A2-2,

$$E = iR + v_C \tag{9}$$

FIGURE A2-2

Since $i = dq/dt$ and since $q = Cv_C$,

$$i = C\frac{dv_C}{dt}$$

Substituting in Equation (9),

$$E = CR\left(\frac{dv_C}{dt}\right) + v_C \tag{10}$$

To solve this differential equation, we rearrange it in the form

$$\frac{dv_C}{dt} = \frac{E - v_C}{CR}$$

$$\frac{dv_C}{E - v_C} = \frac{dt}{CR}$$

Multiplying both sides by -1,

$$\frac{dv_C}{v_C - E} = -\frac{dt}{CR}$$

Integrating gives

$$\ln (v_C - E) \Big]_0^{v_C} = \frac{-t}{CR} \Big]_0^t$$

$$\ln (v_C - E) - \ln (0 - E) = \frac{-t}{CR} + 0$$

which becomes

$$\ln \left(\frac{v_C - E}{-E}\right) = \frac{-t}{CR}$$

Switching from the logarithmic form to the exponential form,

$$\frac{v_C - E}{-E} = e^{-t/CR}$$

Rearranging this to solve for v_C,

$$v_C = E - Ee^{-t/CR}$$

Thus,

$$v_C = E(1 - e^{-x}) \tag{11}$$

where e $= 2.718$ *(the base of natural logarithms) and* $x = t/CR$.

Once the capacitor has charged, if we throw the switch in the circuit of Figure A2-2 to position 2, the applied voltage in the loop consisting of the capacitor and the resistor is zero, and

$$0 = iR + v_C = CR\frac{dv_C}{dt} + v_C$$

If we rewrite the equation as $\dfrac{dv_C}{v_C} = \dfrac{-dt}{CR}$ and integrate,

$$\ln v_C \Big]_0^{v_C} = \frac{-t}{CR} \Big]_0^t$$

$$\ln \left(\frac{v_C}{V_0} \right) = \frac{-t}{CR} \quad \text{or} \quad \frac{v_C}{V_0} = e^{-t/CR}$$

Thus,

$$v_C = V_0 e^{-x} \tag{12}$$

where V_0 is the initial voltage across the capacitor, $e = 2.718$, and $x = t/CR$.

2-3 Energy Stored by a Capacitor

To determine the energy stored by a capacitor, we calculate the area under the instantaneous power graph of Figure 12-19 by establishing the integral

$$w = \int_0^t p \, dt \tag{13}$$

But
$$p = vi \quad \text{and} \quad i = C \frac{dv}{dt}$$

Therefore,
$$w = \int_0^t Cv \frac{dv}{dt} \, dt$$

$$= \int_0^v Cv \, dv$$

and
$$w = \tfrac{1}{2} Cv^2$$

Hence, no matter how a capacitor charges, the energy stored by it at *any* instant in time is proportional to its capacitance and the square of the potential difference between its plates.

2-4 Instantaneous Current in an *LR* Circuit

When we throw the switch to position 1 in the circuit of Fig. A2-3,

$$E = v_R + v_L = iR + L \left(\frac{di}{dt} \right) \tag{15}$$

To solve this differential equation, we rearrange it in the form

$$\frac{di}{i - \dfrac{E}{R}} = -\frac{R}{L} dt$$

FIGURE A2-3

Integrating gives

$$\ln\left(i - \frac{E}{R}\right)\Big]_0^i = -\frac{R}{L}t\Big]_0^t$$

$$\ln\left(i - \frac{E}{R}\right) - \ln\left(0 - \frac{E}{R}\right) = -\frac{R}{L}t + 0$$

which becomes

$$\ln\left(\frac{i - \dfrac{E}{R}}{-\dfrac{E}{R}}\right) = -\frac{R}{L}t$$

Switching from the logarithmic form to the exponential form,

$$\frac{i - \dfrac{E}{R}}{-\dfrac{E}{R}} = e^{-(R/L)t}$$

Rearranging this to solve for i,

$$i = \frac{E}{R} - \frac{E}{R}\left(e^{-(R/L)t}\right)$$

Thus,

$$i = \frac{E}{R}(1 - e^{-x}) \qquad (16)$$

where e = 2.718 (*the base of natural logarithms*) *and* $x = tR/L$.

Once the current has reached its steady-state value of $I = E/R$, if we throw the switch in the circuit of Figure A2-3 to position 2, the applied voltage in the loop consisting of the inductance and the resistor is zero, and

$$0 = v_R + v_L = iR + L\left(\frac{di}{dt}\right)$$

If we rewrite the equation as $di/i = -R\,dt/L$ and integrate,

$$\ln i \Big]_{I_0}^{i} = \frac{-Rt}{L} \Big]_0^t$$

$$\ln\left(\frac{i}{I_0}\right) = \frac{-Rt}{L} \quad \text{or} \quad \frac{i}{I_0} = e^{-Rt/L}$$

Thus,

$$i = I_0 e^{-x} \tag{17}$$

where I_0 is the initial current through the inductance, e = 2.718, and $x = tR/L$.

2-5 Energy Stored by an Inductor

To determine the energy stored in the magnetic field resulting from current through an inductor, we calculate the area under the instantaneous power graph of Figure 15-12 by establishing the integral

$$w = \int_0^t p\,dt \tag{18}$$

But

$$p = iv_L = iL\frac{di}{dt}$$

Hence the integral becomes

$$w = \int_0^i Li\,di$$

and

$$w = \tfrac{1}{2}Li^2 \tag{19}$$

Hence, the energy stored by *any* inductor at *any* instant in time is proportional to its inductance and the square of the current through it at that instant in time.

2-6 Rms and Average Values of a Sine Wave

By using integration techniques to average instantaneous power and instantaneous current waveforms, we can derive expressions for rms and average values for any ac waveform. To find the rms value of an alternating current, we start with the definition that

> *The rms value of an alternating current is equal to the direct current which will produce the same average heating effect in a given resistance.*

To find the average heating effect of an alternating current, we can integrate the instantaneous power over a complete cycle (2π radians) and divide by the number of radians included in the integration. Hence,

$$P_{av} = \frac{1}{2\pi} \int_0^{2\pi} p \, d\phi = \frac{1}{2\pi} \int_0^{2\pi} i^2 R \, d\phi$$

From the foregoing definition,

$$P_{av} = I^2 R$$

where I is the rms value of the alternating current.

Hence,

$$I^2 R = \frac{R}{2\pi} \int_0^{2\pi} i^2 \, d\phi$$

and

$$I = \sqrt{\frac{1}{2\pi} \int_0^{2\pi} i^2 \, d\phi} \tag{20}$$

Equation (20) gives the rms value of an alternating current of *any* waveshape.[†] For a sine wave, we substitute $i = I_m \sin \phi$.

$$I^2 = \frac{1}{2\pi} \int_0^{2\pi} I_m^2 \sin^2 \phi \, d\phi$$

From the relationship, $\sin^2 \phi = \frac{1}{2}(1 - \cos 2\phi)$[‡]

$$I^2 = \frac{I_m^2}{4\pi} \int_0^{2\pi} (1 - \cos 2\phi) \, d\phi$$

$$= \frac{I_m^2}{4\pi} [\phi - \tfrac{1}{2} \sin 2\phi]_0^{2\pi}$$

Since $\sin 0$ and $\sin 4\pi$ are both zero,

$$I^2 = \frac{2\pi I_m^2}{4\pi}$$

[†] Equation (20) shows why we can call this value of an alternating current its **root-mean-square** or **rms** value.

[‡] See footnote on page 445.

and
$$I = \frac{I_m}{\sqrt{2}}$$
(21)

We find the average value of a symmetrical alternating-current waveform from our decision to average the instantaneous current for one half-cycle only. Therefore,

$$I_{av} = \frac{1}{\pi} \int_0^\pi i \, d\phi$$
(22)

Applying Equation (22) to a sine wave,

$$I_{av} = \frac{1}{\pi} \int_0^\pi I_m \sin \phi \, d\phi$$

$$= \frac{I_m}{\pi} [-\cos \phi]_0^\pi$$

$$\therefore I_{av} = \frac{2}{\pi} I_m$$
(23)

2-7 Inductive Reactance

When an ideal inductor is connected across a sine-wave voltage source, the instantaneous voltage drop across the inductance must equal the instantaneous applied voltage. Hence,

$$L \frac{di}{dt} = E_m \sin \omega t$$

from which
$$\frac{di}{dt} = \frac{E_m}{L} \sin \omega t$$

Proceeding with the indefinite integration, we can solve for the instantaneous current in an inductance.

$$i = \frac{E_m}{L} \int \sin \omega t \, dt$$

$$= \frac{E_m}{\omega L} \times -\cos \omega t + C$$

Except when dealing with transients caused by opening or closing a switch in the circuit, the constant of integration C is zero. And if we check the values of sines and cosines for a complete revolution, we find that a *negative* cosine wave is the same as a sine wave *lagging* by 90° or $\pi/2$ radians. Hence,

$$i_L = \frac{E_m}{\omega L} \sin \left(\omega t - \frac{\pi}{2} \right)$$
(24)

The instantaneous current i in Equation (24) will attain its maximum value when $\sin(\omega t - \pi/2) = 1$. Hence,

$$I_m = \frac{E_m}{\omega L}$$

Since I_L and V_L are both 0.707 of their peak values, from the definition of inductive reactance,

$$X_L = \frac{V_L}{I_L} = \frac{E_m}{I_m} = \omega L$$

$$\therefore \quad X_L = \omega L \qquad (25)$$

where $\omega = 2\pi f$.

2-8 Capacitive Reactance

In Equation (17-7) for the instantaneous current in a capacitor,

$$i = C\frac{dv}{dt} \qquad (17\text{-}7)$$

we substitute $E_m \sin \omega t$ for the instantaneous voltage across the capacitor.

$$i = C\frac{d}{dt}(E_m \sin \omega t)$$

$$= \omega C E_m \cos \omega t$$

Since $\cos \phi$ is the same as the sine of an angle *leading* ϕ by 90° or $\pi/2$ radians

$$i_C = \omega C E_m \sin\left(\omega t + \frac{\pi}{2}\right) \qquad (26)$$

By noting that $i = I_m$ when $\sin(\omega t + \pi/2) = 1$ in Equation (26), we obtain

$$I_m = \omega C E_m$$

from which $\qquad\qquad X_C = \dfrac{1}{\omega C} \qquad\qquad (27)$

2-9 General Transformer Equation

We can arrive at the general transformer equation (24-3) by writing Equation (14-1) as it applies to the *instantaneous* voltage induced into a transformer winding. Therefore,

$$e = N\frac{d\phi}{dt} \qquad (14\text{-}1)$$

Since the flux in the core of the transformer must be a sine wave,

$$\phi = \Phi_m \sin \omega t \quad \text{and} \quad \frac{d\phi}{dt} = \omega \Phi_m \cos \omega t$$

Hence,

$$e = \omega N \Phi_m \cos \omega t \tag{28}$$

Because the instantaneous voltage in Equation (28) will be at its peak value when $\cos \omega t = 1$,

$$E_m = 2\pi f N \Phi_m$$

and

$$E = \frac{E_m}{\sqrt{2}} = 4.44 f N \Phi_m \tag{24-3}$$

2-10 Maximum Transformer Efficiency

The efficiency of a transformer is

$$\eta = \frac{S_L \cos \phi}{S_L \cos \phi + I^2 R_e + P_{CL}} \tag{24-14}$$

Substituting $S = IV$ in Equation (24-14) gives

$$\eta = \frac{IV \cos \phi}{IV \cos \phi + I^2 R_e + P_{CL}}$$

If we now consider that the current is the only variable, maximum efficiency will occur when the derivative of η with respect to I is zero.

$$\frac{d\eta}{dI} = \frac{(IV \cos \phi + I^2 R_e + P_{CL})V \cos \phi - IV \cos \phi (V \cos \phi + 2IR_e)}{(IV \cos \phi + I^2 R_e + P_{CL})^2}$$

For the derivative to equal zero, the numerator must equal zero, from which

$$V \cos \phi (I^2 R_e + P_{CL} - 2I^2 R_e) = 0$$

and

$$P_{CL} = I^2 R_e \tag{29}$$

APPENDIX 3   AMERICAN WIRE GAUGE TABLE

Although we have adopted SI units for our pencil and paper analysis of electric circuits, practical electric conductors in North America will still be manufactured for some time using the foot as the unit of length and the mil (one-thousandth of an inch) as the unit of diameter. Before we can use Equation (4-4) to calculate the resistance of a practical conductor of a given AWG size, we must determine its cross-sectional area in square meters using the conversion factor

$$1 \text{ mil} = 0.0254 \text{ mm}$$

In the case of copper conductors, most wire tables spare us the tedious calculations by including a column giving the resistance in ohms per thousand feet at 20°C for each AWG conductor size.

It is worthwhile noting that in setting up the AWG wire table (Table A3-1), the cross-sectional dimensions were chosen so that each *decrease* of one gauge number represents a 25% *increase* in cross-sectional *area*. On this basis, a decrease of *three* gauge numbers represents an increase in cross-sectional area of $1.25 \times 1.25 \times 1.25$ or approximately a 2 : 1 increase. Similarly, a change of 10 wire gauge numbers represents a 10 : 1 change in cross-sectional area. Also, since doubling the cross-sectional area cuts the resistance in half, a decrease of three wire gauge numbers cuts the resistance of the conductor of a given length in half. Therefore, if we memorize the dimensions for any one wire gauge number, we can approximate the dimensions of any other wire gauge number. For this purpose, AWG No. 10 wire is quite convenient.

TABLE A3-1

AMERICAN WIRE GAUGE CONDUCTOR SIZES

AWG Number	Diameter (mils)*	Ohms per 1000 ft†	AWG Number	Diameter (mils)*	Ohms per 1000 ft†
0000	460.0	0.049 01	19	35.89	8.051
000	409.6	0.061 80	20	31.96	10.15
00	364.8	0.077 93	21	28.46	12.80
0	324.9	0.098 27	22	25.35	16.14
1	289.3	0.1239	23	22.57	20.36
2	257.6	0.1563	24	20.10	25.67
3	229.4	0.1970	25	17.90	32.37
4	204.3	0.2485	26	15.94	40.81
5	181.9	0.3133	27	14.20	51.47
6	162.0	0.3951	28	12.64	64.90
7	144.3	0.4982	29	11.26	81.83
8	128.5	0.6282	30	10.03	103.2
9	114.4	0.7921	31	8.928	130.1
10	101.9	0.9989	32	7.950	164.1
11	90.74	1.260	33	7.080	206.9
12	80.81	1.588	34	6.305	260.9
13	71.96	2.003	35	5.615	329.0
14	64.08	2.525	36	5.000	414.8
15	57.07	3.184	37	4.453	523.1
16	50.82	4.016	38	3.965	659.6
17	45.26	5.064	39	3.531	831.8
18	40.30	6.385	40	3.145	1049.0

*1 mil–0.001 inch.
†Copper at 20°C.

AWG gauge number   10   diameter (mils) 100 (approx)
ohms/thousand feet   1
  (copper at 20°C)

EXAMPLE
If the resistance of AWG No. 10 copper wire is 1 Ω per thousand feet at 20°C, what is the resistance at 20°C of 200 ft of AWG No. 14 copper wire?

SOLUTION
Since the cross-sectional area of AWG No. 14 wire is $1/(1.25)^4$ that of AWG No. 10 wire,

$$R = 2.5 \times \frac{200}{1000} \times 1 \ \Omega \approx \mathbf{0.5 \ \Omega}$$

REVIEW GLOSSARY

CHAPTER 1—INTRODUCTION

Ampere The base SI unit of electric current, i.e., the flow of electric charge.

BASIC A symbolic language commonly used in programming personal computers.

Conductor A material, or (more specifically) wires made from a material that has an atomic structure which permits the flow of electric charge.

Dimensional analysis An orderly procedure for ensuring that physical equations balance in terms of their units (dimensions) as well as their numerical values.

Electric circuit An interconnection of electric circuit elements that permits the flow of electric charge.

International System of Units (SI) A universal system of units based on MKSA base units—meter, kilogram, second, ampere (now used almost exclusively in electrical engineering).

Kilogram The base SI unit of mass.

Load A device that converts electric energy to some other form of energy.

Meter The base SI unit of length or distance.

Personal computer (PC) A desk-top digital computer made possible by relatively inexpensive very-large-scale integrated circuits. Their modest cost puts quite sophisticated computers within the realm of personal ownership.

SI Abbreviation for International System of Units (Système Internationale d'Unités).

Schematic diagram A circuit diagram using standard graphic symbols to show the electrical interconnection of a specific electric circuit.

Scientific notation Representation of very large or very small numerical values by using powers of ten.

Second The base SI unit of time.

Source A device that converts some other form of energy into electric energy.

Switch A device for interrupting the continuity of an electric circuit.

Unit prefix A prefix which creates decimal multiples or submultiples of root units (as an engineering alternative to scientific notation).

CHAPTER 2—CURRENT AND VOLTAGE

Ampere The SI unit of electric current. An ampere is the rate of electric charge flow when one coulomb of electric charge passes a given point in an electric circuit in one second.

Atom The smallest particle of an element that can exist alone or in combination.

Battery A group of two or more voltaic cells.

Conventional current direction The direction in which *positive* charge carriers flow in an electric circuit. This convention assures that current is a positive quantity in equations.

Coulomb The SI unit of quantity of electric charge. A coulomb represents the quantity of electric charge possessed by 6.24×10^{18} electrons.

Coulomb's Law Defines the force between two electric charges in terms of the quantities of the charges and the distance between them.

Covalent bonding Formation of a chemical compound by atoms *sharing* valence electrons.

Direct current A unidirectional current, usually of almost constant magnitude.

Dynamo (generator) A device for converting mechanical energy into electric energy.

Electromotive force (emf) The *electron-moving force* created by charge separation in an electric source.

Electron The negatively charged particle which normally orbits the nucleus of an atom.

Electron flow An electric current consisting of the flow of free electrons. The type of charge flow found in metallic conductors.

Electron shell A group of several closely spaced permissible energy levels that can be occupied by orbiting electrons.

Energy level A means of identifying the location of orbiting electrons in an atom by the kinetic energy they possess.

Field force A force which can act at a considerable distance without physical contact.

Free electron The valence electrons of a metal which are free to move about among the positive metal ions forming a crystal lattice structure.

Ion An atom which has a net electric charge due to having a temporary surplus or deficiency of electrons.

Ionic bonding Formation of a chemical compound by one atom *lending* valence electrons to another.

Joule The SI unit of work or energy. A joule is the work accomplished by a force of one newton acting through a distance of one meter.

Metallic bonding The characteristic of positive metal ions taking up positions in crystalline patterns, thus leaving their valence electrons *free* to drift through the structure.

Molecule The smallest particle of a substance that can retain the chemical properties of the substance.

Nucleus The core of an atom composed mainly of protons and neutrons.

Neutron A particle found in the nucleus of all atoms (except hydrogen). Neutrons are slightly heavier than protons but have no electric charge.

Potential difference The difference in potential energy between two points in an electric circuit.

Potential fall (voltage drop) The loss of potential energy as an electric charge moves through a load.

Potential rise (voltage rise) The gain in potential energy as an electric charge is forced to move through a source.

Proton A positively charged particle found in the nucleus of all atoms.

Quantity A measure of the magnitude of an electric charge in coulombs.

Valence electron An orbital electron located in the outermost occupied electron shell of an atom.

Volt The SI unit of potential difference. A volt is the difference in potential between two points in an electric circuit when one joule of energy is required to move an electric charge of one coulomb between those two points.

Voltage A collective term encompassing potential difference, potential rise, and potential fall.

Voltaic cell A chemical cell that converts chemical energy into electric energy.

CHAPTER 3—CONDUCTORS AND INSULATORS, SEMICONDUCTORS, AND VOLTAGE SOURCES

Anode An electrode through which *conventional* current flows from a metallic conductor to a nonmetallic conductor (such as an electrolyte or vacuum).

Barrier potential The potential difference at which electrons cease flowing across a *pn*-junction in search of holes.

Battery Two or more voltaic cells connected for supplying electric energy. (Common usage permits speaking of a single-cell battery.)

Cathode An electrode through which *conventional* current flows from a nonmetallic conductor to a metallic conductor.

Depletion layer An area on both sides of a *pn*-junction where there is a reduced concentration of charge carriers.

Diode A two-element electron tube or semiconductor device.

Dry cell A voltaic cell in which the electrolyte is in the form of a paste or saturated membrane.

Electrode An electric conductor designed to provide electrical contact with a nonmetallic portion of an electric circuit.

Electrolyte A conducting medium in which the flow of electric current takes place by migration of ions.

Electrolytic cell A vessel containing an anode, a cathode, and an electrolyte in which electric current causes an electrochemical reaction.

Faraday's Law Defines the mass of a substance deposited at the cathode of an electrolytic cell in terms of the quantity of electric charge carriers passing through the cell.

Forward bias A potential difference applied to a *pn*-junction with a polarity such as to encourage the flow of charge carriers across the junction.

Fuel cell An electrochemical cell that can continuously convert the chemical energy of a fuel and oxidant into electric energy without involving either electrode in a chemical reaction.

Hole flow Flow of electric charge in the form of mobile vacancies in the electronic valence structure of a semiconductor, which thus appear to behave as "positive electrons."

Insulator A nonconducting material that provides electrical isolation between an electric conductor and other conducting parts or from ground.

***n*-type semiconductor** An intrinsic (pure) semiconductor crystal that has been doped with a pentavalent element to provide some surplus electrons as charge carriers.

Photovoltaic cell (solar battery) A device which produces a potential difference between its two electrodes as the result of the absorption of the photon energy of light waves.

***pn*-Junction** An electron device consisting of a layer of *p*-type semiconductor material fused to a layer of *n*-type semiconductor material.

Polarization A reduction in the current available from a voltaic cell due to the formation of a layer of insulating bubbles on one of the electrodes.

Primary cell A voltaic cell that produces electric current by electrochemical action without regard to reversibility of the chemical reaction.

***p*-type semiconductor** An intrinsic (pure) semiconductor crystal that has been doped with a trivalent element thus creating holes where electrons are missing. These holes behave as positive charge-carriers.

Reverse bias A potential difference applied to a *pn*-junction with a polarity such as to move the charge carriers away from the junction.

Secondary cell (storage battery) A voltaic cell which may be recharged by reversing the electrochemical action when electric current is forced to flow in the opposite direction through the cell.

Semiconductor A material in which the number of charge carriers per unit volume is between that of conductors and insulators. In practical semiconductors, the number and types of charge carriers are carefully controlled in the manufacturing process.

Superconductor A metallic conductor operating near absolute zero temperature, or a special synthetic material, which has negligible resistance to electric current.

Voltaic cell An electrolytic cell designed to convert chemical energy into electric energy. Chemical reaction between dissimilar electrode materials in an electrolyte creates a potential difference between the two electrodes.

CHAPTER 4—RESISTANCE

Constantan A nickel-copper alloy with an extremely low temperature coefficient of resistance. Used for making metallic resistors.

IR drop The voltage drop across a resistor due to electric current through the resistor.

Linear resistor A resistor which maintains a constant V/I ratio with changing values of current.

Nonlinear resistor A resistor whose resistance can vary depending on applied voltage, current, temperature, or incident light.

Ohm The SI unit of resistance. An electric circuit has a resistance of one ohm when a current through the circuit of one ampere causes a voltage drop across the circuit of one volt.

Ohm's Law Defines the constant proportionality between voltage drop and current in a linear resistor.

Ohm meter The SI unit of resistivity.

Photoresistor A nonlinear resistor of semiconductor material whose resistance decreases as incident light increases.

Resistance The opposition of an electric circuit to electric current through that circuit.

Resistivity (specific resistance) The resistance of a unit length and cross-section conductor of a given material.

Resistor An electric circuit component made of metal or carbon and used to insert a specific amount of resistance into an electric circuit.

Resistor color code A method of identifying the resistance and tolerance of tiny resistors by a set of four colored bands.

Temperature coefficient (of resistance) The amount of change in each ohm of resistance of a material for each degree change in temperature from the reference temperature of 20°C.

Thermistor A nonlinear resistor of semiconductor material whose resistance decreases as its temperature increases.

Tolerance The percentage by which the actual resistance of a resistor may vary from the specified value.

Varistor A nonlinear resistor of semiconductor material whose resistance decreases as the applied voltage increases.

Volt-ampere characteristic Graphical representation of the V/I ratio (resistance) of a resistor (particularly a nonlinear resistor) for a range of values of current and *IR* drop.

CHAPTER 5—WORK AND POWER

Efficiency The ratio of useful output energy of a device to total input energy. Efficiency is usually expressed as a percentage.

Energy The ability to do work. Hence energy is numerically equal to work and is measured in joules.

Horsepower A mechanical unit of power which is equivalent to 746 watts in SI units.

Joule The SI unit of work or energy. In electrical systems, one joule of energy is required to raise one coulomb of electric charge through a potential difference of one volt.

Kilowatthour A practical unit of work. One kilowatthour equals 3.6 megajoules.

Power The *rate* of doing work. Power is measured in watts, which is the same as joules per second.

Watt The SI unit of power. One watt equals one joule per second.

Work The accomplishment of motion against the action of a force which tends to oppose the motion. Work is numerically equal to the energy expended in accomplishing the work and is measured in joules.

CHAPTER 6—SERIES AND PARALLEL CIRCUITS

Branch One of several parallel paths for current in an electric circuit.

Conductance The ability of an electric circuit to pass current, hence the reciprocal of resistance.

Conductivity The conductance of a unit length and cross-section conductor of a given material.

Double-subscript notation A procedure for indicating the polarity of the voltage between two points in an electric circuit.

Equivalent resistance The resistance of a single resistor which is equivalent to two or more resistors in parallel in an electric circuit.

Internal resistance The resistance which appears to be in series with a voltage source and its terminals.

Kirchhoff's Current Law At any node in an electric circuit, the algebraic sum of the currents entering the node must equal the algebraic sum of the currents leaving that node.

Kirchhoff's Voltage Law In any complete electric circuit, the algebraic sum of the applied voltages must equal the algebraic sum of the voltage drops.

Maximum power transfer For a given voltage source, maximum load power occurs when the load resistance equals the internal resistance of the source.

Node A junction point common to two or more components in an electric circuit.

Open-circuit voltage The voltage appearing at the terminals of a voltage source when no current is drawn from the source (open circuit). Hence, the *IR* drop across the internal resistance of the source is zero.

Parallel circuit Two or more branches form a parallel circuit when the same (common) voltage appears across all these branches.

Series circuit Two or more components form a series circuit when the same (common) current flows through all these components.

Siemens The SI unit of conductance. The siemens is the reciprocal of the ohm.

Voltage regulation Control of the terminal voltage of a voltage source by increasing the generated emf as the current drain increases.

CHAPTER 7—SERIES-PARALLEL CIRCUITS

Bleeder resistor The part of a voltage divider in parallel with the output terminals.

Current-divider principle In a parallel circuit, the ratio between any two branch currents is the same as the ratio of the two conductances through which these currents flow.

Equivalent circuit A simplified circuit diagram in which two or more components are replaced by a single equivalent component.

Ground A point in an electric circuit which is electrically connected to ground, and thus has the same potential as the earth. Ground then becomes a reference point for measuring voltages in the circuit.

Ladder network An electric circuit in which the layout of the branches in a schematic diagram resemble the rungs of a ladder.

Series dropping resistor The part of a voltage divider in series with the output terminals.

Series-parallel circuit An electric circuit in which some portions of the circuit fit the definition of a series circuit and other fit the definition of a parallel circuit.

Voltage divider A resistance network designed to provide a load with a lower applied voltage than the available source voltage.

Voltage-divider principle In a series circuit, the ratio between any two voltage drops is the same as the ratio of the two resistances across which these voltage drops occur.

Voltage regulation The no-load voltage of a voltage source minus the full-load voltage divided by the full-load voltage, usually expressed as a percentage.

Wheatstone bridge A precision resistance-measuring circuit with four resistors connected as the four sides of a diamond shape, in which a voltage source is connected to one pair of diagonally opposite terminals, and a load (a sensitive balance-indicating milli-ammeter) is connected to the other pair of terminals.

CHAPTER 8—RESISTANCE NETWORKS

Constant-current source An equivalent circuit for a voltage source in which the source appears to supply a constant current to the output terminals and an internal resistance in *parallel*.

Constant-voltage source The usual representation of a voltage source as consisting of a constant source voltage in *series* with the output terminals and an internal resistance.

Loop equations A procedure for writing a set of Kirchhoff's voltage-law equtions for a network based on tracing loop currents around the network.

Mesh analysis A format procedure for writing Kirchhoff's voltage-law equations to solve for various branch currents in a network.

Nodal analysis A format procedure for writing Kirchhoff's current-law equations to solve for various node voltages in a network.

Short-circuit current The current which would be drawn from a constant-voltage source if its output terminals were short-circuited.

Source conversion A procedure for replacing a constant-voltage source in a network diagram with an equivalent constant-current source, and vice versa.

Superposition theorem A procedure for solving for branch currents in a multisource network by considering each source to operate independently.

CHAPTER 9—EQUIVALENT-CIRCUIT THEOREMS

Delta-wye transformation A set of equations for replacing three branches in a network connected in delta (Δ) fashion by an equivalent circuit connected in wye (Y) fashion, or vice versa.

Dependent (controlled) source A voltage or current source whose magnitude is dependent on (or controlled by) a voltage or current appearing in some other branch of the network.

Independent source A voltage or current source whose magnitude is independent of what happens elsewhere in a network containing the source.

Millman's theorem A procedure which combines both Norton's and Thévenin's theorems in simplifying multi-source two-terminal networks.

Norton's theorem A procedure for replacing any two-terminal network of fixed resistances and voltage sources by a single constant-current source and resistance in parallel.

Reciprocity theorem If a voltage source in branch A of a linear network causes a current I in branch B, then moving that voltage source to branch B will cause a current I in branch A equal to the original current in branch B.

Substitution theorem Any branch of a linear network can be replaced by any combination of circuit elements which will produce the same voltage across and current through that branch.

Thévenin's theorem A procedure for replacing any two-terminal network of fixed resistances and voltage sources by a single constant-voltage source and resistance in series.

CHAPTER 10—ELECTRICAL MEASUREMENT

Ammeter An instrument for measuring the magnitude of the electric current at a point in an electric circuit.

Ayrton shunt A multirange ammeter shunt designed to protect the meter movement from an open-circuit shunt as the selected range is changed.

Bridge rectifier Four (semiconductor) diodes arranged in a bridge formation to provide a direct current output from an alternating current input.

Calibrating (swamping) resistor The precision resistor in series with a meter movement to provide a constant, standard resistance between the meter terminals.

Digital meter An electronic measuring instrument which provides a digital readout, rather than the pointer and scale of an analog meter.

Electrodynamometer A meter movement based on the magnetic interaction of both fixed and moving coils.

Galvanometer A meter movement with the zero position of the pointer at center-scale.

Moving-coil meter (D'Arsonval or Weston movement) A meter movement based on the magnetic interaction of a moving coil and a fixed permanent magnet.

Moving-iron meter A meter movement based on the magnetic interaction between a moveable and a fixed soft-iron vane which are temporarily magnetized by electric current in a fixed coil.

Multimeter A measuring instrument which uses a selector switch to permit a single meter movement to measure volts, amperes, and ohms in several ranges.

Ohmmeter An instrument for directly measuring the resistance between two points in an electric circuit which in *not* energized.

Potentiometer A resistor with an adjustable tap. *In measurement,* a precision potentiometer is used to determine an unknown voltage by comparing it with a known voltage.

Shunt resistor A precision resistor connected in parallel with a basic meter movement to provide a higher ammeter full-scale reading.

Torque The moment of force which produces a change in rotary motion.

Voltmeter An instrument for measuring the voltage between two points in an electric circuit.

Voltmeter loading effect The error in voltmeter reading due to the additional current a voltmeter draws when connected to an electric circuit.

Voltmeter sensitivity A measure of the loading effect that a given voltmeter will produce in an electric circuit.

Wattmeter An electrodynamometer instrument that measures the power input to a load as the product of the current through the stationary coil and the voltage applied to the moving coil.

CHAPTER 11—CAPACITANCE

Capacitance The opposition of an electric circuit to any *change* in the voltage across that circuit.

Capacitor A device constructed for the purpose of storing an electric charge with a potential difference between its terminals.

Dielectric An insulating material exposed to an electric field (such as between the conducting plates of a capacitor).

Dielectric constant The ratio between the absolute permittivity of a given dielectric and the absolute permittivity of free space.

Dielectric strength The electric field strength (or voltage gradient) required to break down a dielectric.

Elastance The opposition to the setting up of electric lines of force in a dielectric. (The reciprocal of capacitance.)

Electric field The region in which an electric charge is acted upon by an electric force.

Electric field strength The magnitude of the force acting on a unit charge in an electric field. Electric field strength is numerically equal to the voltage gradient in volts per meter.

Electric flux The total number of electric lines of force in an electric field.

Electric flux density The electric flux per unit cross-sectional area.

Electric line of force The path along which a weightless electrically charged particle will move from one electrically charged plate to another.

Electrostatic induction The separation of electric charges in an isolated conductor due to the proximity of an electric charge.

Equipotential surface An imaginary surface in an electric field created by joining all points that are at the same electric potential.

Farad The base SI unit of capacitance.

Ferroelectric dielectric A dielectric material with a much higher dielectric constant than most other dielectrics.

Permittivity The capacitance between opposite faces of a unit length and cross section of a dielectric material.

Reciprocal farad The SI unit of elastance.

Static electricity Electric charges which are separated by friction between insulating materials.

Voltage gradient The ratio of the potential difference between two charged plates to the distance between the plates. (Voltage gradient is numerically equal to electric field strength.)

Working voltage The maximum voltage that can be applied to a given capacitor without danger of dielectric breakdown.

CHAPTER 12—CAPACITANCE IN DC CIRCUITS

***CR* waveshaping circuit** A four-terminal network in which the instantaneous output voltage differs from the instantaneous input voltage due to the time required for capacitor(s) in the network to charge and discharge.

Instantaneous value A voltage or current whose magnitude is dependent on an exact instant in time.

Rate of change of voltage The rate at which an instantaneous voltage changes in volts per second.

Steady-state value A voltage or current whose magnitude is *not* dependent on an exact instant in time.

Stray capacitance Capacitance between any pair of electric conductors which are at different electric potentials.

Time constant The time it would take the potential difference across a capacitor to rise to the same value as the applied voltage if it were to continue to rise at its initial rate of change for the whole time interval. A time constant (in seconds) is numerically equal to the product of the capacitance and the resistance through which it charges.

Transient A change in the steady-state condition of voltage and/or current.

Universal exponential curve An accurate graph which can be used to determine instantaneous current or voltage during a *CR* transient.

CHAPTER 13—MAGNETIC CIRCUITS

Air gap A small segment of a complete magnetic circuit composed of any nonmagnetic material (such as air).

Ampere The SI unit of magnetomotive force (based on an electric current of one ampere in a one-turn coil).

Coercive force the amount of negative magnetic field strength required to demagnetize a particular sample of a ferromagnetic material.

Eddy current The circular current in an electrically conductive core due to the emf induced into the core by a changing magnetic field.

Electromagnet A solenoid used to produce a substantial magnetic field.

Ferromagnetic material A material with a permeability considerably greater than that of free space. A material that can be magnetized.

Fringing The tendency of magnetic lines of force to spread apart as they cross an air gap in a ferromagnetic circuit.

Hysteresis The lagging of the magnetization of a ferromagnetic material behind the application of the magnetomotive force that produces it.

Leakage flux The magnetic lines of force that complete their circuit outside the designated magnetic circuit.

Linear magnetic circuit A magnetic circuit (usually of nonmagnetic material) in which permeability is independent of the magnetic field strength.

Magnetic circuit The closed loop path followed by magnetic lines of force.

Magnetic domain A group of adjacent atoms of a ferromagnetic material that appear to respond in unison to a magnetizing force.

Magnetic field The region in which a magnetic pole is acted upon by a magnetic force.

Magnetic field strength A measure of the magnetizing force available to create magnetic flux. Also called magnetic field intensity, magnetizing force, and mmf gradient.

Magnetic flux The total number of magnetic lines of force in a magnetic field.

Magnetic flux density The magnetic flux per unit cross-sectional area of a magnetic field.

Magnetic lines of force The paths along which a theoretical isolated magnetic pole will move from one pole of a magnet to another.

Magnetic pole The boundary where magnetic lines of force leave or enter a ferromagnetic material.

Magnetic shielding Protecting a space from a magnetic field by enclosing it in a container made of a high-permeability magnetic material.

Magnetization curve (*BH* graph) A graph which details the magnetic properties of a given ferromagnetic material.

Magnetizing force Magnetomotive force per unit length of a magnetic circuit, also called magnetic field strength or magnetic field intensity.

Nonlinear magnetic circuit A magnetic circuit (usually of ferromagnetic material) in which permeability depends on magnetic field strength.

Permanent magnet A ferromagnetic material in which most of the magnetic domains retain their alignment after a strong external magnetizing force is removed.

Permeability Permeance per unit length and cross-sectional area of a magnetic material.

Permeance A measure of the ability of a given magnetic circuit to permit the creation of magnetic lines of force (hence, the reciprocal of reluctance).

Reciprocal henry The SI unit of reluctance.

Reluctance The opposition of a given magnetic circuit to the establishing of magnetic flux.

Retentivity The ability of a magnetic material to retain some magnetism after an external magnetizing force is removed.

Right-hand rules Procedures for determining the direction of the magnetic lines of force around current-carrying straight conductors and solenoids.

Solenoid An electromagnet created by winding a coil of wire around a cylindrical form.

Tesla The SI unit of magnetic flux density.

Weber The SI unit of magnetic flux.

CHAPTER 14—INDUCTANCE

Counter emf (cemf) The self-induced voltage in an inductor.

Electromagnetic induction Generation of a voltage when an electric conductor cuts across magnetic lines of force.

Faraday's Law The voltage induced in an electric circuit is proportional to the time rate of change of the flux of magnetic induction linked with the circuit.

Henry The SI unit of self inductance.

Inductance See self-inductance.

Inductor A device constructed for the express purpose of displaying the property of self-inductance.

Lenz's Law The polarity of the induced voltage must be such that any current resulting from it will develop a magnetic flux which tends to oppose any change in the original flux.

Self-inductance The property of an electric circuit that opposes any change in the current in that circuit.

Self-induction Generation of a voltage in an electric circuit by a changing current in the same circuit.

CHAPTER 15—INDUCTANCE IN DC CIRCUITS

Discharge resistor The resistor placed in parallel with an inductor to protect the circuit from a high voltage induced by a collapsing magnetic field.

Time constant The time it would take the current to rise to its steady-state value if it were to continue to rise at its initial rate of change for the whole time interval.

Transient A change in the steady-state condition of voltage and/or current.

Universal exponential curve An accurate graph which can be used to determine instantaneous current or voltage during an *LR* transient.

CHAPTER 16—ALTERNATING CURRENT

Alternating current An electric current which reverses at regularly recurring intervals, hence having alternately positive and negative values.

Average value (half-cycle average) The average of instantaneous values of current or voltage over a time interval equal to one half cycle. This non-time-varying value for a sine wave is 0.637 of the peak value.

Cycle One complete set of positive and negative values of an alternating current or voltage.

Effective or equivalent dc value See rms value.

Frequency The number of cycles per unit time.

Hertz The SI unit of frequency (equal to cycles per second).

Instantaneous value A voltage, current, or power whose magnitude is dependent on an exact instant in time.

Peak value A non-time-varying value equal to the highest numerical value of instantaneous voltage, current, or power during a cycle.

Period The time taken for one complete set of positive and negative values of an alternating current, hence the reciprocal of frequency.

Periodic wave Any time-varying function such as voltage, current, or power which continually repeats the same sequence of events after a given time interval.

Phase angle The measure of the progression of a periodic wave in time or space from a chosen instant or position.

Root-mean-square (rms) value The square root of the average value of the squares of instantaneous voltage or current over a full cycle. This non-time-varying value for a sine wave is 0.707 of the peak value.

Sine wave An alternating voltage or current whose instantaneous values are directly proportional to the sine of the phase angle.

CHAPTER 17—REACTANCE

Capacitive reactance The opposition of capacitance to alternating current.

Inductive reactance The opposition of inductance to alternating current.

Ohm The SI unit of capacitive or inductive reactance.

CHAPTER 18—PHASORS

Complex conjugate Of a complex number has the same real component; the imaginary component has the same magnitude but the *opposite* sign.

Complex number A number $(a + jb)$ composed of a real number a and an imaginary number jb, thus existing in the two-dimensional complex plane.

Frequency domain A set of complex numbers representing such phasor quantities as rms voltage, rms current, and impedance.

Imaginary number A number multiplied by $\sqrt{-1}$ (j). Since such a number does not qualify as a "real" number, it was termed "imaginary."

Phase angle A measure of the progression of a periodic wave in time or space from a chosen instant or position (expressed in radians or degrees).

Phasor A complex number associated with a sinusoidally varying electrical quantity (particularly, diagrammatic representation of such a complex number).

Phasor diagram A diagram that shows the magnitudes and phase relationships of two or more sinusoidally varying voltages and/or currents.

Phasor quantity Any electrical quantity (including impedance) that is expressed by a complex number.

Polar coordinates Representation of a complex number (for example, phasor quantity) in the complex plane by a magnitude and an angle.

Pythagoras theorem The square of the hypotenuse of a right-angled triangle is equal to the sum of the squares of the other two sides.

Real number A number between $-\infty$ and $+\infty$ that can be represented geometrically in one dimension (a straight line).

Rectangular coordinates Representation of a complex number (for example, phasor quantity) in the complex plane by a real component and an imaginary component.

Scalar A quantity which possesses magnitude but has no direction or angular information.

Time domain A set of numbers representing the instantaneous values of voltage, current, and power in an electric circuit.

Vector A quantity which possesses both magnitude and direction. Vector quantities are complex numbers, but they do *not* include time-varying functions such as alternating current.

CHAPTER 19—IMPEDANCE

Admittance The overall ability of an electric circuit to *pass* alternating current; hence, the reciprocal of impedance.

Admittance diagram A diagram representing admittance, conductance, and susceptance by the sides of a right-angled triangle.

Conductance The ability of resistance to *pass* electric current (either direct or alternating); hence, the reciprocal of resistance.

Equivalent reactive voltage The difference between the voltages across the capacitance and inductance of a series ac circuit.

Impedance The total opposition to current in an ac circuit.

Impedance diagram A diagram representing impedance, resistance, and reactance by the sides of a right-angled triangle.

Ohm The SI unit of impedance, resistance, and inductive and capacitive reactance.

Practical inductor An impedance representing the resistance and inductive reactance of a practical inductor as acting in series.

Siemens The SI unit of admittance, conductance, and inductive and capacitive susceptance.

Susceptance The ability of inductance and capacitance to *pass* alternating current; hence, the reciprocal of reactance.

CHAPTER 20—POWER IN ALTERNATING-CURRENT CIRCUITS

Active power (real power or true power) The average power input to the resistance of an ac circuit, expressed in watts.

Apparent power The product of the total rms voltage and total rms current in an ac circuit, expressed in voltamperes.

Average power The average of instantaneous power over a complete cycle, expressed in watts. Average power equals active power.

Dielectric hysteresis The conversion of electric energy into heat as the dielectric of a capacitor is stressed in alternate directions many times a second in an ac circuit.

Effective resistance (ac resistance) The resistance of an ac circuit as measured with a wattmeter and ammeter. ($R = P/I^2$)

Lagging power factor The power factor of an inductive load.

Leading power factor The power factor of a capacitive load.

Ohmic resistance The resistance of an electric circuit as measured with an ohmmeter (a dc instrument).

Phasor power Apparent power (which is a phasor quantity).

Power factor The ratio between the active power and the apparent power of a load in an ac circuit.

Power factor correction (improvement) Reduction in the total current drawn from an ac source when capacitive reactance is connected in parallel with an inductive load.

Power triangle A diagram representing apparent power, active power, and reactive power by the sides of a right-angled triangle.

Reactive factor The ratio between the reactive power and the apparent power of an ac load.

Reactive power (quadrature power) The product of the rms voltage and current in an ideal inductor and/or capacitor, expressed in vars.

Skin effect The tendency of high frequency alternating currents to flow at the surface of a conductor due to self-induction within the conductor.

Var (Voltampere reactive) The unit of reactive power.

Varmeter An electrodynamometer movement connected so as to indicate the reactive power of an ac load.

Voltampere The unit of apparent power.

Watt The SI unit of active or average power.

Wattmeter An electrodynamometer movement connected so as to indicate the active or average power of a load.

CHAPTER 21—SERIES AND PARALLEL IMPEDANCES

Circle diagram A diagram showing the effect of varying a single parameter of an ac circuit on the voltage and/or current distribution.

Current-divider principle The ratio between any two parallel branch currents is the same as the ratio of their admittances (or the reciprocal of the ratio of their impedances).

Kirchhoff's Current Law At any node in an ac circuit, the phasor sum of the currents entering the node must equal the phasor sum of the currents leaving that node.

Kirchhoff's Voltage Law In an ac circuit, the phasor sum of the applied voltages must equal the phasor sum of the voltage drops.

Maximum power transfer Occurs in an ac circuit when the load impedance is the complex conjugate of the internal impedance of the source.

Parallel equivalent circuit The rectangular coordinates of an admittance specify an equivalent circuit consisting of a conductance and a susceptance in parallel.

Series equivalent circuit The rectangular coordinates of an impedance specify an equivalent circuit consisting of a resistance and a reactance in series.

Source conversion A procedure for replacing a constant-voltage source in an ac network diagram with an equivalent constant-current source, and vice versa.

Total admittance (of parallel branches) The phasor sum of the branch admittances.

Total impedance (of a series ac circuit) The phasor sum of the individual impedances.

Voltage-divider principle In a series ac circuit, the ratio between any two voltage drops is the same as the ratio of the two impedances across which these voltage drops occur.

CHAPTER 22—IMPEDANCE NETWORKS

Alternating-current bridge A precision device for determining ac circuit parameters based on a diamond-shaped network with an ac voltage source connected to one pair of diagonally opposite terminals and a balance indicator (zero current) connected to the other pair of terminals.

Capacitance comparison bridge An ac bridge used to determine an unknown capacitance by comparing it with a standard capacitor.

Delta-wye transformation A set of equations for replacing three branches in an ac network connected in delta (Δ) fashion by an equivalent circuit connected in wye (Y) fashion, or vice versa.

General bridge equation In a *balanced* ac bridge network, the phasor product of one pair of opposite impedances must equal the phasor product of the other pair of opposite impedances.

Hay bridge An ac bridge used to determine an unknown inductance whose resistance is quite small compared to its inductive reactance.

Loop equations A procedure for writing a set of Kirchhoff's voltage-law equations for an ac network based on tracing loop currents around the network.

Maxwell bridge An ac bridge used to determine an unknown inductance whose resistance is an appreciable fraction of its total impedance.

Mesh equations A format set of Kirchhoff's voltage-law equations used to solve for various branch currents in an ac network.

Nodal analysis A format procedure for writing Kirchhoff's current-law equations to solve for the various node voltages in an ac network.

Norton's theorem A procedure for replacing any two-terminal network of fixed impedances and ac voltage sources by a single constant-current source and impedance in parallel.

Superposition theorem A procedure for solving for branch currents in a multisource ac network by considering each source to operate independently.

Thévenin's theorem A procedure for replacing any two-terminal network of fixed impedances and ac voltage sources by a single constant-voltage source and impedance in series.

CHAPTER 23—RESONANCE

Antiresonance The condition in which the impedance of an *LC* network is maximum. Used in reference to parallel resonant circuits.

Bandwidth The frequency difference between the half-power frequencies of a resonant circuit.

Half-power point A frequency at which the active power input to a tuned circuit is one half of the power input at the resonant frequency.

Logarithmic scale A graph scale laid out so that the linear displacement along the axis is proportional to the logarithm of the numbers represented.

Quality factor (*Q* factor) The ratio between the reactive power of either the inductance or capacitance at resonance and the active power of a resonant circuit.

Resonance The condition in which the impedance of a series *LC* circuit is minimum.

Resonant circuit An ac network containing both inductance and capacitance where the condition of resonance or antiresonance can be achieved by varying circuit parameters.

Resonant frequency The frequency of the voltage source at which resonance occurs in an *LC* network.

Resonant rise of tank current The increase in the ratio of branch current to total source current as a parallel resonant circuit approaches resonance.

Resonant rise of voltage The increase in the ratio of either the capacitive or inductive voltage to source voltage as a series resonant circuit approaches resonance.

Selectivity A measure of the bandwidth of a resonant circuit in terms of resonant frequency and quality factor.

Sensitivity A measure of the resonant rise of voltage in a series resonant circuit in terms of applied voltage and quality factor.

Tank current The current in the inductive and capacitive branches of a parallel resonant circuit.

Tuning The process of varying the parameters of a resonant circuit to achieve the condition of resonance.

Varactor A reverse-biased *pn*-junction whose capacitance depends on the magnitude of the dc reverse bias voltage.

CHAPTER 24—TRANSFORMERS

Audio transformer A transformer designed to have a uniform response over the frequency range of 20 Hz to 20 kHz.

Autotransformer A transformer in which primary and secondary windings have some turns in common.

Copper loss The portion of the power input to a transformer required for the I^2R losses due to the resistance of the windings.

Core loss The portion of the power input to a transformer required for the hysteresis and eddy current losses in the iron core.

Exciting current (magnetizing current) The portion of the primary current of a transformer required to produce a self-induced voltage in the primary winding to oppose the applied voltage.

Funicular phasor diagram A phasor diagram in which phasors are drawn head-to-tail, rather than from a common origin.

General transformer equation The equation describing the voltage induced into a transformer winding in terms of the frequency, number of turns, and maximum magnetic flux permissible in the core.

Impedance matching Using the transformation ratio of a transformer to make a load impedance appear to have some other magnitude when reflected into the primary circuit of the transformer.

Leakage reactance The inductive reactance that appears to act in series with a transformer winding due to some of the primary flux not linking the secondary winding, and vice versa.

Open-circuit test A means of determining the core losses of a transformer.

Primary The winding of a transformer connected to a voltage source.

Reflected impedance The impedance looking into the primary terminals of a transformer with a load connected to the secondary; equal to the square of the transformation ratio times the load impedance.

Secondary The winding of a transformer connected to a load.

Short-circuit test A means of determining the equivalent resistance and reactance of transformer windings.

Step-down transformer A transformer whose secondary voltage is less than the applied primary voltage.

Step-up transformer A transformer whose secondary voltage is greater than the applied primary voltage.

Transformation ratio The ratio of the number of primary turns to the number of secondary turns.

Transformer A static device based on electromagnetic induction which transfers energy from a primary circuit to a secondary circuit, usually with a change in voltage.

Transformer efficiency The ratio of active load power to total active power input to the transformer.

Transformer voltage regulation The percentage change in secondary voltage from no load to full load with respect to full load voltage.

CHAPTER 25—COUPLED CIRCUITS

Band-pass filter A four-terminal coupling network that passes all frequencies between upper and lower cutoff frequencies and rejects all frequencies outside these boundaries.

Bode diagram A graph of the frequency response of a four-terminal coupling network constructed from straight-line asymptotes.

Coefficient of coupling The ratio between mutual flux and primary flux in magnetically coupled circuits.

Coupled circuit A network consisting of a signal source and a load connected through a four-terminal coupling network.

Coupled impedance The impedance which appears to be in *series* with the primary impedance of a four-terminal coupling network when a load is connected to the secondary.

Critical coupling The coefficient of coupling required for maximum power transfer between tuned circuits.

Cutoff frequency The frequency designated as the boundary between the pass band and the reject band of a filter network.

Decibel A logarithmic unit representing the ratio between power output and power input in a four-terminal coupling network.

Double-resonant filter A resonant filter designed to pass one band of frequencies and to reject another.

High-pass filter A four-terminal coupling network that passes all frequencies higher than and rejects all frequencies lower than the cutoff frequency of the network.

Hybrid parameters A combination of impedance parameters for the input terminals and admittance parameters for the output terminals of any four-terminal coupling network.

Low-pass filter A four-terminal coupling network that passes all frequencies lower than and rejects all frequencies higher than the cutoff frequency of the network.

Mutual inductance The inductance (in henrys) of a pair of magnetically coupled coils when a changing current in one coil induces a voltage in the other coil.

Open-circuit impedance parameters A method of representing any four-terminal coupling network by an input (output) impedance in series with a reverse- (forward-) transfer impedance.

Power gain (loss) The ratio between output power and input power in a four-terminal network, usually expressed in decibels.

Resonant filter A four-terminal band-pass or band-reject filter network in which the cutoff frequencies are determined by LC resonant circuits.

Short-circuit admittance parameters A method of representing any four-terminal coupling network by an input (output) admittance in parallel with a reverse- (forward-) transfer admittance.

Tuned transformer A pair of loosely coupled coils whose primary and/or secondary inductances are tuned to a desired pass-band of frequencies by variable capacitors.

CHAPTER 26—THREE-PHASE SYSTEMS

Balanced load A polyphase load which loads each phase with identical load impedances.

Delta-connected load A three-phase load in which the three branches of the load are connected in delta (Δ) fashion. (A three-wire load.)

Delta-connected source A three-phase source in which the three windings are connected in delta (Δ) fashion. (A three-wire source.)

Double subscript notation A procedure for identifying the tracing direction through voltage sources and loads in a polyphase system.

Line current The currents flowing in the three lines of a three-phase system. In a wye-connected system, line current is the same as phase current.

Line voltage The voltage from line to line in a three-phase system. In a delta-connected system, line voltage is the same as phase voltage.

Neutral The common return conductor in a four-wire, three-phase system; usually grounded.

Phase current The currents flowing in the arms of a three-phase source or load.

Phase voltage The voltage across each winding of a three-phase source or each arm of a three-phase load.

Single-phase system A system in which the source provides only a single voltage (which may be center-tapped).

Three-phase system A system in which the

source provides three voltages of equal magnitudes but with 120° phase differences.

Two-phase system A system in which the source provides two voltages of equal magnitudes, but with a 90° phase difference.

Two-wattmeter power measurement A method of connecting two wattmeters to a three-phase load so that they indicate both total active power input and power factor of the load.

Unbalanced load A polyphase load in which the three arms do not have the same impedances.

Wye-connected load A three-phase load in which the three arms are connected in wye (Y) fashion. In a four-wire system (wye-connected source), the junction point of the three arms is connected to the neutral conductor. In a three-wire system (delta-connected source), the junction point is left floating.

Wye-connected source A three-phase source in which the three windings are connected wye (Y) fashion, providing a common neutral conductor.

ANSWERS TO SELECTED PROBLEMS

CHAPTER 1

1-2-1. (a) $l = 0.6$ m; (b) $m = 0.45$ kg

1-3-3. (a) -723; (b) 230.21;
  (c) 6450; (d) 6.25×10^{-14};
  (e) $786.484\ 74$; (f) 195

1-5-5. (a) 4.55×10^2; (b) 5.9×10^4;
  (c) 10^4; (d) 7.65×10^{-2};
  (e) 3.7×10^{-4}; (f) 5.47×10^{-1}

1-5-7. 2.5×10^4 m

1-5-9. 8.999×10^{-28} g

1-5-11. (a) $45\ 700$; (b) 6.917×10^{-3};
  (c) 2.3328;
  (d) $9.143\ 740\ 3 \times 10^{-10}$;
  (e) $495.983\ 87$

1-6-13. (a) 47 kΩ; (b) 45 MHz;
  (c) 1.5 kW; (d) 50.5 ms;
  (e) 0.5 mV or $500\ \mu$V;
  (f) 390 pF

1-6-15. 1.65 kg

1-6-17. 16.67 ms

1-6-19. 22 g

1-6-21. 1257 mm^2

1-7-23. $2144\ \Omega$

1-7-25. 1 ft$^2 = 0.0929$ m^2

1-7-27. 22.35 m/s

1-7-29. $m = 1.2 \times 10^3 \div 0.4536 = 2645.5$ pounds

CHAPTER 2

2-6-1. 5 A

2-6-3. 1.6×10^{-4} A

2-6-5. 18 C

2-6-7. 32 s

2-9-9. 0.25 V

2-9-11. 0.4167 C

2-9-13. 0.3 J

2-9-15. 16.5 J

2-9-17. 9.028 V

2-9-19. 150 mA

CHAPTER 4

4-1-1. $48\ \Omega$

4-1-3. $22\ \Omega$

4-1-5. 5 A

4-1-7.   8.167 A
4-1-9.   45 V
4-1-11. 89.6 V
4-1-13. 2 kΩ
4-1-15. 583.3 Ω
4-1-17. 25 mA
4-1-19. 218 μA
4-1-21. 4.4 V
4-1-23. 3.96 V
4-4-25. 0.86 Ω
4-4-27. 1.719 Ω
4-4-29. 0.86 Ω
4-4-31. 2.252 Ω
4-4-33. 47.9 cm
4-4-35. 0.25 mm
4-4-37. 3.0×10^{-5} Ω·m
4-4-39. 2.82 Ω
4-5-41. 25.5 Ω
4-5-43. 2.535 Ω
4-6-45. 0.002
4-6-47. 8.135 Ω
4-6-49. tungsten
4-8-51. (a) 33 kΩ, ±20%;
         (b) 6.8 kΩ, ±20%;
         (c) 1.0 kΩ, ±10%;
         (d) 0.39 Ω, ±5%;
         (e) 7.5 Ω, ±5%;
         (f) 82 Ω, ±10%
4-8-53. (a) 376 kΩ to 564 kΩ;
         (b) 50 Ω to 62 Ω;
         (c) 2.3 Ω to 2.5 Ω

CHAPTER 5

5-1-1.   2156 J
5-1-3. 5.859×10^5 J
5-2-5.   600 W
5-2-7.   3.78 W
5-2-9.   468.75 W
5-2-11. 40 mW
5-2-13. 1-W rating (0.7258 W)
5-2-15. 5.5 mA
5-2-17. 22 mA
5-2-19. 200 V
5-2-21. 120 V

5-2-23. 320 W
5-3-25. 16.58 A
5-3-27. 88.2%
5-3-29. 82.89%
5-3-31. 2.79 min
5-4-33. 18 kWh
5-4-35. $1.85
5-5-37. 15.81 A
5-5-39. 30 Ω
5-5-41. 7.95 MJ

CHAPTER 6

6-1-1.   30 Ω
6-1-3.   4.08 kΩ
6-1-5.   330 Ω
6-1-7.   0.923 Ω; 39 W
6-1-9.   119.1 Ω
6-4-11. 20 V; 40 V; 60 V
6-4-13. 90 Ω; 100 Ω
6-4-15. 3889 V
6-5-17. 9.1 V; 2.9 V
6-6-19. (a) 4.8 V;   (b) 59.51 W
6-6-21. 20 Ω
6-6-23. 0.5 Ω
6-7-25. (a) 75%;   (b) 180 W
6-7-27. 411.4 W
6-7-29. 0 Ω
6-8-31. 44 A
6-8-33. 113.3 Ω
6-9-35. 800 Ω
6-9-37. 2.5 kΩ; 909.1 Ω
6-9-39. 15.88 mA
6-10-41. 2 S; 0.5 Ω
6-10-43. 30 kΩ
6-12-45. 9.229 mA; 6.153 mA; 4.615 mA

CHAPTER 7

7-2-1.   (a) 36 Ω;   (b) 1800 Ω;
         (c) 1395 Ω
7-2-3.   (a) 250 mA;   (b) 981.5 mA;
         (c) 55.94 μA
7-2-5.   173.3 Ω; 93.33 Ω; 373.3 Ω;
         293.3 Ω; 240 Ω; 360 Ω
7-2-7.   (a) 361.4 Ω;   (b) 354.3 Ω
7-2-9.   (a) 75 Ω;   (b) 77.38 Ω

7-3-17. 10.44 V (or 0.9566 V)

7-3-19. $R_1 = 320\ \Omega$; $R_2 = 533\ \Omega$

7-5-21. 30 kΩ

7-5-23. 47%

7-5-25. $R_S = 11.43\ k\Omega$; 1.839 W; $R_B = 8.889\ k\Omega$, 1.125 W

7-5-27. Point A = +20 V w.r.t. ground; Point B = −4 V w.r.t. ground

7-5-29. $R_s = 1.2\ \Omega$; 7.5 W; $R_1 = 26\ \Omega$, 6.5 W; $R_2 = 31.58\ \Omega$, 4.56 W; $R_3 = 4.858\ \Omega$, 29.64 W

7-6-31. Left = 10 mA; right = 5 mA

7-6-33. 250 Ω

CHAPTER 8

8-2-1. 27 V and 5 Ω

8-3-3. (a) 0.37 μA and 68 kΩ
(b) 333 mA and 225 Ω
(c) 2 mA and 40 kΩ

8-?-5. 6.667 V + at bottom

8-?-7. 4.25 A

8-?-9. 2.434 A

8-?-11. $I_A = 363\ mA$, $I_B = 116.5\ mA$

8-?-13. Left = 85.955 mA, Right = 79.092 mA

8-?-15. Left to right, 147 mA, 117.6 mA, 29.4 mA, 51.46 mA, 22.06 mA

8-?-17. 5.333 mA

8-8-19. Mesh currents, −10.33 mA, −11.5 mA, 6 mA, 6 mA; arm currents, 1.167 mA, zero, 16.33 mA, −17.5 mA

8-?-21. Left to right, −12.38 mA, 2.14 mA, 10.24 mA

8-?-23. 4 V

8-?-25. 2.943 mA

8-10-27. $V_{10} = 80\ V$, $V_{20} = 20\ V$, $V_{30} = 270\ V$

8-?-29. (a) 8.8 V; (b) 11.51 V

8-?-31. Left = 2.53 A, right = 3.84 A

8-?-33. (a) +79.49 V; (b) +19.6 V

8-?-35. 573 V

CHAPTER 9

9-2-1. 160 V and 180 Ω

9-2-3. 80.02 mA

9-2-5. 620 Ω

9-2-7. 35.56 W

9-2-9. 875 mA

9-2-11. 237.08 V and 36.01 kΩ

9-2-13. +264 mV to +15.84 V

9-2-15. 2.944 mA

9-3-17. 888.9 mA and 180 Ω

9-3-19. 235.3 mA

9-3-21. 137.5 mA

9-3-23. 0.2 A

9-4-25. 1.5 A

9-4-27. 20.97 V and 1.0306 kΩ

9-5-29. 10.06 V

9-5-31. 33.86 V

9-5-33. 18 V and 500 Ω

9-5-35. +13.33 V

9-6-37. 200 Ω, 500 Ω, 200 Ω

9-6-39. 117.6 mA

9-6-41. 48.89 V

CHAPTER 10

10-1-1. 0.6 N

10-2-3. 0.050 25 Ω

10-2-5. 0.450 225 Ω; 0.045 022 5 Ω; 0.005 002 5 Ω

10-3-7. 199 kΩ; 800 kΩ; 1 MΩ; 8 MΩ

10-4-9. 10.91 V

10-4-11. 100 kΩ

10-5-13. 0.4948 V

10-6-15. 30 kΩ

10-6-17. 120 kΩ

10-6-19. 53.33 MΩ

10-6-21. 210 kΩ

10-6-23. 40 Ω

10-6-25. 9.09 km

CHAPTER 11

11-2-1. 2.877×10^{-10} N

11-2-3. 8.015×10^{-15} N

11-6-5. 17.81 pF

11-7-7. 13.9 pF

11-7-9. 885 pF
11-9-11. (a) 50 μF; (b) 4 mC, 16 mC
11-10-13. (a) 5.714 μF; (b) 2.286 mC;
(c) 228.6 V, 114.3 V, 57.1 V
11-10-15. 27.8 pF
11-10-17. 12 μF, 240 V, 192 V, 48 V
11-10-19. 680 V

CHAPTER 12

12-3-1. (a) 1.25 mA; (b) 0 V;
(c) 62.5 V/s; (d) 4 s;
(e) 20 s
12-4-3. (a) 175 V; (b) 0.75 mA;
(c) 6.5 s; (d) 2.8 s
12-5-5. (a) 94 s; (b) 1.06 mA
12-5-7. (a) 43 s; (b) 290 V;
(c) 0.394 mA; (d) 26.2 s
12-6-9. 14.3 V, 5.8 V, 19.5 V, 7.9 V,
21.4 V, 8.7 V
12-6-11. (a) +6.549 V
(b) +10 V, +3.451 V,
−6.549 V, 0 V
12-7-13. (a) 178.4 V; (b) 0.7582 mA;
(c) 6.438 s; (d) 2.772 s
12-7-15. (a) 40 μA, 0.8 V/s;
(b) 173.3 s;
(c) 297.4 μA, 5.949 V/s;
(d) 145.36 V
12-7-17. 172.5 V
12-8-19. +188.48 V
12-8-21. (a) −33.58 V; (b) +12.93 V
12-9-23. (a) 0.625 J; (b) 1.28 μJ
12-9-25. 25.6 mJ

CHAPTER 13

13-6-5. 5 × 10^5 At/Wb
13-6-7. 800 turns
13-8-9. 10^{-3} H/m
13-8-11. 4 × 10^{-4} H/m
13-9-13. 4 T
13-10-15. 2000 At/m
13-11-17. 6.908 A
13-12-19. 1.61 T (approx.)

13-14-21. 0.475 A (approx.)
13-16-23. 387 mA (approx.)
13-17-25. 436 mA (approx.)
13-18-27. 1537 turns (approx.)
13-?-29. 144.3 N

CHAPTER 14

14-5-1. 0.25 H
14-5-3. 333.3 V
14-6-5. 7.896 H
14-6-7. 0.506 H
14-6-9. 270 turns
14-7-11. 120 mH
14-8-13. 12.77 mH

CHAPTER 15

15-3-1. 1.5 s
15-3-3. 0.8 V
15-3-5. (a) 0 A; (b) 1.2 A/s;
(c) 1.5 A; (d) 6.25 s
15-4-7. (a) 0.3 s; (b) 0.21 s;
(c) 2.465 A
15-5-9. (a) 1.373 s; (b) 1.197 A
15-6-11. (a) 22.5 J; (b) 36 W
15-7-13. (a) 66.67 Ω; (b) 0.375 s;
(c) 22.5 J
15-8-15. (a) 37.18 V; (b) 1.88 μs
15-9-17. (a) 4.685 V
15-9-19. 1.557 A

CHAPTER 16

16-5-1. 106.07 V
16-5-3. 951 V
16-5-5. −88.17 V
16-6-7. 126.2 V
16-6-9. 200 sin 2513t V
16-7-11. 8.507 A
16-7-13. 0.6 sin 377t A
16-8-15. 90 W
16-10-17. 95.55 V
16-10-19. 240 mA
16-11-21. 106.07 V

16-11-23. 325 V
16-11-25. 15.9 V
16-11-27. 40 mA
16-11-29. (a) 469 mA;   (b) 166 V
16-11-31. 41.67 min
16-11-33. 10.66 V

CHAPTER 17

17-2-1. 29.25 Ω
17-2-3. 14.4 V
17-3-5. 1508 Ω
17-3-7. 636.6 Hz
17-3-9. 469.1 μH
17-3-11. 306.6 mH
17-5-13. 16.67 kΩ
17-5-15. 20 mA
17-6-17. 7.958 kΩ
17-6-19. 124.3 Hz
17-6-21. 7.958 μF
17-6-23. 0.1479 μF
17-7-25. 215 μH

CHAPTER 18

18-2-1. 280 V
18-2-3. $v_T = 216 \sin(1508t + 0.59)$; 240 Hz
18-5-5. 280 V; 100 V
18-5-7. 0 V
18-5-9. 25.9 V$/-5°$
18-7-11. (a) $+19.99 + j15.01$;
         (b) $-27.58 + j23.14$;
         (c) $+89 - j154.15$;
         (d) $-30 - j51.96$;
         (e) $-5.909 + j12.91$;
         (f) $+5.501 + j3.997$
18-8-13. 26.4$/+24.62°$
18-8-15. 198.12$/-133.74°$
18-8-17. 134.39$/+77.28°$
18-9-19. 0.8256$/-16.39°$
18-10-21. (a) 53.3$/-51°$
         (b) 0.3154$/-77°$
18-10-23. 10.75$\Omega/-30°$

CHAPTER 19

19-1-1. (a) 44.72 V;   (b) 12.49 mV
19-1-3. 38.42 V$/+33.69°$
19-2-5. 842.9 $\Omega/+53.62°$
19-2-7. (a) 133.3 $\Omega/+72°$;
         (b) $41.2 + j126.8$ Ω
19-2-9. (a) 409.53 Ω;   (b) $/+56.9°$
19-2-11. $R = 2.078$ kΩ; $L = 0.9549$ H
19-2-13. 48 Ω; 291.7 mH
19-3-15. 42.1 Ω; 983.8 mH
19-4-17. (a) 22.36 V;   (b) 8.718 mV
19-4-19. 1312 $\Omega/-23.85°$
19-4-21. 0.942 A$/+49.9°$
19-4-23. 49.74 Hz
19-4-25. 2.327 μF
19-5-27. 150 $\Omega/+48.18°$
19-5-29. 3.063 μF
19-5-31. 0.9619 μF or 6.531 μF
19-5-33. 1125.4 kHz
19-6-35. (a) 170 mA$/-28.07°$;
         (b) 1.327 A$/+90°$
19-6-37. (a) 2.062 A$/+75.96°$;
         (b) 58.2 $\Omega/-75.96°$
19-7-39. 22.36 $\Omega/-63.43°$
19-7-41. 13.12 mS$/-23.85°$
19-7-43. $G = 433$ mS; $B = 250$ mS (inductive)
19-7-45. (a) 439.17 $\Omega/+49.77°$;
         (b) 675.7 $\Omega/+6.498°$;
         (c) 457.04 $\Omega/-47.76°$
19-7-47. 10.28 k$\Omega/+10.29°$
19-7-49. 166.22 $\Omega/+4.493°$
19-8-51. $R_p = 2.772$ kΩ; $L_p = 3.82$ H
19-8-53. $R = 2.943$ Ω; $L = 31.22$ mH
19-8-55. 4.745 μF

CHAPTER 20

20-1-1. (a) 660 W;   (b) 1320 W
20-1-3. 150 W
20-2-5. (a) 0 W;   (b) 7.54 mvars;
         (c) 7.54 mW
20-3-7. (a) 143.86 vars;   (b) 143.86 VA;
         (c) 143.86 W
20-4-9. (a) 75.78 VA;   (b) 9.571 W;
         (c) 75.173 vars

20-4-11. 30.07 kW
20-4-13. 1.59 MHz
20-6-15. 12.63% lagging
20-6-17. 99.2% lagging
20-6-19. (a) 10.368 Ω; (b) 36.67 mH
20-6-21. (a) 24.24 A; (b) 2667 W
20-6-23. 98%
20-6-25. 147.58 V
20-7-27. (a) 45.9 μF;
          (b) 6.0 A, 3.6 A, 4.8 A
20-7-29. 99.8% leading
20-7-31. 680.79 W
20-9-33. 469.8 Ω
20-9-35. (a) 12 Ω, 156 mH;
          (b) 15.43 Ω, 1.768 H

CHAPTER 21

21-2-1. 1119.17 $\Omega/+45.36°$
21-2-3. (a) 1.071 A$/-47.98°$;
         (b) 228.53 V$/+14.07°$
21-3-5. 3.5 mS$/-29.64°$
21-3-7. 88.7% leading
21-4-9. 287.4 mA$/+69.83°$;
         230 mA$/+106.7°$
21-4-11. 265.6 mA$/+54.9°$
21-4-13. 51.88 mV$/+3.62°$
21-4-15. 533 mV$/-80°$
21-5-17. 519.62 Ω and 0.5305 μF
21-5-19. 416.1 Ω and 0.7026 H
21-5-21. (a) 26.332 kΩ and 0.0532 μF;
          (b) 14.78 kΩ and 0.04218 μF
21-5-23. 315 + j107 Ω
21-5-25. 148.242 − j86.843 Ω
21-6-27. (a) I_x = 666.7 mA$/-10°$ and
              Z_x = 120 $\Omega/+30°$;
          (b) I_x = 13.868 μA$/+41.31°$
              and Z_x = 1.803 k$\Omega/-56.31°$
21-6-29. I_x = 2.2 A$/-98.13°$ and
          Y_x = 12 − j16 mS
21-7-31. 240 μW
21-7-33. 43.3 Ω and 0.0478 μH
21-8-37. 50 $\Omega/-30°$; 50 $\Omega/-30°$; $Z_1 = Z_2$
          with a 120° difference in angle

CHAPTER 22

22-1. 13.183 A$/-21.14°$ and
       12.198 A$/-179.23°$
22-3. 334 mA$/-88.8°$
22-5. 43.27 V$/+4.49°$
22-7. 935 mA$/+124.59°$;
       935 mA$/+124.59°$;
       1.87 A$/+124.59°$
22-9. 1.009 A$/+52.52°$;
       1.434 A$/-21.34°$
22-11. 2.845 A
22-13. 684 mA
22-15. 153.25 V; 173.1 V
22-17. 2.84 A$/64.88°$
22-19. 20.12 V$/+35.66°$
22-21. 930 mA$/+18.22°$
22-23. 18.15 V
22-4-25. 20.55 V$/+31.08°$ and
          17.037 k$\Omega/-58.92°$
22-4-27. 24 V$/0°$, 0.02 μF, and 33 kΩ
22-4-29. 84.62 Ω
22-4-31. 8.36°
22-33. 0.0265 μF
22-7-35. 83.32 k$\Omega/+0.95°$,
          1.658 k$\Omega/-89.05°$,
          8.288 k$\Omega/-89.05°$
22-7-37. 1.625 A
22-8-39. 9.149 MΩ and 940 pF
22-8-41. 2.878 + j1696.5 Ω
22-8-43. 126.7 μH and 10 Ω

CHAPTER 23

23-2-1. 1.989 kHz
23-2-3. 234.3 pF
23-2-5. (a) 238 μH; (b) 1.8835 MHz
23-3-7. 50
23-3-9. 18.753 Ω
23-4-11. 40 mV
23-5-13. 66.315 kHz
23-5-15. 100.66
23-5-17. (a) 466.9 kHz and 461.1 kHz;
          (b) 3.953 Ω and 108.47 μH
23-7-19. 2.652 MHz

23-7-21. $0.3332 \ \mu A$

23-7-23. (a) $23.579 \ \mu H$; (b) $53.05 \ pF$

CHAPTER 24

24-2-1. 819 turns

24-2-3. 91 mA (approximately)

24-2-5. $75.1 \ cm^2$

24-3-7. 17

24-3-9. 107.1 V

24-3-11. 560 turns

24-3-13. 51.2 turns

24-4-15. $5625 \ \Omega$

24-4-17. (a) $34.72 \ \Omega$; (b) $8.68 \ \Omega$

24-6-19. $R_e = 0.1725 \ \Omega$; $X_e = 0.4002 \ \Omega$

24-7-21. 98.3%

24-8-23. 3.846%

24-8-25. 2.118%

CHAPTER 25

25-2-1. $\mathbf{z}_{11} = 643 \ \Omega$, $\mathbf{z}_{12} = \mathbf{z}_{21} = 286 \ \Omega$, $\mathbf{z}_{22} = 322 \ \Omega$

25-2-3. $\mathbf{z}_{11} = \mathbf{z}_{22} = 20 + j0 \ \Omega$, $\mathbf{z}_{12} = \mathbf{z}_{21} = -j800 \ \Omega$

25-2-5. $\mathbf{Z}_m = 400 \ \Omega \underline{/0°}$, $\mathbf{Z}_p = 200 \ \Omega \underline{/0°}$, $\mathbf{Z}_s = 800 \ \Omega \underline{/0°}$

25-2-7. $\mathbf{Z}_m = -j1200 \ \Omega$, $\mathbf{Z}_p = \mathbf{Z}_s = 50 + j1200 \ \Omega$

25-2-9. $59.26 \ \Omega \underline{/0°}$

25-2-11. $500 \ \Omega \underline{/0°}$

25-2-13. $\mathbf{z}_{11} = 643.5 \ \Omega$, $\mathbf{z}_{22} = 321.24 \ \Omega$, $\mathbf{z}_{12} = \mathbf{z}_{21} = 285.09 \ \Omega$

25-2-15. $\mathbf{z}_{11} = \mathbf{z}_{22} = 0$, $\mathbf{z}_{12} = \mathbf{z}_{21} = -j3600 \ \Omega$

25-2-17. $512.7 \ \Omega \underline{/0°}$

25-2-19. $\mathbf{z}_{11} = r_e + r_b$, $\mathbf{z}_{12} = r_b$, $\mathbf{z}_{21} = r_b + r_m$, $\mathbf{z}_{22} = r_b + r_c$

25-3-21. $\mathbf{y}_{11} = 7 \ mS \underline{/0°}$, $\mathbf{y}_{22} = 9 \ mS \underline{/0°}$, $\mathbf{y}_{12} = \mathbf{y}_{21} = 5 \ mS \underline{/180°}$

25-3-23. $\mathbf{Y}_m = 200 \ \mu S \underline{/0°}$, $\mathbf{Y}_p = 200 \ \mu S \underline{/0°}$, $\mathbf{Y}_s = 400 \ \mu S \underline{/0°}$

25-3-25. $\mathbf{y}_{11} = \mathbf{y}_{22} = 2.698 \ mS \underline{/-51.85°}$, $\mathbf{y}_{12} = \mathbf{y}_{21} = 1.667 \ mS \underline{/180°}$

25-4-27. $\mathbf{h}_{11} = 388.98 \ \Omega$, $\mathbf{h}_{12} = 0.888$, $\mathbf{h}_{21} = -0.888$, $\mathbf{h}_{22} = 3.106 \ mS$

25-4-29. Transistor $A = 103.284$, overall $A = 14.775$

25-6-31. 0.07264

25-7-33. $1197.43 \ \Omega \underline{/-72.34°}$

25-8-35. (a) 338.38 pF; (b) 0.026; (c) 11.86 kHz

25-9-37. $0.44 \ \mu F$

25-9-39. $87.95 \ \mu H$ and $521.1 \ \mu H$

25-10-41. (a) 690 Hz; (b) 123.6 V; (d) $6.604 \ \Omega$

25-10-43. (a) 14.47 Hz; (b) $3.535 \ V \underline{/+45°}$; (c) $2.843 \ V \underline{/+55.35°}$

25-10-45. (a) 1.129 nF; (b) 51.95 kHz; (c) 7.5 mV

25-11-47. $+34 \ dB$

25-11-49. 792.4 mW

25-11-51. 7 dB

CHAPTER 26

26-4-1. (b) $12 \ A \underline{/0°}$, $12 \ A \underline{/+120°}$, $12 \ A \underline{/-120°}$; (d) 0 A

26-4-3. (b) $4 \ A \underline{/-30°}$, $3 \ A \underline{/+150°}$, $2.4 \ A \underline{/-30°}$; (d) $3.3995 \ A \underline{/-30°}$

26-5-5. (b) $50 \ A \underline{/0°}$, $50 \ A \underline{/+120°}$, $50 \ A \underline{/-120°}$; (e) $86.6 \ A \underline{/-30°}$, $86.6 \ A \underline{/+90°}$, $86.6 \ A \underline{/-150°}$

26-5-7. (b) $110 \ A \underline{/-45°}$, $50 \ A \underline{/+120°}$, $55 \ A \underline{/-83.1°}$; (e) $158.82 \ A \underline{/-49.67°}$, $102.88 \ A \underline{/+107.88°}$, $74.85 \ A \underline{/+161.96°}$

26-6-9. (c) $208 \ V \underline{/+30°}$, $208 \ V \underline{/+150°}$, $208 \ V \underline{/-90°}$; (d) $18.909 \ A \underline{/+30°}$, $18.909 \ A \underline{/+150°}$, $18.909 \ A \underline{/-90°}$; (e) $32.752 \ A \underline{/0°}$, $32.751 \ A \underline{/+120°}$, $32.751 \ A \underline{/-120°}$

26-6-11. (c) 208 V$\underline{/+30}$, 208 V$\underline{/+150°}$,
208 V$\underline{/-90°}$;

   (d) 41.6 A$\underline{/-15°}$,
    18.909 A$\underline{/+150°}$,
    20.8 A$\underline{/-53.1°}$;

   (e) 60.07 A$\underline{/-19.67°}$,
    38.91 A$\underline{/+137.89°}$,
    28.31 A$\underline{/-168.04°}$

26-7-13. 4.32 kW

26-7-15. 727.44 W

26-8-17. 2.162 kW

26-9-19. 41.25 kW

26-9-21. (a) 3 kW; (b) 0.866;
   (c) 9.616 A;
   (d) 37.47 Ω$\underline{/+30°}$; (e) 60°

26-10-23. CBA

26-10-25. 2.162 kW

26-11-27. (b) 31.754 A$\underline{/-30°}$,
    31.754 A$\underline{/+90°}$,
    31.754 A$\underline{/-150°}$; (d) 0 A

26-11-29. 5.893 A$\underline{/+12.6°}$,
    1.31 A$\underline{/-167.4°}$,
    4.584 A$\underline{/-167.4°}$

ANSWERS TO SELF-CHECK PROBLEMS

CHAPTER 1

1-4:1. Calculator answer = 49.8436 m^2
But the proper answer can have only 3 significant figures therefore:
AREA = 50.0 m^2

1-4:2. Since circumference = $C = 2 \cdot \pi \cdot R$
calculator answer = 113.0973355
Since 2 and π are pure numbers then R is the only measurement, therefore the answer can only have 2 significant figures (the same as R):
$C = 110$ mm. is correct

1-5:1. (a) $f = 5.1 \times 10^3$ Hz, $C = 2.7 \times 10^{-10}$ F, $R = 1.8 \times 10^5 \ \Omega$
(b) $X_c = 115\ 580.9318$ (calculator answer)
$= 120\ 000 \ \Omega$ (proper significant figures)
(c) $Z = 213\ 913.4213$ (calculator answer)
$= 210\ 000 \ \Omega$ (proper significant figures)

1-5:2. (a) $L = 159$ mH; $C = 47 \ \mu$F
(b) $f = 58.22005891$ (using calculator)
$f = 58$ Hz (expressed properly)

1-6:1. $f = 5.1 \times 10^3$ Hz, $C = 270 \times 10^{-12}$ F,
$R = 1.8 \times 10^6 \ \Omega$ and $L = 159 \times 10^{-3}$ H

1-6:2. $f = 5.1$ kHz, $C = 270$ pF, $R = 1.8$ MΩ and $L = 159$ mH

1-7:1. $55\dfrac{\text{mi}}{\text{h}} \cdot \dfrac{1600 \text{ m}}{1.0 \text{ mi}} \cdot \dfrac{1.0 \text{ h}}{3600 \text{ s}} = 24$ m/s

1-7:2. $10\dfrac{\text{km}}{1} \cdot \dfrac{1.0 \text{ mi}}{1.6 \text{ km}} \cdot \dfrac{3.785 \text{ 1}}{1 \text{ gal}} = 24$ mi/gal

```
1-8:1. 100 REM COMPUTER SOLUTION TO SELF-CHECK PROBLEM 1-8.1
       110 INPUT "ENTER F,C,R:F,C,R
       120 LET PI = 3.14159
       130 LET X = 1/(2*PI*F*C)
       140 LET Z = SQR(R^2+X^2)
       150 PRINT "Z = ";Z;" OHMS"
       160 END

1-8:2. 100 REM COMPUTER SOLUTION TO SELF-CHECK PROBLEM 1-8.2
       200 INPUT "ENTER L, C : "; L,C
       300 PI = 3.14159
       400 F = 1/(2*PI*SQR(L*C))
       500 PRINT "F = ";F;" HERTZ"
       600 END
```

CHAPTER 2

2-5:1. $Q = (2.5 \times 10^{-6} \text{ C}) \cdot (6.24 \times 10^{18}\text{e/C})$
$Q = 1.56 \times 10^{13}$ electrons

2-5:2. $Q = \dfrac{(100\ 000 \text{ atoms}) \cdot (92 \text{ e})}{(6.24 \times 10^{18} \text{ e/C})} = 1.474 \times 10^{-12}$
$Q = 1.47 \text{ pC}$

2-6:1. Using Equation (2-2):
$$I = \frac{Q}{t} = \frac{380 \times 10^{19} \text{ e}}{60 \text{ s}} = 6.33 \times 10^{19} \text{ e/s}$$
$I = (6.33 \times 10^{19} \text{ e/s})/(6.24 \times 10^{18} \text{ e/C}) = 10.1 \text{ C/s}$
$I = 10.1 \text{ A}$

2-6:2. (a) From Equation (2-2):
$Q = I \cdot t = (25 \text{ C/s}) \cdot (8 \text{ h}) \cdot (3600 \text{ s/h}) = 720\ 000 \text{ C}$
$Q = 720 \text{ kC}$
(b) $Q = (720 \text{ kC}) \cdot (6.25 \times 10^{18} \text{ e/C})$
$Q = 4.5 \times 10^{24}$ electrons

2-9:1. Using Equation (2-4):
$$E = \frac{W}{Q} = \frac{36 \text{ J}}{4 \text{ C}} = 9.0 \text{ V}$$

2-9:2. From Equation (2-4):
$W = E \cdot Q$
Using Equation (2-2):
$W = E \cdot Q = E \cdot I \cdot t =$
$(120 \text{ J/C}) \cdot (12 \text{ C/s}) \cdot (60 \text{ s}) = 86\ 400 \text{ J}$
$W = 85 \text{ kJ}$

2-10-1. (a) 0 V
(b) 12 V

2-11:1. (a) downward
(b) upward

CHAPTER 4

4-1:1. Using Equation (4-2):

$$R = \frac{V}{I} = \frac{120 \text{ V}}{8 \text{ A}} = 15 \text{ V/A}$$

$$R = 15 \ \Omega$$

4-1:2. From Equation (4-2):

$$I = \frac{V}{R} = \frac{9 \text{ V}}{600 \ \Omega} = 0.015 \ \frac{V}{V/A}$$

$$I = 15 \text{ mA}$$

4-3:1. Using the relationships between resistance, length, and area:

$$\frac{R_B}{R_A} = \frac{L_B}{L_A} \cdot \frac{A_A}{A_B} = \frac{L_B}{L_A} \cdot \frac{\pi d_A^2/4}{\pi d_B^2/4} = \frac{L_B}{L_A} \cdot \frac{d_A^2}{d_B^2}$$

$$\frac{R_B}{R_A} = \frac{50}{100} \cdot \frac{2^2}{3^2} = 0.222$$

$$R_B = 0.222 \cdot R_A = 0.222 \cdot 2 = 0.44 \ \Omega$$

4-4:1. First, find cross-sectional area:

$$A = \pi d^2/4 = \pi \cdot (0.8 \times 10^{-3})^2/4$$

$$A = 2.83 \times 10^{-7} \text{ m},$$

Using Equation (4-4) and using resistivities from Table 4-1:

(a) $$l = \frac{R \cdot A}{\rho} = \frac{(15 \ \Omega) \cdot (2.83 \times 10^{-7} \text{ m}^2)}{1.72 \times 10^{-8} \ \Omega \cdot \text{m}} = 77 \text{ m}$$

(b) $$l = \frac{R \cdot A}{\rho} = \frac{(15 \ \Omega) \cdot (2.83 \times 10^{-7} \text{ m}^2)}{110 \times 10^{-8} \ \Omega \cdot \text{m}} = 3.9 \text{ m}$$

4-4:2. From Equation (4-4):

$$R_{A1} = \frac{\rho_{A1} \cdot l_{A1}}{A_{A1}} = R_{sil} = \frac{\rho_{sil} \cdot l_{sil}}{A_{sil}} =$$

since $A_{A1} = A_{sil}$ then

$$\rho_{A1} \cdot l_{A1} = \rho_{sil} \cdot l_{sil}$$

$$\Rightarrow l_{sil} = \frac{\rho_{A1} \cdot l_{A1}}{\rho_{sil}} = \frac{(2.83 \times 10^{-8} \cdot (100)}{1.64 \times 10^{-8}} = 173 \text{ m}$$

4-5:1. From Equation (4-5) and using information from Table 4-2:

$$R_2 = R_1 \cdot \frac{x + T_2}{x + T_1} = (15 \ \Omega) \cdot \frac{202 + 3600 \text{ C}}{202 + 20 \text{ C}} = 260 \ \Omega$$

4-5:2. Rewriting Equation (4-6):

$$R_T = R_{20} \cdot \frac{x + T}{x + 20}$$

$$\Rightarrow R_{20} = R_T \cdot \frac{x + 20}{x + T} = (15 \ \Omega) \cdot \frac{6250 + 20 \text{ C}}{6250 + 1500 \text{ C}} = 12 \ \Omega$$

4-6:1. Using Equation (4-8) and using information from Table 4-3 and Table 4-1:

$$R = \frac{\rho \cdot l}{A} \cdot (1 + \alpha \Delta T) =$$

$$R = \frac{(2.83 \times 10^{-8} \ \Omega \cdot \text{m}) \cdot (1000 \text{ m})}{0.5 \times 10^{-6} \text{ m}^2} \cdot [1 + 0.0039(100 - 20)]$$

$$R = 56.6 \cdot [1.312] = 74 \ \Omega$$

4-6:2. Rewriting Equation (4-8):

$$R_T = R_{20} \cdot (1 + \alpha \Delta T)$$

$$\Rightarrow \Delta T = \frac{R_T/R_{20} - 1}{\alpha} = \frac{150/50 - 1}{0.0055} = \frac{2}{0.0055} = 363.6$$

therefore

$$T = \Delta T + 20 = 363.6 + 20 = 380°C$$

4-8:1. (a) $10 \times 10^5 = 1.0$ MΩ $\pm 10\%$

(b) $33 \times 10^3 = 33$ kΩ $\pm 20\%$

(c) $62 \times 10^0 = 62$ Ω $\pm 5\%$

4-8:2. By the color code this resistor is nominally 3.9 kΩ $\pm 5\%$ therefore the resistance of this resistor should be between:

$3900 \times 0.95 = 3705$ Ω and

$3900 \times 1.05 = 4095$ Ω

The resistor is out of tolerance.

4-11:1. Using Equation (4-10):

$V = I \cdot R = (0.5$ A$) \cdot (240$ $\Omega) $

$V = 120$ V

4-11:2. Using Equation (4-2):

$R = V/I = (240$ V$)/(5$ A$) = 48$ Ω

Now using this resistance and Equation (4-9):

$I = E/R = (200$ V$)/(48$ $\Omega) = 4.2$ A

CHAPTER 5

5-1:1. From Equation (2-4):

$W = V \cdot Q = (7.5$ V$) \cdot (180$ C$) = 1350$ J

5-2:1. Using Equation (5-4):

$$P = \frac{V^2}{R} = \frac{120^2}{12} = 1200$$

$P = 1.2$ kW

5-2:2. From Equation (5-3):

$$R = \frac{P}{I^2} = \frac{1600}{15^2} = 7.11$$

$R = 7.1$ Ω

5-3:1. (a) Using Equation (5-2):

$P = V \cdot I = (7.5$ V$) \cdot (140$ A$) = 1050$ W $= 1.1$ kW

(b) Using Equation (5-6):

$$\eta = \frac{P_{out}}{P_{in}} = \frac{(1.0 \text{ hp}) \cdot (746 \text{ W/hp})}{1050 \text{ W}} = \frac{746 \text{ W}}{1050 \text{ W}} = 0.7105$$

$\eta = 71\%$

5-3:2. From Equation (5-6):

$P_{out} = \eta \cdot P_{in} = \eta \cdot V \cdot I = (0.8) \cdot (240$ V$) \cdot (10$ A$) = 1920$ W

$P_{out} = 1.9$ kW

5-4:1. From Equation (5-8) and using Equation (5-2):

$$t = \frac{W}{P} = \frac{1.0 \text{ kWh}}{(12 \text{ V}) \cdot (5.0 \text{ A})} = \frac{1.0 \text{ kWh}}{0.060 \text{ kW}} = 17 \text{ hours}$$

5-4:2. (a) Using Equation (5-2):

$P = V \cdot I = (230$ V$) \cdot (30$ A$) = 6.9$ kW

(b) Using Equation (5-8):

$W = P \cdot t = (6.9$ kW$) \cdot (9.0$ hr$) = 62.1$ kWh

$\Rightarrow$ cost $= (62.1$ kWh$) \cdot (\$0.07/\text{kWh}) = \$4.35$

CHAPTER 6

6-1:1. (a) Using Equation (6-1):

$R_T = R_1 + R_2 + R_3 = 1.0 + 2.2 + 1.8 = 5.0$ kΩ

(b) Using Ohm's Law:

$$I = \frac{E}{R_T} = \frac{15 \text{ V}}{5 \text{ k}\Omega} = 3.0 \text{ mA}$$

(c) *All* resistors have the *same* 3.0 mA of current.

6-1:2. First find R_T using Equation (6-1):

$R_T = R_1 + R_2 = 330 + 470 = 800 \text{ }\Omega$

then use Ohm's Law:

$E = I \cdot R_T = (0.050 \text{ A}) \cdot (800 \text{ }\Omega) = 40 \text{ V}$

6-2:1. (a) clockwise around circuit, (b) counterclockwise around circuit, (c) top of R_{eq} is + and bottom is − .

6-4:1. (a) Using Ohm's Law:

$$I = \frac{E}{R_T} = \frac{E}{R_1 + R_2} = \frac{24 \text{ V}}{2.4 + 5.6 \text{ k}\Omega} = \frac{24 \text{ V}}{8.0 \text{ k}\Omega} = 3.0 \text{ mA}$$

(b) Using Ohm's Law for each resistor:

$V_1 = I \cdot R_1 = (3 \text{ mA}) \cdot (2.4 \text{ k}\Omega) = 7.2 \text{ V}$

$V_2 = I \cdot R_2 = (3 \text{ mA}) \cdot (5.6 \text{ k}\Omega) = 16.8 \text{ V}$

(c) Using Equation (6-2):

$E = V_1 + V_2$

$24.0 \stackrel{\text{?}}{=} 7.2 + 16.8$ **CHECKS!**

6-4:2. (a) Using Ohm's Law for each resistor:

$V_1 = I \cdot R_1 = (0.05 \text{ A}) \cdot (330 \text{ }\Omega) = 16.5 \text{ V}$

$V_2 = I \cdot R_2 = (0.05 \text{ A}) \cdot (470 \text{ }\Omega) = 23.5 \text{ V}$

(b) Using Equation (6-2):

$E = V_1 + V_2 = 16.5 + 23.5 = 40 \text{ V}$

6-5:1.

| | Switch CLOSED | | Switch OPEN | |
	Current In	Voltage Across	Current In	Voltage Across
Source	3.0 A	22 V rise	0 A	22 V rise
Switch	3.0 A	0 V drop	0 A	22 V drop
Fuse	3.0 A	1 V drop	0 A	0 V drop
Rheostat	3.0 A	6 V drop	0 A	0 V drop
Lamp	3.0 A	15 V drop	0 A	0 V drop

6-6:1. From Equation (6-3):

$$R_{int} = \frac{E - V_T}{I} = \frac{13.2 - 9.45 \text{ V}}{25 \text{ A}} = 0.15 \text{ }\Omega$$

6-6:2. From Equation (6-3):

$$I = \frac{E - V_T}{R_{int}} = \frac{26 - 24 \text{ V}}{0.05 \text{ }\Omega} = 40 \text{ A}$$

6-7:1. Since for maximum power transfer:

$R_L = R_{int} = 2.0 \text{ }\Omega$

then by Ohm's Law:

$$I = \frac{E}{R_L + R_{int}} = \frac{20 \text{ V}}{2 + 2} = 5 \text{ A}$$

6-7:2. (a) From Equation (6-3) at short-circuit:

$V_T = E - I_{SC} \cdot R_{int} = 0$

$$\Rightarrow R_{int} = \frac{E}{I_{SC}} = \frac{120 \text{ V}}{60 \text{ A}} = 2.0 \text{ }\Omega$$

for maximum power transfer:

(b) $R_L = R_{int} = 2.0 \; \Omega$

$$\eta = \frac{100 \cdot P_L}{P_L + P_{int}} = \frac{100 \cdot I^2 \cdot R_L}{I^2 \cdot R_L + I^2 \cdot R_{int}} = \frac{100 \cdot R_L}{R_L + R_{int}}$$

Solving for R_L:

$$R_L = \frac{\eta \cdot R_{int}}{100 - \eta} = \frac{90 \cdot 2}{100 - 90} = 18 \; \Omega$$

6-8:1. (a) Using Ohm's Law for each branch:

$$I_1 = \frac{E}{R_1} = \frac{12 \; V}{1.2 \; k\Omega} = 10 \; mA$$

Similarly:

$I_2 = 6.7 \; mA$

$I_3 = 3.3 \; mA$

(b) $R_{eq} = \dfrac{E}{I_1 + I_2 + I_3} = \dfrac{12 \; V}{10 + 6.7 + 3.3 \; mA} = \dfrac{12 \; V}{20 \; mA}$

$R_{eq} = 600 \; \Omega$

6-8:2. Using Ohm's Law

$$R_{eq} = \frac{E}{I_1 + I_2} = \frac{E}{E/R_1 + E/R_2} = \frac{1}{1/R_1 + 1/R_2}$$

$$R_{eq} = \frac{1}{1/330 + 1/470} = 190 \; \Omega$$

6-9:1. Using Kirchhoff's Current Law:

$I_T = I_1 + I_2 = 60 + 20 = 80 \; mA$

6-9:2. (a) Using Ohm's Law for each branch:

$$I_1 = \frac{E}{R_1} = \frac{24 \; V}{330 \; \Omega} = 72.72 \; mA$$

Similarly $I_2 = 51.06 \; mA$

(b) Using Kirchhoff's Current Law:

$I_T = I_1 + I_2 = 72.72 + 51.06 = 120 \; mA$

6-10:1. Using Equation (6-10):

$$G_1 = \frac{1}{R_1} = \frac{1}{10 \; \Omega} = 100 \; mS$$

$$G_T = \frac{I_T}{E} = \frac{12 \; A}{50 \; V} = 240 \; mS$$

$G_2 = G_T - G_1 = 240 - 100 = 140 \; mS$

6-10:2. (a) From Equation (6-10):

$$G_1 = \frac{1}{R_1} = \frac{1}{330 \; \Omega} = 3.0 \; mS$$

Similarly $G_2 = 2.1 \; mS$

(b) Rewriting Ohm's Law using conductance:

$$I_1 = \frac{E}{R_2} = E \cdot G_1 = (24 \; V) \cdot (3.0 \; mS) = 73 \; mA$$

Similarly

$I_2 = 51 \; mA$

6-12:1.

	Switch CLOSED		Switch OPEN	
	Current In	*Voltage Across*	*Current In*	*Voltage Across*
Source	18 A	120 V rise	12 A	120 V rise
Fuse	18 A	0 V drop	12 A	0 V drop
Heater	12 A	120 V drop	12 A	120 V drop
Switch	6 A	0 V drop	0 A	120 V drop
Lamp	6 A	120 V drop	0 A	0 V drop

CHAPTER 7

7-2:1.

Component	Resistance	Voltage	Current	Power
R_1	1.8 kΩ	36 V	20 mA	720 mW
R_2	2.4 kΩ	24 V	10 mA	240 mW
R_3	1.2 kΩ	12 V	10 mA	120 mW
Totals	1.2 kΩ	36 V	30 mA	1080 mW

7-3:1. (a) Since R_3 is in series with R_2.

$I_3 = I_2 = 10$ mA

(b) By Kirchhoff's Current Law:

$I_T = = I_1 + I_2 = 5 + 10 = 15$ mA

(c) Since R_1 is in parallel with the source:

$E = V_1 = I_1 \cdot R_1 = (5 \text{ mA}) \cdot (5.6 \text{ k}\Omega) = 28$ V

(d) Using Kirchhoff's Voltage Law for the series combination of R_2 and R_3:

$E = V_2 + V_3$

$\Rightarrow V_3 = E - I_2 \cdot R_2 = 28 - (10 \text{ mA}) \cdot (1 \text{ k}\Omega)$

$V_3 = 18$ V

(e) Using Ohm's Law:

$R_3 = V_3/I_3 = (18 \text{ V})/(10 \text{ mA}) = 1.8$ kΩ

7-4:1. Using Equation (7-1):

$$V_x = E \cdot \frac{R_x}{R_T} = (24) \cdot \frac{330}{(330 + 470)} = 9.9 \text{ V}$$

7-4:2. From Equation (7-1):

$$R_x = R_T \cdot \frac{V_x}{E} = (1000 \ \Omega) \cdot \frac{25 \text{ V}}{60 \text{ V}} = 417 \ \Omega$$

7-5:1. (a) Using Kirchhoff's Current Law and Ohm's Law:

$$R_B = \frac{V_L}{I_T - I_L} = \frac{(6 \text{ mA}) \cdot (3.3 \text{ k}\Omega)}{24 - 6 \text{ mA}} = \frac{19.8 \text{ V}}{18 \text{ mA}} = 1.1 \text{ k}\Omega$$

(b) Using Kirchhoff's Voltage Law and Ohm's Law:

$$R_D = \frac{E - V_L}{I_T} = \frac{75 - 19.8 \text{ V}}{24 \text{ mA}} = 2.3 \text{ k}\Omega$$

7-6:1. (a) Using Equation (7-5):

$$I_1 = I_T \cdot \frac{R_2}{R_1 + R_2} = (60 \text{ mA}) \cdot \frac{3.3 \text{ k}\Omega}{1.1 + 3.3 \text{ k}\Omega} = 45 \text{ mA}$$

(b) Similarly using Equation (7-6):

$$I_2 = I_T \cdot \frac{R_1}{R_1 + R_2} = (60 \text{ mA}) \cdot \frac{1.1 \text{ k}\Omega}{1.1 + 3.3 \text{ k}\Omega} = 15 \text{ mA}$$

7-6:2. Using Equation (7-5):

$$R_1 = R_2 \cdot \left(\frac{I_T}{I_1} - 1 \right) = 1000 \cdot \left(\frac{15 + 25}{15} - 1 \right) =$$

$R_1 = 1000 \cdot (1.667)$

$R_1 = 1.7$ kΩ

CHAPTER 8

8-3:1. (a) Using Equation (8-1):

$$R_{\text{int}} = \frac{V_{\text{OC}}}{I_{\text{SC}}} = \frac{13.2 \text{ V}}{150 \text{ A}} = 0.088 \ \Omega$$

(b)

8-4:1. (a) Using the voltage-divider principle:
$$V_T = E \cdot \frac{R_L}{R_{int} + R_L} =$$
$$\Rightarrow R_{int} = R_L \cdot (E/V_T - 1) = 10 \cdot (24/20 - 1)$$
(b) $I_x = 12$ A in parallel with 2 Ω.

8-5:1. (NOTE: resistances in kΩ and currents in mA)
(a) $2 \cdot (I_1 + I_2) + \quad 1.8 \cdot I_1 \quad + 1.6 \cdot (I_1 + I_2) = 36$
$\quad 2 \cdot (I_1 + I_2) + (1.2 + 2.4) \cdot I_2 + 1.6 \cdot (I_1 + I_2) = 36$
(b)
$$I_1 = \frac{\begin{vmatrix} 36 & 3.6 \\ 36 & 7.2 \end{vmatrix}}{\begin{vmatrix} 5.4 & 3.6 \\ 3.6 & 7.2 \end{vmatrix}} \qquad I_2 = \frac{\begin{vmatrix} 5.4 & 36 \\ 3.6 & 36 \end{vmatrix}}{\begin{vmatrix} 5.4 & 3.6 \\ 3.6 & 7.2 \end{vmatrix}}$$
(c) $I_1 = 5.0$ mA $\quad I_2 = 2.5$ mA

8-7:1. (a) $10 \cdot I_1 + 30 \cdot (I_1 - I_2) = 24 + 20 = 44$
$\quad 20 \cdot I_2 + 30 \cdot (I_2 - I_1) = 75 - 20 = 55$
(b)
$$I_1 = \frac{\begin{vmatrix} 44 & -30 \\ 55 & 50 \end{vmatrix}}{\begin{vmatrix} 40 & -30 \\ -30 & 50 \end{vmatrix}} = 3.5 \text{ A} \qquad I_2 = \frac{\begin{vmatrix} 40 & 44 \\ -30 & 55 \end{vmatrix}}{\begin{vmatrix} 40 & -30 \\ -30 & 50 \end{vmatrix}} = 3.2 \text{ A}$$
(c) current in 30 kΩ = $3.5 - 3.2 = 0.3$ A upward

8-8:1. (NOTE: resistance in kΩ and current in mA)
(a) $\quad 51 \cdot I_A - 33 \cdot I_B + \quad 0 \cdot I_C = 20$
$\quad -33 \cdot I_A + 70 \cdot I_B - 15 \cdot I_C = \quad 0$
$\quad \quad 0 \cdot I_A - 15 \cdot I_B + 45 \cdot I_C = -9$
(b) $I_A = 0.540$ mA, $\quad I_B = 0.238$ mA, $\quad I_C = -0.124$ mA

8-9:1. (a) Kirchhoff's Current Law Equations:
at Node A: $\quad \quad \quad I_1 + I_2 = I_A$
at Node B: $\quad \quad \quad I_3 + I_B = I_2$
at Reference Node: $I_A = I_1 + I_B + I_3$
note that the third equation is redundant information
(b) substituting V/R ratios for the resistor currents:
at Node A: $I_A = \dfrac{V_A}{R_1} + \dfrac{V_A - V_B}{R_2}$
at Node B: $I_B = \dfrac{-V_B}{R_3} + \dfrac{V_A - V_B}{R_2}$

(c) substituting in the circuit values and simplifying:
   at Node A: $100 = 3V_A - V_B$
   at Node B: $150 = 3V_A - 7V_B$

   therefore: $V_B = -8.33$ V
   $ V_A = 30.55$ V

8-10:1. (b) NOTE: currents are in mA and resistances are in kΩ

$$V_1 \cdot \left(\frac{1}{1.0} + \frac{1}{2.2} + \frac{1}{4.7} \right) - V_2 \cdot \left(\frac{1}{4.7} \right) = 20$$

$$V_2 \cdot \left(\frac{1}{4.7} + \frac{1}{1.2} \right) - V_1 \cdot \left(\frac{1}{4.7} \right) = -15$$

   (c) $V_1 = 10.44$ V, $V_2 = -12.21$ V

8-11:1. (a) and (b)

a   b

(c) actual current in the 3 kΩ resistor is 1 mA downward

CHAPTER 9

9-2:1. (a)

$$R_X = \frac{R_S \cdot R_B}{R_S + R_B} = \frac{(143\ \Omega) \cdot (300\ \Omega)}{143 + 300\ \Omega} = 96.84 = 97\ \Omega$$

$$E_X = E \cdot \frac{R_B}{R_S + R_B} = (25\ \text{V}) \cdot \frac{300\ \Omega}{143 + 300\ \Omega} = 16.93 = 17\ \text{V}$$

   (b)

$$I_L = \frac{V_X}{R_X + R_L} = \frac{16.93\ \text{V}}{96.84 + 3000\ \Omega} = 5.5\ \text{mA}$$

9-2:2. (a)

$$R_X = 60 + \frac{(50) \cdot (120 + 80)}{(50) + (120 + 80)} = 100\ \Omega$$

$$E_X = (20\ \text{V}) \cdot \frac{120 + 80\ \Omega}{50 + 120 + 80\ \Omega} = 16\ \text{V}$$

   (b)

$$V_L = E_X \cdot \frac{R_L}{R_X + R_L} = (16\ \text{V}) \cdot \frac{100\ \Omega}{100 + 100\ \Omega} = 8\ \text{V}$$

9-3:1. (a)

$$I_X = \frac{E_X}{R_X} = \frac{16 \text{ V}}{100 \text{ }\Omega} = 0.16 \text{ A}$$

(b)

$$I_L = I_X \cdot \frac{R_X}{R_X + R_L} = (0.16 \text{ A}) \cdot \frac{100 \text{ }\Omega}{100 + 100 \text{ }\Omega} = 80 \text{ mA}$$

9-4:1. (a)

(b)

$$V_L = E_X \cdot \frac{R_L}{R_X + R_L} = (2 \text{ V}) \cdot \frac{60 \text{ }\Omega}{140 + 60 \text{ }\Omega} = 0.6 \text{ V}$$

9-5:1. (a)

$$V_1 \cdot \left(\frac{1}{1.2} + \frac{1}{2.2} + \frac{1}{1.5}\right) - V_2 \cdot \left(\frac{1}{1.5}\right) = -5$$

$$-V_1 \cdot \left(\frac{1}{1.5}\right) + V_2 \cdot \left(\frac{1}{1.5} + \frac{1}{0.82}\right) = 2 \cdot I_1 = 2 \cdot V_1 \cdot \left(\frac{1}{2.2}\right)$$

(b) $V_1 = -3.58 \text{ V}, \quad V_2 = -2.99 \text{ V}$

(c)

$$I_L = \frac{V_2 - V_1}{R_L} = \frac{(-2.99) - (-3.58) \text{ V}}{1.5 \text{ k}\Omega} = 0.3933 \times 10^{-3}$$

$I_L = 0.39 \text{ MA to the left}$

9-6:1. Using Equations (9-6), (9-7), and (9-8):

$$R_X = \frac{2.2 \cdot 1.8 + 1.8 \cdot 4.7 + 4.7 \cdot 2.2}{2.2} = \frac{22.76}{2.2} = 10.35 \text{ k}\Omega$$

Similarly

$R_Y = 12.64 \text{ k}\Omega$

$R_Z = 4.84 \text{ k}\Omega$

CHAPTER 10

10-1:1. From Equation (10-1):
$$I = \frac{F}{B \cdot l} = \frac{4 \times 10^{-3} \text{ N}}{(10^{-2} \text{ T}) \cdot (0.05 \text{ m})} = 8.0 \text{ A}$$

10-1:2. From Equation (10-2):
$$N = \frac{T}{B \cdot I \cdot l \cdot d} = \frac{220 \times 10^{-6} \text{ N} \cdot \text{m}}{(0.25 \text{ T}) \cdot (0.05 \text{ A}) \cdot (0.014 \text{ m})^2} = 89.796$$
$$N = 90 \text{ turns}$$

10-2:1. (a) since the movement path and the shunt are in parallel:
$$V_M = V_{sh} = I_M \cdot R_M = I_{sh} \cdot R_{sh} = (I_T - I_M) \cdot R_{sh}$$
$$\Rightarrow R_{sh} = \frac{I_{sh} \cdot R_{sh}}{I_T - I_M} = \frac{(0.010 \text{ A}) \cdot (70 \text{ }\Omega)}{0.100 - 0.010 \text{ A}} = 7.78 \text{ }\Omega$$
(b) similarly; $R_{sh} = 70.1$ mΩ

10-2:2. Using Ohm's Law:
$$I_M = \frac{V_M}{R_M} = \frac{0.03 \text{ V}}{150 \text{ }\Omega} = 200 \text{ }\mu\text{A} \text{ in the movement path since the movement and the}$$
shunt are in parallel:
$$V_M = V_{sh}$$
$$I_M \cdot R_M = I_{sh} \cdot R_{sh} = (I_T - I_M) \cdot R_{sh}$$
$$(200 \times 10^{-6} \text{ A}) \cdot (150 \text{ }\Omega) = (1.0 - 200 \times 10^{-6} \text{ A}) \cdot (R_{sh})$$
$$R_{sh} = 0.030\ 006 \text{ }\Omega$$

10-3:1. (a) sensitivity $= (200 \times 10^{-6} \text{ A})^{-1} = 5000 \text{ }\Omega/\text{V}$
(b) $R_T = $ (full scale volts) $\cdot$ (sensitivity)
$$R_T = (0.100 \text{ V}) \cdot (5000 \text{ }\Omega/\text{V}) = 500 \text{ }\Omega$$
$$R_M = R_T - 100 = 500 - 100 = 400 \text{ }\Omega$$
(c) similarly; $R_M = 49.9$ kΩ

10-4:1. (a) Using the voltage divider principle:
$$V_L = E \cdot \frac{R_L}{R_{in} + R_L} = 12 \cdot \frac{50}{10 + 50} = 10 \text{ V}$$
(b) first find the resistance of the voltmeter:
$$R_v = (1000 \text{ }\Omega/\text{V}) \cdot (50 \text{ V}) = 50 \text{ k}\Omega$$
now find the "effective" load resistance when the voltmeter is connected across R_L:
$$R_{eff} = \frac{R_L \cdot R_v}{R_L + R_v} = \frac{50 \cdot 50}{50 + 50} = 25 \text{ k}\Omega$$
finally, use the voltage-divider principle:
$$V_L = E \cdot \frac{R_{eff}}{R_{in} + R_{eff}} = 12 \cdot \frac{25}{10 + 25} = 8.6 \text{ V}$$

10-4:2. (a) $V_L = 10$ V (unchanged)
(b) using a procedure identical to previous problem:
$$R_v = 50 \text{ M}\Omega$$
$$R_{eff} = 49950 \text{ }\Omega$$
$$V_L = 9.998 = 10 \text{ V}$$

10-6:1. (a) at half-scale reading:
$$R_M = R_X = 100 \text{ k}\Omega$$
(b) at half-scale reading:
$$I_M = \frac{E}{R_M + R_X} = \frac{5 \text{ V}}{100 + 100 \text{ k}\Omega} = 25 \text{ }\mu\text{A} \quad \text{therefore full-scale } I_M = 50 \text{ }\mu\text{A}$$

10-6:2. Using Equation (10-3):

$$R_X = \frac{R_A \cdot R_Y}{R_B} = \frac{(10\ k) \cdot (65)}{200} = 3.25\ k\Omega$$

$$\text{resistance error} = \frac{R_{nom} - R_X}{R_{nom}} \cdot 100\% = \frac{3300 - 3250}{3300} \cdot 100\% = 1.5\%$$

therefore the resistance is within tolerance.

CHAPTER 11

11-2:1. From Equation (11-2):

$$F = E \cdot Q = (10^6\ \text{N/C}) \cdot (1.6 \times 10^{-19}\ \text{C}) = 1.6 \times 10^{-13}\ \text{N}$$
$$F = 160\ \text{fN}$$

11-2:2. From Equation (11-3):

$$V = E \cdot s = (50\,000\ \text{V/m}) \cdot (10^{-3}\ \text{m}) = 50\ \text{V}$$

11-4:1. (a) Using the definition of dielectric strength, Equation (11-3) and the information from Table 11-1:

$$3\ \text{kV/mm} > E = V/s$$
$$\Rightarrow V < (3\ \text{kV/mm}) \cdot (10\ \text{mm})$$
$$V < 30\ \text{kV}$$

    (b) similarly;

$$V < 600\ \text{kV}$$

11-6:1. (a) From Equation (11-7):

$$V = \frac{Q}{C} = \frac{2 \times 10^{-9}\ \text{C}}{0.01 \times 10^{-6}\ \text{F}} = 0.20\ \text{V}$$

    (b) similarly;

$$V = 200\ \text{V}$$

11-6:2. From Equation (11-7):

$$Q = C \cdot V = (47 \times 10^{-6}\ \text{F}) \cdot (30\ \text{V}) = 1.4\ \text{mC}$$

11-7:1. (a) Using Equation (11-13)

$$C = \frac{8.85 \cdot A}{10^{12} \cdot s} = \frac{8.85 \cdot (10^{-2}\ \text{m}^2)}{10^{12} \cdot (0.1 \times 10^{-3}\ \text{m})} = 885 \times 10^{-12}$$
$$C = 890\ \text{pF}$$

    (b) From Equation (11-7):

$$V = \frac{Q}{C} = \frac{20 \times 10^{-9}\ \text{C}}{885 \times 10^{-12}\ \text{F}} = 22.599$$
$$V = 23\ \text{V}$$

11-7:2. Using the relationships from Equation (11-8):

$$\frac{C_2}{C_1} = \frac{A_2}{A_1} \cdot \frac{s_1}{s_2} = \frac{2}{1} \cdot \frac{1}{2} = 1$$

The capacitance will not change.

11-8:1. Since the capacitors are identical except for the dielectric, from Equation (11-15):

$$\frac{C_2}{C_1} = \frac{k_2}{k_1} = \frac{3}{1} =$$
$$\Rightarrow C_2 = 3 \cdot C_1$$

11-8:2. (a) Combining Equations (11-15) and (11-7):

$$V = \frac{Q}{C} = \frac{Q \cdot 10^{12} \cdot s}{8.85 \cdot k \cdot A} = \frac{(2 \times 10^{-7}) \cdot (10^{12}) \cdot (10^{-4})}{(8.85) \cdot (1) \cdot (10^{-2})} = 225.99$$
$$V = 230\ \text{V with air dielectric}$$

(b) similarly;

    $V = 230/6 = 38$ V with glass dielectric

(c) $V = 230/7500 = 30$ mV with ceramic dielectric

11-9:1. Using Equation (11-17):

$$C_T = C_1 + C_2 + C_3 = 1 + 2.2 + 3.3 = 6.5 \ \mu F$$

11-9:2. Since the capacitors are in parallel:

$$V_1 = V_2 = V_3 = E = 10 \text{ V}$$

From Equation (11-7) for the 1.0 μF capacitor:

$$Q_1 = C_1 \cdot V_1 = (1 \ \mu F) \cdot (10 \text{ V}) = 10 \ \mu C$$

summarizing the results of similar calculations:

Capacitance	voltage across	charge in
$C_1 = 1.0 \ \mu F$	10 V	10 μC
$C_2 = 2.2 \ \mu F$	10 V	22 μC
$C_3 = 3.3 \ \mu F$	10 V	33 μC
$C_T = 6.5 \ \mu F$	10 V	65 μC

11-10:1. Using Equation (11-18):

$$C_{eq} = \frac{1}{1/C_1 + 1/C_2 + 1/C_3} = \frac{1}{1/1 + 1/2.2 + 1/3.3} = 0.5689$$

$$C_{eq} = 0.57 \ \mu F$$

11-10:2. Since the capacitors are in series:

$$Q_1 = Q_2 = Q_3 = Q_T = C_{eq} \cdot V_T = (0.57 \ \mu F) \cdot (10 \text{ V})$$
$$= 5.7 \ \mu C$$

From Equation (11-7) for the 1.0 μF capacitor:

$$V_1 = \frac{Q_1}{C_1} = \frac{5.7 \ \mu C}{1.0 \ \mu F} = 5.7 \text{ V}$$

summarizing the results of similar calculations:

Capacitance	voltage across	charge in
$C_1 = 1.0 \ \ \mu F$	5.7 V	5.7 μC
$C_2 = 2.2 \ \ \mu F$	2.6 V	5.7 μC
$C_3 = 3.3 \ \ \mu F$	1.7 V	5.7 μC
$C_T = 0.57 \ \mu F$	10. V	5.7 μC

CHAPTER 12

12-1:1. Since the maximum current occurs the instant that the switch is closed, using Equation (12-2):

$$R = \frac{E}{(\text{initial } I)} = \frac{100 \text{ V}}{20 \ \mu A} = 5 \text{ M}\Omega$$

12-1:2. (a) Using Equation (12-2):

$$\text{initial } I = \frac{E}{R} = \frac{15 \text{ V}}{120 \text{ k}\Omega} = 125 \ \mu A$$

(b) the current gets smaller and eventually reaches zero when the capacitor is charged to 15 V.

12-2:1. (a) From the graph at $t = 2$ s, $i \approx 20 \ \mu A$;

therefore using Ohm's Law and Kirchhoff's Voltage Law:

$$v_R = i \cdot R = (20 \ \mu A) \cdot (2 \text{ M}\Omega) = 40 \text{ V}$$
$$v_c = E - v_R = 100 - 40 = 60 \text{ V}$$

(b) From the graph at $t = 10$ s, $i \approx 0$; therefore:
$v_R = i \cdot R = 0$ V
$v_c = 100$ V (fully charged)

12-2:2. (a) Using Equation (12-3):
$$\text{initial } \frac{dv}{dt} = \frac{E}{C \cdot R} = \frac{15 \text{ V}}{(0.47 \ \mu\text{F}) \cdot (120 \text{ k}\Omega)} = 270 \text{ V/s}$$

(b) The value of the slope becomes smaller as the potential difference across the charged capacitor approaches its maximum value of 15 V.

12-3:1. (a) Since it takes one time constant to charge to 63% of E:
$t = \tau = R \cdot C = (250 \text{ k}\Omega) \cdot (0.4 \ \mu\text{F}) = 0.1$ s
(b) Since it takes (practically) five time constants to fully charge the capacitor:
$t = 5 \cdot \tau = 5 \cdot R \cdot C = 0.5$ s

12-3:2. The battery voltage affects the charging current but not the charging time which is five time constants:
(a) $t = 5 \cdot \tau = 5 \cdot R \cdot C = 5 \cdot (500 \text{ k}\Omega) \cdot (1.0 \ \mu\text{F}) = 2.5$ s
(b) $t = 5 \cdot \tau = 5 \cdot R \cdot C = 5 \cdot (2.0 \text{ M}\Omega) \cdot (1.0 \ \mu\text{F}) = 10.$ s

12-4:1. (a) Since $(10 \text{ V})/(20 \text{ V}) = 50\%$
from the charging curve $y = 50\%$ at $x \approx 0.7\tau$
therefore
$t = 0.7 \cdot \tau = 0.7 \cdot R \cdot C = 0.7 \cdot (250 \text{ k}\Omega) \cdot (0.4 \ \mu\text{F})$
$t = 70$ ms
(b) similarly; $19/20 = 95\%$
$y = 95\% \Rightarrow x \approx 3.0\tau$
$t = 3 \cdot \tau = 300$ ms

12-4:2. First find x:
$$x = \frac{t}{\tau} = \frac{t}{R \cdot C} = \frac{0.200 \text{ s}}{(250 \text{ k}\Omega) \cdot (0.4 \ \mu\text{F})} = 2.0$$
from the charging curve: $y(2) \approx 87\%$
therefore
$v_C = 0.87 \cdot E = 0.87 \cdot (20 \text{ V}) = 17.4$ V
$v_R = E - v_C = 20 - 17.4 = 2.6$ V

12-5:1. (a) Using Equation (12-5):
$$\text{initial } I = \frac{V_{init}}{R} = \frac{100 \text{ V}}{250 \text{ k}\Omega} = 400 \ \mu\text{A}$$
(b) the time to discharge the capacitor is (practically):
$t = 5\tau = 5 \cdot R \cdot C = 5 \cdot (250 \text{ k}\Omega) \cdot (0.2 \ \mu\text{F}) = 250$ ms.

12-5:2. First find x:
$$x = \frac{t}{\tau} = \frac{t}{R \cdot C} = \frac{0.060 \text{ s}}{(250 \text{ k}\Omega) \cdot (0.2 \ \mu\text{F})} = 1.2$$
from the discharge graph: $y(1.2) \approx 30\%$
therefore
$v_C = 0.30 \cdot E = 0.30 \cdot (100 \text{ V}) = 30$ V
$v_R = E - v_C = 0 - 30 = -30$ V

12-7:1. First determine the Thévenin equivalent charging circuit and the time-constant for the charging circuit:
$R_X = 330/2 = 165 \text{ k}\Omega$
$E_X = 100/2 = 50$ V
$\tau = R_X \cdot C = (165 \text{ k}\Omega) \cdot (4.7 \ \mu\text{F}) = 0.7755$ s
(a) $t = 5\tau = 5 \cdot 0.7755 = 3.88$ s

(b) fully charged $V_C = E_X = 50$ V
(c) From Equation (12-9) and since $x = t/\tau$:
$$v_C = E_X \cdot [1 - \exp(-t/\tau)] = 50 \cdot [1 - \exp(-1/0.7755)]$$
$$v_C = 50 \cdot [1 - 0.2754] = 36.2 \text{ V}$$

12-7:2. First determine the Thévenin equivalent discharging circuit and the time constant for the discharging circuit:
$$R_X = 330 = 330 \text{ k}\Omega$$
$$E_X = 0 \text{ V}; V_o = 50 \text{ V}$$
$$\tau = R_X \cdot C = (330 \text{ k}\Omega) \cdot (4.7 \text{ } \mu\text{F}) = 1.551 \text{ s}$$
(a) discharge time $= 5 \cdot \tau = 5 \cdot (1.551) = 7.8$ s
(b) Solving Equation (12-11) for x:
$$x = -\ln(v_C/V_0) = -\ln(10/50) = 1.609$$
$$t = x \cdot \tau = (1.609) \cdot (1.551) = 2.5 \text{ s}$$

12-8:1. First find the values of the Thévenin equivalent circuit and the time constant:
$$R_X = 160 \text{ }\Omega$$
$$E_X = 100 \text{ V}$$
$$\tau = R_X \cdot C = (160 \text{ k}\Omega) \cdot (47 \text{ } \mu\text{F}) = 7.52 \text{ s}$$
(a) to reach steady-state
$$t = 5\tau = (5) \cdot (7.52) = 37.6 \text{ s}$$
(b) Using Equation (12-13) and since $x = t/\tau$:
$$v_C = V_F + (V_0 - V_F) \cdot \exp(-t/\tau) =$$
$$v_C = 100 + (500 - 100) \cdot \exp(-10/7.52)$$
$$v_C = 100 + (400) \cdot (0.2645) = 205.8$$
$$v_C = 210 \text{ V}$$

12-8:2. By inspection of the circuit diagram
(a) $V_0 = -12$ V
(b) $V_F = +9$ V
(c) $t = 5\tau = 5 \cdot R \cdot C = 5 \cdot (470 \text{ k}\Omega) \cdot (0.1 \text{ } \mu\text{F}) = 235$ ms
(d) solving Equation (12-13) for x:
$$x = -\ln\left(\frac{v_C - V_F}{V_0 - V_F}\right)$$
since $x = t/\tau$
$$t = -\tau \cdot \ln\left(\frac{v_C - V_F}{V_0 - V_F}\right) = (-47 \text{ ms}) \cdot \ln\left(\frac{0 - 9}{-12 - 9}\right)$$
$$t = (-47 \text{ ms}) \cdot (-0.8473) = 39.82$$
$$t = 40 \text{ ms}$$

12-9:1. (a) Using Equation (12-14):
$$W = 0.5 \cdot C \cdot V^2 = 0.5 \cdot (33 \text{ } \mu\text{F}) \cdot (50 \text{ V})^2 = 41 \text{ mJ}$$
(b) similarly;
$$W = 10 \text{ mJ}.$$

12-9:2. From Equation (12-14) since $C_1 = C_2$:
$$\frac{W_2}{W_1} = \frac{V_2^2}{V_1^2} = \frac{(10 \cdot V_1)^2}{(1 \cdot V_1)^2} = 100$$
The higher voltage capacitor has 100 times the stored energy of the lower voltage capacitor.

CHAPTER 13

13-5:1. Using Equation (13-1):
$$F_m = N \cdot I = (50 \text{ turns}) \cdot (2 \text{ A})$$
$$F_m = 100 \text{ ampere} \cdot \text{turns}$$

13-6:1. Combining Equations (13-2) and (13-1):
$$R_m = \frac{F_m}{\Phi} = \frac{N \cdot I}{\Phi} = \frac{(50 \text{ turns}) \cdot (2 \text{ A})}{0.050 \text{ Wb}} = 2000 \frac{\text{A} \cdot \text{t}}{\text{Wb}}$$
$$R_m = 2000 \text{ H}^{-1}$$

13-6:2. From Equation (13-2) and using Equation (13-1):
$$F_m = R_m \cdot \Phi = N \cdot I$$
solving for N:
$$N = \frac{R_m \cdot \Phi}{I} = \frac{(100 \text{ H}^{-1}) \cdot (0.200 \text{ Wb})}{0.050 \text{ A}} = 400 \text{ turns}$$

13-9:1. Using Equation (13-3):
$$P_m = 1/R_m = 1/(10^7 \text{ H}^{-1}) = 10^{-7} \text{ H}$$

13-9:2. Using Equation (13-4):
$$\mu = \frac{l}{R_m \cdot A} = \frac{0.15 \text{ m}}{(10^7 \text{ H}^{-1}) \cdot (0.25 \times 10^{-4} \text{ m}^2)}$$
$$\mu = 6 \times 10^{-4} \text{ H/m}$$

13-10:1. Using Equation (13-5):
$$B = \frac{\Phi}{A} = \frac{2 \times 10^{-6} \text{ Wb}}{0.25 \times 10^{-4} \text{ m}^2} = 80 \text{ mT}$$

13-11:1. From Equation (13-7):
$$H = B/\mu$$
Using the permeability found in Self-Check Problem 13-9:2:
$$\mu = 6 \times 10^{-4} \text{ H/m}$$
and the flux density found in Self-Check Problem 13-10:1:
$$B = 80 \text{ mT}$$
$$H = B/\mu = (80 \times 10^{-3})/(6 \times 10^{-4}) = 130 \text{ A/m}$$

13-15:1. From the "Cast iron" curve for Figure 13-21: when $B = 0.5 \text{ T} \Rightarrow H \approx 2000 \text{ A/m}$
Using Equation (13-7):
$$\mu = \frac{B}{H} = \frac{0.5}{2000} = 2.5 \times 10^{-4} \text{ H/m}$$

13-15:2. (a) Using Equation (13-6) and Equation (13-1):
$$H = \frac{F_m}{l} = \frac{N \cdot I}{l} = \frac{(1200 \text{ turns}) \cdot (2.0 \text{ A})}{0.40 \text{ m}} = 6000 \text{ A/m}$$
(b) by using the "Cast iron" magnetization curve:
$$B = 0.85 \text{ T}$$
(c) From Equation (13-5):
$$\Phi = B \cdot A = (0.85 \text{ Wb/m}^2) \cdot (3 \times 10^{-4} \text{ m}^2) = 2.6 \times 10^{-4} \text{ Wb}$$

CHAPTER 14

14-2:1. Using Equation (14-1):
$$e_T = N \cdot \frac{d\phi}{dt} = (50) \cdot (0.1 \text{ Wb/s}) = 5.0 \text{ V}$$

14-2:2. First find the average rate of change of linked flux:
$$\text{average } \frac{d\phi}{dt} = \frac{\Delta \phi}{\Delta t} = \frac{(15 - 1 \text{ mWb})}{2 \text{ s}} = 7 \text{ mWb/s}$$
Using Equation (14-1):
$$\text{average } e_T = N \cdot \frac{\Delta \phi}{\Delta t} = (250) \cdot (7 \text{ mWb/s}) = 1.75 \text{ V}$$

14-5:1. Using Equation (14-2):

$$L = N \cdot \frac{d\phi}{di} = (50) \cdot (0.1 \text{ Wb/A}) = 5.0 \text{ H}$$

14-5:2. (a) since only *changing* current induces a voltage:

$$e_L = 0 \text{ V}$$

(b) From Equation (14-3):

$$e_L = L \cdot \frac{di}{dt} = (100 \text{ mH}) \cdot (0.5 \text{ A/s}) = 50 \text{ V}$$

14-6:1. From Equation (14-5):

$$N = \sqrt{\frac{L \cdot l}{\mu \cdot A}} = \sqrt{\frac{(0.500 \text{ H}) \cdot (0.424 \text{ m})}{(7.6 \times 10^{-4} \text{ H/m}) \cdot (2.25 \times 10^{-4} \text{ m}^2)}}$$

$$N = 1100$$

14-6:2. From Equation (14-5) and since $\mu_2 A_2 = \mu_1 A_1$:

$$\frac{L_2}{L_1} = \frac{N_2^2}{N_1^2} \cdot \frac{l_1}{l_2} = \frac{2^2}{1^2} \cdot \frac{1}{0.5} = 8$$

the second coil has 8 times the inductance of the first coil

14-7:1. (a) Using Equation (14-6):

$$L_T = L_1 + L_2 + L_3 = 100 + 220 + 330 = 650 \text{ mH}$$

(b) From the definition of the unit henry:

$$e_T = L_T \cdot \frac{di}{dt} = (650 \text{ mH}) \cdot (10 \text{ A/s}) = 6.5 \text{ V}$$

14-8:1. (a) Using Equation (14-7):

$$L_{eq} = \frac{1}{1/L_1 + 1/L_2 + 1/L_3} = \frac{1}{1/100 + 1/220 + 1/330} = 56.897$$

$$L_{eq} = 57 \text{ mH}$$

(b) From the definition of the unit henry:

$$e_T = L_T \cdot \frac{di}{dt} = (57 \text{ mH}) \cdot (10 \text{ A/s}) = 0.57 \text{ V}$$

CHAPTER 15

15-2:1. (a) By inspection of the graph in Figure 15-4:

$$i = 63\% \text{ of } I_m = 0.63 \cdot (10 \text{ A}) = 6.3 \text{ A}$$

(b) Using Ohm's Law:

$$v_R = i \cdot R = (6.3 \text{ A}) \cdot (1 \text{ } \Omega) = 6.3 \text{ V}$$

(c) Using Kirchhoff's Voltage Law:

$$v_L = E - v_R = 10 - 6.3 = 3.7 \text{ V}$$

15-2:2. (a) Using Equation (15-3):

$$I_m = \frac{E}{R} = \frac{20 \text{ V}}{10 \text{ } \Omega} = 2.0 \text{ A}$$

(b) By inspection of the graph in Figure 15-4:

$$i = 0 \text{ A at } t = 0 \text{ s}$$

(c) Using Equation (15-4):

$$\text{initial } \frac{di}{dt} = \frac{E}{L} = \frac{20 \text{ V}}{0.500 \text{ H}} = 40 \text{ A/s}$$

15-3:1. First find the time constant using Equation (15-5):

$$\tau = L/R = (500 \text{ mH})/(10 \text{ } \Omega) = 50 \text{ ms}$$

Since steady-state is reached after five time constants:
$t = 5\tau = (5) \cdot (50 \text{ ms}) = 250 \text{ ms}$

15-3:2. The source voltage has no effect on the response time. Comparing time constants:
$$\frac{t_2}{t_1} = \frac{5\tau_2}{5\tau_1} = \frac{L_2/R_2}{L_1/R_1} = \frac{(1 \text{ H})/(200 \text{ }\Omega)}{(50 \text{ mH})/(10 \text{ }\Omega)} = \frac{5 \text{ ms}}{5 \text{ ms}}$$
the times to reach steady-state are identical.

15-4:1. First determine the time constant and the steady-state current:
$\tau = L/R = (500 \text{ mH})/(10 \text{ }\Omega) = 50 \text{ ms}$
$I_m = E/R = (30 \text{ V})/(10 \text{ }\Omega) = 3.0 \text{ A}$
Now find x:
$x = t/\tau = (80 \text{ ms})/(50 \text{ ms}) = 1.6$
By inspection of the rising-current graph in Figure 15-6:
$y(1.6) = 80\%$
Therefore
$i = 0.80 \cdot I_m = (0.8) \cdot (3.0) = 2.4 \text{ A}$

15-4:2. First find y:
$$y = \frac{i}{I_m} = \frac{0.5 \text{ A}}{3.0 \text{ A}} = 0.1667 = 17\%$$
By inspection of the rising current graph of Figure 15-6:
$x \approx 0.2$ when $y = 17\%$
Use x to find the time:
$t = x \cdot \tau = 0.2 \cdot (50 \text{ ms}) = 10 \text{ ms}$

15-5:1. First find the Thévenin-equivalent circuit as "seen" by the practical inductor with the switch closed.
$R_X = 1.5 \text{ }\Omega$
$E_X = 10 \text{ V}$
Now find the circuit time constant and steady-state current reached after the switch is closed:
$\tau = L/(R_X + R_{in}) = (500 \text{ mH})/(1.5 + 0.5 \text{ }\Omega) = 250 \text{ ms}$
$I_m = E_X/(R_X + R_{in}) = (10 \text{ V})/(1.5 + 0.5 \text{ }\Omega) = 5 \text{ A}$
(a) From Equation (15-6):
$\quad i = I_m \cdot [1 - \exp(-t/\tau)] = (5.0 \text{ A}) \cdot [1 - \exp(-0.1/0.25)]$
$\quad i = (5) \cdot [0.3297] = 1.648$
$\quad i = 1.6 \text{ A}$
(b) Using Kirchhoff's Voltage Law on Thévenin-equivalent circuit:
$\quad v_{ind} = E_X - i \cdot R_X = 10 - (1.648 \text{ A}) \cdot (1.5 \text{ }\Omega) = 7.527$
$\quad v_{ind} = 7.5 \text{ V}$

15-5:2. Referring to the previous solution:
$\tau = 250 \text{ ms}$
$I_m = 5.0 \text{ A}$
From Equation (15-6):
$i = I_m \cdot [1 - \exp(-t/\tau)]$
solving for t:
$t = -\tau \cdot \ln[1 - (i/I_m)] = (-250 \text{ ms}) \cdot \ln[1 - (2/5)]$
$t = (-250 \text{ ms}) \cdot (-0.5108) = 127.7$
$t = 130 \text{ ms}$

15-6:1. From Equation (15-8):
$$I = \sqrt{\frac{2 \cdot W}{L}} = \sqrt{\frac{2 \cdot (10 \text{ J})}{0.5 \text{ H}}} = 6.3 \text{ A}$$

15-6:2. Using Equation (15-8):

$$W_1 = 0.5 \cdot L_1 \cdot I_1^2 = W_2 = 0.5 \cdot L_2 \cdot I_2^2$$

$$\Rightarrow I_2 = \sqrt{\frac{L_1 \cdot I_1^2}{L_2}} = \sqrt{\frac{(0.1 \text{ H}) \cdot (0.05)^2}{0.03 \text{ H}}} = 93 \text{ mA}$$

15-7:1. (a) BEFORE: See solution to Self-Check Problem 15-5:1:

$$I_m = 5 \text{ A}$$
$$v_{ind} = E_x - i \cdot R_x = 10 \text{ V} - (5 \text{ A}) \cdot (1.5 \text{ }\Omega) = 2.5 \text{ V}$$

AFTER: Find the Thévenin-equivalent resistance "seen" by the practical inductor after the switch is opened:

$$R_x = 3 \text{ }\Omega$$
$$v_{ind} = E_x - i \cdot R_x = 0 \text{ V} - (5 \text{ A}) \cdot (3 \text{ }\Omega) = -15 \text{ V}$$

(the negative sign indicates that the polarity of the voltage across the practical coil has reversed)

(b) Find the time constant:

$$\tau = L/(R_x + R_{in}) = (500 \text{ mH})/(3 + 0.5 \text{ }\Omega) = 142.86$$
$$\tau = 140 \text{ ms}$$

For the current to cease:

$$t = 5 \cdot \tau = 5 \cdot (142.86 \text{ ms}) = 710 \text{ ms}$$

15-7:2. (a) First find the current using the discharge graph of Figure 15-6:

$$x = t/\tau = 250/(142.86) = 1.75$$
$$\Rightarrow y \approx 18\%$$
$$\Rightarrow i = 0.18 \cdot (5 \text{ A}) = 0.9 \text{ A}$$

Finally using Ohm's Law:

$$v_{ind} = v_R = i \cdot R_X = (0.9 \text{ A}) \cdot (3 \text{ }\Omega) = 2.7 \text{ V}$$

(b) Using Ohm's Law:

$$i = v/R_x = (1.5 \text{ V})/(3 \text{ }\Omega) = 0.5 \text{ A}$$

Using the discharge graph of Figure 15-6:

$$y = i/I_m = (0.5 \text{ A})/(5 \text{ A}) = 0.1 = 10\%$$
$$\Rightarrow x \approx 2.3$$

therefore:

$$t = 2.3\tau = (2.3) \cdot (142.86 \text{ ms}) = 330 \text{ ms}$$

15-8:1. From the solution of Self-Check Problem 15-7:1

$$\tau = 142.86 \text{ ms}$$
$$I_0 = 5 \text{ A}$$

(a) From Equation (15-10):

$$i = I_0 \cdot \exp(-t/\tau) = (5 \text{ A}) \cdot \exp(-250/142.86) = 0.8689 \text{ A}$$
$$v_{ind} = v_R = i \cdot R_x = (0.8689 \text{ A}) \cdot (3 \text{ }\Omega) = 2.6 \text{ V}$$

(b) Using Ohm's Law:

$$i = v/R_x = (1.5 \text{ V})/(3 \text{ }\Omega) = 0.5 \text{ A}$$

Substituting $x = t/\tau$ into Equation (15-10) and solving for t:

$$t = -\tau \cdot \ln(i/I_0) = (-142.86) \cdot \ln(0.5/5) = 330 \text{ ms}$$

15-9:1. (a) Using information from Example 15-8 with the switch closed:

$$\tau = 2.0 \text{ s}$$
$$I_F = 10 \text{ A}$$
$$I_0 = 2.0 \text{ A}$$

From Equation (15-12):

$$i_L = I_F + (I_0 - I_F) \cdot \exp(-t/\tau)$$
$$i_L = 10 + (2 - 10) \cdot \exp(-1/2)$$
$$i_L = 5.1 \text{ A}$$

(b) ≈ 10 A (new steady-state value since $t > 5\tau$)

(c) $i_L = I_F + (I_0 - I_F) \cdot \exp(-t/\tau) = 10 + (2 - 10) \cdot \exp(-4/2)$
$i_L = 8.917$ A
$v_L = E - i \cdot R = (10\text{ V}) - (8.917\text{ A}) \cdot (1\ \Omega) = 1.1$ V

CHAPTER 16

16-5:1. Solving Equation (16-1) for ϕ:
$e = E_m \cdot \sin \phi$
$\Rightarrow \phi = \sin^{-1}(e/E_m)$
(a) $\phi = \sin^{-1}(0/150) = \sin^{-1}(0)$
$\phi = 0°, 180°, 360°, 540°,$ etc.
(b) $\phi = \sin^{-1}(+26/150) = \sin^{-1}(0.1733)$
$\phi = 10°, 170°, 370°, 530°,$ etc.
(c) $\phi = \sin^{-1}(-75/150) = \sin^{-1}(-0.5)$
$\phi = 210°, 330°, 570°, 690°,$ etc.

16-5:2. Since $e = 150 \cdot \sin(\phi) = E_m \cdot \sin(360° \cdot f \cdot t)$ and since $f = 60$ Hz
$\Rightarrow \phi = (360° \cdot f \cdot t) = 21\,600° \cdot t$
$\Rightarrow t = \phi°/21\,600°$
Using the angles from Self-Check Problem 16-5:1:
(a) $t = 0$ ms, 8.33 ms, 16.67 ms, 25.0 ms, etc.
(b) $t = 0.463$ ms, 7.87 ms, 17.1 ms, 24.5 ms, etc.
(c) $t = 9.72$ ms, 15.3 ms, 26.4 ms, 31.9 ms, etc.

16-6:1. Since $e = 170 \cdot \sin(377t) = 170 \cdot \sin \phi$
$\Rightarrow \phi = \sin^{-1}(e/170) = \sin^{-1}(144/170)$
$\phi = 1.01$ rad, 2.13 rad, 7.29 rad, etc.

16-6:2. (a) Since $e = 100 \cdot \sin 6283t = 100 \cdot \sin \omega t$
then by inspection $\omega = 6283$ rad/s
(b) Since $\omega = 6283 = 2\pi \cdot f$
$\Rightarrow f = 6283/2\pi = 1000$ Hz
$f = 1.0$ kHz
(c) Since $e = 100 \cdot \sin 6283t = 100$
$\Rightarrow 6283t = \sin^{-1}(100/100)$
$= \sin^{-1}(1) = 1.5708$ rad, 7.8540 rad, etc.
$\Rightarrow t = 1.5708/6283$
$t = 250\ \mu$s, 1250 μs, etc.

16-7:1. (a) $I_m = E_m/R = (170\text{ v})/(850\ \Omega) = 0.20$ A $= 200$ mA
(b) $i = 200 \cdot \sin 377t$ mA

16-7:2.

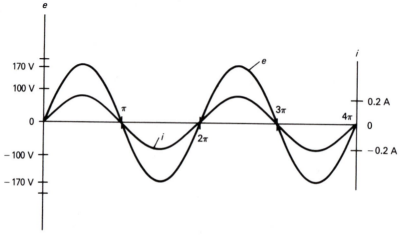

16-8:1. (a) Since $I_m = (170 \text{ V})/(850 \text{ }\Omega) = 0.2 \text{ A}$ (b) $p = P_m \cdot \sin^2 \omega t = \frac{1}{2} P_m \cdot [1 - \cos 2\omega t]$
$P_m = I_m^2 \cdot R$
$= (0.2 \text{ A})^2 \cdot (850 \text{ }\Omega) = 34 \text{ W}$
$\qquad\qquad\qquad\qquad = \frac{1}{2} \cdot 34 \cdot [1 - \cos 2.377t]$
$\qquad\qquad\qquad\qquad = 17 - 17 \cdot \cos 746t \text{ W}$

16-8:2.

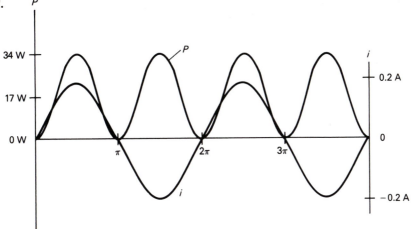

16-10:1. (a) Using Equation (16-17):
$E_{av} = 0.637 \cdot E_m = 0.637 \cdot (170 \text{ V})$
$E_{av} = 108 \text{ V}$
(b) Using Equation (16-16):
$I_{av} = 0.637 \cdot I_m = 0.637 \cdot (E_m/R) = 0.637 \cdot (170 \text{ V})/(850 \text{ }\Omega)$
$I_{av} = 127 \text{ mA}$
(c) The *full*-cycle average of a sine-wave is zero.

16-11:1. (a) Using Equation (16-19):
$I = 0.707 \cdot I_m = 0.707 \cdot (34 \text{ mA}) = 24.0 \text{ mA}$
(b) $P = I^2 \cdot R = (0.024 \text{ A})^2 \cdot (1000 \text{ }\Omega) = 576 \text{ mW}$
(c) From Equation (16-18):
$P_m = 2 P = 2 \cdot (0.576 \text{ mW}) = 1.15 \text{ W}$
(d) solving $P = V \cdot I$ for V:
$V = P/I = (0.576 \text{ mW})/(0.024 \text{ mA}) = 24 \text{ V}$

CHAPTER 17

17-1:1. From Equation (17-3):
$i_L = I_m \cdot \sin(\omega t - \pi/2)$
$i_L = 200 \cdot \sin(377t - \pi/2) \text{ mA}$

17-1:2.

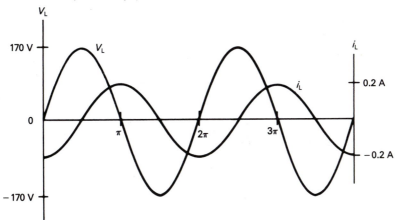

17-2:1. From Equation (17-4) and using the relation between rms and peak values:
$$X_L = \frac{V_L}{I_L} = \frac{0.707 \cdot V_{Lm}}{0.707 \cdot I_{Lm}} = \frac{(170 \text{ V})}{(0.200 \text{ A})} = \frac{170 \text{ V}}{0.2 \text{ A}} = 850 \ \Omega$$

17-3:1. From Equation (17-5); X_L
(a) $X_L = 2\pi \cdot f \cdot L = 2\pi \cdot (400) \cdot (0.010)$
   $X_L = 25 \ \Omega$
(b) $X_L = 2\pi \cdot f \cdot L = 2\pi \cdot (400 \times 10^3) \cdot (0.010)$
   $X_L = 25 \text{ k}\Omega$
(c) Since X_L is proportional to f,
   then if f doubles, X_L doubles

17-3:2. From Equation (17-6) and since $X_L = 850 \ \Omega$ (see Self-Check Problem 17-2:1):
$$L = \frac{X_L}{\omega} = \frac{850 \ \Omega}{377 \text{ rad/s}} = 2.3 \text{ H}$$

17-4:1. From Equation (17-8):
$i_c = I_m \cdot \sin(\omega t + \pi/2) = 25 \cdot \sin(377t + \pi/2) \text{ mA}$

17-4:2.

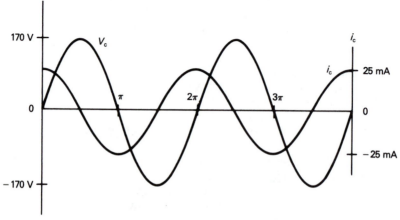

17-5:1. From Equation (17-9) and using the relation between rms and peak values:
$$X_c = \frac{V_c}{I_c} = \frac{0.707 \cdot V_{cm}}{0.707 \cdot I_{cm}} = \frac{(170 \text{ V})}{(0.025 \text{ A})} = 6800$$
$X_c = 6.8 \text{ k}\Omega$

17-6:1. From Equation (17-11):
$X_c = [2\pi \cdot f \cdot C]^{-1}$
(a) $X_c = [2\pi \cdot (400) \cdot (0.22 \times 10^{-6})]^{-1} = 1809$
   $= 1.8 \text{ k}\Omega$
(b) $X_c = [2\pi \cdot (400 \times 10^3) \cdot (0.22 \times 10^{-6})]^{-1} = 1.809$
   $= 1.8 \ \Omega$
(c) Since X_c is inversely proportional to f,
   if f doubles, X_c becomes half of its former value.

17-6:2. From Equation (17-12) and since $X_c = 6.8 \text{ k}\Omega$ (see Self-Check Problem 17-5:1):
$C = [\omega \cdot X_c]^{-1} = [(377) \cdot (6800)]^{-1} = 3.9 \times 10^{-7}$
$C = 0.39 \ \mu\text{F}$

17-7:1. First determine the rms voltage across the load:
$V = 0.707 \cdot V_m = 0.707 \cdot (170 \text{ V}) = 120 \text{ V}$

(a) From Table 17-1:
$$I_R = \frac{V_R}{R} = \frac{120 \text{ V}}{1.2 \text{ k}\Omega} = 0.1 \text{ A}$$
$I_R = 100$ mA in phase with the voltage drop V_R.

(b) From Table 17-1:
$$I_L = \frac{V_L}{\omega \cdot L} = \frac{120 \text{ V}}{(377) \cdot (4.7)} = \frac{120 \text{ V}}{1772 \ \Omega} = 6.77 \times 10^{-2} \text{ A}$$
$I_L = 68$ mA *lagging* the voltage drop V_L by 90°

(c) From Table 17-1:
$$I_c = V_c \cdot [\omega \cdot C] = (120 \text{ V}) \cdot [(377) \cdot (3.3 \times 10^{-6}) \ \Omega^{-1}]$$
$$I_c = 1.493 \times 10^{-1} \text{ A}$$
$$I_c = 150 \text{ mA } \textit{leading} \text{ the voltage drop } V_c \text{ by } 90°$$

CHAPTER 18

18-2:1. Since $I_m = 5$ mA and $\omega = 6283$ rad/s:

(a) Since $v_R = V_{Rm} \cdot \sin \omega t$
$$v_R = I_m \cdot R \cdot \sin \omega t = (5 \text{ mA}) \cdot (4.7 \text{ k}\Omega) \cdot \sin 6283t$$
$$v_R = 23.5 \cdot \sin 6283t \text{ V}$$

(b) Since $X_L = \omega \cdot L = (6283 \text{ rad/s}) \cdot (0.50 \text{ H})$
$$X_L = 3142 \ \Omega$$
then $V_{Lm} = I_m \cdot X_c = (5 \times 10^{-3} \text{ A}) \cdot (3142 \ \Omega)$
$$V_{Lm} = 15.7 \text{ V}$$
also recall that the current lags the voltage (or equivalently the voltage leads the current) by 90° in an ideal inductor
so $v_L = 15.7 \cdot \sin(6283t + 90°)$ V

(c) by Equation (18-2):
$$e_T = v_R + v_L$$
so $e_T = 23.5 \cdot \sin(6283t) + 15.7 \cdot \sin(6283t + 90°)$ V

18-5:1. Phasor addition by geometrical construction.

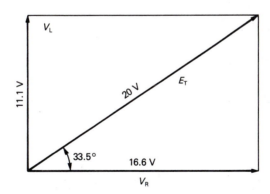

18-6:1. From Equation (18-7):
$$E_T = \sqrt{V_R^2 + V_L^2} = \sqrt{(16.6)^2 + (11.1)^2} = \sqrt{387.77}$$
$$E_T = 20.0 \text{ V}$$
and from Equation (18-4):
$$\phi = \tan^{-1}(V_L/V_R)$$
$$= \tan^{-1}(11.1/16.6) = \tan^{-1}(0.6687)$$

$\phi = 33.77°$

therefore

$E_T = 20.0 \text{ V} \underline{/33.8°}$

18-6:2. (a) From Equation (18-6):

$V_R = E_T \cdot \cos \phi$

$V_R = (120 \text{ V}) \cdot \cos 60° = 120 \cdot 0.5 = 60 \text{ V}$

$\Rightarrow \mathbf{V_R} = 60 \text{ V} \underline{/0°}$

(b) From Equation (18-5):

$V_L = E_T \cdot \sin \phi$

$V_L = (120 \text{ V}) \cdot \sin 60° = 120 \cdot 0.866 = 104 \text{ V}$

$\Rightarrow \mathbf{V_L} = 104 \text{ V} \underline{/90°}$

18-6:3. (a) Since $\mathbf{V_R} = V_R \underline{/0°}$:

$\mathbf{V_R} = 40 \text{ V} \underline{/0°}$:

(b) From Equation (18-7):

$V_L = \sqrt{E_T^2 - V_R^2} = \sqrt{50^2 - 40^2} = \sqrt{900}$

$V_L = 30 \text{ V}$

Since $\mathbf{V_L} = V_L \underline{/90°}$

$\mathbf{V_L} = 30 \text{ V} \underline{/90°}$

(c) From Equation (18-6):

$\cos \phi = V_R / E_T$

$\Rightarrow \phi = \cos^{-1}(V_R/E_T) = \cos^{-1}(40/50)$

$\phi = 36.88°$

therefore $\mathbf{E_T} = 50 \text{ V} \underline{/37°}$

18-7:1. (a) From Equation (18-9) for source voltage phasors:

$\mathbf{E_T} = E \underline{/\phi} = +E \cdot \cos \phi + jE \cdot \sin \phi$

$\mathbf{E_T} = 75 \text{ V} \underline{/60°} = 75 \cdot \cos 60° + j75 \cdot \sin 60°$

$\mathbf{E_T} = 37.5 + j65.0 \text{ V}$

(b) Similarly for voltage drop phasors:

$\mathbf{V} = 16 \text{ V} \underline{/-15°} = 16 \cdot \cos -15° + j16 \cdot \sin -15°$

$\mathbf{V} = 15.5 - j4.1 \text{ V}$

(c) Similarly for current phasors:

$\mathbf{I} = 50 \text{ mA} \underline{/120°} = 50 \cdot \cos 120° + j50 \cdot \sin 120°$

$\mathbf{I} = -25.0 + j43.3 \text{ mA}$

18-7:2 By rewriting Equation (18-8) in general form:

$\mathbf{E_T} = \sqrt{E_x^2 + E_y^2} \underline{/\tan^{-1}(E_y/E_x)}$

(a) $\mathbf{E_T} = 10 + j20 \text{ V} = \sqrt{10^2 + 20^2} \underline{/\tan^{-1}(20/10)}$

$\mathbf{E_T} = 22.4 \text{ V} \underline{/63.4°}$

(b) Similarly for voltage drop phasors:

$\mathbf{V} = \sqrt{V_x^2 + V_y^2} \underline{/\tan^{-1}(V_y/V_x)}$

$\mathbf{V} = -16 - j38 \text{ V} = \sqrt{16^2 + 38^2} \underline{/\tan^{-1}(-38/-16)}$

$\mathbf{V} = 41.2 \text{ V} \underline{/-113°}$

(c) Similarly for current phasors:

$\mathbf{I} = \sqrt{I_x^2 + I_y^2} \underline{/\tan^{-1}(I_y/I_x)}$

$\mathbf{I} = 5.5 - j2.5 \text{ A} = \sqrt{5.5^2 + 2.5^2} \underline{/\tan^{-1}(-2.5/5.5)}$

$\mathbf{I} = 6.04 \text{ V} \underline{/-24.4°}$

18-8:1. First get all the phasors into rectangular form:

$\mathbf{V_3} = 35 \text{ V} \underline{/70°} = 12 + j33 \text{ V}$

$\mathbf{V_4} = 24 \text{ V} \underline{/-135°} = -17 - j17 \text{ V}$

(a) $\mathbf{V_1} + \mathbf{V_2} = (24 + j15) + (30 - j42)$

$= 54 - j27 \text{ V}$

(b) $\mathbf{V_1} + \mathbf{V_3} = (24 + j15) + (12 + j33)$
$= 36 + j48$ V

(c) $\mathbf{V_2} + \mathbf{V_3} = (30 - j42) + (12 + j33)$
$= 42 - j9$ V

(d) $\mathbf{V_3} + \mathbf{V_4} = (12 + j33) + (-17 - j17)$
$= -5 + j16$ V

18-9:1. First get all the phasors into rectangular form:
$\mathbf{V_3} = 12 + j33$ V
$\mathbf{V_4} = -17 - j17$ V

(a) $\mathbf{V_1} - \mathbf{V_2} = (24 + j15) - (30 - j42)$
$= -6 + j57$ V

(b) $\mathbf{V_1} - \mathbf{V_3} = (24 + j15) - (12 + j33)$
$= 12 - j18$ V

(c) $\mathbf{V_2} - \mathbf{V_4} = (30 - j42) - (-17 - j17)$
$= 47 - j25$ V

18-10:1. As a first step, express all phasors in both rectangular and polar forms:
$\mathbf{V_1} = 24 + j15$ V $= 28$ V $\underline{/32°}$
$\mathbf{V_2} = 30 - j42$ V $= 52$ V $\underline{/-54°}$
$\mathbf{V_3} = 35$ V $\underline{/70°}$ $= 12 + j33$ V
$\mathbf{V_4} = 24$ V $\underline{/-135°}$ $= -17 - j17$ V

(a) $\mathbf{V_1} \times \mathbf{V_3} = (28$ V $\underline{/32°}) \times (35$ V $\underline{/70°})$
$= 980$ V $\underline{/102°}$

(b) $\mathbf{V_1} \div \mathbf{V_3} = (28$ V $\underline{/32°}) \div (35$ V $\underline{/70°})$
$= 0.8$ V $\underline{/-38°}$

(c) $\mathbf{V_1} \times (\mathbf{V_3} + \mathbf{V_4}) = (28$ V $\underline{/32°}) \times [(12 + j33) + (-17 - j17)]$
$= (28$ V $\underline{/32°}) \times [-5 + j16$ V$]$
$= (28$ V $\underline{/32°}) \times [16.8$ V $\underline{/107°}]$
$= 470$ V^2 $\underline{/139°}$

(d) $\dfrac{\mathbf{V_3} \cdot \mathbf{V_4}}{(\mathbf{V_1} + \mathbf{V_2})} = \dfrac{(35 \text{ V } \underline{/70°}) \cdot (24 \text{ V } \underline{/-135°})}{[(24 + j15) + (30 - j42)]} = \dfrac{840 \text{ V}^2 \underline{/-65°}}{[54 - j27 \text{ V}]}$

$= \dfrac{840 \text{ V}^2 \underline{/-65°}}{60.4 \text{ V } \underline{/-26.6°}} = 13.9$ V $\underline{/-38.4°}$

CHAPTER 19

19-1:1. (a) Using Equation (19-1):
$\mathbf{E} = V_R + jV_L$
$\mathbf{E} = 100 + j66$ V

(b) Using Equation (19-2):
$\mathbf{E} = \sqrt{V_R^2 + V_L^2} \, \underline{/\tan^{-1} (V_L/V_R)}$
$\mathbf{E} = \sqrt{100^2 + 66^2} \, \underline{/\tan^{-1} (66/100)} = \sqrt{14356} \, \underline{/\tan^{-1} (0.66)}$
$\mathbf{E} = 120$ V $\underline{/33°}$

19-2:1. From Equation (19-3):
$$\mathbf{I} = \dfrac{\mathbf{E}}{\mathbf{Z}} = \dfrac{120 \text{ V } \underline{/0°}}{635 + j300 \ \Omega} = \dfrac{120 \text{ V } \underline{/0°}}{702 \ \Omega \ \underline{/25.3°}}$$
$\mathbf{I} = 171$ mA $\underline{/-25.3°}$

19-2:2. (a) Using equation (19-4):
$\mathbf{Z} = R + jX_L = R + j(2\pi fL)$
$= 1500 + j(2\pi \cdot 50 \cdot 1.8) = 1500 + j570 \ \Omega$
$\mathbf{Z} = 1.6$ kΩ $\underline{/21°}$

(b) From Equation (19-3):
$$I = \frac{E}{Z} = \frac{230 \text{ V } \underline{/0°}}{1.6 \text{ k}\Omega \underline{/21°}} = 140 \text{ mA } \underline{/-21°}$$

19-2:3. (a) Using Equation (19-5):
$$Z = \sqrt{R^2 + X_L^2} \underline{/\tan^{-1} (X_L/R)}$$
$$= \sqrt{R^2 + (2\pi f L)^2} \underline{/\tan^{-1} (2\pi f L/R)}$$
$$= \sqrt{160° + (2\pi \cdot 100 \cdot 0.050)^2} \underline{/\tan^{-1} [(2\pi \cdot 100 \cdot 0.050)/160]}$$
$$= \sqrt{160^2 + (31.4)^2} \underline{/\tan^{-1}[(31.4)/160]}$$
$$Z = 163 \ \Omega \ \underline{/11°}$$
From Equation (19-3):
$$E = Z \cdot I = (163 \ \Omega \ \underline{/11°}) \cdot (2.0 \text{ A } \underline{/0°})$$
$$E = 330 \text{ V } \underline{/11°}$$

(b) Similarly;
$$Z = 160 + j314 \ \Omega = 353 \ \Omega \ \underline{/63°}$$
$$E = Z \cdot I = (353 \ \Omega \ \underline{/63°}) \cdot (2.0 \text{ A } \underline{/0°}) = 710 \text{ V } \underline{/63°}$$

19-3:1. $\mathbf{Z}_{\text{coil}} = R_{\text{coil}} + jX_L = R_{\text{coil}} + j(2\pi f L)$
$$= 50 + j(2\pi \cdot 60 \cdot 0.60) = 50 + j226 \ \Omega$$
$$\mathbf{Z}_T = R + \mathbf{Z}_{\text{coil}} = 220 + (50 + j226) = 270 + j226 \ \Omega$$
$$\mathbf{Z}_T = 352 \ \Omega \ \underline{/40°}$$

19-3:2. (a) Using the impedance from the previous problem:
$$I = \frac{E}{Z_T} = \frac{120 \text{ V } \underline{/0°}}{352 \ \Omega \ \underline{/40°}} = 0.34 \text{ A } \underline{/-40°}$$

(b) Since $V_R = I \cdot R = (0.34 \text{ A } \underline{/-40°}) \cdot 220 \ \Omega$
$$V_R = 75 \text{ V } \underline{/-40°}$$
therefore the voltmeter would read 75 V

(c) Since $V_{\text{coil}} = I \cdot Z_{\text{coil}} = (0.34 \text{ A } \underline{/-40°}) \cdot (50 + j226 \ \Omega)$
$$= (0.34 \text{ A } \underline{/-40°}) \cdot (231 \ \Omega \ \underline{/77.5°})$$
$$V_{\text{coil}} = 79 \text{ V } \underline{/37.5°}$$
therefore the voltmeter would read 79 V

19-4:1. (a) From Equation (19-10):
$$Z = R - jX_c = R - j[(2\pi f C)^{-1}]$$
$$= 2200 - j[(2\pi \cdot 400 \cdot 0.047 \cdot 10^{-6})^{-1}] = 2200 - j8466 \ \Omega$$
$$Z = 8.7 \text{ k}\Omega \ \underline{/-75°}$$

(b) Using Ohm's Law:
$$I = \frac{E}{Z} = \frac{50 \text{ V } \underline{/0°}}{8.7 \text{ k}\Omega \ \underline{/-75°}} = 5.7 \text{ mA } \underline{/+75°}$$

19-4:2. (a) Using Ohm's Law:
$$Z = \frac{E}{I} = \frac{50 \text{ V}}{2 \text{ mA}} = 25 \text{ k}\Omega$$
and
$$X_c = (2\pi f C)^{-1} = (2\pi \cdot 400 \cdot 0.047 \cdot 10^{-6})^{-1} = 8466 \ \Omega$$
therefore:
$$R = \sqrt{Z^2 - X_c^2} = \sqrt{25\,000^2 - 8466^2} = 23\,523$$
$$R = 24 \text{ k}\Omega$$

(b) Using Ohm's Law:
$$I = \frac{E}{Z} = \frac{50 \text{ V } \underline{/0°}}{23523 - j8466} = \frac{50 \text{ V } \underline{/0°}}{25 \text{ k}\Omega \ \underline{/-19.8°}} = 2.0 \text{ mA } \underline{/+19.8°}$$

therefore the current will lead the voltage by 20°

19-5:1. From Equation (19-15):

$$\mathbf{Z} = R + j(X_L - X_c) = R + j[2\pi fL - (2\pi fC)^{-1}]$$
$$= 100 + j[(2\pi \cdot 1000 \cdot 0.030) - (2\pi \cdot 1000 \cdot 1.6 \cdot 10^{-6})^{-1}]$$
$$\mathbf{Z} = 100 + j[188.5 - 99.47]$$
$$\mathbf{Z} = 100 + j89 \; \Omega$$

19-5:2. First find the circuit impedance:

$$\mathbf{Z} = \frac{\mathbf{E}}{\mathbf{I}} = \frac{100 \text{ V } \underline{/0°}}{0.8 \text{ A } \underline{/-28°}} = 125 \; \Omega \; \underline{/28°} = 110 + j59 \; \Omega$$

(a) Since $\mathbf{Z} = R + j(X_L - X_c) = 110 + j59$ by inspection: $R = 100 \; \Omega$

(b) Since $X_c = 100 \; \Omega$

and $j(X_L - X_c) = j59 \Rightarrow X_L = 59 + X_c = 159 \; \Omega$

finally

$$L = X_L/2\pi f = 159/(2\pi \cdot 1000)$$
$$L = 25 \text{ mH}$$

19-6:1. From Equation (19-17) and using Ohm's Law:

$$\mathbf{I}_T = I_R + j(I_c - I_L)$$

$$\mathbf{I}_T = \frac{E}{R} + j \cdot \frac{E}{X_c} - j \cdot \frac{E}{X_L}$$

$$\mathbf{I}_T = \frac{120}{R} + j \cdot \frac{120}{(2\pi fC)^{-1}} - j \cdot \frac{120}{2\pi fL}$$

$$\mathbf{I}_T = \frac{120}{1000} + j \cdot \frac{120}{(2\pi \cdot 800 \cdot 1 \cdot 10^{-6})^{-1}} - j \cdot \frac{120}{2\pi \cdot 800 \cdot 0.1}$$

$$\mathbf{I}_T = 0.120 + j0.603 - j0.239 = 120 + j364 \text{ mA}$$
$$\mathbf{I}_T = 383 \text{ mA } \underline{/72°}$$

(b) By Equation (19-6):

$$\mathbf{Z}_{eq} = \frac{\mathbf{E}}{\mathbf{I}_T} = \frac{120 \text{ V } \underline{/0°}}{383 \text{ mA } \underline{/72°}} = 310 \; \Omega \; \underline{/-72°}$$

19-6:2. (a) by Equation (19-17):

$$\mathbf{I}_T = I_R + j(I_c - I_L) = 4 - j3 \text{ A}$$

by inspection: $I_R = 4$ A and: $R = E/I_R = (120 \text{ V})/(4 \text{ A})$
$$R = 30 \; \Omega$$

(b) since $\mathbf{I}_T = I_R + j(I_c - I_L) = 4 - j3$ A

by inspection: $(I_c - I_L) = -3$ A

since $I_c = \frac{E}{X_c} = \frac{E}{(2\pi fC)^{-1}} = \frac{120}{(2\pi \cdot 60 \cdot 42 \cdot 10^{-6})^{-1}}$

$$I_c = 1.9 \text{ A}$$

then $(1.9 - I_L) = -3$

$$I_L = 4.9 \text{ A}$$

finally solving $X_L = E/I_L = 2\pi fL$ for L:

$$\Rightarrow L = \frac{E}{I_L} \cdot \frac{1}{2\pi f} = \frac{120}{4.9} \cdot \frac{1}{2\pi \cdot 60} = \frac{24.5}{377}$$

$$L = 65 \text{ mH}$$

19-7:1. (a) Using Equation (19-23):

$$B_c = 2\pi fC = \omega C = (377) \cdot (2.2 \cdot 10^{-6}) = 830 \; \mu S$$

(b) Using Equation (19-22):

$$B_L = (2\pi fL)^{-1} = (\omega L)^{-1} = [(377) \cdot (2.0)]^{-1} = 1.3 \text{ mS}$$

(c) Using Equation (19-20):

$$\mathbf{Y} = G + j(B_c - B_L) = 0 + j(830 - 1300) \; \mu S$$
$$\mathbf{Y} = -j470 \; \mu S$$

19-7:2. (a) From Equation (19-20) and using the susceptances from the previous problem:
$$Y = G + j(B_c - B_L) = 1/3300 + j[(830 - 1300) \cdot 10^{-6}] \text{ S}$$
$$= 300 - j470 \ \mu\text{S}$$
$$Y = 560 \ \mu\text{S} \ \underline{/-57°}$$
(b) From Ohm's Law in admittance form:
$$I = E/Z = E \cdot Y = (120 \text{ V} \ \underline{/0°}) \cdot (560 \ \mu\text{S} \ \underline{/-57°})$$
$$= 68 \text{ mA} \ \underline{/-57°}$$

19-7:3. For the original circuit:
$$Y_T = \frac{1}{Z} = \frac{I_T}{E} = \frac{20 \text{ mA} \ \underline{/-20°}}{90 \text{ V} \ \underline{/0°}} = 222 \ \mu\text{S} \ \underline{/-20°}$$
$$Y_T = 209 - j76 \ \mu\text{S}$$
Using Equation (19-20):
$$Y_T = 209 - j76 \ \mu\text{S} = G + j(B_c - B_L)$$
by inspection: $G = 1/R = 209 \ \mu\text{S}$
$$\Rightarrow R = (209 \ \mu\text{S})^{-1} = 4.8 \text{ k}\Omega$$
for the circuit *without* the inductor:
$$Y = \frac{1}{Z} = \frac{I}{E} = \frac{25 \text{ mA}}{90 \text{ V}} = 278 \ \mu\text{S}$$
for this circuit: $Y = \sqrt{G^2 + B_c^2}$
$$\Rightarrow B_c = \sqrt{Y^2 - G^2} = \sqrt{278^2 - 209^2}$$
$$B_c = 183 \ \mu\text{S}$$
returning to the *original* circuit:
$$j(B_c - B_L) = -j76 \ \mu\text{S}$$
$$B_L = B_c + 76 = 183 + 76$$
$$B_L = 262 \ \mu\text{S}$$

19-8:1. $Z_{coil} = R + jX_L = R + j(\omega L) = 100 + j377 \ \Omega$
$$Z_{coil} = 390 \ \Omega \ \underline{/75°}$$
$$Y_{coil} = \frac{1}{Z_{coil}} = \frac{1}{390 \ \Omega \ \underline{/75°}} = 2.6 \text{ mS} \ \underline{/-75°}$$
$$Y_{coil} = G - jB_L = 0.66 - j2.5 \text{ mS}$$
by inspection: $G = 0.66 \text{ mS}$
$$B_L = 2.5 \text{ mS}$$

19-8:2. Find the impedance of the series equivalent:
$$Z_{cap} = \frac{1}{Y_{cap}} = \frac{1}{10 + j300 \ \mu\text{S}} = \frac{1}{300 \ \mu\text{S} \ \underline{/88°}} = 111 - j3330 \ \Omega$$
therefore: $R = 110 \ \Omega$
$$X = 3300 \ \Omega$$

CHAPTER 20

20-1:1. From Equation (20-3):
$$P = V_R^2/R = (120 \text{ V})^2/(360 \ \Omega) = 40 \text{ W}$$

20-1:2. (a) Using the average power from the previous problem and since $P_m = 2 \cdot P$:
$$P_m = 2 \cdot (40 \text{ W}) = 80 \text{ W}$$
(b) since $p = P_m \cdot (\sin^2 \omega t)$
$$\Rightarrow \text{minimum } p = P_m \cdot (0)^2 = 0 \text{ W}$$
This occurs twice per voltage cycle (see Figure 20-1).

20-2:1. (a) From Equation (20-4):
$$p = -0.5 \cdot I_m \cdot V_m \cdot \sin(2\omega t) \text{ W}$$
and since $V_m = V_L \cdot \sqrt{2}$ and $I_m = V_m/X_L$ and $X_L = \omega \cdot L$:

$$p = -0.5 \cdot \left(\frac{V_L \cdot \sqrt{2}}{\omega \cdot L}\right) \cdot (V_L \cdot \sqrt{2}) \cdot \sin(2\omega t)$$

$$p = \frac{-(0.5) \cdot (120 \cdot \sqrt{2})^2}{377 \cdot 1} \cdot \sin(2 \cdot 377 \cdot t)$$

$$p = -38 \cdot \sin 754t \text{ W}$$

(b) average P is average of $38 \cdot \sin(754t) = 38 \cdot 0 = 0$

20-2:2. Using Equation (20-6):
$$Q = \frac{V_L^2}{X_L} = \frac{V_L^2}{\omega \cdot L} = \frac{120^2}{377 \cdot 1} = 38.196 = 38 \text{ vars}$$

20-3:1. (a) Since $V_m = V_C \cdot \sqrt{2}$ and $I_m = I_c \cdot \sqrt{2}$
Using Equation (20-7):
$$p = 0.5 \cdot I_m \cdot V_m \cdot \sin(2\omega t) = I_c \cdot V_c \cdot \sin(2\omega t)$$
$$= (0.1 \text{ A}) \cdot (120 \text{ V}) \cdot \sin(2 \cdot 377 \cdot t)$$
$$p = 12 \cdot \sin 754t \text{ W}$$

(b) average real power is zero watts (see Figure 20-4)

20-3:2. Using Equation (20-8):
$$Q = I_c \cdot V_c = (0.1 \text{ A}) \cdot (120 \text{ V}) = 12 \text{ vars}$$

20-4:1. Using Equation (20-11):
$$S = V_T \cdot I_T = (15 \text{ V}) \cdot (2 \text{ A}) = 30 \text{ VA}$$

20-4:2. (a) from Equation (20-12):
$$S = \frac{V_T^2}{Z} = \frac{(120 \text{ V})^2}{\sqrt{300^2 + 400^2}} = \frac{(120 \text{ V})^2}{500 \ \Omega} = 28.8 \text{ VA}$$

(b) first determine the magnitude of the current in the series circuit:
$$I = V_T/Z = (120 \text{ V})/(500 \ \Omega) = 0.24 \text{ A}$$
then determine the power in the resistor
$$P = I^2 \cdot R = (0.24 \text{ A})^2 \cdot (400 \ \Omega) = 23.0 \text{ W}$$

(c) Using the reactance and the series current:
$$Q_L = I^2 \cdot X_L = 0.24^2 \cdot 300 = 17.3 \text{ vars}$$

20-5:1. Since this circuit has capacitive reactance and by Equation (20-14):
$$S = P + jQ_c = \frac{V_T^2}{R} + j \cdot \frac{V_T^2}{X_c} = \frac{(60 \text{ V})^2}{1200 \ \Omega} + j \cdot \frac{(60 \text{ V})^2}{500 \ \Omega}$$

$$S = 3 + j7.2 \text{ VA}$$

(NOTE: Since this is a *parallel* circuit, the analysis uses the *voltage* which is common to the devices)

20-5:2. First, since this is a *series* circuit, determine the magnitude of the *current* common to all circuit elements:
$$I = \frac{E}{Z} = \frac{120 \text{ V}}{\sqrt{500^2 + (600 - 400)^2}} = \frac{120 \text{ V}}{539 \ \Omega} = 0.223 \text{ A}$$

From Equation (20-14):
$$S = P + jQ_c - jQ_L = I^2 \cdot R + j \cdot I^2 \cdot (X_c - X_L)$$
$$S = (0.223 \text{ A})^2 \cdot (500 \ \Omega) + j \cdot (0.223 \text{ A})^2 \cdot (400 - 600 \ \Omega)$$
$$S = 24.8 - j9.93 \text{ VA}$$

20-6:1. First find the real and apparent power of the circuit using the common voltage:
$$Q_c = \frac{V_c^2}{X_c} = \frac{(120 \text{ V})^2}{500} = 28.8 \text{ vars (capacitive)}$$
$$P = \frac{V_c^2}{R} = \frac{(120 \text{ V})^2}{1200} = 12.0 \text{ W}$$
$$S = \sqrt{P^2 + Q^2} = \sqrt{12^2 + 28.8^2} = 31.2 \text{ VA}$$

Power factor $= \cos \phi = \dfrac{P}{S} = \dfrac{12}{31.2} = 0.3846$ (capacitive)

Power factor $= \cos \phi = 38\%$ leading.

20-6:2. Since this is a series circuit, work with the common current:

$$\cos \phi = \frac{P}{S} = \frac{I^2 \cdot R}{I^2 \cdot Z} = \frac{R}{Z} = \frac{500}{\sqrt{500^2 + (600 - 400)^2}} = \frac{500}{539}$$

$\cos \phi = 0.93$ (inductive)

$\cos \phi = 93\%$ lagging.

20-7:1. First find the Q of the original circuit:

$$Q = \sqrt{S^2 - P^2} = \sqrt{240^2 - 120^2} = 208 \text{ vars inductive}$$

Therefore we need to add a *capacitor in parallel* with the source that will have a $Q = 208$ vars capacitive:

$$Q_c = \frac{V^2}{X_c} \qquad 208 = \frac{(120 \text{ V})^2}{[2\pi \cdot f \cdot C]^{-1}} = (120 \text{ V})^2 \cdot (377 \text{ rad/s}) \cdot C$$

$$\Rightarrow C = \frac{208}{120^2 \cdot 377} = 38 \ \mu\text{F}$$

therefore

add 38 μF in parallel with the source

20-7:2. From the solution to Self-Check Problem 20-6:1 for the original circuit:

$$S_{\text{orig}} = 12 + j28.8 \text{ VA}$$

for the modified circuit:

$$S_{\text{mod}} = \frac{P}{\cos \phi} = \frac{12.0}{0.95} = 12.63 \text{ VA}$$

since the power factor is lagging:

$$Q_{\text{mod}} = \sqrt{S^2 - P^2} = \sqrt{12.63^2 - 12^2} = 3.944 \text{ var (inductive)}$$

therefore the additional component must be an inductor with the following reactive power:

$Q_{\text{add}} = Q_{\text{mod}} - Q_{\text{orig}} = -3.994 - 28.8 = -32.744$ vars

$Q_{\text{add}} = 32.744$ vars inductive

since

$$Q_{\text{add}} = \frac{V^2}{X_L} = \frac{V^2}{\omega \cdot L}$$

$$\Rightarrow L = \frac{V^2}{\omega \cdot Q_{\text{add}}} = \frac{(120 \text{ V})^2}{(377 \text{ rad/s}) \cdot (32.744 \text{ var})} = 1.167$$

$L = 1.2$ H

20-9:1. From Equation (20-19):

$$R = \frac{P}{I^2} = \frac{S \cdot \cos \phi}{I^2} = \frac{E \cdot I \cdot \cos \phi}{I^2}$$

$$R = \frac{(120 \text{ V}) \cdot (0.70)}{(10 \text{ A})} = 8.4 \ \Omega$$

CHAPTER 21

21-2:1. Using Equation (21-2):

$$\begin{aligned}
\mathbf{Z}_T &= \mathbf{Z}_1 + \mathbf{Z}_2 + \mathbf{Z}_3 \\
&= (250 + j500 \ \Omega) + (400 \ \Omega \ \underline{/-40^\circ}) + (-j100 \ \Omega) \\
&= (250 + j500 \ \Omega) + (306 - j257) + (-j100 \ \Omega) \\
&= 556 + j143 \ \Omega = 574 \ \Omega \ \underline{/14.4^\circ}
\end{aligned}$$

21-2:2. Using Equation (21-3):
$$\mathbf{V}_1 = \mathbf{E} \cdot \frac{\mathbf{Z}_1}{\mathbf{Z}_T} = (60 \text{ V } \underline{/0°}) \cdot \frac{(250 + \text{j}500 \ \Omega)}{(574 \ \Omega \ \underline{/14.4°})} = 58.4 \text{ V } \underline{/49.0°}$$
similarly:
$$\mathbf{V}_2 = \mathbf{E} \cdot \frac{\mathbf{Z}_2}{\mathbf{Z}_T} = (60 \text{ V } \underline{/0°}) \cdot \frac{(400 \ \Omega \ \underline{/-40°})}{(574 \ \Omega \ \underline{/14.4°})} = 41.8 \text{ V } \underline{/-54.4°}$$
$$\mathbf{V}_3 = 10.4 \text{ V } \underline{/-104.4°}$$

21-3:1. When dealing with parallel branches, first determine the admittance of each branch:
$$\mathbf{Y}_1 = \mathbf{Z}_1^{-1} = (400 \ \Omega)^{-1} = 2.5 \text{ mS}$$
$$\mathbf{Y}_2 = \mathbf{Z}_2^{-1} = (-\text{j}250 \ \Omega)^{-1} = +\text{j}4.0 \text{ mS}$$
$$\mathbf{Y}_3 = \mathbf{Z}_3^{-1} = (400 + \text{j}500 \ \Omega)^{-1} = 0.9756 - \text{j}1.220 \text{ mS}$$
Using Equation (21-5):
$$\mathbf{Y}_T = \mathbf{Y}_1 + \mathbf{Y}_1 + \mathbf{Y}_1$$
$$= (2.5 \text{ mS}) + (\text{j}4.0 \text{ mS}) + (0.9756 - \text{j}1.220 \text{ mS})$$
$$\mathbf{Y}_T = 3.4756 + \text{j}2.78 \text{ mS} = 4.451 \text{ mS} \ \underline{/38.7°}$$

21-3:2. Use the admittance form of Ohm's Law:
$$\mathbf{I}_T = \mathbf{E} \cdot \mathbf{Y}_T = (60 \text{ V } \underline{/0°}) \cdot (4.451 \text{ mS } \underline{/38.7°})$$
$$\mathbf{I}_T = 267.1 \text{ mA } \underline{/38.7°}$$
Using Equation (21-10):
$$\mathbf{I}_1 = \mathbf{I}_T \cdot \frac{\mathbf{Y}_1}{\mathbf{Y}_T} = (267.1 \text{ mA } \underline{/38.7°}) \cdot \frac{(2.5 \text{ mS})}{(4.451 \text{ mS } \underline{/38.7°})}$$
$$\mathbf{I}_1 = 150 \text{ mA}$$
Similarly:
$$\mathbf{I}_2 = \text{j}240 \text{ mA}$$
$$\mathbf{I}_3 = 93.7 \text{ mA } \underline{/-51.3°}$$

21-4:1. (a) First determine the equivalent impedance for the parallel combination of $\mathbf{Z}_2$ and $\mathbf{Z}_3$ using Equation (21-8):
$$\mathbf{Z}_{23} = \frac{\mathbf{Z}_2 \cdot \mathbf{Z}_3}{\mathbf{Z}_2 + \mathbf{Z}_3} = \frac{(-\text{j}500 \ \Omega) \cdot (400 + \text{j}500 \ \Omega)}{(-\text{j}500 \ \Omega) + (400 + \text{j}500 \ \Omega)}$$
$$\mathbf{Z}_{23} = 625 - \text{j}500 \ \Omega$$
Now, using the series rule:
$$\mathbf{Z}_T = \mathbf{Z}_1 + \mathbf{Z}_{23} = (400 - \text{j}500 \ \Omega) + (625 - \text{j}500 \ \Omega)$$
$$\mathbf{Z}_T = 1025 - \text{j}1000 \ \Omega$$
(b) Using the voltage-divider principle:
$$\mathbf{V}_1 = \mathbf{E} \cdot \frac{\mathbf{Z}_1}{\mathbf{Z}_T} = (100 \text{ V } \underline{/0°}) \cdot \frac{(400 - \text{j}500 \ \Omega)}{(1025 - \text{j}1000 \ \Omega)}$$
$$\mathbf{V}_1 = 44.7 \text{ V } \underline{/-7°}$$

21-4:2. (a) First use the series rule to combine $\mathbf{Z}_2$ and $\mathbf{Z}_3$:
$$\mathbf{Z}_{23} = \mathbf{Z}_2 + \mathbf{Z}_3 = (-\text{j}500 \ \Omega) + (400 + \text{j}500 \ \Omega) = 400 \ \Omega$$
Now use Equation (21-8) on the parallel combination:
$$\mathbf{Z}_T = \frac{\mathbf{Z}_1 \cdot \mathbf{Z}_{23}}{\mathbf{Z}_1 + \mathbf{Z}_{23}} = \frac{(400 - \text{j}500 \ \Omega) \cdot (400 \ \Omega)}{(400 - \text{j}500 \ \Omega) + (400 \ \Omega)}$$
$$\mathbf{Z}_T = 256 - \text{j}89.9 \ \Omega$$
(b) Using the current-divider principle from Equation (21-11):
$$\mathbf{I}_1 = \mathbf{I}_T \cdot \frac{\mathbf{Z}_{23}}{\mathbf{Z}_1 + \mathbf{Z}_{23}}$$
$$\mathbf{I}_1 = (100 \text{ mA } \underline{/0°}) \cdot \frac{(400 \ \Omega)}{(400 - \text{j}500 \ \Omega) + (400 \ \Omega)}$$
$$\mathbf{I}_1 = 42 \text{ mA } \underline{/32°}$$

21-5:1. First determine the 400-Hz series reactance:
$$X_S = 2\pi f L_S = 2\pi \cdot (400 \text{ Hz}) \cdot (0.120 \text{ H}) = 302 \text{ }\Omega$$
Using Equation (21-19):
$$R_P = \frac{R_S^2 + X_S^2}{R_S} = \frac{50^2 + 302^2}{50} = 1869$$
$$R_P = 1.9 \text{ k}\Omega$$
Using Equation (21-20):
$$X_P = \frac{R_S^2 + X_S^2}{X_S} = \frac{50^2 + 302^2}{302} = 309.88$$
$$X_P = 310 \text{ }\Omega$$

21-5:2. First determine the parallel equivalent resistance and reactance:
$$Y_T = Y_1 + Y_2 = 16.7 \text{ mS } \underline{/-60°} + 12.5 \text{ mS } \underline{/45°}$$
$$Y_T = 17.19 - j5.624 \text{ mS}$$
$$\Rightarrow R_P = (17.19 \text{ mS})^{-1} = 58.18 \text{ }\Omega$$
$$X_P = (5.624 \text{ mS})^{-1} = 177.8 \text{ }\Omega \text{ (inductive)}$$
Using Equation (21-21):
$$R_S = \frac{R_P \cdot X_P^2}{R_P^2 + X_P^2} = \frac{(58.18) \cdot (177.8)^2}{58.18^2 + 177.8^2} = 52.55$$
$$R_S = 53 \text{ }\Omega$$
Using Equation (21-22):
$$X_S = \frac{R_P^2 \cdot X_P}{R_P^2 + X_P^2} = \frac{(58.18)^2 \cdot (177.8)}{58.18^2 + 177.8^2} = 17.20$$
$$X_S = 17 \text{ }\Omega$$

21-6:1. The equivalent constant-current source will have a current generator in parallel with the Z_{int}.

The current source is found by using Equation (21-28):
$$I_{SC} = \frac{E}{Z_{int}} = \frac{150 \text{ V } \underline{/0°}}{300 + j400 \text{ }\Omega} = 300 \text{ mA } \underline{/-53°}$$

21-7:1. For maximum power transfer:
$$Z_L = \text{conjugate } Z_{int}$$
$$Z_L = 30 - j40 \text{ }\Omega$$
therefore the load will be the series combination of a resistor and a capacitor with the following values:
$$R = 30 \text{ }\Omega$$
and
$$C = (\omega X_c)^{-1} = [2\pi \cdot (400 \text{ Hz}) \cdot (40)]^{-1} = 9.9 \text{ }\mu\text{F}$$

21-7:2. For maximum power transfer:
$$Z_L = \text{conjugate } Z_{int} = 100 \text{ }\Omega \text{ } \underline{/+25°}$$
current in the maximum-power circuit is:
$$I_L = \frac{E}{Z_{int} + Z_L} = \frac{120 \text{ V } \underline{/0°}}{(100 \text{ }\Omega \text{ } \underline{/-25°}) + (100 \text{ }\Omega \text{ } \underline{/+25°})}$$
$$I_L = \frac{120 \text{ V } \underline{/0°}}{181 \text{ }\Omega \text{ } \underline{/0°}} = 663 \text{ mA}$$
Using the current to find the load power:
$$P_L = I^2 \cdot R_L = (0.663)^2 \cdot (90.6 \text{ }\Omega) = 39.8 \text{ W}$$

CHAPTER 22

22-1:1. (a) $12 \cdot (I_1 + I_2) + (-j5) \cdot (I_1 + I_2) + (+j10) \cdot (I_1) = 25 + j0$
$12 \cdot (I_1 + I_2) + (-j5) \cdot (I_1 + I_2) + (-j15) \cdot (I_2) + (18) \cdot (I_2) = 25 - j15$

(b) Using determinants:

$$\mathbf{I_1} = \frac{\begin{vmatrix} (25 + j0) & (12 - j5) \\ (25 - j15) & (30 - j20) \end{vmatrix}}{\begin{vmatrix} (12 + j5) & (12 - j5) \\ (12 - j5) & (30 - j20) \end{vmatrix}} = 1.636 \text{ A } \underline{/-25.4°}$$

$$\mathbf{I_2} = \frac{\begin{vmatrix} (12 + j5) & (25 + j0) \\ (12 - j5) & (25 - j15) \end{vmatrix}}{D} = 0.300 \text{ A } \underline{/+38.0°}$$

(c) inductor current = $\mathbf{I_1}$ = 1.636 A $\underline{/-25.4°}$ down
    source 1 current = $\mathbf{I_1} + \mathbf{I_2}$ = 1.790 A $\underline{/-16.8°}$ R to L
    source 2 current = $\mathbf{I_2}$ = 0.300 A $\underline{/+38.0°}$ R to L

22-2:1. (a)

$$+(12 + j5) \cdot (\mathbf{I_1}) - (0 + j10) \cdot (\mathbf{I_2}) = 25 + j0$$
$$-(0 + j10) \cdot (\mathbf{I_1}) + (18 - j5) \cdot (\mathbf{I_2}) = 0 - j15$$

(b) Using determinants:

$$\mathbf{I_1} = \frac{\begin{vmatrix} (25 + j0) & (0 - j10) \\ (0 - j15) & (18 - j5) \end{vmatrix}}{\begin{vmatrix} (12 + j5) & (0 - j10) \\ (0 - j10) & (18 - j5) \end{vmatrix}} = 1.790 \text{ A } \underline{/-16.8°}$$

$$\mathbf{I_2} = \frac{\begin{vmatrix} (12 + j5) & (25 + j0) \\ (0 - j10) & (0 - j15) \end{vmatrix}}{D} = 0.300 \text{ A } \underline{/+38.0°}$$

(c) inductor current = $\mathbf{I_1} - \mathbf{I_2}$ = 1.636 A $\underline{/-25.4°}$ down
    source 1 current = $\mathbf{I_1}$ = 1.790 A $\underline{/-16.8°}$ R to L
    source 2 current = $\mathbf{I_2}$ = 0.300 A $\underline{/+38.0°}$ R to L

22-2:2. (a) Writing the mesh equations directly in the matrix format:

$$+(20 + j10) \cdot \mathbf{I_1} - (+j10) \cdot \mathbf{I_2} - (0) \cdot \mathbf{I_3} = 40 + j0$$
$$-(+j10) \cdot \mathbf{I_1} + (+j10) \cdot \mathbf{I_2} - (+j15) \cdot \mathbf{I_3} = 0 + j0$$
$$-(0) \cdot \mathbf{I_1} - (+j15) \cdot \mathbf{I_2} + (30 + j15) \cdot \mathbf{I_3} = 0 - j25$$

(b)

$$\mathbf{I_1} = \frac{\begin{vmatrix} (+40) & (-j10) & (0) \\ (0) & (+j10) & (-j15) \\ (-j25) & (-j15) & (30 + j15) \end{vmatrix}}{\begin{vmatrix} (20 + j10) & (-j10) & (0) \\ (-j10) & (+j10) & (-j15) \\ (0) & (-j15) & (30 + j15) \end{vmatrix}} = 1.912 \text{ A } \underline{/-0.5°}$$

$$\mathbf{I_2} = \frac{\begin{vmatrix} (20 + j10) & (+40) & (0) \\ (-j10) & (0) & (-j15) \\ (0) & (-j25) & (30 + j15) \end{vmatrix}}{D} = 1.887 \text{ A } \underline{/+4.9°}$$

$$I_3 = \frac{\begin{vmatrix} (20 + j10) & (-j10) & (+40) \\ (-j10) & (+j10) & (0) \\ (0) & (-j15) & (-j25) \end{vmatrix}}{D} = 0.119 \text{ A } \underline{/100.3^0}$$

(c) Using Kirchhoff's Voltage Law to find the voltages:

$V_A = E_1 - I_1 \cdot (20 \ \Omega) = (40 \text{ V}) - (1.912 \text{ A } \underline{/-0.5°}) \cdot (20 \ \Omega)$

$V_A = 1.789 \text{ V } \underline{/10.3°}$

$V_B = E_2 + I_3 \cdot (30 \ \Omega)$

$\quad = (j25 \text{ V}) + (0.119 \text{ A } \underline{/100.3°}) \cdot (30 \ \Omega)$

$V_B = 28.53 \text{ V } \underline{/91.3°}$

22-3:1. (a)

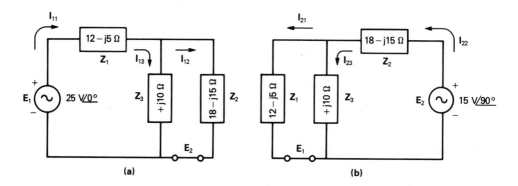

(a)   (b)

(b) Solving for the component currents in the circuit shown in Figure 22-37 (a):

$Z_T = \dfrac{Z_3 \cdot Z_2}{Z_3 + Z_2} + Z_1 = 18.32 \ \Omega \ \underline{/20.6°}$

$I_{11} = \dfrac{E_1}{Z_T} = \dfrac{25 \text{ V } \underline{/0°}}{18.32 \ \Omega \ \underline{/20.6°}} = 1.36 \text{ A } \underline{/-20.6°}$

$I_{13} = I_{11} \cdot \dfrac{Z_2}{Z_3 + Z_2} = (1.36 \text{ A } \underline{/-20.6°}) \cdot \dfrac{18 - j15 \ \Omega}{18 - j5 \ \Omega}$

$I_{13} = 1.71 \text{ A } \underline{/-44.8°}$

$I_{12} = I_{11} \cdot \dfrac{Z_3}{Z_3 + Z_2} = (1.36 \text{ A } \underline{/-20.6°}) \cdot \dfrac{j10 \ \Omega}{18 - j5 \ \Omega}$

$I_{12} = 0.73 \text{ A } \underline{/+85.0°}$

Similarly for the circuit shown in the figure from Self-Check Answer 22-3:1:

$Z_T = \dfrac{Z_3 \cdot Z_1}{Z_3 + Z_1} + Z_2 = 26.33 \ \Omega \ \underline{/-17.6°}$

$I_{22} = \dfrac{E_2}{Z_T} = \dfrac{15 \text{ V } \underline{/90°}}{26.33 \ \Omega \ \underline{/-17.6°}} = 0.57 \text{ A } \underline{/107.6°}$

$I_{23} = I_{22} \cdot \dfrac{Z_1}{Z_3 + Z_1} = 0.57 \text{ A } \underline{/+62.4°}$

$I_{21} = I_{22} \cdot \dfrac{Z_3}{Z_3 + Z_1} = 0.44 \text{ A } \underline{/175.0°}$

(c) inductor current $= I_{13} + I_{23} = 1.636 \text{ A } \underline{/-25.4°}$ down

source 1 current $= I_{11} - I_{21} = 1.790 \text{ A } \underline{/-16.8°}$ R to L

source 2 current $= I_{12} - I_{22} = 0.300 \text{ A } \underline{/+38.0°}$ R to L

22-4:1. (a) Applying Thévenin's theorem to the circuit shown in part a of the figure below:

$$E_{TH} = E_2 \cdot \frac{Z_3}{Z_3 + Z_2} = (j15 \text{ V}) \cdot \frac{j10 \text{ } \Omega}{18 - j5 \text{ } \Omega}$$

$$E_{TH} = 8.03 \text{ V } \underline{/-164.5°}$$

$$Z_{TH} = Z_1 + \frac{Z_3 \cdot Z_2}{Z_3 + Z_2} = 18.32 \text{ } \Omega \underline{/20.6°}$$

Using this information, the Thévenin equivalent circuit is shown in part b of the figure below.

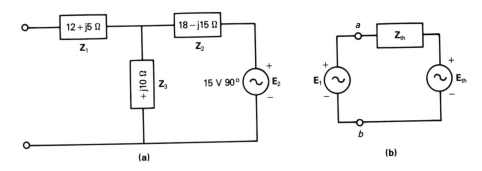

(a)   (b)

(b) Using figure (b) to solve for the source current:

$$I_1 = \frac{E_1 - E_{TH}}{Z_{TH}} = \frac{(25 \text{ V}) - (8.03 \text{ V } \underline{/-164.5°})}{18.32 \text{ } \Omega \underline{/20.6°}}$$

$$I_1 = 1.79 \text{ A } \underline{/-16.8°}$$

22-5:1. (a) Converting the sources:

$$I_1 = \frac{E_1}{Z_1} = \frac{25 \text{ V}}{12 - j5 \text{ } \Omega} = 1.93 \text{ A } \underline{/22.6°}$$

$$I_2 = \frac{E_2}{Z_2} = \frac{j15 \text{ V}}{18 - j15 \text{ } \Omega} = 0.64 \text{ A } \underline{/129.8°}$$

Using this information, the Norton-equivalent circuit is shown in the figure below:

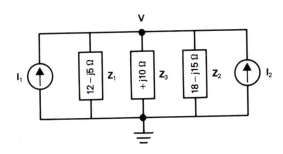

(b) Using the Norton-equivalent circuit:

$$I_L = (I_1 + I_2) \cdot \frac{Y_3}{Y_1 + Y_2 + Y_3}; \text{ where } Y_n = Z_n^{-1}$$

$$I_L = (1.365 + j1.231 \text{ A}) \cdot \frac{-j0.1 \text{ S}}{0.104 - j0.043 \text{ S}}$$

$$I_L = 1.64 \text{ A } \underline{/-25.4°}$$

22-6:1. (a) Using the Norton-equivalent circuit shown in the figure for Self-Check Answer 22-5:1., the nodal equation is written by inspection:

$$\left[\frac{1}{Z_1} + \frac{1}{Z_3} + \frac{1}{Z_2}\right] \cdot V = I_1 + I_2$$

$$\left[\frac{1}{12 - j5} + \frac{1}{j10} + \frac{1}{18 - j15}\right] \cdot V = 1.365 + j1.231$$

(b) Solving the nodal equation gives:

$$V = \frac{1.365 + j1.231 \text{ A}}{0.104 - j0.043 \text{ S}} = 16.36 \text{ V } \underline{/64.6°}$$

$$I_L = \frac{V}{Z_L} = \frac{16.36 \text{ V } \underline{/64.6°}}{j10 \text{ }\Omega} = 1.636 \text{ A } \underline{/-25.4°}$$

22-6:2. (a) Using Norton's theorem to convert the sources:

$$I_1 = \frac{E_1}{R_1} = \frac{40 \text{ V } \underline{/0°}}{20 \text{ }\Omega} = 2.0 \text{ A } \underline{/0°}$$

$$I_2 = \frac{E_2}{R_2} = \frac{25 \text{ V } \underline{/90°}}{30 \text{ }\Omega} = 0.833 \text{ A } \underline{/90°}$$

The equivalent circuit containing current sources is shown in the figure below.

(b) Using the equivalent circuit in the above figure, nodal equations are written directly into matrix format:

$$\left[\frac{1}{20} + \frac{1}{j10} + \frac{1}{-j15}\right] \cdot V_A - \left[\frac{1}{-j15}\right] \cdot V_B = 2 + j0$$

$$-\left[\frac{1}{-j15}\right] \cdot V_A + \left[\frac{1}{j15} + \frac{1}{30} + \frac{1}{-j15}\right] \cdot V_B = 0 + j0.833$$

(c)

$$V_A = \frac{\begin{vmatrix} (2 + j0) & (0 - j0.066) \\ (0 + j0.833) & (0.033 + j0) \end{vmatrix}}{\begin{vmatrix} (0.050 - j0.033) & (0 - j0.066) \\ (0 - j0.066) & (0.033 + j0) \end{vmatrix}} = 1.789 \text{ V } \underline{/10.3°}$$

$$V_B = \frac{\begin{vmatrix} (0.050 - j0.033) & (2 + j0) \\ (0 - j0.066) & (0 + j0.833) \end{vmatrix}}{D} = 28.53 \text{ V } \underline{/91.3°}$$

$$I_C = \frac{V_A - V_B}{Z_c} = \frac{(1.789 \text{ V } \underline{/10.3°}) - (28.53 \text{ V } \underline{/91.3°})}{-j15 \text{ }\Omega}$$

$$I_C = 1.89 \text{ A } \underline{/4.9°}$$

22-7:1. First determine the numerator for the transformation Equations (22-9, -10, and -11):

$N = Z_A \cdot Z_B + Z_B \cdot Z_C + Z_C \cdot Z_A = 5700 + j600$

Solving the transformation equations:

$$Z_X = \frac{N}{Z_A} = 78 - j84 \ \Omega$$

$$Z_Y = \frac{N}{Z_B} = 72.857 + j90 \ \Omega$$

$$Z_Z = \frac{N}{Z_C} = 114 + j12 \ \Omega$$

22-8:1. Using Equations (22-16 and -17):

$$R_X = \frac{R_2 \cdot R_3}{R_1} = \frac{(6.6 \ \Omega) \cdot (1.0 \ k\Omega)}{33 \ k\Omega} = 0.2 \ \Omega$$

$$C_X = \frac{C_S \cdot R_1}{R_3} = \frac{(0.1 \ \mu F) \cdot (33 \ k\Omega)}{1.0 \ k\Omega} = 3.3 \ \mu F$$

CHAPTER 23

23-1:1. Setting the reactances equal:

$$X_C = \frac{1}{\omega C} = X_L = \omega L$$

$$\Rightarrow \omega^2 = \frac{1}{LC} = (2\pi f_r)^2 = \frac{1}{(20 \ mH) \cdot (0.033 \ \mu F)} = 1.515 \times 10^9$$

$$f_r = \frac{1}{2\pi} \cdot \sqrt{1.515 \times 10^9} = 6195$$

$$f_r = 6.2 \ kHz$$

23-2:1. (a) Using Equation (23-1):

$$f_r = \frac{1}{2\pi\sqrt{LC}}$$

It is seen that changing R has no effect on f_r.

(b) From Equation (23-1) and letting $L_2 = 10 \cdot L_1$:

$$\frac{f_{r2}}{f_{r1}} = \frac{2\pi\sqrt{L_1 C_1}}{2\pi\sqrt{L_2 C_2}} = \frac{\sqrt{L_1}}{\sqrt{L_2}} = \frac{1}{\sqrt{10}} = 0.3162$$

Therefore,

the resonant frequency drops to 31% of its former value.

23-2:2. From Equation (23-1):

$$L = \frac{1}{(2\pi f)^2 \cdot C} = \frac{1}{[2\pi \cdot (20 \times 10^3)]^2 \cdot (100 \times 10^{-12})} = 0.6333$$

$$L = 0.63 \ H$$

23-3:1. (a) Using Equation (23-4):

$$Q = \frac{1}{R} \cdot \sqrt{\frac{L}{C}} = \frac{1}{50 \ \Omega} \cdot \sqrt{\frac{0.1 \ H}{0.1 \ \mu F}} = \frac{1000}{50}$$

$$Q = 20$$

(b) Similarly:

$$Q = 10$$

23-3:2. Using Equation (23-3):

$$Q = \frac{X_L}{R} = \frac{1583 \ \Omega}{15 \ \Omega} = 105.53$$

$$Q = 110$$

23-4:1. Using Equation (23-5) and the resonant Q from Self-Check Problem 23-3:1:

(a) $V_C = Q \cdot E = (20) \cdot (100) = 2.0 \text{ kV}$

(b) $V_C = Q \cdot E = (10) \cdot (100) = 1.0 \text{ kV}$

23-4:2. Using Ohm's Law at resonance to determine the common current:

$$I_r = \frac{E}{R} = \frac{100 \text{ V } \underline{/0°}}{100 \text{ }\Omega} = 1.0 \text{ A } \underline{/0°}$$

Using the common current and Ohm's Law across each device:

$$V_R = I \cdot R = (1 \text{ A } \underline{/0°}) \cdot (100 \text{ }\Omega) = 100 \text{ V } \underline{/0°}$$

$$V_L = I \cdot X_L = (1 \text{ A } \underline{/0°}) \cdot (+j2000 \text{ }\Omega) = 2000 \text{ V } \underline{/+90°}$$

$$V_C = I \cdot X_C = (1 \text{ A } \underline{/0°}) \cdot (-j2000 \text{ }\Omega) = 2000 \text{ V } \underline{/-90°}$$

23-5:1. (a) Using Equation (23-7):

$$B = \frac{R}{2\pi L} = \frac{50 \text{ }\Omega}{2\pi \cdot (0.1 \text{ H})} = 79.577$$

$$B = 80 \text{ Hz}$$

(b) Similarly:

$$B = 160 \text{ Hz}$$

23-5:2. Using Equation (23-8) and Q from Self-Check Problem 23-3:2:

$$B = \frac{f_r}{Q} = \frac{840 \text{ kHz}}{105.53} = 7960$$

$$B = 8.0 \text{ kHz}$$

23-6:1. (a) Using Equation (23-1):

$$f_r = \frac{1}{2\pi\sqrt{LC}} = \frac{1}{2\pi \sqrt{(50 \text{ }\mu\text{H}) \cdot (0.33 \text{ }\mu\text{F})}} = 39\,181$$

$$f_r = 39 \text{ kHz}$$

(b) Using Equation (23-9):

$$Q = \frac{R_P}{X_L} = \frac{R_P}{2\pi f_r L} = \frac{2.0 \text{ k}\Omega}{2\pi \cdot (39\,181 \text{ Hz}) \cdot (50 \text{ }\mu\text{H})} = 162.48$$

$$Q = 160$$

23-7:1. (a) Using Equation (23-12):

$$f_r = \frac{1}{2\pi\sqrt{LC}} \cdot \sqrt{1 - \frac{CR_S^2}{L}}$$

$$f_r = \frac{1}{2\pi \cdot \sqrt{(50 \text{ }\mu\text{H}) \cdot (0.33 \text{ }\mu\text{F})}} \cdot \sqrt{1 - \frac{(0.33 \text{ }\mu\text{F}) \cdot (3)^2}{(50 \text{ }\mu\text{H})}}$$

$$f_r = 39\,181 \cdot \sqrt{0.9406} = 37\,999.7$$

$$f_r = 38 \text{ kHz}$$

(b) Since the Q of the circuit is the Q of the coil:

$$Q = \frac{X_S}{R_S} = \frac{2\pi f_r L}{R_S} = \frac{2\pi \cdot (38\,000 \text{ Hz}) \cdot (50 \text{ }\mu\text{H})}{3 \text{ }\Omega} = \frac{11.94}{3}$$

$$Q = 4.0$$

23-7:2. (a) Following the calculations for the previous problem:

$$f_r = \frac{1}{2\pi \cdot \sqrt{(50 \text{ }\mu\text{H}) \cdot (0.33 \text{ }\mu\text{F})}} \cdot \sqrt{1 - \frac{(0.33 \text{ }\mu\text{F}) \cdot (1)^2}{(50 \text{ }\mu\text{H})}}$$

$$f_r = 39\,181 \cdot \sqrt{0.9934} = 39\,051.7$$

$$f_r = 39 \text{ kHz}$$

(b) Similarly:

$$Q = \frac{X_S}{R_S} = \frac{2\pi f_r L}{R_S} = \frac{2\pi \cdot (38\,000 \text{ Hz}) \cdot (50 \text{ }\mu\text{H})}{1 \text{ }\Omega} = 11.94$$

$$Q = 12$$

Note that with a Q of 12 the f_r is virtually identical to the theoretical parallel-resonant circuit analyzed in Self-Check Problem 23-6:1.

23-8:1. To determine bandwidth we need the parallel equivalent resistance "seen" by the capacitor (see Figure 23-15):

$$R_P = \frac{R_S^2 + (2\pi f_r L)^2}{R_S} = \frac{3^2 + [2\pi \cdot (37\,999.7 \text{ Hz}) \cdot (50 \text{ }\mu\text{H})]^2}{3}\ 50.505$$

Since R_P is in parallel with R_{int}:

$$R_{eq} = \frac{R_P \cdot R_{int}}{R_P + R_{int}} = \frac{(50.505) \cdot (1000)}{(50.505) + (1000)}$$

$$R_{eq} = 48.077 \text{ }\Omega$$

Finally:

$$B = f_r \cdot \frac{X_c}{R_{eq}} = (37\,999.7 \text{ Hz}) \cdot \frac{12.692 \text{ }\Omega}{48.077 \text{ }\Omega} = 10\,032$$

$$B = 10 \text{ kHz}$$

23-8:2. By the same analysis as the previous problem:

$$R_P = 151.52 \text{ }\Omega$$
$$R_{eq} = 131.58 \text{ }\Omega$$
$$B = 3665.3 = 3.7 \text{ kHz}$$

CHAPTER 24

24-2:1. (a) From Equation (24-4):

$$\Phi_m = \frac{E_P}{4.44 \cdot f N_P} = \frac{200 \text{ V}}{4.44 \cdot (400 \text{ Hz}) \cdot (500)} = 2.25 \times 10^{-4} \text{ Wb}$$

(b) find the flux density:

$$B = \frac{\Phi}{A} = \frac{2.25 \times 10^{-4} \text{ Wb}}{2.0 \times 10^{-4} \text{ m}^2} = 1.126 \text{ T}$$

Using Figure 13-21 for cast-steel:

$$H_m = 1000 \text{ A} \cdot \text{t}$$

Using Equation (24-6):

$$I = \frac{H_m \cdot l}{\sqrt{2} \cdot N} = \frac{(1000 \text{ A}) \cdot (0.16 \text{ m})}{\sqrt{2} \cdot (500)} = 0.2263 \text{ A}$$

$$I = 230 \text{ mA}$$

24-3:1. From Equation (24-8):

(a) $$E_S = \frac{E_P}{a} = \frac{120 \text{ V}}{0.030} = 4000 = 4.0 \text{ kV}$$

(b) $$N_S = \frac{N_P}{a} = \frac{60}{0.030} = 2000 \text{ turns}$$

24-3:2. (a) From Equation (24-8):

$$E_S = E_P \cdot \frac{N_S}{N_P} = (60 \text{ V}) \cdot \frac{300}{1200} = 15 \text{ V}$$

(b) From Equation (24-10):

$$I_S = I'_P \cdot \frac{N_P}{N_S} = (0.50 \text{ A}) \cdot \frac{1200}{300} = 2.0 \text{ A}$$

24-4:1. Using Equation (24-11):

$$Z_P \approx a^2 \cdot Z_L = (12)^2 \cdot (50) = 7.2 \text{ k}\Omega$$

24-4:2. Use Ohm's Law to determine the primary impedance "seen" by the source:

$$Z_P = \frac{E_P}{I_P} = \frac{60 \text{ V}}{0.5 \text{ A}} = 120 \text{ } \Omega$$

From Equation (24-11):

$$Z_L \approx Z_P \cdot \frac{N_S^2}{N_P^2} = (120 \text{ } \Omega) \cdot \frac{300^2}{1200^2} = 7.5 \text{ } \Omega$$

24-5:1. Using Equation (24-12):

$$\mathbf{Z}_{ref} = a^2(R_S + jX_S + \mathbf{Z}_L) = \frac{N_P^2}{N_S^2} \cdot (\mathbf{Z}_S + \mathbf{Z}_L)$$

$$\mathbf{Z}_{ref} = \frac{1200^2}{400^2} \cdot (3 + j2 + 50) = 477 + j18 \text{ } \Omega = 477 \text{ } \Omega \text{ } \underline{/2.2°}$$

24-6:1. To determine R_e, use the wattmeter and ammeter:

$$R_e = \frac{P}{I^2} = \frac{2.8 \text{ W}}{(2.0 \text{ A})^2} = 0.70 \text{ } \Omega$$

Using the voltmeter and ammeter:

$$Z = \frac{E}{I} = \frac{21 \text{ V}}{2.0 \text{ A}} = 10.5 \text{ } \Omega$$

$$X_e = \sqrt{Z^2 - R_e^2} = \sqrt{(10.5)^2 + (0.70)^2}$$

$$X_e = 10.5 \text{ } \Omega$$

24-6:2. To determine the parallel-equivalent values shown in Figure 24-7:

$$R_{CL} = \frac{V^2}{P} = \frac{(120 \text{ V})^2}{2.4 \text{ W}} = 6.0 \text{ k} \Omega$$

$$\Rightarrow G_{CL} = R_{CL}^{-1} = (6000 \text{ } \Omega)^{-1} = 166.67 \text{ } \mu S$$

$$Y = \frac{I}{V} = \frac{0.1 \text{ A}}{120 \text{ V}} = 833.33 \text{ } \mu S$$

$$B = \sqrt{Y^2 - G^2} = \sqrt{(833.33)^2 - (166.67)^2} = 816.5 \text{ } \mu S$$

Finally

$$L_m = \frac{1}{B \cdot \omega} = \frac{1}{(816.5 \text{ } \mu S) \cdot (377 \text{ rad/s})} = 3.249$$

$$L_m = 3.2 \text{ H}$$

24-7:1. Determining the power to the load:

$$P_L = I^2 R_L = (11 \text{ A})^2 \cdot (2 \text{ } \Omega) = 242 \text{ W}$$

Using Equation (24-13):

$$\eta = \frac{P_L}{P_L + I^2 R_e + P_{CL}} = \frac{242}{242 + 2.8 + 2.4} = 0.979$$

$$\eta = 98\%$$

24-7:2. Since for maximum efficiency:

copper loss = core loss

and since $R_e = 0.70 \text{ } \Omega$ (see Self-Check Problem 24-6:1)

$$I^2 \cdot R_e = P_{CL}$$

$$\Rightarrow I = \sqrt{\frac{P_{CL}}{R_e}} = \sqrt{\frac{2.4 \text{ W}}{0.7}} = 1.85 \text{ A (primary current)}$$

Therefore

$$I_L = a \cdot I_P = 5 \cdot (1.85 \text{ A}) = 9.258$$

$$I_L = 9.3 \text{ A}$$

24-8:1. At no-load, the secondary voltage is

$$V_S = \frac{V_P}{a} = \frac{120 \text{ V}}{5} = 24 \text{ V}$$

Using Equation (24-16):

$$\text{voltage regulation} = \frac{V_{NL} - V_{FL}}{V_{FL}} \cdot 100\% = \frac{24 - 23 \text{ V}}{23 \text{ V}} \cdot 100\%$$

voltage regulation = 4.3%

24-9:1. Since the 200-turn coil is now the secondary, then the 100-turn coil is the primary, thus:

(a) $V_S = \dfrac{200 \text{ turns}}{200 + 100 \text{ turns}} \cdot (120 \text{ V}) = 80 \text{ V}$

(b) Using the load voltage and current magnitudes:

$S_L = V_L \cdot I_L = (80 \text{ V}) \cdot (9.0 \text{ A}) = 720 \text{ VA}$

(c) Of the 9.0 A of load current, 3 A are in the secondary and the remaining 6 A are in the primary:

$S_S = V_L \cdot I_S = (80 \text{ V}) \cdot (3.0 \text{ A}) = 240 \text{ VA}$

CHAPTER 25

25-2:1. From Equations (25-11, -12 and -13):

$\mathbf{z}_{12} = \mathbf{z}_{21} = \mathbf{Z}_m = 100 + j20 \ \Omega$

$\mathbf{z}_{11} = \mathbf{Z}_P + \mathbf{z}_{12} = (-j15 \ \Omega) + (100 + j20 \ \Omega)$
$\mathbf{z}_{11} = 100 + j5 \ \Omega$

$\mathbf{z}_{22} = \mathbf{Z}_S + \mathbf{z}_{21} = (-j12 \ \Omega) + (100 + j20 \ \Omega)$
$\mathbf{z}_{22} = 100 + j8 \ \Omega$

(a) Using Equation (25-17):

$$\text{coupled } \mathbf{Z} = \frac{\mathbf{Z}_m^2}{\mathbf{z}_{22} + \mathbf{Z}_L} = \frac{(100 + j20 \ \Omega)^2}{(100 + j8) + (50 \ \Omega)}$$

coupled $\mathbf{Z} = 69.2 \ \Omega \ \underline{/19.6°}$

(b) From Equation (25-16):

$\mathbf{Z}_{in} = \mathbf{z}_{11} - (\text{coupled } \mathbf{Z})$
$\quad = (100 + j5 \ \Omega) - (69.2 \ \Omega \ \underline{/19.6°})$
$\mathbf{Z}_{in} = 39.2 \ \Omega \ \underline{/-27.7°}$

25-3:1. (a) Using Equations (25-28, -29, and -30):

$\mathbf{Y}_m = -\mathbf{y}_{12} = -\mathbf{y}_{21} = -j150 \text{ mS}$

$\mathbf{Y}_P = \mathbf{y}_{11} + \mathbf{y}_{12} = (+j100 \text{ mS}) + (+j150 \text{ mS})$
$\mathbf{Y}_P = +j250 \text{ mS}$

$\mathbf{Y}_S = \mathbf{y}_{22} + \mathbf{y}_{21} = (+j90 \text{ mS}) + (+j150 \text{ mS})$
$\mathbf{Y}_S = +j240 \text{ mS}$

(b) Comparing Figures (25-9) and (25-7):

$$C_P = \frac{\mathbf{Y}_P}{j\omega} = \frac{+j250 \text{ mS}}{j(40\,000\pi)} = 2.0 \ \mu\text{F}$$

$$C_S = \frac{\mathbf{Y}_S}{j\omega} = \frac{+j240 \text{ mS}}{j(40\,000\pi)} = 1.9 \ \mu\text{F}$$

$$L_m = \frac{-j}{\mathbf{Y}_m \cdot \omega} = \frac{-j}{(-j150 \text{ mS}) \cdot (40\,000\pi)} = 53 \ \mu\text{H}$$

25-4:1. By inspection of Figure 25-12:

load current $= \mathbf{I}_2$

source current $= \mathbf{I}_1$

Therefore the current amplification becomes:

$$A_I = \frac{I_2}{I_1} = h_{21} = -0.95$$

that means that the load current is 95% of the source current and *downward* in R_L.

25-4:2. Since
load $P = V_2 \cdot I_2$ and
source $P = E_1 \cdot I_i$ and from Example 25-6:

$$A = \frac{V_2}{E_1}$$

then

$$A_P = \frac{V_2}{E_1} \cdot \frac{I_2}{I_1} = A \cdot A_i = (40) \cdot (0.95) = 38$$

Note that the negative sign on A_i indicates current direction in R_L and is not used in power calculations.

25-6:1. Using Equation (25-43):
$$M = k \cdot \sqrt{L_P \cdot L_S} = 0.2 \cdot \sqrt{(100 \ \mu H) \cdot (50 \ \mu H)}$$
$$M = 14 \ \mu H$$

25-6:2. From Equation (25-42):

$$E_S = \omega \cdot M \cdot I_P = \omega \cdot M \cdot \frac{E_P}{X_P} = E_P \cdot \frac{M}{L_P}$$

$$E_S = (250 \ \text{mV}) \cdot \frac{14 \ \mu H}{100 \ \mu H} = 35 \ \text{mV}$$

25-7:1. Using Equation (25-50) and the information in Example 25-10:

$$\mathbf{Z}_{in} = \mathbf{Z}_P + \frac{(\omega M)^2}{\mathbf{Z}_S + \mathbf{Z}_L}$$

$$\mathbf{Z}_{in} = (25 + j1885) + \frac{395 \ 000}{(25 + j377 \cdot 0.8) + 50}$$

$$\mathbf{Z}_{in} = (25 + j1885) + (307 - j1233)$$
$$\mathbf{Z}_{in} = 730 \ \Omega \ \underline{/63°}$$

25-9:1. First, C and L_S must be series resonant at 35-kHz to pass the desired signal:

$$X_L = \omega L_S = X_C = \frac{1}{\omega C}$$

$$\Rightarrow L_S = \frac{1}{\omega^2 \cdot C} = \frac{1}{(70\,000\pi)^2 \cdot (3.3 \ \mu F)} = 630 \ \mu H$$

Now find the reactance of the series-resonant combination at the reject frequency of 19 kHz:

$$X_S = \omega L_S - \frac{1}{\omega C}$$

$$= [38\,000\pi \cdot (630 \ \mu H)] - \frac{1}{38\,000\pi \cdot (0.033 \ \mu F)}$$

$X_S = 74.9 - 253.8 = 179 \ \Omega$ capacitive

The component L_P must have the same magnitude reactance to produce parallel resonance and block the 19-kHz signal:

$$L_P = \frac{X_P}{\omega} = \frac{179 \ \Omega}{38000\pi} = 1.5 \ \text{mH}$$

25-10:1. From Equation (23-1):

$$L = \frac{1}{(2\pi f_c)^2 \cdot C} = \frac{1}{(16\,000\pi)^2 \cdot (0.33 \ \mu F)}$$

$L = 1.2$ mH

Therefore the LC filter would look like the π-circuit in Figure 25-23 but with $C = 0.33\ \mu\text{F}$ and $L = 1.2$ mH.

25-10:2. From Equation (25-24):

$$R = \frac{1}{2\pi f_c \cdot C} = \frac{1}{(16\,000\pi) \cdot (0.33\ \mu\text{F})} = 60\ \Omega$$

Therefore the RC filter would look like the series circuit in Figure 25-24(b) but with $C = 0.33\ \mu\text{F}$ and $R = 60\ \Omega$.

25-11:1. Using Equation (25-55):

$$\text{dB} = 10 \cdot \log\left(\frac{P_{\text{out}}}{P_{\text{in}}}\right) = 10 \cdot \log\left(\frac{6}{12}\right) = -3.0\ \text{dB}$$

25-11:2. From Equation (25-55):

$$\text{dB} = 10 \cdot \log\,(P_{\text{out}}/P_{\text{in}})$$
$$23 = 10 \cdot \log\,(P_{\text{out}}/0.020)$$
$$\Rightarrow P_{\text{out}} = 0.020 \cdot 10^{2.3} = 0.020 \cdot 199.52$$
$$P_{\text{out}} = 4.0\ \text{W}$$

25-12:1.

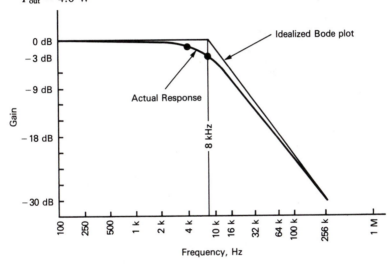

CHAPTER 26

26-4:1. (a) Using Equations (26-3):

$$\mathbf{I_A} = \frac{\mathbf{V_{AN}}}{\mathbf{Z_A}} = \frac{120\ \text{V}\ \underline{/0°}}{30 + j40\ \Omega} = 1.44 - j1.92\ \text{A}$$

$$\mathbf{I_B} = \frac{\mathbf{V_{BN}}}{\mathbf{Z_B}} = \frac{120\ \text{V}\ \underline{/-120°}}{35 - j35\ \Omega} = 0.627 - j2.342\ \text{A}$$

$$\mathbf{I_C} = \frac{\mathbf{V_{CN}}}{\mathbf{Z_C}} = \frac{120\ \text{V}\ \underline{/+120°}}{50\ \Omega} = -1.2 + j2.078\ \text{A}$$

(b) Using Equation (26-4):

$$\mathbf{I_N} = \mathbf{I_A} + \mathbf{I_B} + \mathbf{I_C}$$
$$\mathbf{I_N} = (1.44 - j1.92\ \text{A}) + (0.627 - j2.342\ \text{A}) + (-1.2 + j2.078\ \text{A})$$
$$\mathbf{I_N} = 0.868 - j2.184\ \text{A}$$

26-4:2. First find the neutral current due to the fixed loads. The voltage magnitude will not affect the answer, so use the voltage shown in Figure 26-11 for convenience:

Find the first line current;

$$I_A = \frac{S_A}{V_{AN}} = \frac{1200 \text{ VA } \underline{/-\cos^{-1}(0.8)}}{120 \text{ V } \underline{/0°}} = 8.000 - j6.000 \text{ A}$$

Since load B is resistive:

$$R_B = \frac{V^2}{P} = \frac{120^2}{1450} = 9.931 \ \Omega$$

$$I_B = \frac{V_{BN}}{Z_B} = \frac{120 \text{ V } \underline{/-120°}}{9.931 \ \Omega} = -6.042 - j10.465 \text{ A}$$

To reduce the neutral current to zero:

$$I_C = I_N - I_A - I_B$$
$$= 0 - (8.000 - j6.000 \text{ A}) - (-6.042 - j10.465 \text{ A})$$
$$I_C = -1.958 + j16.465 \text{ A}$$

Solving for the third load:

$$Z_C = \frac{E_{CN}}{I_C} = \frac{120 \text{ V } \underline{/120°}}{-1.958 + j16.465 \text{ A}} = 7.2 \ \Omega \ \underline{/23°}$$

26-5:1. (a) First compute the phase currents using Equations (26-6):

$$I_{AC} = \frac{V_{AC}}{Z_{AC}} = \frac{120 \text{ V } \underline{/0°}}{30 + j40 \ \Omega} = 1.440 - j1.920 \text{ A}$$

$$I_{CB} = \frac{V_{CB}}{Z_{CB}} = \frac{120 \text{ V } \underline{/+120°}}{50 \ \Omega} = -1.200 + j2.078 \text{ A}$$

$$I_{BA} = \frac{V_{BA}}{Z_{BA}} = \frac{120 \text{ V } \underline{/-120°}}{35 - j35 \ \Omega} = 0.627 - j2.342 \text{ A}$$

Now the line currents can be found using Equations (26-7):

$$I_A = I_{AC} - I_{BA}$$
$$I_A = (1.440 - j1.920 \text{ A}) - (0.627 - j2.342 \text{ A})$$
$$I_A = 0.813 + j0.422 = 0.916 \text{ A } \underline{/27.4°}$$

$$I_B = I_{BA} - I_{CB}$$
$$I_B = (0.627 - j2.342 \text{ A}) - (-1.200 + j2.078 \text{ A})$$
$$I_B = 1.828 - j4.420 = 4.783 \text{ A } \underline{/-67.5°}$$

$$I_C = I_{CB} - I_{AC}$$
$$I_C = (-1.200 + j2.078 \text{ A}) - (1.440 - j1.920 \text{ A})$$
$$I_C = -2.640 + j3.999 = 4.791 \text{ A } \underline{/123.4°}$$

(b) To check the line currents apply Equation (26.8):

$$I_A + I_B + I_C = (0.813 + j0.422) + (1.828 - j4.420) + (-2.640 + j3.999)$$
$$I_A + I_B + I_C = 0$$

26-5:2. Determine the phase currents using Equations (26-6):

$$I_{AC} = \frac{V_{AC}}{Z_{AC}} = \frac{120 \text{ V } \underline{/0°}}{12 \ \Omega \ \underline{/25°}} = 10 \text{ A } \underline{/-25°}$$

$$I_{CB} = \frac{V_{CB}}{Z_{CB}} = \frac{120 \text{ V } \underline{/+120°}}{12 \ \Omega \ \underline{/25°}} = 10 \text{ A } \underline{/+95°}$$

$$I_{BA} = \frac{V_{BA}}{Z_{BA}} = \frac{120 \text{ V } \underline{/-120°}}{12 \ \Omega \ \underline{/25°}} = 10 \text{ A } \underline{/-145°}$$

Using Equations (26-7) to obtain line currents:

$$I_A = I_{AC} - I_{BA} = 17.3 \text{ A } \underline{/5°}$$
$$I_B = I_{BA} - I_{CB} = 17.3 \text{ A } \underline{/-115°}$$
$$I_C = I_{CB} - I_{AC} = 17.3 \text{ A } \underline{/125°}$$

26-6:1. Using Equation (26-10):

$V_L = \sqrt{3} \cdot V_P = \sqrt{3} \cdot (120 \text{ V}) = 208 \text{ V}$

$Z = \dfrac{V_L}{I_L} = \dfrac{(208 \text{ V})}{15 \text{ A}} = 13.866 = 13.9 \text{ }\Omega$

26-6:2. Inspection of Figure 26-25 shows the coil currents to be identical to the line currents. Use the line voltages from Equations (26-9) and solve the problem as was done in Self-Check Problem 26-5:1.

First compute the phase currents using Equations (26-6):

$\mathbf{I}_{AC} = \dfrac{\mathbf{V}_{AC}}{\mathbf{Z}_{AC}} = \dfrac{208 \text{ V } \underline{/-30°}}{30 + j40 \text{ }\Omega} = 0.498 - j4.130 \text{ A}$

$\mathbf{I}_{CB} = \dfrac{\mathbf{V}_{CB}}{\mathbf{Z}_{CB}} = \dfrac{208 \text{ V } \underline{/+90°}}{50\Omega} = 0.000 + j4.160 \text{ A}$

$\mathbf{I}_{BA} = \dfrac{\mathbf{V}_{BA}}{\mathbf{Z}_{BA}} = \dfrac{208 \text{ V } \underline{/-150°}}{35 - j35 \text{ }\Omega} = -1.088 - j4.059 \text{ A}$

Now the line currents can be found using Equations (26-7):

$\mathbf{I}_A = \mathbf{I}_{AC} - \mathbf{I}_{BA} = 1.586 - j0.071 \text{ A}$
$\mathbf{I}_A = 1.588 \text{ A } \underline{/-2.6°}$
$\mathbf{I}_B = \mathbf{I}_{BA} - \mathbf{I}_{CB} = -1.088 - j8.219 \text{ A}$
$\mathbf{I}_B = 8.291 \text{ A } \underline{/-97.5°}$
$\mathbf{I}_C = \mathbf{I}_{CB} - \mathbf{I}_{AC} = -0.498 + j8.290 \text{ A}$
$\mathbf{I}_C = 8.305 \text{ A } \underline{/93.4°}$

26-7:1. Using Equation (26-11A):

$P_T = 3 \cdot V_p \cdot I_P \cdot \cos \phi$
$P_T = 3 \cdot (120 \text{ V}) \cdot (10 \text{ A}) \cdot \cos (25°) = 3263$
$P_T = 3.2 \text{ kW}$

OR by using Equation (26-11B):

$P_T = \sqrt{3} \cdot V_L \cdot I_L \cdot \cos \phi$
$P_T = \sqrt{3} \cdot (120 \text{ V}) \cdot (17.3 \text{ A}) \cdot \cos (25°) = 3263$
$P_T = 3.2 \text{ kW}$

26-7:2. Using Equation (26-11B):

$P_T = \sqrt{3} \cdot V_L \cdot I_L \cdot \cos \phi$
$P_T = \sqrt{3} \cdot (208 \text{ V}) \cdot (5.0 \text{ A}) \cdot (0.80) = 1441$
$P_T = 1.4 \text{ kW}$

26-9:1. Since the total power is the sum of the wattmeter readings:

$P_T = P_1 + P_2 = 300 + 500 \text{ W} = 800 \text{ W}$

26-9:2. (a) Identical to Self-Check Problem 26-9:1

$P_T = P_1 + P_2 = 300 + 500 \text{ W} = 800 \text{ W}$

(b) Since the load is balanced, using Equation (26-15):

$\tan \phi = \sqrt{3} \cdot \dfrac{P_2 - P_1}{P_2 + P_1} = \sqrt{3} \cdot \dfrac{500 - 300 \text{ W}}{500 + 300 \text{ W}} = 0.433$

$\Rightarrow \phi = \tan^{-1}(0.433) = 23.4°$

Therefore

$\cos \phi = \cos (23.4°) = 0.92$

26-10:1. Since this is a Δ-load, first compute the phase currents:

$\mathbf{I}_{AB} = \dfrac{\mathbf{V}_{AB}}{\mathbf{Z}_{AB}} = \dfrac{208 \text{ V } \underline{/0°}}{30 + j40 \text{ }\Omega} = 2.496 - j3.328 \text{ A}$

$\mathbf{I}_{BC} = \dfrac{\mathbf{V}_{BC}}{\mathbf{Z}_{BC}} = \dfrac{208 \text{ V } \underline{/-120°}}{35 - j35 \text{ }\Omega} = 1.088 - j4.059 \text{ A}$

$$\mathbf{I_{CA}} = \frac{\mathbf{V_{CA}}}{\mathbf{Z_{CA}}} = \frac{208 \text{ V } \underline{/+120°}}{50 \text{ }\Omega} = -2.080 + j3.603 \text{ A}$$

Now the line currents can be found:

$\mathbf{I_A} = \mathbf{I_{AB}} - \mathbf{I_{CA}} = 4.576 - j6.931 \text{ A}$
$\mathbf{I_A} = 8.305 \text{ A } \underline{/-56.6°}$
$\mathbf{I_B} = \mathbf{I_{BC}} - \mathbf{I_{AB}} = -1.408 - j0.731 \text{ A}$
$\mathbf{I_B} = 1.586 \text{ A } \underline{/-152.6°}$
$\mathbf{I_C} = \mathbf{I_{CA}} - \mathbf{I_{BC}} = -3.168 + j7.662 \text{ A}$
$\mathbf{I_C} = 8.291 \text{ A } \underline{/112.5°}$

26-10:2. The difference between this problem and the previous one is that the phases of $\mathbf{V_{BC}}$ and $\mathbf{V_{CA}}$ have switched:

$$\mathbf{I_{AB}} = \frac{\mathbf{V_{AB}}}{\mathbf{Z_{AB}}} = \frac{208 \text{ V } \underline{/0°}}{30 + j40 \text{ }\Omega} = 2.496 - j3.328 \text{ A}$$

$$\mathbf{I_{BC}} = \frac{\mathbf{V_{BC}}}{\mathbf{Z_{BC}}} = \frac{208 \text{ V } \underline{/+120°}}{35 - j35 \text{ }\Omega} = -4.059 + j1.088 \text{ A}$$

$$\mathbf{I_{CA}} = \frac{\mathbf{V_{CA}}}{\mathbf{Z_{CA}}} = \frac{208 \text{ V } \underline{/-120°}}{50 \text{ }\Omega} = -2.080 - j3.603 \text{ A}$$

Now the line currents can be found:

$\mathbf{I_A} = \mathbf{I_{AB}} - \mathbf{I_{CA}} = 4.576 + j0.275 \text{ A}$
$\mathbf{I_A} = 4.584 \text{ A } \underline{/+3.4°}$
$\mathbf{I_B} = \mathbf{I_{BC}} - \mathbf{I_{AB}} = -6.555 + j4.416 \text{ A}$
$\mathbf{I_B} = 7.904 \text{ A } \underline{/+146.0°}$
$\mathbf{I_C} = \mathbf{I_{CA}} - \mathbf{I_{BC}} = 1.979 - j4.691 \text{ A}$
$\mathbf{I_C} = 5.091 \text{ A } \underline{/-67.1°}$

26-11:1. First determine the numerator for the transformation Equations (22-5, -6, and -7):

$\mathbf{N} = \mathbf{Z_A} \cdot \mathbf{Z_B} + \mathbf{Z_B} \cdot \mathbf{Z_C} + \mathbf{Z_C} \cdot \mathbf{Z_A} = 5700 + j600$

Solving the transformation equations:

$$\mathbf{Z_{AB}} = \frac{\mathbf{N}}{\mathbf{Z_C}} = 114 + j12 \text{ }\Omega$$

$$\mathbf{Z_{BC}} = \frac{\mathbf{N}}{\mathbf{Z_A}} = 78 - j84 \text{ }\Omega$$

$$\mathbf{Z_{CA}} = \frac{\mathbf{N}}{\mathbf{Z_B}} = 72.857 + j90 \text{ }\Omega$$

Solving for phase currents:

$$\mathbf{I_{AB}} = \frac{\mathbf{V_{AB}}}{\mathbf{Z_{AB}}} = \frac{208 \text{ V } \underline{/0°}}{114 + j12 \text{ }\Omega} = 1.805 - j0.190 \text{ A}$$

$$\mathbf{I_{BC}} = \frac{\mathbf{V_{BC}}}{\mathbf{Z_{BC}}} = \frac{208 \text{ V } \underline{/-120°}}{78 - j84 \text{ }\Omega} = 0.534 - j1.734 \text{ A}$$

$$\mathbf{I_{CA}} = \frac{\mathbf{V_{CA}}}{\mathbf{Z_{CA}}} = \frac{208 \text{ V } \underline{/+120°}}{72.857 + j90 \text{ }\Omega} = 0.644 + j1.677 \text{ A}$$

Now the line currents can be found:

$\mathbf{I_A} = \mathbf{I_{AB}} - \mathbf{I_{CA}} = 1.161 - j1.876 \text{ A}$
$\mathbf{I_A} = 2.199 \text{ A } \underline{/-58.1°}$
$\mathbf{I_B} = \mathbf{I_{BC}} - \mathbf{I_{AB}} = -1.271 - j1.544 \text{ A}$
$\mathbf{I_B} = 2.000 \text{ A } \underline{/-129.5°}$
$\mathbf{I_C} = \mathbf{I_{CA}} - \mathbf{I_{BC}} = 0.110 + j3.411 \text{ A}$
$\mathbf{I_C} = 3.413 \text{ A } \underline{/88.2°}$

INDEX